MUIRHEAD • Aspects of Multivariate Statistical Theory
PARZEN • Modern Probability Theory and Its Applications
PURI and SEN • Nonparametric Methods in General Linear Models
PURI and SEN • Nonparametric Methods in Multivariate Analysis
RANDLES and WOLFE • Introduction to the Theory of Nonparametric Statistics
RAO • Linear Statistical Inference and Its Applications, *Second Edition*
RAO and SEDRANSK • W.G. Cochran's Impact on Statistics
ROHATGI • An Introduction to Probability Theory and Mathematical Statistics
ROHATGI • Statistical Inference
ROSS • Stochastic Processes
RUBINSTEIN • Simulation and The Monte Carlo Method
SCHEFFE • The Analysis of Variance
SEBER • Linear Regression Analysis
SEBER • Multivariate Observations
SEN • Sequential Nonparametrics: Invariance Principles and Statistical Inference
SERFLING • Approximation Theorems of Mathematical Statistics
TJUR • Probability Based on Radon Measures
WILLIAMS • Diffusions, Markov Processes, and Martingales, Volume I: Foundations
ZACKS • Theory of Statistical Inference

Applied Probability and Statistics

ABRAHAM and LEDOLTER • Statistical Methods for Forecasting
AGRESTI • Analysis of Ordinal Categorical Data
AICKIN • Linear Statistical Analysis of Discrete Data
ANDERSON, AUQUIER, HAUCK, OAKES, VANDAELE, and WEISBERG • Statistical Methods for Comparative Studies
ARTHANARI and DODGE • Mathematical Programming in Statistics
BAILEY • The Elements of Stochastic Processes with Applications to the Natural Sciences
BAILEY • Mathematics, Statistics and Systems for Health
BARNETT • Interpreting Multivariate Data
BARNETT and LEWIS • Outliers in Statistical Data, *Second Edition*
BARTHOLOMEW • Stochastic Models for Social Processes, *Third Edition*
BARTHOLOMEW and FORBES • Statistical Techniques for Manpower Planning
BECK and ARNOLD • Parameter Estimation in Engineering and Science
BELSLEY, KUH, and WELSCH • Regression Diagnostics: Identifying Influential Data and Sources of Collinearity
BHAT • Elements of Applied Stochastic Processes, *Second Edition*
BLOOMFIELD • Fourier Analysis of Time Series: An Introduction
BOX • R. A. Fisher, The Life of a Scientist
BOX and DRAPER • Evolutionary Operation: A Statistical Method for Process Improvement
BOX, HUNTER, and HUNTER • Statistics for Experimenters: An Introduction to Design, Data Analysis, and Model Building
BROWN and HOLLANDER • Statistics: A Biomedical Introduction
BUNKE and BUNKE • Statistical Inference in Linear Models, Volume I
CHAMBERS • Computational Methods for Data Analysis
CHATTERJEE and PRICE • Regression Analysis by Example
CHOW • Econometric Analysis by Control Methods
CLARKE and DISNEY • Probability and Random Processes: A First Course with Applications
COCHRAN • Sampling Techniques, *Third Edition*
COCHRAN and COX • Experimental Designs, *Second Edition*
CONOVER • Practical Nonparametric Statistics, *Second Edition*

Applied Probability and Statistics (Continued)

CONOVER and IMAN • Introduction to Modern Business Statistics
CORNELL • Experiments with Mixtures: Designs, Models and The Analysis of Mixture Data
COX • Planning of Experiments
DANIEL • Biostatistics: A Foundation for Analysis in the Health Sciences, *Third Edition*
DANIEL • Applications of Statistics to Industrial Experimentation
DANIEL and WOOD • Fitting Equations to Data: Computer Analysis of Multifactor Data, *Second Edition*
DAVID • Order Statistics, *Second Edition*
DAVISON • Multidimensional Scaling
DEMING • Sample Design in Business Research
DILLON and GOLDSTEIN • Multivariate Analysis: Methods and Applications
DODGE • Analysis of Experiments with Missing Data
DODGE and ROMIG • Sampling Inspection Tables, *Second Edition*
DOWDY and WEARDEN • Statistics for Research
DRAPER and SMITH • Applied Regression Analysis, *Second Edition*
DUNN • Basic Statistics: A Primer for the Biomedical Sciences, *Second Edition*
DUNN and CLARK • Applied Statistics: Analysis of Variance and Regression
ELANDT-JOHNSON and JOHNSON • Survival Models and Data Analysis
FLEISS • Statistical Methods for Rates and Proportions, *Second Edition*
FLEISS • The Design and Analysis of Clinical Experiments
FOX • Linear Statistical Models and Related Methods
FRANKEN, KÖNIG, ARNDT, and SCHMIDT • Queues and Point Processes
GALAMBOS • The Asymptotic Theory of Extreme Order Statistics
GIBBONS, OLKIN, and SOBEL • Selecting and Ordering Populations: A New Statistical Methodology
GNANADESIKAN • Methods for Statistical Data Analysis of Multivariate Observations
GOLDBERGER • Econometric Theory
GOLDSTEIN and DILLON • Discrete Discriminant Analysis
GREENBERG and WEBSTER • Advanced Econometrics: A Bridge to the Literature
GROSS and CLARK • Survival Distributions: Reliability Applications in the Biomedical Sciences
GROSS and HARRIS • Fundamentals of Queueing Theory, *Second Edition*
GUPTA and PANCHAPAKESAN • Multiple Decision Procedures: Theory and Methodology of Selecting and Ranking Populations
GUTTMAN, WILKS, and HUNTER • Introductory Engineering Statistics, *Third Edition*
HAHN and SHAPIRO • Statistical Models in Engineering
HALD • Statistical Tables and Formulas
HALD • Statistical Theory with Engineering Applications
HAND • Discrimination and Classification
HILDEBRAND, LAING, and ROSENTHAL • Prediction Analysis of Cross Classifications
HOAGLIN, MOSTELLER and TUKEY • Exploring Data Tables, Trends and Shapes
HOAGLIN, MOSTELLER, and TUKEY • Understanding Robust and Exploratory Data Analysis
HOEL • Elementary Statistics, *Fourth Edition*
HOEL and JESSEN • Basic Statistics for Business and Economics, *Third Edition*
HOGG and KLUGMAN • Loss Distributions

(*continued on back*)

Exploring Data Tables, Trends, and Shapes

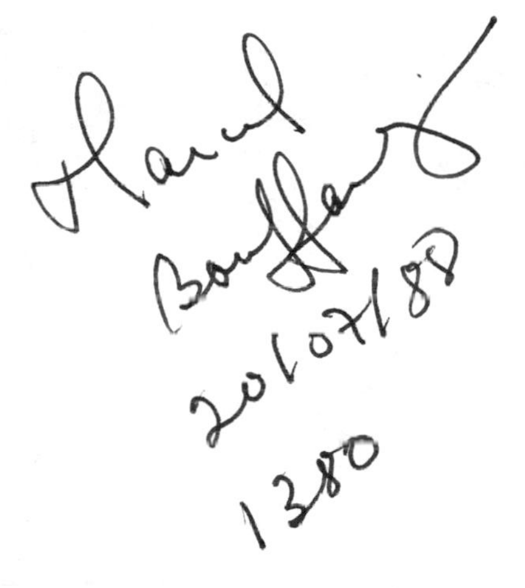

Exploring Data Tables, Trends, and Shapes

Edited by

DAVID C. HOAGLIN
Harvard University

FREDERICK MOSTELLER
Harvard University

JOHN W. TUKEY
Princeton University and AT & T Bell Laboratories

John Wiley & Sons
New York • Chichester • Brisbane • Toronto • Singapore

***Library of Congress Cataloging in Publication Data*:**

Main entry under title:

Exploring data tables, trends, and shapes.

(Wiley series in probability and mathematical statistics. Applied probability and statistics, ISSN 0271-6256)

Includes index.

1. Mathematical statistics. I. Hoaglin, David C. (David Caster), 1944– . II. Mosteller, Frederick, 1916– . III. Tukey, John Wilder, 1915– . IV. Series.

QA276.E97 1985 519.5 84-27144

ISBN 0-471-09776-4

Printed in the United States of America

10 9 8 7 6 5 4 3 2 1

To

Cleo Youtz

who has helped in so many ways

with this book and other projects

Preface

The present book, though largely self-contained, continues our *Understanding Robust and Exploratory Data Analysis* (Wiley, 1983), which we often refer to as UREDA.

Exploratory and robust/resistant techniques are becoming a core component of statistical practice. This book presents a variety of more advanced techniques and extensions of basic exploratory tools, explains why these further developments are valuable, and provides insight into why and how they were devised. In addition to illustrating these techniques, we offer conceptual and logical support for them.

The techniques of *exploratory data analysis*, particularly as embodied in the book of that title by Tukey (Addison-Wesley, 1977), may seem to have sprung from nowhere and to be supported only by anecdote. Many of their purposes parallel those of more conventional techniques. Indeed, we can even express some justifications for particular techniques by using concepts and results from classical statistical theory. This book explains and illustrates such parallelisms and justifications.

Robust and resistant techniques have considerable support in the statistical research literature, both at a highly abstract mathematical level and in extensive Monte Carlo studies. The companion volume, UREDA, primarily treats techniques for estimating location and scale. The present book applies many of the same basic ideas, particularly in multi-way tables and robust regression. Again we provide a basic understanding of the techniques at a reduced level of mathematical sophistication.

By studying this book the user will become more effective in handling the more advanced robust and exploratory techniques, the student better able to understand them, and the teacher better able to explain them.

Most chapters are self-contained, but readers desiring a detailed account of the basic exploratory and robust/resistant techniques should consult the companion volume, UREDA. It deals with the basic operations of displaying and comparing batches, estimating location and scale, fitting resistant straight lines, and summarizing additive structure in two-way tables. In addition, it examines broadly applicable ideas of transformations and residuals. A narrative table of contents for UREDA follows the table of contents of this volume.

The first chapter examines, from a broad philosophical and practical perspective, the role of exploratory and robust techniques in the overall data-analytic enterprise. It helps to integrate these ideas with more traditional statistics and gives some flavor of current and future thrusts.

The remaining ten chapters present a variety of techniques under three general headings: tables, trends, and shapes. To provide a more detailed overview of them, the table of contents includes a brief annotation for each chapter.

For *tables* of measured responses, three chapters discuss a resistant approach for the two-way additive model related to paired comparisons (Chapter 2), a sequence of two-way models that summarize important forms of systematic nonadditivity (Chapter 3), and resistant techniques for three-way layouts (Chapter 4). Chapter 5 illustrates exploratory ideas in diagnosing unusual counts in a sizable contingency table.

For straight-line *trends*, three chapters focus on regression, from eye fitting of simple straight lines (Chapter 6), through applying a resistant line-fitting technique, one variable at a time, in multiple regression (Chapter 7), to advanced techniques of robust regression (Chapter 8).

And for *shapes*, Chapter 9 introduces a systematic approach for checking models associated with discrete frequency distributions, and two further chapters use quantiles from continuous data—first to expose qualitative and comparative features (Chapter 10) and then to summarize distribution shape numerically (Chapter 11).

To a greater degree than in UREDA, the chapters in this volume offer new techniques and perspectives, beyond those described in *Exploratory Data Analysis* (EDA) and other published sources. Chapter 1 takes several new steps toward a theory and philosophy of data analysis. The square-combining-table techniques in Chapter 2 were introduced only in the limited preliminary edition of EDA. Unified resistant approaches to additive-plus-multiplicative description of two-way tables are made available in Chapter 3. Chapter 4 provides an extensive comparison of median polish and least squares for three-way tables. Chapter 5 examines three new approaches for discovering anomalous counts in contingency tables.

Chapter 6 analyzes in more detail the results of an empirical study of straight lines fitted by eye. A general approach for extending any resistant line-fitting procedure to multiple-regression models and to two-way analysis of covariance makes its appearance in Chapter 7. Chapter 8 distills several topics in robust regression and describes them in less technical language than available elsewhere.

Most of Chapter 9's techniques for discrete frequency distributions have been developed recently and have not appeared in print before. Chapter 10 gives a broad account of resistant methods for qualitative study of continuous distributions; and Chapter 11 describes a framework, the *g*-and-*h* distributions, for analyzing distribution shape quantitatively.

As in UREDA, we illustrate most methods in an example or two, using real data. Varying in size and complexity, these examples help introduce the reader to techniques and illustrate why and when one method is preferable to another. Through them we give some evidence of the efficacy of the techniques in concrete applications.

A brief collection of exercises at the end of most chapters encourages the reader to participate directly in applying the techniques to other sets of data, establishing their properties, or extending them to new situations.

Without attempting to be exhaustive, lists of additional literature supplement the references in most chapters to facilitate access to related material.

The level of mathematics varies, both within and among chapters. By allowing this we hope to keep as many readers as possible in touch with each section of the book. This plan requires tolerance on the part of the readers: The quite well prepared need to appreciate that writing for broad understanding does not mean they are being talked down to; the less well prepared must be willing to skip along, omitting parts that require too much background or too much mathematics. Rigor, when available, may sometimes be sacrificed in order to communicate the main idea to more readers.

The mathematical prerequisite for the reader is not high in most chapters. At the same time, mathematical sophistication is matched to the requirement for explaining each technique, and so the level rises occasionally, especially in Chapter 8.

Although the present volume includes a number of new developments, several of the techniques we discuss also appear in *Exploratory Data Analysis* by John W. Tukey, in *Data Analysis and Regression* by Frederick Mosteller and John W. Tukey (Addison-Wesley, 1977), or in *Applications, Basics, and Computing of Exploratory Data Analysis* by Paul F. Velleman and David C. Hoaglin (Duxbury Press, 1981). In the present book, as in UREDA, the emphasis is more on the rationale and development of the methods, and less on illustrating their use, though most chapters offer detailed illustrations.

Together with UREDA, this book completes our originally planned two-volume monograph on exploratory and robust/resistant methods.

DAVID C. HOAGLIN
FREDERICK MOSTELLER
JOHN W. TUKEY

Cambridge, Massachusetts
Murray Hill, New Jersey
February 1985

Acknowledgments

Although the chapters in this book are signed, each has benefited from repeated editing and advice, not only from the three primary editors but also from the members of a working group on exploratory data analysis in the Department of Statistics at Harvard University. Beginning in the spring of 1977, this project has involved students, faculty, academic visitors, and others in its weekly meetings. Those who have participated at one time or another are Mary B. Breckenridge, Nancy Romanowicz Cook, John D. Emerson, Miriam Gasko, John P. Gilbert (deceased), Katherine Godfrey, Colin Goodall, Katherine Taylor Halvorsen, David C. Hoaglin, Boris Iglewicz, Lois Kellerman, Peter J. Kempthorne, Guoying Li, Lillian Lin, James Miller, Frederick Mosteller, Anita Parunak, James L. Rosenberger, Andrew F. Siegel, Keith A. Soper, Michael A. Stoto, Judith F. Strenio, John W. Tukey, Paul F. Velleman, George Y. Wong, and Cleo Youtz. Through their sharing of ideas and friendly criticism, all contributed to the development of the material.

The stimulus for Chapter 5 came from discussions with Laurel Casjens and her husband, Carlcton DeTar. She has kindly allowed us to use her data, and she has given us advice about her interpretations and the general milieu of her archaeological study.

In addition to funding from the National Science Foundation (through grants SES 75-15702 and SES 8023644), these activities received partial support from the Middlebury College Faculty Leave Program and the U.S. Army Research Office (contract DAAG 29-82-K-0085). Anita Parunak's work on Chapter 5 was supported in part by a grant from the Hazel I. Stoll estate to the Highway Safety Research Institute, The University of Michigan. John W. Tukey's activities at Princeton also received partial support from the U.S. Army Research Office (Durham), and his work on Chapter 9 was supported in part by Grant DE-AC02-81ER10841 from the U.S. Department of Energy to Princeton University and in part by AT & T Bell Laboratories.

A. S. C. Ehrenberg, Jerome H. Friedman, Peter J. Huber, Andrew B. Jonas, Amos Tversky, Paul F. Velleman, and Roy E. Welsch generously provided comments on various draft chapters.

Cleo Youtz helped unstintingly with the numerous steps of assembling, checking, and coordinating the manuscript.

Peter J. Kempthorne rendered special assistance by preparing Figure 8-1.

Edwin B. Newman generously supplied background information on the Massachusetts mental health data set in Chapter 7.

For both this volume and UREDA, Beatrice Shube, our editor at John Wiley & Sons, provided frequent encouragement and many forms of advice. We have gained much from her wise counsel.

Gail Borden, Holly Grano, Marjorie Olson, and Cleo Youtz prepared the manuscript with great care and efficiency.

D. C. H.
F. M.
J. W. T.

Contents

conveniently using median differences. These medians are then combined, in a way related to usual analyses of paired-comparisons data, to yield row effects and column effects. This approach may be preferable to median polish, particularly when the two-way table contains holes and has at most 20% bad data values.

Two-way tables of data may exhibit structure that goes beyond that describable in a simple additive fit. One approach to such nonadditive structure (discussed by Mandel and others) adds a general multiplicative term to produce an additive-plus-multiplicative model. A generalization of median polish yields a resistant fit of this model. New resistant strategies supplement existing ones in assessing various fits and their residuals.

Tables with a measured response and three or more factors extend the ideas of the two-way table. To gain protection against the adverse effects of isolated unusual data values, median polish is extended to higher-way tables. The analyses parallel the more traditional analysis based on means, to which they are compared. Generalizations of the

5G. Illustrations of the New Standardization 217
5H. Summary 221
5I. Conclusion 223

6. FITTING STRAIGHT LINES BY EYE
FREDERICK MOSTELLER, ANDREW F. SIEGEL,
EDWARD TRAPIDO, AND CLEO YOUTZ 225

Because investigators frequently fit lines by eye, it is useful to know something about the properties of such a procedure. Students fitted lines by eye to each of four sets of points presented in an experimental design. Their pooled slope was closer to the slope of the major axis than to the slope of the least-squares regression line, and the median efficiency of fitting was about 63 percent.

6A. Method 226
6B. Results 229
6C. Summary 238

7. RESISTANT MULTIPLE REGRESSION, ONE VARIABLE AT A TIME
JOHN D. EMERSON AND DAVID C. HOAGLIN 241

When regression data come with more than one carrier (or predictor), repeated application of the exploratory resistant line guards against the effects of isolated wild data values. Taking the carriers in a specified order, the analysis sweeps the effect of each carrier out of the response and out of all later carriers. A related approach provides a resistant analog of two-way analysis of covariance.

7A. Resistant Lines 242
7B. Sweeping Out 246
7C. Example 250
7D. When Carriers Come in Blocks 263
7E. Summary 273
Exercises 275

8. ROBUST REGRESSION
GUOYING LI 281

Robust regression methods, besides providing estimates that are often more useful, can call attention to unusual data in a regression data set.

These methods, including those based on M-estimators and W-estimators as well as bounded-influence regression, protect against distortion by anomalous data and have good efficiency over a wide range of possibilities for the error structure.

New graphical techniques help to assess how closely an observed discrete frequency distribution follows a member of one of the common families of discrete distributions, such as the binomial or the Poisson. Used carefully, these techniques prevent unusual counts from distorting the overall assessment. The slope of a straight-line pattern in the plot identifies the main parameter of the family of distributions. Other plots help to show whether a cell is discrepant, as well as to assess whether different parts of the data indicate different values of the main parameter.

Instead of the classical measures based on the third and fourth moments, numerical and graphical techniques based on quantiles serve to describe skewness and elongation. Variants of quantile–quantile plots aid in comparing distributions.

This chapter approaches the study of distribution shape more quantitatively. Selected monotonic functions of a standard Gaussian random variable can be described by constants g and h that indicate skewness and elongation, respectively. The resulting family of g-and-h distributions offers resistance and flexibility in summarizing distribution shapes quantitatively.

Companion Volume

Understanding Robust and Exploratory Data Analysis

Narrative Table of Contents

variability. A spread-versus-level plot facilitates choice of a suitable power transformation.

4. TRANSFORMING DATA

JOHN D. EMERSON AND MICHAEL A. STOTO

Using transformed instead of raw data values often yields more effective displays, summaries, and analyses. Gains from such a change in the basic scale of measurement include more nearly symmetric shapes in batches, elimination of apparent outliers, better comparability of spread among batches at different levels, and less systematic residuals from fitting simple models. Matched transformations aid interpretation by mimicking the scale of the raw data. A few guidelines help in judging when transformation is likely to be worthwhile.

5. RESISTANT LINES FOR y VERSUS x

JOHN D. EMERSON AND DAVID C. HOAGLIN

The exploratory resistant line offers an alternative to the least-squares regression line for y-versus-x data. The method divides the data into three groups and achieves resistance to unusual data points by using medians within the groups. A straightforward iteration, in terms of the residuals, then yields the fitted line. Long-known techniques based on groups and summaries provide a background for the three-group resistant line. The ideas of influence and leverage aid understanding of this and other alternatives to least-squares regression.

6. ANALYSIS OF TWO-WAY TABLES BY MEDIANS

JOHN D. EMERSON AND DAVID C. HOAGLIN

In a two-way table of responses, each of two factors separately takes on several levels; the table contains a value of the response variable for each possible combination of levels of the factors. Median polish uses the median in an iterative fashion to provide a resistant analysis. Insights about the method come from comparing its results with those of other methods of analysis, including least squares and least absolute residuals. The suitability of median polish for data exploration derives in part from its nonzero breakdown bound and from its ability to cope with missing data values.

7. EXAMINING RESIDUALS

COLIN GOODALL

Exploratory data analysis emphasizes looking at and analyzing residuals to help understand data and to investigate models. A fit describes the data, and examining residuals often gives guidance for improving that description. Strong patterns usually indicate that further refinement of the fit will be a worthwhile aid to understanding the structure in the raw data. A regression setting illustrates the use of residuals and some associated plots.

8. MATHEMATICAL ASPECTS OF TRANSFORMATION
JOHN D. EMERSON

Four desirable features of data are symmetry in single batches, equal spread in several batches at different levels, straightness in y-versus-x data, and additive structure in two-way tables. Each property serves as a primary objective of transforming data in the associated data structure. The chapter provides a variety of justifications for the assertions and techniques introduced in earlier chapters. For each of the four data structures, a special plot aids in finding a suitable power transformation.

9. INTRODUCTION TO MORE REFINED ESTIMATORS
DAVID C. HOAGLIN, FREDERICK MOSTELLER, AND JOHN W. TUKEY

This chapter sets the stage for the three chapters that follow it and for Chapters 7 and 8 of the companion volume. To consolidate messages in data, one often assumes a model and estimates parameters, such as location and scale. Robust estimators, which perform well over a wide range of assumptions about the behavior of the fluctuations in the data, are desirable. The chapter introduces a variety of methods of studying location estimators in symmetric situations; and it explores concepts of efficiency, the role of sample size in shifting attention from variability to bias, and further criteria useful in assessing robust estimators.

10. COMPARING LOCATION ESTIMATORS: TRIMMED MEANS, MEDIANS, AND TRIMEAN
JAMES L. ROSENBERGER AND MIRIAM GASKO

Many simple location estimators for the center of a symmetric distribution belong to a class called L-estimators: linear combinations of order statistics. In the absence of precise knowledge about the true distribution underlying a data set, it is customary to consider the behavior of estimators for alternative distributions that range from the Gaussian to the very heavy-tailed Cauchy. The concept of efficiency provides a framework for assessing the performance of several L-estimators over the set of distributions considered.

11. *M*-ESTIMATORS OF LOCATION: AN OUTLINE OF THE THEORY
COLIN GOODALL

Three classes of estimators—L-estimators, R-estimators, and M-estimators—have been prominent in research on robust estimation. The best robust estimators, at least for small to moderate samples, come from the class of M-estimators. These estimators minimize functions of the deviations of the observations from the estimate: the sum of squared deviations and the sum of absolute deviations are the two best-known examples. M-estimation can be viewed as generalizing maximum-likelihood estimation. A suitably chosen

M-estimator has good robustness of efficiency in large samples and high efficiency in small samples taken from a variety of realistic distributions.

12. ROBUST SCALE ESTIMATORS AND CONFIDENCE INTERVALS FOR LOCATION

BORIS IGLEWICZ

This chapter describes and compares robust estimators for the scale of a symmetric distribution. Robust alternatives to the standard deviation include the mean absolute deviation from the sample median, the median absolute deviation from the sample median, the fourth-spread, and other more intricate estimators. The Gaussian distribution and two symmetric distributions having heavier tails serve as the setting for comparing performance. These estimators lead to robust confidence intervals for location parameters and help to produce tests for departure from Gaussian shape.

Exploring Data Tables, Trends, and Shapes

CHAPTER 1

Theories of Data Analysis: From Magical Thinking Through Classical Statistics

Persi Diaconis
Stanford University

Exploratory data analysis (EDA) seeks to reveal structure, or simple descriptions, in data. We look at numbers or graphs and try to find patterns. We pursue leads suggested by background information, imagination, patterns perceived, and experience with other data analyses.

Magical Thinking

Magical thinking involves our inclination to seek and interpret connections between the events around us, together with our disinclination to revise belief after further observation. In one manifestation it may be believed that the gods are sending signs, and that a particular ritual holds the key to understanding. This belief persists despite the facts. Scholars have assembled a rich collection of facts and theories to describe human myths and rituals. Cassirer (1972) offers an inspiring presentation of the role of magical thinking in modern life. Shweder (1977) describes other anthropological and psychological studies.

The South Sea cargo cults provide a clear example of magical thinking. During World War II airplanes landed and unloaded food and materials. To bring this about again, the natives build fires along the sides of makeshift runways. They have a ceremonial controller who sits in a hut wearing a wooden helmet complete with bamboo bar antennas. They recreate the pattern observed in the past and wait for the planes to land. No airplanes land, yet the ritual continues. Worsley (1968) offers documentation and an anthropological interpretation.

Several studies of human learning indicate that such behavior is common in our own activities. In a typical study, an experimenter generates a pattern of zeros and ones by flipping a coin with chance of heads .7. Subjects are asked to predict the outcomes, not knowing the chance generating mechanism. The subjects seek and see patterns (e.g., alternation or a head following two tails), often including patterns that do not appear in the actual data. They often report a complex working system even after hundreds of trials. Subjects do not converge to and adopt the right strategy (always guess heads).

At one extreme, we can view the techniques of EDA as a ritual designed to reveal patterns in a data set. Thus, we may believe that naturally occurring data sets contain structure, that EDA is a useful vehicle for revealing the structure, and that revealed structure can sometimes be interpreted in the language of the subject matter that produced the data. If we make no attempt to check whether the structure could have arisen by chance, and tend to accept the findings as gospel, then the ritual comes close to magical thinking.

None of this argues that exploration is useless or wrong. Consulting statisticians cannot be universal subject matter experts, and data often present striking patterns. Such explorations have been, and continue to be, an important part of science. The final section of this chapter argues that a controlled form of magical thinking—in the guise of "working hypothesis"—is a basic ingredient of scientific progress.

Classical Mathematical Statistics

Classical mathematical statistics offers a different description of what we do (or should do) when we examine data. It seeks to interpret patterns as chance fluctuations. In its most rigid formulation, we decide upon models and hypotheses *before* seeing the data. Then we compute estimates and carry out tests of our assumptions. Classical statistics offers an antidote for some of the problems of magical thinking. However, as a description of what a real scientist does when confronting a real, rich data base, it seems as far off as a primitive ritual.

In some situations the standard errors and P-values of classical statistics are useful, even though the interpretation of "wrong once in 20 times under chance conditions" is not valid or interesting. Long experience has given practitioners a common understanding of the use and usefulness of classical procedures. Some of the surprising things a young statistician learns when working with an experienced applied statistician come in the form of statements like "oh, the overall F test is always significant in problems like this; that's nothing to get excited about. It's the relative size of the t tests for the contrasts that you have to watch." In these situations, standard errors and P-values have evolved as a useful way of communicating. Such uses can be studied; they are most worthy of respect.

Many areas of science and technology—the latter including medicine and agriculture, for example—are cooperative joint ventures of many people and many groups. One of the most important functions that formalized schemes of data analysis can play is to facilitate and deepen communication among these groups and individuals. Once the statistician gives up the hubris of being "the decider," he or she can seize a vital role, helping to make science and technology function better, because groups and people now understand the quantitative aspects of the strengths of each other's results more clearly. The qualitative aspects, frequently even more important, still have to be judged by those who know the field.

In one way classical statistics facilitates magical thinking: sometimes, the "person with the data" looks to "the statistician" as someone who will give the answers and take away the uncertainties, just as the primitive tribesman looks to the local shaman as one who will practice the rituals and take away the sickness. Tukey (1969) discusses the use of classical statistics as a ritual for sanctification.

Scientific Thinking

People have varied views of science. Is it a magnificent structure, erected to last forever, made up of irrefutable results? Work in modern philosophy and history of science argues differently. Thomas Kuhn (1970) argues that scientific revolutions can drastically change the focus of normal science, that old problems and inadequacies are forgotten when the paradigm shifts.

As one extreme, Paul Feyerabend (1978) argues that there are no canons of rationality or privileged class of good reasons. This attitude views science as a thing of tissue and string, somehow limping along despite itself. Surely this is too extreme—science has taken huge steps forward in the past few hundred years and, even though it errs and we cannot neatly say just what it is, clearly has substance.

As the readable overview of Hacking (1983) demonstrates, many occupy positions between the extremes. John Tukey has said that "all the laws of physics are wrong, at least in some ultimate detail, though many are awfully good approximations." Some such middle ground must be appropriate.

In much of science, collection of new data is both possible and practical. Such science proceeds by combining exploration with fresh attempts at confirmation. Repetition by distinct experimenters or observers, gathering independent data, is a hallmark that distinguishes science from magical thinking, although both share a substantial dependence on exploration.

Much of science also falls under John Tukey's label "uncomfortable science," because real repetition is not feasible or practical. Geology and geophysics are often based on nonreplicable observation—fossil plate boundaries seem inadequately preserved to give us good replication of today's plate boundaries. Fortunately, macroeconomics cannot find a second example of the Great Depression. Astronomical observation is often nonreplicable.

Scientific thinking for uncomfortable science is not easy or simple to describe. It depends heavily, when practiced in the best style, on borrowing concepts, insights, and quantities from situations judged to be parallel, at least in specific aspects, to the situation at hand. (Some examples are in remedy 4 of Section 1C.) As a result, changes in theory, which can involve new concepts or insights as well as new mathematical models, can have great impact on the conclusions drawn from a fixed set of observations. A frankly exploratory attitude seems mandatory when working with such data.

This chapter examines some theories of data analysis that fall between the extremes of mathematical statistics and magical thinking. The chapter reviews the empirical work on intuitive statistics, suggests practical remedies for the most common problems, and goes on to review some of the frameworks erected especially for EDA. It then discusses uses for mathematics in EDA, and in conclusion it emphasizes the necessity of exploratory analysis and argues for the usefulness of a controlled form of magical thinking.

1A. INTUITIVE STATISTICS—SOME INFERENTIAL PROBLEMS

The skills, background information, and biases of a data analyst may affect the conclusions drawn from a body of data. People can imagine patterns in data. Sampling fluctuations can produce an appearance of structure. Without a standardized ritual, different investigators may come to different conclusions from the same evidence. This section surveys some studies that quantify these claims.

Drawing Scatterplots

Studies of human perceptual abilities and human judgment shed some light on the subjective element in an exploratory analysis. Let us begin with a study that investigates how choices in drawing a scatterplot affect perception of the association between the two variables being plotted. Cleveland, Diaconis, and McGill (1982) compared two styles of plotting the same data. The first draws the scatterplot so that the points occupy a small part of the

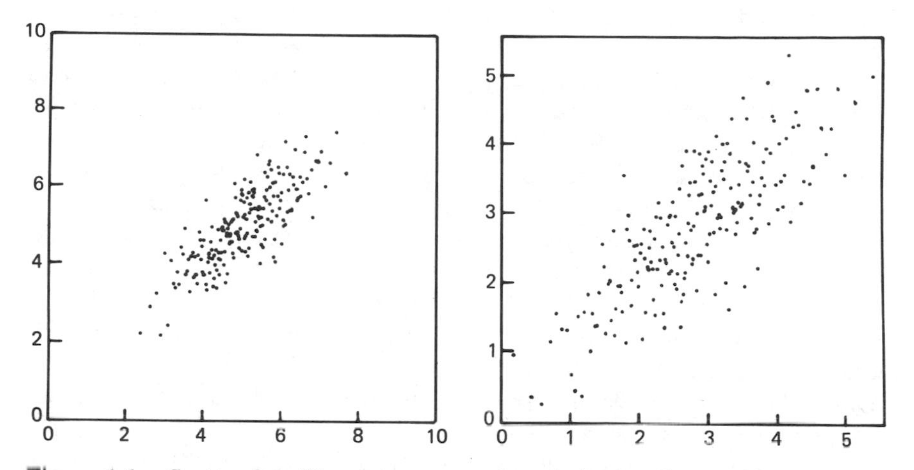

Figure 1-1. Scatterplots illustrating two styles of plotting data. Both scatterplots have $r = .8$. *Source*: William S. Cleveland, Persi Diaconis, and Robert McGill (1982). "Variables on scatterplots look more highly correlated when the scales are increased," *Science*, **216**, 1138–1141 (Figure 1, p. 1139). Copyright 1982 by the AAAS.

total space available. In the second version the points essentially fill the space available.

An example appears in Figure 1-1. Professional statisticians and other scientists with statistical training were asked to judge "how associated the two variables are." Most of the subjects judged a small plot as more associated than a big plot of the same points. The small plots are simply rescaled versions of the big plots; any of the standard measures of association (e.g., the correlation coefficient) is unchanged by rescaling. The subjective effects of rescaling are most pronounced for data sets with correlation between .3 and .8. Rescaling can shift the perceived association by 10–15%. As studies like this become more frequent, we can look forward to a better understanding of our psychological reaction to commonly used graphical procedures and comparisons between competing procedures.

Anchoring and Experimenter Bias

A different branch of psychology investigates how people make judgments under uncertainty. The book by Kahneman, Slovic, and Tversky (1982) assembles the basic experimental papers with extensive discussion. The book by Nisbett and Ross (1980) presents the basic research in a unified fashion, describing many replications and follow-up studies. Here we briefly describe some experiments that illustrate the degree to which preconceptions affect inference.

The anchoring phenomenon, described in Chapter 1 of Kahneman, Slovic, and Tversky (1982), exemplifies the effect of an individual's preliminary judgments. Subjects were asked to estimate various percentages (e.g., the percentage of African countries in the United Nations). For each quantity, a number between 0 and 100 was determined by spinning a wheel in view of the subject. The subjects were instructed first to indicate whether that number was higher or lower than the actual value of the quantity and then to estimate that actual value by moving up or down from the given number. The starting number had a marked effect on estimates. For example, the median estimates of the percentage of African countries in the United Nations were 25 and 45 for groups of subjects that received 10 and 65, respectively, as starting points.

The anchoring phenomenon shows that an irrelevant random start can markedly affect the outcome. When people have a *theory* before encountering evidence, the theory takes precedence. Studies discussed in Chapter 8 of Nisbett and Ross suggest that data (whether they support the theory, oppose the theory, or are mixed) will tend to result in more belief in the original theory than seems normatively warranted.

In one study, Ross and Lepper (1980) presented Stanford University students with two purportedly authentic studies on the deterrent effects of capital punishment. The students had previously indicated that they either strongly believed capital punishment to be a deterrent to potential murderers or strongly believed it to be worthless as a deterrent. In a balanced design, each student read about the results and methods of an empirical study that demonstrated the effectiveness of capital punishment and about the results and methods of a study that demonstrated the lack of effect of capital punishment (the order was randomized).

Subjects found whichever study supported their own position to be significantly "more convincing" and "better conducted" than the study opposing their position. Subjects were asked about their beliefs after reading about only one study. Belief in initial position was strengthened if the study supported the subject's initial position. But belief in initial position changed relatively little if the study opposed the subject's initial position—subjects found flaws in the study design and conclusions in this circumstance. After reading about *both* studies—one that supported their initial position and one that opposed their initial position—the subjects were more convinced of the correctness of their initial position than they were before reading about *any* evidence. The evidence polarized students with differing views instead of bringing them closer together.

Another circle of studies quantifies the effect of experimenter bias. Many experiments have now shown that when experimenters expect certain responses from research subjects, they are substantially more likely to observe what they expect. Rosenthal (1981) summarizes this work. Barber (1976) also provides many quantifications of the effect of experimenter bias.

The literature we have been discussing documents many other pitfalls that can trap intuitive statisticians. Even experienced analysts often believe in the "law of small numbers"; they think that a sample randomly drawn from a population is highly representative—similar to the population in all its essential characteristics [Chapter 2 of Kahneman, Slovic, and Tversky (1982)]. People seem not to pay much attention to sample size, and the difference between a random sample and one selected to prove a point seems unimportant to them [Chapter 4 of Nisbett and Ross (1980)].

Availability and Representativeness

The psychologists who have undertaken these studies have done much more than point to normative aberrations. Often they can explain the behavior of subjects on the basis of a few heuristics that we all seem to use. The two main heuristics, availability and representativeness, each deserve an example.

When using the availability heuristic, we judge population frequencies by the ease with which we can think of an example. For instance, Kahneman and Tversky showed that subjects asked to read lists of well-known personalities of both sexes subsequently overestimated the proportion of the sex with the most famous names.

When using the representativeness heuristic, we attribute great weight to the resemblance or representativeness between the sample in hand and the population. For example, if subjects are asked to assess the relative likelihood of the following three birth-order sequences of boys (B) and girls (G)

$$(1)\ BBBBBB \qquad (2)\ GBBGGB \qquad (3)\ GGGBBB$$

they think that (2) is most likely, followed by (3), followed by (1). If anything, (1) is actually most likely because of the slight preponderance of male births, and (2) and (3) are equally likely. But (2) seems more representative of a random sequence.

Magical Thinking Again

One key difference between thinking and magical thinking is the change in belief brought about by experience. If a theory or hypothesis does not hold up under replication, it should fall by the wayside. The evidence from psychology just presented also shows that learning from experience can be very difficult. Subjects with years of statistical training (including the present author) often err in much the same way as beginning students. This is particularly true if the problem comes up in a real-life situation. The human learning studies cited in the introduction and the Ross–Lepper study on capital punishment make the same point: Once we note a pattern or take hold of an idea, it can be harder than seems normatively warranted to revise our belief after further observation.

Data-Analytic Examples

The studies described above are not so directly related to the daily work of the data analyst. Although suggestive, they are often conducted in carefully controlled circumstances on small, specific tasks. The studies involve a variety of subjects from beginning students to seasoned experimenters. It has been argued that no serious scientist will be misled, in a large way, in a real study. Not with all the checks and balances of normal scientific protocol! Unfortunately, examples of faulty conclusions drawn from snooping about in data are all too easy to find. To illustrate, we mention three major examples.

Uncontrolled observations in medical trials. A doctor or medical team notes a pattern in available patient records and starts making medical recommendations on that basis. Studies surveyed in Bunker, Barnes, and Mosteller (1977) show the importance of—and need for—the elaborate machinery of randomized clinical trials and tests. Without it, optimistic results are reported far too often, and useless or harmful treatments become widely used.

Without such enthusiasms about treatments not yet really tested, however, it seems almost certain that we would be worse off in an important way, since we would not have as many good candidates for randomized trials. Gilbert, McPeek, and Mosteller (1977) have shown that in surgery the fraction of randomized trials declaring the novel technique a success seems to be roughly one-half. Perhaps this fraction is about what it ought to be. (If all randomized trials found successes, these trials would be providing only a ritual stamp of approval!)

Expert testimony on legal cases. Often two groups of statisticians acting as expert witnesses will come to very different conclusions by choosing to focus on different aspects of the same data base or by choosing different techniques.

Modern research in extrasensory perception (ESP). According to both its staunchest advocates and its critics, ESP has yet to produce a replicable experiment in over 100 years of effort. [See Jahn (1982, p. 139) or Diaconis

(1978) for further discussion.] Yet this flourishing field supports half a dozen journals, several dating back 40 years. What can all the discussion be about?

There can be no simple answer, but one recurring theme is snooping about in data. In describing a particularly clear example, Gardner (1981, Chapter 18) details some experiments performed by Charles Tart, a parapsychologist at the University of California at Davis. Tart's experiments involved subjects guessing at 1 of 10 symbols. Some subjects guessed correctly far more often than chance predictions. Statisticians looking at Tart's data discovered a faulty random number generator and a fatally flawed protocol. Furthermore, it was not clear just what aspects of the data were to be looked at—in modern ESP testing, subjects can get credit for correct guesses when their guessing sequences are shifted forward (or back). Clearly replication was called for.

The attempted replication showed no paranormal effect. *However*, Tart spent many pages in the article that described the replication doing data analysis by new and imaginative methods, testing a variety of hypotheses formed after the data were collected. Sure enough, some of these tests *were* significant. Most of the parties involved agree that the new tests are merely suggestive, calling for further tests and data analysis, presumably null results but new patterns. Many similar cases of active data snooping can be found in the ESP journal literature.

In summary, things can go wrong in real studies. The psychologists' work already provides a useful list of potential problems. Experiments show that we can often be trained to recognize and correct these problems.

1B. MULTIPLICITY—A PERVASIVE PROBLEM

Multiplicity is one of the most prominent difficulties with data-analytic procedures. Roughly speaking, if enough different statistics are computed, some of them will be sure to show structure. With the computer, many statistics are forced on us. A few examples will be helpful.

EXAMPLE: PARAPSYCHOLOGY

In a "Ganzfeld" experiment for ESP, a "sending" subject concentrates on one of four pictures. A "receiving" subject tries to discern which of the four pictures is being thought of. The four pictures are then given to the receiving subject, who rank orders them—the picture that best fits the impressions received gets rank 1, and the picture that fits worst gets rank 4.

A simple way of testing for ESP is to count the number of times that the picture ranked 1 was the picture actually used. The chance of being correct is 1 in 4, if chance alone is operating. A typical experiment generates 30 permutations. From binomial tables, 12 or more direct hits constitute a significant result at the 5% level.

Other tests could be suggested. For example, one can count the number of times the correct picture was ranked 1 or 2. Another test can be based on the sum of the ranks given to correct pictures (summed over the 30 trials). There are many other possibilities.

For any *single* statistic, it is easy to compute a cutoff that would be appropriate if that statistic were used alone. In the Ganzfeld literature, cutoffs are chosen so that the chance of a significant result on a *single* test is 5%. Often, however, many such tests are tried on the same set of 30 rankings. Proponents of ESP have counted a study significant if any one of the tests is significant. This can be very misleading.

Hyman (1982) performed a Monte Carlo experiment to assess the effect of multiple testing. Using the three tests described above (direct hit, top two, sum of ranks), Hyman showed that the chance of at least one test significant at 5% is about 12.5%. Taking account of all the tests actually performed, Hyman concluded that the chance of a significant study was about 25%. Hyman examined all 42 of the available Ganzfeld studies and concluded that the proportion of significant studies roughly corresponds to chance expectations. This contrasts sharply with the claims of the proponents of these studies—they compared the proportion of significant studies with 5% and did not allow for multiplicity produced by the number of comparisons in each study.

EXAMPLE: DISPLAY PSYCHOLOGY

Problems of multiplicity are likely to be rampant in any really thorough data analysis. For example, in their study of scatterplots, Cleveland, Diaconis, and McGill (1982) computed a difference between two averages for each of 78 subjects in the study. This gave 78 numbers. If the numbers were "about zero," a certain variable in the study would be judged to have no effect. In analyzing these numbers, many "central values" (e.g., mean, median, biweight) were tried. The final published averages used 10%-trimmed means. For any single average, it is straightforward to derive a significance test. It is less simple, but sometimes feasible, when there are many averages. In this example, replication on new data provided the defense against the problems of multiplicity.

EXAMPLE: LINEAR REGRESSION

Freedman (1983) presents a clear example showing how formal tests can lead us astray if they are applied to data that have been selected as the result of a preliminary analysis. The example involves fitting a linear model. In one simulated case, a matrix was created with 100 rows (data points) and 51 columns (variables). All the entries in this matrix were independent observations drawn from a standard normal distribution, so that, in fact, the columns were independent. The 51st column was taken as the dependent variable Y in a regression equation, and the first 50 columns were taken as the independent variables $X_1, X_2, \ldots, X_{50}$. By construction, Y was independent of the Xs. Indeed the whole matrix consists of "noise."

Freedman analyzed these data in two successive multiple linear regressions. The first phase produced the best linear predictor of Y using $X_1, X_2, \ldots, X_{50}$. It turned out that 15 coefficients out of 50 were significant at the 25% level and one coefficient out of 50 was significant at the 5% level. As a crude model of the results of a preliminary analysis, the 15 variables whose coefficients were significant at the 25% level were used in the second stage, which obtained the best linear predictor of Y using these 15 variables. This time 14 of the 15 variables were significant at the 25% level, and 6 of the 15 were significant at the 5% level. The results from the second pass seem to demonstrate a definite relationship between Y and the Xs. That is, between noise and independent noise.

Freedman verifies that the results reported are about what standard theory predicts. In this stark light, the conclusions seem clear: standard statistical procedures do not behave the way the books describe, when applied to data selected in a preliminary analysis. Yet, if we were shown one such trial in which the data had names and a story accompanied the final variables, making their inclusion seem reasonable, many of us might find this conclusion less clear.

EXAMPLE: MULTIPLE PEEKS AT THE DATA

Multiplicity can be a problem when data come in sequentially, or when a large rich data base is sampled repeatedly or examined in a piecemeal fashion.

For a clear example, consider testing whether the mean of a normal distribution is zero, when the variance is known to be one. The usual 5% test rejects if $|\bar{X}| > 1.96/\sqrt{n}$. Suppose that the n observations are available sequentially and that the usual test is run on the first observation, then again with the first two observations, then again with the first three observations, and so on. The testing stops if any of the tests are significant or if the cutoff of n is reached.

For any single test, the chance of (falsely) rejecting the null hypothesis of zero mean is 5%. When $n = 2$, the chance that at least one of two tests rejects goes up to about 8%. When $n = 5$, the chance that at least one of the five rejects is about 14%; and when $n = 10$, the chance is about 19%. Multiple tests increase the chance of a mistake. The numbers used here are based on approximations derived by Siegmund (1977).

Sequential analysts study ways of adjusting standard tests to account for multiple looks. Pocock (1977) shows that, when $n = 2$, the test that rejects if $|X_1|$ or $|X_1 + X_2|/\sqrt{2} > 2.18$ has about 5% chance of error. In general, one recalculates the sample mean $\overline{X}_i = (X_1 + \cdots + X_i)/i$ after the arrival of each observation and rejects as soon as the standardized absolute value $Y_i = \sqrt{i}\,|\overline{X}_i|$ is too large. The appropriate cutoff is 2.41 when $n = 5$ and 2.56 when $n = 10$.

Again, when spelled out, the conclusion seems clear. In a real scientific situation the problem might arise as follows: On a hunch, we look at an average using a small part of our data. Nothing happens, so we look at more data. If the hunch is strong, we may well look again at all the data.

Beyond the four given above, examples of multiplicity include:

Preliminary data selection and screening.

Comparison of many "treatments," either with a standard or with each other.

Transformation of variables.

Not much work has been done on the effect of the first of these. The second is the problem of multiple comparisons; thorough discussion appears in Miller (1981). Some discussion of the effect of transformation of variables is in Bickel and Doksum (1981).

We next review some of the standard cures for these problems.

1C. SOME REMEDIES

Statistical practice has developed a number of useful ways of dealing with the problems posed by exploratory analysis. This section describes some practical remedies. Section 1D describes some more theoretical approaches.

Remedy 1: Publish without *P*-Values

A number of fine exploratory analyses have been published without a single probability computation or *P*-value. In publishing such a study, an investigator is distinguishing between confirmatory analysis and exploratory analysis. Ideally, the paper would call attention to the potential subjective elements and might even discuss how much data snooping contributed to the final analysis. If confirmatory and exploratory analysis appear in the same document, the two efforts should be clearly distinguished.

Calling attention to the problem causes much of the controversy to go away. Of course, such a study will not be published unless the results are exceptionally informative descriptions of interesting data or are striking and clearly worthy of followup. We now briefly describe some of these studies, both as fine examples and to show that journals do accept interesting data analyses without *P*-values.

EXAMPLE: AIR POLLUTION

In a series of papers in *Science* and other journals, Cleveland and co-workers (1974, 1976, 1976, 1979) and Bruntz et al. (1974) have investigated air pollution in Eastern cities in the United States. Their results are striking. To give one example: ozone is a secondary pollutant, believed to be produced by two primary pollutants "cooking" in the atmosphere. The primary pollutants are lower on weekends than on weekdays. Cleveland et al. (1974) found that, on average, ozone was slightly *higher* on weekends. This suggests that we do not yet understand how ozone is produced. These papers are particularly noteworthy because of the notorious difficulty of working with air pollution data. The available data base is huge. At best, pollution data show great spatial and temporal variability. Worse yet, poor methods of measurement and recording often render such data unreliable. Many groups around the country are trying to fit more-or-less standard statistical models to air pollution data. I think it is fair to say that the Bell Labs group, using exploratory techniques, has triumphed where classical techniques have faltered.

EXAMPLE: ECONOMICS

Economic data often involve many variables, some of them imprecisely measured. Chen, Gnanadesikan, and Kettenring (1974) studied a large number of American corporations. Their main purpose was to determine a fair rate of return on investments for American Telephone and Telegraph Corporation by finding other companies that experience "comparable risk." They considered variables thought to be related to risk, such as variability of stock price, debt ratio, and other standard financial variables, and looked for companies most like AT & T in the values of these variables. The data base contained 60 variables for each of 10 years on each of 4200 companies. Preliminary considerations reduced these to 14 variables for 10 years on each of about 90 companies. The investigators then attempted to find structure, or clusters, for a single year in the 14-dimensional space of companies. Typical conclusions of this phase of the study were that:

1. AT & T falls in the center of "industrials" but in the extremes of "utilities." (This was considered important because in some previous rate cases AT & T had been compared with utilities.)
2. The companies most similar to AT & T generally had a much higher rate of return than AT & T.

The investigators then demonstrated that their preliminary conclusions were reasonably stable over the 10 years of available data.

The AT & T study involved a huge data base and wholesale use of large computers. Slater (1974, 1975) uses graphs and other EDA tools to explore economic data on a smaller scale. Leamer (1978) discusses all of econometric model building from a data-analysis perspective. His book contains numerous other examples of data analysis in economic settings.

EXAMPLE: MEDICINE

The main conclusions of Reaven and Miller (1979) flow from a single picture. Their study examines the relationship between chemical diabetes and overt diabetes. Diabetes has been considered a homogeneous disorder, caused primarily by the body's failure to secrete enough insulin. This view suggests that the data should show a smooth continuum ranging from patients with the mildest discernible degree of glucose intolerance through patients who depend on insulin from outside sources to prevent death. For each of the 145 adult subjects in the study, five variables were measured. The variables, somewhat crudely described, are (1) relative weight, (2) a measure of glucose tolerance, (3) a second measure of glucose tolerance called *glucose area*, (4) a measure of insulin secretion called *insulin area*, and (5) a measure of how glucose and insulin interact, abbreviated *SSPG*.

The data were put onto the PRIM-9 graphics system at the Stanford Linear Accelerator Center. The system makes "scatterplots" of three variables showing the points on a video screen; when the display is "rotated," the point cloud moves in such a way that points closest to the viewer move in one direction, while those furthest away move in the opposite direction. Parallax fools the eye into seeing three dimensions. The system also allows investigators to change the three-dimensional space being viewed into any projection involving three orthogonal linear combinations of the five variables.

Figure 1-2 shows an artist's rendition of one of the views found. The picture shows a fat middle and two wings. The middle contains normal patients, and the two wings contain patients with chemical diabetes and

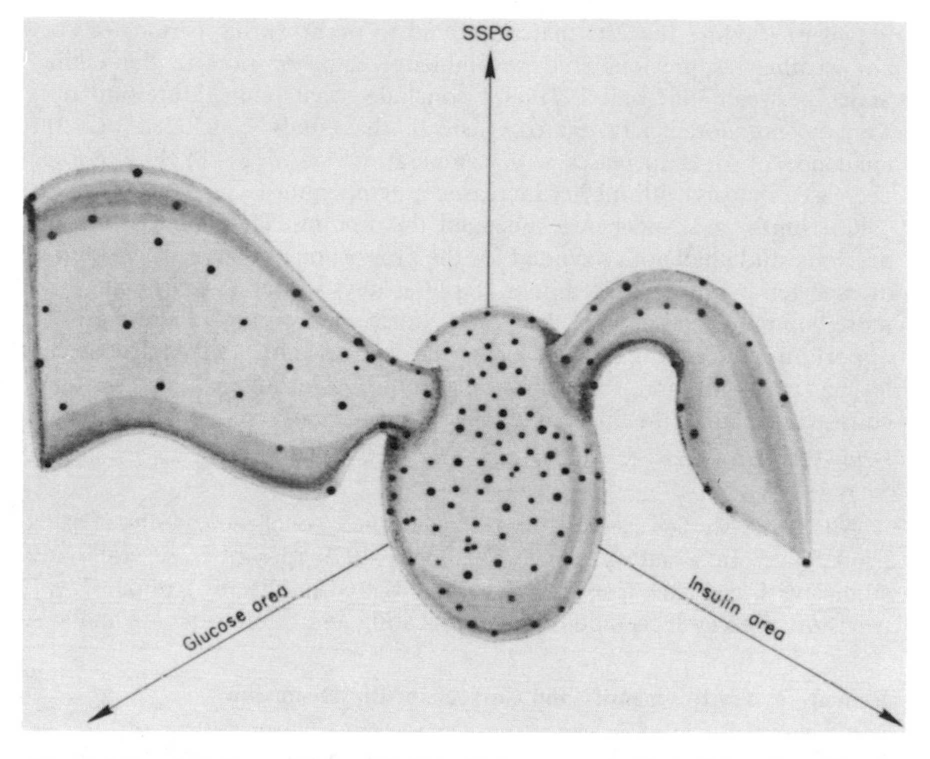

Figure 1-2. Artist's rendition of diabetes data as seen in three dimensions. View is approximately along the 45° line as seen through the PRIM-9 program on the computer; coordinate axes are in the background. *Source*: G. Reaven and R. Miller (1979). "An attempt to define the nature of chemical diabetes using multidimensional analyses," *Diabetologia*, **16**, 17–24 (used with permission of Springer-Verlag).

overt diabetes, respectively. No occupied path leads from one form of diabetes to the other, except through the region occupied by the normal people. This suggests that the usual "smooth transition" model is wrong; indeed, patients with chemical diabetes seldom develop overt diabetes. Reaven and Miller draw further conclusions and suggest followup studies that are now under way. In further graphical analysis of this data set, Diaconis and Friedman (1983) show that a method of drawing four-dimensional scatterplots reveals some additional structure that relates the three variables in Figure 1-2 to relative weight.

EXAMPLE: PSYCHOLOGY

Carlsmith and Anderson (1979) give a nice example of exploratory techniques (and common sense) overturning a finding bolstered by *P*-values.

Prevalent folklore suggests that riots tend to occur during periods of very hot weather. A previous study examined 102 major riots in the United States between 1967 and 1971 and concluded that temperature and riots were *not* monotonically related. Instead, that study concluded that the incidence of riots increases with temperature up to 81–85°F and then decreases sharply with further increases in temperature.

Carlsmith and Anderson challenged this finding. They argued that the previous study had not accounted for the proportion of days during summer in each temperature range. For example, if days in the 81–85°F range are more common than days in the 91–95° temperature range, there are more opportunities for riots in the former range. Carlsmith and Anderson collected fresh data and, using a variety of EDA techniques, demonstrated convincingly that the likelihood of a riot increases monotonically with temperature.

All of the studies sketched above reach their conclusions without using probability. In a survey article, Mallows (1979) describes several other studies with the same features. We conclude that publishing without P-values is often a viable method of communicating an exploratory data analysis.

Remedy 2: Try to Quantify and Correct for the Distortion

There are several techniques for analyzing how a standard statistical procedure behaves on data selected as "interesting" by an exploratory data analysis.

Specific Allowance for Multiplicity

Problems of multiplicity are reviewed in Section 1B. Miller (1981) describes a broad range of techniques for dealing with multiple comparisons. One of the most useful techniques has turned out to be a simple system of inequalities called the Bonferroni inequalities. These generalize the fact that the probability of a union is no greater than the sum of the individual probabilities; thus the chance that at least one test is significant cannot exceed the sum of the chances that the individual tests are significant.

Should we correct for multiplicity? After all, we have been practicing data analysis with P-values for many years. If the consequence of the average amount of distortion introduced by skilled workers had been too bad, we can expect that practices would gradually have changed, just as the comparison of the variability of two samples by an F test appears only in books and papers by innocent authors. This line of thinking suggests a dichotomy:

> For routine analysis in established fields, we want to keep the impact of multiplicity close to where it has traditionally been in the practice of each field's leaders.
>
> With no tradition, or with multiplicity far in excess of tradition, adjustment seems mandatory.

Bayesian Quantification and Bode's Law

Many of the problems addressed in this chapter can be approached by using tools developed by Bayesian statisticians. A book-length review of this literature is given by Leamer (1978), who also presents a theoretical framework for evaluating data-analytic studies on nonexperimental data. We review this framework in Section 1D.

To illustrate the Bayesian approach to the problems associated with data analysis, we describe one example: the astronomical relation known as Bode's law. Leamer discusses this in Section 9.5. Bayesian calculations for Bode's law have also been described by Good (1969) and Efron (1971). The following account draws on all three sources.

In 1772 J. E. Bode gave a simple rule for the mean distance of a planet from the sun as a function of the planet order. Let d_n be the distance from the sun to the nth nearest planet. Bode's law predicts that

$$d_n = 4 + 3 \times 2^n.$$

For the first eight planets, it gives: 4, 7, 10, 16, 28, 52, 100, 196 (using $n = -\infty, 0, 1, 2, \ldots, 6$). The seven planets known in 1800 had mean distances of 3.9, 7.2, 10, 15.2, 52, 95, 192—the units have been chosen so that the distance of the earth to the sun is 10. The law certainly seems to do well, aside from a missing planet 28 units from the sun. This led a group of astronomers to search the heavens at roughly 28 units from the sun. They found a "planet," actually the asteroid series. Clearly the predictive success of Bode's law adds to its believability.

The obvious question is whether Bode's law is real—in the sense that planets around other stars show a similar geometric relationship. An alternative to "reality" is that Bode's law is a numerological artifact, the result of fooling around with simple progressions. After all, the comparison involves only eight numbers. A sequence like $a + b \times c^n$ has three free parameters. If the planetary distances were approximately in the proportions 1, 2, 5, 10, 17, ..., it seems certain that Bode or one of his colleagues would have noted this and proposed a law of the form $a + b \times n^2$. There are many further possibilities.

Although difficult and different, the problem of testing "the reality of Bode's law" is not completely intractable. The basic idea is to describe the steps in the data analysis, set up a way of generating data, and see how often a "law" that fits as well can be found. All the steps are difficult, but perhaps not impossible. Quantifying the steps in the data analysis is perhaps the most difficult phase. Still, even thinking of alternatives is a useful exercise. As to generating more data, the simplest approach is to find a new sample (e.g., data from another solar system). Failing this, a simple mathematical generating mechanism may be tried.

Good suggested the following mechanism for testing B: "Bode's law is true" versus $\bar{B}$: "Bode's law is not true."

Under B the logarithms of the planetary distances are independent Gaussian with mean $\log(a + b \times 2^n)$ and common variance. Thus, on a log scale, the data approximately obey Bode's law.
Under $\bar{B}$ the logarithms of the planetary distances are like the ordered lengths between points dropped at random into an interval.

Good carried out the testing in a Bayesian framework, putting prior distributions on the parameters involved. He concludes that the odds are 30^4 to 1 in favor of Bode's law. Efron offers a criticism of Good's formulation and various alternatives. Using different models for B and $\bar{B}$, he concludes that the odds for "the reality of Bode's law" are roughly even. This is a big shift from Good's 30^4 to 1, but it still suggests that Bode's law should be taken seriously. Both authors discuss how similar analyses might be carried out for testing the validity of other simple laws.

The ideas introduced by Good, Efron, and Leamer are fresh and deserve further trials on more down-to-earth examples. They offer one potential route through the difficult maze of quantifying the results of a truly exploratory analysis.

Remedy 3: Try It Out on Fresh Data

Replication on fresh data, preferably by another group of experimenters, is a mainstay of "the scientific method." If done skillfully, replication can eliminate all difficulties with extensive exploration.

EXAMPLE: BIRTHDAY AND DEATHDAY

Here is an example of a published, purely data-analytic finding that seems nonreplicable. In "Deathday and Birthday" David Phillips (1978) found that people's birthdays and time of death are associated, more people dying just after their birthday than just before. The effect seemed more pronounced among famous people. The evidence consisted of some averages and graphs.

I put Phillips' findings to a test during a course on sample surveys at Stanford University during the winter quarter of 1980. Thirteen students in the course tested Phillips' claim on new data as part of their final project. The students read Phillips' article, each designed a test statistic, and then each took a sample from a different source, such as an almanac. Without exception, each student's conclusion disputed Phillips' hypothesis. Some further analysis of the birthday–deathday problem, again rejecting Phillips' hypothesis, is in Schulz and Bazerman (1980).

Although it copes with exploration, replication may not cure other, nonstatistical, ills that science occasionally suffers. Blondlot's N rays [see Klotz (1980) for a modern account] is a well-known case of a widely repeated experiment that turned out to be a complete artifact. Some further examples are described in Vogt and Hyman (1979, Chapter 5).

As emphasized in the discussion of scientific thinking in the introduction to this section, there are many branches of science, but few of technology, where "new data" are impossible to obtain. Remedies 4, 5, and 6 give some forms of partial replication that can be used when new data are hard to come by.

Remedy 4: Borrowing Strength

When direct repetition is not feasible, we often turn to more or less parallel situations. The results of their analyses can then be *borrowed*, to add *strength* to the result of analyzing the data of our prime concern. Clearly such borrowing has to be done carefully and in the light of the best subject-matter insight available. Equally, it often is the best available way to support—or challenge—the apparent results of our prime analysis. Less obviously, as discussed in Section 1E, it can play an important part in protecting us from real dangers of a still more subtle kind.

A familiar example arises in the analysis of variance, where different groups are assumed to have different average values, but common variability about each group's average. Under this assumption we can use all the data to estimate the common variability—thus borrowing strength—and wind up with narrower confidence intervals for each group's average.

A second example of borrowing strength comes from the literature on combining tests of significance. Sometimes studies indicate a trend, but the indication is not striking enough to be judged significant. If several such studies show a trend in the same direction, the significance tests can be combined and may result in a significant judgment. A convenient reference to this literature is Rosenthal (1978).

A third example, illustrating the difficulties of borrowing strength, arises in consideration of animal test data for problems like human sensitivity to radiation or chemicals. Such extrapolation is widely used and often controversial in these difficult problems. Scientists argue about the suitability of data taken on specially bred, supersensitive, hairless mice. In some instances changing the test animal can change estimates of "safe levels" by several orders of magnitude. Animals are different from humans, and they may react when most humans would not.

Systematic study of these issues is actively under way. As one example, Bartsch et al. (1980) compared the results of quick, cheap, tests for mutagenicity of chemicals (on bacteria and hamster cells) with more extensive animal trials of the same chemicals. The bacterial tests (on 89 chemicals) correctly identified 76% of the known carcinogens as mutagenic and correctly identified 57% of the known noncarcinogens as nonmutagenic. Of the chemicals identified as mutagens, 95% were known carcinogens. Although encouraging these results still leave us worried about missing a wonder drug or failing to detect a truly lethal substance.

A final example of borrowing strength involves the informal judgments we make all the time, as when we use someone's past performance on a variety of tasks to help judge whether his or her future work will be excellent. Tools for quantifying these more subjective judgments can be found in the literature on subjective probability under the label "combination of evidence." Diaconis and Zabell (1982) and Shafer (1976) provide references and discussion.

Remedy 5: Cross-Validation

In many problems, such as Bode's law, replication is not a practical possibility. One compromise, which works for large data sets, is what statisticians call cross-validation. The idea is to take a (random) sample from a data set, use exploratory techniques on the sample, and then try the result on the rest of the data. Mosteller and Tukey (1977) and Efron (1982) contain further discussion. For a careful application in the field of geomagnetism, see Macdonald and Ward (1963).

Remedy 6: Bootstrapping the Exploration

Efron (1979) describes a simple, widely applicable method called the "bootstrap" for assessing variability in a data analysis. The idea is to use the available sample as a model of the population and draw further samples from the available sample to assess variability. Efron and Gong (1983) and Gong (1982) put the bootstrap to work at the exploratory data analysis level.

In one example a medical investigator had a "multiple discrimination" problem. He had studied 180 patients with a liver disease. For each patient, he measured about 20 variables—age, blood pressure, etc. He also knew whether the patient had lived through the disease or died during the course of the disease. His goal was to investigate the association of the measured variables with the end result. He was seeking a simple rule to predict the outcome from the measured variables. Through a rich data-analytic process he found a rule that seemed to work—would it work on new data?

The investigator first performed preliminary exploratory analysis to reduce 20 measurements per patient to 5. He then used a standard technique called logistic regression to get a rule for predicting outcomes from the five measured variables. It turned out that only three of the five variables really mattered in the prediction rule. The investigator was working at understanding the chemistry of these variables and the effect of changing the levels of some of them in the hope of saving patients.

To investigate the stability of the analysis, Efron and Gong applied the bootstrap. They regarded the set of 180 patients as a population, and they drew a new sample of 180 from this population *with replacement*. Then, by carefully quizzing the investigator and his assistants, they replicated the entire data analysis: the preliminary data screening, initial variable selection, logistic discrimination, and final selection of variables. They went through this procedure 500 times. Among many other findings, they noted that no single measured variable appeared in more than half of the 500 final sets of predictor variables. This amount of instability suggests caution in taking any single predictor very seriously.

The exercise just described took a massive amount of computing and months of work. It probably results only in a very crude approximation to the effect of rooting about in the data. Nevertheless, it represents a significant breakthrough in an extremely hard problem.

Remedy 7: Remedies to Come

As the techniques of EDA become more widely recognized, and used, we can look forward to more research aimed at aiding the analyst in avoiding self-deception. Here are two suggestions. In the middle of an interactive data-analysis session it might be useful to have available a display of random, unstructured noise. This should come in a form close to the data being examined. It is easy to imagine structure, dividing lines, and outliers in scatter plots of samples from uniform distributions.

A second suggestion for research is systematically to study adaptations or specializations of the heuristics and biases described by Tversky and Kahneman (1974) along with Nisbett and Ross (1980) to the situations encountered in routine data analysis.

1D. THEORIES FOR DATA ANALYSIS

Suitable theories of inference and the mathematics of probability underlie many, nowadays routine, applications of statistics. These include randomized clinical trials, sample surveys, least-squares fits to underlying models, quality control, and many other examples. Yet, none of the classical theories of statistics comes close to capturing what a real scientist does when exploring new data in a real scientific problem. All formal theories—Neyman–Pearson, decision-theoretic, Bayesian, Fisherian, and others—work with prespecified probability models. Typically, independent and identically distributed observations come from a distribution supposed known up to a few parameters. In practice, of course, hypotheses often emerge after the data have been examined; patterns seen in the data combine with subject-matter knowledge in a mix that has so far defied description.

This section surveys some alternatives to the more usual theories of inference. These alternative frameworks have been created as a step toward foundations for exploratory data analysis. Two useful categories emerge: theories that explicitly avoid probability and theories that depend on subjective interpretation of probability.

Probability-Free Theories

The classical approaches to comparing and interpreting statistical techniques depend on probability. Finch (1979) and Mallows (1983) have started to develop a theory of data description that does not depend on assumptions such as random samples or stochastic errors. They offer a framework for comparing different summaries and methods for assessing how much of the data a given description captures.

An example adapted from Mallows (1983) treats inaccuracy in a nonstochastic setting. We begin with a descriptor δ that takes data sets x in a space $\mathscr{X}$ into descriptions in a space $\mathscr{D}$. Informally, the inaccuracy of δ measures how closely the description approximates the data. Formal treatment requires a measure of distance $d(x, y)$ between two data sets x and y. The *inaccuracy* $i(\delta, x)$ of the descriptor δ of the data set x is defined as the distance between the most distant data sets y and z with the same description as x:

$$i(\delta, x) = \max_{y,z} \{ d(y, z) | \delta(y) = \delta(z) = \delta(x) \}.$$

Observe that, with this definition, $i(\delta, x) = \infty$ means that some data sets have the same description as x and yet are arbitrarily different from x, whereas $i(\delta, x) = 0$ implies that x is uniquely described by δ.

For an example, consider the inaccuracy of a description of the batch $x_1, x_2, \ldots, x_n$ of real numbers. Let $x_{(1)} \leq x_{(2)} \leq \cdots \leq x_{(n)}$ denote the ordered values. One reasonable choice of the distance between two batches uses the sum of squared differences between the corresponding ordered values:

$$d^2(x, y) = \frac{1}{n} \sum (x_{(i)} - y_{(i)})^2.$$

Bickel and Freedman (1981, Appendix) discuss some properties of this measure of distance. It is the Mallows metric between the empirical distribution functions of x and y: If F and G are any two distribution functions, the Mallows metric between them is

$$d^2(F, G) = \int_0^1 [F^{-1}(t) - G^{-1}(t)]^2 \, dt.$$

Bickel and Freedman also discuss appropriate generalizations (Vassershtein metrics) for multivariate data. Table 1-1 gives the squared inaccuracy $i^2(\delta, x)$ for several common descriptors of a batch based on this distance. The entries in the table are straightforward to derive. For example, the inaccuracy for the median is ∞ because there are data sets with the same median but arbitrarily different order statistics.

For the endpoints descriptor, the maximum i^2 is achieved by choosing one data set with $n - 1$ values equal to $x_{(1)}$ and one value equal to $x_{(n)}$ and the second data set with $n - 1$ values equal to $x_{(n)}$ and one value equal to $x_{(1)}$. The derivation of the inaccuracy for the other descriptors is similar.

Among the summaries based on order statistics, the median is least accurate, followed by the 3-number and the 5-number summaries, followed by the empirical cumulative distribution function (ECDF). For roughly symmetric batches and large n, the 3-number summary is almost four times as accurate as the endpoints summary. Under similar assumptions, the 5-number summary is about four times as accurate as the 3-number summary.

TABLE 1-1. Squared inaccuracy i^2 of common batch descriptors; the distance measure is the average squared difference between the corresponding order statistics.

Descriptor δ	Description $\delta(x)$	Squared Inaccuracy $i^2(\delta, x)$
Median	Q_2	∞
Endpoints	$(x_{(1)}, x_{(n)})$	$(x_{(n)} - x_{(1)})^2\left(\frac{n-2}{n}\right)$
3-number summary[a]	$(x_{(1)}, Q_2, x_{(n)})$	$\{(x_{(n)} - Q_2)^2 + (Q_2 - x_{(1)})^2\}\left(\frac{n-3}{2n}\right)$
5-number summary[a]	$(x_{(1)}, Q_1, Q_2, Q_3, x_{(n)})$	$\{(Q_1 - x_{(1)})^2 + (Q_2 - Q_1)^2 + (Q_3 - Q_2)^2 + (x_{(n)} - Q_3)^2\}\left(\frac{n-5}{4n}\right)$
Mean	$\bar{x}$	∞
Mean and standard deviation	$(\bar{x}, s)$	$s^2\frac{2(n-2)}{n}$
ECDF	F_n	0

[a] For simplicity the expression for $i^2(\delta, x)$ assumes that Q_2, Q_1, and Q_3 are single order statistics (e.g., for the 3-number summary, n is odd).

Source: Adapted in part from Mallows (1983, Table 2, p. 142).

TABLE 1-2. Numbers of men and women accepted and rejected among applicants to a large department at the University of California–Berkeley, for graduate study in academic year 1973–1974.

	Accepted	Rejected	Total
Men	54	137	191
Women	94	299	393
Total	148	436	584

Source: D. A. Freedman and D. Lane (1983). "Significance testing in a nonstochastic setting." In P. J. Bickel, K. A. Doksum, and J. L. Hodges, Jr. (Eds.), *A Festschrift for Erich L. Lehmann*. Belmont, CA: Wadsworth International Group (Table 1, p. 187). Reprinted with permission of Wadsworth, Inc.

Another property of a description that does not depend on probability is the *breakdown point*. Informally, the breakdown point of a description is β if we can change the description an arbitrary amount by changing $n\beta$ data points (but not fewer). Thus, in large samples, the mean has breakdown point 0, the median has breakdown point $\frac{1}{2}$, and γ-trimmed means have breakdown point γ. It is possible to develop a reasonable amount of theory from this idea, as in Mallows (1983) and Donoho and Huber (1983).

The examples just described are a beginning of a probability-free language for descriptions. Finch (1979) has applied this language to a kind of inference. Working with a finite set Δ of descriptors, he defines the *descriptive power* of δ at x as the proportion of descriptors in Δ that describe x less accurately than δ does. Finch applies this approach to two-sample problems and 2×2 tables. The ideas are closely related to the nonstochastic interpretation of significance testing developed by Freedman and Lane (1983a, 1983b).

To illustrate, we use data collected as part of an investigation into the possibility of sex bias in graduate admissions at the University of California–Berkeley. Table 1-2 gives the numbers of men and women accepted and rejected among all applicants for admission to one of Berkeley's largest departments for the 1973–1974 academic year. The acceptance rate for men applicants was 28%. For women, it was 24%. Is the difference between 24% and 28% real? One approach is to compute a chi-squared test for independence. For this table, the significance probability is about .26, suggesting that the difference could easily be accidental. The usual interpretation of the chi-squared test depends on assumptions such as random sampling from larger populations. These are at best imprecise here—the data derive from the inherently nonrepeatable graduate admissions process of 1973–1974—they are not a sample from any reasonably clearly defined population.

Freedman and Lane (1983a) offer an interpretation for the significance probability .26: suppose we divide the pool of applicants into two groups according to whether they are right- or left-handed, or whether their last name begins with A–L or M–Z, or just "at random" (so that no name like "sex" or "handedness" can be associated to the division). For each such division of the applicants into two groups, we could form a table, as in Table 1-2, by counting those admitted and those not admitted in each group. For each table, a chi-squared test can be performed. Freedman and Lane prove that the proportion of divisions for which the observed chi-squared statistic is larger than the statistic for the original division by sex is about .26.

Their result gives a way to interpret the significance probability .26—it is the proportion of divisions showing an inequity at least as large as the original division. Clearly most of these divisions must be regarded as irrelevant to the admissions process. Thus the 4% difference between the proportions of male and female applicants accepted for admission is not unusual. For a fascinating analysis of the full set of data underlying Table 1-2, see Bickel, Hammel, and O'Connell (1975). If the division into two groups is taken as a description (e.g., percent males, percent left-handed), their approach is similar to the approach of Finch—they count the proportion of descriptions with a more extreme value of the goodness-of-fit criterion. Freedman and Lane also developed a similar rationale for the usual tests in multiple linear regression.

Martin-Löf (1974) developed a model-free interpretation for some of the same test statistics based on counting. The motivation and final results are similar to those of Mallows, Finch, and Freedman and Lane. We can look forward to a more unified theory as these accounts expand and merge.

Bayesian Theories

Subjective judgments and background knowledge underlie many of the steps taken during an exploratory analysis. The subjective Bayesian approach to statistics seems a natural tool to apply to EDA. In the subjective approach, probability represents the investigator's state of knowledge. Probability changes, after contact with data, via Bayes' theorem. Connections between Bayesian statistics and EDA are discussed by Box (1980),

Dempster (1983), Good (1983), and Leamer (1978). This section discusses some implications of their work. It ends with an explanation of why the Bayesian approach has had little impact on the practice of EDA.

Dempster (1983) suggests that the difference between EDA and statistical modeling is often exaggerated. Although statistical model building makes use of formal probability calculations, the probabilities usually have no sharply defined interpretation (either frequentist or Bayesian), and the whole model-building process is simply a form of exploratory analysis. Dempster notes that both EDA and statistical modeling cycle back and forth between fitting curves and looking at residuals. EDA generally starts with summaries and displays, whereas the modeling approach starts by fitting a curve. If the cycling back and forth is carried out diligently and skillfully, the starting place might not have a great effect on the final reported results. An empirical test of this point would be fascinating.

This leads Dempster to suggest fitting truly large models in the exploratory phases—"a model as big as an elephant" as Jimmie Savage is reported to have said. Such a model might well have as many parameters as data points; and Bayesian methods represent a possible way to make parameters identifiable. Dempster goes on to compare the approach just outlined (using probability) with nonprobabilistic approaches.

Good (1983) does not attempt to present a full-blown theory of data analysis. He describes some of the things that such a theory must cover [Mallows and Walley (1981) do this provocatively as well]. Good emphasizes that such a theory will involve psychology, in that techniques of description and display must be designed to match salient features of a data set to human cognitive abilities. For Good, EDA is concerned mainly with the *encouragement* of hypothesis formulation. Good discusses the problem of distinguishing a pattern in the data from a mere coincidence. He argues that it is possible to judge that a pattern has an underlying explanation *even if we are unable to specify the explanation sharply*. This suggests that patterns with extremely small prior probability of being potentially explicable (given the context) will be discarded. Patterns with an appreciable prior probability of being potentially incorporated in a useful hypothesis for explaining the data will be displayed and labeled "salient features of the data."

One quantification of "explicativity" can be based on subjective probability. For example, if H is a hypothesis about the data ("this drug really has an effect" or "the experimenter was cheating"), then $P(H|G)$ denotes the investigator's prior probability of H based only on the background information G. If E is an event or pattern ("80 of 100 patients recovered" or "the successful trials are at prime numbers"), then $P(E|H,G)$ denotes the likelihood that E occurs, given G and H.

Informally, a good explanation of an event or pattern E is a hypothesis H such that $P(E|H,G) \gg P(E|G)$ but $P(H|G)$ is not too small. Good offers a more precise quantification of explicativity as

$$\log P(E|H,G) - \log P(E|G) + \tfrac{1}{2}\log P(H|G).$$

The difference between the first two terms measures how much H increases the likelihood of E. The third term will be a large negative number, thus decreasing explicativity, if the prior probability of H is very small. Although the factor $\frac{1}{2}$ could properly be replaced by any other small constant, Good gives an argument to justify it. Such numerical quantification, however crude, seems mandatory if data analysis is to be automated.

Good's reflections lead to a number of useful practical suggestions. He considers techniques for reducing dimensionality—such as principal components, factor analysis, and multidimensional scaling; these have generally been used to represent high-dimensional data in two dimensions. Good suggests using these techniques to reduce to six or seven dimensions and then using devices like projection pursuit or color graphics to explore the reduced version of the data.

The book by Leamer (1978) represents an attempt to deal systematically with the problems encountered in exploratory analysis. The subtitle "ad hoc inference with nonexperimental data" suggests the book's two themes. First, the book attempts to deal with the "data mining, fishing, number crunching" side of statistics. Second, the techniques are intended for application to data collected in a wide variety of ways, not necessarily as part of a designed, randomized trial. Leamer realizes that the classical version of Bayesian statistics, which requires fully specified probabilistic assumptions, is not particularly relevant to guiding an exploratory analysis. He suggests that the Bayesian approach is sufficiently flexible to permit the main ingredients of an exploratory analysis to be legitimate or at least understandable in a Bayesian framework.

Most of Leamer's examples involve regression. Topics covered in detail include "data selection searches," in which the data in hand are used to detect and investigate deviations from initial models (or simple descriptions); "post data model construction," in which models, or theory, are worked up after seeing data; and "simplification searches," in which a larger model is "pruned down" to allow findings to be summarized and communicated. Many of the problems are illustrated on econometric data. Leamer is quite modest about the accomplishments of the theory he sets forth. Nevertheless, he has constructed a reasonably complete framework to model the meanderings of a real data analyst. It offers a baseline for comparison and improvement.

To conclude this section, we consider two challenges that EDA poses to Bayesian techniques. First, exploration is important in problems where nothing is unknown. Consider a complete list of a population—like the records of all employees in a company or a list of features of the world's major rivers. Here nothing is unknown or random—a complete, accurate list is assumed available. Still, a great deal of useful data-analytic work is possible in the form of simple descriptions and summaries. Because there are no unknown quantities (or parameters), no guidance for the data-analytic investigation can be expected from the usual models of Bayesian (or frequentist) statistics.

To discuss the second challenge, let us review the subjective Bayesian approach to inference. First, probability represents an investigator's state of knowledge. Second, probability changes, after contact with data, via Bayes' theorem. Tukey (1972, pp. 61–63), Mallows (1970), and Savage (1970) have discussed the limitations of this approach. Here are two of Tukey's objections: "the discovery of the irrelevance of past knowledge to the data before us can be one of the great triumphs of science." Furthermore, "Bayesian techniques assume we know all the alternate possible states of nature."

Both of these objections seem cogent. They link up with recent work in Bayesian statistics that suggests that Bayes' theorem is not the only way that probability shifts in response to new evidence. In surveying this recent work, Diaconis and Zabell (1982) describe several other methods of changing an initial probability assessment. In addition to Bayes' rule and complete quantification, Jeffreys' rule—a generalization of Bayes' rule—permits part of the previously quantified probability to be retained.

Bayes' rule will most often be the route to probability change in situations where a lot of experience has dictated the relevant variables. In exploratory situations such experience is usually not available. The impact of new data may well be to remind us of numerous variables not previously contemplated. Then Bayes' rule does not apply, and one of the alternative methods of changing probability must be used. Faced with many complex shifts in probability, the Bayesian data analyst may well decide to abandon the usual Bayesian machinery until the situation has stabilized long enough to make such calculations cost effective.

1E. USES FOR MATHEMATICS

Most of the discussion in this chapter has pointed to the difference between exploratory techniques and the techniques of probability and mathematical statistics. This section shows how the tools developed for classical statistics can lead to insight in EDA. Of course, much of the material in UREDA and in other chapters of this volume has this emphasis.

An important teaching of EDA is to try out several summaries, possibly some new ones. This process necessarily leads, in due course, to a wealth of alternative techniques. (Many of these techniques will not hold much interest outside the specific context for which they were created.) This diversity leads some to say that mathematics is not useful for EDA (because they recognize that the formative, data-specific stages are the heart of EDA).

As shown in UREDA and this volume, many exploratory techniques are broadly useful and merit closer study. Mathematical methods can be useful at this stage. Mathematics can help in fine tuning, finding pitfalls, and choosing the better of several exploratory devices.

Careful studies of EDA techniques have made heavy use of the Monte Carlo technique in the spirit of "try it in a problem where we know the answer and see what happens." Experimental sampling on a computer is a mathematical activity, whether it involves simple repetition or a sophisticated combination of importance sampling and other variance-reduction techniques.

Computational mathematics often tells us "less about more," in that the results are specific and approximate, although we have great flexibility as to what we may study. Theorem-proving mathematics usually tells us "more about less," in that we come to exact results, or remarkable approximations, but often with severe restrictions on what they apply to (populations may have to be Gaussian, or sample sizes may have to be unrealistically large).

The two approaches can be combined with great effect. It seems mandatory to check asymptotic theorems by computational methods to see whether their conclusions hold in cases of practical interest. Similarly, simulation studies, even if the examples span a broad range, are limited by the imagination of the investigator. It seems mandatory to have some theoretical backup to help organize and verify the computational results. (Many a Monte Carlo study has all numbers "off" through a programming error.) The following example shows how theorem-proving mathematics can add insight to a technique that has been very thoroughly studied by simulation.

Tukey's biweight (discussed in Chapter 11 of UREDA) is a popular robust estimator of the center of a batch of data. The estimator works with data standardized by a "tuning constant" times a scale estimate (usually the MAD—the median of the absolute deviations about the median). A natural question is: Which tuning constants are good, or does it matter? Monte Carlo work on this question proceeds by trying the estimator in situations where the correct answer and form of error terms are known; the tuning constant is chosen so that the estimator behaves well in these situations. Based on such studies, tuning constants between 4.8 and 9.1 are in use as of this writing.

Freedman and Diaconis (1982) asked a sharp mathematical question of the biweight: Is it consistent? That is, suppose a large sample is taken from a population that is symmetric about zero (say). Suppose too that the population density is positive and smooth about zero. Compute the biweight estimate of the center of symmetry. Will this estimate be close to zero? For most location estimators, like the median or trimmed means, the answer is yes. For the biweight, the answer depends on the tuning constant. Freedman and Diaconis show that, for constants smaller than 5.4, the answer is no. The results generalize: In the language of Chapter 11 of UREDA, any M-estimator based on a redescending ψ-function can be inconsistent if the tuning constant is too small.

Consistency and inconsistency are tied to increasing sample size. The practical statistician may wonder what these results say about sample size 20 or 100. A rule of thumb is that, if a procedure behaves badly in large samples, it will have similar, objectionable features for moderate sample sizes. In the case of the biweight, the moderate-sample phenomenon is that, for small tuning constants, the biweight is quite tricky to compute numerically. It may oscillate among several answers for a given data set; a slight change in numerical technique can produce a different answer.

The Monte Carlo investigation of the biweight has been going on for about 10 years. It ranks high as a model of thoroughness for the computational approach. Nonetheless, the theorem-proving investigation pointed to a new direction: the counterexamples to consistency are symmetric densities that are multimodal. None of the numerical trials has yet considered such examples. Presumably such cases are rare in practice.

The moral is clear—"In union there is strength." We need to use *both* computational mathematics *and* theorem-proving mathematics to guide the growth of our new techniques and insights. This conclusion runs through much of the best statistical practice. Indeed, Student's first derivation of a distribution for Student's t combined experimental sampling with the method of moments to fit one of Karl Pearson's frequency curves. Nearly a decade elapsed before R. A. Fisher provided a formal proof. Empirical and theoretical work have continued through the years. Their union provides a very thorough understanding of the usefulness and validity of Student's t.

Many other novel techniques have yielded to mathematical analysis. In addition to robust techniques and other examples in EDA, the bootstrap [Efron (1979), Bickel and Freedman (1981)], cross validation [Efron (1982)], projection pursuit [Friedman and Tukey (1974), Diaconis and Freedman (1984)], and recursive partitioning [Breiman et al. (1983)] are worth mentioning. Each of these began as a novel technique which did not fit neatly within standard statistics. As the techniques became more widely used, they have been successfully analyzed using both kinds of mathematics.

1F. IN DEFENSE OF CONTROLLED MAGICAL THINKING

Magical thinking couples the inclination to seek meaningful connections and interpretations with the disinclination to learn from experience. Our task is to enhance the first and control the second.

For the first, the scientist has an obligation to explore—to seek meaningful connections. Edison tried many candidates for a filament to provide the first incandescent lamp. A critical part of the discovery of continental drift involved staring at a map and noticing how the continents could fit together. In both cases, exploration was vital, and formal significance was absent or unnoticed. In many areas, progress can only come from exploration.

The second aspect of magical thinking can also be of enormous use: In order to make progress in a complex situation we often assume a model, paradigm, or working hypothesis and proceed as if it were true. This often involves ignoring data and competing theories. Thomas Kuhn (1970, Chapter 2; 1977, Chapter 9) argues that many of the discoveries of science were made because of the unswerving belief in a working hypothesis. Such beliefs are rewarded when:

The belief allows us to forget about philosophical controversy and get down to "honest work" which will be interpretable from most philosophical perspectives.

The beliefs are longshots that prove to be correct.

The beliefs are wrong, but to defend them, we undertake a detailed series of experiments that ultimately leads to a convincing refutation.

Perhaps Francis Bacon, quoted by Kuhn (1970, p. 18), put it best when he wrote

> Truth emerges more readily from error than from confusion.

Little theory is available to guide our decisions on how long to cling to a belief unsupported by data. Kuhn makes a clear case for the eminent practicality of magical thinking continued about the right length of time and documents how revolutions, great and small, occur when such beliefs cease. Let us consider the desire and skill for exploration together with an awareness of the uses and pitfalls of working hypotheses as a controlled form of magical thinking.

Where do we stand? The new exploratory techniques seem to be a mandatory supplement to more classical statistical procedures. The argument is two-pronged.

First, exploratory techniques find useful structure where classical techniques fall flat. This is shown in the examples described in remedy 1 of Section 1C. Second, in many situations, the "usual assumptions" are not even approximately valid. Therefore, the resulting P-values or levels of significance do not mean what they say.

Sometimes we may be able to supply more appropriate P-values, as indicated in remedies 2 and 4 of Section 1C. None of the special theories outlined in Section 1D are ready for routine use. The computer and mathematics can combine to give a reasonable way of checking some more widely used procedures, but again under stringent restrictions.

Despite its lack of sophistication, a purely data-analytic approach has something to offer. It is an empirical fact that using the tools of EDA to root about in large rich collections of data leads to useful results. One need not always be a subject matter expert or be able to explain the patterns in order to proceed usefully.

Of course, we try hard to control the tendency to see patterns in noise and label our findings as exploratory. Despite the cost of false leads, it seems that controlled magical thinking is here to stay.

REFERENCES

Barber, T. X. (1976). *Pitfalls in Human Research: Ten Pivotal Points*. New York: Permagon Press.

Bartsch, H., Malaveille, C., Camus, A.-M., Martel-Planche, G., Brun, G., Hautefeuille, A., Sabadie, N., Barbin, A., Kuroki, T., Drevon, C., Piccoli, C., and Montesano, R. (1980). "Bacterial and mammalian mutagenicity tests: validation and comparative studies on 180 chemicals." In R. Montesano, H. Bartsch, and L. Tomatis (Eds.), *Molecular and Cellular Aspects of Carcinogen Screening Tests* (IARC Scientific Publications No. 27). Lyon, France: International Agency for Research on Cancer, pp. 179–241.

Bickel, P. J. and Doksum, K. A. (1981). "An analysis of transformations revisited," *Journal of the American Statistical Association*, **76**, 296–311.

Bickel, P. J. and Freedman, D. A. (1981). "Some asymptotic theory for the bootstrap," *Annals of Statistics*, **9**, 1196–1217.

Bickel, P. J., Hammel, E. A., and O'Connell, J. W. (1975). "Sex bias in graduate admissions: data from Berkeley," *Science*, **187**, 398–404. Reprinted in W. B. Fairley and F. Mosteller (Eds.) (1977), *Statistics and Public Policy*. Reading, MA: Addison-Wesley, pp. 113–130.

Box, G. E. P. (1980). "Sampling and Bayes' inference in scientific modeling and robustness," *Journal of the Royal Statistical Society, Series A*, **143**, 383–430 (with discussion).

Breiman, L., Friedman, J. H., Olshen, R., and Stone, C. (1983). *Classification and Regression Trees*. Belmont, CA: Wadsworth.

Bruntz, S. M., Cleveland, W. S., Graedel, T. E., Kleiner, B., and Warner, J. L. (1974). "Ozone concentration in New Jersey and New York: statistical association with related variables," *Science*, **186**, 257–259.

Bunker, J. P., Barnes, B. A., and Mosteller, F. (Eds.) (1977). *Costs, Risks, and Benefits of Surgery*. New York: Oxford University Press.

Carlsmith, J. M. and Anderson, C. A. (1979). "Ambient temperature and the occurrence of collective violence: a new analysis," *Journal of Personal and Social Psychology*, **37**, 337–344.

Cassirer, E. (1972). *An Essay on Man*. New Haven: Yale University Press.

Chen, H.-J., Gnanadesikan, R., and Kettenring, J. R. (1974). "Statistical methods for grouping corporations," *Sankhyā, Series B*, **36**, 1–28.

Cleveland, W. S., Diaconis, P., and McGill, R. (1982). "Variables on scatterplots look more highly correlated when the scales are increased," *Science*, **216**, 1138–1141.

Cleveland, W. S. and Graedel, T. E. (1979). "Photochemical air pollution in the Northeast United States," *Science*, **204**, 1273–1278.

Cleveland, W. S., Graedel, T. E., Kleiner, B., and Warner, J. L. (1974). "Sunday and workday variations in photochemical air pollutants in New Jersey and New York," *Science*, **186**, 1037–1038.

Cleveland, W. S., Kleiner, B., McRae, J. E., and Warner, J. L. (1976). "Photochemical air pollution: transport from the New York City area into Connecticut and Massachussets," *Science*, **191**, 179–181.

Cleveland, W. S., Kleiner, B., and Warner, J. L. (1976). "Robust statistical methods and photochemical air pollution data," *Journal of the Air Pollution Control Association*, **26**, 36–38.

Dempster, A. P. (1983). "Purposes and limitations of data analysis." In G. E. P. Box, T. Leonard, and C.-F. Wu (Eds.), *Scientific Inference, Data Analysis, and Robustness*. New York: Academic, pp. 117–133.

Diaconis, P. (1978). "Statistical problems in ESP research," *Science*, **201**, 131–136; **201**, 1145–1146 (related correspondence).

Diaconis, P. (1981). "Magical thinking in the analysis of scientific data," *Annals of the New York Academy of Sciences*, **364**, 236–244.

Diaconis, P. and Freedman, D. (1984). "Asymptotics of graphical projection pursuit," *Annals of Statistics*, **12**, 793–815.

Diaconis, P. and Friedman, J. H. (1983). "M and N plots." In H. Rizvi, J. Rustagi, and D. Siegmund (Eds.), *Recent Advances in Statistics: Papers in Honor of Herman Chernoff on His Sixtieth Birthday*. New York: Academic, pp. 425–447.

Diaconis, P. and Zabell, S. L. (1982). "Updating subjective probability," *Journal of the American Statistical Association*, **77**, 822–830.

Donoho, D. L. and Huber, P. J. (1983). "The notion of breakdown point." In P. J. Bickel, K. A. Doksum, and J. L. Hodges, Jr. (Eds.), *A Festschrift for Erich L. Lehmann*. Belmont, CA: Wadsworth International Group, pp. 157–184.

Efron, B. (1971). "Does an observed sequence of numbers follow a simple rule? (Another look at Bode's law)," *Journal of the American Statistical Association*, **66**, 552–559.

Efron, B. (1979). "Bootstrap methods: another look at the jackknife," *Annals of Statistics*, **7**, 1–26.

Efron, B. (1982). *The Jackknife, the Bootstrap, and Other Resampling Plans*. Philadelphia: SIAM.

Efron, B. and Gong, G. (1983). "A leisurely look at the bootstrap, the jackknife, and cross-validation," *The American Statistician*, **37**, 36–48.

Feyerabend, P. (1978). *Against Method: Outline of an Anarchistic Theory of Knowledge*. London: Verso.

Finch, P. D. (1979). "Description and analogy in the practice of statistics," *Biometrika*, **66**, 195–208 (with discussion).

Freedman, D. A. (1983). "A note on screening regression equations," *The American Statistician*, **37**, 152–155.

Freedman, D. A. and Diaconis, P. (1982). "On inconsistent M-estimators," *Annals of Statistics*, **10**, 454–461.

Freedman, D. A. and Lane, D. (1983a). "Significance testing in a nonstochastic setting." In P. J. Bickel, K. A. Doksum, and J. L. Hodges, Jr. (Eds.), *A Festschrift for Erich L. Lehmann*. Belmont, CA: Wadsworth International Group, pp. 185–208.

Freedman, D. A. and Lane, D. (1983b). "A nonstochastic interpretation of reported significance levels," *Journal of Business & Economic Statistics*, **1**, 292–298.

Friedman, J. H. and Tukey, J. W. (1974). "A projection pursuit algorithm for exploratory data analysis," *IEEE Transactions on Computers*, **C-23**, 881–890.

Gabbe, J. D., Wilk, M. B., and Brown, W. L. (1967). "Statistical analysis and modeling of the high-energy proton data from the *Telstar* satellite," *Bell System Technical Journal*, **46**, 1301–1450.

Gardner, M. (1981). *Science: Good, Bad, and Bogus*. New York: Prometheus.

Gilbert, J. P., McPeek, B., and Mosteller, F. (1977). "Progress in surgery and anesthesia: benefits and risks of innovative therapy." In J. P. Bunker, B. A. Barnes, and F. Mosteller (Eds.), *Costs, Risks, and Benefits of Surgery*. New York: Oxford University Press, pp. 124–169.

Gnanadesikan, R. (1977). *Methods for Statistical Data Analysis of Multivariate Observations*. New York: Wiley.

Gong, G. (1982). "Cross-validation, the jackknife, and the bootstrap: excess error estimation in forward logistic regression." Ph.D. Dissertation, Department of Statistics, Stanford University.

Good, I. J. (1969). "A subjective evaluation of Bode's law and an 'objective' test for approximate numerical rationality," *Journal of the American Statistical Association*, **64**, 23–66 (with discussion).

Good, I. J. (1983). "The philosophy of exploratory data analysis," *Philosophy of Science*, **50**, 283–295.

Hacking, I. (1983). *Representing and Intervening: Introductory Topics in the Philosophy of Natural Science*. Cambridge: Cambridge University Press.

Hyman, R. (1982). "Does the Ganzfeld experiment answer the critics' objection?" Presented at the combined meeting of The Parapsychological Association and the Society for Psychical Research, Cambridge, England. To appear in *Journal of Parapsychology*, March 1985.

Jahn, R. G. (1982). "The persistent paradox of psychic phenomena: an engineering perspective," *Proceedings of the IEEE*, **70**, 136–170

Kahneman, D., Slovic, P., and Tversky, A. (Eds.) (1982). *Judgment under Uncertainty: Heuristics and Biases*. Cambridge: Cambridge University Press.

Klotz, I. M. (1980). "The *N*-ray affair," *Scientific American*, **242**, No. 5, 168–175.

Kuhn, T. S. (1970). *The Structure of Scientific Revolutions*, 2nd ed. Chicago: University of Chicago Press.

Kuhn, T. S. (1977). *The Essential Tension: Selected Studies in Scientific Tradition and Change*. Chicago: University of Chicago Press.

Leamer, E.E. (1978). *Specification Searches: Ad Hoc Inference with Nonexperimental Data*. New York: Wiley.

Macdonald, N. J. and Ward, F. (1963). "The prediction of geomagnetic disturbance indices. 1. The elimination of internally predictable variations," *Journal of Geophysical Research*, **68**, 3351–3373.

Mallows, C. L. (1970). "Some comments on Bayesian methods." In D. L. Meyer and R. O. Collier, Jr. (Eds.), *Bayesian Statistics*. Itasca, IL: Peacock, pp. 71–84.

Mallows, C. L. (1979). "Robust methods—some examples of their use," *The American Statistician*, **33**, 179–184.

Mallows, C. L. (1983). "Data description." In G. E. P. Box, T. Leonard, and C.-F. Wu (Eds.), *Scientific Inference, Data Analysis, and Robustness*. New York: Academic, pp. 135–151.

Mallows, C. L. and Walley, P. (1981). "A theory of data analysis." In *1980 Proceedings of the Business and Economic Statistics Section*. Washington, DC: American Statistical Association, pp. 8–14.

Martin-Löf, P. (1974). "The notion of redundancy and its use as a quantitative measure of the discrepancy between a statistical hypothesis and a set of observational data," *Scandinavian Journal of Statistics*, **1**, 3–18.

Miller, R. G., Jr. (1981). *Simultaneous Statistical Inference*, 2nd ed. New York: Springer-Verlag.

Mosteller, F. and Tukey, J. W. (1977). *Data Analysis and Regression: A Second Course in Statistics*. Reading, MA: Addison-Wesley.

Nisbett, R. and Ross, L. (1980). *Human Inference: Strategies and Shortcomings of Social Judgment*. Englewood Cliffs, NJ: Prentice-Hall.

Phillips, D. P. (1978). "Deathday and birthday: an unexpected connection." In J. M. Tanur, F. Mosteller, W. H. Kruskal, R. F. Link, R. S. Pieters, G. R. Rising, and E. L. Lehmann (Eds.), *Statistics: A Guide to the Unknown*, 2nd ed. San Francisco, CA: Holden-Day, pp. 71–85.

Pocock, S. J. (1977). "Group sequential methods in the design and analysis of clinical trials," *Biometrika*, **64**, 191–199.

Reaven, G. and Miller, R. (1979). "An attempt to define the nature of chemical diabetes using multidimensional analyses," *Diabetologia*, **16**, 17–24.

Rosenthal, R. (1978). "Combining results of independent studies," *Psychological Bulletin*, **85**, 185–193.

Rosenthal, R. (1981). "Pavlov's mice, Pfungst's horse, and Pygmalion's PONS: some models for the study of interpersonal expectancy effects," *Annals of the New York Academy of Sciences*, **364**, 182–198.

Ross, L. and Lepper, M. R. (1980). "The perseverance of beliefs: empirical and normative considerations." In R. A. Shweder (Ed.), *New Directions for Methodology of Behavioral Sciences: Fallible Judgement in Behavioral Research*. San Francisco: Jossey-Bass.

Savage, L. J. (1970). "The shifting foundations of statistics." In R. Colodny (Ed.), *Logic, Laws and Life*. Pittsburgh: University of Pittsburgh Press, pp. 3–18. Reprinted in *The Writings of Leonard Jimmie Savage—A Memorial Selection*. Washington, D.C.: The American Statistical Association, 1981, pp. 721–736.

Schulz, R. and Bazerman, M. (1980). "Ceremonial occasions and mortality: a second look," *American Psychologist*, **35**, 253–261.

Shafer, G. (1976). *A Mathematical Theory of Evidence*. Princeton, NJ: Princeton University Press.

Shweder, R. A. (1977). "Likeness and likelihood in everyday thought: magical thinking in judgments about personality," *Current Anthropology*, **18**, 637–658.

Siegmund, D. (1977). "Repeated significance tests for a normal mean," *Biometrika*, **64**, 177–189.

Slater, P. B. (1974). "Exploratory analyses of trip distribution data," *Journal of Regional Science*, **14**, 377–388.

Slater, P. B. (1975). "Petroleum trade in 1970: an exploratory analysis," *IEEE Transactions on Systems, Man, and Cybernetics*, **SMC-5**, 278–283.

Tukey, J. W. (1969). "Analyzing data: sanctification or detective work," *American Psychologist*, **24**, 83–91.

Tukey, J. W. (1972). "Data analysis, computation, and mathematics," *Quarterly of Applied Mathematics*, **30**, 51–65.

Tukey, J. W. (1977). *Exploratory Data Analysis*. Reading, MA: Addison-Wesley.

Tversky, A. and Kahneman, D. (1974). "Judgment under uncertainty: heuristics and biases," *Science*, **185**, 1124–1131.

Vogt, E. Z. and Hyman, R. (1979). *Water Witching U.S.A.*, 2nd ed. Chicago: University of Chicago Press.

Worsley, P. (1968). *The Trumpet Shall Sound: A Study of "Cargo Cults" in Melanesia*, 2nd ed. New York: Schocken.

CHAPTER 2

Fitting by Organized Comparisons: The Square Combining Table

Katherine Godfrey
Harvard University

Data for analysis often come in the form of a two-way table, involving values for each combination of several levels of two different variables. For example, a medical investigator may collect data on remission rates for different types of tumors in patients treated by various therapies. One natural question in analyzing such data is whether the therapies differ in performance. The classical method for investigating this question uses analysis of variance. One exploratory method uses median polish (UREDA, Chapter 6). In this chapter we discuss a third method, the square combining table.

In Section 2A we look at a simple example of combining comparisons, the underlying motivation to the square combining table. Next we extend the discussion to the general two-way table, with a demonstration of how to calculate a square-combining-table fit (Section 2B). Section 2C deals with the similarity between the square-combining-table technique and the method of paired comparisons. Finally, Section 2D discusses the analysis of two-way tables when the data are incomplete—that is, two-way tables with holes.

2A. COMBINING COMPARISONS

The underlying concept of the *square combining table* extends the long-standing method of paired comparisons used in psychological experiments. In such designs, a subject considers a pair of items (called *stimuli*) and ranks the two. For example, he or she may lift two weights and select the heavier, or eat two brands of ice cream and choose the one he or she prefers.

The square combining table generalizes this idea. When someone simply ranks two objects, he or she gives no information as to how great he or she judges the difference between them to be. If we do have a means for gathering that information, we can not only form a picture of the average order of the different stimuli, but attach a measurement scale to that ordering as well.

Fleckenstein, Freund, and Jackson (1958) used this scheme in an experiment to compare brands of carbon paper. Individual typists each tested two brands of carbon paper. After typing the same material with each brand, the subjects reported their preference for one brand over the other on a seven-point scale (coded for analysis from -3 to $+3$). The experimenters decided that giving a typist more than two brands to compare would yield less precise results. On the other hand, if each typist rated only one brand of paper, differences from typist to typist in what each ranking meant could bias the results—or greatly increase their variability.

Thirty typists gave preference scores for each pair of brands. We look now at the result for three brands, labeled C, D, and E. Summing over all 30 typists, the total of the preference scores for brand C over brand D was $+61$. The total for brand C over brand E was $+13$. The total for brand D over brand E was -40; that is, the typists as a whole preferred brand E to brand D.

Now that we have a comparative score for each pair, we want to combine them to get estimates for the location of the individual brands along an ordered scale of total scores. To do this we need to make one arbitrary choice—the location of the zero point on that scale. One convenient approach makes the scores for the individual brands add to zero. Once we have made that choice, our method of estimating the scores for the individual brands uses the square combining table.

The term "square combining table" emphasizes the two main features. The table is square because it has as many rows and as many columns as there are items, and it combines the comparisons between pairs of items to yield values for the individual items in the pairs. We set up such a table in Table 2-1. The first brand in each pair appears at the head of the column, and the second brand at the head of the row. Thus the value for the total preference score for brand C over brand E is given in the first column and the third row: $+13$.

We fill in the cells of the table below the main diagonal from the results of the experiment. The main diagonal of the table receives zeros when we make the natural assumption that a typist would show no preference between two sheets of the same brand of carbon paper. (In fact, the typists might show some preferences by chance.) To complete the table, we fill in the upper right half according to the property of antisymmetry: if the sum

TABLE 2-1. The square combining table applied to observed preferences (first brand over second brand) among brands of carbon paper.

	First Brand		
Second Brand	*C*	*D*	*E*
C	0	−61	−13
D	+61	0	+40
E	+13	−40	0
Column total	+74	−101	+27
Column mean	+24.7	−33.7	+9.0

Source: Mary Fleckenstein, Richard A. Freund, and J. Edward Jackson (1958). "A paired comparison test of typewriter carbon papers," *Tappi*, **41**, 128–130 (data from Table III, p. 130).

of the preference scores for brand C over brand E is $+13$, then the sum of the preference scores for brand E over brand C would be -13. (We assume either that the order of presentation within a pair has no effect or that the orders were balanced.)

We take the means of the columns in the square combining table to obtain the scores for the individual brands. Brand C, with a combined score of $+24.7$, is the brand preferred by the subjects as a whole, and brand D, at -33.7, was judged the poorest. The scaling indicates that the typists found brand D much worse than either of the other brands, and that the difference between brands D and E was over twice as large as the difference between brands C and E. As noted earlier, the sum of the scores for the three brands is zero, a consequence of using the square combining table to fit the scores.

2B. TWO-WAY TABLES

In the carbon paper experiment, we considered only one variable—brand of carbon paper. We can extend our method to estimate row effects and column effects in a two-way table by calculating differences between pairs of rows *and* differences between pairs of columns and combining each set of differences in a square combining table.

We think first of an additive description of the table in which each data value y_{ij} is the sum of four components. The first component is constant for all values in the table. We denote this by m. The second component is constant for all data values in the same row. This is a_i, the row effect for row i. The third component is constant for all data values in the same column. This is b_j, the column effect for column j. The final component is the residual, e_{ij}. The equation

$$y_{ij} = m + a_i + b_j + e_{ij} \tag{1a}$$

summarizes this description. In terms of the estimands, we also write

$$y_{ij} = \mu + \alpha_i + \beta_j + \varepsilon_{ij}, \tag{1b}$$

as in the standard formulation of two-way analysis of variance. These two equations remind us that, although our calculations work with the y_{ij}, we are almost always concerned with analyzing what lies behind the data we have. Accordingly we often regard a_i as an estimate of α_i and similarly for m, b_j, and e_{ij}.

Equations (1a) and (1b) each cover a variety of decompositions of the y_{ij}. To fix a specific decomposition, we impose constraints. It is classical to require that the mean of the α_i be zero, but often we do better to make their median zero, and so on in both equations.

Neither equation (1a) nor equation (1b) provides separately for interaction—nonadditivity—between the row factor and column factor. We cannot distinguish meaningful interaction from other components of the ε_{ij}.

Generally we think of μ as an overall typical value of what the data y_{ij} represent, to which the α_i and the β_j are adjustments reflecting the contributions of row i and column j, respectively. For this reason, we often require that the row effects sum to zero, and that the column effects also sum to zero. This is one motivation for using means to center the row effects and column effects derived from the square combining table. Here we require the *median* of the e_{ij} to be zero as a way of making m an overall typical value of the data. This is another arbitrary choice, and other techniques of estimating row and column effects for a two-way table may make different choices. For example, two-way analysis of variance produces residuals whose *mean* is zero. In a median-polish fit, the overall median of the residuals may not be zero, but the row medians and column medians of the residuals are all zero.

TABLE 2-2. Infant mortality rates in the United States, all races, 1964–1966, by region and father's education. (Entries are numbers of deaths per 1000 live births.)

	Education of Father (in years)				
Region	≤ 8	9–11	12	13–15	≥ 16
Northeast	25.3	25.3	18.2	18.3	16.3
North Central	32.1	29.0	18.8	24.3	19.0
South	38.8	31.0	19.3	15.7	16.8
West	25.4	21.1	20.3	24.0	17.5

Source: U.S. Department of Health, Education, and Welfare, National Center for Health Statistics, *Infant Mortality Rates: Socioeconomic Factors, United States*, Vital and Health Statistics, Series 22, No. 14, Rockville, MD, 1972. DHEW publication number (HSM)72-1045 (data from Table 8, p. 21).

Fitting a Two-Way Table

The square combining table handles differences between pairs. To make a square-combining-table fit to a two-way table, we form the differences of all pairs of rows and, as well, the differences of all pairs of columns. These differences give us two square combining tables, one for the row differences and one for the column differences. We derive our row effects from the former and our column effects separately from the latter.

Table 2-2 gives infant mortality rates per 1000 live births in the United States for the period 1964–1966. The rates are broken down by region of the country and by the education of the father. We now use the square combining table to calculate column (education) effects and row (region) effects, starting with columns.

First we calculate the pairwise column differences. We do this by taking the difference in each row between each pair of columns. Because there are four rows, we have four differences for each column pair. Table 2-3 shows the differences for the first and second columns as (25.3 − 25.3), (32.1 − 29.0), (38.8 − 31.0), and (25.4 − 21.1).

Now that we have the differences, we summarize the four differences for each pair. To obtain resistance, we use medians. Thus our estimate for the difference between columns 1 and 2 is the median of the four differences, 3.7. We use these medians of differences to fill in the square combining table

TABLE 2-3. Column differences from Table 2-2 (first named column minus second named).

Row	Column Pair 1,2	1,3	1,4	1,5	2,3
1	0	7.1	7.0	9.0	7.1
2	3.1	13.3	7.8	13.1	10.2
3	7.8	19.5	23.1	22.0	11.7
4	4.3	5.1	1.4	7.9	0.8
Median	3.7	10.2	7.4	11.05	8.65

Row	Column Pair 2,4	2,5	3,4	3,5	4,5
1	7.0	9.0	−0.1	1.9	2.0
2	4.7	10.0	−5.5	−0.2	5.3
3	15.3	14.2	3.6	2.5	−1.1
4	−2.9	3.6	−3.7	2.8	6.5
Median	5.85	9.5	−1.9	2.2	3.65

in Table 2-4. Taking means of the values in each column of the square combining table yields estimates of the column effects, the β_j. Because these are estimates, we denote them by b_1, b_2, through b_5.

Working with row differences, we obtain a second square combining table for the rows (Table 2-5). Taking column means of the values in this table gives the estimates for the row effects, a_1 through a_4.

TABLE 2-4. Square combining table for the columns of Table 2-2.

Second Column	First Column 1	2	3	4	5
1	0	−3.7	−10.2	−7.4	−11.05
2	3.7	0	−8.65	−5.85	−9.5
3	10.2	8.65	0	1.9	−2.2
4	7.4	5.85	−1.9	0	−3.65
5	11.05	9.5	2.2	3.65	0
Column mean	6.5	4.1	−3.7	−1.5	−5.3
Column effect	b_1	b_2	b_3	b_4	b_5

TABLE 2-5. Square combining table for the rows of Table 2-2.

Second Row	First Row 1	2	3	4
1	0	3.7	1.1	1.2
2	−3.7	0	0.5	−1.5
3	−1.1	−0.5	0	0.7
4	−1.2	1.5	−0.7	0
Column mean	−1.5	1.2	0.2	0.1
Row effect	a_1	a_2	a_3	a_4

Once we have the fitted row and column effects, we can estimate the overall typical value μ. First we subtract from each value in the original table its corresponding row effect and column effect. This gives the table shown in Table 2-6. For example, in row 1, column 1, $20.3 = 25.3 - 6.5 - (-1.5)$. In the notation of equation (1a), each value in the main body of Table 2-6 has the form $m + e_{ij}$. To meet the condition that the median of the residuals be zero, we take as our estimate of μ the overall median of the values in Table 2-6. This gives the final square-combining-table fit to the data of Table 2-2, as displayed in Table 2-7.

Looking at the residuals in Table 2-7, we see that the infant mortality rate is particularly high for southern children whose fathers did not attend high school, low for western babies whose fathers did not finish high school, and low for the children of southerners who attended but did not finish college. Generally, the more education the father has, the lower the infant mortality rate is, although the observed effect is not monotonic. The safest

TABLE 2-6. The data of Table 2-2 with row and column effects removed.[a]

Region	Education of Father (in years) ≤ 8	9–11	12	13–15	≥ 16	Row Effect
Northeast	20.3	22.7	23.4	21.3	23.1	−1.5
North Central	24.4	23.7	21.3	24.6	23.1	1.2
South	32.1	26.7	22.8	17.0	21.9	0.2
West	18.8	16.9	23.9	25.4	22.7	0.1
Column effect	6.5	4.1	−3.7	−1.5	−5.3	

[a] The overall median of the 20 cells in the main body of the table is 22.9.

TABLE 2-7. Square-combining-table fit, with residuals, for the infant mortality data of Table 2-2.

	Education of Father (in years)					Row
Region	≤ 8	9–11	12	13–15	≥ 16	Effect
Northeast	−2.6	−0.2	0.5	−1.6	0.2	−1.5
North Central	1.5	0.8	−1.6	1.7	0.2	1.2
South	9.2	3.8	−0.1	−5.9	−1.0	0.2
West	−4.1	−6.0	1.0	2.5	−0.2	0.1
Column effect	6.5	4.1	−3.7	−1.5	−5.3	22.9

TABLE 2-8. Stem-and-leaf display of the residuals from Table 2-7.

(1|2 represents 1.2)

Stem	Leaf
9	2
8	
7	
6	
5	
4	
3	8
2	5
1	570
+0	5282
−0	122
−1	660
−2	6
−3	
−4	1
−5	9
−6	0
−7	

Median = 0.05
Upper fourth = 1.25
Lower fourth = −1.6
Fourth-spread = 2.85

region is the Northeast; the least safe is the North Central. The effects of education are larger than those of region.

A stem-and-leaf display (UREDA, Chapter 1) of the residuals from the square-combining-table fit shows the pattern of the residuals (Table 2-8). Most of the residuals are small, and the overall median is 0.05. The residual 9.2 for southern infants whose fathers have no more than 8 years of education is a noticeably high outlier.

The General Two-Way Table

To make a square-combining-table fit to a general two-way table with I rows and J columns, we calculate the differences for each pair of rows and each pair of columns, as we did in Table 2-3 for the columns of Table 2-2. By taking medians in Table 2-3, we estimated the separations between the columns. The column separations take the form

$$\Delta_{kj} = \underset{i=1,\ldots,I}{\text{med}} \{y_{ij} - y_{ik}\}. \tag{2}$$

The number Δ_{kj} estimates the separation between columns j and k, and is our estimate for the difference between the corresponding column effects $(\beta_j - \beta_k)$. We call this separation Δ_{kj} rather than Δ_{jk} to produce a square combining table in which we estimate effects as the means of the columns of the square combining table. We do this because people usually find it easier to sum down columns rather than across rows.

We calculate the separations for the rows δ_{st} according to the formula

$$\delta_{st} = \underset{j=1,\ldots,J}{\text{med}} \{y_{tj} - y_{sj}\}. \tag{3}$$

The number δ_{st} estimates the separation between rows s and t, and is our estimate for the difference $\alpha_t - \alpha_s$ between the row effects α_s and α_t.

As we saw before, the square combining table is antisymmetric. That is, $\Delta_{kj} = -\Delta_{jk}$, and $\delta_{st} = -\delta_{ts}$. Both Δ_{kk} and δ_{ss} are zero, as these numbers lie along the main diagonals of the square combining table for columns and the square combining table for rows. Therefore, although there are I^2 pairs of rows and J^2 pairs of columns, we need calculate only $I(I-1)/2$ and $J(J-1)/2$ different medians of differences in order to find all the Δ_{kj} and δ_{st}.

Once we have the square combining table of Δ_{kj} for columns and the square combining table of δ_{st} for rows, we calculate the estimates b_j for column effects and a_i and row effects. The b_j are the column means of the table of Δ_{kj}. The a_i are the column means of the table of δ_{st}. Taking means centers the a_i and the b_j so that they sum to zero. Equations (4) and (5) yield the a_i and the b_j:

$$a_i = \frac{1}{I} \sum_{s=1}^{I} \delta_{si}, \tag{4}$$

$$b_j = \frac{1}{J} \sum_{k=1}^{J} \Delta_{kj}. \tag{5}$$

To complete the fit to the two-way table, we need to calculate an estimate of μ, the typical value. We choose our estimate $\hat{\mu}$ so that the overall median of the residuals is zero. That is,

$$\hat{\mu} = \operatorname*{med}_{i,j} \{ y_{ij} - a_i - b_j \}. \tag{6}$$

TABLE 2-9. Column effects and row effects from a two-way table by using the square combining table.

Y (4 × 5)

y_{11}	y_{12}	y_{13}	y_{14}	y_{15}
y_{21}	y_{22}	y_{23}	y_{24}	y_{25}
y_{31}	y_{32}	y_{33}	y_{34}	y_{35}
y_{41}	y_{42}	y_{43}	y_{44}	y_{45}

Row differences →

4 × 4 square combining table for rows

δ_{11}	δ_{12}	δ_{13}	δ_{14}
δ_{21}	δ_{22}	δ_{23}	δ_{24}
δ_{31}	δ_{32}	δ_{33}	δ_{34}
δ_{41}	δ_{42}	δ_{43}	δ_{44}
a_1	a_2	a_3	a_4

Column means

Column differences ↓

5 × 5 square combining table for columns

Δ_{11}	Δ_{12}	Δ_{13}	Δ_{14}	Δ_{15}
Δ_{21}	Δ_{22}	Δ_{23}	Δ_{24}	Δ_{25}
Δ_{31}	Δ_{32}	Δ_{33}	Δ_{34}	Δ_{35}
Δ_{41}	Δ_{42}	Δ_{43}	Δ_{44}	Δ_{45}
Δ_{51}	Δ_{52}	Δ_{53}	Δ_{54}	Δ_{55}
b_1	b_2	b_3	b_4	b_5

Column means

Once we subtract the fitted effects and typical value from the original data values, we are left with the residuals e_{ij}, as in Table 2-7.

Table 2-9 shows schematically the process of calculating the a_i and the b_j. For this exhibit, we set I equal to 4 and J equal to 5, following the infant mortality example.

Equations (2) and (3) use medians of the pairwise differences to obtain resistant estimates of the separations between rows and between columns. The median is not the only possible resistant summary, and we could use another such summary to calculate the Δ_{kj} and δ_{st} for the square combining tables. Chapter 10 of UREDA discusses several resistant summaries. The square-combining-table fit is resistant because the Δ_{kj} and δ_{st} are resistant summaries of the column differences and row differences. If we were to take means instead of medians, and if we chose m to be the overall mean instead of the overall median, we would obtain a nonresistant fit. In fact, the fit would be identical to the usual two-way analysis-of-variance fit. Whatever summary we use in equations (2) and (3), we take means in equations (4) and (5), so that the row effects and column effects sum to zero.

2C. PAIRED COMPARISONS

We mentioned earlier that the motivation behind the square combining table parallels that for the method of paired comparisons. In both methods, we think of comparing two things at a time (two columns or two stimuli), with the aim of calculating a separation between them that represents how different they are.

In a paired-comparison experiment, the subject judges a pair of stimuli and selects the "better" of the two, such as the heavier of two weights. With n stimuli, there are $\binom{n}{2} = n(n-1)/2$ possible pairs. Since the number of possible pairs increases as n^2, an individual subject in a large experiment will not see all possible pairs, but the experiment will be balanced so that each pair receives the same number of judgments.

Once the experiment is complete, the $n \times n$ proportion matrix $\{P_{ij}\}$, $i, j = 1, \ldots, n$ gives the results across all subjects. P_{ij} is the fraction of subjects who compared stimulus i and stimulus j and chose stimulus j over stimulus i. P_{ji} will then be $1 - P_{ij}$. P_{ii} should be 0.5, or nearly so. In an experiment that does not expose subjects to pairs of identical stimuli, the P_{ii} are taken to be 0.5.

After we have the proportion matrix, we need to choose a method for scaling the proportions. The standard method assumes that the underlying separations have a Gaussian distribution, an assumption known as Thurstone's law of comparative judgment (Thurstone, 1927). If there is no

TABLE 2-10. Proportion matrix for four vegetables in a paired-comparison experiment: proportion preferring first vegetable to second vegetable.

Second	First Vegetable			
Vegetable	1	2	3	4
1. Turnips	.500	.818	.811	.892
2. Cabbage	.182	.500	.723	.845
3. Asparagus	.189	.277	.500	.601
4. Peas	.108	.155	.399	.500

Source: J. P. Guilford (1954). *Psychometric Methods*, 2nd ed. New York: McGraw-Hill. (The data come from Table 7.4, p. 160.)

correlation between the responses to any pair of stimuli, and if the variances of the underlying stimulus effects (the β_j of our model) are equal, this assumption leads to a very simple method of scaling: we replace each proportion P_{ij} by the corresponding standard Gaussian deviate S_{ij}. If Z is a Gaussian random variable with mean 0 and variance 1, then S_{ij} obeys the following rule:

$$P_{ij} = P\{Z \leq S_{ij} | Z \text{ unit Gaussian}\}. \tag{7}$$

By the symmetry of the Gaussian distribution, $S_{ji} = -S_{ij}$, and if $P_{ii} = 0.5$, $S_{ii} = 0$. The resulting matrix $\{S_{ij}\}$, known as the *scale separation matrix*, is antisymmetric and has a zero main diagonal, just as the square combining table does. As with the square combining table, we estimate column (stimulus) effects by taking column means in the scale separation matrix. For a paired-comparison study, these column means are then usually rescaled, but the relative position of the stimuli remains the same. For example, the column means (known as *scale separation factors*) are often translated so that the smallest scale separation factor has the value zero.

We take an example from Guilford (1954), who discusses the method in detail. Table 2-10 shows the proportion matrix for four vegetables judged in terms of preferences as foods. From this Guilford derives the scale separation matrix $\{S_{ij}\}$, which appears in Table 2-11.

The column means of the scale separation matrix are the estimates for the vegetable effects. Turnips are the least favorite vegetable of the four,

TABLE 2-11. Scale separation matrix for the four vegetables in a paired-comparison experiment.

Second Vegetable	First Vegetable 1	2	3	4
1. Turnips	.000	.908	.882	1.237
2. Cabbage	−.908	.000	.592	1.015
3. Asparagus	−.882	−.592	.000	.256
4. Peas	−1.237	−1.015	−.256	.000
Column mean	−.757	−.175	.305	.627
Vegetable effect[a]	−0.8	−0.2	0.3	0.6
	b_1	b_2	b_3	b_4

[a]Rounded value of column mean.

Source: J. P. Guilford (1954). *Psychometric Methods*, 2nd ed. New York: McGraw-Hill. (The values in the table come from Table 7.5, p. 162.)

peas are the best liked, and the four vegetables seem reasonably well separated.

Other methods of scaling the proportions will produce different scale separation matrices, and thus different stimulus effects. David (1963) discusses alternative scaling techniques and treats the topic of paired comparisons in depth.

Because nothing is resistant about the calculation of the scale separation matrix, it may need a resistant analysis. Exercise 1 pursues this possibility via the square combining table.

2D. ANALYZING TABLES CONTAINING HOLES

A two-way table may not have information to fill all its cells. For example, in a two-way table with age as the row factor and type of treatment for cancer as the column factor, not every treatment may have been given to people in one of the age groups, or one treatment may not be appropriate for an age group. The two-way table would contain a "hole" in the cell defined by that age group and the unused treatment. Alternatively, a slip in transcribing or recording data may produce a clearly anomalous value, say a value larger than 1 when all values are proportions (between 0 and 1).

Unless the error can be reliably corrected, such values are best replaced by holes. In these situations, the investigator faces a choice: (1) use only the original data, or (2) somehow estimate values to fill in the holes.

The first option generally sacrifices the simplicity of the intended analysis. For example, the least-squares analysis of variance would have to be reformulated as a multiple regression problem or handled by one of several alternative methods, all of which involve nonstandard computation. These techniques would naturally produce estimates for the missing cells as a byproduct.

To retain the usual least-squares calculations under the second option, various workers have devised schemes that begin by filling the holes, calculate residuals, adjust, and iterate until the residuals corresponding to the holes are zero. Seber (1977, Section 10.2) discusses both approaches.

Filling the holes maintains balance in the table and thus allows the contributions of the row factor and the column factor to be assessed without the qualification that regression models require. Such balance also produces formal *orthogonality*, which leads to the simplest relationship among sums

TABLE 2-12. Percentage of respondents indicating "a great deal" or "quite a lot" of confidence in American institutions, as recorded by the Gallup poll.

	Year				
Institution	1980	1979	1977	1975	1973
Organized religion	66	65	64	68	66
Banks	60	60			
The military	52	54	57	58	
Public schools	51	53	54		58
U.S. Supreme Court	47	45	46	49	44
Newspapers	42	51			39
Organized labor	35	36	39	38	30
Congress	34	34	40	40	42
Television	33	38			37
Big business	29	32	22	23	26

Source: Michael W. Kirst (1981). "Loss of support for public secondary schools: some causes and solutions," *Daedalus*, Summer, 45–68 (Table 3, p. 49). His source: *San Jose Mercury*, November 6, 1980. Reprinted by permission of DAEDALUS, the Journal of the American Academy of Arts and Sciences, Summer 1981, Cambridge, MA.

TABLE 2-13. Square-combining-table fit to the Gallup poll data.

	Year					
Institution	1980	1979	1977	1975	1973	Row Effect
Organized religion	1	0	−2	0	0	19
Banks	0	0				14
The military	−2	0	2	1		8
Public schools	−2	0	0		4	7
U.S. Supreme Court	2	0	0	1	−2	−1
Newspapers	−5	4			−9	1
Organized labor	0	1	3	0	−6	−11
Congress	−3	−3	2	0	4	−9
Television	−3	2			0	−10
Big business	1	4	−7	−8	−3	−18
Column effect	1	−1	0	2	0	47

of squares in the analysis-of-variance table. The investigator most easily achieves formal orthogonality by having each possible row and column combination appear an equal number of times (often once) in the data.

The square-combining-table method of analyzing a two-way table does not require balance, and the calculations proceed in the usual way when balance is absent. We take column and row differences of the available data. Some differences will be missing. We take medians of the available differences to obtain the Δ_{kj} and δ_{st} and then use these numbers to construct the square combining tables. These tables yield the row effects and column effects.

Table 2-12 gives the results of Gallup polls from 1973 to 1980 studying the confidence Americans have in various institutions. Because the poll did not include all the institutions every year, the table has holes. We can, nevertheless, use a square combining table to calculate effects for the rows and the columns. The final results are in Table 2-13.

The overall confidence is close to 50% for the institutions analyzed. Organized religion, banks, the military, and public schools enjoyed more confidence than the average for institutions, while Americans polled had less than average confidence in organized labor, Congress, television, and big business. Newspapers and the Supreme Court enjoyed about the average confidence. The bottom line of Table 2-13 displays no discernible trend across time, and the year effects are small compared to the institution effects. In fact, the year effects are the same order of magnitude as the residuals.

A few large residuals in the body of the table deserve attention. In both 1975 and 1977, confidence in big business was low, but in 1979, confidence in big business was high. In 1973, American confidence in organized labor was low. The residual for newspapers is very low in 1973, high in 1979, and then low in 1980.

Further Fitting and Display

When we look at the residuals in Table 2-13, we find a suggestion of some more-or-less linear time trends. If we fit such trends, even roughly, the residuals get even smaller.

Table 2-14 shows the result of such a fit, displayed as effectively as we know how. We have recombined

the common term, m, and

the new institution effect, a_i, with

each row's linear trend, l_i,

and specified the result for each row by its values at the first and last years for which we have data. We have also inserted blank columns for the missing years.

Beyond the rising and falling appearances summarized in Table 2-14*b*, the residuals for big business suggest lower public confidence in 1975 and 1977. Once this appearance attracts our attention, we may also note a complementary pattern in the residuals for organized labor: higher in 1975 and 1977 than in 1973 and 1980. We mention these possibilities without trying to pursue them. A further investigation would look for explanations of these and other apparent patterns. It might also ask whether some rows of residuals in Table 2-13 reflect discrete shifts of public opinion rather than steady trends.

The Effect of Holes on a Square-Combining-Table Fit

How does a fit to an incomplete table compare to the fit for the complete data? For square-combining-table techniques, we can follow the distortion a single missing value creates. Consider the 4×5 table of Table 2-15, and blank out one value (y_{23}). We compare the two square combining tables for rows and for columns for the table with and without the missing value. The absence of y_{ij} affects only the separations Δ_{lj} and Δ_{jl} ($l \neq j$) and δ_{ki} and δ_{ik} ($k \neq i$). This concentrates the impact of the missing value on the fitted effects for row i and column j of the original table. Although the hole

TABLE 2-14. Further analyses of the entries in Table 2-13 to include linear time trends.

a. Year effects

Year	1980	1979	1978	1977	1976	1975	1974	1973
Effect	−1	−1		0		2		0

b. Time trends

	1980	1979	1978	1977	1976	1975	1974	1973
Organized religion	66	————	steady at		————	————	————	66
Banks	61	61						
Military	54	————	falling from		——→	57		
Public schools	52	————	falling from		————	————	————	57
Supreme Court	47	————	rising from		————	————	————	44
Newspapers	48	————	rising from		————	————	————	41
Organized labor	38	————	rising from		————	————	————	34
Congress	35	————	falling from		————	————	————	42
Television	37	————	steady at		————	————	————	37
Big business	33	————	rising from		————	————	————	25

c. Residuals

	1980	1979	1978	1977	1976	1975	1974	1973
Organized religion	1	0		2		0		0
Banks	0	0						
Military	−1	1		2		−1		
Public schools	0	1		0				1
Supreme Court	1	−1		0		2		0
Newspapers	−5	5						−2
Organized labor	−2	0		3		1		−4
Congress	0	−1		2		−2		0
Television	−3	2						0
Big business	−3	1		−8		−6		1

affects all the fitted values to some extent, the change will be less for the fits to the other row effects and column effects.

This tolerance of holes makes the square combining table particularly useful in fitting tables with missing data. The median-polish method (UREDA, Chapter 6) can also handle two-way tables with missing values, but the presence of holes can greatly increase the number of steps needed to

TABLE 2-15. Square-combining-table analysis of a two-way table, complete and with one value missing (X).[a]

Complete		(2, 3) Value Missing
y_{11} y_{12} y_{13} y_{14} y_{15}		y_{11} y_{12} y_{13} y_{14} y_{15}
y_{21} y_{22} y_{23} y_{24} y_{25}	Data	y_{21} y_{22} X y_{24} y_{25}
y_{31} y_{32} y_{33} y_{34} y_{35}		y_{31} y_{32} y_{33} y_{34} y_{35}
y_{41} y_{42} y_{43} y_{44} y_{45}		y_{41} y_{42} y_{43} y_{44} y_{45}
0 Δ_{12} Δ_{13} Δ_{14} Δ_{15}		0 Δ_{12} Δ^*_{13} Δ_{14} Δ_{15}
Δ_{21} 0 Δ_{23} Δ_{24} Δ_{25}	Square combining	Δ_{21} 0 Δ^*_{23} Δ_{24} Δ_{25}
Δ_{31} Δ_{32} 0 Δ_{34} Δ_{35}	table for columns	Δ^*_{31} Δ^*_{32} 0 Δ^*_{34} Δ^*_{35}
Δ_{41} Δ_{42} Δ_{43} 0 Δ_{45}		Δ_{41} Δ_{42} Δ^*_{43} 0 Δ_{45}
Δ_{51} Δ_{52} Δ_{53} Δ_{54} 0		Δ_{51} Δ_{52} Δ^*_{53} Δ_{54} 0
b_1 b_2 b_3 b_4 b_5	Column effects	b^*_1 b^*_2 b^*_3 b^*_4 b^*_5
0 δ_{12} δ_{13} δ_{14}		0 δ^*_{12} δ_{13} δ_{14}
δ_{21} 0 δ_{23} δ_{24}	Square combining	δ^*_{21} 0 δ^*_{23} δ^*_{24}
δ_{31} δ_{32} 0 δ_{34}	table for rows	δ_{31} δ^*_{32} 0 δ_{34}
δ_{41} δ_{42} δ_{43} 0		δ_{41} δ^*_{42} δ_{43} 0
a_1 a_2 a_3 a_4	Row effects	a^*_1 a^*_2 a^*_3 a^*_4

[a]An asterisk indicates that a value for a calculated separation, row effect, or column effect may differ from the corresponding value for the complete table.

iterate to a satisfactory fit. Also, the final fit can depend on whether we polish first by rows or by columns. As with the square combining table, the iterative process of polish may spread the distortion from the missing value across most of the residuals.

Median polish and the square combining table are both resistant methods of fitting to a two-way table. Both will provide fairly similar fits to a given table, identical when the original table is perfectly additive (i.e., all residuals can be zero). The patterns of the residuals from the two fits will also be similar; both will tend to tag the same values as outliers, although the actual numerical values of the residuals will seldom be exactly the same. Because median polish (when it converges) forces all rows and columns in the table of residuals to have median zero, whereas the square-combining-table fit requires only the overall median of all residuals to be zero, we can

TABLE 2-16. Median-polish fit to the infant mortality data of Table 2-2 (after two iterations).

	Education of Father (in years)					Row
Region	≤ 8	9–11	12	13–15	≥ 16	Effect
Northeast	−1.1	0.5	0	−0.7	0.7	−1.6
North Central	1.1	−0.4	−4.0	0.7	−1.2	3.0
South	11.2	5.0	−0.1	−4.5	0	−0.4
West	−2.9	−5.6	0.2	3.1	0	0.3
Column effect	7.4	5.8	−0.8	0.0	−3.4	20.6

Source: John D. Emerson and David C. Hoaglin (1983). "Analysis of two-way tables by medians." In D. C. Hoaglin, F. Mosteller, and J. W. Tukey (Eds.), *Understanding Robust and Exploratory Data Analysis*. New York: Wiley (Table 6-9 on p. 175). Copyright © 1983 John Wiley & Sons, Inc. Reprinted by permission of John Wiley & Sons, Inc.

expect to see some greater clustering of residual values about zero for the median-polish fit, particularly when the number of rows or the number of columns in the two-way table is small.

If we look at a side-by-side display of the residuals from the square-combining-table and median-polish fits to the infant mortality data of Table 2-2, we see that the median-polish residuals have a slightly larger range and cluster more tightly about zero in comparison to the square-combining-table residuals. Table 2-16 gives the median-polish fit to the data of Table 2-2. Table 2-17 shows back-to-back stem-and-leaf displays of residuals from the two fits.

Figure 2-1 plots the median-polish residuals (on the vertical axis) against the square-combining-table residuals (on the horizontal axis). Agreement between the two sets of residuals is close. A fitted line would have slope close to 1 and intercept close to zero.

Table 2-18 gives the median-polish fit to the Gallup poll data of Table 2-12. Table 2-13 gives the square-combining-table fit. Like the two fits to Table 2-2, the two fits to Table 2-12 give similar pictures of the data. The ordering of the institutions is almost identical, and the year effects are small for both fits. The point plot in Figure 2-2 shows that agreement between the two sets of residuals is not as good as it was for the two fits to Table 2-2. One reason for this is the disagreement between the two tables of residuals in the rows representing newspapers and big business. Both fits produce a substantial number of residuals at or near zero.

TABLE 2-17. Back-to-back stem-and-leaf display of residuals from the square-combining-table and median-polish fits to the infant mortality data.

Square combining table		Median polish
	11	2
	10	
2	9	
	8	
	7	
	6	
	5	0
	4	
8	3	1
5	2	
750	1	1
8522	0	002577
221	−0	0147
660	−1	12
6	−2	9
	−3	
1	−4	05
9	−5	6
0	−6	

Patterns of Holes

The number of holes in Table 2-12 is relatively small in comparison with the number of cells in the table (9 versus 50), so we would not expect a great deal of difference between the two fits. When holes become more numerous, the pattern of missing values can guide the choice of an appropriate method of analysis. We choose the method that will recover the underlying row effects and column effects without excessive calculations. To simplify matters, we restrict our detailed attention to perfectly additive tables.

Consider a *perfectly* additive two-way table. If we remove any single value from the table, both the square-combining-table and median-polish methods will recover the original fit. The differences between values in the same row of any two given columns will be constant across rows in a perfectly additive table. When we form the square combining table for

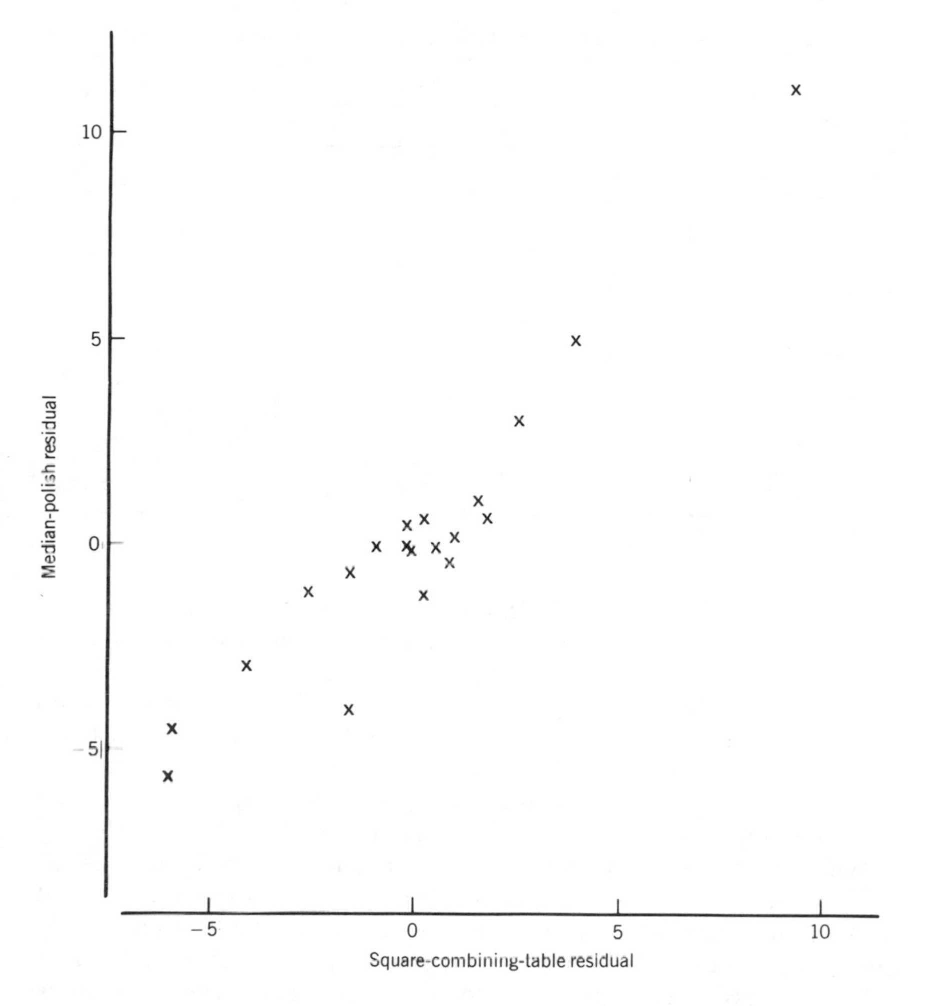

Figure 2-1. Median-polish residuals versus square-combining-table residuals for the infant mortality rate data.

columns, the Δ_{kj} are all medians of identical numbers. Thus, even if one value is missing, the median remains the same, and so does the final fit. In fact, the fit will remain the same as long as there is even a single column difference for each pair of columns. This will always be true if fewer than I values are missing, where I is the number of rows. Similarly, the δ_{st} will all be present so long as fewer than J values are missing. So for a perfectly additive table, the square-combining-table fit remains the same as long as

TABLE 2-18. Median-polish fit to the Gallup poll data (after two iterations).

	Year					Row
Institution	1980	1979	1977	1975	1973	Effect
Organized religion	0	−1	−2	0	0	22
Banks	0	0				16
The military	−4	−2	1	0		12
Public schools	−3	−1	0		4	10
U.S. Supreme Court	1	−1	0	1	−2	2
Newspapers	0	9			−3	−2
Organized labor	−1	0	3	0	−6	−8
Congress	−4	−4	2	0	4	−6
Television	−4	1			0	−7
Big business	3	6	−4	−5	0	−18
Column effect	0	0	0	2	0	44

the number of missing values is less than $\min(I, J)$. The effort required to calculate the square-combining-table fit does not increase. With fewer available differences, the fit actually involves less calculation.

The median-polish method also will eventually recover the original fit to a perfectly additive table with fewer than $\min(I, J)$ missing values. However, when more than one data value is missing, the number of iterations to convergence can more than double. The square combining table provides a more convenient method of analysis in this situation.

When the number of holes is not less than $\min(I, J)$, the location of the holes within the table can indicate which method would give the better fit (better in the sense of closer to the fit for the complete table). Consider the case of a perfectly additive table with $\min(I, J)$ missing values. Table 2-19 shows two possible patterns for the holes in a 3×3 table. The first pattern, Table 2-19*a*, has one hole in each row and column. The Δ_{kj} and δ_{st} are all available, although they are medians of single differences. The median-polish fit requires four iterations, twice the number needed for the original table. The pattern of holes favors using the square combining table.

In Table 2-19*b* the complementary pattern of holes between the first and third rows makes it impossible to calculate a value for δ_{13}, the separation between those two rows. To proceed with a square-combining-table analysis, we must choose a value for δ_{13}. One possible choice is to treat δ_{13} as zero. This assumes that, if we cannot observe a difference directly, none exists. The resulting square-combining-table fit in Table 2-19*d* does not recover the row effects.

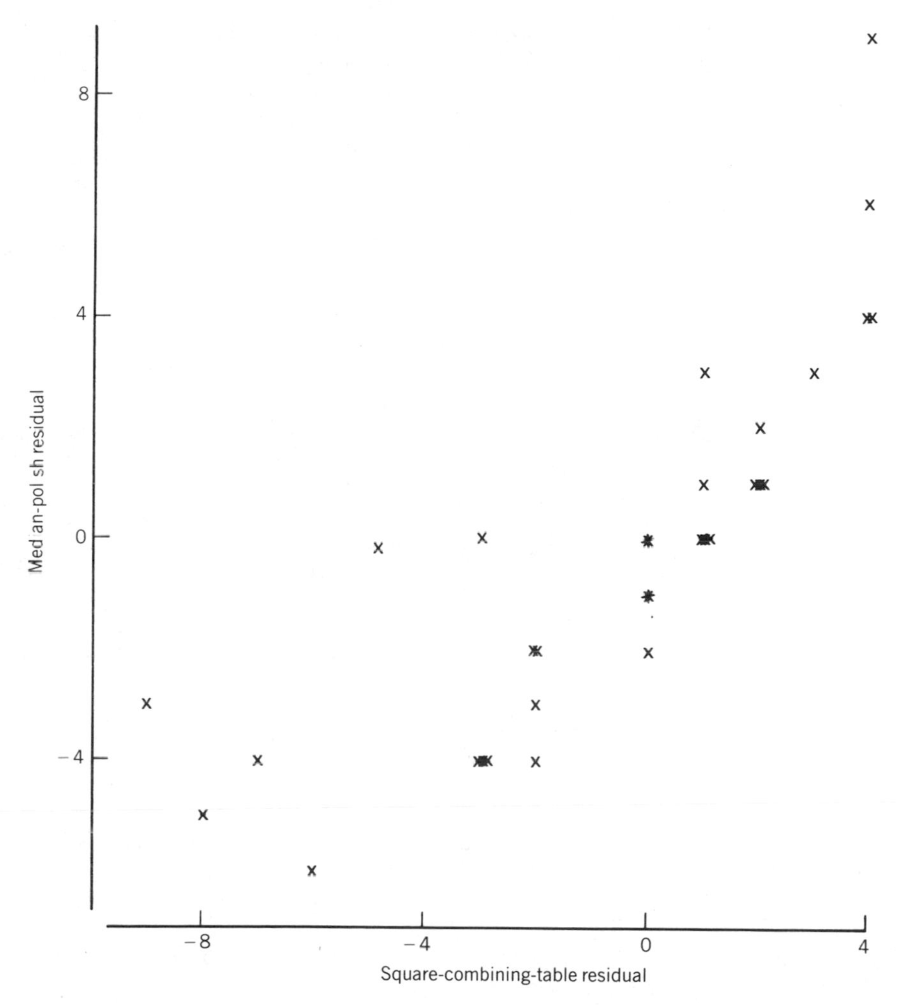

Figure 2-2. Median-polish residuals versus square-combining-table residuals for the Gallup poll data.

Other choices are available, as long as the data table does not separate into two (or more) disconnected tables with no overlap between either rows or columns. These involve indirect comparisons and the use of further combining tables. For example, applying a second combining table to the rows of the square combining table for rows of Table 2-19*b* estimates δ_{13} from δ_{12} and δ_{32}.

TABLE 2-19. Two missing-value patterns in a 3 × 3 table; *X* indicates a missing value.

a. A pattern with one hole in each row and column

X	2	3
4	5	X
7	X	9

b. An alternating or zigzag pattern of holes

X	2	X
4	5	6
7	X	9

c. Median-polish fit to pattern b

0	0	0	−3
0	0	0	0
0	0	0	3
−1	0	1	5

d. Square-combining-table fit to pattern b (assigning $\delta_{13} = 0$)

−2	−2	−2	−1
0	0	0	0
2	2	2	1
−1	0	1	5

e. The complete table

1	2	3
4	5	6
7	8	9

Although median polish takes eight iterations to converge for the data of Table 2-19*b*, it does produce the original additive fit, as shown in Table 2-19*c*.

Technically, in deriving estimates from two-way tables with holes, we could weight the Δ_{kj} and δ_{st} according to the number of differences on which each is based. A proper unbalanced analysis of variance incorporates a similar weighting. The square-combining-table technique sacrifices weighting information for the sake of simplicity.

We offer the following general guidelines for a perfectly additive table with I rows and J columns and H missing values.

1. If $H < \min(I, J)$, the median-polish and square-combining-table methods will both produce the true fit to the data, but median polish is likely to take longer than usual to converge.
2. If $H \geq \min(I, J)$, median polish is preferable when the pattern of missing values produces a missing entry in one of the square combining tables—that is, if one of the Δ_{kj} or δ_{st} cannot be calculated. Median polish will take longer than usual to converge. If all Δ_{kj} and δ_{st} are available, the square-combining-table method will be faster.

These results apply only to perfectly additive tables, but if a table is close to being additive, the guidelines should not be far wrong.

At the other extreme, suppose that a substantial fraction f of the entries in a data table are sour. Then, applying this fraction to each column (or row), the larger fraction $1 - (1 - f)^2 = 2f - f^2$ of the differences will be sour, too. The value of this expression reaches $\frac{1}{2}$ when $f = .293$. Because the median breaks down when half the observations are sour, the square combining tables will also be sour when $f \geq .293$. In choosing between median polish and the square combining table, $f = .2$ seems to be a reasonable changeover point. As long as no more than 20% of the data actually present are sour, the square-combining-table method will work and, in the presence of holes, will reduce effort. This range of f covers the vast majority of data sets. When the percentage of sour data exceeds 20%, we may have to use median polish, no matter how slow it may be.

If the data table is nonadditive in an orderly way—if, for example, it has a multiplicative interaction between the row factor and the column factor—a transformation of the original data may be in order before the analysis proceeds (UREDA, Sections 6H and 8F).

2E. SUMMARY

The square combining table uses measurements on differences between pairs of items to produce estimates of the effects for the items, locating them on an ordered scale. The rationale resembles that behind the method of paired comparisons.

To estimate row effects or column effects for a two-way table, we treat the levels of one factor as the items, form all possible pairwise differences of these within levels of the other factor, calculate the median of these differences, and place these separations in a square combining table. The column means of the square combining table are the estimates of the effects. We can obtain these estimates whenever we can form the square combining table, even if all the data take the form of comparisons.

The square combining table is especially suited to fitting tables with missing data values. As long as the square combining table can be formed, the analysis requires no greater investment of time or effort than the analysis of a complete table. It offers an alternative simple resistant method to median polish, which can take many iterations to converge to a final fit when the table has missing values.

REFERENCES

David, H. A. (1963). *The Method of Paired Comparisons*. New York: Hafner Publishing Company.

Fleckenstein, M., Freund, R. A., and Jackson, J. E. (1958). "A paired comparison test of typewriter carbon papers," *Tappi*, **41**, 128–130.

Guilford, J. P. (1954). *Psychometric Methods*, 2nd ed. New York: McGraw-Hill.

Seber, G. A. F. (1977). *Linear Regression Analysis*. New York: Wiley.

Thurstone, L. L. (1927). "A law of comparative judgment," *Psychological Review*, **34**, 273–286.

Additional Literature

Bradley, R. A. (1976). "Science, statistics, and paired comparisons," *Biometrics*, **32**, 213–239 (with discussion).

Davidson, R. R. and Farquhar, P. H. (1976). "A bibliography on the method of paired comparisons," *Biometrics*, **32**, 241–252.

EXERCISES

1. (a) Apply the square-combining-table technique of Section 2B to the scale separation matrix in Table 2-11. (Note that, because this matrix is antisymmetric, each row effect will be the negative of the corresponding column effect.)

(b) Construct the fit and calculate residuals. How closely do the differences between the resistant vegetable effects reproduce the scale separations?

(c) Use the column means in Table 2-11 to construct a nonresistant fit, and calculate its residuals. Compare them to the ones obtained in part (b).

2. Fill in the calculations for the square combining tables for rows and columns for the analysis in Table 2-13 (the Gallup Poll data).

3. Gulliksen and Tukey (1958) report on the confirmatory analysis of data consisting of paired comparisons among nine specimens of handwriting. Table 2-20 summarizes the preference judgments made by two groups of 100 students each. One group was at Chicago in the late 1930s, and the other was at Princeton in the late 1940s. Analyze the two sets of data by means of the square combining table to obtain effects for the nine specimens.

TABLE 2-20. Preferences between handwriting specimens (entries are percentage favored minus percentage not favored).

Minus Specimen	Plus Specimen								
	50a	50b	50c	70a	70b	70c	80a	80b	80c
a. Chicago data									
50a	0	4	34	90	98	96	98	94	88
50b	−4	0	20	70	90	92	96	96	90
50c	−34	−20	0	52	56	84	82	72	92
70a	−90	−70	52	0	52	74	90	58	56
70b	−98	−90	−56	−52	0	48	60	4	42
70c	−96	−92	−84	−74	−48	0	18	−48	12
80a	−98	−96	−82	90	−60	−18	0	−70	−38
80b	−94	−96	−72	−58	−4	48	70	0	22
80c	−88	−90	−92	−56	−42	−12	38	−22	0
b. Princeton data									
50a	0	4	32	76	96	96	94	66	72
50b	−4	0	20	38	94	92	86	88	82
50c	−32	−20	0	40	64	88	84	68	86
70a	−76	−38	−40	0	56	68	82	40	66
70b	−96	−94	−64	−56	0	28	56	−26	22
70c	−96	−92	−88	−68	−28	0	42	−40	16
80a	−94	−86	−84	−82	−56	−42	0	−70	−24
80b	−66	−88	−68	−40	26	40	70	0	40
80c	−72	−82	−86	−66	−22	−16	24	−40	0

Source: Harold Gulliksen and John W. Tukey (1958). "Reliability for the law of comparative judgment," *Psychometrika*, **23**, 95–110 (entries calculated from Table 1 on page 99).

TABLE 2-21. Fraction of games won between individual American League baseball teams in 1948.

Losing	Winning Team							
Team	Cleveland	Boston	New York	Philadelphia	Detroit	St. Louis	Washington	Chicago
Cleveland	—	.478	.545	.273	.409	.364	.273	.273
Boston	.522	—	.364	.455	.318	.318	.318	.364
New York	.455	.636	—	.455	.409	.273	.227	.273
Philadelphia	.727	.545	.545	—	.545	.182	.364	.273
Detroit	.591	.682	.591	.455	—	.500	.273	.364
St. Louis	.636	.682	.727	.818	.500	—	.545	.381
Washington	.727	.682	.773	.636	.727	.455	—	.429
Chicago	.727	.636	.727	.727	.636	.619	.571	—

Source: Frederick Mosteller (1951). "Remarks on the method of paired comparisons: III. A test of significance for paired comparisons when equal standard deviations and equal correlations are assumed," *Psychometrika*, **16**, 207–218 (data from p'_{ij} Table, p. 210). Reprinted with permission.

TABLE 2-22. Annual premium (in dollars) for a $30,000 "Homeowners 2" Insurance Policy in 1974.

	City				
Company	Marietta, GA	Mill Valley, CA	New Rochelle, NY	St. Cloud, MN	Webster Groves, MO
A	127	105	125	131	140
F	114	138	129	124	143
G	95	106	105	121	95
I	122	121	124	112	150
N	95	116	130	abs[a]	abs[a]
S	99	114	113	98	92

[a] "abs" = absent—Company *N* did not sell insurance in these two cities.
Source: *Money*, February 1974, p. 76.

4. Repeat the analysis in Exercise 3, but omit all entries in the two tables that are larger in absolute value than 80. Contrast the results with those for the analyses of the complete tables.

5. Table 2-21 presents data from Mosteller (1951) on the proportion of games that each American League baseball team won against each of the seven other teams in 1948. What operations are necessary before applying a square combining table to these data? Manipulate the data as necessary and perform the analysis.

6. Table 2-22 displays annual premiums for a particular homeowner's insurance policy in five different cities for six different companies. Make a square-combining-table fit to these data and comment.

7. Eden and Fisher (1927) reported the yield of straw from a crop of oats under eight fertilizer treatments in eight blocks. Table 2-23 gives the data after removing three-eighths of the values in a balanced systematic pattern. Analyze by means of a square combining table. Are there appreciable block effects as well as treatment effects?

8. One approach to residuals (mentioned in UREDA, Section 5A) asks that they produce a zero fit when put through the fitting procedure in place of the original data values. Do the residuals from a square-combining-table fit to a two-way table satisfy this definition of residuals?

TABLE 2-23. Yields of straw under eight different manurial treatments.

	Treatment							
Block	*A*	*B*	*C*	*D*	*E*	*F*	*G*	*H*
1	242	—	—	—	322	200	260	203
2	321	382	—	—	—	261	318	275
3	261	201	298	—	—	—	266	207
4	317	316	381	255	—	—	—	331
5	255	280	300	238	232	—	—	—
6	—	285	294	309	393	258	—	—
7	—	—	256	283	351	306	276	—
8	—	—	—	324	363	376	385	328

Source: T. Eden and R. A. Fisher (1927). "Studies in crop variation. IV. The experimental determination of the value of top dressings with cereals," *Journal of Agricultural Science*, **17**, 548–562 (data from Table II, p. 556).

9. Assuming that the fluctuations ε_{ij} in equation (1b) are distributed as independent Gaussian variables with zero mean and identical variance σ^2, and using the asymptotic result that the variance of the median of a Gaussian sample is $(\pi/2)(\sigma^2/n)$, produce an estimate for the variance of the column separations Δ_{kj} and row separations δ_{st}. From these results, give an estimate of the variance of the fitted row and column effects, a_i and b_j, defined by equations (4) and (5).

CHAPTER 3

Resistant Nonadditive Fits for Two-Way Tables

John D. Emerson
Middlebury College

George Y. Wong
Memorial Sloan-Kettering Cancer Center

In the data structure known as a two-way table, each of two factors varies separately from the other, creating a rectangular array of cells. Each cell contains a numerical value of the response variable. For example, median family income may vary with both the ethnic group and the metropolitan area where the family lives.

A simple additive model represents the response as a sum of a common term, a main effect for each row, a main effect for each column, and a fluctuation term for each cell. This model resembles the two-way analysis-of-variance model which includes in the error term any interaction between rows and columns. However, the analysis-of-variance model ordinarily assumes that the errors have independent Gaussian distributions with mean zero and with common variance. This chapter's models do not rely on such assumptions.

This chapter explores fits for various models. A fit $\hat{y}_{ij}$ for a two-way table describes the data through the equation

$$y_{ij} = \hat{y}_{ij} + e_{ij}.$$

Although we may sometimes regard the components of the fit—for example, the common term and the main effects—as estimates of similar components of a model, in a data-analysis setting we more often emphasize the fitted parameter values and especially the residuals from the fit.

Many considerations may guide choices among fits and among techniques for obtaining them; these include simplicity of description, resistance, and robustness. Median polish is a resistant way of providing a simple additive fit to a two-way table. Other fitting techniques (for example, least squares and least sum of absolute residuals) usually give different simple additive fits. Furthermore, median polish can give slightly different fits that depend on whether the polish begins with rows or with columns and on how many steps of iteration are used.

When the residuals from a simple additive fit reveal systematic departures from additivity, fitting a more complex model to the table can sometimes help us understand and summarize the nonadditive structure. This chapter discusses resistant methods for analyzing the data of a two-way table according to models that have more terms than the simple additive model. We refer to any fit for such a model as a nonadditive fit.

Section 3A reviews the structure of the two-way table and the method of median polish for fitting the simple additive model resistantly. Section 3B reviews the diagnostic plot for examining departures from additivity and for fitting a simple multiplicative term, and Section 3C gives ways to compare the results of the fits. Section 3D describes a method for fitting a multiplicative model to a table of positive values, and Section 3E provides some technical details that help extend this method to tables containing nonpositive entries. Section 3F then assembles these techniques to provide a resistant method that fits a more general model having both additive and multiplicative terms. Section 3G gives some historical background, especially for least-squares fits of multiplicative models, and it provides some further references. Section 3H summarizes the various models and fits discussed in this chapter.

3A. THE SIMPLE ADDITIVE MODEL AND MEDIAN POLISH

A *two-way table* with one observation per cell is denoted by

$$y_{ij}, \qquad i = 1, \ldots, I; \quad j = 1, \ldots, J; \tag{1}$$

and often displayed in a rectangular array with I rows and J columns. The rows correspond to I levels of one factor, and the columns correspond to J levels of a second factor. We refer to the intersection of row i and column j as the (i, j) *cell*.

We want to summarize the observed data y_{ij} and to describe their underlying structure. If each y_{ij} is related to the pair (i, j) by a function $g(i, j)$ and a fluctuation term ε_{ij}, a formal and general statistical model is

$$y_{ij} = g(i, j) + \varepsilon_{ij}. \tag{2}$$

We attempt to discover the form of a suitable function g by analyzing the observed data; in choosing g, we emphasize simplicity where possible. As a first step, we try the *simple additive model*, where

$$g(i, j) = \mu + \alpha_i + \beta_j \tag{3}$$

expresses y_{ij} as the sum of a constant term μ, a row effect α_i for level i of the row factor, a column effect β_j for level j of the column factor, and a fluctuation term. The simple additive model is also called a *main-effects model*, and it has a simple interpretation because the separate contributions of the factors are added together. The customary centerings of the α_i and the β_j, so that their means (or perhaps their medians) are zero, aid this interpretation and offer technical conveniences.

Median polish iteratively obtains a resistant fit,

$$\hat{y}_{ij} = m + a_i + b_j, \tag{4a}$$

and a table of *residuals*,

$$e_{ij} = y_{ij} - \hat{y}_{ij}, \tag{4b}$$

corresponding to the additive model

$$y_{ij} = \mu + \alpha_i + \beta_j + \varepsilon_{ij}. \tag{5}$$

We briefly review this method of analysis here, and we refer to Chapter 6 of UREDA for a detailed discussion of median polish and of other methods.

Unlike least-squares fitting, which uses arithmetic means to summarize and remove the main effects, median polish provides a more resistant analysis. We summarize this technique as a sequence of steps aimed at producing the residuals e_{ij}. Chapter 6 of UREDA gives a formal algebraic treatment of these steps.

Steps for Median Polish

1. Calculate the median for each row in the table, record these values in a column at the right-hand margin, and replace each entry of the table with the difference between that entry and the median for its row.
2. Calculate the median for each column of the current table, including the column of row medians. Record these values in a row at the lower

margin of the table, and replace each entry of the table (and of the column of row medians) with the difference between that entry and its associated column median. These steps comprise one *full step* of median polish.

3. Calculate the median for each row in the current table (including the row of column medians), subtract this value from each of the current row entries, and add it to the corresponding entry at the right margin.
4. Calculate the median value for each column (including the extra column at the right), subtract this value from each entry in its column, and add it to the corresponding entry of the row at the lower margin of the table. We have now completed two full steps of median polish.
5. Repeat steps 3 and 4 until further adjustments to row medians and column medians are negligible. Although two full steps often suffice, in rare instances as many as 10 full steps are needed.
6. The final entry in the lower right corner is m, the a_i are the other entries in the right margin, and the b_j are the entries in the lower margin. The values that remain in the body of the table are the residuals, e_{ij}, from the additive fit.

We illustrate median polish in the examples of this chapter. Median polish can begin with columns instead of rows; although the results are often not identical to those obtained by beginning with rows, the two fits are very similar and equally useful in most applications.

Unlike the corresponding analysis of a two-way table using means, median polish does not ensure that any criterion of mathematical optimality is met. For example, whereas analysis by means minimizes the sum of squared residuals, median polish need not minimize the sum of absolute residuals, even though no step of the process can increase this sum (UREDA, Section 6D). But in practice, a lack of theoretical focus upon median polish does not reduce its effectiveness in resistantly exploring two-way tables. We have found that, for many two-way tables of data, the median-polish fit does achieve the minimum value of the sum of absolute residuals. Sections 6C, 6D, and 6F of UREDA examine the strengths and limitations of median polish in some detail.

We performed nearly all calculations for this chapter using the Minitab statistical package (Ryan, Joiner, and Ryan, 1982). Apparent discrepancies in the results of calculations presented in the text and in tables sometimes arise because, although we present rounded results, we retained many decimal places when doing the computer calculations.

3B. ONE STEP BEYOND AN ADDITIVE FIT

The residuals e_{ij} from median polish will exhibit systematic patterns when the additive fit of equation (4a) cannot capture all the structure of the table. We then look for a more appropriate form for $g(i, j)$ than that of equation (3)—one that can describe *interaction* between the row factor and the column factor.

For a two-way table with multiple observations (or *replications*) in each cell, classical two-way analysis of variance tests for interaction by separating nonadditivity among the cell means from variability within the cells (Scheffé, 1959, Chapter 4). Snedecor and Cochran (1980, Chapters 14 and 20) describe and illustrate the least-squares analysis both with equal and unequal numbers of replications within cells. Our interest here, however, is in systematic patterns of interaction that can be detected even when each cell has a single observation.

An Extended Fit

To test for systematic interaction in a two-way table without replication, Tukey (1949) proposed the model

$$y_{ij} = \mu + \alpha_i + \beta_j + \kappa\alpha_i\beta_j + \varepsilon_{ij}. \tag{6}$$

This *extended model* describes a relatively uncomplicated form of nonadditivity. We write the fit associated with equation (6) as

$$\hat{y}_{ij} = m + a_i + b_j + ka_ib_j \tag{7}$$

and call it an "extended fit" or a "fit with one degree of freedom for nonadditivity." Although we use median polish here to provide the estimates m, a_i, and b_j, the classical approach to fitting model (6) uses means.

One measure of the complexity of a model for an $I \times J$ table is the number of parameters that it uses. Where possible we prefer to use relatively few parameters. The simple additive model of equation (5) uses $1 + I + J$ parameters, and the extended model has one additional parameter, κ. For this reason, Tukey's model is said to have "one degree of freedom for nonadditivity." We use the expression "degree of freedom" loosely to describe the number of parameters of a model minus the number of constraints on these parameters. Because the α_i and the β_j each satisfy one constraint by being centered at 0, we think of model (5) as having $I + J - 1$ degrees of freedom. Similarly, model (6) has $I + J$ degrees of freedom.

Later in this chapter we consider models involving additional degrees of freedom.

The Diagnostic Plot

Although we can often discover nonadditive structure by examining the table of residuals from an additive fit, the *diagnostic plot* of these residuals, e_{ij}, against the *comparison values*, $a_i b_j / m$, more effectively reveals systematic departures from additivity. If this specialized plot shows a systematic pattern or trend, we need to account for row-by-column interaction in the table, either by transforming or by adopting a model with an additional term (such as the extended model).

In principle, a power transformation of y, say $z = y^p$ for a suitable value of p, can approximately remove the form of nonadditivity in equation (6). The linear regression of e_{ij} on the product $a_i b_j$ (which estimates $\alpha_i \beta_j$) has approximate slope $k = (1 - p)/m$, provided that the main effects a_i and b_j are not too large relative to the common term m. Thus $p = 1 - mk$ is the approximate power. Section 6H of UREDA discusses and illustrates transformation to an additive scale, and Section 8F of UREDA gives some related mathematical background.

If the diagnostic plot is linear, it suggests an extended fit according to model (6). Then we interpret the slope as km and solve for k as an estimate of κ. The residuals from this extended fit, r_{ij}, are related to the residuals from the additive fit, e_{ij}, through

$$e_{ij} = k a_i b_j + r_{ij}. \tag{8}$$

EXAMPLE: SPECIFIC VOLUME OF RUBBER

Table 3-1, analyzed by Mandel (1969a), displays the specific volume of peroxide-cured rubber at four temperatures and six pressures. As the results of careful laboratory measurements, these data have higher precision than many data sets. A statistical look at this table of data aids in understanding its structure, and the greater precision allows us to probe deeper. We have used median polish and the diagnostic plot to give a resistant fit to this table, according to equation (7).

To begin analyzing the structure of Table 3-1, we fit a simple additive model to the data, obtaining the main-effects analysis shown in Table 3-2. Although the residuals from the additive fit are relatively small, their systematic sign pattern (+ in NW and SE corners and − in SW and NE corners) indicates the type of additional structure, beyond a main-effects model, that an extended fit could help to describe. The diagnostic plot in

TABLE 3-1. Specific volume (in cubic centimeters per gram) of peroxide-cured rubber at four temperatures (in degrees Celsius) and six pressures (in kilograms per square centimeter above atmospheric pressure).

	Pressure					
Temperature	500	400	300	200	100	0
0	1.0637	1.0678	1.0719	1.0763	1.0807	1.0857
10	1.0697	1.0739	1.0782	1.0828	1.0876	1.0927
20	1.0756	1.0801	1.0846	1.0894	1.0944	1.0998
25	1.0786	1.0830	1.0877	1.0926	1.0977	1.1032

Source: L. A. Wood and G. M. Martin (1964). "Compressibility of natural rubber at pressures below 500 kg/cm^2," *Journal of Research of the National Bureau of Standards—A. Physics and Chemistry*, **68**A, 259–268 (data from Table 1, p. 260).

Figure 3-1 suggests that a model with one degree of freedom for nonadditivity will improve the fit.

Fitting a resistant line (UREDA, Chapter 5) to this diagnostic plot gives a slope estimate of 7.81. This estimate, denoted by km in the discussion above, leads to $k = km/m = 7.21$. The new residuals, $r_{ij} = y_{ij} - (m + a_i + b_j + ka_ib_j)$, and the values of the multiplicative term, ka_ib_j, give the entire extended fit to these data, as shown in Table 3-3. For this example, the indicated power transformation would have an approximate power of $p = 1 - 7.81 = -6.81$; we much prefer to use an extended fit than to adopt such an unwieldy transformation.

TABLE 3-2. Simple additive fit by median polish for the rubber data (unit: 10^{-4} cm^3 / g).

	Pressure						
Temperature	500	400	300	200	100	0	a_i
0	7.0	4.5	1.5	−1.5	−6.5	−9.0	−96.5
10	3.0	1.5	0.5	−0.5	−1.5	−3.0	−32.5
20	−3.0	−1.5	−0.5	0.5	1.5	3.0	32.5
25	−4.5	−4.0	−1.0	1.0	3.0	5.5	64.0
b_j	−111.0	−67.5	−23.5	23.5	72.5	125.0	10837.5 = m

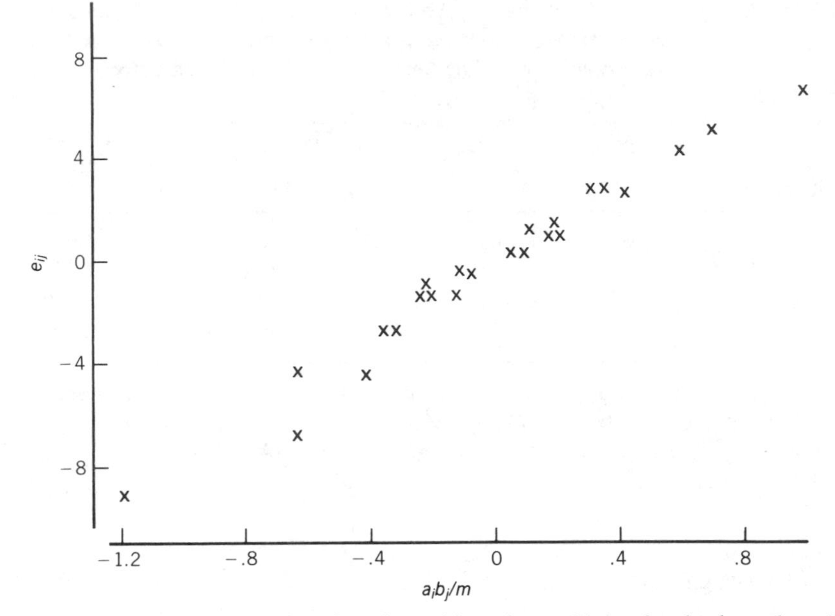

Figure 3-1. Diagnostic plot for the rubber data. (Units for both scales: 10^{-4} cm^3/g.)

TABLE 3-3. Extended fit for rubber data (unit: 10^{-4} cm^3/g).

	Values for ka_ib_j						
Temperature	500	400	300	200	100	0	a_i
0	7.7	4.7	1.6	−1.6	−5.0	−8.7	−96.5
10	2.6	1.6	0.6	−0.6	−1.7	−2.9	−32.5
20	−2.6	−1.6	−0.6	0.6	1.7	2.9	32.5
25	−5.1	−3.1	−1.1	1.1	3.3	5.8	64.0
b_j	−111.0	−67.5	−23.5	23.5	72.5	125.0	10837.5 = m
	Residuals (r_{ij})						
0	−0.73	−0.20	−0.14	0.14	−1.45	−0.30	
10	0.40	−0.08	−0.05	0.05	0.20	−0.07	
20	−0.40	0.08	0.05	−0.05	−0.20	0.07	
25	0.63	−0.88	0.08	−0.08	−0.35	−0.27	

The strong linear trend in the diagnostic plot of Figure 3-1 ensures that an extended fit to this data table will offer a substantial improvement over the simple additive fit; that is, we expect the residuals from the extended fit to be smaller in magnitude and to have less pattern than the residuals from the simple additive fit. A comparison of the residuals in Tables 3-2 and 3-3 quickly confirms that this happens. The residuals from Table 3-3 lack the strong pattern in Table 3-2, and they are only about one-tenth as large. Note that the original data are given to the nearest unit in this scale and that only four of the 24 residuals in Table 3-3 would round to a value other than zero. In the next section we extend the comparison of these sets of residuals and discuss additional ways to compare fits to a two-way table.

Like median polish, the algorithm that gives an extended fit should be regarded as iterative. Although the residuals e_{ij} from median polish of a two-way table have median 0 for all rows and all columns (at least, in principle), the residuals r_{ij} from the extended fit may no longer have this property. Thus, we can repolish the table of r_{ij} and combine the new common term, row effects, and column effects with those obtained from the original median polish for the two-way table we started with. Once these adjustments are made, we can revise the value of k, now using the adjusted values of m, a_i, and b_j, and we can recalculate the extended part of the fit—the term ka_ib_j. A new table of residuals, which we continue to denote by r_{ij}, can thus be given. This process of median polish followed by calculation of a new extended term may, in principle, be repeated indefinitely. In practice, we seldom need to repeat it more than once before (for all practical purposes) convergence is achieved. Exercise 7 invites the reader to try the second step (which clearly suffices) for the rubber data.

Internal versus External Structure

Most of the discussion of this chapter focuses on what one might call the *internal* structure of a two-way table. That is, we ask what patterns and trends lie hidden in the numbers that make up the table. With some tables, like the rubber data in Table 3-1, it may be possible to go a step further and try to discover relationships between a table's internal structure and variables that are external to, but related to, the table. With the rubber data, for example, we can ask how the main effects are related to the levels of temperature and pressure.

EXAMPLE: SPECIFIC VOLUME OF RUBBER (CONTINUED)

Table 3-2 shows that the volume of rubber increases monotonically with increasing temperature, at all levels of pressure. Similarly, the volume

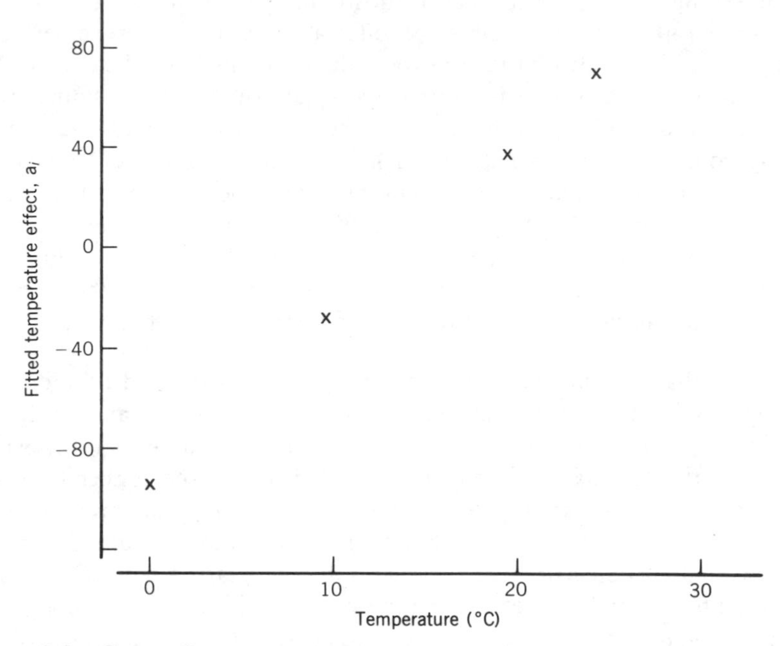

Figure 3-2. Row effects versus temperature in the rubber data. Least-squares regression line: $a(t) = .000643t - .00966$. [Units of a_i: 10^{-4} cm^3/g. Note that $a(t)$ is in cm^3/g, the units of the original data.]

increases with decreasing pressure, at each level of temperature. To quantify these trends, we plot the row effects a_i against temperature, and we plot the column effects b_j against pressure. Figures 3-2 and 3-3 give these plots and the equations of fitted least-squares regression lines; we use least-squares fits instead of the resistant line. Here resistance is not a concern because these effects have no outliers and very little scatter, and because one plot contains only four points.

The regression lines provide good fits of the effects against temperature and pressure, respectively. For temperatures ranging from 0 to 25°C and for pressure in kilograms per square centimeter ranging from 0 to 500, we can conclude that volume depends linearly on temperature at a given pressure and nearly linearly on pressure at a given temperature. Note that the effects tell about this only on the average. (Figure 3-3 suggests slight curvature. As a refinement, we could include a quadratic term in pressure.) How then does volume depend on temperature and pressure together? We approach this question with the aid of the extended fit.

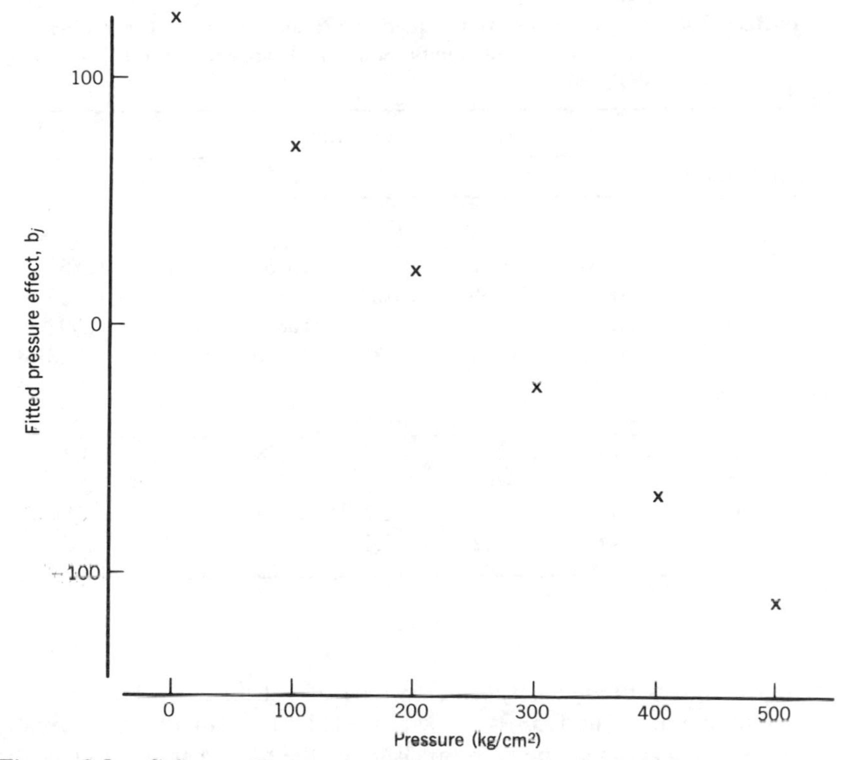

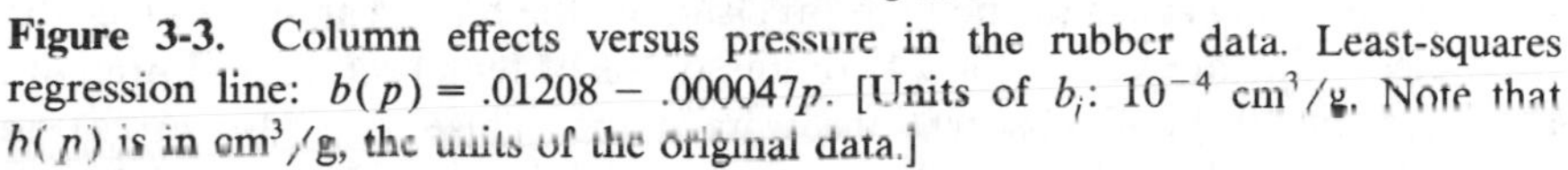

Figure 3-3. Column effects versus pressure in the rubber data. Least-squares regression line: $b(p) = .01208 - .000047p$. [Units of b_j: 10^{-4} cm^3/g. Note that $b(p)$ is in cm^3/g, the units of the original data.]

The form of the extended fit is

$$\hat{y}_{ij} = m + a_i + b_j + ka_ib_j,$$

and we obtained $m = 1.084$ and $k = 7.21$. If we let $y(t, p)$ denote volume when the temperature is t and the pressure is p, the fitted lines $a(t)$ and $b(p)$ given in Figures 3-2 and 3-3 lead us to the equation

$$\begin{aligned}\hat{y}(t, p) &= 1.08375 + (.000643222t - .00965688) \\ &\quad + (-.0000470571p + .0120809) \\ &\quad + 7.2135(.000643222t - .00965688) \\ &\quad \times (-.0000470571p + .0120809) \\ &= 10^{-4}(10853. + 6.9928t - .43779p - .0021834tp). \qquad (9)\end{aligned}$$

TABLE 3-4. Fitted values from equation (9) and residuals for rubber data, based on temperature and pressure (unit: 10^{-4} cm^3/g).

	Pressure					
Temperature	500	400	300	200	100	0
	Predicted values					
0	10634	10678	10722	10766	10810	10853
10	10693	10739	10785	10831	10877	10923
20	10752	10801	10849	10897	10945	10993
25	10782	10831	10880	10930	10979	11028
	Residuals					
0	2.6	−0.2	−3.0	−2.8	−2.5	3.7
10	3.6	−0.4	−3.7	−3.3	−1.3	3.7
20	3.5	0.4	−2.7	−2.9	−1.0	4.8
25	4.0	−1.2	−3.4	−3.7	−1.9	3.9

Equation (9) gives predicted specific volume of vulcanized rubber at each value of temperature and pressure; we needed to retain unusually many decimal places because of the high precision of the data. Table 3-4 gives the predicted value at each of the 24 combinations of temperature and pressure at which the original observations were made. This table also gives the residuals from these predicted values. As we expect because of the constraints imposed by modeling row effects as a linear function of temperature and column effects as a linear function of pressure, the residuals from this fit are neither as small nor as free of structure as those from the extended fit, given in Table 3-3. Indeed, an analysis not presented here reveals that allowing $b(p)$ to be quadratic in p produces further improvements in the fit given in Table 3-4. However, the residuals in Table 3-4 do compare favorably with those from the simple additive fit, given in Table 3-2.

Although the fit given by equation (9) is not as good as that provided by the extended fit of Table 3-3, it does offer the advantage of providing predicted volumes at combinations of temperature and pressure different from those at which observations were made. Thus, we can use the equation to predict volume when the temperature is 15°C and the pressure is 225 kg/cm^2, and we can reasonably expect the prediction to be correct to within 0.0005 cm^3/g. We would be more cautious in making predictions when either temperature or pressure is outside the ranges studied in the experiment that produced these data; such caution is reinforced by an apparent slight departure from linearity at a pressure of 500 kg/cm^2.

Equation (9) gives a fit that ultimately uses only four constants, whereas the extended fit uses 10 degrees of freedom; thus the fit of equation (9) is more parsimonious. Furthermore, now that the two-way table structure has led us to discover a 4-carrier model, we could fit that model (by least squares or resistantly) to obtain improved estimates of the coefficients. We could then study the residuals from the multiple regression fit against t, p, tp, and perhaps other carriers, to consider further expansion or refinement of the model. Chapters 7 and 8 discuss resistant and robust methods for fitting multiple regression models.

In the development that led to equation (9), we have used only the information in the data set, including the values of temperature and pressure. Although we recognize (and appreciate) that scientific laws govern the relationships among volume, temperature, and pressure, our desire here is to explore an interesting data set without the assistance of such information. We refer readers to a science-based discussion of this data set by Mandel (1969a). His exploratory analysis uses least-squares techniques rather than the resistant techniques described and illustrated in this chapter. The residuals from his extended fit are larger than most of ours but not as large as the largest residuals shown in Table 3-3, and his fit using temperature and pressure is better than that given in Table 3-4 because he retains terms that are quadratic in pressure.

3C. ASSESSING AND COMPARING FITS

To explore the structure of a two-way table, we often try several different fits associated with different models. The residuals from these fits tell how good each fit is and something about the nature of departures from a fit. To assess fits, or to compare various fits to a table, we must examine residuals in a systematic way. In this section, we review and illustrate existing techniques, and we develop an additional technique.

Graphical Displays

A stem-and-leaf display of residuals can help us to see whether the residuals are centered at zero, whether and by how much they are skewed, and which residuals, if any, are outliers. Table 3-5 gives stem-and-leaf displays of the residuals from two fits to the rubber data: the simple additive fit and the extended fit. Although these displays tell us about each of the batches of residuals, other forms of display are convenient for comparing residuals.

TABLE 3-5. Parallel stem-and-leaf displays of residuals from two fits to the rubber data. (leaf digit: 10^{-5}; the residuals from Table 3-3 have been rounded.)

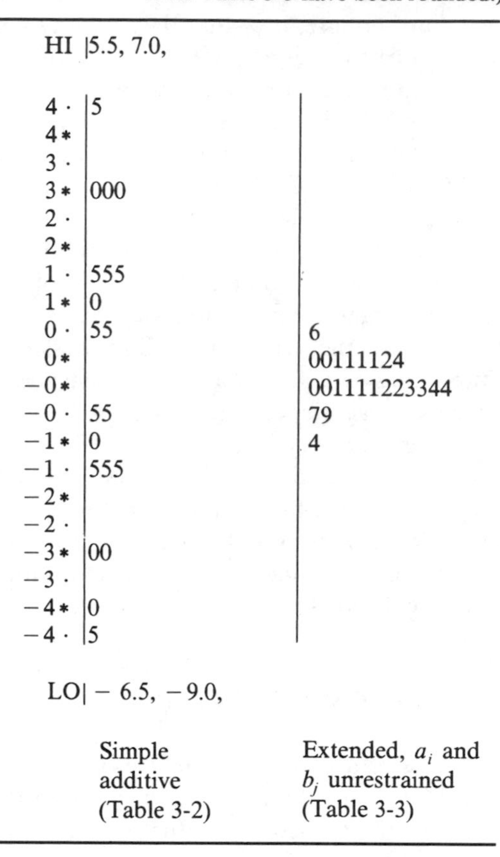

	Simple additive (Table 3-2)	Extended, a_i and b_j unrestrained (Table 3-3)
HI	5.5, 7.0,	
4·	5	
4*		
3·		
3*	000	
2·		
2*		
1·	555	
1*	0	
0·	55	6
0*		00111124
−0*		001111223344
−0·	55	79
−1*	0	4
−1·	555	
−2*		
−2·		
−3*	00	
−3·		
−4*	0	
−4·	5	
LO	− 6.5, −9.0,	

Parallel boxplots offer visual summaries of the shapes of the batches of residuals and therefore facilitate making comparisons. Figure 3-4 shows the parallel boxplots for the residuals from the same two fits to the rubber data. We can see at a glance the vast improvement in fit offered by the extended model.

Reduction in Total Absolute Variation

Although displays can present batches in varying degrees of detail, we often find it useful to have a single numerical summary for a batch of residuals.

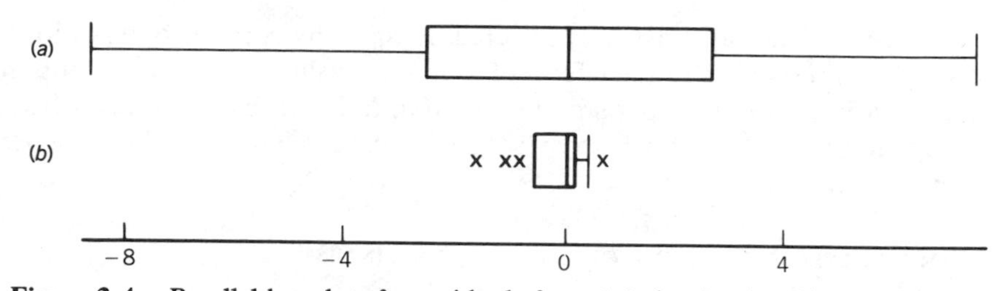

Figure 3-4. Parallel boxplots for residuals from (a) the simple additive fit and (b) the extended fit to the rubber data. (Units: 10^{-4} cm^3/g) Note: The rectangles contain the middle 50% of the residuals. The lines extending from the ends of the rectangles indicate the range of data within 1.5 $\times$ (fourth-spread) of the ends of the rectangles. The $\times$s represent outliers.

We let $\hat{y}_{ij}$ denote the fitted values obtained from any fitting technique, so that

$$y_{ij} = \hat{y}_{ij} + r_{ij}. \tag{10}$$

Two simple measures of the sizes of residuals are the sum of squared residuals, $\Sigma\Sigma r_{ij}^2$, and the sum of absolute residuals, $\Sigma\Sigma|r_{ij}|$.

Two-way analysis of variance with a single replication [see Snedecor and Cochran (1980, Chapter 14)] sometimes uses "the fraction of the sum of squared variation explained by the fit," which may be written

$$R^2 = 1 - \frac{\Sigma\Sigma r_{ij}^2}{\Sigma\Sigma(y_{ij} - \bar{y})^2}.$$

This expression, which uses the mean of the observations $y_{..}$ (or $\bar{y}$) and arises quite naturally in a least-squares framework, is sometimes converted to a percent. The user of these descriptive techniques may report that the two factors "explain" 97% (say) of the total squared variation.

A somewhat more suitable measure for exploratory applications is the percent reduction in total absolute variation achieved by the fit. We define

$$P = \left(1 - \frac{\Sigma\Sigma|r_{ij}|}{\Sigma\Sigma|y_{ij} - \operatorname*{med}_{i,j}\{y_{ij}\}|}\right) \times 100\%. \tag{11}$$

Note that the ratio of double sums gives the fraction of absolute variation of all data values about the median that remains after fitting to obtain the $\hat{y}_{ij}$.

Although P is not as drastically affected as R^2 is by extreme observations, it still lacks resistance to the effects of wild data values. We further caution that squaring $P/100$ (or perhaps $1 - P/100$, followed by subtraction from 1) does not give R^2; and, whereas $P/100$ has the same units as the data, R^2 is in squared units.

EXAMPLE: SPECIFIC VOLUME OF RUBBER (CONTINUED)

For the simple additive fit, based on median polish, to the specific volume of rubber, we calculate $\text{med}_{i,j}\{y_{ij}\} = 1.0829$, $\Sigma\Sigma|y_{ij} - 1.0829| = .1991$, and $\Sigma\Sigma|e_{ij}| = .0069$. Thus, we obtain the percent reduction in total absolute variation

$$P = \left(1 - \frac{.0069}{.1991}\right) \times 100\% = 96.5\%.$$

Evidently much of the systematic structure in this data set is described by additive main effects. However, the corresponding reduction in total absolute variation achieved by the extended fit, for which the sum of absolute residuals is only one-tenth as large, is

$$P = \left(1 - \frac{.00069}{.1991}\right) \times 100\% = 99.7\%.$$

Thus the additional degree of freedom associated with the extended fit allows us to account for most of the sum of absolute values of residuals, 3.2% out of the 3.5% not already described by the main-effects model.

An Aid to Choosing Among Fits

A fit with more constants should fit more closely, even if only by absorbing random variation. We need a simple guide that can help us choose among several fits which use different numbers of constants.

When working with sums of squared residuals SS, it is usual to find a mean square MS by dividing SS by the number of degrees of freedom DF not used in estimating parameters of a model. (For example, a 4×6 table takes $I + J - 1 = 9$ parameters in the simple additive fit, leaving $24 - 9 = 15$ degrees of freedom.) Thus $MS = SS/DF$ provides a measure of the average size of the residual variation.

To properly account for degrees of freedom in assessing a fit, Anscombe (1967) proposed choosing fits that give the minimum of a certain criterion. Tukey (1967) showed that this criterion often has the simple form MS/DF. One analog to MS/DF (or to its square root) for the sum of (unsquared) absolute deviations is

$$\frac{\Sigma\Sigma|r_{ij}|}{DF}, \tag{12a}$$

which is proportional to

$$\frac{1}{DF} \times \frac{\Sigma\Sigma|r_{ij}|}{\Sigma\Sigma|y_{ij} - \text{med}\{y_{ij}\}|} \times 100\%. \tag{12b}$$

A good fit is one which makes this number small. Equivalently, equation (11) invites us to say, after taking into account the number of parameters used, that a fit is good if it makes the expression

$$\frac{100\% - P}{DF} \tag{12c}$$

small.

Because we often first calculate the value of P for a particular fit, we prefer to use expression (12c) as a measure of fit when we are comparing two or more models. In summary, because expressions (12c) and (12a) are proportional, a fit that makes expression (12c) relatively small also makes expression (12a) small.

For the simple additive fit to the rubber data, expression (12c) gives

$$\frac{100\% - 96.5\%}{15} = .23\% \text{ per } DF,$$

and for the extended fit it gives

$$\frac{100\% - 99.7\%}{14} = .021\% \text{ per } DF.$$

The extended fit thus offers great improvement over the simpler fit, even after considering its increased degrees of freedom.

3D. MULTIPLICATIVE FITS

When two-way tables of data do not have a simple additive structure, a logarithmic transformation often provides a scale more appropriate for additivity. The logarithmic transformation is often effective when the model

$$y_{ij} = \nu\gamma_i\delta_j + \varepsilon_{ij} \tag{13}$$

describes the structure of a two-way table of positive values; we refer to this model as a *simple multiplicative model*. Although a multiplicative fluctuation term for equation (13), perhaps of the form $(1 + \varepsilon'_{ij})$, may well be appropriate, we choose the form given because of our use of more general models with both additive and multiplicative structure.

By first taking logs of the data, we can use median polish to give a resistant fit associated with the model of equation (13). Once we obtain a simple additive fit to log y_{ij}, it is easy to transform back to the original scale of measurement and obtain a fit to the data table:

$$\hat{y}_{ij} = hc_i d_j. \tag{14}$$

Thus, median polish can provide a simple multiplicative fit to a table almost as easily as it provides a simple additive fit. This simplest approach faces difficulty, however, if some y_{ij} are zero or negative.

Like the simple additive model, the multiplicative model uses $I + J - 1$ degrees of freedom. A more general model devotes one additional degree of freedom to an additive constant:

$$y_{ij} = \mu + \nu\gamma_i\delta_j + \varepsilon_{ij}. \tag{15}$$

This model describes situations where, after translation of the origin by μ units, a two-way table has multiplicative structure. We write a fit for model (15) as

$$\hat{y}_{ij} = q + hc_i d_j, \tag{16}$$

and we refer to it as a *multiplicative fit*. If we knew, or could estimate, the value of μ, then we could easily obtain the estimates h, c_i, and d_j by the methods already described. We can avoid the need to obtain q if we exploit the perhaps surprising fact that model (16) is equivalent to an extended model (Tukey, 1977, Chapter 12). We state this important result as a theorem about the fits associated with these models.

THEOREM 1: The set of fits of the form

$$\hat{y}_{ij} = q + hc_i d_j$$

and the set of fits of the form

$$\hat{y}_{ij} = m + a_i + b_j + ka_i b_j$$

($k \neq 0$) are the same sets, differently described.

PROOF: Suppose that the values of q, h, c_i, and d_j are given. We define $m = q + h, a_i = h(c_i - 1), b_j = h(d_j - 1)$, and $k = 1/h$. (This step assumes h is nonzero.) Then

$$
\begin{aligned}
m + a_i + b_j + ka_ib_j &= (q + h) + h(c_i - 1) + h(d_j - 1) \\
&\quad + \frac{1}{h}h(c_i - 1)h(d_j - 1) \\
&= q + h + hc_i - h + hd_j - h + hc_id_j - hc_i - hd_j + h \\
&= q + hc_id_j.
\end{aligned}
$$

If $h = 0$, we set $m = q, a_i = 0, b_j = 0$, and $k = 1$, and we obtain

$$m + a_i + b_j + ka_ib_j = m = q.$$

Thus, in every situation, any fit of the first form can be written in the second form, with the given conversion from one set of parameters to the other.

Conversely, suppose that the values of m, a_i, b_j, and k are given: We define $q = m - 1/k, c_i = 1 + ka_i, d_j = 1 + kb_j$, and $h = 1/k$. (This step assumes that k is nonzero.) Then

$$
\begin{aligned}
q + hc_id_j &= \left(m - \frac{1}{k}\right) + \frac{1}{k}(1 + ka_i)(1 + kb_j) \\
&= m - \frac{1}{k} + \frac{1}{k} + a_i + b_j + ka_ib_j \\
&= m + a_i + b_j + ka_ib_j.
\end{aligned}
$$

Thus any fit of the second form can be written in the first form. So the two sets contain exactly the same collection of potential fits.

It is customary to center the row effects and the column effects in the resistant extended fit, as we have done for the simple additive fit, by specifying that $\mathrm{med}\{a_i\} = 0$ and $\mathrm{med}\{b_j\} = 0$. Because of the correspondence described in the theorem, these restrictions lead to constraints in the multiplicative fit—namely that $\mathrm{med}\{c_i\} = 1$ and $\mathrm{med}\{d_j\} = 1$. This result follows from the relationships $c_i = 1 + ka_i$ and $d_j = 1 + kb_j$ and from the fact that a linear function transforms a median to a median. Thus the multiplicative fit of equation (16), like the extended fit, has $I + J$ "independent parameters."

Even with the restrictions on a_i and b_j, and on c_i and d_j, the fits discussed in the theorem are quite general. We have seen that centering the c_i and d_j at 1 is equivalent to centering the a_i and b_j at 0. Inclusion of the parameter h in the multiplicative fit permits this centering of the c_i and d_j, just as the parameter m in the additive fit permits centering the a_i and b_j. Furthermore, the centering of the row effects and column effects could use the mean, or another location measure, instead of the median.

Two complications make it difficult to fit directly the parameters in the multiplicative model of equation (15). First, it is not clear how to estimate ν; in particular, we cannot take q to be an estimate of the center of the batch of y_{ij} because the terms $hc_i d_j$ are not centered at 0. But even if we knew the value of q and subtracted it from the y_{ij}, we could not approach the problem of finding c_i and d_j by taking logs, if any of the terms $y_{ij} - q$ were negative. The theorem often allows us to circumvent these difficulties by fitting the extended model and then reparametrizing to obtain values for q, h, c_i and d_j in the multiplicative fit.

We describe below a psychological investigation of the strengths of adjectives and adverbs in which a multiplicative model of the form given by equation (15) arises from subject-matter considerations. To give a resistant fit to a two-way table that is described by this model, we use the indirect approach. We first use median polish to obtain a simple additive fit. We then construct the diagnostic plot of the residuals from the additive fit, and use it as we described in Section 3B to obtain a resistant slope estimate km for $\kappa\mu$ and thus a value of k for the extended fit. (For some tables, even if not for this one, we may choose to iterate these steps.) Finally, we use the relationships between parameters in the two ways of writing the fit, specified in the proof of the theorem, to give the estimates q, h, c_i, and d_j.

EXAMPLE: ADVERBS AND ADJECTIVES

Cliff (1959) proposed a "law of word mixture" for adverb–adjective combinations that rests on five postulates:

1. A number is associated with each adjective.
2. A number is associated with each adverb.
3. The intensity of an adverb–adjective pair is the product of these two numbers.
4. The intensity of the adjective used alone is the number associated with it when used in combination.

5. A set of adjectives can be chosen that may be scaled on a single dimension for which all adjectives have the same zero point when they are unmodified.

From these postulates, Cliff derived the algebraic formula

$$x_{ij} = c_i s_j + K, \tag{17}$$

where x_{ij} is the scale value of adverb i in combination with adjective j, K is the difference between the arbitrary zero point of the obtained scale values and the psychological zero point of the scale, c_i is the multiplying value of adverb i, and s_j is the psychological scale position of adjective j.

By administering an elaborate questionnaire to hundreds of college students at three universities, Cliff derived psychological scaled values for 150 adverb–adjective pairs. Table 3-6 gives these values obtained from the data for 186 male students at Princeton University. For convenience of display, we place the 15 adjectives in the left-hand margin for the rows, and the 10 adverbs in the top margin for the columns, so that we have a 15 × 10 table of scaled values. We denote these values by y_{ij} for $i = 1, \ldots, 15$ and $j = 1, \ldots, 10$, and we reformulate Cliff's algebraic expression in the notation of equation (16):

$$\hat{y}_{ij} = q + hc_i d_j.$$

Note that Cliff's equation (17) differs from this equation of fit in two important ways. First, we use i to index adjectives instead of adverbs so that the larger number of adjectives will correspond to rows; this change facilitates the displays of data and fits. Second, Cliff's formula absorbs the value of the constant h in the main effects, whereas our use of h permits both the c_i and the d_j to be scaled in a convenient way. Such differences are notational rather than substantive.

A simple additive fit to Cliff's data, using 10 iterations of median polish, gives the same common term m, row effects a_i, and column effects b_j as does the extended fit displayed in Table 3-7. The simple additive fit explains 78% of the total absolute variation. The corresponding residuals, denoted by e_{ij} (shown a little later in Table 3-9 but with signs changed in some rows and columns), lead to the diagnostic plot in Figure 3-5. The slope of a resistant line fit to that plot, -3.98, gives the value of km, which determines the value of k ($= -3.98/1.86 = -2.14$) for the extended fit,

$$\hat{y}_{ij} = m + a_i + b_j + ka_i b_j.$$

Table 3-7 includes the residuals obtained from this fit.

TABLE 3-6. Scale values of adverb–adjective combinations as determined by male students at Princeton University (unit: 10^{-3}).

	Adverb									
Adjective	(Unmodified)	Slightly	Somewhat	Rather	Pretty	Quite	Decidedly	Unusually	Very	Extremely
Evil	739	1394	1289	1115	1090	892	764	702	698	504
Wicked	771	1395	1204	1112	929	882	756	669	632	515
Contemptible	858	1342	1217	995	1046	884	742	753	744	556
Immoral	956	1393	1232	1113	1034	893	791	701	765	494
Disgusting	836	1273	1042	965	822	736	718	652	593	473
Bad	1080	1546	1365	1257	1225	1012	889	808	834	666
Inferior	1274	1548	1486	1396	1183	974	986	926	899	758
Ordinary	1869	1928	1943	1913	1902	1871	1763	1660	1785	1806
Average	2053	2008	2050	2056	2052	1951	1828	1877	1872	1800
Nice	2736	2407	2511	2685	2711	2778	3009	3206	3042	3282
Good	2910	2449	2490	2821	2727	2990	3095	3304	3250	3443
Pleasant	2816	2457	2572	2793	2738	2920	3033	3240	3172	3437
Charming	2993	2564	2716	2939	2869	3058	3178	3262	3244	3387
Admirable	3095	2547	2656	2865	2893	3110	3250	3373	3385	3579
Lovable	3024	2604	2689	2885	2932	3059	3147	3287	3243	3431

Source: Norman Cliff (1959). "Adverbs as multipliers," *Psychological Review*, **66**, 27–44 (data from Table 9, p. 36). Copyright 1959 by the American Psychological Association. Reprinted by permission of the author.

TABLE 3-7. Extended fit to adverb–adjective data (unit: 10^{-3}).

	Residuals (r_{ij})										
	(Unmodified)	Slightly	Somewhat	Rather	Pretty	Quite	Decidedly	Unusually	Very	Extremely	a_i
Evil	−91.7	318.5	166.8	5.3	95.6	−19.7	202.8	198.5	161.8	192.2	−991
Wicked	−50.4	327.3	89.2	9.8	−57.1	−21.0	205.7	176.8	106.8	215.6	−1000
Contemptible	27.5	266.8	95.0	−114.4	51.8	−27.5	181.1	249.8	208.1	244.5	−991
Immoral	65.1	267.2	61.4	−45.6	−13.9	−75.6	158.9	124.3	156.8	101.3	−933
Disgusting	78.5	258.8	−21.1	−85.0	−107.1	−106.5	243.2	237.7	144.4	259.7	−1063
Bad	69.6	320.2	98.1	1.1	70.6	−69.6	116.0	85.7	82.9	112.4	−817
Inferior	193.2	263.3	162.4	82.8	−34.1	−174.2	129.9	117.9	63.7	109.7	−748
Ordinary	19.5	0.0	0.0	−26.0	0.0	−4.5	0.0	−84.5	30.0	123.0	0
Average	83.2	−20.6	10.1	19.1	42.8	−38.2	−76.8	−13.9	−26.8	−44.8	117
Nice	−32.8	−290.3	−172.7	−2.3	−10.0	32.8	161.4	341.5	187.1	361.6	89
Good	−57.1	−414.3	−353.5	−27.8	−170.7	57.1	13.4	197.9	157.8	255.6	1088
Pleasant	−66.9	−335.8	−203.6	12.7	−84.6	66.9	50.8	236.6	180.6	363.0	1006
Charming	−51.7	−364.3	−190.0	27.0	−97.9	51.7	4.8	61.4	59.0	95.1	1163
Admirable	−28.9	−447.5	−313.8	−111.4	−144.4	28.9	−16.4	76.0	105.3	180.6	1240
Lovable	−37.2	−338.0	−230.3	−40.3	−49.5	37.2	−45.5	66.4	38.3	117.0	1179
b_j	−13.0	65.5	80.5	76.5	39.5	13.0	−99.5	−118.0	−107.5	−179.5	1863

Notes: The fit is $\hat{y}_{ij} = m + a_i + b_j + ka_ib_j$ with $m = 1863$ and $k = -2.14$. Approximately 86% of total absolute variation is "explained" by this fit. After converting parameters to a multiplicative fit, we may think of these residuals as being those from the multiplicative fit of equation (16).

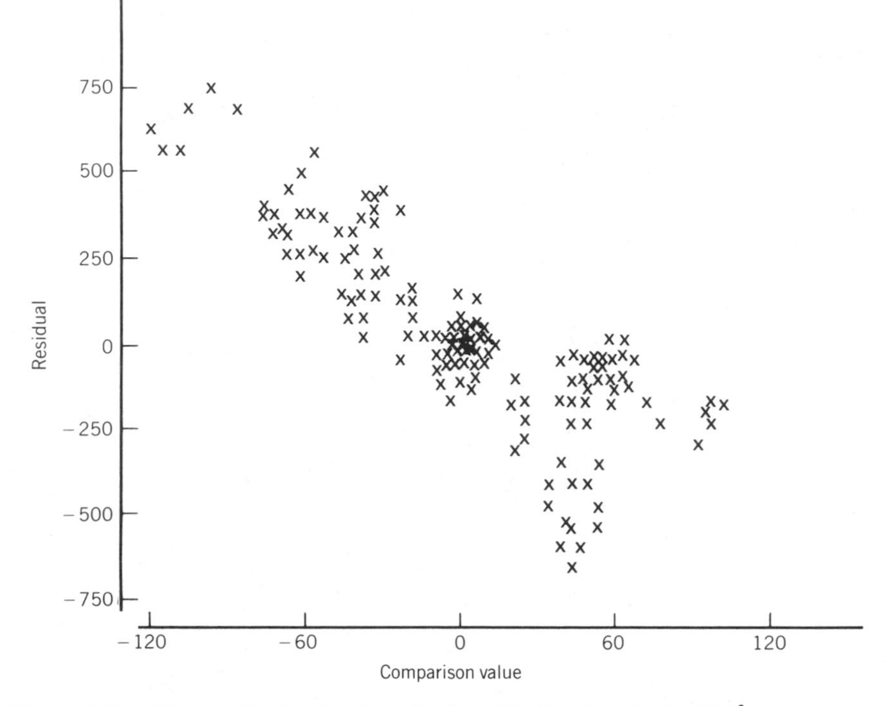

Figure 3-5. Diagnostic plot for the adverb–adjective data (unit: 10^{-3}).

The previous theorem states that the extended fit is equivalent to a multiplicative fit of the form

$$\hat{y}_{ij} = q + hc_i d_j.$$

Table 3-8 provides the details of the conversion to the parameters of this multiplicative fit, which corresponds to the model proposed for these data by Cliff. Thus we have provided a resistant fit to the data table, corresponding to Cliff's model in equation (17). However, the somewhat nonlinear pattern in the diagnostic plot of Figure 3-5, especially at the lower right, hints that the multiplicative fit cannot capture all the systematic structure in this table. Although the residuals from this fit display somewhat less pattern and have smaller magnitudes than those of the simple additive fit, they clearly retain some structure. For example, most negative values are in the first six columns, and especially in the lower left-hand corner of Table 3-7. (When interpreting these values, note that h is negative.)

TABLE 3-8. Conversion from extended fit to multiplicative fit for adverb–adjective data.

		Extended Fit		Multiplicative Fit	
		$m = 1.86$		$q = m - \frac{1}{k} = 2.33$	
		$k = -2.14$		$h = \frac{1}{k} = -.472$	
Adjective	Adverb	a_i	b_j	$c_i = 1 + ka_i$	$d_j = 1 + kb_j$
1 Evil	1 (None)	−.991	−.013	3.12	1.03
2 Wicked	2 Slightly	−1.000	.066	3.14	0.86
3 Contemptible	3 Somewhat	−.991	.081	3.12	0.83
4 Immoral	4 Rather	−.933	.077	2.99	0.84
5 Disgusting	5 Pretty	−1.063	.040	3.27	0.91
6 Bad	6 Quite	−.817	.013	2.75	0.97
7 Inferior	7 Decidedly	.748	−.100	2.60	1.21
8 Ordinary	8 Unusually	0.000	−.118	1.00	1.25
9 Average	9 Very	.117	−.108	0.75	1.23
10 Nice	10 Extremely	.895	−.180	0.91	1.38
11 Good		1.089		−1.32	
12 Pleasant		1.006		−1.15	
13 Charming		1.163		−1.48	
14 Admirable		1.240		−1.65	
15 Lovable		1.179		−1.52	

The limited amount of effort needed to reformulate the extended fit as a multiplicative fit is justified by the advantages gained in interpretation. The largest values of c_i correspond to the most critical adjectives like "wicked" ($i = 1$) and "disgusting" ($i = 5$). Values of c_i near 1 (for "ordinary" and "average") correspond to neutral adjectives, and negative values of c_i are associated with complimentary adjectives like "admirable" ($i = 14$). A graphical display of the 15 adjectives on the number line could help to show spacing, as well as order, and to identify groupings of adjectives having similar connotations.

The d_j for the adverbs are also centered at 1, but range from .83 to 1.38. Adverbs like "somewhat" ($j = 3$) which decrease the level of intensity of an adjective have values less than 1, and those like "extremely" which increase the level of intensity of an adjective have values greater than 1. Note that the strongest adverb–adjective combination, as measured by the resistant multiplicative fit, is "extremely disgusting" ($i = 5, j = 10$). One should display the numerical factors associated with the 10 adverbs on a coordinate axis to identify adverbs that appear nearly similar in strength, and to evaluate the ordering of adverbs provided by the d_j.

The percentage of total absolute variation explained by the extended fit (equivalently, the multiplicative fit) is 86%, an appreciable, but not dramatic, increase over the 78% figure for the simple additive fit. Because the extended fit is equivalent to a multiplicative fit, our analysis shows that a resistant fit based on Cliff's multiplicative model is somewhat better than the fit based on the simple additive model. Of course, any choice of a model and fit should take into account available outside evidence, based either on theoretical considerations or on other data sets. We return to this data set in Section 3E, with the aid of more sophisticated models and fitting techniques.

3E. TECHNIQUES FOR OBTAINING SIMPLE MULTIPLICATIVE FITS

The techniques described in Section 3D for obtaining a resistant multiplicative fit are not always useful. For example, some two-way tables have zero additive structure so that m, the a_i, and the b_j are all 0. Although such tables could have perfect multiplicative structure as given in model (13), the fitting techniques already described cannot obtain a multiplicative fit. The difficulty arises because, when all a_i and b_j are 0, then all comparison values are 0 and we cannot obtain an extended fit which would translate to a multiplicative fit.

These difficulties may seem minor and of little practical importance. Our goal for the remainder of this chapter, however, is to develop and illustrate fitting techniques that can extract whatever multiplicative structure exists in the residuals from a simple additive fit. Although the term ka_ib_j is sometimes successful in doing just this, at other times a more general multiplicative term of the form kc_id_j describes the structure much better. Thus we want to explore fits of the form

$$\hat{y}_{ij} = m + a_i + b_j + kc_id_j; \tag{18}$$

we call these *additive-plus-multiplicative fits*.

We aim to fit the residuals e_{ij} from a simple additive fit according to the equation

$$\hat{e}_{ij} = kc_id_j. \tag{19}$$

Because the e_{ij} have row medians and column medians all equal to 0, further median polish of these residuals does nothing. Therefore, we cannot fit multiplicatively by first obtaining a diagnostic plot and an extended fit, and then reparametrizing. Because nearly half of the e_{ij} are negative and some are likely to be zero, we also cannot take logs and fit a simple additive model in the log scale which might then give a fit as in equation (19). To obtain an additive-plus-multiplicative fit, we need another approach.

Changing Signs and Removing Zeros

We now consider a two-way table of numbers (often, residuals from a previous fit), some of which may be zero or negative. If these data had perfect multiplicative structure, we could write

$$e_{ij} = \kappa\gamma_i\delta_j, \tag{20}$$

where now γ_i or δ_j or both can be negative. By changing the sign on row i whenever γ_i is negative, and by changing the sign on column j whenever δ_j is negative, we could produce a table without negative entries.

In practice, the e_{ij} often lack perfect multiplicative structure, so that we cannot make sign changes on rows and columns to eliminate all negative signs; we write the model as

$$e_{ij} = \kappa\gamma_i\delta_j + \varepsilon_{ij}. \tag{21}$$

We now develop an algorithm for changing a given table to one without negative entries, when possible, or with as few negative entries as possible; this algorithm uses row and column sign changes.

For an $I \times J$ two-way table of numbers $\{e_{ij}\}$, there are 2^I tables that can be produced by choosing sign changes for the rows and 2^J tables that result from choosing column sign changes. When both types of sign changes are used, 2^{I+J} (not all distinct) tables result. In principle, we could consider all these tables and choose to work with one having the fewest remaining negative entries. Some negative entries may well remain, and the table may also contain zeros and missing values. It is essential that we develop a technique for handling these difficulties. One often practical approach replaces negative or zero entries with a small positive entry, but not so small as to "underflow" on the computer. To fit a multiplicative model to such a table of positive entries, we take logs and then use median polish. Although the logs of numbers very close to 0 are negative numbers with large magnitudes, the resistance of median polish ensures that these entries have a suitably small impact on the multiplicative fit.

EXAMPLE: ADVERBS AND ADJECTIVES (CONTINUED)

We found an extended fit to the adverb–adjective data, and Table 3-7 gives the residuals. Because some pattern, both in signs and in magnitudes, remains in the table of residuals, we can hope to improve the fit to the original table by using an additive-plus-multiplicative term from the extended fit and pursue a new, unconstrained multiplicative fit. To do this, we discard the multiplicative fit to the residuals e_{ij} from a simple additive fit. (One can obtain these residuals from Table 3-9.)

Although nearly half of the 150 residuals, e_{ij}, are negative, sign changes on the last six rows and the last five columns reduce the number of negative entries to 14. Table 3-9 gives the residuals after the sign changes. No row or column of Table 3-9 has as many as half of its entries negative, and most negative entries are relatively close to 0. But, to proceed with our multiplicative fitting strategy, we must do something about remaining negative entries and zeros. For subsequent computations, both types of entries are replaced with $+10^{-6}$, so that the resulting adjusted table has only positive entries, and we can take logs. An alternative approach would treat negative entries as missing; unless either I or J is small, the resistance of medians to unusual data values suggests that the choice will not greatly influence the fit.

Table 3-10 gives a median-polish fit to the logs of the adjusted residuals from Table 3-9. We next transform back to the scale of the original data using the exponential function (10^x for logarithms to base 10 and e^x when natural logarithms are employed). Restoration of signs to the last six rows and the last five columns yields a simple multiplicative fit to the e_{ij}, as presented in Table 3-11. Specifying the c_i and the d_j involves an arbitrary choice because $(-c_i) \times (-d_j) = c_i \times d_j$; we made the choices so that positive signs on the d_j correspond to the strongest adverbs. In the next section, we explore the residuals,

$$r_{ij} = y_{ij} - (m + a_i + b_j + kc_i d_j), \tag{22}$$

and we assemble and polish the additive-plus-multiplicative fit.

A Sign-Change Algorithm

An $I \times J$ table has 2^{I+J} associated tables with various sign changes for rows and columns. Because 2^{I+J} is a large number for many tables we encounter (for a 15×10 table, it is 33,554,432), a computer implementation of the technique for finding a multiplicative fit to a table of residuals requires a systematic way to perform sign changes on rows and columns. The goal of such an algorithm is to produce a new table with a minimum number of negative entries. It is easy to reduce the procedure to 2^J trials.

TABLE 3-9. Residuals from additive fit after sign changes on rows 10 through 15 and on columns 6 through 10 (unit: 10^{-3}). The lower right corner value is m.

	j										
i	1	2	3	4	5	6	7	8	9	10	a_i
1	−119.2	457.2	337.2	167.2	179.2	−7.7	7.7	51.2	65.7	187.7	−991.2
2	−78.2	467.2	261.2	173.2	27.2	−6.7	6.7	75.2	122.7	167.7	−1000.2
3	0.0	405.5	265.5	47.5	135.5	0.0	29.5	0.0	19.5	135.5	−991.5
4	39.2	397.7	221.7	106.7	64.7	49.7	39.2	110.7	57.2	256.2	−932.7
5	49.0	407.5	161.5	83.5	−17.5	77.0	−17.5	30.0	99.5	147.5	−1062.5
6	47.0	434.5	238.5	134.5	139.5	47.0	57.5	120.0	104.5	200.5	−816.5
7	172.5	368.0	291.0	205.0	29.0	153.5	29.0	70.5	108.0	177.0	−748.0
8	19.5	0.0	0.0	−26.0	0.0	4.5	0.0	84.5	−30.0	−123.0	0.0
9	86.5	−37.0	−10.0	0.0	33.0	41.5	52.0	−15.5	0.0	0.0	117.0
10	8.0	415.5	326.5	148.5	85.5	8.0	351.5	567.0	392.5	704.5	894.5
11	27.0	566.5	540.5	205.5	262.5	27.0	244.5	472.0	407.5	672.5	1087.5
12	39.0	476.5	376.5	151.5	169.5	39.0	264.5	490.0	411.5	748.5	1005.5
13	19.5	527.0	390.0	163.0	196.0	19.5	252.0	354.5	326.0	541.0	1163.0
14	−5.5	621.0	527.0	314.0	249.0	−5.5	247.0	388.5	390.0	656.0	1240.0
15	4.5	503.0	433.0	233.0	149.0	4.5	205.0	363.5	309.0	569.0	1179.0
b_j	−13.0	65.5	80.5	76.5	39.5	13.0	−99.5	−118.0	−107.5	−179.5	1862.5

Note: We start with the fit $\hat{y}_{ij} = m + a_i + b_j$ and the residuals e_{ij} from this fit. This table shows the residuals after signs of entire rows and columns are changed to reduce the number of negative entries.

TABLE 3-10. Median-polish fit to natural logarithms of adjusted residuals u_{ij}. Corner value is m'.

	Residuals (e'_{ij})										
	j										
i	1	2	3	4	5	6	7	8	9	10	a'_i
1	−15.642	1.052	0.988	0.860	1.114	−15.642	−1.955	−0.323	−0.191	0.191	−0.570
2	−15.711	1.004	0.663	0.825	−0.838	−15.711	−2.163	−0.008	0.363	0.009	−0.500
3	−15.133	1.441	1.258	0.110	1.343	−15.133	−0.110	−17.566	−0.897	0.374	−1.079
4	1.368	0.438	0.094	−0.063	−0.378	1.605	−0.808	−0.027	−0.804	0.027	−0.095
5	1.842	0.713	0.029	0.000	−18.113	2.294	−18.042	−1.082	0.000	−0.273	−0.346
6	1.381	0.359	0.000	0.000	0.221	1.381	−0.593	−0.114	−0.369	−0.385	0.071
7	2.753	0.264	0.270	0.493	−1.277	2.637	−1.206	−0.574	−0.264	−0.437	0.000
8	19.125	−0.906	−0.665	−0.093	0.092	17.659	0.163	18.158	−0.210	−0.877	−18.552
9	20.709	−0.813	−0.572	0.000	17.497	19.974	18.023	0.000	−0.116	−0.783	−18.645
10	−0.703	0.000	−0.000	−0.215	−0.582	−0.703	0.902	1.124	0.639	0.557	0.386
11	0.000	−0.202	−0.009	−0.403	0.026	0.000	0.026	0.427	0.163	−0.002	0.899
12	0.262	−0.480	−0.475	−0.813	−0.515	0.262	0.000	0.360	0.068	0.000	1.004
13	−0.055	−0.005	−0.065	−0.365	0.004	−0.055	0.326	0.411	0.210	0.050	0.629
14	−17.085	−0.084	−0.008	0.046	0.000	−17.085	0.062	0.259	0.146	−0.000	0.873
15	−1.537	−0.067	0.023	−0.023	−0.285	−1.537	0.104	0.420	0.141	0.085	0.644
b'_j	−2.433	0.813	0.573	−0.000	−0.185	−2.433	−0.256	0.000	0.117	0.784	4.830

Note: Fit is $\widehat{\log}(u_{ij} \times 10^3) = m' + a'_i + b'_j$ [see equation (23)]. The u_{ij} arise from the e_{ij} by changing signs on rows 10 through 15 and on columns 6 through 10, and then replacing all nonpositive entries with 10^{-6}.

TABLE 3-11. Multiplicative fit to residuals e_{ij} from additive fit, with signs restored (unit: 10^{-3}).

	Residuals (r_{ij})										
	j										
i	1	2	3	4	5	6	7	8	9	10	c_i
1	−125.46	297.58	211.75	96.47	120.44	13.96	47.04	19.56	13.85	−32.70	−0.565
2	−84.91	296.09	126.71	97.38	−35.79	13.41	51.99	0.66	−37.42	−1.54	−0.606
3	−3.73	309.52	190.06	4.95	100.15	3.73	3.44	42.57	28.35	−42.30	−0.339
4	29.26	141.11	20.02	−7.01	−29.78	−39.76	48.82	3.07	70.69	−7.03	−0.908
5	41.23	207.90	4.61	0.02	−91.02	−69.23	85.99	58.52	0.00	46.33	−0.706
6	35.19	131.08	−0.00	−0.00	27.74	−35.19	46.62	14.57	46.76	94.15	−1.074
7	161.51	85.64	69.05	79.83	−75.00	−142.51	67.89	54.73	32.76	97.20	−1.000
8	19.50	−0.00	−0.00	−26.00	0.00	−4.50	0.00	−84.50	30.00	123.00	0.000
9	86.50	−37.00	−10.00	−0.00	33.00	−41.50	−52.00	15.50	0.00	0.00	0.000
10	8.17	0.00	0.10	35.68	67.54	−8.17	208.92	382.73	185.37	301.01	1.471
11	0.00	127.43	4.95	102.10	−6.92	−0.00	6.38	164.25	61.85	−1.36	2.457
12	−9.01	294.29	229.36	190.17	114.39	9.01	0.00	148.16	27.26	0.00	2.729
13	1.12	2.85	26.48	71.87	−0.85	−1.12	70.18	119.52	61.87	26.47	1.876
14	31.80	55.05	4.40	−14.32	−0.00	−31.80	15.01	88.68	52.99	−0.50	2.394
15	16.44	35.10	−10.03	5.53	49.19	−16.44	20.35	124.86	40.75	46.46	1.905
d_j	−0.087	−2.255	−1.772	−0.999	−0.830	0.087	0.773	1.000	1.124	2.190	125.2

Note: Signs have been restored on rows 10 through 15 and on columns 6 through 10. Fit is $\hat{e}_{ij} = kc_i d_j$.

We assume that $J \leq I$, so that there are fewer columns than rows when the table is not square. We now enumerate the steps of an algorithm for changing signs.

1. Consider all tables produced from the given table by using any possible combination of sign changes for the J columns; there are 2^J such tables.
2. For each of the 2^J tables considered in step 1, make a sign change for each row that has more minus signs than plus signs.
3. For each of 2^J tables formed at step 2, count the number of minus signs still remaining.
4. Consider the collection of tables from step 3 having the minimum number of remaining negative entries, and remove from further consideration all other tables.
5. If one table remains, we use it. When more than one table remains, we must choose one to use. If possible, we choose a table which places the maximum number of outliers in those cells whose signs are negative. Otherwise we may choose a table which leaves the remaining minus signs attached to entries whose magnitudes are closest to 0. (Note that the replacement of these negative entries with very small positive numbers should minimally disturb whatever multiplicative structure underlies the table.)
6. If a single table is not reached in step 5, we may randomly select the final table. (Our experience suggests that choices made in steps 5 and 6 have only modest impact on the fit for the original two-way table, provided I and J are not too small and provided the proportion of zeros is not excessive.)
7. If the chosen table has nonpositive entries, replace all zeros with a small positive number, preferably at least several orders of magnitude smaller than most of the positive residuals. We sometimes treat the negative entries as missing, but for tables with I and J not too small it may be more convenient to recode the negative entries just as we recoded the zeros. (For the adverb–adjective data, we coded all 14 remaining nonpositive entries as 10^{-6} before taking logs.) We refer to the final table as the *adjusted table*.

Once we have made necessary sign changes and other adjustments that lead to a table whose entries are all positive, we can use the techniques outlined early in Section 3D and illustrated in this section to give a simple multiplicative fit to the adjusted table. If the entries of the adjusted table are denoted by u_{ij}, we may write

$$\widehat{\log(u_{ij})} = m' + a'_i + b'_j \tag{23}$$

as the fit given by median polish in the log scale. We then use exponentials (or antilogs) to obtain

$$\begin{aligned}\hat{u}_{ij} &= \exp\left(m' + a'_i + b'_j\right) \\ &= \exp(m')\exp(a'_i)\exp(b'_j).\end{aligned} \tag{24}$$

By letting $k = \exp(m')$, $c'_i = \exp(a'_i)$, and $d'_j = \exp(b'_j)$, we produce a fit of the form

$$\hat{u}_{ij} = kc'_i d'_j.$$

Geometric Median of Multiplicative Parameters

We noted in the theoretical discussion of Section 3D that centering the additive effects at 0 in an extended fit corresponds roughly to centering the multiplicative effects at 1 in a multiplicative fit. Similarly (but not equivalently), the conditions that $\text{med}\{a'_i\} = 0$ and $\text{med}\{b'_j\} = 0$ translate roughly to conditions that $\text{med}\{c'_i\} = 1$ and $\text{med}\{d'_j\} = 1$ for the simple multiplicative fit to the u_{ij}

If I is odd, then $\text{med}\{\log c'_i\} = \log \text{med}\{c'_i\}$ is 0 if and only if $\text{med}\{c'_i\} = 1$. We write out an expression for $\text{med} \log\{c'_i\}$, using parentheses around a subscript to denote an order statistic. If I is even, then the expression

$$\begin{aligned}\text{med}\{\log c'_i\} &= \tfrac{1}{2}\left(\log c'_{(I/2)} + \log c'_{(I/2+1)}\right) \\ &= \log\left(c'_{(I/2)} \times c'_{(I/2+1)}\right)^{1/2}\end{aligned}$$

is 0 if and only if

$$\sqrt{c'_{(I/2)} c'_{(I/2+1)}} = 1.$$

A similar statement holds for the d'_j. We consolidate the conditions on the parameters for the multiplicative fit for u_{ij} by using the concept of the geometric median.

DEFINITION: Let $S = \{x_1, x_2, \ldots, x_n\}$ be a set of numbers with $0 < x_1 \leq x_2 \leq \cdots \leq x_n$. The *geometric median* of S is

$$\text{gmed}(S) = \begin{cases} \text{med}\{x_i\} & \text{for } n \text{ odd} \\ \sqrt{x_{n/2} \cdot x_{n/2+1}} & \text{for } n \text{ even.} \end{cases} \tag{25}$$

We now see that our fitting procedure for the u_{ij} leads to the following conditions on c_i' and d_j':

$$\text{gmed}\{c_i'\} = 1 \quad \text{and} \quad \text{gmed}\{d_j'\} = 1.$$

For an alternative approach that involves only a single order statistic, regardless of whether n is odd or even, we could turn to the "low median" (Siegel, 1983), both in the conditions on the parameters and throughout the analysis.

3F. ADDITIVE-PLUS-MULTIPLICATIVE FITS

The previous sections of this chapter together enable us to fit a rather general model to a two-way table using resistant procedures for the various stages of the fitting process. This *additive-plus-multiplicative* model subsumes all models discussed previously in this chapter as special cases; we write it as

$$y_{ij} = \mu + \alpha_i + \beta_j + \kappa\gamma_i\delta_j + \varepsilon_{ij}. \tag{26}$$

When $\kappa = 0$, it gives the main-effects model of Section 3A; and when $\gamma_i = \alpha_i$ and $\delta_j = \beta_j$, it gives the extended model of Section 3B. It also gives the multiplicative models, both simple and otherwise, discussed in Section 3D.

The additive-plus-multiplicative models include two other important special cases, the *columns-linear model*

$$y_{ij} = \mu + \alpha_i + \beta_j + \kappa\alpha_i\delta_j + \varepsilon_{ij} \tag{27}$$

and the *rows-linear model*

$$y_{ij} = \mu + \alpha_i + \beta_j + \kappa\gamma_i\beta_j + \varepsilon_{ij}. \tag{28}$$

The columns-linear model is suitable if, for each fixed column j, a plot of

$e_{ij} = y_{ij} - (m + a_i + b_j)$ against a_i gives very nearly a straight line. Similarly, when a plot of e_{ij} against b_j is nearly linear for each fixed row i, a rows-linear model may be useful. These models are intermediate in complexity between the extended model and the more general additive-plus-multiplicative model of equation (26); we refer to Mandel (1969a) for further discussion and for illustration with some physical-science data for which a rows-linear model brings useful insights.

In addition to having common constant terms μ and κ for the additive $(\mu + \alpha_i + \beta_j)$ and multiplicative $(\kappa\gamma_i\delta_j)$ components, model (26) provides $2I$ parameters (the α_i and γ_i) for the rows, and $2J$ parameters (the β_j and δ_j) for the columns of an $I \times J$ table. The four side conditions $\text{med}\{\alpha_i\} = \text{med}\{\beta_j\} = 0$ and $\text{gmed}\{\gamma_i\} = \text{gmed}\{\delta_j\} = 1$ imply that the effective number of parameters is $2I + 2J - 2$. The number of degrees of freedom associated with the residuals is $DF = IJ - (2I + 2J - 2)$. [For a more sophisticated approach to degrees of freedom, we refer to Mandel (1971).]

Steps for Obtaining an Additive-Plus-Multiplicative Fit

We now combine the steps described in several previous sections to obtain a resistant additive-plus-multiplicative fit corresponding to model (26).

1. For a two-way table of data, y_{ij}, we first use median polish to obtain a simple additive fit, $\widehat{y_{ij}} = m + a_i + b_j$. (At this stage, we should explore the residuals e_{ij}, construct a diagnostic plot, and consider whether to try a power transformation or an extended fit.)
2. To prepare to fit multiplicatively the residuals from the simple additive fit, we perform sign changes on rows and columns in order to reduce the number of negative entries. We may use the algorithm described in Section 3E, or we may be content to make sign changes on rows and on columns with more minuses than pluses.
3. We replace remaining entries that are not positive with a very small positive number, to obtain the adjusted residuals from the additive fit, u_{ij}. (As noted in the sign-change algorithm, we may sometimes prefer to treat some or all remaining negative entries as missing.)
4. We next take logs and use median polish to construct a fit

$$\widehat{\log(u_{ij})} = m' + a_i' + b_j'.$$

5. By taking exponentials (or antilogs), we return to the scale of the original data and obtain a fit

$$\hat{u}_{ij} = kc_i'd_j'.$$

6. After restoring signs to rows and columns whose signs were changed in step 2, we obtain

$$\hat{e}_{ij} = kc_i d_j.$$

We now have an additive-plus-multiplicative fit

$$\hat{y}_{ij} = m + a_i + b_j + kc_i d_j.$$

7. Finally we repolish the $\hat{e}_{ij}$ to recenter the additive components of the fit at 0.
8. If we desire, we may iterate the fit further.

Iterating the Fit

When enough steps of iteration are used in the median polish that provides the initial simple additive fit, the table of residuals e_{ij} from that fit will have row medians and column medians very nearly equal to 0. After removing $kc_i d_j$ from the residuals of the additive fit, we may find nonzero values among the row medians and column medians of the r_{ij}, suggesting adjustments to some a_i or b_j or perhaps to m. We can make the adjustments by adding to m and to the a_i and the b_j the main effects resulting from median polish of the r_{ij}. We can then recompute the term $kc_i d_j$, and repeat the entire process until the convergence is adequate. McNeil and Tukey (1975) apply such an iterative process to another resistant fitting technique, for which they include computer programs; J. W. Tukey (personal communication) has pointed out that those programs need some correction.

Remarks on Fitting Strategies

We have described a strategy that uses median polish as a cornerstone for providing a resistant additive-plus-multiplicative fit to a two-way table; we devote much of the rest of this chapter to illustrating the strategy with the adverb–adjective data and interpreting the results. For this 15 × 10 table, the strategy works reasonably well; in particular, the general results and conclusions do not change dramatically if we make different choices in the sign-change algorithm (as noted in Section 3E). Some cautionary remarks are in order, however, for guiding the general application of these fitting techniques.

Chapter 6 of UREDA shows that the resistance of median polish to the influence of unusual data values can be measured by its *breakdown value*. This value is, very roughly, min(I/2, $J/2$). The precise value depends on whether I and J are even or odd, but the details are unimportant for this discussion. If either I or J is small—perhaps 3 or 4 or 5—the appearance of two or more values that remain nonpositive after using sign changes can sometimes lead to difficulties in fitting the table of adjusted residuals derived from a resistant additive fit. In particular, the specific sign changes and the choices made in dealing with remaining negative entries and zeros can influence the final fit. We therefore caution that the algorithm we have described may be fully effective for fitting tables with few rows or few columns only when the residuals from a simple additive fit show an unambiguous multiplicative structure. This restriction on the fitting method may not be too important, however, because for small tables we seldom use the relatively many parameters associated with an additive-plus-multiplicative fit.

Although we have used median polish as the cornerstone of our fitting technique, any resistant method for obtaining a simple additive fit to a two-way table could serve as well. Thus a technique for giving an additive fit that minimizes the sum of absolute values could replace median polish as a basic element of our algorithm for obtaining an additive-plus-multiplicative fit. Similarly, trimmed means could replace medians in the basic fitting process for the simple additive model. We refer to Chapter 6 of UREDA for amplification of these ideas, and to Gabriel and Odoroff (1983) and Gabriel (1983) for a discussion of some recent related work.

EXAMPLE: ADVERBS AND ADJECTIVES DATA (CONTINUED)

The residuals r_{ij} given in Table 3-11 are far from centered; in fact, some rows and columns have only a single negative entry. (Column 2 has a median of 127.43, and several rows and columns have medians over 20.) Table 3-12 gives the adjusted additive-plus-multiplicative fit after the main effects are recentered; we do not recalculate the multiplicative term for this example because we found that the further improvement in fit does not justify the additional computational effort. Thus, the fit of Table 3-12 is our final additive-plus-multiplicative fit.

Before going on to look at residuals from our fits in some detail, our first task is to gain insights about what we have fitted in Table 3-12. At the outset we caution that the parameters of the additive-plus-multiplicative fit do not correspond to the parameters of Cliff's model in equation (17).

We begin by thinking of the revised value of m, 1844, as a typical score for an adverb–adjective combination. The a_i and the b_j represent additive effects for the adjectives and adverbs, and both of these sets of parameters are centered at 0. Positive a_i correspond to the most favorable adjectives, negative a_i to the least favorable adjectives, and a_i with small magnitudes

TABLE 3-12. **Final additive-plus-multiplicative fit to adverb–adjective data from Princeton (unit: 10^{-3}).**

	Residuals											
	(Unmodified)	Slightly	Somewhat	Rather	Pretty	Quite	Decidedly	Unusually	Very	Extremely	a_i	c_i
Evil	−165.42	152.02	169.51	62.22	78.21	6.86	−6.86	−52.30	−47.39	−71.09	−949	−0.57
Wicked	−103.94	171.46	105.41	84.05	−57.08	27.24	19.00	−50.27	−77.73	−19.00	−978	−0.61
Contemptible	−14.40	193.26	177.12	0.00	87.22	25.93	−21.17	0.00	−3.59	−51.39	−978	−0.34
Immoral	21.03	27.29	9.53	−9.52	−40.26	−15.12	26.64	−37.05	41.19	−13.69	−922	−0.91
Disgusting	25.05	86.13	−13.83	−10.44	−109.45	−52.54	55.87	10.44	−37.44	31.71	−1044	−0.71
Bad	9.70	0.00	−27.74	−19.77	0.00	−27.80	7.19	−42.82	0.00	70.22	−788	−1.07
Inferior	123.12	−58.34	28.40	47.15	−115.64	−148.03	15.56	−15.56	−26.90	60.37	−707	−1.00
Ordinary	21.76	−103.32	0.00	−18.02	0.00	30.63	−11.68	−114.13	10.99	126.82	0	0.00
Average	96.94	−132.14	−1.81	16.16	41.19	1.81	−55.49	−5.95	−10.82	12.01	108	0.00
Nice	−45.16	−158.92	−55.50	−11.94	11.94	−28.63	141.63	297.49	110.76	249.24	950	1.47
Good	−12.15	9.57	−9.57	95.55	−21.43	20.61	−19.82	120.09	28.04	−12.05	1102	2.46
Pleasant	−86.01	111.68	150.09	118.87	35.12	−35.12	−90.95	39.25	−71.02	−75.44	1084	2.73
Charming	−28.77	−132.63	−5.68	47.68	−33.00	1.85	26.33	57.72	10.70	−1.85	1195	1.88
Admirable	30.73	−51.60	1.06	−9.67	−3.32	0.00	0.00	55.71	30.65	0.00	1243	2.39
Lovable	0.00	−86.92	−28.73	−5.19	30.49	−0.00	−10.03	76.52	3.04	31.58	1197	1.91
b_j	3.4	187.5	99.1	87.2	58.1	−3.4	−69.1	−69.6	−69.7	−164.6	$1843.8 = m$	
d_j	−0.09	−2.26	−1.77	−1.00	−0.83	0.09	0.77	1.00	1.12	2.19	$k = 125.2$	

Notes: The fit is $\hat{y}_{ij} = m + a_i + b_j + kc_i d_j$. Approximately 94% of the total linear variation is "explained."

correspond to adjectives that are nearly neutral. Negative values of b_j are associated with adverbs like "extremely" that strengthen the impact of the modified adjective. The reader should plot the a_i and the b_j on separate coordinate axes and label each point with the appropriate word in order to begin to identify clusters of similar adjectives or adverbs.

Our next task is to relate the multiplicative parameters to the a_i and the b_j. A plot of d_j against b_j (not shown) is reasonably close to a straight line, but a plot of the c_i against the a_i is very far from linear. These observations suggest that, because the d_j and the b_j are nearly linearly related, introducing separate d_j instead of taking $d_j = b_j$ gains little; however, this is not true of the c_i and the a_i. Substituting $d_j = b_j$ yields the fit

$$\hat{y}_{ij} = m + a_i + b_j + kc_ib_j, \tag{29}$$

which corresponds to the rows-linear model of equation (28) and uses $2I + J - 1$ independent parameters. We leave the exploration of this fit to Exercise 11.

The residuals for our final additive-plus-multiplicative fit are smaller in magnitude and have less systematic structure than do either the residuals from a simple additive fit (see Table 3-9) or those from an extended fit (Table 3-7). To facilitate comparisons among batches of residuals, Figure 3-6 shows parallel boxplots for the residuals from each of five fits for the adverb–adjective data. Besides the residuals from the three fits just mentioned, we also give batches of residuals from a simple multiplicative fit and from the initial additive-plus-multiplicative fit (see Table 3-11). For convenience, we give the equations of the residuals from each fit, the percentage of total absolute (linear) variation accounted for by each fit, and the value of a measure of fit that is adjusted for degrees of freedom.

The boxplots and the other measures of total absolute variation all show that a simple additive fit describes the structure of the adverb–adjective data better than a simple multiplicative fit. The extended fit, with one additional degree of freedom, offers substantial improvement over the additive fit, and an additive-plus-multiplicative fit gives considerable further improvement. A slight additional gain is achieved by recentering the residuals from the initial additive-plus-multiplicative fit; interestingly, this final step of median polish converged so very slowly that we used 10 iterations of polish. The effects of the repolishing lead to more points being subsequently identified as outside; 19 such points appear in the final table of residuals.

The percentage reduction in total absolute deviation describes numerically what we have already observed qualitatively in the parallel boxplot display. This measure verifies the superiority of the additive-plus-multiplicative fits. Note, however, that this fit uses $2 + 2 \cdot 9 + 2 \cdot 14 = 48$ degrees of

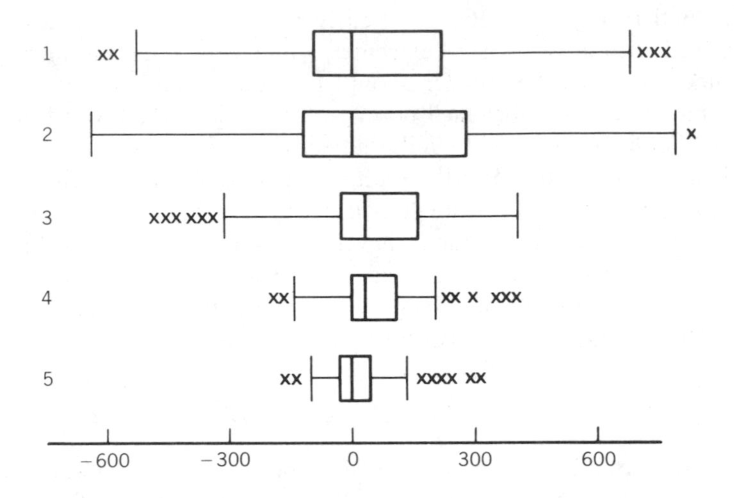

Figure 3-6. Boxplots of residuals from five fits for adverb–adjective da
10^{-3}).

Fit	Residuals	P	DF	(100%
1. Simple additive	$e_{ij} = y_{ij} - (m + a_i + b_j)$	78%	126	.1
2. Simple multiplicative	$r_{ij} = y_{ij} - (hc_i d_j)$	74%	126	.2
3. Extended	$r_{ij} = y_{ij} - (m + a_i + b_j + ka_i b_j)$	86%	125	.1
4. Additive-plus-multiplicative (initial)	$r_{ij} = y_{ij} - (m + a_i + b_j + kc_i d_j)$	94%	102	.0
5. Additive-plus-multiplicative (final)	Same as 4, with a_i and b_j recentered	95%	102	.0

Note: P is the percentage of total absolute variation "explained" by the quantity $(100\% - P)/DF$ is a criterion for choosing among fits; it is desc Section 3C.

freedom (leaving 102 degrees of freedom for the residuals); the exte uses only $2 + 9 + 14 = 25$ degrees of freedom (leaving 125 de freedom for the residuals). When we look at the values of (100% – we see that the fuller description of the additive-plus-multiplicati worthwhile in comparison with the extended fit.

Table 3-13 gives stem-and-leaf displays of the residuals from tended fit (Table 3-7) and from the additive-plus-multiplicative fi

TABLE 3-13. Stem-and-leaf displays of residuals from fits to the adverb–adjective data (unit: 10^{-2}, so that 1|2 represents .12).

a. Residuals from extended fit

stem	leaves
−4*	41
−3 ·	65
−3*	331
−2 ·	9
−2*	30
−1 ·	9777
−1*	41100
−0 ·	9988877665555
−0*	44444333322222222111110000
+0*	0000001111122233334
+0 ·	5555566666667778888889999
1*	000011112224
1 ·	5556666788889999
2*	000133444
2 ·	555666
3*	1224
3 ·	66

b. Residuals from additive-plus-multiplicative fit

stem	leaves
−1 ·	65
−1*	43311000
−0 ·	9887777555555555
−0*	4444333322222222111111111111111100000000000000000
+0*	000000000000000111111112222222222223333333334444
+0 ·	5556677788899
1*	01112224
1 ·	556779
2*	4
2 ·	9

3-12). For the first display, the seven smallest residuals are outliers, and all of them are associated with the adverbs "slightly" and "somewhat" in combination with positive adjectives like "good" and "charming." The second display shows residuals of smaller magnitude and less skewness, although two extreme outliers stand out. They are associated with "extremely nice" and "unusually nice." Unusual adverb–adjective combinations, like "slightly contemptible" and "slightly wicked," tend to have relatively large residuals.

We used Tables 3-7 and 3-12 to calculate the sum of absolute residuals (SAR) separately for each row and for each column of the residuals tables for the two fits. In each table, the SAR for "slightly" stands out as being nearly twice as large as any of those for the other nine adverbs. With the extended fit, the SARs for the adjectives are relatively homogeneous, except for very small values associated with "ordinary" and "average." The largest SAR for adjectives corresponds to "good" with "pleasant" and "nice" not far behind. These commonly used adjectives do not follow Cliff's multiplicative model quite as well as most other adjectives. Using the additive-plus-multiplicative fit, "nice" stands apart among the 15 adjectives as having the largest SAR; we view it as an over-used and perhaps innocuous word. Again, "ordinary" and "average" have small SARs, but now the SARs for "bad" and "admirable" are even smaller. Finally, in going from the extended fit to the more complex fit, the improvement in fit for "admirable" is quite striking. The interested reader can calculate the SARs for the rows and columns of Tables 3-7 and 3-12, and explore the observations we have made here.

Ranking Adverb–Adjective Combinations by Predicted Values

The predicted scores for the 150 adverb–adjective combinations offer a basis for ordering these combinations from least favorable to most favorable. We can use any fit for doing this; here we use the additive-plus-multiplicative fit. We prefer using the fits rather than the data values for this ordering, so that the ordering reflects systematic structure in the scores and not fluctuations that may be partly random. The values predicted by this fit range from 441 to 3579. The values in the raw data range from 473 to 3579. Table 3-14 lists the combinations in order of increasing predicted values under the additive-plus-multiplicative fit.

The adjectives "evil," "wicked," "contemptible," "immoral," "disgusting," "bad," and "inferior" separate entirely from the other eight adjectives; their rankings are below those for all other adverb–adjective combinations. Examination of the fitted effects a_i in Table 3-12 confirms this observation. These seven negative adjectives do tend to mix among them-

TABLE 3-14. Ranking of adverb–adjective combinations by fitted values from the additive-plus-multiplicative fit.

Rank	Fitted Value	Observed Value	Adverb	Adjective
1	441	473	10. Extremely	5. Disgusting
2	508	494	10. Extremely	4. Immoral
3	534	515	10. Extremely	2. Wicked
4	575	504	10. Extremely	1. Evil
5	596	666	10. Extremely	6. Bad
6	607	556	10. Extremely	3. Contemptible
7	630	593	9. Very	5. Disgusting
8	642	652	8. Unusually	5. Disgusting
9	662	718	7. Decidedly	5. Disgusting
10	698	758	10. Extremely	7. Inferior
11	710	632	9. Very	2. Wicked
12	719	669	8. Unusually	2. Wicked
13	724	765	9. Very	4. Immoral
14	737	756	7. Decidedly	2. Wicked
15	738	701	8. Unusually	4. Immoral
16	745	698	9. Very	1. Evil
17	748	744	9. Very	3. Contemptible
18	753	753	8. Unusually	3. Contemptible
19	754	702	8. Unusually	1. Evil
20	763	742	7. Decidedly	3. Contemptible
21	764	791	7. Decidedly	4. Immoral
22	771	764	7. Decidedly	1. Evil
23	789	736	6. Quite	5. Disgusting
24	811	836	1. (Unmodified)	5. Disgusting
25	834	834	9. Very	6. Bad
26	851	808	8. Unusually	6. Bad
27	855	882	6. Quite	2. Wicked
28	858	884	6. Quite	3. Contemptible
29	872	858	1. (Unmodified)	3. Contemptible
30	875	771	1. (Unmodified)	2. Wicked
31	882	889	7. (Decidedly)	6. Bad
32	885	892	6. Quite	1. Evil
33	904	739	1. (Unmodified)	1. Evil
34	908	893	6. Quite	4. Immoral
35	926	899	9. Very	7. Inferior

TABLE 3-14. (Continued)

Rank	Fitted Value	Observed Value	Adverb	Adjective
36	931	822	5. Pretty	5. Disgusting
37	935	956	1. (Unmodified)	4. Immoral
38	942	926	8. Unusually	7. Inferior
39	959	1046	5. Pretty	3. Contemptible
40	970	986	7. Decidedly	7. Inferior
41	975	965	4. Rather	5. Disgusting
42	986	929	5. Pretty	2. Wicked
43	1012	995	4. Rather	3. Contemptible
44	1028	1090	5. Pretty	1. Evil
45	1040	1112	4. Rather	2. Wicked
46	1040	1012	6. Quite	6. Bad
47	1040	1217	3. Somewhat	3. Contemptible
48	1053	1115	4. Rather	1. Evil
49	1056	1042	3. Somewhat	5. Disgusting
50	1070	1080	1. (Unmodified)	6. Bad
51	1074	1034	5. Pretty	4. Immoral
52	1099	1204	3. Somewhat	2. Wicked
53	1119	1289	3. Somewhat	1. Evil
54	1122	974	6. Quite	7. Inferior
55	1123	1113	4. Rather	4. Immoral
56	1149	1342	2. Slightly	3. Contemptible
57	1151	1274	1. (Unmodified)	7. Inferior
58	1187	1273	2. Slightly	5. Disgusting
59	1222	1232	3. Somewhat	4. Immoral
60	1224	1395	2. Slightly	2. Wicked
61	1225	1225	5. Pretty	6. Bad
62	1242	1394	2. Slightly	1. Evil
63	1277	1257	4. Rather	6. Bad
64	1299	1183	5. Pretty	7. Inferior
65	1349	1396	4. Rather	7. Inferior
66	1366	1393	2. Slightly	4. Immoral
67	1393	1365	3. Somewhat	6. Bad
68	1458	1486	3. Somewhat	7. Inferior
69	1546	1546	2. Slightly	6. Bad
70	1606	1548	2. Slightly	7. Inferior
71	1679	1806	10. Extremely	8. Ordinary
72	1774	1785	9. Very	8. Ordinary
73	1774	1660	8. Unusually	8. Ordinary

TABLE 3-14. (Continued)

Rank	Fitted Value	Observed Value	Adverb	Adjective
74	1775	1763	7. Decidedly	8. Ordinary
75	1788	1800	10. Extremely	9. Average
76	1840	1871	6. Quite	8. Ordinary
77	1847	1869	1. (Unmodified)	8. Ordinary
78	1883	1872	9. Very	9. Average
79	1883	1877	8. Decidedly	9. Average
80	1884	1828	7. Pretty	9. Average
81	1902	1902	5. Unusually	8. Ordinary
82	1931	1913	4. Rather	8. Ordinary
83	1943	1943	3. Somewhat	8. Ordinary
84	1949	1951	6. Quite	9. Average
85	1956	2053	1. (Unmodified)	9. Average
86	2011	2052	5. Pretty	9. Average
87	2031	1928	2. Slightly	8. Ordinary
88	2040	2056	4. Rather	9. Average
89	2052	2050	3. Somewhat	9. Average
90	2140	2008	2. Slightly	9. Average
91	2345	2457	2. Slightly	12. Pleasant
92	2422	2572	3. Somewhat	12. Pleasant
93	2439	2449	2. Slightly	11. Good
94	2500	2490	3. Somewhat	11. Good
95	2566	2407	2. Slightly	10. Nice
96	2564	2511	3. Somewhat	10. Nice
97	2598	2547	2. Slightly	14. Admirable
98	2658	2656	3. Somewhat	14. Admirable
99	2686	2793	4. Rather	12. Pleasant
100	2692	2604	2. Slightly	15. Lovable
101	2697	2564	2. Slightly	13. Charming
102	2697	2685	4. Rather	10. Nice
103	2699	2711	5. Pretty	10. Nice
104	2703	2738	5. Pretty	12. Pleasant
105	2718	2689	3. Somewhat	15. Lovable
106	2725	2716	3. Somewhat	13. Charming
107	2736	2821	4. Rather	11. Good
108	2747	2727	5. Pretty	11. Good
109	2774	2736	1. (Unmodified)	10. Nice
110	2799	2778	6. Quite	10. Nice
111	2867	3009	7. Decidedly	10. Nice
112	2875	2865	4. Rather	14. Admirable

TABLE 3-14. (Continued)

Rank	Fitted Value	Observed Value	Adverb	Adjective
113	2890	2885	4. Rather	15. Lovable
114	2891	2939	4. Rather	13. Charming
115	2896	2893	5. Pretty	14. Admirable
116	2902	2932	5. Pretty	15. Lovable
117	2902	2869	5. Pretty	13. Charming
118	2902	2816	1. (Unmodified)	12. Pleasant
119	2908	3206	8. Unusually	10. Nice
120	2922	2910	1. (Unmodified)	11. Good
121	2931	3042	9. Very	10. Nice
122	2955	2920	6. Quite	12. Pleasant
123	2969	2990	6. Quite	11. Good
124	3022	2993	1. (Unmodified)	13. Charming
125	3024	3024	1. (Unmodified)	15. Lovable
126	3033	3282	10. Extremely	10. Nice
127	3056	3058	6. Quite	13. Charming
128	3059	3059	6. Quite	15. Lovable
129	3064	3095	1. (Unmodified)	14. Admirable
130	3110	3110	6. Quite	14. Admirable
131	3115	3095	7. Decidedly	11. Good
132	3124	3033	7. Decidedly	12. Pleasant
133	3152	3178	7. Decidedly	13. Charming
134	3157	3147	7. Decidedly	15. Lovable
135	3184	3304	8. Unusually	11. Good
136	3201	3240	8. Unusually	12. Pleasant
137	3204	3262	8. Unusually	13. Charming
138	3210	3287	8. Unusually	15. Lovable
139	3222	3250	9. Very	11. Good
140	3233	3244	9. Very	13. Charming
141	3240	3243	9. Very	15. Lovable
142	3243	3172	9. Very	12. Pleasant
143	3250	3250	7. Decidedly	14. Admirable
144	3317	3373	8. Unusually	14. Admirable
145	3354	3385	9. Very	14. Admirable
146	3389	3387	10. Extremely	13. Charming
147	3399	3431	10. Extremely	15. Lovable
148	3455	3443	10. Extremely	11. Good
149	3512	3437	10. Extremely	12. Pleasant
150	3579	3579	10. Extremely	14. Admirable

selves, depending on the adverb they are paired with. On the other hand, "ordinary" and "average" tend to cluster together in the middle of the list, with "ordinary" most often ranked lower than "average."

Among the adverbs, "extremely" dominates the very beginning and the end of each list. Of course the strongest adverbs appear at both ends of the ranking, whereas weak adverbs like "pretty," "quite," and "rather" do not appear near either end of the list.

3G. SOME BACKGROUND FOR NONADDITIVE FITS

Mandel (1969a, 1969b, 1971) was among the first to consider explicitly models for two-way tables in which, unlike the extended model of equation (6), the interaction term need not be a function of the main effects for rows and columns. To model the residuals from an additive fit, he proposed

$$e_{ij} = \kappa^{(1)}\gamma_i^{(1)}\delta_j^{(1)} + \kappa^{(2)}\gamma_i^{(2)}\delta_j^{(2)} + \cdots + \kappa^{(l)}\gamma_i^{(l)}\delta_j^{(l)} + \varepsilon_{ij},$$

where l is less than or equal to the rank of the matrix of residuals, $\mathbf{E}$. When Mandel's model is used to generalize our additive-plus-multiplicative model (26), it provides a very general representation for data of a two-way table:

$$y_{ij} = \mu + \alpha_i + \beta_j + \sum_{n=1}^{l} \kappa^{(n)}\gamma_i^{(n)}\delta_j^{(n)} + \varepsilon_{ij}. \tag{30}$$

In his published illustrations, Mandel used either one or two multiplicative terms. In the psychometric literature, Gollob (1968) proposed similar models which combine aspects of analysis-of-variance and factor analytic models.

To give uniqueness to the parameters of model (30), Mandel imposed several constraints:

$$\sum \alpha_i = 0 \quad \text{and} \quad \sum \beta_j = 0;$$

$$\sum_i \gamma_i^{(n)} = 0 \quad \text{and} \quad \sum_j \delta_j^{(n)} = 0 \quad \text{for all } n = 1, \ldots, l;$$

and

$$\sum_i \left(\gamma_i^{(n)}\right)^2 = 1 \quad \text{and} \quad \sum_j \left(\delta_j^{(n)}\right)^2 = 1 \quad \text{for all } n = 1, \ldots, l.$$

These constraints center the various row and column parameters at 0, and for each n they give the vectors

$$\boldsymbol{\gamma}^{(n)} = \left[\gamma_1^{(n)}, \ldots, \gamma_I^{(n)}\right]^T$$

and

$$\boldsymbol{\delta}^{(n)} = \left[\delta_1^{(n)}, \ldots, \delta_J^{(n)}\right]^T$$

unit length.

The Singular Value Decomposition

Mandel estimated the various parameters of model (30) by solving the least-squares problem of minimizing the sum of squares of the residuals from this model. He made use of the *singular value decomposition* of a matrix (Eckart and Young, 1939), a theoretical result from linear algebra that is analogous to the spectral decomposition for square matrices. [See Good (1969) for a discussion of the role of the SVD in statistics.] If $\mathbf{E} = (e_{ij})$ is the $I \times J$ matrix of residuals from a least-squares additive fit to (y_{ij}), consider the $I \times I$ matrix $\mathbf{EE}^T$. Theory ensures that, for each value of $n = 1, \ldots, l$, the coefficients $\kappa^{(n)}$ in equation (30) satisfy the matrix equation

$$\mathbf{EE}^T\boldsymbol{\gamma}^{(n)} = \left(\kappa^{(n)}\right)^2\boldsymbol{\gamma}^{(n)}. \tag{31a}$$

Thus, $\boldsymbol{\gamma}^{(n)}$ is an eigenvector of the square symmetric matrix $\mathbf{EE}^T$, associated with the eigenvalue $(\kappa^{(n)})^2$. Similarly, the equations

$$\mathbf{E}^T\mathbf{E}\boldsymbol{\delta}^{(n)} = \left(\kappa^{(n)}\right)^2\boldsymbol{\delta}^{(n)}, \quad n = 1, \ldots, l, \tag{31b}$$

must hold. The singular value decomposition of $\mathbf{E}$ is then

$$\mathbf{E} = \kappa^{(1)}\boldsymbol{\gamma}^{(1)}\left(\boldsymbol{\delta}^{(1)}\right)^T + \cdots + \kappa^{(r)}\boldsymbol{\gamma}^{(r)}\left(\boldsymbol{\delta}^{(r)}\right)^T, \tag{32}$$

where r, the rank of $\mathbf{E}$, cannot exceed the smaller of I and J.

The application of this decomposition to Mandel's model (30) is more evident if we rewrite the model as

$$y_{ij} - \mu - \alpha_i - \beta_j = \sum_{n=1}^{l} \kappa^{(n)}\gamma_i^{(n)}\delta_j^{(n)} + \varepsilon_{ij}, \tag{33}$$

with $l \le r$ and with $\kappa^{(1)} \ge \kappa^{(2)} \ge \cdots \ge \kappa^{(l)} \ge \cdots \ge \kappa^{(r)}$. The first l terms on the right-hand sides of equations (32) and (33) coincide. Thus the sum of the last $r - l$ terms in equation (32) corresponds to the matrix of ε_{ij} in the model of equation (33). A major result in the theory of least squares says that *the sum of the first l terms of the singular value decomposition of* **E** *gives the matrix of rank l, denoted by* $\mathbf{N}(l)$, *that best approximates* **E** *in the least-squares sense*. That is, the matrix $\mathbf{N}(l) = (n_{ij})$ is the rank-l matrix that minimizes the sum of squares

$$SS = \sum_{i=1}^{I} \sum_{j=1}^{J} (e_{ij} - n_{ij})^2. \tag{34}$$

Furthermore, the minimum value of this sum is

$$\sum_{n=l+1}^{r} (\kappa^{(n)})^2,$$

so that we can account for the fraction

$$\frac{(\kappa^{(1)})^2 + \cdots + (\kappa^{(l)})^2}{(\kappa^{(1)})^2 + \cdots + (\kappa^{(l)})^2 + \cdots + (\kappa^{(r)})^2} \tag{35}$$

of the total sample variation (above and beyond the simple additive model) using the rank-l matrix

$$\mathbf{N}(l) + \kappa^{(1)}\boldsymbol{\gamma}^{(1)}(\boldsymbol{\delta}^{(1)})^T + \cdots + \kappa^{(l)}\boldsymbol{\gamma}^{(l)}(\boldsymbol{\delta}^{(l)})^T. \tag{36}$$

Mandel's model does just this by using a sum of l multiplicative terms.

Householder and Young (1938) first studied the singular value decomposition of a matrix and showed its role in providing a least-squares approximation to a given matrix. Good (1969) and Bradu and Gabriel (1978) discuss the role of the singular value decomposition in analyzing the multiplicative structure of a two-way table. Gabriel (1978) discusses the least-squares theory for additive-plus-multiplicative fits, where the multiplicative term can have any rank. Lowerre (1982) gives a summary of the role of this decomposition in modern statistics and provides some related theory. Gabriel (1971, 1982) and Bradu and Gabriel (1978) have focused on matrix approximations of rank 2 or less to provide a geometrical technique, called the *biplot*, for exploring the structure of a two-way table. Johnson and Graybill (1972) extend Mandel's study of model (30), in the special case with $l = 1$ and independent identically distributed Gaussian errors, by finding likelihood-ratio tests and maximum-likelihood estimates for all parameters.

TABLE 3-15. Summary of models and resistant fitting procedures for two-way tables.

Step[a]	Description of Model	Equation of Model[b]	Fit	Fitting Method	Comments
A	Simple additive model	$y_{ij} = \mu + \alpha_i + \beta_j + \varepsilon_{ij}$ equation (5)	$m + a_i + b_j$	Median polish	Iterative procedure
B	Extended model	$y_{ij} = \mu + \alpha_i + \beta_j + \kappa\alpha_i\beta_j + \varepsilon_{ij}$ equation (6)	$m + a_i + b_j + ka_ib_j$	Diagnostic plot	May iterate; equivalent to fit of form $\hat{y}_{ij} = q + hc_id_j$; equation (16)
D	Simple multiplicative model (positive data)	$y_{ij} = \nu\gamma_i\delta_j + \varepsilon_{ij}$ equation (13)	hc_id_j	Median polish in log scale	Assumes all $y_{ij} > 0$
E	Simple multiplicative model (any data)	$e_{ij} = \kappa\gamma_i\delta_j + \varepsilon_{ij}$ equation (21)	kc_id_j	Sign-change procedure followed by median polish in log scale	Minimizes the number of negative entries, then polishes the rest of the table
F	Additive-plus-multiplicative model	$y_{ij} = \mu + \alpha_i + \beta_j + \kappa\gamma_i\delta_j + \varepsilon_{ij}$ equation (26)	$m + a_i + b_i + kc_id_j$	Step A followed by step E on residuals	May iterate steps A and E

[a] The steps are lettered according to the section of this chapter that discusses them.
[b] The equation numbers identify the location of each model in the text.

Much of the second half of this chapter provides and illustrates resistant techniques for fitting the data of a two-way table according to a model much like Mandel's model (30) when $l = 1$. Our techniques, by relying heavily on medians rather than means, avoid the tendency of least-squares fitting procedures to spread the impact of an unusual data value over other cells in the table. Emerson, Hoaglin, and Kempthorne (1984) show that this impact can be very substantial in an additive-plus-multiplicative fit. Although the techniques of this chapter do not satisfy any optimality criterion, they are often suitable for resistant analysis of two-way tables that exhibit highly structured interaction between rows and columns, particularly when the analysis is exploratory.

3H. SUMMARY

We have described and illustrated fitting techniques for five models for two-way tables, using median polish as the cornerstone of a resistant approach. Table 3-15 summarizes the models and our fitting technique for each and gives very brief comments on the fitting techniques. We also give a few references to the relevant sections of this chapter and to selected equation numbers.

REFERENCES

Anscombe, F. J. (1967). "Topics in the investigation of linear relations fitted by the method of least squares," *Journal of the Royal Statistical Society, Series B*, **29**, 1–29.

Bradu, D. and Gabriel, K. R. (1978). "The biplot as a diagnostic tool for models of two-way tables," *Technometrics*, **20**, 47–68.

Carter, O. R., Collier, B. L., and David, F. L. (1951). "Blast furnace slags as agricultural liming materials," *Agronomy Journal*, **43**, 430–433.

Cliff, N. (1959). "Adverbs as multipliers," *Psychological Review*, **66**, 27–44.

Eckart, C. and Young, G. (1939). "A principal axis transformation for non-Hermitian matrices," *Bulletin of the American Mathematical Society*, **45**, 118–121.

Emerson, J. D., Hoaglin, D. C., and Kempthorne, P. J. (1984). "Leverage in least squares additive-plus-multiplicative fits for two-way tables," *Journal of the American Statistical Association*, **79**, 329–335.

Gabriel, K. R. (1971). "The biplot graphic display of matrices with application to principal components analysis," *Biometrika*, **58**, 453–467.

Gabriel, K. R. (1978). "Least squares approximation of matrices by additive and multiplicative models," *Journal of the Royal Statistical Society, Series B*, **40**, 186–196.

Gabriel, K. R. (1982). "Biplot." In S. Kotz and N. L. Johnson (Eds.), *Encyclopedia of Statistical Sciences*, Vol. 1. New York: Wiley, pp. 263–271.

Gabriel, K. R. (1983). "An alternative computation of median fits in two-way tables," Technical Report 83/01, Department of Statistics and Division of Biostatistics, University of Rochester.

Gabriel, K. R. and Odoroff, C. L. (1983). "Resistant lower rank approximation of matrices," Technical Report 83/02, Department of Statistics and Division of Biostatistics, University of Rochester.

Gollob, H. F. (1968). "A statistical model which combines features of factor analytic and analysis of variance techniques," *Psychometrika*, **33**, 73–116.

Good, I. J. (1969). "Some applications of the singular decomposition of a matrix," *Technometrics*, **11**, 823–831.

Householder, A. S. and Young, G. (1938). "Matrix approximation and latent roots," *American Mathematical Monthly*, **45**, 165–171.

Johnson, D. E. and Graybill, F. A. (1972). "An analysis of a two-way model with interaction and no replication," *Journal of the American Statistical Association*, **67**, 862–868.

Lowerre, J. M. (1982). "An introduction to modern matrix methods and statistics," *The American Statistician*, **36**, 113–115.

Mandel, J. (1969a). "A method for fitting empirical surfaces to physical or chemical data," *Technometrics*, **11**, 411–429.

Mandel, J. (1969b). "The partitioning of interaction in analysis of variance," *Journal of Research of the National Bureau of Standards—B. Mathematical Sciences*, **73B**, 309–328.

Mandel, J. (1971). "A new analysis of variance model for non-additive data," *Technometrics*, **13**, 1–18.

McNeil, D. R. and Tukey, J. W. (1975). "Higher-order diagnosis of two-way tables, illustrated on two sets of demographic empirical distributions," *Biometrics*, **31**, 487–510.

Ryan, T. A., Joiner, B. L., and Ryan, B. F. (1982). *Minitab Reference Manual*. University Park, PA: The Pennsylvania State University.

Scheffé, H. (1959). *The Analysis of Variance*. New York: Wiley.

Siegel, A. F. (1983). "Low median and least absolute residual analysis of two-way tables," *Journal of the American Statistical Association*, **78**, 371–374.

Snedecor, G. W. and Cochran, W. G. (1980). *Statistical Methods*, 7th ed. Ames, IA: The Iowa State University Press.

Tukey, J. W. (1949). "One degree of freedom for non-additivity," *Biometrics*, **5**, 232–242.

Tukey, J. W. (1967). "Discussion of Anscombe's paper," *Journal of the Royal Statistical Society, Series B*, **29**, 47–48.

Tukey, J. W. (1977). *Exploratory Data Analysis*. Reading, MA: Addison-Wesley.

Wood, L. A. and Martin, G. M. (1964). "Compressibility of natural rubber at pressures below 500 kg/cm^2," *Journal of Research of the National Bureau of Standards—A. Physics and Chemistry*, **68**A, 259–268.

Yates, F. (1970). *Experimental Design: Selected Papers*. Darien, CT: Hafner.

Additional Literature

Breckenridge, M. B. (1983). *Age, Time, and Fertility: Applications of Exploratory Data Analysis*. New York: Academic Press.

Cox, D. R. (1984). "Interaction," *International Statistical Review*, **52**, 1–31.

Gabriel, K. R. and Zamir, S. (1979). "Lower rank approximation of matrices by least squares with any choice of weights," *Technometrics*, **21**, 489–498.

Kemperman, J. H. B. (1984). "Least absolute value and median polish." In Y. L. Tong (Ed.), *Inequalities in Statistics and Probability* (IMS Lecture Notes—Monograph Series, Vol. 5). Hayward, CA: Institute of Mathematical Statistics, pp. 84–103.

Kettenring, J. R. (1983). "Chapter 5. A case study in data analysis," *Proceedings of Symposia in Applied Mathematics*, **28**, 105–139.

TABLE 3-16. Specific volume in cm^3/g for unvulcanized rubber.

Temperature in degrees Celsius	Pressure in kg/cm^2 500	400	300	200	100	0
0	1.0554	1.0593	1.0636	1.0679	1.0725	1.0772
20	1.0673[a]	1.0718	1.0764	1.0814	1.0864	1.0917
25	1.0702	1.0748	1.0795	1.0845	1.0896	1.0951

Source: Wood and Martin (1964), as reproduced by Mandel (1969a).
[a] The value is a statistical estimate for the actual value, which was missing. Mandel (1969a) describes the estimation procedure underlying this value.

Kruskal, J. B. (1983). "Chapter 4. Multilinear Methods," *Proceedings of Symposia in Applied Mathematics*, **28**, 75–104.

Snee, R. D. (1982). "Nonadditivity in a two-way classification: is it interaction or nonhomogeneous variance?" *Journal of the American Statistical Association*, **77**, 515–519.

Tukey, J. W. (1962). "The future of data analysis," *Annals of Mathematical Statistics*, **33**, 1–67, 812.

Williams, E. J. (1952). "The interpretation of interactions in factorial experiments," *Biometrika*, **39**, 65–81.

EXERCISES

1. In addition to the data presented in the example of Section 3B, Wood and Martin (1964) presented data for unvulcanized rubber. Table 3-16 gives the specific volume of unvulcanized rubber at three temperatures and six pressures.

(a) Use median polish to provide a simple additive fit to this table of data.

(b) Construct the diagnostic plot for this table and use it to determine the value of k for an extended fit.

(c) Give the table of residuals from this extended fit.

2. Compare the two fits you have obtained in Exercise 1 for the unvulcanized rubber data, using:

(a) the two tables of residuals,

(b) parallel boxplots,

(c) the percent of total sum of absolute residuals explained,

(d) a measure of fit that incorporates degrees of freedom.

Do the residuals suggest that a complex fit using more parameters better describes the structure within this table?

TABLE 3-17. Yields of corn in bushels per acre.

			Soil Type		
Liming Treatment	Pounds/Acre	Elements Added	Very Fine Sandy Loam	Sandy Clay Loam	Loamy Sand
No lime	0	None	11.1	32.6	63.3
Coarse slag	4,000	None	15.3	40.8	65.0
Medium slag	4,000	None	22.7	52.1	58.8
Agricultural slag	4,000	None	23.8	52.8	61.4
Agricultural limestone	4,000	None	25.6	63.1	41.1
Agricultural slag	4,000	B, Zn, Mn	31.2	59.5	78.1
Agricultural limestone	4,000	B, Zn, Mn	25.8	55.3	60.2

Source: Carter, Collier, and David (1951), as cited by Johnson and Graybill (1972).

3. Do the fits obtained in Section 3B for vulcanized rubber adequately describe the data for unvulcanized rubber? Compare your analysis of the data set in Exercises 1 and 2 with that presented in Section 3B. Does a three-way table provide a more appropriate framework for viewing the two data sets together? (Chapter 4 explores resistant analyses for three-way tables.)

4. Table 3-17 presents data on the yields of corn in bushels per acre [from Carter, Collier, and David (1951), as cited by Johnson and Graybill (1972)]. Using a least-squares analysis for Tukey's extended model, Johnson and Graybill found that Tukey's test for systematic nonadditivity (mentioned in Section 3B) did not indicate significant departures from additive structure. Use the more resistant and exploratory methods of Section 3B to decide whether you think that the corn data have nonadditive structure. If so, does a resistant fit to Tukey's extended model adequately describe the nonadditivity? What further exploration of this table might be productive?

5. Bradu and Gabriel (1978) present data from Yates (1970) on the yields of four varieties of cotton in seven centers. They mention that Yates analyzed these data in the log scale. Give a multiplicative fit to the data of Table 3-18 and use the residuals to judge the adequacy of your fit.

TABLE 3-18. Yields of cotton.

	Variety			
Center	1	2	3	4
1	1.55	1.26	1.41	1.78
2	3.39	3.47	2.82	3.89
3	1.95	1.91	1.74	2.29
4	10.47	9.12	9.55	17.78
5	1.45	1.51	1.41	1.70
6	3.72	3.55	3.09	4.27
7	4.47	4.07	3.98	4.47

Source: Yates (1970), as cited by Bradu and Gabriel (1978).

6. Reanalyze the data of Exercise 5 by providing a fit of the form $\hat{y}_{ij} = m + kc_i d_j$. Why is this fit almost certain to be at least somewhat better than that explored in Exercise 5?

7. After reviewing the extended fit for the rubber data (Tables 3-1 and 3-3), provide a second complete step of the iterative fitting process described, by

(a) performing median polish on the residuals r_{ij},
(b) adjusting the fitted values of m, a_i, and b_j,
(c) recalculating the extended fit by basing the value k on the adjusted common term and main effects, and
(d) providing a new table of residuals from this extended fit.

8. Use the methods of Section 3C to compare the fit you found in Exercise 7 with that described in Section 3B for the rubber data.

(a) Are there qualitative improvements in your new residuals?
(b) Has the step of iteration that you performed increased the percentage of total sum of absolute deviations "explained" by the extended model?

9. Consider the table of residuals from the simple additive fit to the rubber data, given in Section 3B. Use sign changes and median polish in the log scale to give a multiplicative fit to the residuals from the additive fit.

10. Compare your residuals in Exercise 9 to those for the extended fit. Do you think that the improvements, if any, gained by the more complex

TABLE 3-19. Scale values of adverb–adjective combinations as determined by (a) students at Wayne State University and (b) male students at Dartmouth College (unit = .001).

	Adverb									
Adjective	(Unmodified)	Slightly	Somewhat	Rather	Pretty	Quite	Decidedly	Unusually	Very	Extremely
a. Wayne State University										
Evil	607	1419	1283	1084	914	752	576	553	465	107
Wicked	650	1328	1124	978	964	786	558	466	446	88
Contemptible	793	1324	1134	1047	867	832	609	661	633	327
Immoral	793	1176	1133	954	884	726	528	451	465	−85
Disgusting	828	1327	963	990	753	687	636	664	502	395
Bad	1024	1497	1323	1232	1018	924	797	662	639	470
Inferior	1323	1520	1516	1295	1180	1127	1013	963	927	705
Ordinary	2074	1980	2038	2034	2026	2023	1949	1875	2073	1936
Average	2145	2023	2080	2172	2094	2101	2020	2062	2039	2052
Nice	2636	2286	2488	2568	2767	2738	2969	3155	3016	3351
Good	2712	2417	2462	2755	2622	2880	3024	3243	3250	3449
Pleasant	2770	2440	2505	2743	2738	2849	3028	3107	3174	3490
Charming	2912	2557	2667	2881	2860	2955	3153	3223	3182	3372
Admirable	2972	2542	2682	2853	2867	3031	3263	3231	3305	3561
Lovable	3054	2626	2705	2829	3062	3025	3157	3250	3327	3462

b. Dartmouth College

Evil	843	1420	1305	1085	1056	891	721	713	640	450
Wicked	782	1412	1132	1038	918	845	692	611	614	375
Contemptible	801	1336	1148	1032	1015	755	647	663	652	465
Immoral	1002	1328	1227	1099	974	839	754	653	690	372
Disgusting	792	1270	1055	935	749	661	625	571	538	325
Bad	1125	1466	1343	1266	1065	1004	920	837	828	725
Inferior	1329	1487	1452	1347	1123	985	970	893	900	703
Ordinary	2043	2005	2006	1964	2025	1920	1969	1824	1903	1790
Average	1984	2052	2003	2048	2044	1950	1864	1989	1977	1799
Nice	2669	2366	2483	2594	2739	2760	2968	3170	3080	3326
Good	2770	2411	2521	2778	2686	2946	3052	3262	3228	3410
Pleasant	2830	2402	2601	2811	2726	2887	3034	3123	3103	3465
Charming	2938	2468	2681	2874	2860	3042	3128	3244	3235	3372
Admirable	2977	2479	2674	2887	2874	3150	3292	3272	3339	3549
Lovable	2991	2593	2732	2883	3090	3060	3192	3330	3339	3493

Source: Norman Cliff (1959). "Adverbs as multipliers," *Psychological Review*, **66**, 27–44 (data from Table 9, p. 36). Copyright 1959 by the American Psychological Association. Reprinted by permission of the author.

procedure justify the additional parameters used in an additive-plus-multiplicative fit?

11. Using some of the results from the analysis of the adverb–adjective data, give a resistant rows-linear fit of the form specified by equation (29).

(a) A plot of e_{ij} against b_j for fixed i should be nearly linear, and a resistant line fit to this plot may aid in determining kc_i.

(b) Compare the quality of your fit with that of the fits examined in Figure 3-6. Be certain to use a measure that takes into account the degrees of freedom associated with your residuals.

12. Table 3-19 gives the scale values for the 150 adverb–adjective combinations as determined by Cliff (1959) from questionnaire responses at Wayne State University and Dartmouth College, respectively.

(a) How well does the final additive-plus-multiplicative fit of Tables 3-12 and 3-14 (based on the data in Table 3-6) predict the scale values for the other two schools? Give particular attention to systematic lack of fit.

(b) Obtain a new resistant additive-plus-multiplicative fit for each of these additional sets of data. Discuss the relative quality of fit for the three tables, and compare the respective values of m, the a_i, the b_j, k, the c_i, and the d_j.

CHAPTER 4

Three-Way Analyses

Nancy Romanowicz Cook
Harvard Medical School

Tables with a measured response and three or more factors extend the ideas of the two-way table. To gain protection against adverse effects of isolated unusual observations, we analyze these data by resistant techniques similar to two-way median polish (described in Section 3A and in Chapter 6 of UREDA). The present chapter concentrates on the three-way case as a leading illustration.

Section 4A examines the structure of the three-way table and establishes notation for the rest of the chapter. As Section 4B describes, it is often advantageous to summarize the behavior of the response in terms of separate additive contributions for the three factors, immediately extending the basic additive model for the two-way table to a main-effects-only analysis. Section 4C develops the corresponding extension of median polish.

When the residuals from a simple additive fit to a two-way table reveal systematic nonadditivity, we consider re-expressing the data to promote additivity or, alternatively, fitting a multiplicative interaction term (Chapter 3). Section 4D derives the basic diagnostic plot to aid in choosing a re-expression for a three-way table.

We also have available the further possibility of fitting a general set of interaction effects for each combination of two factors from among the three. Section 4F applies median polish to this full-effects model, which is a second immediate extension of the basic additive model for the two-way table. Section 4G shows how to use the two-factor interactions in diagnosing a re-expression.

Although nonresistant, the convenient analyses based on means also apply to three-way tables. Sections 4E and 4H briefly review these techniques for fitting and compare them to the median-polish analyses.

Other issues that arise for two-way tables carry over into the three-way setting as well. Section 4I touches on computational techniques and the least-absolute-deviations criterion, polishing approaches based on location summaries other than the mean and the median, and missing values.

4A. STRUCTURE OF THE THREE-WAY TABLE

In a three-way table of measurements, three factors (say A, B, and C) may each contribute to the value of the response or outcome variable. If we arrange the data array in three-dimensional space, we can visualize these contributions as row, column, and layer effects.

Suppose factor A has I levels, factor B has J levels, and factor C has K levels. The possible combinations of one level from each factor form $I \times J \times K$ cells that contain the observations, often one value per cell. In

TABLE 4-1. Two of the six possible two-way arrangements of a three-way table.

a. Factor A alone and factors B and C combined, with one block for each level of B

	B:	1			2			···	J		
A	C:	1	···	K	1	···	K	···	1	···	K
1		y_{111}	···	y_{11K}	y_{121}	···	y_{12K}	···	y_{1J1}	···	y_{1JK}
⋮		⋮		⋮	⋮		⋮		⋮		⋮
I		y_{I11}	···	y_{I1K}	y_{I21}	···	y_{I2K}	···	y_{IJ1}	···	y_{IJK}

b. Factor C alone and factors A and B combined, with one block for each level of A

A	B	C: 1	2	···	K
	1	y_{111}	y_{112}	···	y_{11K}
1	⋮	⋮	⋮		⋮
	J	y_{1J1}	y_{1J2}	···	y_{1JK}
	1	y_{211}	y_{212}	···	y_{21K}
2	⋮	⋮	⋮		⋮
	J	y_{2J1}	y_{2J2}	···	y_{2JK}
⋮	⋮	⋮	⋮		⋮
	1	y_{I11}	y_{I12}	···	y_{I1K}
I	⋮	⋮	⋮		⋮
	J	y_{IJ1}	y_{IJ2}	···	y_{IJK}

this, the simplest case, we represent the measurement in a cell by y_{ijk}, where i, j, and k stand for the levels of factors A, B, and C, respectively.

We can arrange the three-way array in two dimensions in six ways. Table 4-1 shows two possibilities, but any of the four other arrangements of this type may be more desirable. In practice, we often determine the most convenient groupings of the factors by the relative sizes of I, J, and K—putting the factor with most levels by itself—or by how the arrangement fits on a page.

EXAMPLE: WEIGHT GAIN IN PIGS

At the Iowa Agricultural Experiment Station, investigators conducted an experiment to estimate the effects of three food supplements to a diet of corn on weight gains in male pigs. The supplements and their percentages were:

Methionine (3 levels: 0%, .25%, .50%)
Protein (in soybean meal—2 levels: 12%, 14%)
Lysine (4 levels: 0%, .05%, .10%, .15%).

Each combination of supplements was given to two pigs. (Thus the data come from a total of 48 pigs.) The response measured for each pig was the *average daily weight gain*. From Snedecor and Cochran (1980, p. 319) we

TABLE 4-2. Treatment totals of average daily weight gains (unit = .01 lb) of male pigs fed three food supplements. (Layout follows Table 4-1b.)

Methionine (in .01%)	Protein (in %)	Lysine (in .01%) 0	5	10	15
0	12	208	230	235	222
	14	297	308	246	209
25	12	208	224	275	252
	14	249	258	261	281
50	12	206	208	253	249
	14	291	303	285	289

Source: George W. Snedecor and William G. Cochran (1980). *Statistical Methods*, 7th ed. Ames, IA: The Iowa State University Press (data from Table 16.13.2, p. 319).

take the total of the average gains for the two pigs that received the same supplements. These totals serve as the single measurements (per cell) y_{ijk}. Table 4-2 presents one arrangement of the table, corresponding to the format in Table 4-1*b*.

4B. DECOMPOSITIONS AND MODELS FOR THREE-WAY ANALYSIS

We can break down data such as those in Table 4-2 in ways similar to the ways used for two-way tables. These decompositions are often reminiscent of corresponding "models," many of which are special cases of the general form

$$y_{ijk} = f(i, j, k) + \varepsilon_{ijk}, \tag{1}$$

where f is some function of the levels of the row, column, and layer factors (here, the fractions of methionine, protein, and lysine), and the ε_{ijk} are fluctuations arising from natural variability and from the measurement process.

If the three factors exert their influences separately and additively, the model takes the simple form

$$y_{ijk} = \mu + \alpha_i + \beta_j + \gamma_k + \varepsilon_{ijk}, \tag{2}$$

where α_i is the effect of level i of factor A, β_j is the effect of level j of factor B, γ_k is the effect of level k of factor C, and μ is an overall value or common term. Because it treats the effect of each factor as separate from the effects of other factors, we call equation (2) a *main-effects-only* or *simple additive* model. In Section 4C we show how to estimate these effects resistantly by a technique similar to median polish for two-way tables.

In preparation for the three-way median-polish procedure we impose certain side conditions on the effects:

$$\operatorname*{med}_i \{\alpha_i\} = 0,$$

$$\operatorname*{med}_j \{\beta_j\} = 0,$$

$$\operatorname*{med}_k \{\gamma_k\} = 0,$$

where $\operatorname{med}_i\{\alpha_i\}$ denotes the median of the values $\alpha_1, \ldots, \alpha_I$. These conditions are analogous to the usual conditions in analysis of variance that

$$\sum_i \alpha_i = \sum_j \beta_j = \sum_k \gamma_k = 0,$$

or, equivalently, that the average of each set of effects is zero.

We find the main-effects model in equation (2) both simple and attractive, but frequently it cannot adequately describe the data. Although additivity of effects may often be a good initial model, we require interaction terms when the factors fail to act separately. The most general additive decomposition of this form for a three-way table (with more than one observation per cell, indicated by the additional subscript r) is the *full-effects* model:

$$y_{ijkr} = \mu + \alpha_i + \beta_j + \gamma_k + (\alpha\beta)_{ij} + (\alpha\gamma)_{ik} + (\beta\gamma)_{jk} + (\alpha\beta\gamma)_{ijk} + \varepsilon_{ijkr}, \tag{3}$$

where $(\alpha\beta)_{ij}$ is the two-factor interaction effect for level i of factor A and level j of factor B, with similar interpretations for $(\alpha\gamma)_{ik}$ and $(\beta\gamma)_{jk}$, and where $(\alpha\beta\gamma)_{ijk}$ is a three-factor interaction effect. [If we have only one observation per cell, as in the pig feeding example of Table 4-2, we cannot explicitly consider patterns of interaction as general as $(\alpha\beta\gamma)_{ijk}$, because we have no basis for separating $(\alpha\beta\gamma)_{ijk}$ from ε_{ijk}, to which ε_{ijkr} reduces when r takes only one value.]

The interaction terms indicate, in one common notation, joint contributions of two or more factors, above and beyond the contributions represented by their main effects (or by interactions of lower order). These are not necessarily anything like products of main effects; the notation $(\alpha\beta)_{ij}$ simply denotes the joint contribution of the two factors whose main effects are α_i and β_j, described as a supplement to these factors' main effects, and it does this without introducing any new Greek letters. It generalizes easily to more complicated combinations of factors. For model (3) we find it convenient to impose the additional side conditions

$$\operatorname*{med}_i (\alpha\beta)_{ij} = 0 \text{ for each } j; \qquad \operatorname*{med}_j (\alpha\beta)_{ij} = 0 \text{ for each } i;$$

$$\operatorname*{med}_i (\alpha\gamma)_{ik} = 0 \text{ for each } k; \qquad \operatorname*{med}_k (\alpha\gamma)_{ik} = 0 \text{ for each } i;$$

$$\operatorname*{med}_j (\beta\gamma)_{jk} = 0 \text{ for each } k; \qquad \operatorname*{med}_k (\beta\gamma)_{jk} = 0 \text{ for each } j;$$

and

$$\operatorname*{med}_{i}(\alpha\beta\gamma)_{ijk} = 0 \text{ for each } j \text{ and } k;$$

$$\operatorname*{med}_{j}(\alpha\beta\gamma)_{ijk} = 0 \text{ for each } i \text{ and } k;$$

$$\operatorname*{med}_{k}(\alpha\beta\gamma)_{ijk} = 0 \text{ for each } i \text{ and } j;$$

where $\text{med}_i(\alpha\beta)_{ij}$ denotes the median of the $(\alpha\beta)_{ij}$ over the I values of i, holding j fixed, for example.

A decomposition which includes only some of the interaction effects in equation (3), such as

$$y_{ijk} = \mu + \alpha_i + \beta_j + \gamma_k + (\alpha\beta)_{ij} + \varepsilon_{ijk}, \tag{4}$$

may also be appropriate in a particular situation. Three-way analyses frequently omit the three-factor interaction term $(\alpha\beta\gamma)_{ijk}$—what might otherwise be three-factor interaction is absorbed into the ε_{ijk}.

4C. MEDIAN-POLISH ANALYSIS FOR THE MAIN-EFFECTS-ONLY CASE

Many methods can be used for fitting three-way models, just as in the two-way case. These include taking out means, medians, midmeans, or midextremes, and polishing on these by iteration. Because median polish is straightforward and resistant, we illustrate its use here. The present section considers the main-effects-only model, and Section 4F takes up the full-effects model.

In fitting the main-effects-only model (2), we need to estimate the effect of each factor alone, without introducing any interaction terms. To estimate the effect of each level of factor A (that is, the α_i), we first take the median of the JK observations for each level of A (ignoring levels of factors B and C). We subtract the median of these I medians from each of them to get our initial estimates of the α_i. The median we subtract is our first estimate of μ, the overall value.

In turn we obtain estimates for effects of levels of factors B and C in a similar way. First we subtract the estimates of μ and α_i from the appropriate cells. We then take medians of these adjusted data for each level of factor B, say, to estimate the β_j. (Again we should subtract the median of these to center them. This β median is added to our first estimate of μ.)

Once we have subtracted the estimated β_j, we estimate the γ_k in the same way. This concludes the first cycle in our process. We now have estimates for μ, the α_i, the β_j, and the γ_k.

We usually repeat the process until the medians we take out are all near zero. This procedure works fairly well for complete tables, but much less well, though still usefully, for tables with some empty cells.

In practice, we may want to defer centering our estimates of the α_i, β_j, and γ_k until the last step in the iterative process. This saves some computation and leads to the same fit.

Algorithmic Representation

We now illustrate the procedure algebraically. First, to establish notation, we let a parenthesized superscript denote the number of the iterative cycle. Thus $a_i^{(q)}$, $b_j^{(q)}$, and $c_k^{(q)}$ denote the estimated effects of the various levels of factors A, B, and C, respectively, removed in cycle q. That is, $a_i^{(q)}$ is the median over level i of factor A removed in cycle q. (Later we add these up over cycles and take care of any necessary centering.) At the start of the procedure we may assign

$$a_i^{(0)} = b_j^{(0)} = c_k^{(0)} = 0 \quad \text{for all } i, j, k. \tag{5}$$

If the $r_{ijk}^{(q)}$ are the residuals after taking out the row, column, and layer effects in cycle q, then we start with

$$r_{ijk}^{(0)} = y_{ijk}, \tag{6}$$

the original data.

We now examine the details of the iteration, removing medians in the order: factor A, factor B, factor C. The first cycle proceeds as follows: First we take the median for each level of factor A,

$$a_i^{(1)} = \operatorname*{med}_{jk} \left\{ r_{ijk}^{(0)} \right\}, \qquad i = 1, \ldots, I, \tag{7}$$

the median over all j and k, holding i fixed. Next, we subtract these medians from the $r_{ijk}^{(0)}$ and take new medians for each level of factor B,

$$b_j^{(1)} = \operatorname*{med}_{ik} \left\{ r_{ijk}^{(0)} - a_i^{(1)} \right\}, \qquad j = 1, \ldots, J. \tag{8}$$

We then subtract these $b_j^{(1)}$ and take the median for each level of factor C,

$$c_k^{(1)} = \underset{ij}{\text{med}} \left\{ r_{ijk}^{(0)} - a_i^{(1)} - b_j^{(1)} \right\}, \qquad k = 1, \ldots, K. \tag{9}$$

Subtracting these $c_k^{(1)}$ yields the residuals

$$r_{ijk}^{(1)} = r_{ijk}^{(0)} - a_i^{(1)} - b_j^{(1)} - c_k^{(1)}. \tag{10}$$

If we wish to center the effect estimates at this stage, we take the median of each set of effects,

$$\begin{aligned} m_a^{(1)} &= \underset{i}{\text{med}} \left\{ a_i^{(1)} \right\}, \\ m_b^{(1)} &= \underset{j}{\text{med}} \left\{ b_j^{(1)} \right\}, \\ m_c^{(1)} &= \underset{k}{\text{med}} \left\{ c_k^{(1)} \right\}, \end{aligned} \tag{11}$$

so that we get the effect estimates

$$\begin{aligned} \hat{\alpha}_i^{(1)} &= a_i^{(1)} - m_a^{(1)}, \\ \hat{\beta}_j^{(1)} &= b_j^{(1)} - m_b^{(1)}, \\ \hat{\gamma}_k^{(1)} &= c_k^{(1)} - m_c^{(1)}, \\ \hat{\mu}^{(1)} &= m_a^{(1)} + m_b^{(1)} + m_c^{(1)}. \end{aligned} \tag{12}$$

In this instance we computed the medians in the order A, then B, then C, but in practice any order would do. Sometimes the results depend a little on the order chosen.

We continue with more cycles, this time saving the centering until the last iteration. In cycle n we have the following:

$$\begin{aligned} a_i^{(n)} &= \underset{jk}{\text{med}} \left\{ r_{ijk}^{(n-1)} \right\}, \\ b_j^{(n)} &= \underset{ik}{\text{med}} \left\{ r_{ijk}^{(n-1)} - a_i^{(n)} \right\}, \\ c_k^{(n)} &= \underset{ij}{\text{med}} \left\{ r_{ijk}^{(n-1)} - a_i^{(n)} - b_j^{(n)} \right\}, \end{aligned} \tag{13}$$

$$r_{ijk}^{(n)} = r_{ijk}^{(n-1)} - a_i^{(n)} - b_j^{(n)} - c_k^{(n)}, \tag{14}$$

and

$$\begin{aligned} A_i^{(n)} &= a_i^{(1)} + a_i^{(2)} + \cdots + a_i^{(n)}, \\ B_j^{(n)} &= b_j^{(1)} + b_j^{(2)} + \cdots + b_j^{(n)}, \\ C_k^{(n)} &= c_k^{(1)} + c_k^{(2)} + \cdots + c_k^{(n)}, \end{aligned} \tag{15}$$

where $A_i^{(n)}$, $B_j^{(n)}$, and $C_k^{(n)}$ are the uncentered effect estimates after cycle n.

As n, the number of cycles, increases, the adjustments $a_i^{(n)}$, $b_j^{(n)}$, and $c_k^{(n)}$ usually approach zero, and we stop the iteration process. When this happens, say at cycle t, we calculate the following centered effect estimates:

$$\begin{aligned} \hat{\alpha}_i &= A_i^{(t)} - \operatorname*{med}_i \left\{ A_i^{(t)} \right\}, \qquad i = 1, \ldots, I, \\ \hat{\beta}_j &= B_j^{(t)} - \operatorname*{med}_j \left\{ B_j^{(t)} \right\}, \qquad j = 1, \ldots, J, \\ \hat{\gamma}_k &= C_k^{(t)} - \operatorname*{med}_k \left\{ C_k^{(t)} \right\}, \qquad k = 1, \ldots, K, \\ \hat{\mu} &= \operatorname*{med}_i \left\{ A_i^{(t)} \right\} + \operatorname*{med}_j \left\{ B_j^{(t)} \right\} + \operatorname*{med}_k \left\{ C_k^{(t)} \right\}. \end{aligned} \tag{16}$$

These final estimates depend somewhat on the order used in taking medians, but typically not enough to be of much practical importance. In practice, we usually carry out a fixed number of iterations, at least two or three, and use this as our standard analysis. In this three-way median-polish procedure, the calculations are guided to approach the side conditions or "complete removal" conditions

$$\begin{aligned} &\operatorname*{med}_{jk} \left\{ r_{ijk} \right\} = 0 \quad \text{for each } i, \\ &\operatorname*{med}_{ik} \left\{ r_{ijk} \right\} = 0 \quad \text{for each } j, \\ &\operatorname*{med}_{ij} \left\{ r_{ijk} \right\} = 0 \quad \text{for each } k, \\ &\operatorname*{med}_i \left\{ \hat{\alpha}_i \right\} = 0, \\ &\operatorname*{med}_j \left\{ \hat{\beta}_j \right\} = 0, \\ &\operatorname*{med}_k \left\{ \hat{\gamma}_k \right\} = 0. \end{aligned} \tag{17}$$

These show that nothing remains to take out, at least in median terms. This algorithm, of course, is best programmed into a computer, eliminating tedious hand calculation and many possible errors. We return to computation issues briefly in Section 4I.

EXAMPLE: WEIGHT GAIN IN PIGS (MAIN-EFFECTS-ONLY BY MEDIANS)

The pig feeding data provide a straightforward example. For convenience we refer to M-effects for the levels of methionine, P-effects for the levels of protein, and L-effects for the levels of lysine. The arrangement of Table 4-2 makes it easiest to take the median for each of the four levels of lysine. We show these column medians in Table 4-3. We then subtract each median from the numbers in its column to obtain a first set of residuals. These appear in Table 4-4. We express the numbers as halves or whole numbers for simplicity, throughout the rest of the calculation (up to Table 4-12).

TABLE 4-3. The pig feeding data accompanied by column medians.

Methionine (in .01%)	Protein (in %)	Lysine (in .01%)			
		0	5	10	15
0	12	208	230	235	222
	14	297	308	246	209
25	12	208	224	275	252
	14	249	258	261	281
50	12	206	208	253	249
	14	291	303	285	289
Column median		228.5	244	257	250.5

TABLE 4-4. The pig feeding data after removal of the median value for each level of lysine.

Methionine (in .01%)	Protein (in %)	Lysine (in .01%)			
		0	5	10	15
0	12	−20.5	−14	−22	−28.5
	14	68.5	64	−11	−41.5
25	12	−20.5	−20	18	1.5
	14	20.5	14	4	30.5
50	12	−22.5	−36	−4	−1.5
	14	62.5	59	28	38.5
L-effects		228.5	244	257	250.5

TABLE 4-5. Rearrangement of Table 4-4 to make methionine the column factor. Column medians come from these new columns.

Lysine (in .01%)	Protein (in %)	Methionine (in .01%)			
		0	25	50	*L*-Effects
0	12	−20.5	−20.5	−22.5	228.5
	14	68.5	20.5	62.5	228.5
5	12	−14	−20	−36	244
	14	64	14	59	244
10	12	−22	18	−4	257
	14	−11	4	28	257
15	12	−28.5	1.5	−1.5	250.5
	14	−41.5	30.5	38.5	250.5
Column median		−17	9	13	

TABLE 4-6. Rearrangement of Table 4-5 to make protein the column factor. Column medians now correspond to the levels of protein.

Lysine (in .01%)	Methionine (in .01%)	Protein (in %)			
		12	14	*L*-Effects	*M*-Effects
0	0	−3.5	85.5	228.5	−17
	25	−29.5	11.5	228.5	9
	50	−35.5	49.5	228.5	13
5	0	3	81	244	−17
	25	−29	5	244	9
	50	−49	46	244	13
10	0	−5	6	257	−17
	25	9	−5	257	9
	50	−17	15	257	13
15	0	−11.5	−24.5	250.5	−17
	25	−7.5	21.5	250.5	9
	50	−14.5	25.5	250.5	13
Column median		−13	18		

Next, we want to take the medians for one of the other factors, either methionine or protein, but the arrangement of Table 4-4 is somewhat awkward because the two factors are combined along the row dimension. To overcome this, we rearrange the table so that one of the other factors stands by itself. Thus, in Table 4-5, methionine becomes the column factor, and protein and lysine combine as the row factor. We then take column medians to obtain estimates of the methionine effects. Once the residuals

TABLE 4-7. The pig feeding data after taking out lysine, methionine, and protein effects.

Lysine (in .01%)	Methionine (in .01%)	Protein (in %) 12	14	*L*-Effects	*M*-Effects
0	0	9.5	67.5	228.5	−17
	25	−16.5	−6.5	228.5	9
	50	−22.5	31.5	228.5	13
5	0	16	63	244	−17
	25	−16	−13	244	9
	50	−36	28	244	13
10	0	8	−12	257	−17
	25	22	−23	257	9
	50	−4	−3	257	13
15	0	1.5	−42.5	250.5	−17
	25	5.5	3.5	250.5	9
	50	−1.5	7.5	250.5	13
P-effects		−13	18		

TABLE 4-8. Main-effects analysis of the pig feeding data after one cycle.

Methionine (in .01%)	Protein (in %)	Lysine (in .01%) 0	5	10	15	*M*-Effects	*P*-Effects
0	12	9.5	16	8	1.5	−26	−15.5
	14	67.5	63	−12	−42.5	−26	15.5
25	12	−16.5	−16	22	5.5	0	−15.5
	14	−6.5	−13	−23	3.5	0	15.5
50	12	−22.5	−36	−4	−1.5	4	−15.5
	14	31.5	28	−3	7.5	4	15.5
L-effects		−18.5	−3	10	3.5	Overall	258.5

from this step have been calculated, a further rearrangement (in Table 4-6) brings out protein as the single factor so that we can easily estimate its effects. Table 4-7 shows the results after we take out the median over each level of protein.

This concludes the first cycle. We now center our effect estimates and produce an estimate of the overall value. Table 4-8 gives these results. The residuals appear in stem-and-leaf in Table 4-9, which shows two very large positive residuals. Continuing with polishing, the effect adjustments finally become zero after five cycles, when we have

$$\operatorname*{med}_{jk}\{r_{ijk}\} = 0 \quad \text{for each } i,$$

$$\operatorname*{med}_{ik}\{r_{ijk}\} = 0 \quad \text{for each } j,$$

$$\operatorname*{med}_{ij}\{r_{ijk}\} = 0 \quad \text{for each } k.$$

The final effect analysis appears in Table 4-10, and Table 4-11 shows a stem-and-leaf display of the corresponding residuals. We note one large negative residual (-64, from 0% methionine, 14% protein, and .15% lysine) and three large positive residuals (47, 44.5, 40).

TABLE 4-9. Stem-and-leaf display of residuals after one cycle in fitting a main-effects model to the pig feeding data (median at 0, fourths at -14.5 and 12.75).

(1 | 2 represents 0.12 lb)

1	−4	2
2	−3	6
4	−2	23
8	−1	2366
12	−0	1346
12	0	135789
6	1	6
5	2	28
3	3	1
	4	
	5	
2	6	37

TABLE 4-10. Final main-effects analysis of pig feeding data by median polish.

Methionine (in .01%)	Protein (in %)	Lysine (in .01%) 0	5	10	15	*M*-Effects	*P*-Effects
0	12	−2	2	3	−11	0	−20
	14	47	40	−26	−64	0	20
25	12	−0.5	−2.5	44.5	20.5	−1.5	−20
	14	0.5	−8.5	−9.5	9.5	−1.5	20
50	12	−20	−36	5	0	16	−20
	14	25	19	−3	0	16	20
L-Effects		−20	−2	2	3	Overall	250

TABLE 4-11. Stem-and-leaf display of residuals from final main-effects analysis of the pig feeding data by median polish (median at 0, fourths at −9 and 14.25).

(1 | 2 represents 0.12 lb)

1	−6	4
	−5	
	−4	
2	−3	6
4	−2	06
5	−1	1
12	−0	0022389
12	0	002359
6	1	9
5	2	05
	3	
3	4	047

To follow the progress of successive cycles, Table 4-12 tracks the *changes* in the centered effect estimates at each cycle. At the bottom of this table several summary measures of the changes facilitate comparison of the impact of each cycle. The sum of sizes (omitting the overall value) decreases from 96 to 0.5, showing that the changes do become smaller from one cycle to the next. Considering the individual changes in comparison with the mean size of residual, we see that, except perhaps for those in the *M*-effects,

TABLE 4-12. Change of effect estimates at each iterative cycle in median-polish main-effects analysis of the pig feeding data.

	Cycle					
Effect	1	2	3	4	5	Sum
*M*0	−26	21	5	0	0	0
*M*25	0	0	−0.5	−1	0	−1.5
*M*50	4	11	2	−1	0	16
*P*12	−15.5	−3.5	−1	0	0	−20
*P*14	15.5	3.5	1	0	0	20
*L*0	−18.5	−0.5	−1	0	0	−20
*L*5	−3	−0.5	1	0.5	0	−2
*L*10	10	−5.5	−1.5	−0.5	−0.5	2
*L*15	3.5	0.5	−1	0	0	3
Overall value	258.5	−8.5	−1.5	2	−0.5	250
Sum of effect sizes	354.5	54.5	15.5	5	1	
Sum of sizes[a]	96	46	14	3	0.5	
Mean size	10.7	5.1	1.6	0.3	0.1	
Largest size	26	21	5	1	0.5	
Mean size of residual	19.2	17	16.6	16.6	16.6	

[a] Omitting overall value.

the changes are quite minor after cycle two. Thus, a two-cycle or three-cycle analysis would be adequate for this set of data.

Figure 4-1 presents parallel plots of the effect estimates and a boxplot of the residuals. (Chapter 3 of UREDA discusses boxplots.) Because these all use the same scale, we can readily compare the relative contributions of the different factors. In this analysis the effect for the percentage of protein seems to be largest, whereas methionine and lysine supplements contribute smaller effects. The most noticeable feature, though, is the size of the residuals. Their fourth-spread (represented by the length of the box) is comparable in size to the effect estimates. We could take this as suggesting that the three dietary supplements make modest contributions at best when compared to routine variation, as long as the residuals do not conceal substantial two-factor interaction effects which increase their spread.

This way of comparing the various pieces of the fit (the effects and the residuals) serves the same general purpose as comparing mean squares in

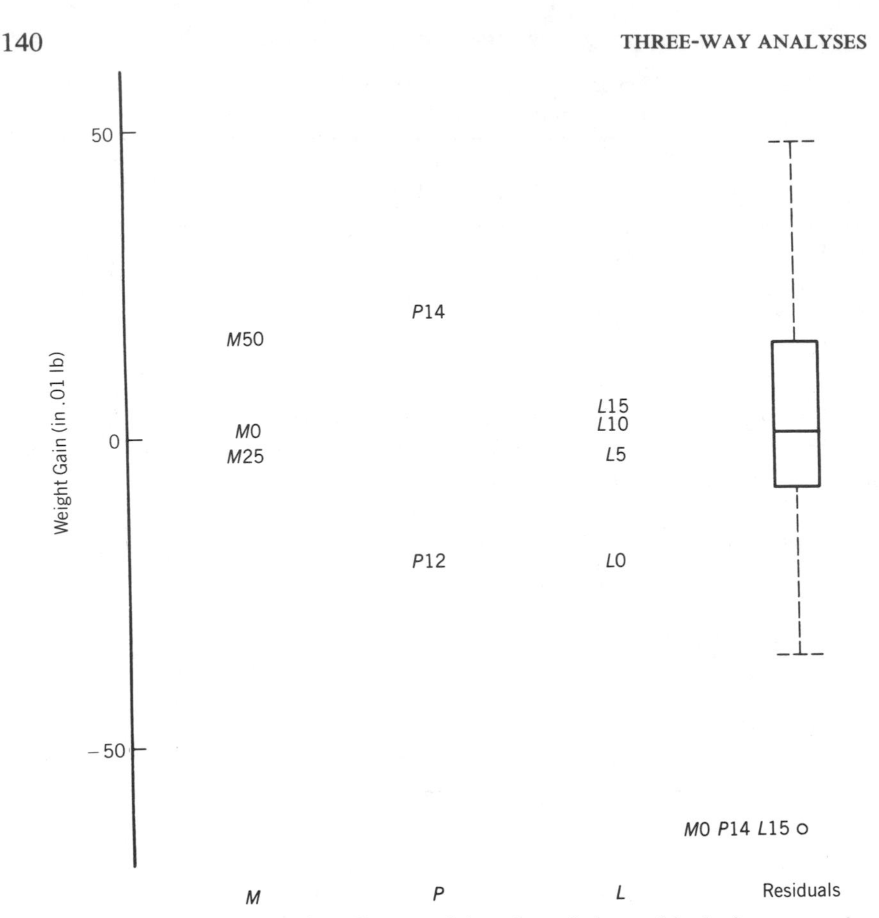

Figure 4-1. Dot plots of the effects and boxplot of the residuals from a main-effects-only analysis of the pig feeding data by median polish.

the analysis of variance. The graphical comparison, however, provides more detail on the effects and permits more judgment in interpreting the contributions of different factors.

For more refined judgments about the overall significance of each set of effects and about the possible significance of differences between individual effects for the same factor, one would need rules based on the sizes of the residuals and on the dimensions of the three-way table. Such confirmatory matters are properly part of procedures for robust analysis of variance. At an exploratory level we derive considerable benefit from graphical displays that bring together the residuals and the various components of the fit, both in the present situation and in more complex ones.

When such a graphical comparison indicates that the fit is not close, we can try to improve the fit. Two ways of doing this are (1) to fit some interaction terms (as we demonstrate in Section 4F) or (2) to apply a transformation to the data. In practice, we often check whether a transformation is useful before adopting more complex models. In a two-way table we examine the need for a transformation by means of the diagnostic plot (discussed in Section 3B and in Sections 6H and 8F of UREDA). This tool may easily be extended to three-way tables in a main-effects analysis, as we shall see in Section 4D.

Another Way to Do Main-Effects-Only Decomposition

Another method of analyzing a three-way table for main effects will lead to similar results. Readers who do not want to examine another set of details should skip ahead to Section 4D.

We take the pig feeding data as presented in Table 4-2 and analyze them as if they were simply a two-way table; that is, with lysine as one factor and with methionine and protein combined ($M \times P$) as the other. Thus we treat methionine and protein as one factor with six levels. Once we get the combined $M \times P$-effects, we treat them as a two-way table to separate the methionine and protein effects. Although less direct than the median-polish analysis for main effects, this procedure can easily use a computer program for median polish of a two-way table, first on a 4×6 table and then on a 2×3 table. The results of the two analyses will probably be somewhat different, but the difference should usually be small. This algorithm can also be adapted to analyze more complex tables (e.g., four-way) in main-effects-only terms; and, as we will see, we get some of the two-factor interaction effects automatically. A disadvantage of this method for the main-effects-only decomposition is that often some side conditions do not hold; that is, the medians of the residuals over factor-levels [as in equation (17)] are not always zero. With only a little further effort, however, we can arrange to satisfy all the side conditions.

EXAMPLE: WEIGHT GAIN IN PIGS (MAIN-EFFECTS-ONLY BY MEDIANS AGAIN)

We now try the new algorithm on the pig feeding data. Starting at Table 4-4, which already has the first L-effects removed, we take out medians across rows. The result appears in Table 4-13. We continue to treat the data as a two-way table and polish until all row and column medians are zero. We then center our effect estimates by taking out their medians as parts of the overall value. Table 4-14 shows this result.

TABLE 4-13. The pig feeding data after subtracting row medians from the residuals of Table 4-4.

Methionine (in .01%)	Protein (in %)	Lysine (in .01%) 0	5	10	15	$M \times P$-Effects
0	12	0.5	7	−1	−7.5	−21
	14	42	37.5	−37.5	−68	26.5
25	12	−11.5	−11	27	10.5	−9
	14	3.5	−3	−13	13.5	17
50	12	−9.5	−23	9	11.5	−13
	14	13.5	10	−21	−10.5	49
L-effects		228.5	244	257	250.5	

TABLE 4-14. Analysis of the pig feeding data by two-way median polish with methionine and protein as the combined factor.

Methionine (in .01%)	Protein (in %)	Lysine (in .01%) 0	5	10	15	$M \times P$-Effects
0	12	−3.5	3	4	−11.5	−21
	14	37.5	33	−33	−72.5	27
25	12	−11.5	−11	36	10.5	−13
	14	3.5	−3	−4	13.5	13
50	12	−10.5	−24	17	10.5	−16
	14	13.5	10	−12	−10.5	45
L-effects		−17.5	−2	2	4.5	250

The numbers in the rightmost column are estimates of the combined $M \times P$-effects. They combine main effects for M, main effects for P, and $M \times P$-interaction effects. In our main-effects-only analysis, though, we want the M-effects and P-effects alone. To get these, we rearrange the last column as another, smaller two-way table and median-polish that. Table 4-15 shows this two-way table and its polished result.

At this stage we have estimates of the L-effects from Table 4-14, estimates of 'the M-effects and P-effects from Table 4-15, estimates of the

TABLE 4-15. Further analysis of the $M \times P$-effects from Table 4-14.

a. The $M \times P$-effects as a two-way table

Methionine (in .01%)	Protein (in %)	
	12	14
0	−21	27
25	−13	13
50	−16	45

b. Analysis by median polish

Methionine (in .01%)	Protein (in %)		
	12	14	*M*-Effects
0	0	0	0
25	11	−11	−3
50	−6.5	6.5	11.5
P-effects	−24	24	3

$M \times P$-interaction effects (the residuals in Table 4-15*b*), and the residuals left in Table 4-14. If we are interested only in the main effects, we can add the $M \times P$-interactions back to the residuals in Table 4-14. The recombined numbers could then serve as our final main-effects-only residuals. We can also try to balance things a little more by re-fitting the *L*-effects, that is, taking medians down columns again. This is done in Table 4-16. To polish a little more, we also take row medians again (in Table 4-16*b*) and re-estimate the *M*-effects and *P*-effects. The final analysis appears in Table 4-17, along with the results of the original analysis from Table 4-10.

The results from the new analysis differ slightly from those of our first analysis, although most entries are of the same order of magnitude. As mentioned earlier, however, the side conditions do not hold in the new analysis. For instance, the median over all residuals for 14% protein is −4, not 0. This algorithm, then, should be used when computational convenience is desired, at the expense of some side conditions not holding. We could, however, take a step or two of further adjustment toward making those medians zero.

TABLE 4-16. Adding back the $M \times P$-interaction effects and re-fitting the L-effects.

a. Adding back the interactions (the residuals from Table 4-15)

		Table 4-14 Residuals				Table 4-15	Both Recombined			
M	*P*	*L*: 0	5	10	15	Residuals	*L*: 0	5	10	15
0	12	−3.5	3	4	−11.5	0	−3.5	3	4	−11.5
	14	37.5	33	−33	−72.5	0	37.5	33	−33	−72.5
25	12	−11.5	−11	36	10.5	11	−0.5	0	47	21.5
	14	3.5	−3	−4	13.5	−11	−7.5	−14	−15	2.5
50	12	−10.5	−24	17	10.5	−6.5	−17	−30.5	10.5	4
	14	13.5	10	−12	−10.5	6.5	20	16.5	−5.5	−4
						Column median	−2	1.5	−1	−1

b. Taking out an adjustment to the L-effects (column medians)

M	*P*	*L*: 0	5	10	15	Row Median
0	12	−1.5	1.5	5	−10.5	0
	14	39.5	31.5	−32	−71.5	0
25	12	1.5	−1.5	48	22.5	12
	14	−5.5	−15.5	−14	3.5	−10
50	12	−15	−32	11.5	5	−5
	14	22	15	−4.5	−3	6
L-effects		−2	1.5	−1	−1	

TABLE 4-17. Final alternative main-effects-only analysis of the pig feeding data. The original results from Table 4-10 appear as the lower entries in italics.

Methionine (in .01%)	Protein (in %)	Lysine (in .01%) 0	5	10	15	*M*-Effects	*P*-Effects
0	12	−1.5	1.5	5	−10.5	0	−24
		−2	*2*	*3*	*−11*	*0*	*−20*
	14	39.5	31.5	−32	−71.5	0	24
		47	*40*	*−26*	*−64*	*0*	*20*
25	12	0.5	−2.5	47	21.5	−2	−24
		−0.5	*−2.5*	*44.5*	*20.5*	*−1.5*	*−20*
	14	−6.5	−16.5	−15	2.5	−2	24
		0.5	*−8.5*	*−9.5*	*9.5*	*−1.5*	*20*
50	12	−15.5	−32.5	11	4.5	12	−24
		−20	*−36*	*5*	*0*	*16*	*−20*
	14	21.5	14.5	−5	−3.5	12	24
		25	*19*	*−3*	*0*	*16*	*20*
		−19.5	−0.5	1	3.5	Overall	253
L-effects		*−20*	*−2*	*2*	*3*	value	*250*

4D. NONADDITIVITY AND A DIAGNOSTIC PLOT IN MAIN-EFFECTS-ONLY ANALYSIS

Whenever we might find a pattern of systematic nonadditivity in the residuals from a main-effects-only analysis, we should plot the residuals against a suitable comparison value, chosen so that the slope indicates roughly which power transformations might promote additivity. In the two-way diagnostic plot (Section 3B and Sections 6H and 8F of UREDA), the slope of the line of the residuals against the comparison values is $1 - p$, where p is the exponent in the suggested power transformation. In this section, we demonstrate that the same relationship holds for the three-way main-effects model and show how to define the appropriate comparison values.

Suppose that

$$x_{ijk} = y_{ijk}^{p} \tag{18}$$

is a transformation of the original data y_{ijk} such that an additive main-effects model fits the x_{ijk} exactly:

$$x_{ijk} = m + a_i + b_j + c_k. \tag{19}$$

Then

$$\begin{aligned} y_{ijk} = x_{ijk}^{1/p} &= \left(m + a_i + b_j + c_k\right)^{1/p} \\ &= m^{1/p}\left(1 + A_i + B_j + C_k\right)^{1/p}, \end{aligned} \tag{20}$$

where $A_i = a_i/m$, $B_j = b_j/m$, and $C_k = c_k/m$. If the row, column, and layer effects are small relative to the overall value, then A_i, B_j, and C_k are also small. We then use the expansion of $(1 + t)^{1/p}$ about $t = 0$, with $t = A_i + B_j + C_k$. Through second-order terms this yields

$$\begin{aligned} y_{ijk} &\approx m^{1/p}\left[1 + \frac{1}{p}\left(A_i + B_j + C_k\right) + \frac{1-p}{2p^2}\left(A_i + B_j + C_k\right)^2\right] \\ &\approx m^{1/p}\left[1 + \left(\frac{1}{p}A_i + \frac{1-p}{2p^2}A_i^2\right) + \left(\frac{1}{p}B_j + \frac{1-p}{2p^2}B_j^2\right)\right. \\ &\quad \left. + \left(\frac{1}{p}C_k + \frac{1-p}{2p^2}C_k^2\right) + \frac{1-p}{p^2}\left(A_iB_j + A_iC_k + B_jC_k\right)\right]. \end{aligned} \tag{21}$$

The leading nonadditive term, when we replace A_i, B_j, and C_k by a_i/m, b_j/m, and c_k/m, respectively, becomes

$$\frac{1-p}{p^2} m^{(1/p)-2}\left(a_ib_j + a_ic_k + b_jc_k\right). \tag{22}$$

If we bring the additive terms that involve only one factor over to the left side of equation (21), this nonadditive term should approximately equal the residual from the main-effects analysis of the y_{ijk}.

Because we do not have estimates for m, a_i, b_j, and c_k, we need to replace these by quantities we do have. We have already fitted the original observed data y_{ijk} and obtained

$$\hat{y}_{ijk} = \hat{\mu} + \hat{\alpha}_i + \hat{\beta}_j + \hat{\gamma}_k. \tag{23}$$

Expanding $\hat{x}_{ijk} = \hat{y}_{ijk}^{p}$ about $\hat{\mu} = \text{med}_{i,j,k}\{y_{ijk}\}$, we have, approximately,

$$\hat{x}_{ijk} \approx \hat{\mu}^{p} + p\hat{\mu}^{p-1}(\hat{y}_{ijk} - \hat{\mu}). \tag{24}$$

From this we obtain, for example,

$$\underset{i,j,k}{\text{med}}\{\hat{x}_{ijk}\} = \hat{m} \approx \hat{\mu}^{p} + p\hat{\mu}^{p-1}(\hat{\mu} - \hat{\mu}) = \hat{\mu}^{p}$$

and (25)

$$\underset{j,k}{\text{med}}\{\hat{x}_{ijk}\} = \hat{m} + \hat{a}_i \approx \hat{\mu}^{p} + p\hat{\mu}^{p-1}(\hat{\mu} + \hat{\alpha}_i - \hat{\mu})$$

$$= \hat{\mu}^{p} + p\hat{\mu}^{p-1}\hat{\alpha}_i.$$

This implies

$$\begin{aligned} \hat{a}_i &\approx p\hat{\mu}^{p-1}\hat{\alpha}_i, \\ \hat{b}_j &\approx p\hat{\mu}^{p-1}\hat{\beta}_j, \\ \hat{c}_k &\approx p\hat{\mu}^{p-1}\hat{\gamma}_k. \end{aligned} \tag{26}$$

After substitution of these results in (22), the leading nonadditive term becomes

$$\frac{1-p}{\hat{\mu}}(\hat{\alpha}_i\hat{\beta}_j + \hat{\alpha}_i\hat{\gamma}_k + \hat{\beta}_j\hat{\gamma}_k). \tag{27}$$

Therefore, a plot of the residuals

$$r_{ijk} = y_{ijk} - \hat{\mu} - \hat{\alpha}_i - \hat{\beta}_j - \hat{\gamma}_k \tag{28}$$

against the comparison values

$$CV_{ijk} = \frac{\hat{\alpha}_i\hat{\beta}_j + \hat{\alpha}_i\hat{\gamma}_k + \hat{\beta}_j\hat{\gamma}_k}{\hat{\mu}} \tag{29}$$

might well approximately yield a line with slope $1 - p$, indicating that a transformation to the pth power could be useful in removing nonadditivity.

TABLE 4-18. Comparison values for the pig feeding data.

Methionine (in .01%)	Protein (in %)	Lysine (in .01%) 0	5	10	15
0	12	1.60	0.16	−0.16	−0.24
	14	−1.60	−0.16	0.16	0.24
25	12	1.84	0.29	−0.05	−0.14
	14	−1.60	−0.27	0.03	0.10
50	12	−0.96	−1.25	−1.31	−1.33
	14	−1.60	0.99	1.57	1.71

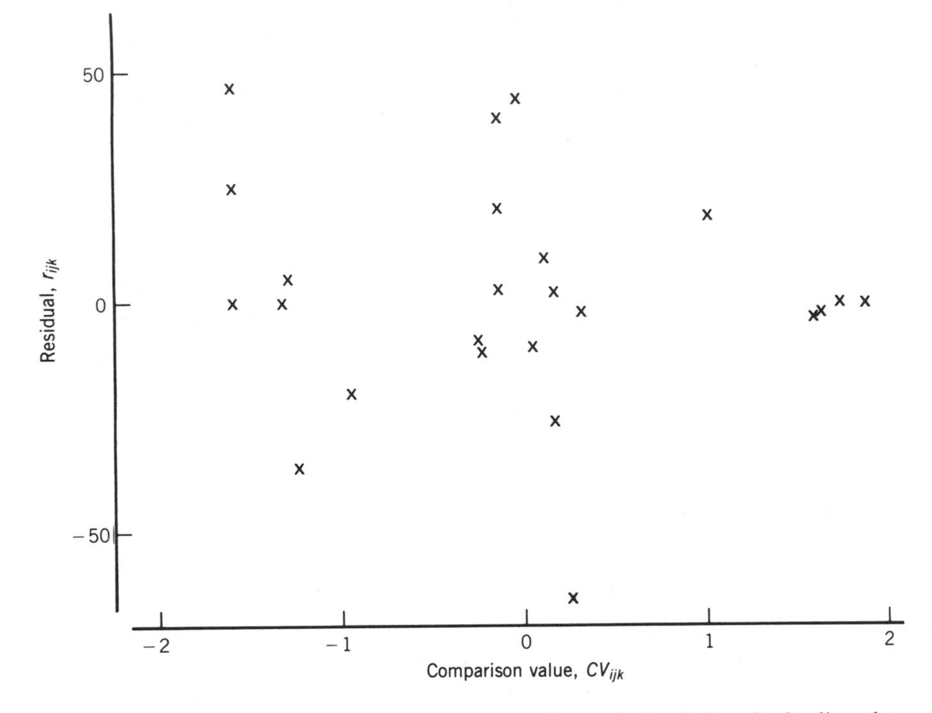

Figure 4-2. Diagnostic plot for main-effects-only analysis of the pig feeding data by median polish.

If the plot does not yield a reasonably straight-line pattern, no power transformation is likely to do a really good job. Note that the diagnostic plot is appropriate whatever the method we use to fit the model. It requires only that $\hat{\mu}$ be neither zero nor very small.

EXAMPLE: WEIGHT GAIN IN PIGS

A diagnostic plot will indicate whether the residuals in the pig feeding data contain systematic interactions that can be removed by a power transformation. Table 4-18 gives the comparison values, and Figure 4-2 plots the residuals from Table 4-10 against these comparison values. In this figure the absence of any clear relationship suggests that $1 - p = 0$ or $p = 1$. We therefore anticipate no benefit in transformation of these data.

EXAMPLE: YARN BREAKAGE (MAIN-EFFECTS-ONLY BY MEDIANS)

Box and Cox (1964) give data from a textile experiment that repeatedly subjected specimens of worsted yarn to a load until they broke. The original investigators wanted to see how the number of cycles to failure was affected

TABLE 4-19. Yarn breakage data. Number of cycles of repeated loading of worsted yarn until failure under different loading conditions.

A: Length (mm)	B: Amplitude (mm)	C: Load (grams)		
		40	45	50
250	8	674	370	292
	9	338	266	210
	10	170	118	90
300	8	1414	1198	634
	9	1022	620	438
	10	442	332	220
350	8	3636	3184	2000
	9	1568	1070	566
	10	1140	884	360

Source: G. E. P. Box and D. R. Cox (1964). "An analysis of transformations," *Journal of the Royal Statistical Society, Series B*, **26**, 211–252 (data from Table 4, p. 223).

TABLE 4-20. Final main-effects-only analysis by median polish of yarn breakage data.

A	B	*C* 40	45	50	*A*-Effects	*B*-Effects
250	8	−220	−166	−62	−520	436
	9	−120	166	292	−520	0
	10	0	306	460	−520	−288
300	8	0	142	−240	0	436
	9	44	0	0	0	0
	10	−248	0	70	0	−288
350	8	1670	1576	574	552	436
	9	38	−102	−424	552	0
	10	−102	0	−342	552	−288
C-effects		358	0	−182	Overall	620

by the three factors:

A: length of test specimen (3 levels: 250, 300, 350 mm).

B: amplitude of loading cycle (3 levels: 8, 9, 10 mm).

C: load (3 levels: 40, 45, 50 g).

Table 4-19 presents the original data, the number of cycles to failure. A main-effects-only median-polish analysis of the data yields the results in Table 4-20. As we see in the plot of main effects and residuals in Figure 4-3, two residuals are very high, and otherwise the three factors make moderate contributions, relative to the fourth-spread of the residuals. To judge the quality of the fit, however, we must look further at the residuals. Because these show some systematic structure (most easily seen within each level of factor *A*), we ask whether a transformation is needed. Table 4-21 gives the comparison values, and Figure 4-4 shows the diagnostic plot.

The diagnostic plot shows a clear relationship. A resistant line through the origin has slope 1.00, indicating a p of 0—that is, a log transformation. Thus we try the (base-10) log transformation, which yields the transformed data in Table 4-22. (All entries have been multiplied by 100 to simplify the arithmetic.) A median-polish analysis of the transformed data produces the decomposition in Table 4-23 and the comparison values in Table 4-24. The diagnostic plot in Figure 4-5 indicates that the log transformation seems satisfactory. There is no longer any pattern in the plot.

The plots of effects and residuals in Figure 4-6 also indicate that the fit is better. Here the residuals are much smaller, and the effect estimates are

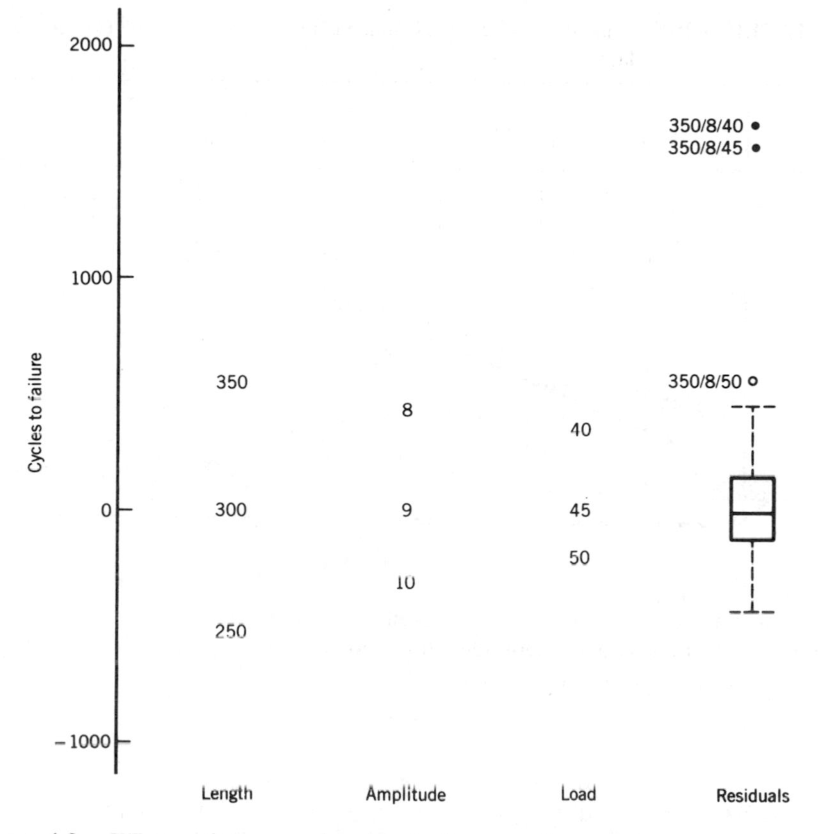

Figure 4-3. Effect estimates and residuals from median-polish main-effects-only analysis of the yarn breakage data.

TABLE 4-21. Comparison values for the yarn breakage data.

		C		
A	*B*	40	45	50
250	8	−414.2	−365.7	−341.0
	9	−300.3	0	152.6
	10	−225.0	241.5	478.7
300	8	251.8	0	−128.0
	9	0	0	0
	10	166.3	0	84.5
350	8	958.7	388.2	354.1
	9	318.7	0	−162.0
	10	−104.0	−256.4	−333.9

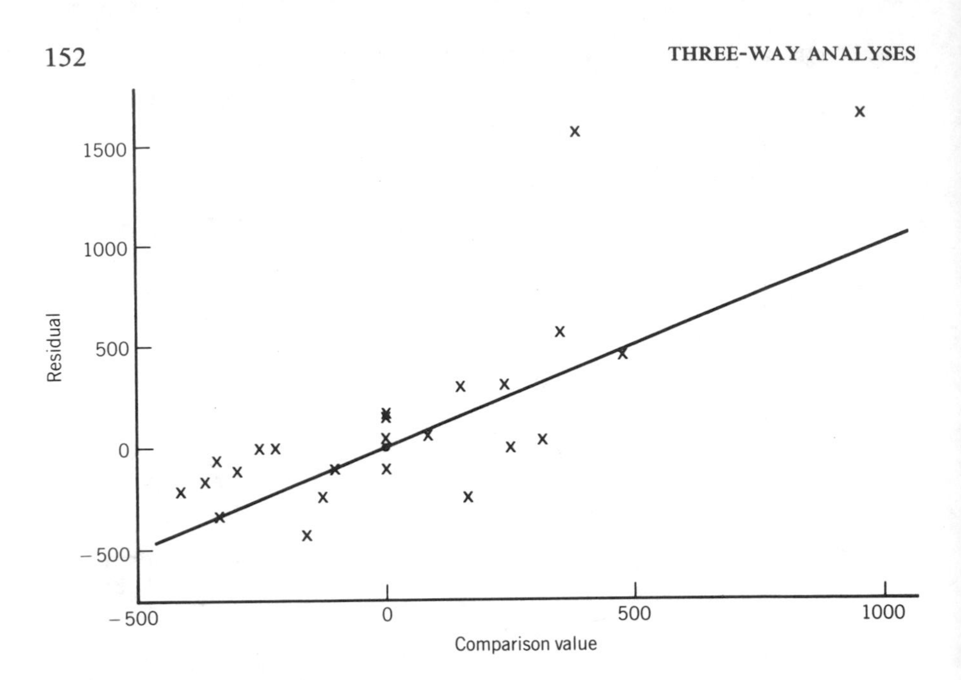

Figure 4-4. Diagnostic plot for main-effects-only analysis of the yarn breakage data. The dot at the origin represents three points. The line (through the origin) has slope 1.00 (calculated as the median of the slopes of the lines joining each point and the origin).

TABLE 4-22. Common logarithms (× 100) of the yarn breakage data.

			C	
A	B	40	45	50
250	8	282.9	256.8	246.5
	9	252.9	242.5	232.2
	10	223.0	207.2	195.4
300	8	315.0	307.8	280.2
	9	301.0	279.2	264.1
	10	264.5	252.1	234.2
350	8	356.1	350.3	330.1
	9	319.5	302.9	275.3
	10	305.7	294.6	255.6

TABLE 4-23. Main-effects-only analysis for logs (×100) of yarn breakage data.

		C				
A	*B*	40	45	50	*A*-Effects	*B*-Effects
250	8	2.3	−11.5	−3.6	−38.6	27.7
	9	0	1.9	9.8	−38.6	0
	10	−2.9	−6.4	0	−38.6	−27.0
300	8	−4.2	1.0	−8.5	0	27.7
	9	9.5	0	3.1	0	0
	10	0	−0.1	0.2	0	−27.0
350	8	0	6.5	4.5	36.9	27.7
	9	−8.9	−13.2	−22.6	36.9	0
	10	4.3	5.5	−15.3	36.9	−27.0
C-effects		12.3	0	−18.2	Overall	279.2

TABLE 4-24. Comparison values for logs (×100) of yarn breakage data.

		C		
A	*B*	40	45	50
250	8	−4.31	−3.83	−3.12
	9	−1.70	0	2.52
	10	0.84	3.73	8.01
300	8	1.22	0	−1.81
	9	0	0	0
	10	−1.19	0	1.76
350	8	6.51	3.66	−0.55
	9	1.63	0	−2.41
	10	−3.13	−3.57	−4.21

comparatively larger. The length of the specimen has the greatest effect on cycles to failure, followed by amplitude, and the effect of load is smaller and does not differ much from the residuals in magnitude.

After studying the original diagnostic plot in Figure 4-4, we may want also to consider a smaller slope. In this direction the simplest alternative to the log transformation is the square root, corresponding to a slope of 0.5. When we reanalyze the data in the square root scale and make the new

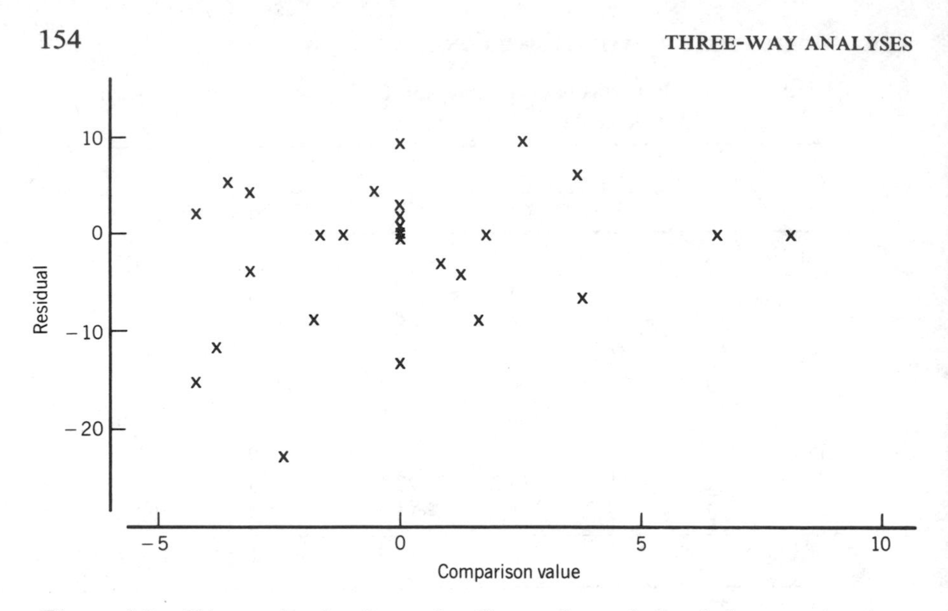

Figure 4-5. Diagnostic plot for main-effects-only analysis of the yarn breakage data in the log scale.

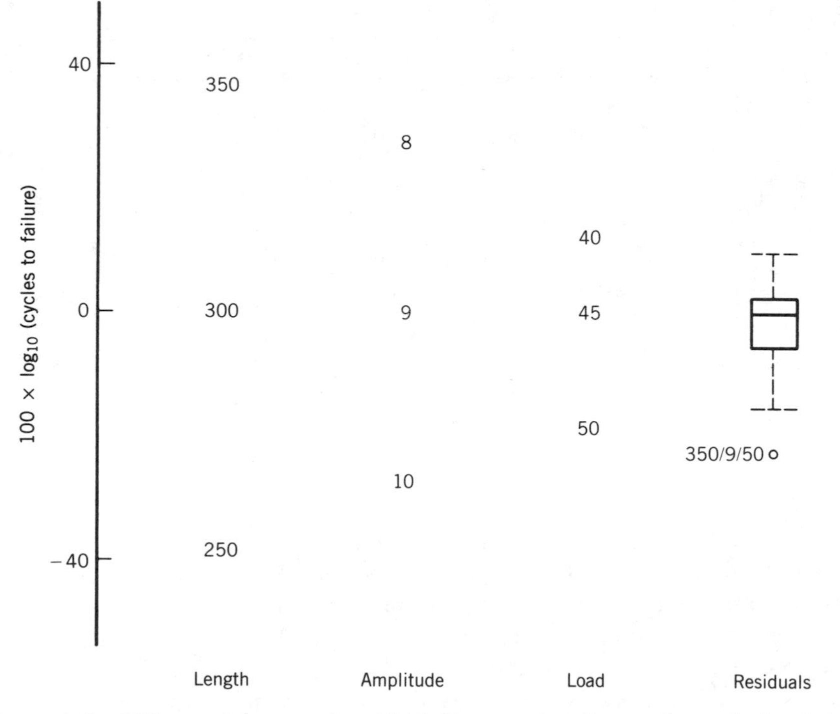

Figure 4-6. Effect estimates and residuals from main-effects-only analysis of the yarn breakage data in the log scale.

diagnostic plot (not shown), a positive slope suggests that the square root transformation is considerably less effective in reducing the nonadditivity than the log transformation. As Box and Cox point out, the behavior of yarn breakage data in relation to the three factors makes the logarithm a promising choice.

EXAMPLE: CONSTRUCTED DATA (MAIN-EFFECTS-ONLY BY MEDIANS)

It is useful also to see what message the diagnostic plot gives when we know the answer. For a further example we construct a three-way table in which the main-effects model fits exactly. Then we square each entry and proceed with our analysis.

In this constructed table each of the three factors has three levels, and the common value and main effects are as follows:

$$\mu = 20,$$

$$(\alpha_1, \alpha_2, \alpha_3) = (-1, 0, 3),$$

$$(\beta_1, \beta_2, \beta_3) = (-1, 0, 1),$$

$$(\gamma_1, \gamma_2, \gamma_3) = (-1, 0, 2),$$

TABLE 4-25. Constructed data, perfectly additive in the square root scale.

		C		
A	B	1	2	3
	1	289	324	400
1	2	324	361	441
	3	361	400	484
	1	324	361	441
2	2	361	400	484
	3	400	441	529
	1	441	484	576
3	2	484	529	625
	3	529	576	676

so that

$$x_{ijk} = \mu + \alpha_i + \beta_j + \gamma_k,$$

$$y_{ijk} = x_{ijk}^2.$$

Table 4-25 gives the "data" y_{ijk}, Table 4-26 shows the median-polish analysis, and Table 4-27 records the comparison values. In the diagnostic plot, Figure 4-7, the points show excellent agreement with the reference line, whose slope is $\frac{1}{2}$. Thus, to a close approximation, the diagnostic plot

TABLE 4-26. Main-effects analysis of the constructed data by median polish.

		C				
A	*B*	1	2	3	*A*-Effects	*B*-Effects
	1	8	4	−4		−39
1	2	4	2	−2	−41	0
	3	0	0	0		41
	1	2	0	−4		−39
2	2	0	0	0	0	0
	3	−2	0	4		41
	1	−10	−6	2		−39
3	2	−6	0	12	129	0
	3	−2	6	22		41
C-effects		−39	0	84	Overall	400

TABLE 4-27. Comparison values for the constructed data.

		C		
A	*B*	1	2	3
	1	11.80	4.00	−12.80
1	2	4.00	0	−8.61
	3	−4.20	−4.20	−4.20
	1	3.80	0	−8.19
2	2	0	0	0
	3	−4.00	0	8.61
	1	−21.35	−12.58	6.32
3	2	−12.58	0	27.09
	3	−3.35	13.22	48.92

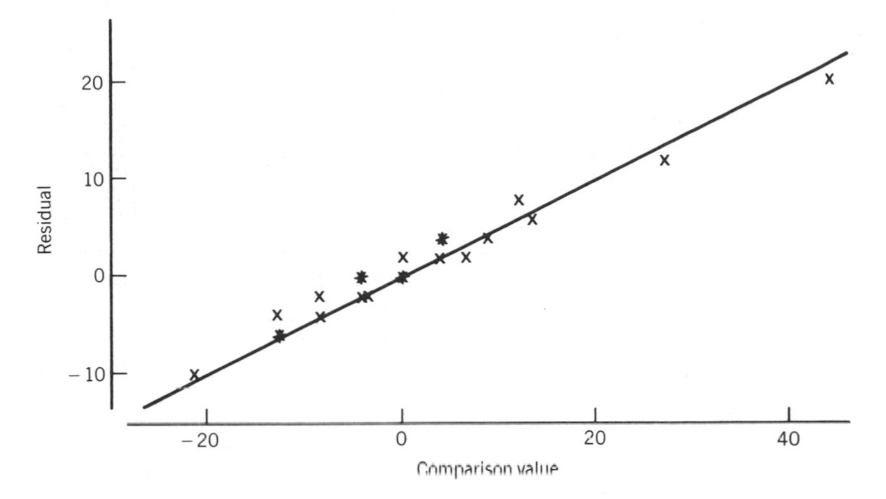

Figure 4-7. Diagnostic plot for the constructed data, perfectly additive in the square root scale. The reference line has slope 0.5.

suggests the square root transformation, which will exactly remove the nonadditivity in these data.

Because we have constructed this example to require a square root transformation, we may wonder why the points in the diagnostic plot did not all lie on the line with slope $\frac{1}{2}$. The answer lies in the theory on which the diagnostic plot is based. According to our construction, y_{ijk}, the data for analysis in equation (18), must be transformed by the square root $(p - \frac{1}{2})$ to achieve perfect additivity. Although in this special case the expansion of equation (21) is exact, those underlying equations (25) and (26) are not, and the differences are enough to displace some points from the line.

Because actual data almost always involve random fluctuations, we continue the analysis of constructed examples with a rather extreme case. We select 27 standard Gaussian random deviates from a random number table, round these to the nearest whole number, and add them to the x_{ijk}. We again square the numbers, perform a median-polish analysis, and calculate comparison values. The diagnostic plot, Figure 4-8, does not suggest any nonzero slope, although it is compatible with positive slopes as large as 1. This outcome should hardly be surprising, however, because the perturbations are essentially the same size as the main effects, and only 9 cells out of 27 remained unchanged.

In the face of only a few isolated perturbations (even of substantial size), however, the median-polish analysis would have tended to minimize their impact on the diagnostic plot, leaving us a clear indication.

4E. ANALYSIS USING MEANS

More traditional analyses of three-way and more-way tables use the mean instead of the median. Wherever we took medians along rows, columns, or layers, we would take means instead, as we do in ordinary least squares. The use of means is not resistant, but it may be desired in a particular setting. Although secondary to resistance, one great computational convenience is that repeated polishing is unnecessary when the table contains no holes; the procedure is noniterative because, except for the effects of rounding, the estimates "converge" after only one cycle. We give the computational formulas below.

For convenience, we introduce the dot notation here: a dot in place of a subscript signifies averaging over that subscript; for example, $y_{\cdot jk} =$

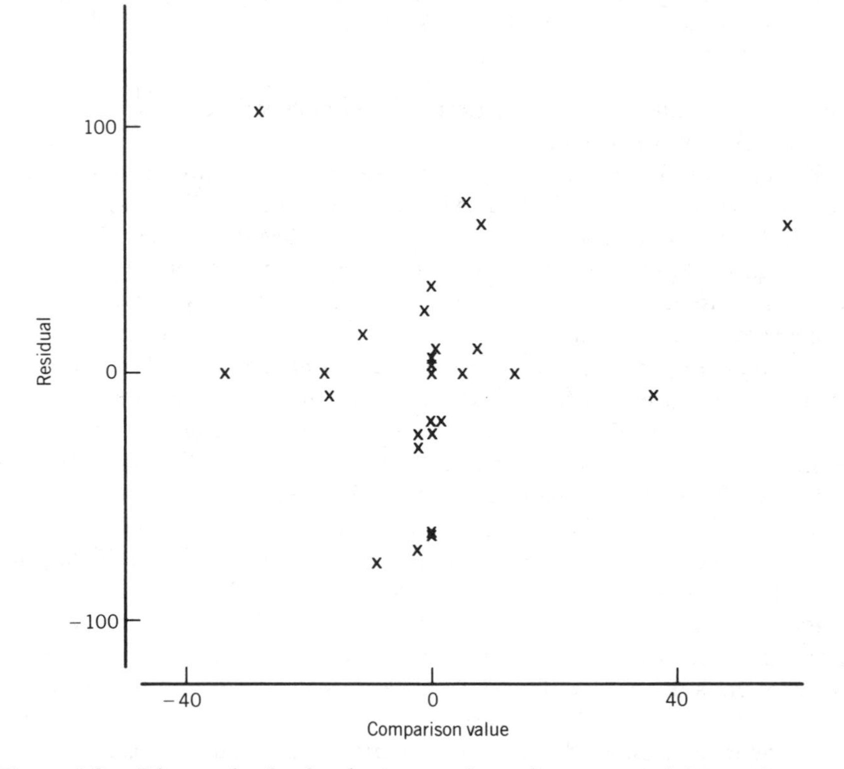

Figure 4-8. Diagnostic plot for the (untransformed) constructed data with rounded unit Gaussian perturbations added.

$\sum_i y_{ijk}/I$. Although we could easily mimic the median-polish procedure, now using means, the direct formulas are simpler to compute. Again let y_{ijk} be the observed data ($i = 1, \ldots, I$; $j = 1, \ldots, J$; and $k = 1, \ldots, K$) and let a_i, b_j, c_k be the respective row, column, and layer means. Then the analysis proceeds as follows:

$$
\begin{aligned}
a_i &= y_{i\cdot\cdot} = \sum_j \sum_k y_{ijk}/JK, \\
b_j &= y_{\cdot j \cdot} = \sum_i \sum_k y_{ijk}/IK, \\
c_k &= y_{\cdot\cdot k} = \sum_i \sum_j y_{ijk}/IJ, \\
m &= a_{\cdot} = b_{\cdot} = c_{\cdot} = y_{\cdot\cdot\cdot}.
\end{aligned}
\tag{30}
$$

The effect estimates are

$$
\begin{aligned}
\hat{\alpha}_i &= a_i - m, \\
\hat{\beta}_j &= b_j - m, \\
\hat{\gamma}_k &= c_k - m, \\
\hat{\mu} &- m,
\end{aligned}
\tag{31}
$$

with the residuals

$$
\begin{aligned}
r_{ijk} &= y_{ijk} - a_i - b_j - c_k + 2m \\
&= y_{ijk} - \hat{\alpha}_i - \hat{\beta}_j - \hat{\gamma}_k - \hat{\mu}.
\end{aligned}
\tag{32}
$$

The estimates are the same as those obtained for the three-way main-effects-only analysis-of-variance model corresponding to equation (2).

EXAMPLE: WEIGHT GAIN IN PIGS (MAIN-EFFECTS-ONLY BY MEANS)

For the pig feeding data, Table 4-28 shows the results, rounding to one decimal place. We can compare these with the results of the median-polish analysis from Table 4-10. Figure 4-9, the plot of the effect estimates and the residuals, facilitates comparisons with Figure 4-1. When we look at the

TABLE 4-28. Main-effects-only mean analysis of pig feeding data.

Methionine (in .01%)	Protein (in %)	Lysine (in .01%) 0	5	10	15	*M*-Effects	*P*-Effects
0	12	−6.5	3.5	4.5	0.4	−7.6	−21.1
	14	40.3	39.3	−26.7	−54.8	−7.6	21.1
25	12	−13.1	−9.1	37.9	23.8	−1.0	−21.1
	14	−14.3	−17.3	−18.3	10.6	−1.0	21.1
50	12	−24.6	−34.6	6.4	11.3	8.5	−21.1
	14	18.2	18.2	−3.8	9.1	8.5	21.1
L-effects		−8.8	3.2	7.2	−1.7	Overall	252.0

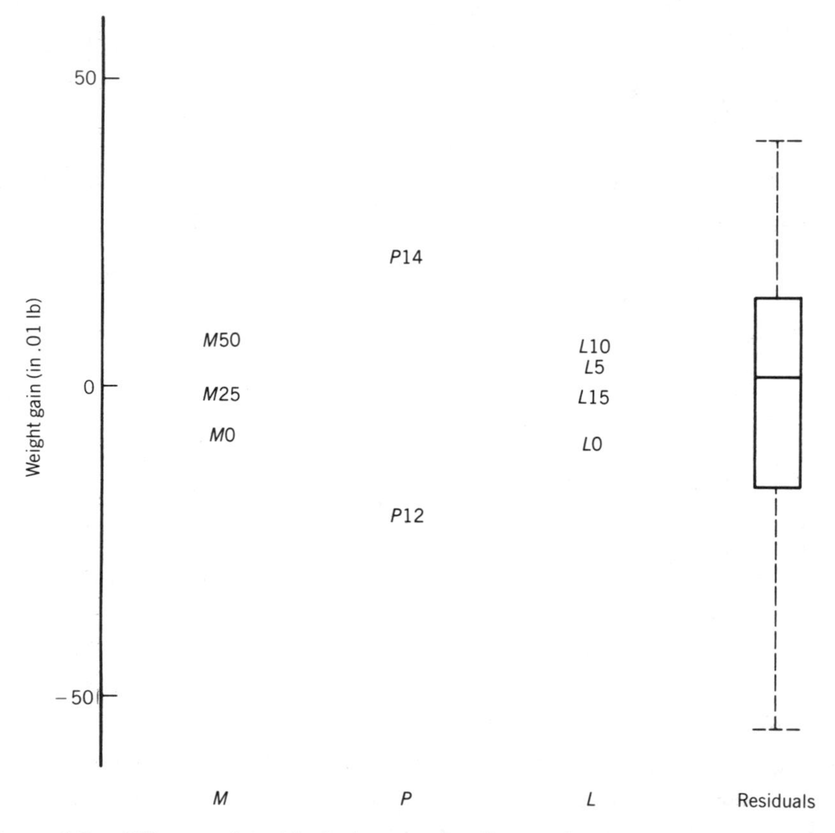

Figure 4-9. Effects and residuals from main-effects-only mean analysis of the pig feeding data.

TABLE 4-29. Pig feeding data with introduced outliers (the entries 0 and 1000).

Methionine (in .01%)	Protein (in %)	Lysine (in .01%) 0	5	10	15
0	12	208	230	235	222
	14	297	1000	246	209
25	12	208	224	275	252
	14	249	258	261	281
50	12	0	208	253	249
	14	291	303	285	289

results of the main-effects-only analyses using means and medians, we see many similarities, something that will not always happen. The estimates for the P-effects are almost the same, and the greatest residuals appear in the same cells. Slight discrepancies show up in the estimates for the L-effects and M-effects, both in sign and magnitude, and it is not apparent which set of estimates is more satisfactory in this example.

EXAMPLE: INTRODUCED OUTLIERS

To show how median polish can be preferable to using means, we introduce two extreme outliers into the pig feeding data, as shown in Table 4-29. We have changed the smallest entry to 0 and the largest to 1000. So few aberrant values should not distort a resistant analysis seriously. Table 4-30 shows the median-polish analysis and includes in each cell the corresponding value for the median-polish analysis of the unmodified data from Table 4-10. The effect estimates are unchanged, except for rounding error, leaving the residuals the same for the 22 numbers that were not changed. The two outliers have produced extreme residuals.

Table 4-31 gives the main-effects-only mean analysis for the modified data. Compared to Table 4-30, the effect estimates and residuals show that the outliers have had a dramatic effect on all aspects of the mean analysis. The sizes of the effect estimates are much greater in the mean analysis, and some signs have been reversed among the L-effects. The mean analysis gives equal weight to the outliers in calculating means to estimate the effects and thus permits distortion of the estimates. The analysis using medians is much more resistant to the outliers.

TABLE 4-30. Median-polish analysis of pig feeding data with introduced outliers. The lower entry in each cell is the result for the unmodified data (from Table 4-10).

Methionine (in .01%)	Protein (in %)	Lysine (in .01%) 0	5	10	15	*M*-Effects	*P*-Effects
0	12	−2	2	3	−11	0	−20
		−2	2	3	−11	0	−20
	14	46.5	731.5	−26.5	−63.5	0	20.5
		47	40	−26	−64	0	20
25	12	0	−2	45	21	−2	−20
		−0.5	−2.5	44.5	20.5	−1.5	−20
	14	0.5	−8.5	−9.5	9.5	−2	20.5
		0.5	−8.5	−9.5	9.5	−1.5	20
50	12	−226	−36	5	0	16	−20
		−20	−36	5	0	16	−20
	14	24.5	18.5	−3.5	−0.5	16	20.5
		25	19	−3	0	16	20
L-effects		−20	−2	2	3	Overall	250
		−20	−2	2	3	value	250

TABLE 4-31. Main-effects-only mean analysis of pig feeding data with introduced outliers.

Methionine (in .01%)	Protein (in %)	Lysine (in .01%) 0	5	10	15	M-Effects	P-Effects
0	12	−1.0	−140.6	−24.3	−28.5	58.67	−58.54
	14	−29.0	512.3	−130.4	−158.5	58.67	58.54
25	12	78.9	−66.8	95.6	81.4	−21.21	−58.54
	14	2.8	−149.8	−35.5	−6.7	−21.21	58.54
50	12	−112.8	−66.5	89.8	94.7	−37.46	−58.54
	14	61.1	−88.6	4.8	17.6	−37.46	58.54
L-effects		−63.38	98.29	−13.04	−21.88	Overall	272.21

4F. MEDIAN-POLISH ANALYSIS FOR THE FULL-EFFECTS CASE

Up to this point we have discussed main-effects-only analyses of three-way data. We looked at median polish, mean analysis, and diagnostic plots. In some examples, such as the pig feeding data, the fit of the main-effects-only model did not seem satisfactory. The residuals were large, but the diagnostic plot gave no indication that a transformation would improve the fit. We now turn to fitting the full-effects model, which includes all three two-factor interaction terms. When we have only one observation per cell, this is often the most general model that we can consider for a specific three-way table.

As introduced in Section 4B, the full-effects model for a three-way table is

$$y_{ijk} = \mu + \alpha_i + \beta_j + \gamma_k + (\alpha\beta)_{ij} + (\alpha\gamma)_{ik} + (\beta\gamma)_{jk} + \varepsilon_{ijk}.$$

We can again apply median polish to fit this, although the algebra is somewhat more complicated. (The arithmetic is, if anything, easier.) Now we take the medians along one dimension or subscript, holding the other two fixed, rather than over all combinations of two subscripts while holding the third fixed.

If we imagine a three-way table of measurements as a cubic structure, with rows, columns, and layers, the fitting procedure becomes clearer. In the main-effects-only analysis, we take the median for, say, the first row over all columns and layers. That is, we take a median of all data in a two-way table, the top slice in our cubic structure, to get the first row effect. In a full-effects analysis we take the median for the first row and first layer over all columns to get the combined effect for the first row and first layer. We do the same for all rows and layers, for all rows and columns, and for all columns and layers. To get the main effects for rows, we take medians of the combined effects involving rows. We get the overall value by summing the medians of each set of main effects. We therefore produce estimates for all two-factor interaction effects, as well as for the main effects and overall value.

In terms of the cubic structure, we take medians parallel to each axis (separately) in order to sweep information from the data into the three sets of two-factor effects. We then use medians along both directions of each table of two-factor effects to extract information that can go into the main effects of those two factors. Finally, we use the median to center each of the three sets of main effects and transfer a contribution to the common term.

We now describe the method in algebraic terms. In the first cycle (for one possible order of factors), the steps, using the same notation as in equations (6) through (14), go as follows:

$$r_{ijk}^{(0)} = y_{ijk},$$

$$(ab)_{ij}^{(1)} = \operatorname*{med}_{k}\left\{ r_{ijk}^{(0)} \right\} \quad \text{for each } i, j,$$

$$(ac)_{ik}^{(1)} = \operatorname*{med}_{j}\left\{ r_{ijk}^{(0)} - (ab)_{ij}^{(1)} \right\} \quad \text{for each } i, k,$$

$$a_{i}^{(1)} = \operatorname*{med}_{j}\left\{ (ab)_{ij}^{(1)} \right\} \quad \text{for each } i,$$

$$(bc)_{jk}^{(1)} = \operatorname*{med}_{i}\left\{ r_{ijk}^{(0)} - (ab)_{ij}^{(1)} - (ac)_{ik}^{(1)} \right\} \quad \text{for each } j, k,$$

$$c_{k}^{(1)} = \operatorname*{med}_{i}\left\{ (ac)_{ik}^{(1)} \right\} \quad \text{for each } k,$$

$$b_{j}^{(1)} = \operatorname*{med}_{i}\left\{ (ab)_{ij}^{(1)} - a_{i}^{(1)} \right\} \quad \text{for each } j,$$

$$m_{a}^{(1)} = \operatorname*{med}_{i}\left\{ a_{i}^{(1)} \right\}.$$

We can think of this cycle in terms of three steps, each involving a different subscript over whose values we take the median(s): first k, to get the $(ab)_{ij}^{(1)}$; second j, to get the $(ac)_{ik}^{(1)}$ and the $a_{i}^{(1)}$; and third i, to get the $(bc)_{jk}^{(1)}$, the $c_{k}^{(1)}$, the $b_{j}^{(1)}$, and $m_{a}^{(1)}$. At the end of the first cycle we have the residuals

$$\begin{aligned} r_{ijk}^{(1)} &= r_{ijk}^{(0)} - (ab)_{ij}^{(1)} - (ac)_{ik}^{(1)} - (bc)_{jk}^{(1)} \\ &= r_{ijk}^{(0)} - \hat{\mu}^{(1)} - \hat{\alpha}_{i}^{(1)} - \hat{\beta}_{j}^{(1)} - \hat{\gamma}_{k}^{(1)} \\ &\quad - \left(\widehat{\alpha\beta}\right)_{ij}^{(1)} - \left(\widehat{\alpha\gamma}\right)_{ik}^{(1)} - \left(\widehat{\beta\gamma}\right)_{jk}^{(1)} \end{aligned}$$

and the estimates

$$\widehat{(\alpha\beta)}_{ij}^{(1)} = (ab)_{ij}^{(1)} - a_i^{(1)} - b_j^{(1)},$$

$$\widehat{(\alpha\gamma)}_{ik}^{(1)} = (ac)_{ik}^{(1)} - c_k^{(1)},$$

$$\widehat{(\beta\gamma)}_{jk}^{(1)} = (bc)_{jk}^{(1)},$$

$$\hat{\alpha}_i^{(1)} = a_i^{(1)} - m_a^{(1)},$$

$$\hat{\beta}_j^{(1)} = b_j^{(1)},$$

$$\hat{\gamma}_k^{(1)} = c_k^{(1)},$$

$$\hat{\mu}^{(1)} = m_a^{(1)}.$$

In practice we often record the steps in a cycle by enlarging upon a two-way arrangement of the three-way table. Table 4-32 uses the factors and

TABLE 4-32. One two-way arrangement of the eight pieces in the decomposition of a three-way table according to the full-effects model.

Methionine	Protein	Lysine (in .01%)				
(in .01%)	(in %)	0	5	10	15	
0	12					
	14					
25	12		Residuals			$M \times P$-effects
	14		r_{ijk}			$\widehat{(\alpha\gamma)}_{ik}$
50	12					
	14					
	12		$P \times L$-effects			P-effects
	14		$\widehat{(\alpha\beta)}_{ij}$			$\hat{\alpha}_i$
0			$M \times L$-effects			M-effects
25			$\widehat{(\beta\gamma)}_{jk}$			$\hat{\gamma}_k$
50						
			L-effects $\hat{\beta}_j$			Overall value
						$\hat{\mu}$

TABLE 4-33. The pig feeding data (from Table 4-2) and medians over levels of methionine for each combination of protein and lysine.

Methionine (in .01%)	Protein (in %)	Lysine (in .01%) 0	5	10	15
0	12	208	230	235	222
	14	297	308	246	209
25	12	208	224	275	252
	14	249	258	261	281
50	12	206	208	253	249
	14	291	303	285	289
Medians over M					
		0	5	10	15
	12	208	224	253	249
	14	291	303	261	281

levels in the pig feeding data to illustrate one convenient format for organizing the eight pieces of the decomposition according to the full-effects model: the overall value (or common term), the three sets of main effects, the three sets of two-factor interactions, and the table of residuals.

In this arrangement, protein is factor A, lysine is factor B, and methionine is factor C. Where we began with the data y_{ijk}, we now place the residuals r_{ijk}. The interaction effects $\widehat{(\alpha\gamma)}_{ik}$ for the combined factor appear as a column to the right of the residuals, and the other sets of interaction effects, $\widehat{(\alpha\beta)}_{ij}$ and $\widehat{(\beta\gamma)}_{jk}$, appear as two-way tables beneath the residuals. Because columns in this two-way arrangement correspond to levels of B, the main effects $\hat{\beta}_j$ appear in a row at the bottom. The main effects $\hat{\alpha}_i$ and $\hat{\gamma}_k$ form columns to the right of the interactions $\widehat{(\alpha\beta)}_{ij}$ and $\widehat{(\beta\gamma)}_{jk}$, respectively, and $\hat{\mu}$ occupies a natural position at the lower right.

EXAMPLE: WEIGHT GAIN IN PIGS (FULL-EFFECTS BY MEDIANS)

To fit the full-effects model to the pig feeding data (Table 4-2), we begin by calculating the $(ab)_{ij}^{(1)}$ in Table 4-33. (Our ultimate objective is a decomposition of the form shown in Table 4-32.) Table 4-34 shows the result of subtracting out these $(ab)_{ij}^{(1)}$ and continues with the $(ac)_{ik}^{(1)}$ and the $a_i^{(1)}$. Table 4-35 proceeds to the third step of the first cycle by subtracting both

TABLE 4-34. **The second step in the first cycle of fitting the full-effects model to the pig feeding data: medians over levels of lysine.**

Methionine (in .01%)	Protein (in %)	Lysine (in .01%) 0	5	10	15	Medians over L
0	12	0	6	−18	−27	−9
	14	6	5	−15	−72	−5
25	12	0	0	22	3	1.5
	14	−42	−45	0	0	−21
50	12	−2	−16	0	0	−1
	14	0	0	24	8	4
	12	208	224	253	249	236.5
	14	291	303	261	281	286

TABLE 4-35. **The third step in the first cycle of fitting the full-effects model to the pig feeding data: medians over levels of protein.**

Methionine (in .01%)	Protein (in %)	Lysine (in .01%) 0	5	10	15	
0	12	9	15	−9	−18	−9
	14	11	10	−10	−67	−5
25	12	−1.5	−1.5	20.5	1.5	1.5
	14	−21	−24	21	21	−21
50	12	−1	−15	1	1	−1
	14	−4	−4	20	4	4
	12	−28.5	−12.5	16.5	12.5	236.5
	14	5	17	−25	−5	286
Medians over P						
		0	5	10	15	
	0	10	12.5	−9.5	−42.5	−7
	25	−11	−13	21	11	−10
	50	−2.5	−9.5	10.5	2.5	1.5
		−12	2	−4	4	261

the $(ac)_{ik}^{(1)}$ and the $a_i^{(1)}$ and calculating the $(bc)_{jk}^{(1)}$, the $c_k^{(1)}$, the $b_j^{(1)}$, and $m_a^{(1)}$. Subtracting these four components takes us to Table 4-36 and fills in the format of Table 4-32.

In later cycles, we often take two medians which contribute to the same effect during the course of a cycle. In particular, we have two contributions to each of the main effects. These come from medians of the two two-factor interaction terms that involve that factor. For instance, the two contributions to the a_i come from the median of the $(ab)_{ij}$ over j and the median of the $(ac)_{ik}$ over k, respectively. We distinguish the two by including the symbols $*$ and $**$ in the superscript: a_i^* and a_i^{**}, for example.

The steps in cycle n, then, go as follows for $n \geq 2$:

$$(ab)_{ij}^{(n)} = \operatorname*{med}_{k} \left\{ r_{ijk}^{(n-1)} \right\},$$

$$a_i^{*(n)} = \operatorname*{med}_{k} \left\{ \left(\widehat{\alpha\gamma}\right)_{ik}^{(n-1)} \right\},$$

$$b_j^{*(n)} = \operatorname*{med}_{k} \left\{ \left(\widehat{\beta\gamma}\right)_{jk}^{(n-1)} \right\},$$

$$m_c^{(n)} = \operatorname*{med}_{k} \left\{ \hat{\gamma}_k^{(n-1)} \right\},$$

$$(ac)_{ik}^{(n)} = \operatorname*{med}_{j} \left\{ r_{ijk}^{(n-1)} - (ab)_{ij}^{(n)} \right\},$$

$$a_i^{**(n)} = \operatorname*{med}_{j} \left\{ \left(\widehat{\alpha\beta}\right)_{ij}^{(n-1)} + (ab)_{ij}^{(n)} \right\},$$

$$c_k^{*(n)} = \operatorname*{med}_{j} \left\{ \left(\widehat{\beta\gamma}\right)_{jk}^{(n-1)} - b_j^{*(n)} \right\},$$

$$m_b^{(n)} = \operatorname*{med}_{j} \left\{ \hat{\beta}_j^{(n-1)} + b_j^{*(n)} \right\},$$

$$(bc)_{jk}^{(n)} = \operatorname*{med}_{i} \left\{ r_{ijk}^{(n-1)} - (ab)_{ij}^{(n)} - (ac)_{ik}^{(n)} \right\},$$

$$c_k^{**(n)} = \operatorname*{med}_{i} \left\{ \left(\widehat{\alpha\gamma}\right)_{ik}^{(n-1)} + (ac)_{ik}^{(n)} - a_i^{*(n)} \right\},$$

$$b_j^{**(n)} = \operatorname*{med}_{i} \left\{ \left(\widehat{\alpha\beta}\right)_{ij}^{(n-1)} + (ab)_{ij}^{(n)} - a_i^{**(n)} \right\},$$

$$m_a^{(n)} = \operatorname*{med}_{i} \left\{ \hat{\alpha}_i^{(n-1)} + a_i^{*(n)} + a_i^{**(n)} \right\}.$$

TABLE 4-36. Effect estimates and residuals at the end of the first cycle of fitting the full-effects model to the pig feeding data.

Methionine in (.01%)	Protein (in %)	Lysine (in .01%)				
		0	5	10	15	
0	12	−1	2.5	.5	24.5	−2
	14	1	−2.5	−.5	−24.5	2
25	12	9.5	11.5	−.5	−9.5	11.5
	14	−10	−11	0	10	−11
50	12	1.5	−5.5	−9.5	−1.5	−2.5
	14	−1.5	5.5	9.5	1.5	2.5
	12	−16.5	−14.5	20.5	8.5	−24.5
	14	17	15	−21	−9	25
0		10	12.5	−9.5	−42.5	−7
25		−11	−13	21	11	−10
50		−2.5	−9.5	10.5	2.5	1.5
		−12	2	−4	4	261

The residuals for cycle n are

$$\begin{aligned} r_{ijk}^{(n)} &= r_{ijk}^{(n-1)} - (ab)_{ij}^{(n)} - (ac)_{ik}^{(n)} - (bc)_{jk}^{(n)} \\ &= r_{ijk}^{(0)} - \hat{\mu}^{(n)} - \hat{\alpha}_i^{(n)} - \hat{\beta}_j^{(n)} - \hat{\gamma}_k^{(n)} - \left(\widehat{\alpha\beta}\right)_{ij}^{(n)} \\ &\quad - \left(\widehat{\alpha\gamma}\right)_{ik}^{(n)} - \left(\widehat{\beta\gamma}\right)_{jk}^{(n)} \end{aligned}$$

with the effect estimates

$$\begin{aligned} \left(\widehat{\alpha\beta}\right)_{ij}^{(n)} &= \left(\widehat{\alpha\beta}\right)_{ij}^{(n-1)} + (ab)_{ij}^{(n)} - a_i^{**(n)} - b_j^{**(n)}, \\ \left(\widehat{\alpha\gamma}\right)_{ik}^{(n)} &= \left(\widehat{\alpha\gamma}\right)_{ik}^{(n-1)} + (ac)_{ik}^{(n)} - a_i^{*(n)} - c_k^{**(n)}, \\ \left(\widehat{\beta\gamma}\right)_{jk}^{(n)} &= \left(\widehat{\beta\gamma}\right)_{jk}^{(n-1)} + (bc)_{jk}^{(n)} - b_j^{*(n)} - c_k^{*(n)}, \\ \hat{\alpha}_i^{(n)} &= \hat{\alpha}_i^{(n-1)} + a_i^{*(n)} + a_i^{**(n)} - m_a^{(n)}, \\ \hat{\beta}_j^{(n)} &= \hat{\beta}_j^{(n-1)} + b_j^{*(n)} + b_j^{**(n)} - m_b^{(n)}, \\ \hat{\gamma}_k^{(n)} &= \hat{\gamma}_k^{(n-1)} + c_k^{*(n)} + c_k^{**(n)} - m_c^{(n)}, \\ \hat{\mu}^{(n)} &= \hat{\mu}^{(n-1)} + m_a^{(n)} + m_b^{(n)} + m_c^{(n)}. \end{aligned}$$

Although these calculations for cycle n may seem to be a torrent of algebra and might be more appealing if embedded in a computer program, they are quite straightforward to interpret if we focus on the grouping according to whether the medians involve values of i, j, or k. It then becomes clearer that they perform precisely the sweeping-out operations that we need to get from the version of Table 4-32 at the end of cycle $n - 1$ to the version at the end of cycle n. Organizing the calculations in this way does recenter each set of effects at each step, instead of leaving all recentering to a final step.

Interpreting the steps briefly, we see that the four equations involving medians over k transfer $(ab)_{ij}$ from the r_{ijk} to $(\widehat{\alpha\beta})_{ij}$, a_i^* from the $(\widehat{\alpha\gamma})_{ik}$ to $\hat{\alpha}_i$, b_j^* from the $(\widehat{\beta\gamma})_{jk}$ to $\hat{\beta}_j$, and m_c from the $\hat{\gamma}_k$ to $\hat{\mu}$. By "transfer" we mean that what we subtract from one term of the table we add into a lower-order term. Thus our books remain balanced, and at any stage will reproduce the data y_{ijk}. Continuing with the medians over j, we see that these four equations then transfer $(ac)_{ik}$ from the r_{ijk} to $\widehat{(\alpha\gamma)}_{ik}$, a_i^{**} from the $\widehat{(\alpha\beta)}_{ij}$ to $\hat{\alpha}_i$, c_k^* from the $\widehat{(\beta\gamma)}_{jk}$ to $\hat{\gamma}_k$, and m_b from the $\hat{\beta}_j$ to $\hat{\mu}$. Finally, after these adjustments, the four equations involving medians over i transfer $(bc)_{jk}$ from the r_{ijk} to $\widehat{(\beta\gamma)}_{jk}$, c_k^{**} from the $\widehat{(\alpha\gamma)}_{ik}$ to $\hat{\gamma}_k$, b_j^{**} from the $\widehat{(\alpha\beta)}_{ij}$ to $\hat{\beta}_j$, and m_a from the $\hat{\alpha}_i$ to $\hat{\mu}$.

The order of the steps implies that some sets of effects may not be completely centered at the close of a given cycle. For example, both $b_j^{*(n)}$ and $c_k^{*(n)}$ are subtracted from $\widehat{(\beta\gamma)}_{jk}^{(n-1)}$ before $(bc)_{jk}^{(n)}$ is added in, and so

TABLE 4-37. The first step in the second cycle of fitting the full-effects model to the pig feeding data by median polish: medians over levels of methionine.

Methionine (in .01%)	Protein (in %)	Lysine (in .01%) 0	5	10	15	
0	12	−1	2.5	.5	24.5	−2
	14	1	−2.5	−.5	−24.5	2
25	12	9.5	11.5	−.5	−9.5	11.5
	14	−10	−11	0	10	−11
50	12	1.5	−5.5	−9.5	−1.5	−2.5
	14	−1.5	5.5	9.5	1.5	2.5
	12	−16.5	−14.5	20.5	8.5	−24.5
	14	17	15	−21	−9	25
0		10	12.5	−9.5	−42.5	−7
25		−11	−13	21	11	−10
50		−2.5	−9.5	10.5	2.5	1.5
		−12	2	−4	4	261
		Medians over M of residuals and effects				
		0	5	10	15	
	12	1.5	2.5	−.5	−1.5	−2
	14	−1.5	−2.5	0	1.5	2
		−2.5	−9.5	10.5	2.5	−7

TABLE 4-38. The second step in the second cycle of fitting the full-effects model to the pig feeding data by median polish: medians over levels of lysine.

Methionine (in .01%)	Protein (in %)	Lysine (in .01%) 0	5	10	15		Medians over L
0	12	−2.5	0	1	26	0	.5
	14	2.5	0	−.5	−26	0	0
25	12	8	9	0	−8	13.5	4
	14	−8.5	−8.5	0	8.5	−13	−4
50	12	0	−8	−9	0	−.5	−4
	14	0	8	9.5	0	.5	4
	12	−15	−12	20	7	−26.5	−2.5
	14	15.5	12.5	−21	−7.5	27	2.5
0		12.5	22	20	45	0	4
25		−8.5	−3.5	10.5	8.5	−3	2.5
50		0	0	0	0	8.5	0
		−14.5	−7.5	6.5	6.5	254	−.5

the $\widehat{(\beta\gamma)}_{jk}^{(n)}$ may need recentering in both directions. Ordinarily, the final cycle of the fitting process is concerned with centering these sets of effects and makes no change in the r_{ijk}.

EXAMPLE: WEIGHT GAIN IN PIGS (FULL-EFFECTS BY MEDIANS CONTINUED)

To illustrate the steps in the general formula, Tables 4-37 through 4-39 show the second cycle for the pig feeding data. We give the final analysis (after five iterations and one further recentering step) in Table 4-40. The residuals and effect estimates are plotted as parallel batches in Figure 4-10. In comparison with the main-effects analysis by median polish in Table 4-10 and Figure 4-1, the residuals are much smaller. This shows that interaction effects have accounted for much of the variability in the earlier residuals (although at a cost of increasing the effective number of parameters fitted from 7 to 18).

What does this full-effects fit seem to be telling us? As in the main-effects-only fit, the percentage of protein added (as soybean meal) shows the largest main-effect contribution, followed by lysine and methionine. The

TABLE 4-39. The third step in the second cycle of fitting the full-effects model to the pig feeding data by median polish: medians over levels of protein.

Methionine (in .01%)	Protein (in %)	Lysine (in .01%) 0	5	10	15	
0	12	−3	−.5	.5	25.5	.5
	14	2.5	0	−.5	−26	0
25	12	4	5	−4	−12	17.5
	14	−4.5	−4.5	4	12.5	−17
50	12	4	−4	−5	4	−4.5
	14	−4	4	5.5	−4	4.5
	12	−12.5	−9.5	22.5	9.5	−29
	14	13	10	−23.5	−10	29.5
0		16.5	26	−16	−41	−4
25		−11	−6	8	6	−.5
50		0	0	0	0	8.5
		−14	−7	7	7	253.5
		Medians over P of residuals and effects				
		0	5	10	15	
0		0	0	0	0	0
25		0	0	0	0	0
50		0	0	0	0	0
		0	0	−.5	0	0

methionine effect may be increasing more rapidly than linearly with the percentage of methionine added. A similar description holds for the lysine effect up through .10%, but then .15% appears to have the same effect as .10%. The $M \times P$ interactions suggest that increased protein makes an additional positive contribution at .50% methionine but has nearly twice as large a negative increment at .25% methionine. Although the observed $M \times L$ interactions include the largest entries (both positive and negative) in the whole decomposition, they do not follow any especially simple

TABLE 4-40. The final full-effects median-polish analysis of the pig feeding data.

Methionine (in .01%)	Protein (in %)	Lysine (in .01%) 0	5	10	15	
0	12	−11	−1	.5	17	0
	14	11	1	−.5	−17	0
25	12	0	8	0	−17	13
	14	0	−8	0	17	−13
50	12	.5	0	−.5	0	−9.5
	14	−.5	0	.5	0	9.5
	12	−11.5	−16	16	11.5	−22
	14	11.5	16	−16	−11.5	22
0		16.5	26	−16	−41	−4
25		−11.5	−6	7.5	6	0
50		0	0	0	0	8.5
		−13.5	−6.5	7	7	253

pattern. All are zero when methionine is at .50%. Otherwise, when no methionine is added, lower levels of lysine seem to help, but higher levels seem to hinder; and at .25% methionine the higher levels of lysine seem to provide a moderate increment. Finally, the $P \times L$ interactions suggest that more protein helps at the lower levels of lysine but hinders at the higher levels. Perhaps this represents some type of substitution effect.

In practice, if the measurement errors were small enough, one would try to use these results in choosing a good combination (or combinations) of the three dietary supplements, as well as in starting to understand the typical daily weight gains that one might expect for other percentages of methionine, protein, and lysine. For this latter purpose it would be customary to move from the effects and interactions to simple functions (e.g., linear and quadratic) of the three percentages and then to develop a suitable regression model. (In a regression analog to the full-effects model, the quadratic part would need to include cross-product terms.) Fitting a full-effects model or, perhaps, a main-effects-only model to the three-way table generally provides a valuable springboard for such simplifications.

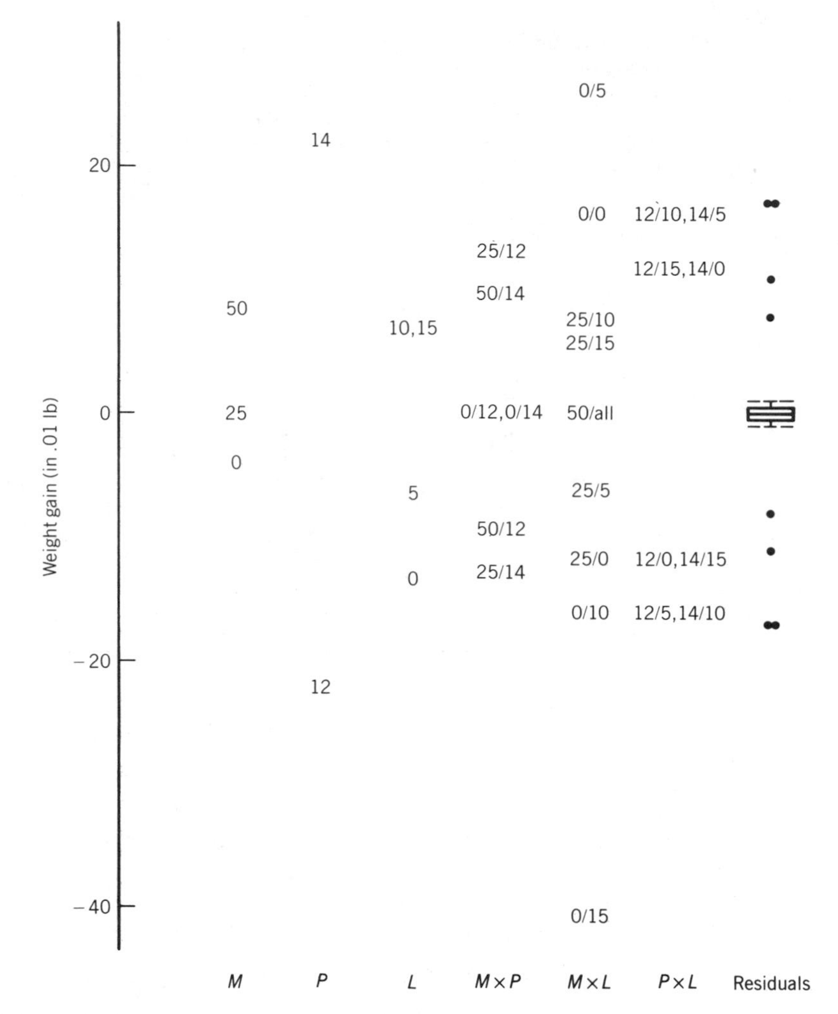

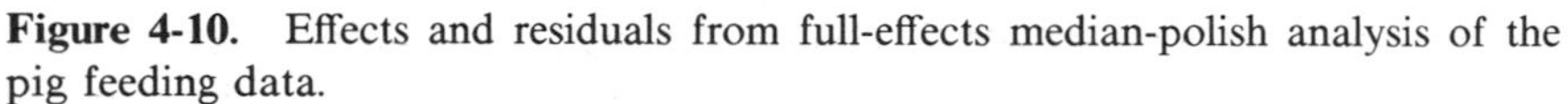

Figure 4-10. Effects and residuals from full-effects median-polish analysis of the pig feeding data.

4G. DIAGNOSTIC PLOTS FOR THE FULL-EFFECTS CASE

In Section 4D we concluded that the leading nonadditive term, which ought to account for most of the structure in the residuals from a main-effects-only fit, was (when rewritten slightly)

$$(1-p)\frac{\hat{\alpha}_i\hat{\beta}_j}{\hat{\mu}} + (1-p)\frac{\hat{\alpha}_i\hat{\gamma}_k}{\hat{\mu}} + (1-p)\frac{\hat{\beta}_j\hat{\gamma}_k}{\hat{\mu}}.$$

In this sum each term takes the form of a two-factor interaction. Thus we might expect each product term to account for part (conceivably, nearly all) of the corresponding two-factor interaction that we calculate in a full-effects analysis of the data in the untransformed scale. To assess p from the full-effects analysis, we plot

$$\widehat{(\alpha\beta)}_{ij} \quad \text{against} \quad \frac{\hat{\alpha}_i \hat{\beta}_j}{\hat{\mu}},$$

$$\widehat{(\alpha\gamma)}_{ik} \quad \text{against} \quad \frac{\hat{\alpha}_i \hat{\gamma}_k}{\hat{\mu}},$$

$$\widehat{(\beta\gamma)}_{jk} \quad \text{against} \quad \frac{\hat{\beta}_j \hat{\gamma}_k}{\hat{\mu}},$$

either separately or in one combined display. As before, we interpret a consistent slope as $1 - p$ and consider re-expression by the pth power.

TABLE 4-41. The full-effects median-polish analysis of the yarn breakage data.

A: Length (mm)	B: Amplitude (mm)	C: Load (g) 40	45	50	
250	8	0	−88	272	−386
	9	222	0	0	0
	10	0	0	−42	140
300	8	−58	0	0	0
	9	50	0	0	0
	10	0	0	0	0
350	8	32	0	−208	1434
	9	0	0	90	−102
	10	0	0	0	0
	8	0	0	−382	578
	9	78	0	0	0
	10	−164	0	70	−288
250		−58	0	126	−354
300		0	0	0	0
350		146	0	−412	552
		274	0	−182	620

TABLE 4-42. Comparison values and corresponding interaction effects for the full-effects analysis of the yarn breakage data.

Comparison Values				Interaction Effects			
$\frac{\hat{\alpha}_i \hat{\beta}_j}{\hat{\mu}}$	$j=1$	2	3	$\widehat{(\alpha\beta)}_{ij}$	$j=1$	2	3
$i=1$	-330.02	0	164.44	$i=1$	-386	0	140
2	0	0	0	2	0	0	0
3	514.61	0	-256.41	3	1434	-102	0
$\frac{\hat{\alpha}_i \hat{\gamma}_k}{\hat{\mu}}$	$k=1$	2	3	$\widehat{(\alpha\gamma)}_{ik}$	$k=1$	2	3
$i=1$	-156.45	0	103.92	$i=1$	-58	0	126
2	0	0	0	2	0	0	0
3	243.95	0	-162.04	3	146	0	-412
$\frac{\hat{\beta}_j \hat{\gamma}_k}{\hat{\mu}}$	$k=1$	2	3	$\widehat{(\beta\gamma)}_{jk}$	$k=1$	2	3
$j=1$	255.44	0	-169.67	$j=1$	0	0	-382
2	0	0	0	2	78	0	0
3	-127.28	0	84.54	3	-164	0	70

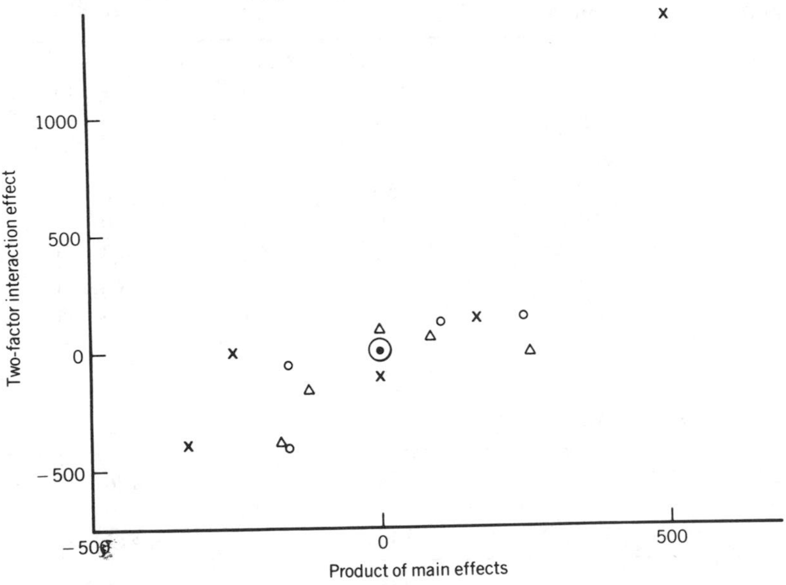

Figure 4-11. Two-factor interaction effects versus main-effect products from full-effects median-polish analysis of the yarn breakage data. The plotting symbols are $\times$ for the $\widehat{(\alpha\beta)}_{ij}$, O for the $\widehat{(\alpha\gamma)}_{ik}$, and $\triangle$ for the $\widehat{(\beta\gamma)}_{jk}$. The circled dot at the origin represents 13 points.

TABLE 4-43. Full-effects median-polish analysis of the logs of the yarn breakage data.

A: Length (mm)	B: Amplitude (mm)	C: Load (g) 40	45	50	
250	8	9	−24	0	0
	9	−26	0	6	1
	10	0	0	0	−9
300	8	0	2	0	−3
	9	0	−9	0	0
	10	0	0	0	0
350	8	−3	0	34	0
	9	0	1	0	−52
	10	0	0	−28	0
	8	0	10	−8	48
	9	13	0	0	0
	10	0	0	2	−71
250		0	−9	6	−86
300		0	0	0	0
350		0	2	−17	95
		28	0	−44	652

EXAMPLE: YARN BREAKAGE (FULL-EFFECTS BY MEDIANS, DIAGNOSED)

For the untransformed yarn breakage data, the full-effects analysis appears in Table 4-41. We show the comparison values described above with their corresponding interaction effects in Table 4-42. Finally, all three sets of numbers are plotted together (using different symbols to distinguish them) in Figure 4-11.

As in the main-effects-only case, the graph shows a clear relationship between the interaction effects and the corresponding comparison values. A three-group resistant line (UREDA, Chapter 5) yields slope 1.26. Thus p, the exponent of y_{ijk} in the suggested transformation, has a value of $-.26$. This is close enough to zero for us to try logs, the same transformation indicated by the main-effects-only analysis. In the full-effects analysis of the logs of the yarn breakage data (Table 4-43) the two-factor interactions generally seem comparable in size to the residuals, with $\widehat{(\alpha\beta)}_{32} = -52$ as an isolated exception. Thus a main-effects-only model should adequately summarize the effects of length, amplitude, and load on the log of the number of cycles to failure, if accompanied by this single two-factor-interaction value.

4H. FITTING THE FULL-EFFECTS MODEL BY MEANS

We may fit the full-effects model by using means, just as we used means in the main-effects-only analysis. The procedure resembles median polish, except that now (a) we take means instead of medians and (b) no iteration is necessary. Rather than mimic the steps in the median-polish analysis, we list the more direct computational formulas below, using the same dot notation as in equation (30):

$$\hat{\mu} = y_{\cdots},$$

$$\hat{\alpha}_i = y_{i\cdot\cdot} - y_{\cdots} = y_{i\cdot\cdot} - \hat{\mu},$$

$$\hat{\beta}_j = y_{\cdot j\cdot} - y_{\cdots} = y_{\cdot j\cdot} - \hat{\mu},$$

$$\hat{\gamma}_k = y_{\cdot\cdot k} - y_{\cdots} = y_{\cdot\cdot k} - \hat{\mu},$$

$$\widehat{(\alpha\beta)}_{ij} = y_{ij\cdot} - y_{i\cdot\cdot} - y_{\cdot j\cdot} + y_{\cdots} = y_{ij\cdot} - \hat{\alpha}_i - \hat{\beta}_j - \hat{\mu},$$

$$\widehat{(\alpha\gamma)}_{ik} = y_{i\cdot k} - y_{i\cdot\cdot} - y_{\cdot\cdot k} + y_{\cdots} = y_{i\cdot k} - \hat{\alpha}_i - \hat{\gamma}_k - \hat{\mu},$$

$$\widehat{(\beta\gamma)}_{jk} = y_{\cdot jk} - y_{\cdot j\cdot} - y_{\cdot\cdot k} + y_{\cdots} = y_{\cdot jk} - \hat{\beta}_j - \hat{\gamma}_k - \hat{\mu},$$

$$r_{ijk} = y_{ijk} - y_{ij\cdot} - y_{i\cdot k} - y_{\cdot jk} + y_{i\cdot\cdot} + y_{\cdot j\cdot} + y_{\cdot\cdot k} - y_{\cdots}$$

$$= y_{ijk} - \hat{\mu} - \hat{\alpha}_i - \hat{\beta}_j - \hat{\gamma}_k - \widehat{(\alpha\beta)}_{ij} - \widehat{(\alpha\gamma)}_{ik} - \widehat{(\beta\gamma)}_{jk}.$$

TABLE 4-44. Full-effects analysis by means for the pig feeding data.

Methionine (in .01%)	Protein (in %)	Lysine (in .01%) 0	5	10	15	
0	12	−9.2	−5.0	−1.2	15.3	0.5
	14	9.2	5.0	1.2	−15.3	−0.5
25	12	5.5	7.6	2.0	−15.0	9.9
	14	−5.5	−7.6	−2.0	15.0	−9.9
50	12	3.7	−2.6	−0.8	−0.3	−10.4
	14	−3.7	2.6	0.8	0.3	10.4
	12	−14.7	−13.4	16.3	11.8	−21.1
	14	14.7	13.4	−16.3	−11.8	21.1
0		16.9	21.4	−11.1	−27.2	−7.6
25		−13.7	−13.2	9.8	17.1	−1.0
50		−3.2	−8.2	1.3	10.1	8.5
		−8.8	3.2	7.2	−1.6	252.0

EXAMPLE: WEIGHT GAIN IN PIGS (FULL-EFFECTS BY MEANS)

For the pig feeding data Table 4-44 shows the full-effects analysis by means, and Figure 4-12 plots the effect estimates and residuals. The results differ little from those found in the median-polish analysis, which appear in Table 4-40 and Figure 4-10. Most of the effect estimates from the two analyses are close in both sign and magnitude, something that will not always happen. We also find the largest residuals in the same cells. The spread of the residuals, however, differs between the two analyses. In the mean analysis the fourth-spread is 8.7, whereas in the median-polish analysis it is only 1.0. The median-polish analysis has produced more very small residuals.

The substantial similarity between these two sets of residuals and, especially, the exact symmetry of each set about zero follow from the fact that one factor (protein) has only two levels. When all factors have more than two levels, we get the full benefit of the resistance of the median-polish

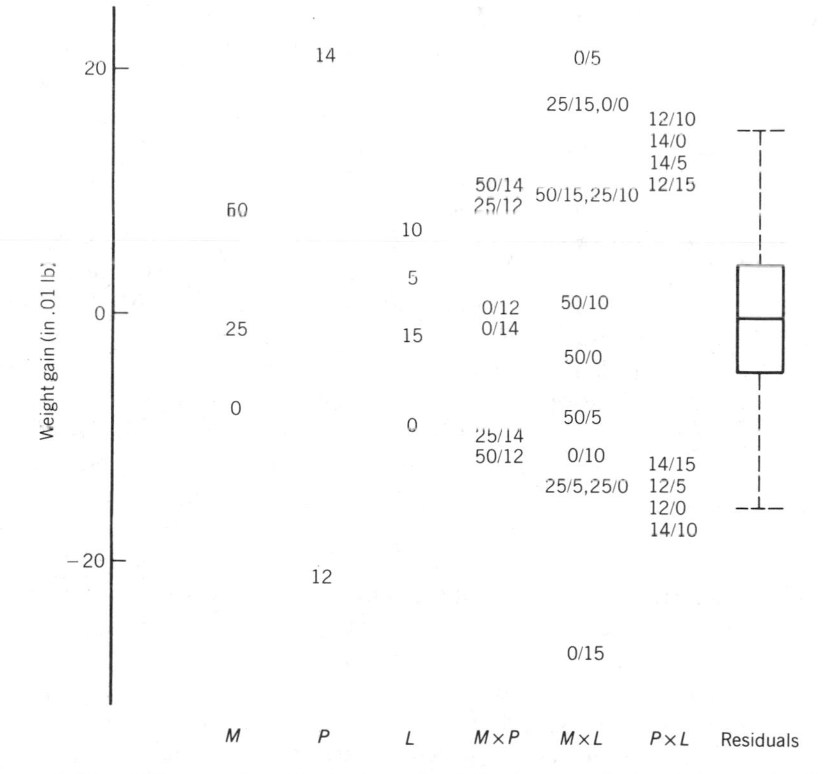

Figure 4-12. Effects and residuals from the full-effects analysis of the pig feeding data by means.

analysis. A single isolated large interaction effect will not disturb the other effects in its interaction term or the main effects. The analysis based on means, however, often makes it difficult to distinguish such an isolated interaction effect from moderate to large effects throughout the interaction term.

4I. COMPUTATION, OTHER POLISHES, AND MISSING VALUES

A number of practical issues that arise in making resistant analyses of three-way tables deserve brief mention.

Computational Approaches

The detailed bookkeeping and iterations of either the main-effects-only analysis or the full-effects analysis by median polish strongly invite the use of a computer. Although one could readily program the algorithms of Sections 4C and 4F, it may be helpful to draw upon the two-way median-polish programs of Velleman and Hoaglin (1981). Similarly one could readily construct macros in the Minitab statistical system (Ryan, Joiner, and Ryan, 1981) that handle the three-way table by applying two-way median polish and other operations.

Alternatively, software for least-absolute-residuals regression can fit models such as (2) and (3): each individual parameter in the model requires a separate indicator variable as its carrier. Thus the carrier for α_1 takes the value 1 for each element of $\mathbf{y}$ at level 1 of factor A and the value 0 otherwise; and the regression equation for the full-effects model contains such a carrier for each of the other α_i, for all the β_j, all the γ_k, all the $(\alpha\beta)_{ij}$, and so on. To impose constraints such as $\text{med}\{\alpha_i\} = 0$, one uses the corresponding linear constraint (e.g., $\Sigma\alpha_i = 0$) and recenters the computed effect estimates. Computer algorithms for least-absolute-residuals fitting such as the one by Bartels and Conn (1980a, 1980b) accept such linear constraints as input. Otherwise one can introduce the constraints by removing suitable carriers (such as the one for α_I) and making compensating changes in the values of the remaining carriers.

As in two-way tables (Chapter 6 of UREDA), the median-polish fits may not necessarily be least-absolute-residuals fits, and even a least-absolute-residuals fit may not be unique. These modest complications seem a satisfactory price to pay for resistance.

Low-Median Polish

To avoid all division (and hence any fractions) in analyzing three-way tables, one can define *low-median polish* in exactly the same way that Sections 4C and 4F describe median polish. The *low median* of an odd number of values is the ordinary median, whereas for an even number of values it is the lower of the two central values (whose average is the ordinary median). Siegel (1983) discusses the low median and its use in two-way tables. Among its further advantages, low-median polish facilitates rapid convergence and production of residual tables with many zeros.

Other Polishes and Related Techniques

Where the algorithms of Sections 4C and 4F calculate a median, it would be straightforward to substitute another measure of location. In effect, Sections 4E and 4H do this with the mean, but the resulting calculations are special because no iteration is required. Section 6E of UREDA discusses other polishes for two-way tables.

Writing a higher-way table as a two-way table with composite factors and applying median polish often lead to useful analyses. Seheult and Tukey (1982) illustrate this approach in an example.

For the special class of higher-way tables known as 2^n factorial experiments, in which each of n factors has exactly two levels, Seheult and Tukey (1982) describe an exploratory approach based on applying a resistant location estimator to the set of comparisons (differences) associated with each main effect and interaction.

Missing Values

To deal with missing values or holes, we can follow much the same strategy as in two-way tables. If we have only a few holes, we can continue to use the algorithms for complete tables described in this chapter; we should expect to need a few more iterations. In the face of numerous holes, we would apply a generalization of the square-combining-table technique (Chapter 2) and pay some penalty in reduced resistance.

4J. SUMMARY

After reviewing the structure of the three-way table and the customary decompositions and models, we have described and illustrated median polish, which generalizes the simple steps that have formed the basis for resistant analyses of two-way tables. The main-effects-only model and the full-effects model serve as leading cases.

Discussions of the corresponding analyses using means point out that, in comparison with the traditional analyses based on means, the exploratory techniques trade additional computational effort for resistance—in the form of protection against leakage of the effects of an isolated wild value from its cell into the whole table or those of an isolated large interaction effect into the rest of the interaction term. When the data happen to be well behaved, the two approaches yield rather similar results.

When the data contain systematic patterns of nonadditivity, we show that generalizations of the two-way diagnostic plot make it easy to examine these patterns and to learn whether a power transformation would help to promote additivity.

A brief concluding section reviews computational issues, other techniques that use polishing, and treatment of missing values.

REFERENCES

Bartels, R. H. and Conn, A. R. (1980a). "Linearly constrained discrete l_1 problems," *ACM Transactions on Mathematical Software*, **6**, 594–608.

Bartels, R. H. and Conn, A. R. (1980b). "Algorithm 563. A program for linearly constrained discrete l_1 problems," *ACM Transactions on Mathematical Software*, **6**, 609–614.

Box, G. E. P. and Cox, D. R. (1964). "An analysis of transformations," *Journal of the Royal Statistical Society, Series B*, **26**, 211–252 (with discussion).

Brown, M. B. (1975). "Exploring interaction effects in the ANOVA," *Applied Statistics*, **24**, 288–298.

Ryan, T. A., Joiner, B. L., and Ryan, B. F. (1981). *Minitab Reference Manual*. University Park, PA: Minitab Project, The Pennsylvania State University.

Salthouse, T. A. (1984). "The skill of typing," *Scientific American*, **250**, 2 (February), 128–135.

Seheult, A. and Tukey, J. W. (1982). "Some resistant procedures for analyzing 2^n factorial experiments," *Utilitas Mathematica*, **21B**, 57–98.

Siegel, A. F. (1983). "Low median and least absolute residual analysis of two-way tables," *Journal of the American Statistical Association*, **78**, 371–374.

Snedecor, G. W. and Cochran, W. G. (1980). *Statistical Methods*, 7th ed. Ames, IA: The Iowa State University Press.

Velleman, P. F. and Hoaglin, D. C. (1981). *Applications, Basics, and Computing of Exploratory Data Analysis*. Boston, MA: Duxbury Press.

Additional Literature

Bloomfield, P. and Steiger, W. L. (1983). *Least Absolute Deviations: Theory, Applications, and Algorithms*. Boston: Birkhäuser.

Cox, D. R. (1984). "Interaction," *International Statistical Review*, **52**, 1–31 (with discussion).

McKean, J. W. and Schrader, R. M. (1980). "The geometry of robust procedures in linear

models," *Journal of the Royal Statistical Society, Series B*, **42**, 366–371.
Schrader, R. M. and Hettmansperger, T. P. (1980). "Robust analysis of variance based on a likelihood ratio criterion," *Biometrika*, **67**, 93–101.

EXERCISES

1. Try other orders of the factors in the pig feeding data (Table 4-2) to explore dependence of final median-polish main-effects-only fit on order. (In Section 4C an iteration calculates the median for each level of lysine, then the median for each level of methionine, and finally the median for each level of protein.)

(a) Take medians in the order methionine, lysine, protein.

(b) Take medians in the order protein, methionine, lysine.

2. A discussion of accountants' salary levels reported the median annual salary in the United States (Table 4-45) for each of the groups formed by three classification variables: the individual's position in the company, the size of the company, and whether the company is a firm of Certified Public Accountants or a corporation. Among CPA firms, size was defined as follows: large = national firm, medium = regional firm, small = local firm. Among corporations the divisions were based on annual sales: large = over $1 billion, medium = $50–200 million,

TABLE 4-45. **Median annual compensation (in $1000) of accountants in the United States, 1980, by position, size of company, and type of company.**

Position within Company	Small		Medium		Large	
	CPA Firm	Corporation	CPA Firm	Corporation	CPA Firm	Corporation
Junior	15.5	14.1	16.9	16.1	18.0	18.2
Senior	19.5	18.5	21.1	21.0	24.8	24.6
Manager	28.4	23.0	32.6	29.4	38.7	47.6
Partner or controller	37.0	28.2	54.7	40.6	87.7	98.2
Senior partner or chief financial officer	64.3	39.5	94.6	70.5	196.4	177.6

Source: *Savvy*, June 1980, p. 20 (adapted from Source Finance Personnel Services' *Accounting and Finance Salary Survey and Career Planning Guide for 1980*).

small = under $10 million. Use a three-way analysis to summarize the contributions of the three factors to median annual salary. Discuss the roles of choice of model and re-expression in the analysis.

3. An experiment to investigate the effects of the type of gold material and the method of condensation on the hardness of gold fillings produced the data in Table 4-46, analyzed by Brown (1975). Each of five dentists (D) used three methods of condensation (C) and eight types of gold material (G). Fit a full-effects model by median polish, and plot the sets of effects and the residuals (as in Figure 4-10). Examine any unusual two-factor interaction effects and discuss their impact on the full-effects fit by means. Would a main-effects-only fit by median polish seem likely to be satisfactory?

TABLE 4-46. Hardness of gold fillings prepared from eight types of gold by five dentists using three methods of condensation.

	Method of	Type of Gold							
Dentist	Condensation	1	2	3	4	5	6	7	8
1	1	792	824	813	792	792	907	792	835
	2	772	772	782	698	665	1115	835	870
	3	782	803	752	620	835	847	560	585
2	1	803	803	715	803	813	858	907	882
	2	752	772	772	782	743	933	792	824
	3	715	707	835	715	673	698	734	681
3	1	715	724	743	627	752	858	762	724
	2	792	715	813	743	613	824	847	782
	3	762	606	743	681	743	715	824	681
4	1	673	946	792	743	762	894	792	649
	2	657	743	690	882	772	813	870	858
	3	690	245	493	707	289	715	813	312
5	1	634	715	707	698	715	772	1048	870
	2	649	724	803	665	752	824	933	835
	3	724	627	421	483	405	536	405	312

Source: Morton B. Brown (1975). "Exploring interaction effects in the ANOVA," *Applied Statistics*, **24**, 288–298 (data from Table 1 on p. 290). His source: F. Xhonga (1971). "Direct gold alloys—Part II," *Journal of the American Academy of Gold Foil Operators*, **14**, 5–15. Reproduced by permission of the Royal Statistical Society.

4. Re-express the yarn breakage data (Table 4-19) in the square root scale, make a main-effects-only analysis by medians, calculate the comparison values according to equation (29), and make the diagnostic plot. What range of exponents seems plausible for a further re-expression?

5. For each of the three factors (length, amplitude, and load) in the yarn breakage data, relate the main effects from the main-effects-only analysis in the log scale (Table 4-23) to the numerical values of the levels. Do these levels also need re-expression? Interpret the relationships in terms of the physical situation. Do they suggest a regression model that would involve fewer parameters than the main-effects-only model?

6. Salthouse (1984) gives the average time (in milliseconds) that 10 skilled typists required to make each of 30 keystrokes on a standard typewriter keyboard (Table 4-47). Analyze these data by medians and discuss the effect of hand, row (on the keyboard), and finger on the keystroke time. Do interactions between the factors seem important? Optional: Use a table of the relative frequency of occurrence of the 26 letters in English prose to investigate whether keystroke time seems related to frequency

TABLE 4-47. Average time (in milliseconds) for typewriter keystrokes. The arrangement follows the layout of a standard keyboard. The time for the space bar was 225.2 msec.

	Upper		Middle		Lower	
Left hand						
	Q	221.0	A	202.7	Z	279.2
	W	211.4	S	226.3	X	226.5
	E	191.3	D	220.6	C	232.7
	R	194.5	F	194.4	V	259.3
	T	199.9	G	208.7	B	262.9
Right hand						
	Y	205.2	H	168.7	N	202.3
	U	196.1	J	183.5	M	226.2
	I	191.7	K	191.5	,	254.0
	O	190.5	L	209.1	.	298.0
	P	236.2	;	542.0		

Source: Timothy A. Salthouse (1984). "The skill of typing," *Scientific American*, **250**, 2 (February), 128–135 (data from p. 128).

of occurrence, once the effects of hand, row, and finger have been cleared away.

7. The *worst-case breakdown bound* (discussed for two-way tables in Section 6F of UREDA) gives the largest fraction of observations that can be turned into arbitrarily bad values without causing the estimate or fit to change greatly. Investigate this breakdown bound in a general three-way table for
   (a) the main-effects-only fit by median polish;
   (b) the main-effects-only fit by means;
   (c) the full-effects fit by median polish;
   (d) the full-effects fit by means.

8. Investigate the extent to which the full-effects median-polish fit for the pig feeding data (Table 4-2) depends on the choice of factors as A, B, and C in applying the algorithm of Section 4F. (Taking A = protein, B = lysine, and C = methionine yields the result in Table 4-40.) Determine the full-effects analysis by median polish for these orders of factors:
   (a) A = methionine, B = lysine, C = protein;
   (b) A = lysine, B = methionine, C = protein.

   Compare the results with Table 4-40.

9. Derive the expression for comparison values in a diagnostic plot that would aid in determining a re-expression to remove systematic three-factor interaction in a three-way table. (Assume that two-factor interactions remain.)

10. Assemble the three two-way tables giving perceived favorableness of adverb–adjective combinations (Tables 3-6 and 3-19) into a three-way table and analyze. Take guidance as appropriate from the fits discussed in Section 3F.

CHAPTER 5

Identifying Extreme Cells in a Sizable Contingency Table: Probabilistic and Exploratory Approaches

Frederick Mosteller
Harvard University

Anita Parunak
University of Michigan

To discover outliers and interaction in two-way tables of measurements, usually we study residuals that become available after applying some process of fitting to the data in the cells. When studying the same matters in contingency tables—tables of counts—we often face more features that need attention. We discuss here three methods of analyzing tables of counts, all oriented primarily toward identifying outliers.

Probabilistic Simulation

The first method builds on the theoretical variation in the cell entries we would have if the hypergeometric distribution were appropriate, as in Fisher's exact test, except possibly for a few outliers—if, in other words, the individual cases entered the table (i) individually at random, and (ii) without influence of row chosen upon column chosen or *vice versa*. Using that assumption, we develop a method of simulating contingency tables that allows us to see where the largest observed standardized residual from the actual data falls in comparison with a simulated approximation to the theoretical distribution, under independence, of such a largest residual. This approach readily generalizes to test other values such as the smallest, or the second largest, or jth largest or jth smallest values.

This first method employs a new standardization that avoids difficulties that classical standardizations show for cells with small expectations, difficulties common to instances of fitting hypergeometric, binomial, and Poisson distributions, as well as to more general contingency tables, as illustrated in Section 5G.

An Exploratory Analysis

A second method of identifying outliers, more nearly in tune with exploratory data analysis, allows the departures of the cell counts from proportionality to be approximately adjusted for the variability anticipated for hypergeometric distributions, and then uses a general rule of thumb discussed in UREDA (Section 2C) to identify extreme observations by their size in comparison with all residuals considered as a single batch.

This method may prove appropriate for data that depart substantially from the hypergeometric model, not just in a few outliers, but more generally. Like the first method, this method neglects the perturbation of its row and column totals by an outlier.

Allowing for Perturbed Totals: The Single-Outlier Case

Templeton (1956) examines the problem of looking for a single outlier in a two-way contingency table. He recognizes that the outlier will have perturbed the row and column totals. To avoid the resulting difficulty, he treats the count in the cell as if it were a missing entry. The formulas become fairly complicated, and he does not pursue the possibility of two or more outliers.

Allowing for Perturbed Totals: A Resistant Analysis

The third method (given in Section 5F) employs a resistant analysis that does not require the fitted cell entries to add up to the observed row and column totals. Releasing these constraints gives us a more flexible way to study residuals and look for patterns in the data, one even more in tune with exploratory data analysis. Like the second method, it compares suspiciously large residuals with the general size of all residuals taken as a single batch.

TABLE 5-1. Contingency table giving the number of artifacts by distance to permanent water.

	Distance to Permanent Water						
Artifact	$j=1$ Immediate Vicinity	2 Within $\frac{1}{4}$ Mile[a]	3 $\frac{1}{4}$ to $\frac{1}{2}$ Mile	4 $\frac{1}{2}$ to 1 Mile	5 1 to 3 Miles	6 Over 3 Miles	Row Sum x_{i+}
i							
1 Specialized unifaces	20	102	54	38	29	3	246
2 Unifaces 2 or more edges retouched	33	136	86	58	56	7	376
3 Unifaces 1 edge retouched	27	122	68	51	53	0	321
4 Limited bifacial retouched	2	10	8	5	4	0	29
5 Large, heavy tools	11	82	34	35	30	2	194
6 Whole biface	10	53	25	17	17	3	125
7 Round biface, base snapped, side notched	39	185	88	100	58	13	483
8 Pointed biface, base snapped, side notched	34	179	70	78	60	11	432
9 Rectangular biface, base snapped, side notched	26	78	24	26	14	6	174
10 Biface midsection	24	88	32	41	26	3	214
11 Humboldt Pinto Northern (projectile points)	8	44	16	28	39	3	138
12 Elko gypsum (projectile points)	15	75	30	35	27	8	190
13 Eastgate and Rose Spring (projectile points)	11	32	5	11	21	2	82
14 Cottonwood Desert, side notched	12	28	5	18	7	0	70
15 Drills	2	10	4	2	6	0	24
16 Pots	3	8	4	6	8	0	29
17 Hammerstones	1	2	0	3	3	0	9
18 Grinding stones	13	5	3	9	7	0	37
19 Point fragments	20	36	19	20	28	1	124
Column sum, x_{+j}	311	1275	575	581	493	62	3297
Number of sites	42	174	90	99	121	11	537

[a] These counts exclude those in the Immediate Vicinity column.

Source: Laurel Casjens (1974). "The Prehistoric Human Ecology of Southern Ruby Valley, Nevada." Doctoral dissertation, Department of Anthropology, Harvard University. Reproduced here with the permission of Laurel Casjens.

An Archaeological Example

Table 5-1 shows a large (19×6) contingency table giving the counts of 19 kinds of artifacts found at various distances from permanent water in southern Ruby Valley, Nevada. The anthropologist, Laurel Casjens (1974), wanted to know whether the counts in some cells departed extremely from the results that the theory of independence would predict. We do not plan to use the ordering of the distances in our analysis here, since we are looking for outliers scattered around the table, rather than for trends. Although classical tests for independence do not form part of our discussion, the reader may be interested to know that the usual chi-squared statistic on 90 ($= 18 \times 5$) degrees of freedom is 190, corresponding to a unit Gaussian deviate of roughly 7.5.

Dr. Casjens (personal communication) discusses the reason for choosing distance from permanent water as a dimension of the table:

> The purpose of this study was to determine the association between the locations of certain natural resources and the types of sites utilized by the prehistoric hunter–gatherers who inhabited the area. Water is a vital resource, one that is difficult to transport, and one that is relatively scarce and localized. We chose permanent water sources because ephemeral sources are dry most of the year and cannot be relied on. While ephemeral sources would have been used, we are unable to determine if sites near them were in use during wet or dry periods.
>
> We expected to find a correlation between proximity to water and certain site types (as defined by kinds of artifacts present). Results of the study show that, in most cases, tool assemblages were associated with one or more micro-environments, some of which involve water.

Casjens has additional information about the sites, and so we asked her to comment here and later in this chapter.

5A. THE HYPERGEOMETRIC DISTRIBUTION

For a contingency table with r rows and c columns, we denote the observed count in a cell by

$$x_{ij}, \qquad i = 1, 2, \ldots, r, \qquad j = 1, 2, \ldots, c.$$

For the row, column, and grand totals, we use the notation

$$x_{i+} = \sum_{j=1}^{c} x_{ij}, \qquad x_{+j} = \sum_{i=1}^{r} x_{ij}, \qquad x_{++} = \sum_{i=1}^{r} \sum_{j=1}^{c} x_{ij}.$$

At times we also use N for x_{++}. Because we focus on the distribution when the row chosen and the column chosen are independent, we can collapse the data to form a 2×2 table with x_{ij} in the upper left-hand corner and totals for the other rows and columns in the other three cells, as shown in Table 5-2.

Most analyses of contingency tables treat the counts as if the marginal totals were fixed. Under random sampling with fixed marginal totals x_{i+} and x_{+j}, the probability that the random variable X_{ij} equals x_{ij}, $P\{X_{ij} = x_{ij}\}$, is given by either equivalent form of the hypergeometric probability distribution:

$$\frac{\binom{x_{+j}}{x_{ij}}\binom{x_{++} - x_{+j}}{x_{i+} \quad x_{ij}}}{\binom{x_{++}}{x_{i+}}} \quad \text{and} \quad \frac{\binom{x_{i+}}{x_{ij}}\binom{x_{++} - x_{i+}}{x_{+j} - x_{ij}}}{\binom{x_{++}}{x_{+j}}}, \tag{1}$$

where

$$\binom{n}{k} - \frac{n!}{k!\,(n-k)!}$$

is the usual binomial coefficient. The mean and variance of X_{ij} are

$$E(X_{ij}) = \frac{x_{i+}x_{+j}}{x_{++}} = \mu_{ij} \tag{2}$$

TABLE 5-2. The 2×2 table formed from cell (i, j) and the totals for other rows and other columns.

	Column j	Other Columns	Totals
Row i	x_{ij}	$x_{i+} - x_{ij}$	x_{i+}
Other rows	$x_{+j} - x_{ij}$	$x_{++} - x_{i+} - x_{+j} + x_{ij}$	$x_{++} - x_{i+}$
Totals	x_{+j}	$x_{++} - x_{+j}$	x_{++}

and

$$\operatorname{var}(X_{ij}) = \frac{x_{i+}(x_{++} - x_{i+})x_{+j}(x_{++} - x_{+j})}{x_{++}^2(x_{++} - 1)} = \sigma_{ij}^2. \tag{3}$$

Consequently, when we check whether a particular cell, chosen in advance, has an unusual size, we refer it to the hypergeometric distribution to get a P-value. For computational convenience, we often substitute for the hypergeometric one of its approximations such as the normal approximation or the chi statistic for the 2×2 table. For independence and random sampling, the joint distribution of all X_{ij} is known but complicated, and we need not write it here.

Omitting an Entry

Now that we have a notation, we can consider what entry we might obtain if we treated a cell as missing. Essentially we want to reduce row, column, and grand totals by the cell entry x_{ij}. We then insert in the cell and add to the reduced row, column, and grand totals a quantity w which would be reproduced under the assumption of independence. If we attach an * to a symbol to indicate that x_{ij} has been subtracted from it, then the new row, column, and grand totals are $x_{i+}^* + w$, $x_{+j}^* + w$, and $x_{++}^* + w$. The independence assumption leads to

$$w = \frac{(x_{i+}^* + w)(x_{+j}^* + w)}{x_{++}^* + w}.$$

Solving this equation for w gives, for the imputed value of x_{ij}

$$w = \frac{x_{i+}^* x_{+j}^*}{x_{++}^* - x_{i+}^* - x_{+j}^*}.$$

For example, if we omit the upper left entry in the table

10	100	110
100	200	300
110	300	410

then $x_{11} = 10$, $x_{1+}^* = 100$, $x_{+1}^* = 100$, $x_{++}^* = 400$, so that independence would suggest the value

$$w = \frac{(100)(100)}{200} = 50$$

in place of the original 10. Now that this approach has shown us how crucial removal can be, we shall not use this missing-value approach further. In due course, we will replace it with the method described in Section 5F.

Random Sampling Models

The probabilistic simulation or "random sampling" approach treats every possible allocation of the x_{+j} individuals in column j and of the $x_{++} - x_{+j}$ individuals in the "other columns" to the two rows (labeled "Row i" and "Other rows" in Table 5-2) as equally likely. In an alternative interpretation, it treats every possible allocation of the x_{i+} individuals in row i and of the $x_{++} - x_{i+}$ individuals in the "other rows" to the two columns (labeled "Column j" and "Other columns" in Table 5-2) as equally likely. Both interpretations reflect the constraints that row and column sums are fixed.

The two forms in equation (1) correspond to these two interpretations. We speak of this situation as random sampling or as the situation "under independence." This language flows from the idea that, in a 2 × 2 probability table with events A and not-A versus events B and not-B, we can compute the probability that both A and B occur as $P\{A\}P\{B\}$ when A and B are independent. The mean proportion of elements in cell (i, j) under random sampling is just

$$\frac{x_{i+}}{N}\frac{x_{+j}}{N},$$

which corresponds to the product of the two probabilities. Because independence is our null assumption in the original $r \times c$ table, we get independence, as a null assumption, in each collapsed 2 × 2 table.

5B. ASSESSING OUTLIERS

To search for outliers in a two-way contingency table, we need to handle a variety of matters. Some cell counts tend to be large or small because their row and column totals are large or small. Second, we might want to use average cell counts for a reasonable probability model as a base line against which to measure residuals. Third, even with such a base line, large numbers usually tend to vary more than small ones, and we should also allow for that. Fourth, the discreteness of the counts in a contingency table complicates comparisons among distributions of cell values. Fifth, the shapes of the distributions may create trouble. Sixth, the problem of multiplicity created by the obligation to focus on extreme values causes further trouble. Thus we must take several preliminary steps before we can discuss outliers in a contingency table.

In our problem, far from picking out a cell in advance, we plan to scan all the cells, looking for the more extreme ones. Partly for this reason, standard approaches, some of which we discuss below, will not alone deal with our problem.

Naive Extremeness

How shall we measure the extent of departure from independence? One method could express the difference of x_{ij} from $E(X_{ij})$ in standard-deviation units. This tends to bring comparability to the distributions of the entries in large and small cells because it adjusts for both expected size and variability. We expect cells with large average counts to have larger variances. We can see this from

$$\frac{\operatorname{var}(X_{ij})}{E(X_{ij})} = \frac{(x_{++} - x_{i+})(x_{++} - x_{+j})}{x_{++}(x_{++} - 1)},$$

in which the right-hand side is less than (but often close to) 1.

Standardization of the entry would then allow both for the expected size of the count, by subtracting it, and for the variability, by using its square root as a unit of measurement. The usual standardization is the standard score or standard deviate

$$z_{ij} = \frac{x_{ij} - E(X_{ij})}{\sqrt{\operatorname{var}(X_{ij})}} = \frac{x_{ij} - \mu_{ij}}{\sigma_{ij}}. \tag{4}$$

Then the random variable Z_{ij} has mean 0 and variance 1.

When μ_{ij} is good-sized, but not a large fraction of either x_{i+} or x_{+j}, then Z_{ij} approximately follows the standard normal distribution—subject, of course, to the usual awkwardness that the discreteness of x_{ij} presents. To the extent that (a) this approximate normality holds and that (b) we may ignore discreteness, equal zs for different cells correspond to equal probabilities in the tails beyond the observed z.

When we want to approximate the probability that $X_{ij} \geq x_{ij}$ (a common calculation for x_{ij} in the upper tail), we usually subtract $\frac{1}{2}$ from the numerator of the rightmost expression of equation (4) before entering the standard normal table. This adjustment is the familiar "half-correction."

In discussing a value of z, we want to consider its extremeness in the distribution. When all cells have identical distributions for Z, we can make such interpretations easily. When the row totals are equal to one another and the column totals are similarly equal among themselves, and the table is randomly constructed, all cell entries have the same distribution. Ordinarily the totals for rows are not equal, nor are the totals for columns. The usual standardization, including the $\frac{1}{2}$ correction, thus takes three steps in the direction of producing equivalent distributions by adjusting for mean and standard deviation and helping with discreteness, although it does nothing for shape of the distribution or for multiplicity.

Other Standardizations

Commercial statistical packages such as the BMDP programs (Dixon et al., 1981) give a variety of ways of standardizing residuals in contingency tables, as do Bishop, Fienberg, and Holland (1975). The methods that are close to the one presented here are the chi statistic (square root of the contribution to chi-squared) defined as

$$(x_{ij} - \mu_{ij})/\mu_{ij}^{1/2}$$

and the standardized deviates proposed by Haberman (1973). These latter use the variance of the hypergeometric given in expression (3) except that they neglect the -1 in the denominator. (Haberman derived the formula from another approach.) This change makes it possible to write the residual as

$$(x_{ij} - \mu_{ij})\Big/\big[\mu_{ij}(1 - x_{i+}/x_{++})(1 - x_{+j}/x_{++})\big]^{1/2}.$$

For large x_{++} these residuals will be much the same as the standardization for the hypergeometric

$$(x_{ij} - \mu_{ij})/\sigma_{ij},$$

which gives zero mean and unit variance.

Another popular standardization uses the Freeman–Tukey deviate (Freeman and Tukey, 1950; Bishop, Fienberg, and Holland, 1975, p. 137)

$$\sqrt{x_{ij}} + \sqrt{x_{ij} + 1} - \sqrt{4\mu_{ij} + 1},$$

which behaves similarly to the standard score when μ_{ij} is not too small. (For computing normal approximations to $P\{X_{ij} \geq x_{ij}\}$, replacing x_{ij} by $x_{ij} - .5$ in the above formula should improve the accuracy when $x_{ij} \neq 0$.)

Problems with Normal Approximations

Thus far we have discussed extremeness within a cell. We propose to scan all the cells, looking for the most extreme cell among many. Lack of independence of the cells in a given row or column now afflicts us. If one cell is larger than its expected value, then other cells in its row and column must tend to be smaller because the row and column totals are fixed.

If the expected value of X_{ij} is small, say 1, then the likely values of x_{ij} are few: $0, 1, 2, 3, 4$, perhaps, and some individual values will have to have large probabilities. Whereas, when the expected value of X_{ij} is large, say 50 (but not close to either its row total or its column total), we might expect the values of x_{ij} that are likely to occur to range over 30 or 40 numbers, say from 30 to 70. Now, no individual value need have a large probability. Consequently, no standardization, including expression (4), can achieve equivalence of distribution. For some cells the distributions have to be more lumpy than for others.

We are not going to get all our distributions alike, but we do need to avoid the greatest difficulties. A few trials easily convince us that the greatest difficulty with the classical standardization arises in the upper tail when μ_{ij} is small.

As part of a study of improved standardizations for Poisson and binomial distributions, as well as for the hypergeometric distribution, Parunak (1983) has introduced what we here call the G-standardization, which uses

$$\frac{x_{ij} - \mu_{ij}}{G_{ij}\sigma_{ij}}, \tag{5}$$

where

$$G_{ij} = 1 + e^{-\mu_{ij}}.$$

We illustrate some of the advantages of the G-standardization in Section 5G.

Having chosen our standardization—our measure of deviation of x_{ij} from μ_{ij} for each cell—we have to face up to the difficulties associated with

1. looking at the largest such measure, and
2. drawing a random sample, under independence, of contingency tables with the same marginal totals as the observed table.

Once we have fixed our measure of deviation, and agreed to look at its largest values—and once we know how to conduct our simulation, generating random contingency tables with the observed margins—we can compare our single observed largest with the observed distribution of the random largest, one such for each randomly generated table. This comparison is a legitimate way to test our observed largest against the hypothesis of independence, no matter how we choose our standardization, no matter how well or poorly the null distributions of the standardized residuals for the different cells agree. We will make more sensitive tests, however, and reduce the number of unreasonable answers, by making the distributions more nearly alike.

We were careful about out choice of standardization because we wanted better sensitivity and fewer unreasonable answers. We relied on the hypergeometric distribution in two quite different ways: (a) in choosing a standardization to make the cell distributions more nearly alike, and (b) in doing our simulations.

A Naive Approach to the Problem of Multiplicity

One approach by formula to handling the extremeness is the Bonferroni method. To assess how extreme the largest cell is out of 114, we multiply its observed naive tail area from the normal table by 114 (or more generally rc), or something close to this. If the 114 observations were independent and were drawn from an approximately standard normal distribution, this would be a reasonable move. In the actual problem, we have some correlation because of the fixed sums in the rows and the columns. Possibly multiplying by $(r - 1)(c - 1)$ would be more appropriate to take some account of this correlation.

5C. THE SIMULATION APPROACH

To exemplify the actual situation under random sampling, we ran 500 simulated tables having the same row totals and column totals as in Table 5-1. We turn now to that actual simulation.

To decide in the archaeological (or any other) example whether an observed largest standardized residual is big compared to what independence (random sampling) would predict, we use the following steps in a simulation algorithm:

1. *Generate tables.* We randomly generate 500 contingency tables from the same fixed margins as the archaeological table, under the assumption of independence.
2. *Apply an appropriate standardization.* We transform the entries in each of the 500 tables to appropriate standard scores, perhaps using $(x_{ij} - \mu_{ij})/\sigma_{ij}(1 + e^{-\mu_{ij}})$, which is formula (5).
3. *Record largest entry.* We call the largest standard score in each new table the largest order statistic. We record that largest value from each table.
4. *Form distribution of largest entries.* These largest values, 500 of them, form an empirical distribution, the simulated distribution of the largest order statistic. For a random table with the same fixed margins as the original table, that distribution gives us a way to estimate the probability that the largest standard score would be larger than a given value. Similarly, the second largest values in each of the 500 tables form an empirical distribution; as do the third largest, and so on.
5. *Compare observed value with simulated distribution.* Then we bring out the observed largest standard score in the archaeological table. We look at it against the background of the simulated distribution of the largest order statistic. If it falls either entirely above, or quite high in the tail of, the simulated distribution, we conclude that it is clearly and demonstrably unusually large.

We chose 500 simulations as a compromise between cost and accuracy. The standard deviation of the proportion observed to the right of the true upper 5% point of a distribution is a little less than $\sqrt{0.05/500} = 0.01$, or 1%, and so the estimated 5% point is likely to lie between the true 3% and 7% points. To get a 95% probability of being within 1% in either direction would require 2000 simulations.

Generating Random Tables

Just how to construct a random contingency table that winds up with fixed row and column totals may not leap to mind. A small numerical example gives the thread of the method we used, and it generalizes easily to larger tables.

EXAMPLE:

Suppose that we want to generate a table with the fixed margins (totals):

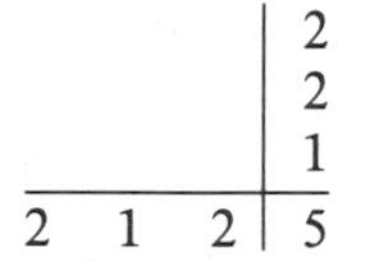

			2
			2
			1
2	1	2	5

We convert the cumulations, 2, 3 (and 5) of the column totals, 2, 1, 2, to numbers ranging between 0 and 1, because uniform random number generators produce numbers in this interval. We first cumulate and then divide by the overall total. So the column margins yield $2/5$, $(2 + 1)/5$, $(2 + 1 + 2)/5$ or .4, .6, 1.0, respectively. Then, if the first random number, u, were less than or equal to .4, the first column would receive one count; if u were greater than .4 but less than or equal to .6, the second column would receive one count; and if u were greater than .6, the third column would receive one count.

Now we might use the same type of cumulative proportions to choose the row that will receive the count. However, we are able to make a simplification. (We do not present here the algebra justifying the simplification.) We can allow the first row to get counts until it is full—that is, until the number of counts generated in the row equals the original first-row total. Then we move down and fill up the second row. The column proportions change after each step, as do the column and row counts.

Thus if the first four random numbers in our example are .57, .61, .32, .41, the steps, beginning at the empty table with fixed margins, look like:

				2
				2
				1
	2	1	2	5
Cumulative proportions	.4	.6	1.0	

(i) Random number .57 chooses second column (of first row) and leads to

		1		1
				2
				1
	2	0	2	4
Cumulative proportions	.5	.5	1.0	

Total in row 1 reduced by 1. Total in column 2 reduced by 1.

Note that, at the bottom of the table, we now have a new set of cumulative proportions, to which we compare the second random number, in order to place the corresponding count.

(ii) Random number .61 chooses third column (of first row)

	0	1	1	0
				2
				1
	2	0	1	3
Cumulative proportions	$.6\dot{6}$	$.6\dot{6}$	1.0	

Row 1 emptied.
Column 3 reduced by 1.

(The dot over a final 6 means that they go on forever.)

(iii) Random number .32 chooses first column (of second row)

	0	1	1	0
	1			1
				1
	1	0	1	2
Cumulative proportions	.5	.5	1.0	

Row 2 reduced by 1.
Column 1 reduced by 1.

(iv) Random number .41 chooses first column (of second row)

	0	1	1	0
	2	0	0	0
				1
	0	0	1	1
Cumulative proportions	.0	.0	1.0	

Row 2 emptied.
Column 1 emptied.

And the last row must be 0, 0, 1, whatever the uniform random number chosen.

Generalization. To give a precise statement of how one generates a single random $r \times c$ table **X1** whose margins coincide with those of a given table **X**, we proceed algorithmically in a sequence of steps. (The symbol $\leftarrow$ denotes the operation of assigning the value on the right of $\leftarrow$ to the variable on the left of $\leftarrow$.)

ALGORITHM T (RANDOM TABLE WITH FIXED MARGINS): Given the nonzero marginal totals x_{i+} $(i = 1, \ldots, r)$, x_{+j} $(j = 1, \ldots, c)$, and x_{++}, this algorithm fills each row in turn, updating the column margins CM_j and overall margin OM after placing each random observation. The row margin RM_i serves as a counter to indicate when row i is full.

T1. [Initialize.] Set $RM_i \leftarrow x_{i+}$ for $1 \leq i \leq r$, $CM_j \leftarrow x_{+j}$ for $1 \leq j \leq c$, and $OM \leftarrow x_{++}$. Set $\mathbf{X1}(i, j) \leftarrow 0$ for $1 \leq i \leq r$ and $1 \leq j \leq c$. Set $i \leftarrow 1$.

T2. [Calculate probabilities corresponding to current column margins.] Set $p_0 \leftarrow 0$ and

$$p_j \leftarrow \sum_{k=1}^{j} CM_k/OM \quad \text{for } 1 \leq j \leq c.$$

T3. [Choose column randomly according to current probabilities.] Generate a random number u from the uniform distribution on $(0, 1)$. Find the integer t such that $p_{t-1} < u \leq p_t$, and set $\mathbf{X1}(i, t) \leftarrow \mathbf{X1}(i, t) + 1$.

T4. [Reduce margins.] Set $RM_i \leftarrow RM_i - 1$, $CM_t \leftarrow CM_t - 1$, and $OM \leftarrow OM - 1$.

T5. [Current row filled?] If $RM_i > 0$, go back to step **T2**.

T6. [Ready for last row?] If $i < r - 1$, set $i \leftarrow i + 1$ and go back to step **T2**.

T7. [Fill last row.] Set $\mathbf{X1}(r, j) \leftarrow CM_j$ for $1 \leq j \leq c$.

T8. [Output.] Deliver the simulated table **X1**.

As the 3×3 example presented earlier illustrates, the last row can be placed in **X1** without generating any random numbers.

Simplified Notation

As we generalize our approach, because we examine the largest of the $n = r \times c$ values, it will be convenient if we do not have to carry the double subscripts throughout. We plan to use x_i as the cell entry and number the cells from left to right and from top to bottom so that we have $x_1, \ldots, x_n$. Of course, the detailed calculations use the double subscripts, but our exposition need not.

The Probabilistic Simulation

1. *Random generation of tables.* We use the marginal totals from the original table, **X**, for the r rows and for the c columns. Then we randomly generate a table, **X1**, of counts with the fixed margins.†

†As the source of uniform pseudorandom numbers, we used the multiplicative congruential generator of Lewis, Goodman, and Miller (1969) recommended to us by David C. Hoaglin, who has examined the theoretical properties of several congruential random-number generators (Hoaglin, 1976).

2. *Standardization.* We make the distributions in the cells more nearly comparable by transforming the entries within each table with the new standardization. For the original table, **X**, the new standardized values are

$$y_i = \frac{x_i - \text{expected}}{(\text{standard deviation})(1 + e^{-\text{expected}})}$$
$$= \frac{x_i - \mu_i}{\sigma_i(1 + e^{-\mu_i})} = \frac{x_i - \mu_i}{\sigma_i G_i}, \qquad (6)$$

where G_i is $1 + e^{-\mu_i}$. (The constant e is 2.718..., the base of the natural logarithms.) These new standardized values appear in table **Y**. We also transform the counts in simulated table **X1** to standardized values in table **Y1**. Similarly, **X2** is transformed to **Y2**, and we continue this to get 500 tables, **Y1** to **Y500**. Note that the expectations and the standard deviations [from equation (3)] of the cells need not be recomputed for each table, because the margins are fixed.

3. *Order statistics.* We consider the $n = r \times c$ standardized values $y_1, y_2, \ldots, y_n$ in table **Y**. When ordered from least to greatest, the values $y_1, y_2, \ldots, y_n$ may be relabeled according to their ranks with parentheses in the subscripts to indicate order statistics:

$$y_{(1)} \leq y_{(2)} \leq \cdots \leq y_{(n)}.$$

4. *Distribution of the order statistics from the simulated data.* Likewise, we can order the entries in each of the 500 simulated tables from least to greatest. For example, in simulated table **Y1**, we label these entries as:

$$y^1_{(1)} \leq y^1_{(2)} \leq \cdots \leq y^1_{(n)},$$

with the superscript indicating the number of the simulation, not a power.

If we record the largest values, $y^1_{(n)}$ from simulated table **Y1**, $y^2_{(n)}$ from table **Y2**,..., and $y^{500}_{(n)}$ from table **Y500**, we can construct the simulated empirical distribution of the largest order statistic under independence. In the same manner, the simulated distribution of the kth order statistic is made up of the 500 values, $y^1_{(k)}, y^2_{(k)}, \ldots, y^{500}_{(k)}$, one from each of the 500 tables.

We want to know whether the kth order statistic, $y_{(k)}$, in the original table, **Y**, is unusually large (or small) among all possible kth order statistics

TABLE 5-3. Three hypothetical distributions of the 39 Humboldt Pinto Northern artifacts "1 to 3 miles" from permanent water among the 121 sites (from Table 5-1).

	Number of Sites		
Number of Artifacts	*A*	*B*	*C*
2	0	13	19
1	39	13	1
0	82	95	101
Total	39	39	39
SSD	26.4	52.4	64.4
$\frac{\text{Estimated variance}}{\text{Poisson variance}}$	68%	134%	165%

in $r \times c$ tables with the same fixed margins. So we compare the observed $y_{(k)}$ with the distribution of the kth order statistic under our standard assumptions. For example, if the largest order statistic, $y_{(n)}$, in the original table is above an upper 2.5% point of the distribution of simulated values, then $y_{(n)}$ is considered to be unusually large relative to an estimated one-sided 2.5% level.

One Way That Independence May Fail

Although we are not in a position to treat the matter here, we should point out that, in the archaeological example, the distribution of counts over sites within cells could have a substantial effect on an analysis. For example, the column "1 to 3 miles" to permanent water in Table 5-1 comes from counts at 121 sites. If we choose a particular cell in this column, say that for Humboldt Pinto Northern, we find a total of 39 artifacts. These 39 could be clustered among some of the 121 sites in many ways. Table 5-3 shows three such possible ways. "Common sense" suggests that the last of the three is much more likely than the first, but we need to be prepared for any of them.

Such different clusterings can lead to different estimates of the variance of the count associated with the cell. If the distribution were Poisson, which is approximately what our main analysis uses, then the variance associated with the cell would be estimated by the count itself. The corresponding variance estimates for the sum of 121 observations from the three hypothetical distributions appear on the line labeled SSD. (The estimated variance is the sum of squares of the deviations because the variance of a sum of independent items is the sum of the variances.) Finally, from the ratio of the estimated variance to the Poisson value, we see that this ratio varies among these three examples by a factor of about 2.5. Thus the clustering by sites could make a substantial difference in the analysis. Except for noting its existence, we do not deal with this problem here.

Note also, however, that when we come to consider Dr. Casjens's explanations for possible unusual cells, in all but one case her preferred explanations involve nonrandom behavior of sites. (The other case involves nonrandom behavior— across sites—of arrowhead collectors.)

Such "clustering" is often of importance in understanding a contingency table.

5D. APPLYING THE SIMULATION APPROACH TO THE TABLE OF ARCHAEOLOGICAL DATA

We now apply our method of identifying unusually large values to the 19×6 archaeological table, Table 5-1. Table 5-4 shows the G-standardized values $y_i = (x_i - \mu_i)/\sigma_i G_i$ that result from applying equation (5) or (6) to the counts in Table 5-1. The largest value, $y_{(n)}$, is 5.22 in the cell in row 18, column 1.

Illustrative Calculation for Row $i = 18$, Column $j = 1$

$$x_{ij} = 13,$$

$$x_{i+} = 37, \qquad x_{++} - x_{i+} = 3260,$$

$$x_{+j} = 311, \qquad x_{++} - x_{+j} = 2986,$$

$$x_{++} = 3297,$$

$$\mu_{ij} = 37(311)/3297 = 3.49,$$

$$\sigma_{ij} = \sqrt{\frac{37(3260)(311)(2986)}{3297^2(3296)}} = 1.77,$$

$$G_{ij} = 1 + e^{-\mu_{ij}} = 1.03,$$

$$\text{new standardized residual} = \frac{13.0 - 3.49}{1.77(1.03)} = 5.22.$$

TABLE 5-4. G-standardized residuals from the model of unclustered independence for the archaeological data in Table 5-1.

Artifact	Immediate Vicinity	Within $\frac{1}{4}$ Mile	$\frac{1}{4}$ to $\frac{1}{2}$ Mile	$\frac{1}{2}$ to 1 Mile	1 to 3 Miles	Over 3 Miles	Row Sum
1	−.73	.94	1.94	−.93	−1.45	−.79	246
2	−.46	−1.06	2.95	−1.19	−.03	−.03	376
3	−.66	−.26	1.86	−.86	.82	−2.60	321
4	−.44	−.46	1.44	−.05	−.18	−.47	29
5	−1.85	1.06	.03	.16	.21	−.88	194
6	−.56	.87	.77	−1.20	−.43	.40	125
7	−1.10	−.18	.49	1.92	−1.96	1.42	483
8	−1.19	1.26	−.73	.25	−.66	1.09	432
9	2.55	1.71	−1.30	−.95	−2.62	1.51	174
10	.92	.76	−.99	.61	−1.19	−.52	214
11	−1.49	−1.67	−1.85	.84	4.48	.24	138
12	−.75	.23	−.62	.30	−.30	2.37	190
13	1.25	.07	−2.74	−1.01	2.74	.31	82
14	2.23	.23	−2.30	1.80	−1.17	−.92	70
15	−.17	.30	−.10	−1.18	1.39	−.42	24
16	.16	−1.23	−.52	.43	1.92	−.47	29
17	.12	−.98	−1.14	1.03	1.55	−.22	9
18	5.22	−3.16	−1.50	1.07	.68	−.56	37
19	2.60	−2.25	−.63	−.44	2.43	−.82	124
Column sum	311	1275	575	581	493	62	3297

Note: The standardized values are

$$y_{ij} = \frac{x_{ij} - \mu_{ij}}{\sigma_{ij} G_{ij}},$$

where $G_{ij} = 1 + e^{-\mu_{ij}}$, in each cell. x_{ij} are the counts from Table 5-1; μ_{ij} and σ_{ij} come from formulas (2) and (3), respectively.

Of course, 5.22 is a very large value for a standard normal observation. However, we want to find out whether it is a larger standardized residual than we can expect any cell to have in any table with these margins. It is a question of comparing multiple large values. These would follow the non-normal distribution of the largest order statistic of a sample from the standard normal, if the standardization were perfect. In practice, matters may be somewhat more complicated. But simulation will take us through without trouble.

We generated 500 tables from the margins given in the observed archaeological table, and we standardized the counts within each simulated table

TABLE 5-5. The simulated distribution of the largest order statistic in simulated tables, Y1, Y2, . . . , Y500. Stem-and-leaf display.

Depth	Stem	Leaves (unit .01)
2	16	37
5	17	359
8	18	899
30	19	0013334577777788889999
42	20	001444678888
66	21	111223333333334455555888
108	22	000000111122222234555566666777788888899999
148	23	1122224444455555555567778888888888888999
200	24	0000011111111112233333333333444444555556666677778888
(64)	25	0001111111222222222222333334444455555555666666666667777888888888
236	26	000011111111112233344444444445577778888888999
192	27	1111223333333445568899
170	28	0001112222333334555555666666777777889999
131	29	00000000012445555566778888899999
99	30	000001122245555666688889
75	31	001122555555666799
57	32	111123333344468
42	33	00557779
34	34	046778899
25	35	003778
19	36	044567
13	37	125
10	38	4
9	39	067
6	40	5
5	41	26
3	42	3
2	43	2
1	44	7

according to formula (5). Tabulating the *largest* standardized value from each table creates a simulated distribution of the largest order statistic. Table 5-5 shows this distribution in a stem-and-leaf display. We briefly explain the stem-and-leaf display, which UREDA (Chapter 1) describes in detail.

Stem-and-Leaf Display

The stem-and-leaf display is like a histogram on its side (a histogram turned clockwise through 90°). It has three main column types. The third "column" gives digits (leaves) at a suitable decimal place in the data; here the leaves are multiples of .01. The second column gives the stems. Together, one stem and one leaf reproduce the initial digits of one measurement. Thus in line 1, the 3 at 16| means 1.63, and in line 2 the 5 at 17| means 1.75. The first column counts the number of observations or leaves from an end of the display. Thus in the first line, 2 is the number of leaves on stem 16; and in the second line, 5 is the number of leaves up to and including the 17 stem. The median of the batch of 500 occurs at stem 25, 50.5 digits to the right. Thus the median is 2.56. The parentheses around the number of leaves (64) on the 25 stem mark both that the median occurs in that line and that the number 64 is a count for the line and not a depth from an end.

The simulated distribution is formed from the standardized variables in expression (5), which are approximately standard normal but are not independent. If we did not have a G-standardization to help handle the large values in distributions with small means, this distribution of the first order statistic would be markedly lumpy where it ought not to be lumpy.

Outside the Distribution

Now we are in a position to find out whether the largest order statistic, $y_{(114)} = 5.22$, in cell (18, 1) of table **Y** (Table 5-4), from the archaeological data is unusually large, in comparison with all possible largest values in $r \times c$ tables with the same fixed margins. Comparing $y_{(114)}$ with the simulated distribution of the largest order statistic in Table 5-5, we find that $y_{(114)}$ is substantially larger than all the values of the simulated distribution. The largest of the 500 simulated values was 4.47.

Percent Points

We illustrate a second use of Table 5-5 by choosing the 97.5 percent point in the simulated distribution as a cutoff point.

For 500 points, the 97.5 percent point will be 12.5 (= .025 × 500) values in from the high side. The 12th value in Table 5-5 for the first order statistic is 3.72 and the 13th value is 3.71, and so the 97.5 percent point is 3.715. Table 5-6 lists this, along with the observed value 5.22.

A similar table corresponding to Table 5-5 was constructed for each of the next three order statistics after the largest. Table 5-6 gives the 97.5 percent points from the distributions of the four largest order statistics and the observed order statistics from the archaeological data (Table 5-4). We expect to focus attention on a small number of high-value order statistics, because we are looking for a few high outliers. The second most extreme is at $k = 113$, cell (11, 5). Its observed deviation of 4.48 also exceeds 4.47, the largest value from any of the simulations, as well as the 97.5 percent point of 3.715 for the largest deviation. Thus it too ought to be taken as almost certainly unusually large.

Once we have seen that a few, in our case the largest and second largest, observations almost certainly are outliers and cannot be regarded as merely the highest two observations of 114 from the hypergeometric distribution, we are no longer as well prepared to work with the $y_{(k)}$ further down the list. One thing we might do is withdraw the cells containing these two large observations from Table 5-1 and then rerun the simulation based on the

TABLE 5-6. The 97.5 percent points of randomly generated distributions of the four largest order statistics (k = 114, 113, 112, and 111). The kth standardized value from the actual archaeological data is $y_{(k)}$ (Table 5-4).

k	Largest for 500 Tables	Simulated 97.5 Percent Point	$y_{(k)}$	Cell
114	4.47	3.715[a]	5.22	(18, 1)
113	4.32	2.965	4.48	(11, 5)
112	4.23	2.67	2.95	(2, 3)
111	4.16	2.425	2.74	(13, 5)

[a] The simulated 95 percent point, corresponding to a one-sided 5 percent approach, falls at 3.50, the 25th entry from above in Table 5-5.

largest standardized deviate found among 112 cells instead of 114. This would be a considerable further effort, which we would like to avoid.

Because we doubt that the distribution of the largest observation among 112 behaves very differently from the largest among 114, we also plan to use the upper bound, 97.5 percent point, and 95 percent point for the largest of all 114 as if they applied to the largest of the remaining 112. Thus, we compare the third largest value, 2.95, with the original 97.5 percent point (and 95 percent point) of the largest of all 114, which is 3.715 (and 3.50, respectively). It is considerably smaller. Because of using 114 for 112, we expect it to be smaller, but only a bit smaller, and so we do not have any reason to regard 2.95 as demonstrated to be large. (It actually falls at the 77 percent point of the simulated largest values.)

By storing, for the half-dozen or so largest deviations in each of the 500 tables, not only their sizes but what cells they came from, it would be easy to rework our stored data and get distributions for the largest, second largest, etc., when any few specified cells were withdrawn from the competition. This may well be the method of choice for this approach. (We used this approach as part of exploring the need for improved standardization—in particular, the special problems associated with cells having small expected values.)

Similarly, since we have not rejected the third largest, we could compare the fourth largest observed value (now regarded as the second largest of 112) with the second largest among the 114, thus comparing 2.74 with 2.965. Again it is substantially smaller, not just a bit smaller.

And so we conclude that, on the basis of this analysis, we have no clearly and demonstrably extreme values left among these top observations after we reject the first two of the 114.

Interpretations for Extreme Cells

Laurel Casjens (personal communication) interprets cell (18, 1), which represents grinding stones in the immediate vicinity of water, as follows:

> Grinding stones are tools which cannot be easily transported. Their high count near water sources is due to sites in one specific area, a marsh. Their abundance thus reflects usage of marsh resources and/or grinding of grass seeds carried into marsh-side sites from the surrounding valley floor.

In interpreting the unusually large value in cell (11, 5), Casjens notes that these artifacts (row 11) are early projectile points. She also mentions that cell (12, 6), whose standardized residual, 2.37, is not so extreme, involves middle period projectile points. She continues,

> These two cases, in which projectile points are relatively more abundant at large distances from water, reflect the lack of correlation between water sources and hunting activities. Hunting activities can be assumed to have taken place almost everywhere, without regard to campsite or water location.
>
> Unfortunately, this distribution may also reflect a modern bias introduced by "arrowhead collectors," who collect projectile points exclusively, and who collect more heavily in areas near water sources because of their higher site densities.

Of cell (2, 3), unifaces with two or more edges retouched located $\frac{1}{4}$ to $\frac{1}{2}$ mile from water, which although large may not be an outlier, Casjens writes,

> The relatively large number of heavily used unifaces (generally scraping or heavy cutting tools) $\frac{1}{4}$ to $\frac{1}{2}$ mile from water is due to two very large sites in this location. These two sites share their topographic situation (slight rise in the valley floor) with a number of sites having a similar artifact content, but which are not as near water. Thus, it would appear that their association with water is spurious, and that water is not the significant determinant of this assemblage's location.

If we had intended to look for relation to distance from water, rather than to look for the occurrence of isolated outliers, we might well have combined columns 1 and 2, set aside column 3, and combined columns 4, 5, and 6, to obtain a 19×2 contingency table focused on (a) types of objects and (b) relationship to water. We would then have needed to look at only 19 2×2 tables, one for each type of object. Row 11 would then lead to a standardized residual of -3.58, which would probably be shown to be significant by a simulation.

5E. AN EXPLORATORY APPROACH, BASED ON DEVIATIONS FROM INDEPENDENCE

An approach different from that of the simulation effort given above might follow the lines of straightforward exploratory data analysis. It would not assume that the hypergeometric model was necessarily correct. Instead, it might suppose that there is variation in terrain and in hunting and counting artifacts, and so it would allow for these other sources of variation. Its goal would be to identify cells whose standardized residuals are unusually large in magnitude relative to the rest of the cells. We are still using independence as a reference point, since our residuals are based on that idea. But we are *not* assuming independence as a basis for assumed amounts of variability. (We are also not studying the distribution of any one of the order statistics themselves, only how large each of their observed values is when compared with other observed values.)

This approach would still adjust for marginal totals and get residuals by standardizing the difference between observed and fitted, again only as a device for roughly equating scales. We might, and did, use σ_{ij} from equation (3) for the scale. That is, we simply used equation (4) and did not divide further by $G_{ij} = 1 + e^{-\mu_{ij}}$. The results are not much different from those in Table 5-4, the largest difference being 0.40 for the item in row 17 in the "1 to 3 Miles" column.

Then we analyze the observed standardized values as a batch. For example, as in UREDA (Section 2C), to find outliers we first determine the upper and lower fourths (or quartiles), F_U and F_L, of the observed distribution. Then we lay off upper and lower outlier limits

$$\text{upper limit} = F_U + 1.5(F_U - F_L),$$

$$\text{lower limit} = F_L - 1.5(F_U - F_L).$$

This approach bases the estimate of variability, $F_U - F_L$ (essentially the interquartile range), on the central portion of the distribution and thus largely insulates it from any outliers that may be present in the table. Observations falling outside the limits are regarded as outliers. In the present instance only 5.38 and 4.48 (corresponding to 5.22 and 4.48, respectively, in Table 5-4) would qualify as outliers.

For a normal distribution, the interquartile range would be about 4/3 of a standard deviation. The "old" standardized scores from equation (4) have a unit standard deviation according to theory when the null hypothesis of independence is true. For the archaeological data, the interquartile range is 1.82. Multiplying by 3/4 gives 1.36, and so the estimate of the standard deviation with a reduced effect of outliers is about 1.36. This means that the variance estimate based on the interquartile range is greater by roughly 85% ($= 100 \times 1.36^2 - 100$). Thus the observed variability in the standardized scores is nearly double that provided by the hypergeometric distribution. We discuss the meaning of this excess further in Section 5F. The two outliers seem firmly identified even when the additional sources of variability are allowed to contribute to the measure of variation.

5F. A LOGARITHMIC EXPLORATORY APPROACH

The following analysis brings a resistant approach to the log-linear method of analyzing contingency tables. David Hoaglin encouraged us to apply it, John Emerson did the initial fitting, and Cleo Youtz worked on one form of residual analysis.

As we discussed earlier, an outlier, created by an excessively large count, enlarges the row, column, and grand totals, and thus misrepresents the expected value and variance of the cell entry. One way around this would fit the table entries x_{ij} by an estimate having the multiplicative form

$$\hat{x}_{ij} = MA_iB_j$$

where M, the A_i, and the B_j are resistantly fitted and replace the observed grand total and the row and column proportions, respectively. Thus, for an undisturbed table (no outliers), M would estimate x_{++}, A_i would estimate x_{i+}/x_{++}, and B_j would estimate x_{+j}/x_{++}. This approach differs from what we usually do in fitting cell values, as a basis for finding residuals, by releasing M, the A_i, and the B_j from being fixed as the observed total and observed marginal proportions and allowing them to be estimated robustly.

Taking logarithms gives us

$$\log(\hat{x}_{ij}) = m + a_i + b_j,$$

where the lower-case letters on the right are the logarithms of the corresponding capital letters introduced above. By using this additive model and fitting the $\log x_{ij}$ table resistantly, we can release the constraints on row, column, and grand totals and thus mitigate the outlier effect on the estimated cell mean. By rescaling so that $\Sigma A_i = \Sigma B_j = 1$, MA_i and MB_j will give new estimates of what the row and column totals might have been, outliers aside.

In executing the analysis, some special treatment is required for the cells with zero entries. Usually some number between 0 and $\frac{1}{2}$ is added to such a cell (or to all cells) to avoid taking the logarithm of zero. In our problem the analysis was made twice, once after adding the small number 10^{-6} to the zero cells and once after adding $\frac{1}{2}$ to the zero cells. Although second-decimal-place differences between the two analyses occurred in standard scores, they are too small to warrant discussion. We report the 10^{-6} analysis made by John Emerson. The resistant analysis applied median polish (UREDA, Chapter 6) to the two-way table of $\log x_{ij}$. After the median polish was carried out, the cell means were estimated by taking antilogarithms of $m + a_i + b_j$. This gives us counts on the original scale. (In this discussion, we neglect possible bias introduced by the transforming.) It turned out, for example, that the estimate of x_{++} was 3226 instead of the observed 3297.

We now wish to examine the residuals and see whether we have any blatant outliers. As in Section 5E, we need to choose a scale for variability. Our method of fitting acts as if typical values (means, medians, etc.) for the distributions generating the random counts in the cells could be represented by a multiplicative model. We are still not bound by the hypergeometric probability distribution, although it seems to be the natural model. Again, careful workers might like to use a standardization that would be more accurate in the hypergeometric case, but we may also be much less careful.

In choosing a scale factor, much experience with counts leads us to think that the variance of a cell random variable is approximately proportional to its mean. (For a Poisson random variable the variance exactly equals the mean, but we need not suppose the data are as "pure" as a Poisson variable would imply.) We can use $\sqrt{\text{estimated cell mean}}$ to scale and ask what multiplier is required to get the standard deviation. Of course, we can also look into the actual relation between the deviations and the means by plotting the residuals $x_{ij} - MA_iB_j$ against MA_iB_j, after excluding any definitive outliers. Furthermore, we can look at the variance of residuals for collections of cells having about the same estimated mean, and see whether these variances are proportional to the mean. It is likely to be more effective to try such ways of examining the residuals *after* we have chosen a fairly good standardization.

We standardized by dividing each residual by $\sqrt{\text{estimated cell mean}}$, and we looked at the distribution of these standardized values. The stem-and-leaf display in Table 5-7 identifies four outliers using the fourth-spread approach described and used in Section 5E. They are associated with cells (18, 1), (11, 5), (2, 3), and (12, 6) in that order. These four cells also have the largest standardized residuals in Tables 5-4 and 5-6 in the same order.

In addition, we looked at the fourth-spread. If the Poisson variability were a good approximation, then the standardized residuals screened of demonstrated outliers would have a fourth-spread of about $\frac{4}{3} = 1.33$. The actual fourth-spread is 1.5, slightly larger. Thus our estimate of the multiplier for the standard deviation is $1.5/1.33 = 1.12$, and so the excess variation over what would be associated with the Poisson is about 25% $[= 100(1.12)^2 - 100]$. Using this resistant approach, we have reduced the apparent excess variability considerably; recall that we found 85% excess using the nonresistant approach of Section 5E. Considering likely amounts of additional variability introduced by cluster sampling (existence of "sites"), this excess of 25% seems small, rather than large. If we were able to allow for sites in our analysis, we would not be surprised to find all this variability accounted for.

TABLE 5-7. Stem-and-leaf display for the standardized residuals resulting from the resistantly fitted cell means scaled by $\sqrt{\text{estimated cell mean}}$.

Depth		Leaf unit = 0.1
1	−2•	7
4	−2*	320
7	−1•	955
19	−1*	433111110000
36	−0•	99888887766655555
57	−0*	444433322221000000000
57	+0*	0000000111222233444
37	+0•	566789999
28	1*	0011222444
18	1•	6779
14	2*	03444
9	2•	679
6	3*	00

Hi| 34, 35, 41, 63

Depth of median = (114 + 1)/2 = 57.5
Depth of fourth = (57 + 1)/2 = 29
$F_U = .9$, $F_L = -.6$, $d_F = .9 - (-.6) = 1.5$, $1.5(d_F) = 2.25$
Cutoffs: Lower $-.6 - 2.25 = -2.85$ (−28.5 in leaf units)
Upper $.9 + 2.25 = 3.15$ (31.5 in leaf units)

In standardizing we might have used a variance like Haberman's with x_{i+}/x_{++} and x_{+j}/x_{++} replaced by A_i and B_j, respectively. Emerson carried out this analysis with similar results. Or we might use the Freeman–Tukey standardization. Cleo Youtz carried this out with similar results as well.

Possibly, we could get an improved hypergeometric simulation by using the MA_i and MB_j, obtained from the resistant analysis, as fixed row and column totals for the simulated tables, and then analyze the order statistics as we did in Section 5C. We have not explored the merits of such an approach, although it is certainly not disturbed as much by high (or low) outliers for large cells as the approach of Section 5C alone is. (A high or low outlier for a small cell has only small effects on the row, column, and grand totals.)

Hoaglin, Iglewicz, and Tukey (1981) find the expected number of observations that fall outside the limits $F_L - 1.5(F_U - F_L)$ and $F_U + 1.5(F_U - F_L)$ in samples from a normal distribution to be no greater than approximately $.007n + .4$, for $n \geq 5$. In our problem with $n = 114$, this expected number "outside" is about 1.2. Thus the third and fourth values, 34 and 35, both only just outside, seem to be about what might be expected; and we return to the position that at most two cells appear definitely unusual.

5G. ILLUSTRATIONS OF THE NEW STANDARDIZATION

The new standardization is useful for special distributions, aside from its application to contingency tables. Without attempting to analyze the effectiveness of the new standardization here, we illustrate it for a few examples and discuss some properties.

EXAMPLE: HYPERGEOMETRIC DISTRIBUTION

Let us begin with the hypergeometric random variable X with row and column totals 9 and 311 and grand total 3297. Then $\mu = 0.85$ and $\sigma = 0.88$. The distribution is skewed; indeed, the probabilities are monotonically

TABLE 5-8. Relations among cumulative probabilities and standardizations for the hypergeometric distribution with margins 9 and 311 and grand sum 3297. Expected = 0.85, standard deviation = 0.88.[a]

	Hypergeometric Probabilities					
x	$P\{X = x\}$	$P\{X \geq x\}$	z	$\dfrac{x - \mu}{\sigma}$	$\dfrac{x - \mu}{\sigma G}$	$\dfrac{x - \mu - 0.5}{\sigma}$
0	.40949	1.00000	$-\infty$	$-.97$	$-.68$	-1.54
1	.38488	.59051	.23	.17	.12	$-.40$
2	.16020	.20563	.82	1.31	.92	.74
3	.03876	.04543	1.69	2.46	1.72	1.89
4	.00601	.00667	2.47	3.60	2.52	3.03
5	.000618	.000663	3.21	4.74	3.32	4.17
6	.0000423	.0000442	3.92	5.88	4.12	5.31
7	1.85×10^{-6}	1.90×10^{-6}	4.62	7.02	4.92	6.45
8	4.72×10^{-8}	4.77×10^{-8}	5.34	8.17	5.72	7.59
Maximum error ($x > 0$):				2.83	0.38	2.26

[a] z = standard normal deviate corresponding to $P\{X \geq x\}$. $G = 1 + e^{-\mu}$.
Editor's suggestion: Try $(x - \mu - 0.5)/\sigma G$ in this and the next two tables (J.W.T.).

decreasing (J-shaped). Table 5-8 shows the probabilities for $x = 0$ to $x = 8$ and the cumulative from the right.

To compare standardizations, we let z be the normal score that gives the correct value of $P\{X \geq x\}$ obtained from the hypergeometric distribution. Then we offer three standardizations: $(x - \mu)/\sigma$, $(x - \mu)/\sigma(1 + e^{-\mu})$, and $(x - \mu - .5)/\sigma$. Essentially, the approximation that gives the value closest to the correct z is the best one. In the example, the half correction is closest for $x = 0$, 1, and 2. Thereafter the new standardization is closest. Over the range $1 \leq x \leq 8$ its maximum error is 0.38, much smaller than the 2.26 for the half correction. In general, the new standardization seems to do its best work in the right-hand tail of a distribution skewed to the right. Because we are especially studying largest values, this approximation is helpful just where we need it. As we illustrate below, Poissons with small means (Table 5-9) and binomials with small p (Table 5-10) also find close approximations.

If the distribution is approximately symmetric with large mean (such as the binomial with $p = \frac{1}{2}$ and n large, or the Poisson with large mean, or the hypergeometric with a large μ which does not fall near the upper or lower limit of the random variable), the standardized distributions will be nearly symmetrical. The correction factor will be about 1 because $e^{-\mu}$ is near 0.

TABLE 5-9. Illustration of the new standardization for the Poisson distribution with $\mu = 1$ (and hence $\sigma = 1$).[a]

x	Poisson Probabilities $P\{X \geq x\}$	z	$\frac{x-\mu}{\sigma}$	$\frac{x-\mu}{\sigma G}$	$\frac{x-\mu-0.5}{\sigma}$
0	1.0000	$-\infty$	−1	−.73	−1.5
1	.6321	−.34	0	.00	−.5
2	.2642	.63	1	.73	.5
3	.0803	1.40	2	1.46	1.5
4	.0190	2.08	3	2.19	2.5
5	.00366	2.68	4	2.92	3.5
6	.000594	3.24	5	3.66	4.5
7	.0000832	3.77	6	4.39	5.5
8	.0000102	4.26	7	5.12	6.5
Maximum error ($x > 0$):			2.74	0.86	2.24

[a] z = standard normal deviate corresponding to $P\{X \geq x\}$. $G = 1 + e^{-\mu}$.

Then we might expect the half correction to do better than the new standardization because we are in the standard symmetric situation.

EXAMPLE: POISSON DISTRIBUTION

As a second example, not related to our contingency table analysis, we consider the Poisson distribution with $\mu = 1$. The probabilities are given in Table 5-9. We note that, among the three standardizations, the half correction is closest for $x = 0, 1$, and after that the new standardization does well. (Although we have not included it, the Freeman–Tukey standardization is also close in the tail.)

EXAMPLE: BINOMIAL DISTRIBUTION

As a third example also not related to our contingency table analysis, we consider a binomial with $\mu = np$ larger than 1, but not so large that $e^{-\mu}$ is negligible. For the choice $n = 10$ and $p = .15$, Table 5-10 gives the results in the same pattern as in the two previous tables. Here $\sigma = \sqrt{10(.15)(.85)} = 1.129$. In this example, the half correction is closest for $x = 0, 1, 2, 3, 4$, and the new standardization for $x = 5, 6, 7, 8$.

TABLE 5-10. Illustration of new standardization for the binomial distribution with $n = 10$ and $p = .15$ (so that $\mu = 1.5$ and $\sigma = 1.129$).[a]

x	Binomial Probabilities $P\{X \geq x\}$	z	$\frac{x-\mu}{\sigma}$	$\frac{x-\mu}{\sigma G}$	$\frac{x-\mu-0.5}{\sigma}$
0	1.0000	$-\infty$	1.33	−1.09	−1.77
1	.8031	−.85	−.44	−.36	−.89
2	.4557	.11	.44	.36	.00
3	.1798	.92	1.33	1.09	.89
4	.0500	1.65	2.21	1.81	1.77
5	.00987	2.33	3.10	2.53	2.66
6	.00138	2.99	3.99	3.26	3.54
7	1.35×10^{-4}	3.64	4.87	3.98	4.43
8	8.66×10^{-6}	4.30	5.76	4.71	5.31
Maximum error ($x > 0$):			1.46	0.49	1.02

[a] z = standard normal deviate corresponding to $P\{X \geq x\}$. $G = 1 + e^{-\mu}$.

Alternative Tail Areas

Each of the standardizations transforms the distribution of X so that the result more closely resembles a standard normal distribution. To compare such standardizations at a given distribution of X, we choose a way of transforming the values of X into standard normal values, and then we ask how close each standardization comes to reproducing those standard normal values. Because the distributions of X are discrete, we have some leeway in constructing this transformation. The examples in Tables 5-8 to 5-10 define z corresponding to x as the standard normal deviate for which $P\{Z \geq z\} = P\{X \geq x\}$; that is, they use the right-hand tail area.

When we are concerned with significance, John W. Tukey has suggested a correspondence between z and x that provides a more balanced treatment of both left-hand and right-hand tails:

1. *Left-hand tail.* If $P\{X \leq x\} \leq \frac{1}{2}$, take z such that $P\{Z \leq z\} = P\{X \leq x\}$.
2. *Right-hand tail.* If $P\{X \geq x\} \leq \frac{1}{2}$, take z such that $P\{Z \geq z\} = P\{X \geq x\}$.

Because $P\{X \leq x\} + P\{X \geq x\} = 1 + P\{X = x\}$, we also need the following:

3. *Transitional region.* If both $P\{X \leq x\}$ and $P\{X \geq x\}$ are greater than $\frac{1}{2}$, let

$$Q = P\{X > x\} + \tfrac{1}{2}P\{X = x\}$$

and determine z so that $P\{Z \geq z\} = Q$, where z will sometimes be > 0 and sometimes < 0.

For example, in Table 5-10, for $x = 0$ we have $P\{X \leq 0\} = .197$, leading to $z = -.85$. And for $x = 1$ we have $P\{X \leq 1\} = .544$ and $P\{X \geq 1\} = .803$, so that $Q = .456 + \frac{1}{2}(.347) = .629$. Thus we regard $x = 1$ as being in the left-hand tail, and we would use $z = -.33$. For $x \geq 2$ we would still have $P\{X \geq x\} \leq \frac{1}{2}$ and assign z-values from the right-hand tail as before. With this alternative way of defining the z-values, the G-standardization is closest for $x = 0, 1, 5, 6, 7, 8$ and the half correction for $x = 2, 3, 4$.

5H. SUMMARY

In a large two-way contingency table, we want to detect cell entries that deviate extremely from results based on independence. This requires a scale for measuring deviance. Each of the three methods proposed can use a variety of scales. (In fact, we used different scales in different methods.) We apply the methods to a 19×6 archaeological example described by Laurel Casjens.

1. *Probabilistic simulation.* Using the given margins of the observed contingency table and assuming independence between rows and columns, we randomly fill in the cells by the method described in Section 5C. Cells of the simulated distribution have a hypergeometric distribution, and the deviation within each cell could be measured by the probability of an entry in the cell being as large as or larger than the entry observed. These probabilities are hard to compute, and so for high values of $x_{ij} - \mu_{ij}$ we refer to a standard normal distribution z_{ij} based on Parunak's G adjustment for skewed distributions as follows:

$$z_{ij} = (x_{ij} - \mu_{ij})/\sigma_{ij}G_{ij}$$

where

x_{ij} = entry in row i and column j;

$\mu_{ij} = x_{i+}x_{+j}/x_{++}$, the usual expected value;

$$\sigma_{ij}^2 = \frac{x_{i+}(x_{++} - x_{i+})x_{+j}(x_{++} - x_{+j})}{x_{++}^2(x_{++} - 1)}, \text{ the hypergeometric variance;}$$

$G_{ij} = 1 + e^{-\mu_{ij}}$;

$$x_{i+} = \sum_j x_{ij}, \qquad x_{+j} = \sum_i x_{ij}, \qquad x_{++} = \sum_{i,j} x_{ij}.$$

We repeated the simulation 500 times, found the largest z_{ij} of the 114 $(= 19 \times 6)$ in each simulation, and found the empirical distribution of this maximum. The largest cell observed in the archaeological data fell above the 97.5 percent point of the simulation, as did the second largest, and so we regarded these points as high outliers with this method. Once these points were rejected, the next point seemed not excessively high, regarded as the highest among 112.

In this method we judge extremeness relative to empirical distributions of the order statistics from the tables of standardized values. Outlying counts, if present, affect the μ_{ij} and σ_{ij} and hence the standardized values in other cells. We found that two cells of the archaeology table did stand out.

2. *Exploratory approach.* Using the usual standardized score $z'_{ij} = (x_{ij} - \mu_{ij})/\sigma_{ij}$ for the observed entries x_{ij}, we applied the rule of thumb for outliers. We made a stem-and-leaf display for the z'_{ij}, calculated the quartiles, and regarded those z'_{ij} falling more than 1.5 times the interquartile range above the upper quartile as outliers. Here we judge extremeness relative to the sample of standardized values for the one observed table of counts. We found the same two outliers as by method 1. We found no low outliers.

3. *Logarithmic exploratory approach.* When the bulk of the data follow a multiplicative model $\mu_{ij} = MA_iB_j$, we can take logarithms and fit $\log x_{ij}$ by $m + a_i + b_j$, where the lower-case m, a_i, and b_j are the logarithms of M, A_i, and B_j. (If $x_{ij} = 0$, we add a small number; adding 10^{-6} or .5 gave equivalent results.) We used median polish to make the fit. By taking antilogarithms, we estimated M, A_i, and B_j and thus got an estimate of MA_iB_j, and used it to get residuals. Then we standardized by scaling each residual by the square root of the estimated cell mean. Here, as in the exploratory approach, we judge extremeness relative to the sample of standardized values for the one observed table of counts.

Using the rule of thumb for outliers, the distribution of the standardized residuals produced four outliers, the largest three of which came from cells with the largest residuals in method 1. Study of the behavior of the rule of thumb in simpler situations suggests that two of the four would be likely to appear by chance. Again two outliers stand out.

4. *Naive approach.* Allowing for the number of cells to be examined separately—but not allowing for the correlations among deviations in different cells, which tend to compensate for some of this multiplicity—by using $.05/rc$ or $.025/rc$ quantiles of the unit normal (instead of .05 or .025 quantiles) as a basis for judging extremeness of any individual cell deviation should work well for large contingency tables (we do not yet know which tables are large).

All four methods could be used to find low outliers as well as high, but were not used in this example. Method 3 does not lean nearly so much on the hypergeometric assumption as do the other methods, and it is much less affected by outliers.

The new standardization is not limited to contingency tables, and for high observed values (half-corrected deviate $> +1$) it gives generally better normal approximations for hypergeometric, binomial, and Poisson distributions which are skewed to the right than does the usual half-corrected standardization. This is illustrated in Section 5G.

5I. CONCLUSION

If most cells of the contingency table are not outliers and follow the hypergeometric distribution, then the simulation approach has attractions, although it is expensive.

The exploratory approach is cheapest and nearest to the classical chi-squared approach, but it is affected by outliers.

The logarithmic exploratory approach is intermediate in cost, is resistant to outliers, and considerably loosens the distributional assumption. For most large tables it seems preferable.

The naive approach is probably sound for "large" tables, although we do not know when a table is "large."

To relate these methods to the usual chi-squared analysis, we note that the terms that compose the chi-squared statistic are the squares of

$$(x_{ij} - \mu_{ij})/\sqrt{\mu_{ij}}$$

and that $\sqrt{\mu_{ij}}$ often approximately equals σ_{ij}. Thus the square roots of single terms of the chi-squared statistic could be used in the straightforward exploratory approach. [If we allow for the usual number of degrees of freedom—number of cells minus number of fitted constants—say $rc - k$, then the role of σ_{ij} is more nearly played by $\sqrt{(rc - k)\mu_{ij}/rc}$.]

REFERENCES

Bishop, Y. M. M., Fienberg, S. E., and Holland, P. W. (1975). *Discrete Multivariate Analysis: Theory and Practice*. Cambridge, MA: MIT Press.

Casjens, L. (1974). "The prehistoric human ecology of southern Ruby Valley, Nevada." Doctoral dissertation, Harvard University.

Dixon, W. J., Brown, M. B., Engelman, L., Frane, J. W., Hill, M. A., Jennrich, R. I., and Toporek, J. D. (1981). *BMDP Statistical Software 1981*. Berkeley: University of California Press.

Freeman, M. F. and Tukey, J. W. (1950). "Transformations related to the angular and the square root," *Annals of Mathematical Statistics*, **21**, 607–611.

Haberman, S. J. (1973). "The analysis of residuals in cross-classified tables," *Biometrics*, **29**, 205–220.

Hoaglin, D. C. (1976). "Theoretical properties of congruential random-number generators: An empirical view." Memorandum NS-340, Department of Statistics, Harvard University.

Hoaglin, D. C., Iglewicz, B., and Tukey, J. W. (1981). "Small-sample performance of a resistant rule for outlier detection," *1980 Proceedings of the Statistical Computing Section*. Washington, D. C.: American Statistical Association, pp. 148–152.

Lewis, P. A. W., Goodman, A. S., and Miller, J. M. (1969). "A pseudo-random number generator for the System/360," *IBM Systems Journal*, **8**, 136–148.

Parunak, A. (1983). "A new standardization." Unpublished manuscript, University of Michigan.

Templeton, J. G. C. (1956). "A test for detecting single-cell disturbances in contingency tables." Doctoral dissertation, Princeton University.

Additional Literature

Brown, M. B. (1974). "Identification of sources of significance in two-way contingency tables," *Applied Statistics*, **23**, 405–413.

Dallal, G. E. (1982). "A simple interaction model for two-dimensional contingency tables," *Journal of the American Statistical Association*, **77**, 425–432.

Cochran, W. G. (1952). "The χ^2 test of goodness of fit," *Annals of Mathematical Statistics*, **23**, 315–345. [Also Paper 49 in William G. Cochran, *Contributions to Statistics*. New York: Wiley, 1982.]

Cochran, W. G. (1954). "Some methods for strengthening the common χ^2 tests," *Biometrics*, **10**, 417–451. [Also Paper 59 in William G. Cochran, *Contributions to Statistics*. New York: Wiley, 1982.]

Fienberg, S. E. (1979). "The use of chi-squared statistics for categorical data problems," *Journal of the Royal Statistical Society*, *Series B*, **41**, 54–64.

Fienberg, S. E. (1980). *The Analysis of Cross-Classified Categorical Data*, 2nd ed. Cambridge, MA: MIT Press.

Fuchs, C. and Kenett, R. (1980). "A test for detecting outlying cells in the multinomial distribution and two-way contingency tables," *Journal of the American Statistical Association*, **75**, 395–398.

Grizzle, J. E., Starmer, C. F., and Koch, G. G. (1969). "Analysis of categorical data by linear models," *Biometrics*, **25**, 489–504.

Haberman, S. J. (1978). *Analysis of Qualitative Data*, Volume 1 (*Introductory Topics*). New York: Academic.

Killion, R. A. and Zahn, D. A. (1976). "A bibliography of contingency table literature: 1900 to 1974," *International Statistical Review*, **44**, 71–112.

Ling, R. F. and Pratt, J. W. (1984). "The accuracy of Peizer approximations to the hypergeometric distribution, with comparisons to some other approximations," *Journal of the American Statistical Association*, **79**, 49–60.

Moore, D. S. (1978). "Chi-square tests." In R. V. Hogg (Ed.), *Studies in Statistics*. Washington, D. C.: Mathematical Association of America, pp. 66–106.

Patefield, W. M. (1981). "Algorithm AS 159. An efficient method of generating random $R \times C$ tables with given row and column totals," *Applied Statistics*, **30**, 91–97.

Plackett, R. L. (1974). *Analysis of Categorical Data*. New York: Hafner.

Plackett, R. L. (1983). "Karl Pearson and the chi-squared test," *International Statistical Review*, **51**, 59–72.

CHAPTER 6

Fitting Straight Lines by Eye†

Frederick Mosteller
Harvard University

Andrew F. Siegel
University of Washington

Edward Trapido
Comprehensive Cancer Center for the State of Florida

Cleo Youtz
Harvard University

The properties of least-squares and other computed lines are well understood, but surprisingly little is known about the commonly used method of fitting by eye. This method involves maneuvering a string, black thread, ruler, or line-bearing transparency until the fit seems satisfactory, and then drawing the line. We report one systematic investigation of straight lines fitted by eye. Because investigators frequently fit lines by eye, we need to know something of the properties of such a procedure.

Students fitted lines by eye to four sets of points given in an experimental design to help us discover the properties of their fitted lines and whether order of presentation or practice made a difference. Other populations of subjects may produce different results. These sets of data were not unusual in curvature or in having outlying points or patterns. Thus additional populations of data sets could profitably be investigated.

---

†A shorter version of this material appeared as "Eye Fitting Straight Lines" in *The American Statistician*, **35**, 1981, pp. 150–152. The common material appears here with the permission of the American Statistical Association.

The principal quantitative reference on fitting straight lines by eye is Finney (1951). He found that one mathematical iteration starting with slopes from eye-fitting by scientists, inexperienced with probit analysis, gave satisfactory approximations to the relative potency in a bioassay, which is usually achieved only after several iterations.

6A. METHOD

We conducted this investigation of eye-fitting in a class of graduate and postdoctoral students in introductory biostatistics. Most students had not studied statistics before and had not yet been shown formal methods for fitting lines in the course. In a previous class session we illustrated the idea of using a regression line fitted to a set of points to estimate the vertical value, y, from the horizontal value, x. And just before the students began their fitting, the instructor demonstrated the placement of a line inscribed on a transparency onto a set of points that fell exactly on a straight line.

Each student received a set of four scatter diagrams and an $8\frac{1}{2} \times 11$-in. transparency with a straight line etched completely across the middle, parallel to the shorter sides. Students moved the transparency over the scatter diagram until satisfied with the fit of the etched line, and then marked a cross ($\times$) on the scatter diagram at each end of the line. This transparency method is preferable to the black-thread method, which requires three hands.

The four scatter diagrams were labeled S for standard, F for fat, V for vertical, and N for negative; these are shown in Figure 6-1. Data sets S, F, and V are linear transformations of each other, so that F has more vertical variation than S, and V has a steeper slope than S. Data sets S, F, and V come from a table of pseudorandom normal deviates in Beyer (1971), whereas data set N is a linear transformation of the fiber strength data in Dunn and Clark (1974, p. 224).

We constructed the values of x and y for S, F, and V from pseudorandom normal deviates (Beyer 1971, p. 202, column 3, 7th through 46th numbers). After constructing the 20 points for S, we obtained those for F and V by linear transformations on x and y. Essentially S has least-squares regression line

$$S\colon\ y = \tfrac{2}{3}x,$$

and F also has least-squares regression line

$$F\colon\ y = \tfrac{2}{3}x,$$

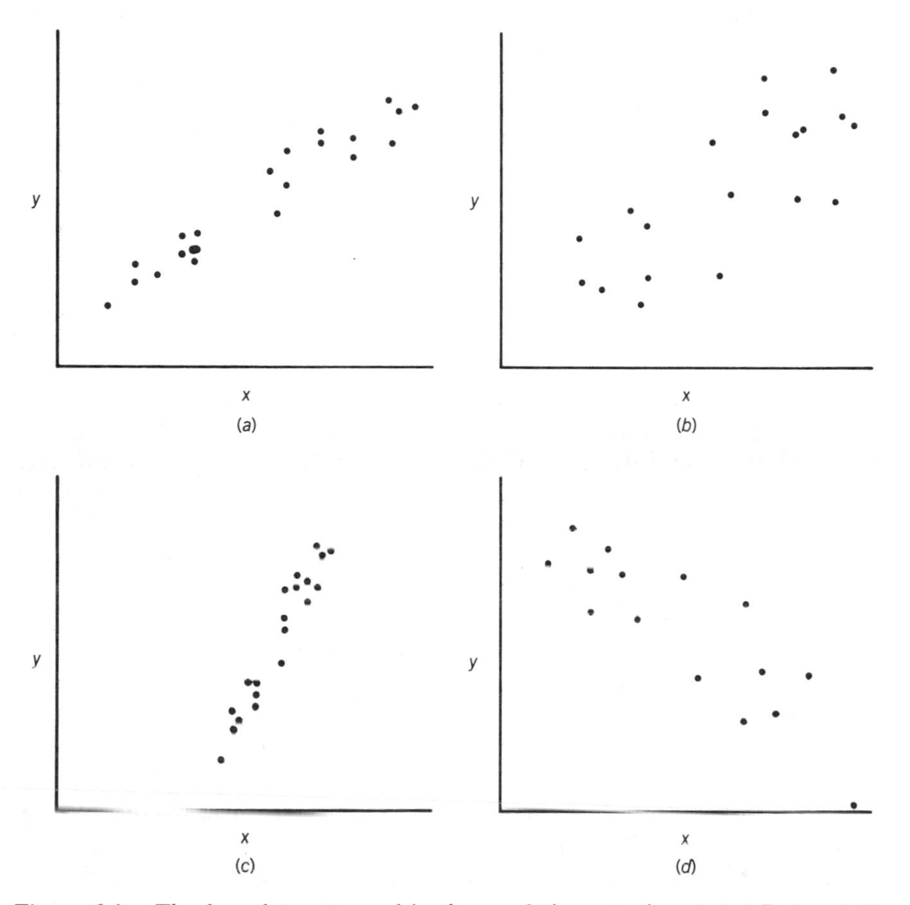

Figure 6-1. The four data sets used in the eye-fitting experiment. (a) Data set S. (b) Data set F. (c) Data set V. (d) Data set N. *Source*: Frederick Mosteller, Andrew F. Siegel, Edward Trapido, and Cleo Youtz (1981). "Eye fitting straight lines," *The American Statistician*, **35**, 150–152 (Figure 1, p. 151).

but its vertical deviations vary three times as much as those for S, by construction. The data points for V are identical with those of S except that the x coordinate has been divided by 3. The least-squares regression line is then

$$V\colon\ y = 2x.$$

The 15 data points for N have a different source. They are diameters (D) of 15 fibers and the logarithm to the base 10 of their breaking strengths (B). To obtain a scale similar to those of the other figures, we transformed these data as follows:

$$x = \frac{D - \overline{D}}{3},$$

$$y = -4(B - \overline{B}).$$

After this transformation, the least-squares line became

$$N\colon\ y = -.70x.$$

The coordinates were plotted as shown in Figures 6-1(a), 6-1(b), 6-1(c), and 6-1(d). In all plots, the scale was 2 in. to 1 unit. For data sets S, F, and V, the plotted axes intersect at $(-2.5, -2)$. For data set N, the plotted axes intersect at $(-2, -2)$. However, the students received no information about the scales or origins.

To assess the effect of the order of presentation, we used a Latin square design with packets stapled in four different orders: I, *SNFV*; II, *NSVF*; III, *FVSN*; and IV, *VFNS*. We distributed the packets systematically in that sequence. Students were thus in four groups, one corresponding to each packet. Students sitting side by side had packets with different orders. We laid out on desks before class 175 packets and were able to collect 153 at the end of the hour.

Four of the scatter diagrams in the 153 packets were not marked—one student did not mark N in packet I, and one student did not mark V, S, and N in packet III. No reason was given for the omissions. The orders and sample sizes of the packets appear in Table 6-1.

TABLE 6-1. Orders of presentation and number of students.

	Order of Scatter Diagram in Packet					Number of Students Responding to Diagrams in Packet			
Packet	1	2	3	4	Packet	1	2	3	4
I	*S*	*N*	*F*	*V*	I	39	38	39	39
II	*N*	*S*	*V*	*F*	II	43	43	43	43
III	*F*	*V*	*S*	*N*	III	38	37	37	37
IV	*V*	*F*	*N*	*S*	IV	33	33	33	33
					Total	153	151	152	152

TABLE 6-2. Coordinates of data points as read from scatter diagrams.

S		*F*		*V*		*N*	
x	*y*	*x*	*y*	*x*	*y*	*x*	*y*
1.20	1.16	1.13	.51	3.85	1.14	1.10	5.62
1.85	2.27	1.80	1.92	4.06	1.89	1.76	6.39
1.88	1.88	1.79	2.92	4.07	2.24	2.08	4.54
2.33	2.06	2.29	1.75	4.26	2.07	2.11	5.47
2.95	2.94	2.92	3.54	4.43	2.93	2.48	5.88
3.20	2.33	3.18	1.44	4.55	2.34	2.77	5.38
3.28	2.97	3.28	2.04	4.57	2.61	3.15	4.37
3.31	2.59	3.27	3.23	4.56	2.97	4.17	5.37
4.86	4.32	4.82	5.13	5.17	3.38	4.53	3.06
5.06	3.39	5.01	2.10	5.15	4.33	5.56	4.76
5.22	4.06	5.18	3.97	5.20	4.11	5.54	2.08
5.22	4.89	5.20	6.44	5.19	4.94	5.89	3.22
5.98	5.27	5.97	6.53	5.48	5.04	6.24	2.29
6.03	5.01	5.96	5.73	5.48	5.27	6.93	3.06
6.70	5.14	6.67	5.24	5.70	4.71	7.95	0.21
6.72	4.70	6.70	3.82	5.71	5.17		
7.50	6.01	7.49	6.73	5.97	6.01		
7.60	5.03	7.58	3.76	5.98	5.06		
7.78	5.76	7.74	5.67	6.05	5.78		
8.02	5.81	8.00	5.43	6.16	5.83		

After the students handed in their scatter diagrams, we used a transparent graph to read the coordinates of the two crosses that each student had marked on each diagram. The transparent graph was scaled 1 in. = 1 unit, and the origin was placed at the intersection of the two axes on the scatter diagram.

The coordinates of the data points, which appear in Table 6-2, as well as the points marked by the students, were measured by the same transparent graph, and thus they differ from the coordinates originally used to plot the scatter diagrams.

6B. RESULTS

We computed the slopes and intercepts of each of the lines joining the two crosses that the students marked. As a summary measure of location, we used the median and, as a measure of variability, the interquartile range. We

TABLE 6-3. A stem-and-leaf display for slopes of students' lines for data set *S* (all orders of presentation combined).

	Stem	Leaf
	.5*	
1	t	2
	f	
3	s	67
	.5•	
	.6*	
4	t	3
9	f	55555
28	s	6666666677777777777
63	.6•	88888888888999999999999999999999999
(33)	.7*	000000000000000001111111111111111
56	t	2222222222222222222233333333333
25	f	444455
19	s	666666777777
7	.7•	88888
	.8*	
2	t	22

$n = 152$ Median = .70 Interquartile range = .04

TABLE 6-4. A stem-and-leaf display of slopes of students' lines for data set *F* (all orders of presentation combined).

	Stem	Leaf
2	.4*	24
	.4•	
	.5*	
4	.5•	66
11	.6*	3344444
13	.6•	68
24	.7*	01122333444
51	.7•	555556666667788899999999999
(32)	.8*	00000001122222223334444444444444
70	.8•	55555556666666777889
50	.9*	00000011112222222333444444
24	.9•	6667777788889999
8	1.0*	00
6	1.0•	68
4	1.1*	0234

$n = 153$ Median = .84 Interquartile range = .14

TABLE 6-5. A stem-and-leaf display for slopes of students' lines for data set V (all orders of presentation combined).

	Stem	Leaf
1	LO	.57
2	1.4*	2
4	1.4•	68
	1.5*	
7	1.5•	568
8	1.6*	3
	1.6•	
	1.7*	
9	1.7•	6
11	1.8*	02
13	1.8•	56
20	1.9*	0223444
33	1.9•	5566778888889
62	2.0*	00000111111112222223333344444
(33)	2.0•	555566666666677788888888889999999
57	2.1*	0000011122223334444
38	2.1•	5555566666777888899
19	2.2*	0111233
12	2.2•	588
9	2.3*	0000123
2	2.3•	69

$n = 152$ Median = 2.07 Interquartile range = .14

chose these measures because they are more resistant to outliers than are the mean and standard deviation.

Tables 6-3 through 6-6 show, in stem-and-leaf displays, the slopes of the students' lines for the four sets of data.

Table 6-7 shows the medians and the interquartile ranges of the slopes. The numbers are arranged vertically according to the order of presentation of the scatter diagrams in the packets. We will not treat the intercepts here.

In Table 6-7 we also give the computed slope of the least-squares line for the data points in each scatter diagram (using the coordinates read from the diagram). The noteworthy departures are for F. The eye-fitted slopes of F are high compared with the least-squares value. Some students may have felt that the line for F should pass through the intersection of the plotted axes.

TABLE 6-6. A stem-and-leaf display for slopes of students' lines for data set *N* (all orders of presentation combined).

	Stem	Leaf
	−1.0•	
5	−1.0*	41000
13	−.9•	99988776
25	−.9*	443322111110
41	−.8•	9999999998877555
49	−.8*	44322111
70	−.7•	998887777777666665555
(22)	−.7*	4433333322222221000000
59	−.6•	999999998888887777666666665
32	−.6*	444443333221110000
14	−.5•	99999998
6	−.5*	4442
2	−.4•	8
	−.4*	
1	−.3•	5

$n = 151$ Median = −.73 Interquartile range = .20

TABLE 6-7. Medians and interquartile ranges of slopes, arranged in order of presentation.

Order of Presentation	*S*	*F*	*V*	*N*
		Medians		
1st	.70	.82	2.12	−.77
2nd	.70	.90	2.06	−.72
3rd	.71	.80	2.09	−.70
4th	.71	.84	2.02	−.72
Least-squares slope from scatter diagrams	.66	.66	1.98	−.70
Slope of major axis	.68	.82	2.11	−.79
		Interquartile ranges		
1st	.04	.15	.10	.20
2nd	.03	.11	.13	.23
3rd	.04	.12	.12	.13
4th	.04	.13	.14	.17

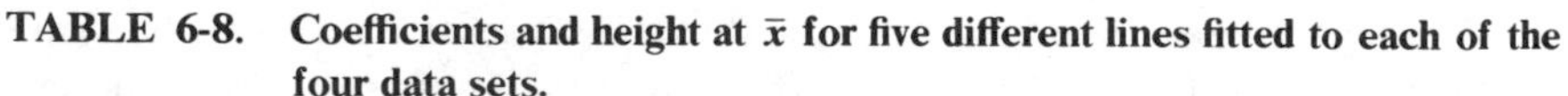

TABLE 6-8. Coefficients and height at $\bar{x}$ for five different lines fitted to each of the four data sets.

Data Set		Least-Squares	L_1	Three-Group Resistant	Major-Axis	Eye-Fitted
S	Slope	0.66	0.66	0.66	0.68	0.70
	Intercept	0.69	0.64	0.72	0.61	0.50
	Height at $\bar{x}$	3.88	3.83	3.91	3.88	3.90
F	Slope	0.66	0.63	0.63	0.82	0.84
	Intercept	0.75	0.78	0.97	−0.02	−0.16
	Height at $\bar{x}$	3.89	3.80	3.99	3.89	3.86
V	Slope	1.98	1.96	1.97	2.11	2.07
	Intercept	−6.18	−6.05	−6.11	−6.81	−6.54
	Height at $\bar{x}$	3.89	3.91	3.90	3.89	3.97
N	Slope	0.70	−0.77	−0.60	−0.79	−0.73
	Intercept	7.03	7.10	6.57	7.40	7.06
	Height at $\bar{x}$	4.11	3.90	4.08	4.11	4.03

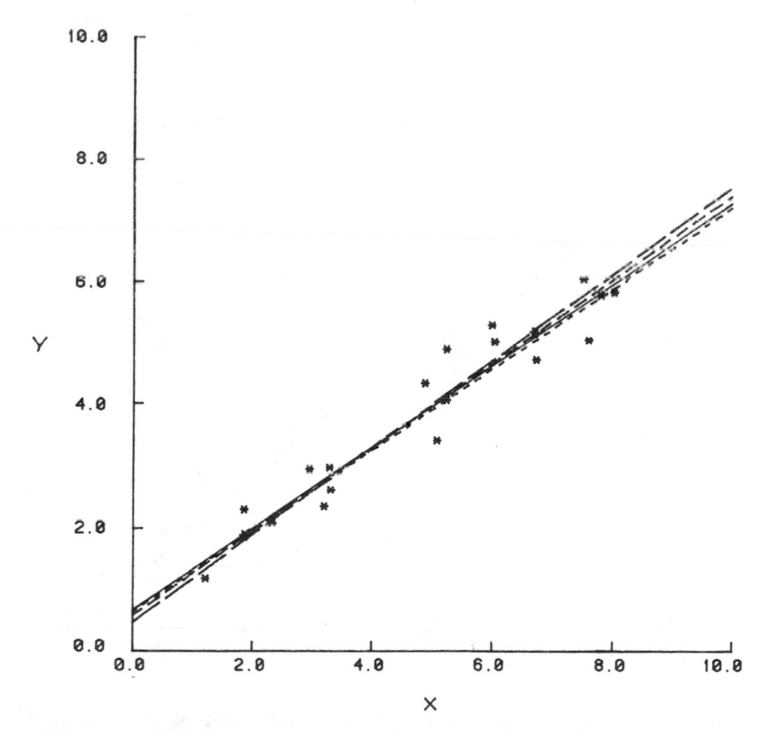

Figure 6-2. Plot of the 20 points of data set *S* together with four different fitted lines: least-squares (——), L_1 (---), major axis (– – –), and an eye-fitted (— — —). The three-group resistant line is not shown because it nearly coincides with the least-square line.

An alternative explanation might be that the students tend to fit the slope of the major axis (the line that minimizes the sum of squares of perpendicular distances). The slope of the major axis would give a slightly poorer fit for the eye-fitted slope of N, a slightly better fit to the eye-fitted slope for V, and a much better fit to the eye-fitted slope of F and S than the least-squares slope.

In Table 6-8 we give the coefficients in the equations for five different lines fitted to each of the four data sets (using the coordinates in Table 6-2):

1. Least-squares line.
2. L_1 line. This minimizes the sum of absolute deviations.
3. Three-group resistant line using medians (Section 7A and UREDA, Chapter 5).

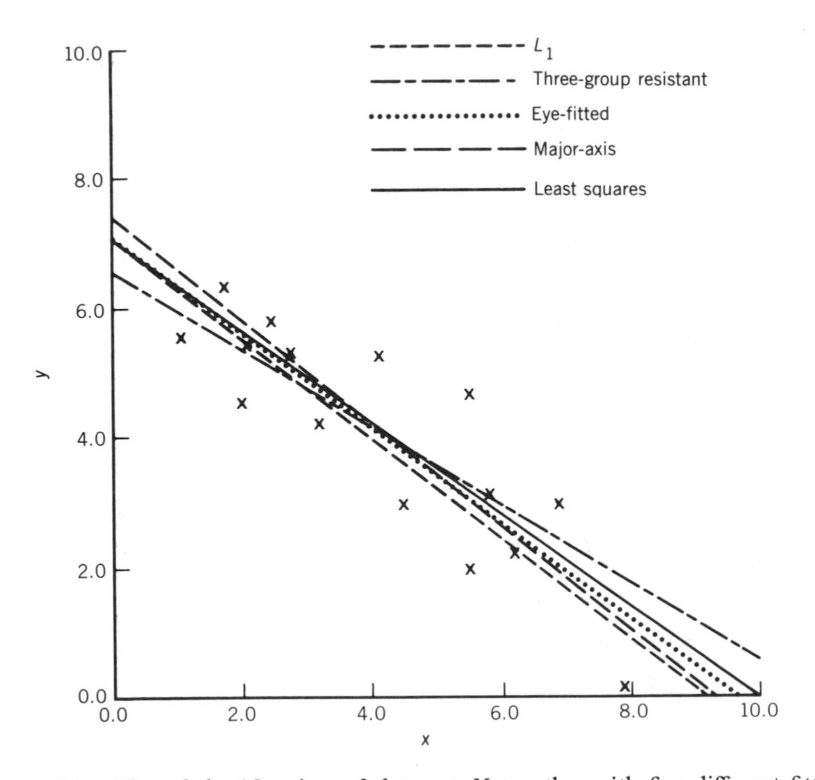

Figure 6-3. Plot of the 15 points of data set N together with five different fitted lines: least-squares (——), L_1 (---), three-group resistant (–---), major-axis (–––), and an eye-fitted (······).

4. Major-axis line. This minimizes the sum of squares of the perpendicular distances.
5. Eye-fitted line. The coefficients are overall medians of students' eye-fitted lines.

Figures 6-2 and 6-3 plot the data set together with the various fitted lines for data sets S and N, respectively.

Order of Presentation

To hunt for an effect from the order of presentation of the scatter diagrams, we look down the columns of Table 6-7. We might predict that experience would reduce the variability, so that the interquartile ranges might get smaller from top to bottom. Using the values for Table 6-7 before rounding to two decimals, we find the average rank correlation between order and size of interquartile range for slopes is -0.15, and so we get no support for this hypothesis. If anything, the students are getting a little worse with practice. As far as we could tell, they were not rushed, and all claimed to have finished before they were asked to hand in the papers (even though a few returned blanks). An alternative explanation is that they became bored or careless after their first experiences.

Predicted Value at $\bar{x}$

To see how the predicted values of y at $\bar{x}$ based on the students' lines compared with $\bar{y}$, we computed the position of $\hat{y}$ at $\bar{x}$ for each student's lines. Table 6-9 shows the medians and interquartile ranges for the height of the students' lines at $\bar{x}$. They compare closely with the actual values of $\bar{y}$. For S, three medians of the four orders fall within the interval provided by the four fitted lines (Table 6-8), for F all four fall in the interval, and for N two fall in the interval. For V, two orders give medians above this interval and two below. Thus the eye-fitting seems to put the line near the centroid of the cloud of points in these examples.

This analysis seems more useful than an analysis of intercepts. The two aspects "slope b" and "y_0 at $\bar{x}$" should be more nearly independently determined. ("Independently" in the statistical sense, or the psychological sense, or both.) But the interpretation of

$$\text{intercept} = (y \text{ at } \bar{x}) - (\text{slope})(\bar{x})$$

combines and confounds these two.

TABLE 6-9. Medians and interquartile ranges of $\hat{y}$ at $\bar{x}$, arranged in order of presentation.

Order of Presentation	S	F	V	N
		Medians		
1st	3.86	3.88	3.87	4.12
2nd	3.93	3.88	4.03	3.95
3rd	3.84	3.82	3.87	4.19
4th	3.89	3.90	4.07	3.92
(mean)	(3.88)	(3.87)	(3.96)	(4.04)
(median)	(3.88)	(3.86)	(3.95)	(4.04)
Value of $\bar{y}$ for the scatter diagram	3.88	3.90	3.89	4.11
		Interquartile ranges		
1st	.10	.15	.11	.16
2nd	.08	.15	.10	.19
3rd	.07	.16	.08	.19
4th	.09	.22	.08	.21

Standard Error of the Least-Squares Slopes, of Intercepts, and of Heights at $\bar{x}$

Using the original data points of the four scatter diagrams, we computed the standard error of the slopes, of the intercepts, and of heights at $\bar{x}$ for the least-squares lines according to the usual formulas. These are given in Table 6-10. We compare them with $\frac{3}{4}$ times the interquartile ranges for the approximately 151 lines of the students for each of the four scatter diagrams, also given in Table 6-10. Because an interquartile range for a normal distribution is about $\frac{4}{3}$ times the true standard deviation, this multiple of the interquartile range gives a rough estimate of the standard deviation of a fitted normal distribution. It somewhat favors the overall student performance, because it makes no allowance for the occasional stray value so clearly present in Tables 6-3, 6-4, and 6-5.

Note that these two quantities estimate different kinds of variability, namely:

How much the same procedure (least squares) would vary from one sample to another.

How much different people vary for the same sample.

TABLE 6-10. Standard errors of slopes, of intercepts, and of heights at $\bar{x}$.

	S	*F*	*V*	*N*	Median
		Slopes			
Sampling s.e.	.04	.13	.13	.12	
Eye-fitting s.e.[a]	.03	.10	.10	.15	
Efficiency[b]	64%	63%	63%	39%	63%
		Intercepts			
Sampling s.e.	.23	.68	.67	.54	
Eye-fitting s.e.[a]	.19	.49	.57	.52	
Efficiency[b]	59%	66%	58%	52%	58%
		Heights at x			
Sampling s.e.	.09	.28	.09	.24	
Eye-fitting s.e.[a]	.07	.13	.13	.19	
Efficiency[b]	62%	82%	32%	61%	62%

[a] Estimated from the interquartile ranges of the students' values.

[b] $\text{Efficiency} = \dfrac{(\text{sampling s.e.})^2}{(\text{sampling s.e.})^2 + (\text{eye fitting s.e.})^2}.$

We may reasonably assume lack of correlation, so it is natural to write

$$\text{total variance} = \text{sampling variance} + \text{eye-fitting variance}.$$

We could define the relative efficiency of eye-fitting, compared with that of least-squares fitting, as the ratio of the variance for least squares (which deals only with sampling variance) to the total variance from eye-fitting (which includes the variability contributed by a random individual doing the fitting). When the individual variation and sampling variation have equal variance, the relative efficiency would be one-half, using the ratio

$$\frac{\text{sampling variance}}{\text{total variance}}.$$

Consider the possibility that the uncertainty expressed by the observed variability of the eye-fitted coefficients is proportional to the standard error of the least-squares estimates of the slope and intercept. (We have no good theory to encourage belief in such a possibility, but it seems plausible.) Then the observed data in this study show a median efficiency of 62–63%. The

TABLE 6-11. Correlations between slopes (151 lines for each).

	Data Set		
	F	*V*	*N*
S	.18	.14	−.14
F		.28	−.08
V			−.05

Source: Frederick Mosteller, Andrew F. Siegel, Edward Trapido, and Cleo Youtz (1981). "Eye fitting straight lines," *The American Statistician*, **35**, 150–152 (Table 2, p. 152).

poorest apparent efficiency is for the V height at $\bar{x}$, 32%. In these data all 20 points for S, F, V, and all 15 for N are well behaved, and so the agreement might not hold if the data contained outliers or patterns other than a single linear relationship. For example, eye-fitting might be better able to avoid the adverse effects of outliers than least squares.

Perhaps more important, the students' standard deviation of the height at $\bar{x}$ for F is not triple, as we might expect from its construction, but only double that for S. As a result, the apparent efficiency rises to 82%. Furthermore, the height at $\bar{x}$ for N, which has less least-squares variability than F, is, for the students, more variable than the height at $\bar{x}$ for F.

Correlations

We eliminated the data for the two students who failed to mark all four of their scatter diagrams. This left 151 sets of data for each of the four plots. The correlations between S and F, S and V, S and N, F and V, F and N, and V and N for the slopes of the students' lines appear in Table 6-11.

The relation seems slight but definite. Those who gave steep slopes on one figure tended to give steep ones on another, but the strength of the correlation is less than we might expect.

6C. SUMMARY

1. Students, not trained in regression, fitted lines by eye to clouds of points. Their pooled slope was closer to the slope of the major axis than to the slope of the least-squares regression line.

2. The median efficiency of fitting was about 63%, with values as high as 82% and as low as 32%. [To estimate the standard deviation among individuals, we used $\frac{3}{4}$ (interquartile range).]

3. At $\bar{x}$, the height of the students' fitted regression lines averaged extremely close to the true value of $\bar{y}$. Thus the students tended to pass their line close to the centroid of the points.

4. The correlation between individual's slopes on various figures was slight but positive among positively sloped clouds and negative between those for positively sloped clouds and the N plot. Thus the steepness (absolute value of slope) was correlated among all pairs of figures.

5. Perhaps new sets of data and different subjects would produce different results. The findings are worth pursuing with further investigations.

ACKNOWLEDGMENTS

The authors wish to express their appreciation to the students who participated. Nina Leech measured the coordinates of the crosses provided by the students. J. W. Tukey suggested the etched line on the transparency. Valuable advice was provided by John Emerson, Katherine Godfrey, Colin Goodall, David Hoaglin, Anita Parunak, Nancy Romanowicz Cook, and Michael Stoto. This work was facilitated by National Science Foundation Grants SES75-15702 and SES 8023644 and Public Health Service Grant No. 5-DO4-AH01698-02, all to Harvard University, and U.S. Army Research Contract DAAG29-79-C-0205 with Princeton University.

REFERENCES

Beyer, W. H. (Ed.) (1971). *Basic Statistical Tables*. Cleveland, Ohio: Chemical Rubber Co.

Dunn, O. J. and Clark, V. A. (1974). *Applied Statistics: Analysis of Variance and Regression*. New York: Wiley.

Finney, D. J. (1951). "Subjective judgment in statistical analysis: An experimental study," *Journal of the Royal Statistical Society, Series B*, **13**, 284–297.

Mosteller, F., Siegel, A. F., Trapido, E., and Youtz, C. (1981). "Eye fitting straight lines," *The American Statistician*, **35**, 150–152.

CHAPTER 7

Resistant Multiple Regression, One Variable at a Time

John D. Emerson
Middlebury College

David C. Hoaglin
Harvard University

We sometimes find it convenient to apply a resistant line-fitting procedure repeatedly to fit complicated regression models. Opportunities to do this arise when:

We seek resistance, and a resistant line-fitting technique is readily available.

We have used a resistant technique for y versus x and we decide to consider a further explanatory variable.

We need a resistant fit as a starting point for a robust multiple regression procedure.

In these situations our strategy is to apply the resistant line-fitting procedure one variable at a time. That is, we reduce the fitting of the more complicated model to a sequence of simple y-versus-x fitting problems.

Providing a resistant initial fit for a multiple regression model offers an important application of resistant techniques for y versus x. The more familiar techniques for obtaining an initial fit—least squares and least absolute residuals—do not guard against adverse effects of influential observations, and a robust regression procedure may have difficulty recovering from a poor initial fit. Starting with a suitable resistant fit thus offers a way to improve the overall procedure of arriving at a good robust fit. Because some of the useful resistant procedures do not generalize readily beyond a single explanatory variable, we proceed one variable at a time.

The basic strategy applies equally well when part of the overall model has a structure of cells or groups. For example, we can analyze a two-way table and decide to incorporate a covariate. Then we usually treat the table as a single unit and fit it by an appropriate technique, such as median polish, before proceeding to the regression component of the model. If we later decided to fit that regression component by using a more robust technique (as discussed in Chapter 8), we would go back and handle the rest of the fit more robustly as well.

7A. RESISTANT LINES

In fitting a line of the form y versus x to data, we often seek to avoid the difficulties that arise when a relatively small number of anomalous data points excessively influence the usual least-squares regression line. A number of line-fitting procedures provide the desired resistance. Their fitted slope and intercept (or slope and central value) change little in response to arbitrary changes in only a small fraction of the data—in the values of y, of x, or of x and y jointly. Chapter 5 of UREDA discusses a three-group resistant line and several alternative procedures, all of which offer some degree of resistance. For convenience the present chapter uses the three-group resistant line, although any other resistant line-fitting procedure could be used in the same way.

The Three-Group Resistant Line

For fitting a line of the form

$$\hat{y} = a + bx$$

to the data $(x_1, y_1), \ldots, (x_n, y_n)$, one simple, moderately efficient technique gains resistance by dividing the data points into three groups on the basis of the x-values and then taking medians within the groups. More specifically, if we assume that the x-values are already in increasing order,

$$x_1 \leq x_2 \leq \cdots \leq x_n,$$

then we attempt to form three groups

$$\text{Left:} \quad (x_1, y_1), \ldots, (x_{n_L}, y_{n_L})$$

$$\text{Middle:} \quad (x_{n_L+1}, y_{n_L+1}), \ldots, (x_{n-n_R}, y_{n-n_R})$$

$$\text{Right:} \quad (x_{n+1-n_R}, y_{n+1-n_R}), \ldots, (x_n, y_n)$$

choosing n_L and n_R so that the groups are as nearly equal in size as possible. [We begin with n_L and n_R equal to the largest integer not exceeding $(n + 1)/3$ and revise them if ties among the x-values require it.] Within the left group and the right group we define the summary points (x_L, y_L) and (x_R, y_R) by taking the median x-value and, separately, the median y-value; for example,

$$x_L = \text{med}\{x_1, \ldots, x_{n_L}\}$$

$$y_L = \text{med}\{y_1, \ldots, y_{n_L}\}.$$

Neither of these summary points need be a data point.

For an initial slope estimate we take

$$b_0 = \frac{y_R - y_L}{x_R - x_L}. \tag{1}$$

Then, after calculating the residuals

$$r_i^{(0)} = y_i - b_0 x_i, \qquad i = 1, \ldots, n,$$

we use $r_i^{(0)}$ in place of y_i to obtain the summary points $(x_L, r_L^{(0)})$ and $(x_R, r_R^{(0)})$ and from these the slope adjustment

$$\delta_1 = \frac{r_R^{(0)} - r_L^{(0)}}{x_R - x_L}$$

and thus the adjusted slope $b_1 = b_0 + \delta_1$.

Ultimately, the slope of the three-group resistant line equals, by definition, the value of b for which the residuals produce zero adjustment. It is not entirely satisfactory to attempt to calculate the slope by continuing to iterate the slope adjustments as described above, because the sequence of slopes may not converge. Instead, we define the residual at data point i as a function of b,

$$r_i(b) = y_i - bx_i, \tag{2}$$

and then define the median residuals in the left and right groups according to

$$r_L(b) = \text{med}\{r_1(b), \ldots, r_{n_L}(b)\} \tag{3a}$$

and

$$r_R(b) = \text{med}\{r_{n+1-n_R}(b), \dots, r_n(b)\}. \tag{3b}$$

The difference between the median residual in the left group and the median residual in the right group is

$$\Delta r(b) = r_R(b) - r_L(b), \tag{4}$$

and the desired value of b, which we denote by $b^{\#}$ for the present, yields $\Delta r(b^{\#}) = 0$. This approach ensures convergence because $\Delta r(b)$ has only a single zero.

Starting from b_0 and $b_1 = b_0 + \delta_1$, we obtain $b^{\#}$ by iteration. Ordinarily $\Delta r(b_0)$ and $\Delta r(b_1)$ have opposite signs, so that $b^{\#}$ lies between b_0 and b_1. [Otherwise we try larger adjustments of the same sign as δ_1 until we have a b_1 for which $\Delta r(b_1)$ does have the opposite sign from $\Delta r(b_0)$.] Once we have localized $b^{\#}$ in an interval, we prepare for the iterative steps by setting b_1^- equal to the smaller of b_0 and b_1 and b_1^+ equal to the larger. We continue by linear interpolation according to

$$b_{j+1} = b_j^- - \Delta r(b_j^-) \frac{b_j^+ - b_j^-}{\Delta r(b_j^+) - \Delta r(b_j^-)}, \qquad j = 1, 2, \dots, \tag{5}$$

until $\Delta r(b_{j+1}) = 0$ (to within some specified tolerance). For a further iteration, b_{j+1} replaces b_j^- if $\Delta r(b_{j+1}) > 0$ and replaces b_j^+ if $\Delta r(b_{j+1}) < 0$. In each case, the other endpoint remains unchanged. Thus, we progressively narrow the interval containing $b^{\#}$, and convergence usually occurs quite rapidly.

To determine the intercept a, we calculate the median (over all the data points) of the residuals after removing the slope:

$$a = \text{med}\{y_i - bx_i\}. \tag{6}$$

Alternatively, we may write the line in terms of slope and central value (of y at some $x = x_0$, which need not be an x_i) rather than slope and intercept. In this form

$$\hat{y} = a^* + b(x - x_0),$$

so that the central value is

$$a^* = \text{med}\{\, y_i - b(x_i - x_0)\}. \tag{7}$$

This representation of the resistant line is often more convenient, both for calculation and for interpretation.

This brief description of the three-group resistant line omits many interesting theoretical details associated with the definition of $b^{\#}$ and the iterative algorithm for obtaining it. Section 5B of UREDA demonstrates that Δr is piecewise linear and strictly decreasing (that is, the graph of Δr as a function of b consists of downward sloping line segments, as in Figure 7-1) and thus that the equation $\Delta r(b) = 0$ has a unique solution. The computer programs in Velleman and Hoaglin (1981) implement this algorithm and also contain one strategy for handling all the combinations of ties among the x-values that may arise in dividing the data into the three groups. This resistant line is also available in the Minitab statistical package (Ryan, Joiner, and Ryan, 1982).

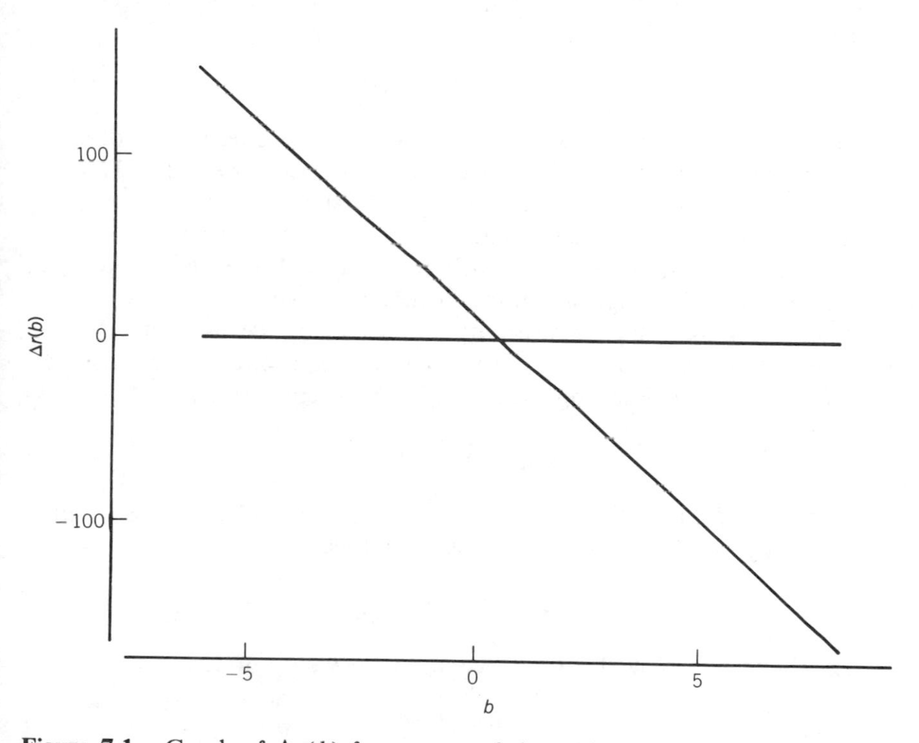

Figure 7-1. Graph of $\Delta r(b)$ for one set of data. The graph consists of 15 line segments.

7B. SWEEPING OUT

We seek a resistant multiple-regression fit

$$\hat{y} = b_0 + b_1 X_1 + \cdots + b_p X_p, \tag{8}$$

which summarizes the linearly describable dependence of the *response variable* y on the *carriers* $X_1, \ldots, X_p$ (and on the constant carrier, whose value is always 1). For an individual data point, we write this fit as

$$\hat{y}_i = b_0 + b_1 x_{i1} + \cdots + b_p x_{ip}. \tag{9}$$

To discuss the process of arriving at such a fit by applying a resistant line one variable at a time, it suffices to take $p = 2$. Models with more carriers merely entail more bookkeeping of a straightforward nature.

The basic operation consists of using a resistant line to summarize a y-versus-x relationship and then removing the fitted slope, leaving a set of residuals. This amounts to sweeping x out of y. When applied to y and X_1, it yields

$$y_{\cdot 1} = y - c_1 X_1. \tag{10}$$

The subscript notation shows a dot and then a list of integers identifying a list of carriers to indicate that those carriers have been swept out. Thus $y_{\cdot 1}$ is what remains of y after removing the fitted slope against X_1, and $X_{2\cdot 1}$ indicates that X_1 has been swept out of X_2. Also, we introduce the coefficient c_1 for the slope of y against X_1 because this slope is usually not the same as b_1, which appears in the fit that also involves X_2.

After the initial step of sweeping X_1 out of y, we ask what further contribution X_2 can make to the fit. To do this properly, we must first adjust X_2 for a possible slope against X_1. This step of sweeping out yields

$$X_{2\cdot 1} = X_2 - d_1 X_1, \tag{11}$$

and we then calculate the resistant slope of $y_{\cdot 1}$ against $X_{2\cdot 1}$. If we denote this slope by c_2, this last step of sweeping out leaves

$$y_{\cdot 12} = y_{\cdot 1} - c_2 X_{2\cdot 1}. \tag{12}$$

For the intercept, we take the median of the values of $y_{\cdot 12}$:

$$b_0 = \operatorname{med}\{(y_{\cdot 12})_i\}. \tag{13}$$

Subtracting b_0 then leaves the residuals, with median 0.

We must still assemble the pieces of the fit from these separate steps in order to obtain the combined fit in terms of the original (unswept) carriers, as in equation (8). From equations (10), (11), (12), and (13) we can construct

$$\hat{y} = b_0 + c_1 X_1 + c_2 (X_2 - d_1 X_1),$$

which is algebraically equivalent to

$$\hat{y} = b_0 + (c_1 - c_2 d_1) X_1 + c_2 X_2.$$

Comparing this second form to

$$\hat{y} = b_0 + b_1 X_1 + b_2 X_2,$$

we see that

$$b_1 = c_1 - c_2 d_1, \qquad b_2 = c_2. \tag{14}$$

Thus we have resistantly fitted the two (nonconstant) carriers X_1 and X_2 to y by using three y-versus-x fitting steps: y versus X_1, X_2 versus X_1, and $y_{\cdot 1}$ versus $X_{2\cdot 1}$.

With more carriers ($X_3, X_4, \ldots$) we would proceed in much the same way. The steps involving y would be y versus X_1, $y_{\cdot 1}$ versus $X_{2\cdot 1}$, $y_{\cdot 12}$ versus $X_{3\cdot 12}$, $y_{\cdot 123}$ versus $X_{4\cdot 123}$, and so on. Usually we would handle the carriers by sweeping each carrier out of all the higher numbered carriers, and for this we would require a more general notation for the coefficients in these adjustments than the d_1 we used in equation (11). Thus we could write

$$X_{2\cdot 1} = X_2 - d_{21} X_1,$$

$$X_{3\cdot 12} = X_3 - d_{31} X_1 - d_{32} X_{2\cdot 1},$$

$$X_{4\cdot 123} = X_4 - d_{41} X_1 - d_{42} X_{2\cdot 1} - d_{43} X_{3\cdot 12},$$

and so on.

Interpreting Coefficients

Although we do not undertake a thorough discussion of the meaning of the regression coefficients $b_1, \ldots, b_p$ in equation (8), it is appropriate to recall that simple interpretations are often misleading. In general, we must think of b_1 as indicating the typical change in y corresponding to a change of one unit in X_1, when we adjust for simultaneous linear change in the other X_j in the data at hand.

Thus the meaning and the numerical value of b_1 depend on what other carriers are in the fit. This qualification applies even when we make only linear adjustments among a set of original carriers without changing the fitted values at all. For example, X_1 has the coefficient b_1 in

$$\hat{y} = b_0 + b_1 X_1 + b_2 X_2,$$

but it has the coefficient c_1 in

$$\hat{y} = b_0 + c_1 X_1 + c_2 X_{2 \cdot 1},$$

because the other carrier is X_2 in the first of these fits and $X_{2 \cdot 1}$ in the second. Equation (14) relates b_1 and c_1 and makes explicit the difference between (a) adjusting for simultaneous linear change in X_2 and (b) adjusting for simultaneous linear change in $X_{2 \cdot 1}$.

At times it may be tempting to use the simpler interpretation of b_1 as the change in y corresponding to a unit change in X_1 when all the other carriers are held fixed. Unfortunately, this simplification is often misleading. Even if, in principle, X_1 could be varied while holding all the other carriers constant, the data may not reflect what happens when this is done. In other situations X_1 and one or more of the other carriers are, mathematically, functions of the same basic variable (e.g., $X_1 = t$ and $X_2 = t^2$), so that we cannot think of varying one carrier while holding the others fixed.

These aspects of regression coefficients imply that it is often much safer and easier to use equation (9) in predicting the value of y corresponding to a new set of x-values similar to those in the data than it is to seize upon a single b_j and interpret it. The blunder of calculating the coefficients, extracting b_j while losing sight of both the other carriers in the model and the way the data arose, and using b_j to decide how much to change X_j seems especially egregious.

The two difficulties in interpreting regression coefficients that this discussion has mentioned do not begin to exhaust the problems of regression coefficients (whether obtained resistantly or by least squares). Mosteller and Tukey (1977, Chapter 13) examine a number of other aspects in some detail.

We also recall that even the meaning of changing X_1 depends on the way that the data were collected. A controlled experiment that actually varied X_1 would tell us about the impact of this manipulation on y. However, from an observational study that merely collected data on each individual's value of X_1, we would learn only how the typical value of y changes in the population as we shift our attention from the individuals who have a given value of X_1 to the other individuals whose value of X_1 is one unit greater. We can learn nothing about the result of actually changing X_1 for any individual. The regression equation does not remind us about this distinction.

Iterative Improvement

A final set of residuals should have zero slope against X_1 and $X_{2\cdot 1}$. To approach this requirement more closely, we generally need to apply further steps of adjustment to $y_{\cdot 12}$, as defined by equation (12). The iterative process of improvement uses equations (10) and (12) alternately. After obtaining initial values of c_1 and c_2, we return to equation (10) with $y_{\cdot 12}$ in place of y and thus calculate an adjustment to c_1. Using the residuals for this new step in place of $y_{\cdot 1}$ in equation (12) then yields an adjustment to c_2, and the process continues until one of these adjustments is small enough to be ignored.

An Algorithm

If we reexamine the whole process—sweeping X_1 out of y and X_2, sweeping $X_{2\cdot 1}$ out of $y_{\cdot 1}$, calculating b_0, and making iterative improvements—from a more algorithmic point of view, we discover that sweeping X_1 out of X_2 does not involve y and so can precede the iterative improvement cycle. Similarly, we may choose to calculate b_0 once, after the improvement cycle has concluded, rather than making some small adjustment in it at each iteration. Thus, assuming that the model contains both X_1 and X_2 from the start (as opposed to our deciding to add X_2 later), the steps of the algorithm would go as follows:

1. Sweep X_1 out of X_2 to obtain

$$X_{2\cdot 1} = X_2 - d_1 X_1.$$

2. Sweep X_1 out of y, leaving

$$y_{\cdot 1} = y - c_1 X_1.$$

3. Sweep $X_{2\cdot 1}$ out of $y_{\cdot 1}$, leaving

$$y_{\cdot 12} = y_{\cdot 1} - c_2 X_{2\cdot 1}.$$

4. If the slope of $y_{\cdot 12}$ against X_1 is not small enough, remove this additional multiple of X_1 from $y_{\cdot 12}$ and adjust c_1; otherwise go to step 6.
5. If the slope of the adjusted $y_{\cdot 12}$ against $X_{2\cdot 1}$ is not small enough, remove this additional multiple of $X_{2\cdot 1}$, adjust c_2, and return to step 4; otherwise go to step 6.
6. Using the final c_1 and c_2, calculate

$$b_1 = c_1 - c_2 d_1,$$

$$b_2 = c_2,$$

$$b_0 = \text{med}\{ y_i - b_1 x_{i1} - b_2 x_{i2} \}.$$

7. Form the residuals

$$y - \hat{y} = y - (b_0 + b_1 X_1 + b_2 X_2).$$

In a model with more carriers, we would need a separate cycle of improvement iterations for each additional carrier. For example, the process of obtaining $X_{3\cdot 12}$ would parallel that for $y_{\cdot 12}$ in steps 2 through 5 of the algorithm; and the calculations for $X_{4\cdot 123}$ would then cycle through steps of sweeping out X_1, $X_{2\cdot 1}$, and $X_{3\cdot 12}$.

This process of developing a resistant fit one variable at a time may appear vastly more laborious than fitting the same model by least squares, but that is largely illusory. Many of the good algorithms for calculating least-squares fits actually sweep out one carrier at a time, and some of them incorporate iterative improvement to increase numerical accuracy. (The iterative improvement would be unnecessary if a computer could perform arithmetic with infinite precision.) Sources for further discussion of these computational aspects of least-squares regression include Golub (1969), Kennedy and Gentle (1980, Section 8.1), and Lawson and Hanson (1974).

7C. EXAMPLE

A 1974 study of funding in the Massachusetts Department of Mental Health examined the relationship between the numbers of admissions to state hospitals and five demographic variables. From a master file of all persons admitted to state hospitals in 1973 it was possible to determine how many patients had come from each of 39 geographic areas in Massachusetts. The response variable is the number of admissions per 100,000 population in the area. We label this variable "RESP" as a continuing reminder of its special role in the analysis. Table 7-1 summarizes this variable and each of the five carriers considered in this exploratory analysis. Table 7-2 presents the raw data on each of the six variables for all 39 geographic areas.

Figures 7-2*a* through 7-2*e* plot the response against each of the five carriers, and Figure 7-2*f* shows the plot of reciprocal median family income against poverty. The equations of the associated resistant lines are indicated for each of the plots, and the line is drawn on each plot. High-leverage points are easy to spot in several of the plots; although such points would affect considerably the equations of least-squares lines, they do not unduly influence the resistant lines. The last plot, reciprocal median family income

TABLE 7-1. Description of variables for the Massachusetts mental health funding data set.

Variable Name	Description	Units
RESP	Number of hospital admissions as a fraction of population	Number per 100,000 population
PVRTY	Percentage of families whose incomes are below the federally defined poverty level	Tenths of a percent
SING	Fraction of population that is single (including divorced or widowed) and over 14 years of age	Fraction as a percent
RINC	The reciprocal of median family income	100 over median income in tens of thousands of dollars
INFM	Infant mortality rate	Deaths per 10,000 births
WARD	Number of child wards of the state as a fraction	Number per 10,000 children 18 years or younger

Source: Edwin B. Newman (1974). "Memorandum to Advisory Council on Mental Health and Mental Retardation re. 1974 Addendum to State Plan." Department of Psychology and Social Relations, Harvard University, June 12, 1974.

TABLE 7-2. Raw data for study of mental health admissions in Massachusetts.

Area	Number	X_1 PVRTY	X_2 SING	X_3 RINC	X_4 INFM	X_5 WARD	y RESP
Berkshire	1	61	38	127	144	50	92
Franklin–Hampshire	2	54	47	213	152	37	163
Holyoke–Chicopee	3	70	40	134	149	40	215
Springfield	4	81	40	131	219	67	210
Westfield	5	46	36	116	174	48	128
Fitchburg	6	63	37	127	171	50	239
Gardner–Athol	7	54	37	128	178	85	317
Blackstone Valley	8	44	35	114	125	58	195
Southbridge	9	53	36	123	213	78	108
Worcester	10	60	42	130	191	52	158
Cambridge–Somerville	11	80	52	177	194	59	263
Concord	12	25	34	104	96	19	101
Lowell	13	59	37	113	167	45	266
Metropolitan–Beaverbrook	14	42	42	112	118	32	290
Mystic Valley	15	34	37	91	165	31	169
Danvers	16	46	37	108	135	41	168
Haverhill–Newburyport	17	63	37	126	178	57	146
Lawrence	18	62	39	122	243	48	145
Lynn	19	67	39	121	107	64	195
Tri-Cities	20	54	41	122	176	63	254
Eastern Middlesex	21	31	37	96	109	46	99
Cape Ann	22	54	38	118	146	41	146
Westwood–Norwood	23	31	37	88	104	31	132
Newton–Weston–Wellesley	24	26	43	95	70	15	105
South Shore	25	41	39	101	156	38	131
Framingham	26	34	34	90	157	49	198
Westborough	27	42	35	101	136	49	237
Boston State	28	82	44	125	257	108	202
Boston University	29	202	57	231	276	153	461
Lindemann Mental Health Center	30	89	46	145	141	124	350
Massachusetts Mental Health Center	31	84	61	214	177	138	275
Tufts Mental Health Center	32	133	50	158	140	88	467
Cape Cod and Islands	33	85	35	148	198	54	201
Brockton	34	51	37	114	154	72	169
Fall River	35	91	37	144	152	70	253
Foxboro	36	45	37	113	119	40	202
New Bedford	37	102	38	151	171	57	195
Plymouth	38	48	33	106	151	84	140
Taunton	39	60	37	120	204	42	179

Source: Edwin B. Newman (1974). "Memorandum to Advisory Council on Mental Health and Mental Retardation re. 1974 Addendum to State Plan." Department of Psychology and Social Relations, Harvard University, June 12, 1974. His source: Massachusetts Department of Mental Health.

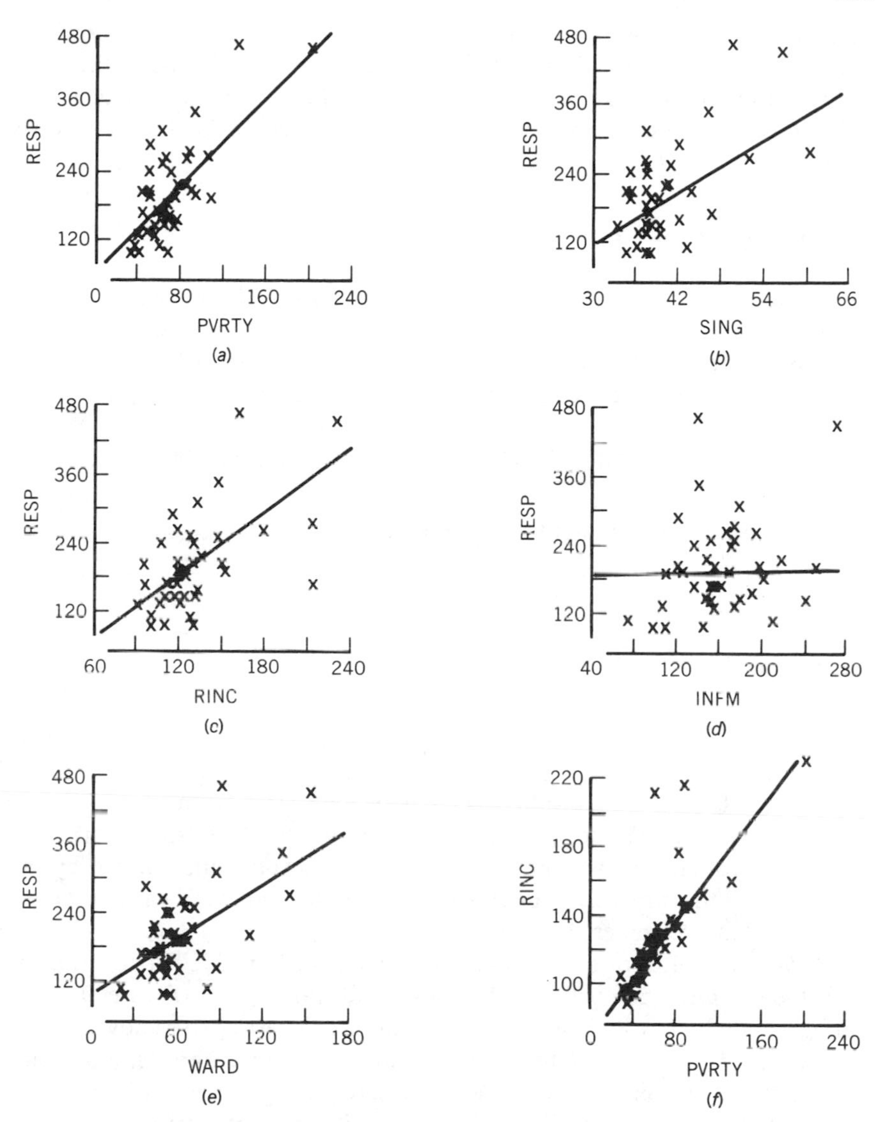

Figure 7-2. Plots of response against each of five carriers, and of reciprocal income against poverty. Equations of the resistant lines are: (*a*) RESP = 73.09 + 1.88PVRTY. (*b*) RESP = −93.08 + 7.08SING. (*c*) RESP = −26.87 + 1.80RINC. (*d*) RESP = 180.95 + 0.082INFM. (*e*) RESP = 105.03 + 1.57WARD. (*f*) RINC = 72.08 + .822PVRTY.

against poverty, illustrates well that carriers can be strongly correlated among themselves, and thus that two carriers can contain much the same information about the response. The scatter in the right-hand side of Figure 7-2*f* shows considerable variation in reciprocal income for higher poverty levels.

Removing the First Carrier

To decide which of the five carriers to remove from the response first, we compare the scatter of the five plots about the resistant lines. Because the response depends on poverty more strongly than on any other carrier, we remove poverty first. This step gives the resistant-line fit

$$\text{RESP}{\cdot}1 = \text{RESP} - (m_1 + c_1\text{PVRTY})$$

or, in the notation of the previous section,

$$y_{\cdot 1} = y - (m_1 + c_1 X_1).$$

Numerically, this equation may be rewritten as

$$y = 73.09 + 1.88X_1 + y_{\cdot 1}, \tag{15}$$

where $y_{\cdot 1}$ denotes the residual. For convenience in using the Minitab statistical package (Ryan, Joiner, and Ryan, 1976, 1982) to perform this step, we have removed not only poverty but also a constant, denoted by m_1. We continue this practice throughout the example, thereby always obtaining residuals that are centered at zero.

Because the plots of the response against each of the five carriers showed that the relationship between response and poverty in Figure 7-2*a* is strongest, we had little difficulty in deciding which carrier to remove first. In subsequent exploration of this example, and in other data sets, additional steps can aid in determining an order for removing carriers when several carriers have nearly equivalent importance. For example, we replaced data values for all six variables by their ranks and then considered both a least-squares multiple regression on the ranks and the rank correlation of the response with each of the five carriers. Using ranks instead of raw data values helps reduce the impact of wild data values on the selection process. (Other standard patterns of values, such as the expected values of the order statistics in a Gaussian sample of the same size, would serve the same purpose.)

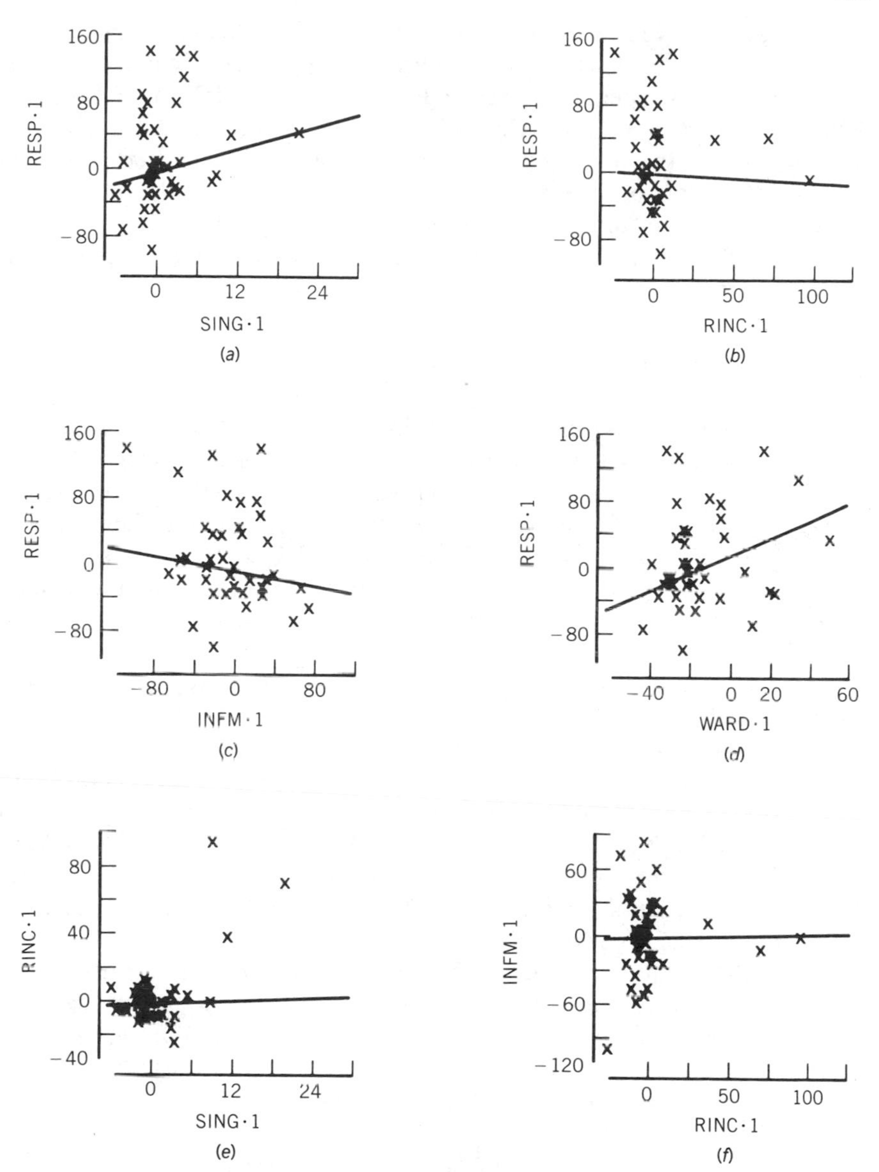

Figure 7-3. Plots of variables after resistant line for each variable against PVRTY has been removed. Equations of the resistant lines for the plots shown are: (*a*) RESP•1 = −3.38 + 2.38SING•1. (*b*) RESP•1 = −0.492 − 0.116RINC•1. (*c*) RESP•1 = −4.58 − 0.283INFM•1. (*d*) RESP•1 = −4.96 + 0.690WARD•1. (*e*) RINC•1 = −0.60 + 0.265SING•1. (*f*) INFM•1 = −1.41 + 0.138RINC•1.

We now remove poverty (X_1) from each of the other four carriers (X_2 through X_5) as well as from the response (y), thus obtaining sets of residuals which we denote by SING•1, RINC•1, INFM•1, WARD•1, and RESP•1. Figure 7-3 shows plots of the response (adjusted for poverty) against each of the other four carriers (also adjusted for poverty). It also includes two plots of selected carriers against each other; these plots show some clear outliers that affect the least-squares regression lines far more than they affect resistant lines, whose equations are provided in the caption for the figure.

Choosing and Removing a Second Carrier

From the plots in the first four panels of Figure 7-3, it is not easy to determine what carrier, if any, should be removed next. Larger values of RESP•1 tend to correspond to larger values of SING•1 and to smaller values of INFM•1, for example. The rank correlation for RESP•1 and SING•1 is .25, and that for RESP•1 and INFM•1 is .24. (A least-squares multiple regression based on the ranks of the five variables also indicated a tie between SING•1 and INFM•1 as the next variable to consider.) After exploring the data further by separately removing each of these variables, our analysis of the residuals suggested that INFM•1 is, by a small margin, the better choice for next removal.

We remove INFM•1 from SING•1, RINC•1, WARD•1, and RESP•1, obtaining residuals with the linear effects of both poverty and infant mortality removed. We denote the resulting adjusted variables by SING•14 (or $X_{2\cdot14}$), RINC•14 (or $X_{3\cdot14}$), WARD•14 (or $X_{5\cdot14}$), and RESP•14 (or $y_{\cdot14}$). At this stage we polish the coefficients by removing PVRTY (X_1) and INFM•1 ($X_{4\cdot1}$) iteratively from RESP•14. A total of four cycles are entirely adequate to achieve convergence.

Using the notation of Section 7B, we now assemble the results of our removal of poverty and infant mortality from the response variable. To facilitate later assembly of components into the completed fit, we show $y_{\cdot1}$ and $X_{4\cdot1}$ at the position in the equation usually occupied by the residual. Removing poverty (X_1) from the response y gives

$$y = m_1 + c_1 X_1 + y_{\cdot1},$$

$$y = (73.09 + 14.676 - 0.223) + (1.88 - 0.168 + 0.004) X_1 + y_{\cdot1}.$$

The numbers enclosed in parentheses represent initial values for intercept and slope and subsequent changes that result from refinement iterations.

Removing X_1 from X_4 gives

$$X_4 = 89.82 + 1.15X_1 + X_{4\cdot1},$$

and removal of $X_{4\cdot1}$ from $y_{\cdot1}$ gives, with subsequent refinement:

$$y_{\cdot1} = (-4.58 + 0.244 - 0.003)$$

$$+(-0.285 - 0.0447 + 0.0005)X_{4\cdot1} + y_{\cdot14}.$$

After incorporating the adjustments, we have three equations:

$$y = 87.54 + 1.72X_1 + y_{\cdot1},$$

$$X_4 = 89.82 + 1.15X_1 + X_{4\cdot1},$$

and

$$y_{\cdot1} = -4.34 - .329X_{4\cdot1} + y_{\cdot14}.$$

Assembling these equations to give y in terms of X_1, X_4, and the residuals, $y_{\cdot14}$, leads to

$$y = 112.8 + 2.10X_1 - .329X_4 + y_{\cdot14}$$

or, by subtracting $y_{\cdot14}$ from each side,

$$\hat{y} = 112.8 + 2.10X_1 - .329X_4. \tag{16}$$

Choosing and Removing a Third Carrier

At this stage of the analysis, we have variables SING•14 ($X_{2\cdot14}$ in the more general notation), RINC•14 ($X_{3\cdot14}$), WARD•14 ($X_{5\cdot14}$), and RESP•14 ($y_{\cdot14}$). We have polished each of these in the iterative removal of X_1 and $X_{4\cdot1}$. Plots of RESP•14 against each of the three adjusted carriers reveal very little relationship; we omit the plots here. Although it is unclear that we should add a further carrier to the fit, we proceed to remove the linear effect of SING through SING•14. This allows us to further illustrate the steps of the resistant-line-one-at-a-time algorithm. Thus, our last phase of analysis begins with the removal of SING•14 from RINC•14, WARD•14, and RESP•14, the last giving residuals denoted by $y_{\cdot142}$ with poverty, infant mortality, and singleness removed. The equation for $y_{\cdot142}$ may be expressed as

$$y_{\cdot 142} = y_{\cdot 14} - (m_{2\cdot 14} + c_{2\cdot 14} X_{2\cdot 14}),$$

where the more detailed subscript notation for the slope and intercept coefficients shows which carriers have previously been removed, as well as the carrier to which they apply.

Numerically, the coefficients from the resistant fit of $X_{2\cdot 14}$ to $y_{\cdot 14}$ give

$$y_{\cdot 142} = y_{\cdot 14} - (-1.35 + 2.71 X_{2\cdot 14}).$$

To polish this result, we iteratively remove X_1, $X_{4\cdot 1}$, and $X_{2\cdot 14}$ from $y_{\cdot 142}$ until the changes to slope and intercept are negligible; once again, four full steps of this iteration are more than adequate for convergence. After making the iterative adjustments to the slope and intercept and using considerable algebraic manipulation, we obtain as our fit using three carriers

$$\hat{y} = 7.88 + 1.81\text{PVRTY} + 3.72\text{SING} - .458\text{INFM}$$

or

$$\hat{y} = 7.88 + 1.81 X_1 + 3.72 X_2 - .458 X_4. \tag{17}$$

Assessing the Fits

To examine the fit given by equation (17), we plot the residuals, $y_{\cdot 142}$, against each of the carriers in the fit. Figure 7-4 provides these plots as well as a plot of the residuals against the fitted values. Each of the plots shows negligible relationship between the residuals and the variable plotted against.

Figure 7-5 displays parallel boxplots for the three resistant fits developed in this section. For comparison, it also includes a boxplot of differences between the response and the median response—that is, the residuals from a constant fit. As we expect, the residuals shrink as carriers are added. The fourth-spreads for the four sets of residuals (100.5, 70.7, 69.2, and 66.5, respectively) show the same monotonic decrease. The sums of absolute residuals for the four fits are (*a*) 2421, (*b*) 1738, (*c*) 1673, and (*d*) 1607. Each of the measures of assessment we have adopted suggests that, once poverty has been added as a carrier, adding further carriers produces only modest improvements in the fit.

One may ask whether the order in which we add carriers in our resistant-fitting algorithm affects the overall fit. Our experience with this

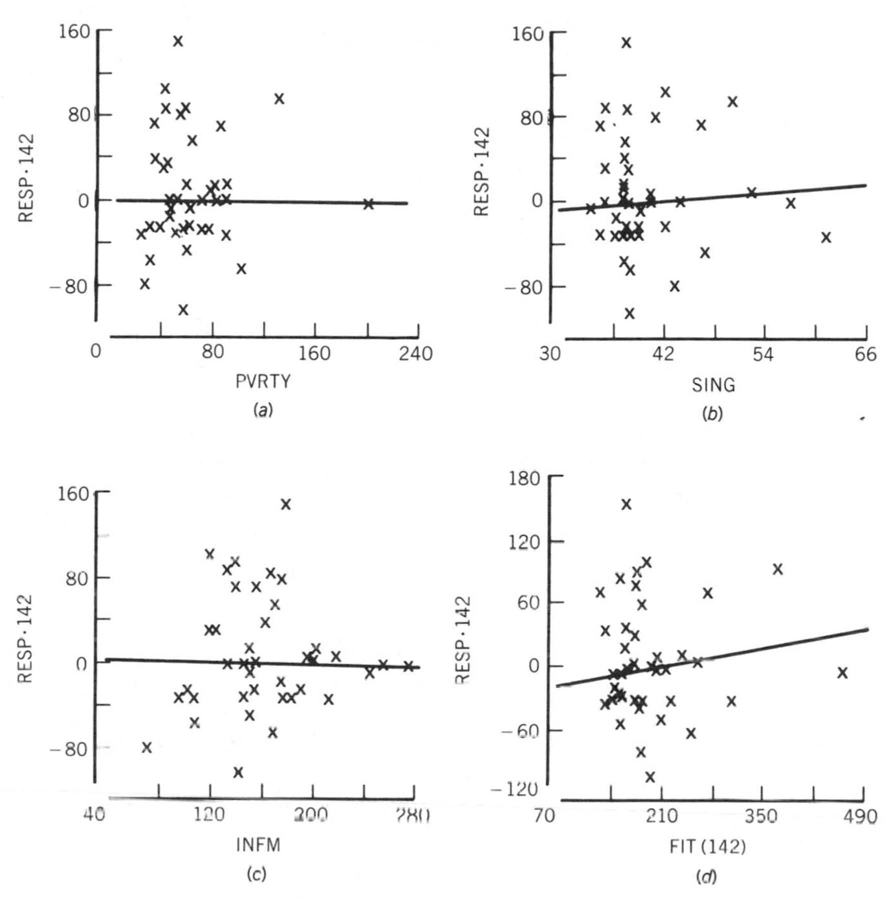

Figure 7-4. Residuals (RESP·142) plotted against each carrier used in fit, and against the fitted values. The equations of the resistant lines are: (a) $y_{\cdot 142} = -.001X_1 + .087$. ($b$) $y_{\cdot 142} = .560X_2 - 23.2$. ($c$) $y_{\cdot 142} = -.011X_4 + 2.58$. ($d$) $y_{\cdot 142} = .143\hat{y} - 29.4$.

data set suggests that order can matter, but to a rather modest extent. For example, with PVRTY already a carrier, adding SING as a carrier before adding INFM reduces the sum of absolute values of the residuals, when all three carriers are used, from 1607 to 1604. We cannot guarantee, however, that more substantial changes will not occur when the order of entry of carriers is altered; thus we recommend that, when two carriers are of nearly equal importance, both orders for adding them be explored.

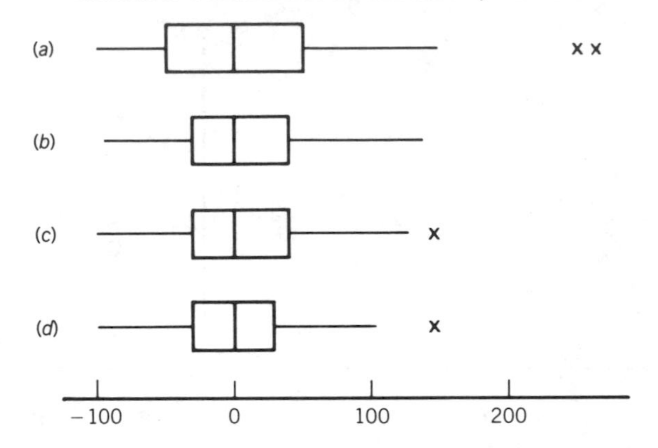

Figure 7-5. Parallel boxplots of residuals from four resistant fits to the Massachusetts mental health admissions data. (a) $\hat{y} = 195$. (b) $\hat{y} = 73.09 + 1.88X_1$. ($c$) $\hat{y} = 112.8 + 2.10X_1 - .329X_4$. ($d$) $\hat{y} = 7.88 + 1.81X_1 + 3.72X_2 - .458X_4$.

Comparing with Least Squares

Both this volume and UREDA have emphasized the differences between classical least-squares analysis and resistant techniques. When, as usually happens, we explore data with the aid of a statistical package like Minitab, we typically compare the results of a resistant analysis with those from a corresponding least-squares fit. For the mental health data, we have performed least-squares regression of the response against the following sets of carriers:

1. PVRTY.
2. PVRTY, INFM.
3. PVRTY, INFM, SING.
4. All five carriers.

Our exploration reveals that the data do contain a few high-leverage points and outliers (see Chapter 8); the exact numbers of such points depend on which carriers are used in the analysis. Even so, in this data set these points do not cause dramatic differences between the residuals from least-squares fits and the residuals from the corresponding resistant fits. There are some modest differences, and we now report on these.

Table 7-3 presents four different ways of comparing the resistant fits with the corresponding least-squares fit: the equation of fit, the sum of squared residuals, the sum of absolute residuals, and the fourth-spread of the

TABLE 7-3. Comparison of least-squares fits and residuals with the corresponding resistant fits and residuals. The first entries are for resistant fits, and the second entries for least-squares fits.

Carriers in the Fit	Equation of Fit	Sum of Squared Residuals	Sum of Absolute Residuals	Fourth-Spread of Residuals
PVRTY	$\hat{y} = 73.09 + 1.88 X_1$	136,127	1738.4	70.7
	$\hat{y}_{LS} = 77.31 + 2.02 X_1$	128,742	1815.7	71.8
PVRTY, INFM	$\hat{y} = 112.8 + 2.10 X_1 - .329 X_4$	127,604	1673.8	69.2
	$\hat{y}_{LS} = 128.6 + 2.37 X_1 - .452 X_4$	118,946	1739.7	70.3
PVRTY, INFM, SING	$\hat{y} = 7.88 + 1.81 X_1 + 3.72 X_2 - .458 X_4$	117,664	1607.2	66.5
	$\hat{y}_{LS} = 41.1 + 2.03 X_1 + 2.59 X_2 - .418 X_4$	113,210	1670.4	65.2
All five carriers[a]	$\hat{y}_{LS} = 49.6 + 2.04 X_1 + 3.80 X_2 - .680 X_3 - 4.51 X_4 + .581 X_5$	103,381	1637.1	62.2

[a] We did not attempt a resistant fit that used all five carriers.

residuals. We present this variety of ways to compare fits because no single objective function for assessing fits is yet considered most appropriate for exploratory analysis. We provide the numerical results needed for these comparisons, for each of three sets of carriers in the fit, and we also give the results for a least-squares fit using all five carriers.

For each of the sets of carriers where we obtained both a resistant fit and a least-squares fit, the corresponding coefficients show differences between the two fits. For example, when PVRTY and INFM are used as carriers, the resistant-fit coefficient for INFM is −.329, whereas the least-squares coefficient of INFM is −.452. In no situation, however, is a coefficient in a resistant fit more than one (least-squares) standard deviation away from the corresponding coefficient produced by least squares. As least-squares theory predicts, the sums of squared residuals for least-squares fits are never larger than those from the resistant fits. But the sums of absolute residuals for the resistant fits are always smaller than the corresponding sums for least squares. Differences between the fourth-spreads for each of the three pairs of fits appear negligible. Overall, we believe that, for exploring these data, our resistant fits are at least as good as those from least squares. We have also gained the assurance that our fits have substantial resistance to a few unusual data values, and this feature is, of course, the primary advantage of using resistant techniques for exploratory analyses.

Another way to compare the residuals from two fitting strategies plots the difference in residuals against one set of residuals. Figure 7-6 plots the differences between the residual from a least-squares fit using three carriers and the residual from the resistant fit against the residual from the resistant fit. If the two fits coincided, all points would lie on a line of slope 0 through the origin. Because the fits differ, this plot exhibits considerable scatter. Most differences are negative, because least-squares fitting centers residuals at the mean, whereas the resistant fitting centers residuals at the median. The least-squares residuals have a median of −11.9, whereas the residuals from the resistant fit have zero median; the median of the differences in the residuals is −8.33. Figure 7-6 includes a horizontal line with intercept at −8.33; the plot exhibits little systematic pattern about this reference line.

Although this example has served primarily to illustrate the process of resistant fitting, one variable at a time, we now return briefly to the original objective: rough prediction of prospective use of state mental health facilities. The percentage of families below the poverty level in an area emerged as the best single predictor of the area's rate of admissions to state hospitals. Furthermore, none of the other variables seemed to improve the prediction enough to justify including them in the fit. The use of poverty alone has an appealing simplicity, but we would want to try it for other years with cleaner data on the response variable (more than a third of the patients in

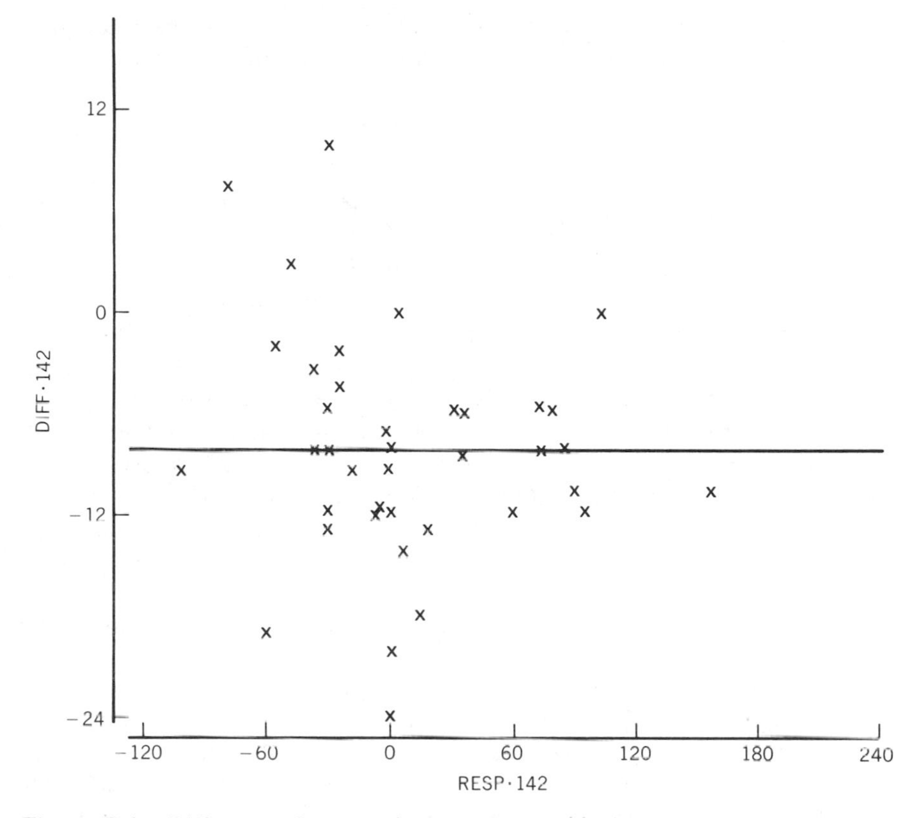

Figure 7-6. Differences between least-squares residuals and residuals from the resistant fit plotted against residuals from resistant fit. Each fit uses three carriers. The line has slope 0 and intercept −8.33, the median of the differences.

1973 could not be counted in the admission rate because no record of an address was made).

7D. WHEN CARRIERS COME IN BLOCKS

In a number of data structures the values of the carriers follow patterns that allow us to streamline the fitting process by working with carriers in blocks, rather than one at a time. Often, many of the carriers take only two values, conventionally 0 and 1.

One simple example is the two-way table, whose customary decomposition into common value, row effects, column effects, and residuals,

$$y_{ij} = m + a_i + b_j + e_{ij} \tag{18}$$

($i = 1, \ldots, I$; $j = 1, \ldots, J$), does not show the carriers explicitly. To write equation (18) in terms of carriers, we would use indicator variables, whose only values are 0 and 1. In the form of equation (8), we have one carrier for each of the coefficients $m, a_1, \ldots, a_I, b_1, \ldots, b_J$. (In actual fitting we discard one of these carriers in centering the a_i and another in centering the b_j, so that $p = I + J - 2$.) Thus, for example, m is associated with the constant carrier and plays the same role as b_0 in equation (8), the carrier for a_i takes the value 1 for all observations in row i of the table and the value 0 otherwise, and the carrier for b_j takes the value 1 for all observations in column j of the table and 0 otherwise. Here y and the carriers each have IJ observations. Ordinarily it is convenient to work with the carriers for the a_i as one block and those for the b_j as another. The specialized resistant technique known as median polish for two-way tables (UREDA, Section 6B) operates in precisely this way by sweeping out medians and alternating between the rows and the columns. In this way the carriers in a block are essentially handled in parallel.

Although the simplicity and familiarity of the two-way table make it a convenient illustration, we would seldom consider analyzing it by applying a resistant line one carrier at a time. Exploiting the special nature of the carriers, as median polish does, requires far less effort. In another family of data structures, however, the carriers include both indicator variables and explanatory variables of the sort that usually serve as x in y versus x. Among the simplest is the structure often analyzed (by least squares) by one-way analysis of covariance, for which we write the fit

$$\hat{y}_{ij} = m + a_i + cz_{ij}$$

($i = 1, \ldots, I$; $j = 1, \ldots, n_i$) in terms of a common value (m), an effect (a_i) for group i, and a slope (c) against the covariate or concomitant variable (z). Here the carriers for the group effects form a natural block, to be swept out before calculating the slope against z. A simple resistant procedure could remove group medians from both y and z and then apply a resistant line to centered y versus centered z.

To examine the treatment of blocks in more detail, we turn to the two-way table in which each cell contains one observation on the response variable (y) and one observation on the concomitant variable (z). The decomposition into fit and residuals can take the form

$$y_{ij} = m + a_i + b_j + cz_{ij} + e_{ij} \tag{19}$$

($i = 1, \ldots, I$; $j = 1, \ldots, J$). Taking the entire additive fit [given by m, the a_i, and the b_j as equation (18) shows] as a block of carriers, we rewrite equation (19) with this block swept out of z:

$$\begin{aligned} y_{ij} = m^*(y) &+ a_i^*(y) + b_j^*(y) \\ &+ c\left[z_{ij} - m(z) - a_i(z) - b_j(z)\right] \\ &+ e_{ij}. \end{aligned} \tag{20}$$

Rather than introduce a flood of new symbols to distinguish the two-way additive fit for z from that for y, we simply attach a parenthesized y or z to the common value, row effects, and column effects. Equation (20) is entirely analogous to

$$\hat{y} = b_0 + c_1 X_1 + b_2 X_{2 \cdot 1}$$

as discussed in Section 7B, and the starred coefficients for y are related to the unstarred ones in the same way as c_1 is related to b_1 [for example, through equation (14)].

Guided by equation (20), we proceed algorithmically as follows:

1. Sweep the additive fit out of z to get z-residuals,

$$e_{ij}(z) = z_{ij} - m(z) - a_i(z) - b_j(z).$$

2. Sweep the additive fit out of y to obtain y-residuals

$$e_{ij}(y) = y_{ij} - m^*(y) - a_i^*(y) - b_j^*(y).$$

3. Calculate and remove a resistant slope from $e_{ij}(y)$ versus $e_{ij}(z)$:

$$e_{ij}(y) - ce_{ij}(z).$$

4. If the value of c is small enough, stop; otherwise return to step 2 with the current residuals from step 3 in place of y, in order to calculate adjustments to $m^*(y)$, the $a_i^*(y)$, and the $b_j^*(y)$ in step 2 and an adjustment to c in step 3.

To recover the coefficients in equation (19) from those fitted according to equation (20) requires only simple algebra:

$$
\begin{aligned}
m &= m^*(y) - cm(z), \\
a_i &= a_i^*(y) - ca_i(z), \qquad i = 1, \ldots, I, \\
b_j &= b_j^*(y) - cb_j(z), \qquad j = 1, \ldots, J.
\end{aligned} \tag{21}
$$

These a_i and b_j are analogous to the adjusted effects in the corresponding analysis of covariance.

In implementing the algorithm above, we generally find it convenient to obtain the additive two-way fits in steps 1 and 2 by median polish and to use the three-group resistant line in step 3. It is straightforward to set up these steps in the Minitab statistical package (Ryan, Joiner, and Ryan 1976, 1982).

We now apply this resistant approach in an example and compare the results to those obtained by least squares.

EXAMPLE: CORN YIELDS

Snedecor and Cochran (1980) present a two-way analysis of covariance involving the yields of six varieties of corn. Table 7-4 shows the data. The number of plants (referred to as the "stand") varied somewhat from plot to plot, and this variable may account for some of the variation in yield, because higher yields may reflect (in part) larger numbers of plants. Thus the analysis needs to adjust for this contribution in order to make comparisons among the varieties more precise. One approach would take yield per plant as the response variable instead of yield and then bring in stand as a concomitant variable if needed. To maintain comparability with the customary least-squares analysis given by Snedecor and Cochran, however, we proceed to analyze yield, with stand as a concomitant variable. Exercise 6 considers analyses of yield per plant.

In situations where it is appropriate to consider dividing the original response variable by the concomitant variable, a further alternative approach would re-express both variables in the log scale. Then (a) analyzing log(yield) with log(stand) as covariate and (b) analyzing log(yield/plant) with log(stand) as covariate would give equivalent results. We do not pursue this possibility in the present example.

Tables 7-5 and 7-6 show the results of using median polish to sweep out the two-way additive fit for stand (z) and yield (y), respectively. In each, the row effects, column effects, and common value border the table of

TABLE 7-4. Data on yield of six varieties of corn in four randomized blocks. The upper entry in each cell is the yield y (in pounds field weight of ear corn), and the lower entry is the number of plants z (the stand).

	Block			
Variety	1	2	3	4
A	202 28	165 22	191 27	134 19
B	145 23	201 26	203 28	180 24
C	188 27	185 24	185 27	220 28
D	201 24	231 28	238 30	261 30
E	202 30	178 26	198 26	226 29
F	228 30	221 25	207 27	204 24

Source: G. W. Snedecor and W. G. Cochran (1980). *Statistical Methods*, 7th ed. Ames, IA: Iowa State University Press (data from Table 18.5.2, p. 378).

residuals. Before fitting any resistant slope we plot the residual of yield against the residual of stand. Figure 7-7 shows a fairly strong linear relationship between $e_{ij}(y)$ and $e_{ij}(z)$, and it also reveals that two points lie substantially above any reasonable line that would summarize this relationship. Thus the resistant analysis has an opportunity to avoid the adverse effects of these two points.

Removing the resistant slope against $e_{ij}(z)$ from $e_{ij}(y)$ and continuing with a few cycles of adjustment yield the fit [expressed in terms of adjusted effects—see equation (21)] in Table 7-7. The large positive residuals for variety *D* in block 1 and variety *E* in block 3 stand out clearly. Figure 7-8 shows three sets of residuals of yield: the initial $e_{ij}(y)$ in Table 7-6, after

TABLE 7-5. Analysis of data on stand (z) by median polish.[a] The variety effects, block effects, and common value border the table of residuals.

	Block				
Variety	1	2	3	4	Effect
A	1.58	−1.25	1.25	−5.25	−1.25
B	−4.42	1.75	1.25	−1.25	−0.25
C	−0.42	−0.25	0.25	2.75	−0.25
D	−6.92	0.25	−0.25	1.25	3.25
E	0.42	−0.42	−2.92	1.58	1.92
F	2.08	0.25	−0.25	−1.75	0.25
Effect	1.42	−1.75	0.75	−0.75	26.25

[a] Using 10 iterations (specified by the subcommand ITERATIONS = 20 after the MPOLISH command in Minitab).

TABLE 7-6. Analysis of data on yield (y) by median polish.[a] The variety effects, block effects, and common value border the table of residuals.

	Block				
Variety	1	2	3	4	Effect
A	20.5	−10.8	10.8	−54.7	−13.5
B	−43.6	18.1	15.8	−15.8	−6.4
C	−1.3	1.3	−3.0	23.4	−5.7
D	−37.0	−1.3	1.3	15.8	43.0
E	1.3	−17.0	−1.3	18.1	5.7
F	10.5	9.2	−9.2	−20.7	22.5
Effect	0.7	−5.0	−0.7	7.9	194.3

[a] Using 10 iterations.

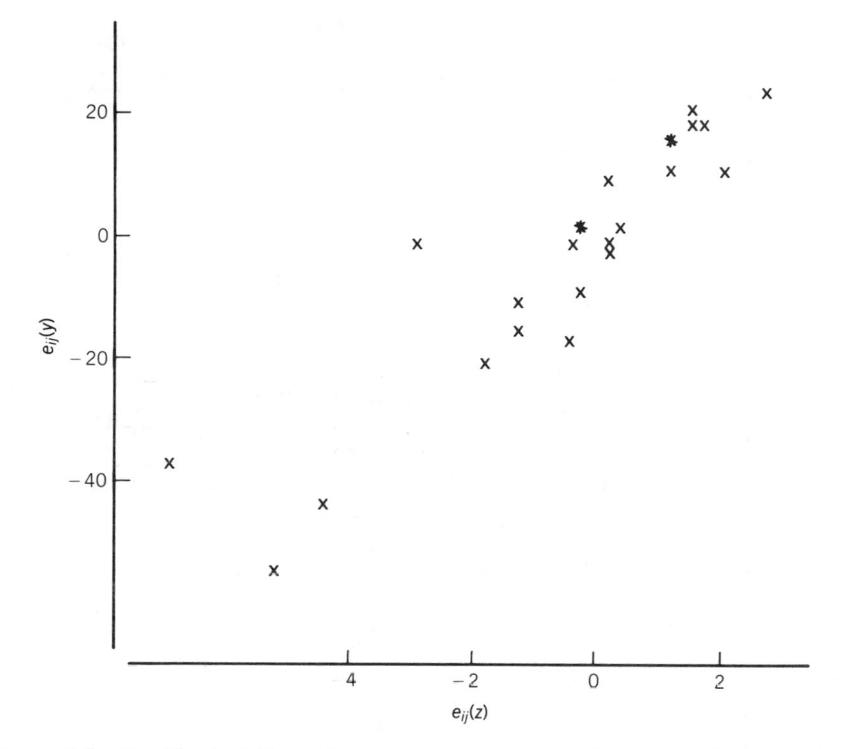

Figure 7-7. Residual yield, $e_{ij}(y)$, versus residual stand, $e_{ij}(z)$, after removing variety effects, block effects, and common value by median polish.

TABLE 7-7. Resistant fit of equation (19) to the data on yield. The value of c is 10.03, and the adjusted effects and common value border the table of residuals.

	Block				
Variety	1	2	3	4	Effect
A	1.6	−0.0	−0.1	−0.1	0.3
B	−1.1	0.0	6.1	−0.0	−3.9
C	1.0	3.3	−2.7	−1.0	−3.0
D	24.0	−10.9	0.1	−0.1	17.0
E	−5.0	−13.7	30.3	5.1	−13.1
F	−10.3	8.0	−2.0	1.9	18.2
Effect	−10.6	14.2	−9.8	13.4	−70.1

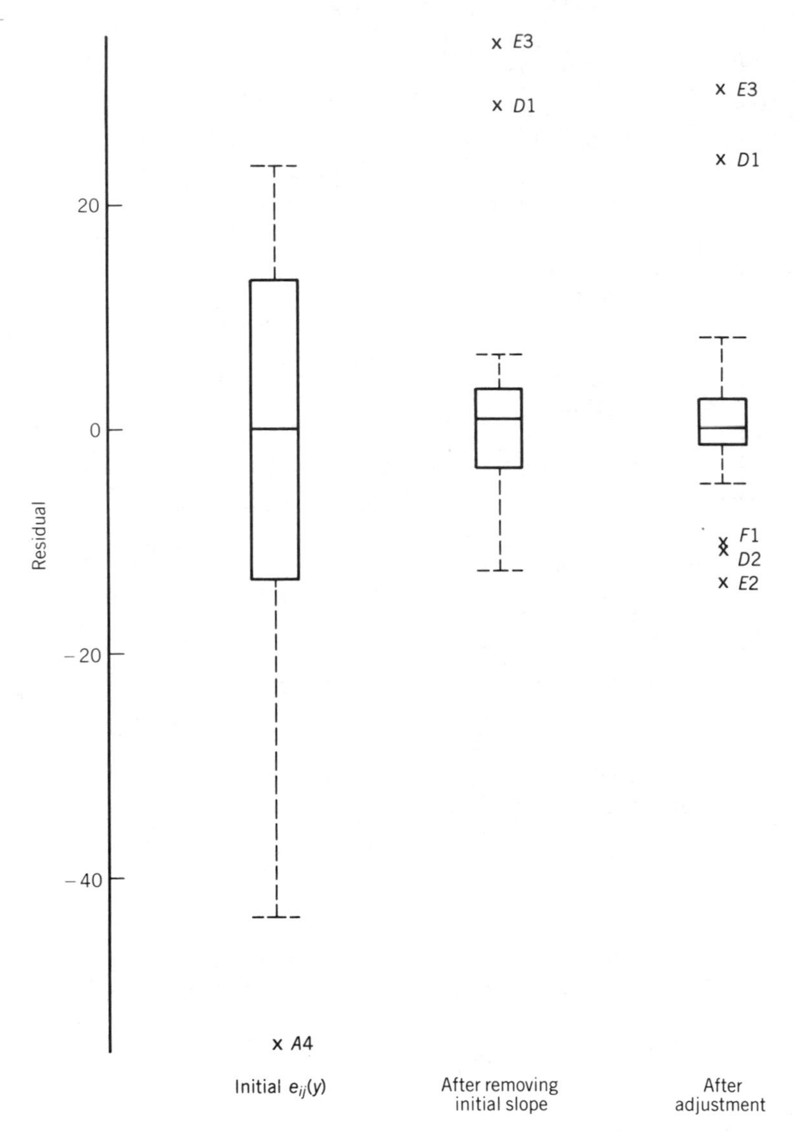

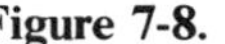

Figure 7-8. Parallel boxplots for three sets of residuals of yield.

removing the initial slope against $e_{ij}(z)$, and after all the cycles of adjustment. Taking the covariate into account has allowed us to summarize a substantial part of the variation in yield that remained after removing effects for variety and block. Among other things, we would now have a much smaller estimate of the residual variability in y.

An interesting second message in Figure 7-8 is that the behavior of variety A in block 4, which seemed unusual after removing only the additive two-way fit, is no longer extreme; the value of the covariate in that cell brought it back into line. Now cells $D1$ and $E3$ have appeared as outlying on the positive side (these were the two points that stood out in Figure 7-7), and $F1$, $D2$, and $E2$ on the negative side.

It is informative to compare the resistant fit and the least-squares fit, which we record in Table 7-8. The most important difference appears in the value of c, which is 10.03 in the resistant fit but only 8.06 in the least-squares fit, which is too sensitive to the two discrepant points in Figure 7-7.

The large difference between the two common values, -70.1 versus -12.4, comes primarily from the difference between the values of c; the value of $m(z)$ is 26.25 in the resistant analysis and 26.33 in the least-squares analysis.

Qualitatively the patterns of effects, shown in Figure 7-9, are not dramatically different, although the orders have changed somewhat and the spacings are not the same. The different conventions for centering the effects mean that we must look at differences among effects rather than at the individual numerical values, but this is just what we would do anyway in

TABLE 7-8. Least-squares fit of equation (19) to the data on yield. The value of c is 8.06, and the adjusted effects and common value border the table of residuals.

	Block				
Variety	1	2	3	4	Effect
A	7.6	1.6	−0.7	−8.5	−7.9
B	−8.3	6.2	4.1	−2.0	−8.8
C	0.3	4.2	−8.0	3.6	−6.6
D	11.3	−8.2	−5.4	2.3	19.6
E	−6.3	−15.4	16.6	5.1	−10.2
F	−4.4	11.6	−6.5	−0.6	13.9
Effect	−10.8	6.5	−5.5	9.8	−12.4

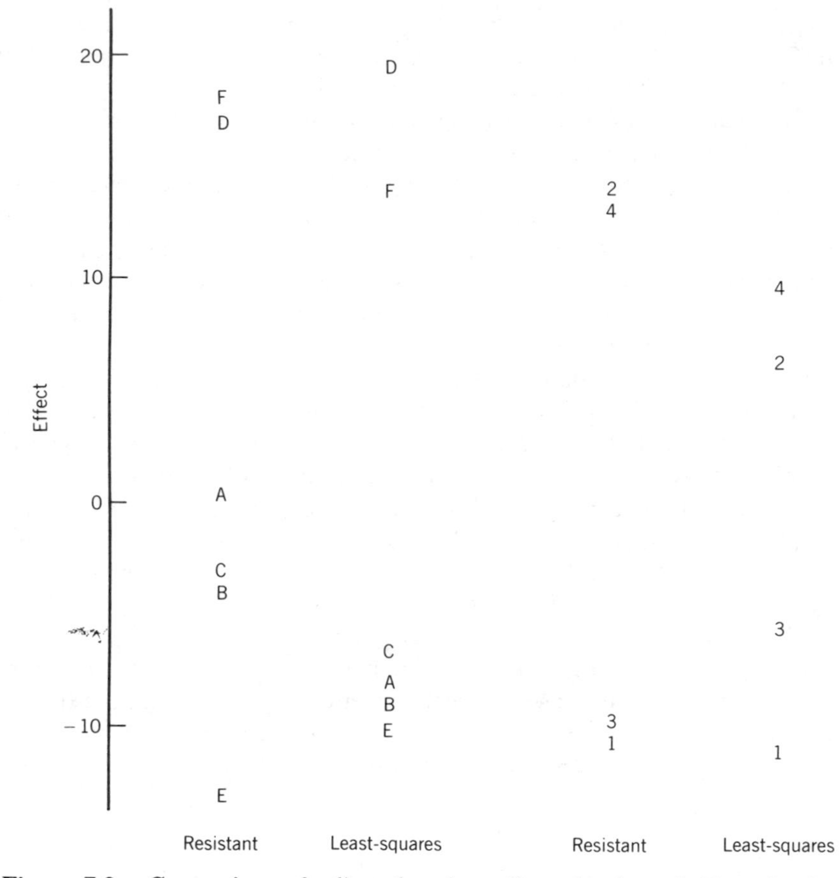

Figure 7-9. Comparison of adjusted variety effects (A through F) and adjusted block effects (1 through 4) obtained from the resistant fit and the least-squares fit.

comparing varieties or blocks. We would still tend to feel that varieties *D* and *F* have higher yield than the other four and that blocks 2 and 4 have higher fertility.

To compare the two sets of residuals, Figure 7-10 presents boxplots. The difference is striking. By allowing a few data values to produce large residuals, the resistant analysis has arrived at a fit whose residuals generally are considerably smaller than those from the least-squares fit. Indeed, the interquartile range of the resistant residuals is less than half that for the least-squares residuals.

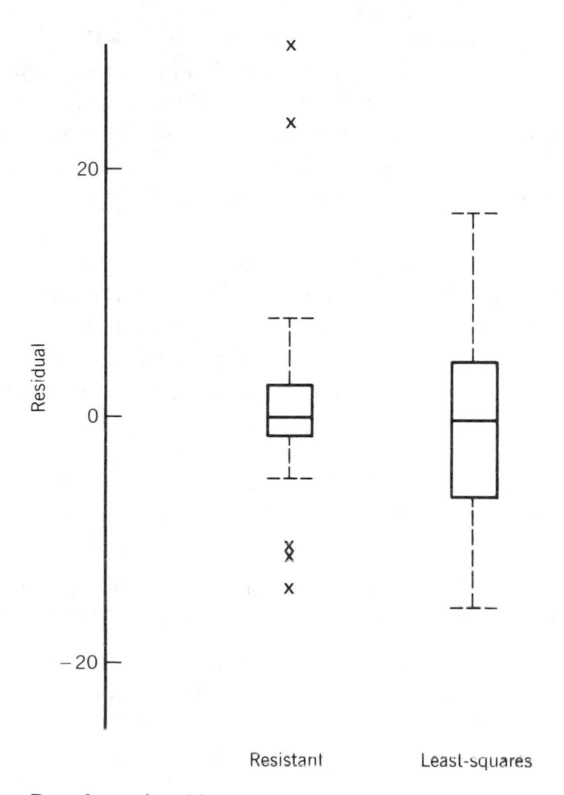

Figure 7-10. Boxplots of residuals from the resistant fit and the least-squares fit of equation (19) to the corn yield data.

7E. SUMMARY

Without aiming to produce a refined fit, the strategy of applying a resistant line one carrier at a time allows any line-fitting technique for y versus x to be extended to multiple regression models. The three-group resistant line serves as the basis for concrete examples.

In algorithmic terms the idea is to reduce the multicarrier fitting problem to a sequence of single-carrier problems by successive sweeping out. Thus, for example, fitting y in terms of the carriers 1, X_1, X_2, and X_3 would initially change the carriers to 1, X_1, $X_{2\cdot 1}$, and $X_{3\cdot 12}$ and then work with y versus X_1, $y_{\cdot 1}$ versus $X_{2\cdot 1}$, and $y_{\cdot 12}$ versus $X_{3\cdot 12}$. Iterative refinement concludes the process.

Although this approach may be most convenient in adding a carrier to an existing resistant fit, one of its most important applications comes in providing a starting point for the iterative calculations of a robust multiple-regression M-estimator or W-estimator. In this role the resistant technique generally avoids adverse effects associated with using least squares or least absolute residuals when the data contain anomalous y-values or points of high leverage.

Some data structures, such as those usually handled by analysis of covariance, combine a systematic, often balanced, structure (e.g., a two-way table) and one or more unstructured regression carriers. For these it is usually more convenient to use a special-purpose fitting technique for sweeping out the systematic part and then repeated resistant lines for the remaining carriers.

REFERENCES

Eden, T. and Fisher, R. A. (1927). "Studies in crop variation. IV. The experimental determination of the value of top dressings with cereals," *Journal of Agricultural Science*, **17**, 548–562.

Golub, G. H. (1969). "Matrix decompositions and statistical calculations." In R. C. Milton and J. A. Nelder (Eds.), *Statistical Computation*. New York: Academic Press, pp., 365–397.

Kennedy, W. J., Jr., and Gentle, J. E. (1980). *Statistical Computing*. New York: Marcel Dekker, Inc.

Lawson, C. L. and Hanson, R. J. (1974). *Solving Least Squares Problems*. Englewood Cliffs, NJ: Prentice-Hall.

Mosteller, F. and Tukey, J. W. (1977). *Data Analysis and Regression*. Reading MA: Addison-Wesley.

Ryan, T. A., Joiner, B. L., and Ryan, B. F. (1976). *Minitab Student Handbook*. North Scituate, MA: Duxbury Press.

Ryan, T. A., Joiner, B. L., and Ryan, B. F. (1982). *Minitab Reference Manual*. University Park, PA: Minitab Project, The Pennsylvania State University.

Snedecor, G. W. (1946). *Statistical Methods*, 4th ed. Ames, IA: Iowa State College Press.

Snedecor, G. W. and Cochran, W. G. (1980). *Statistical Methods*, 7th ed. Ames, IA: Iowa State University Press.

Velleman, P. F. and Hoaglin, D. C. (1981). *Applications, Basics, and Computing of Exploratory Data Analysis*. Boston: Duxbury Press.

Additional Literature

Allen, D. M. and Cady, F. B. (1982). *Analyzing Experimental Data by Regression*. Belmont, CA: Lifetime Learning Publications.

Andrews, D. F. (1974). "A robust method for multiple linear regression," *Technometrics*, **16**, 523–531.

Beaton, A. E. and Tukey, J. W. (1974). "The fitting of power series, meaning polynomials, illustrated on band-spectroscopic data," *Technometrics*, **16**, 147–192 (with discussion).

Birch, J. B. (1983). "On the power of robust tests in analysis of covariance," *Communications in Statistics—Simulation and Computation*, **12**, 159–182.

Birch, J. B. and Myers, R. H. (1982). "Robust analysis of covariance," *Biometrics*, **38**, 699–713.

Bloomfield, P. and Steiger, W. L. (1983). *Least Absolute Deviations: Theory, Applications, and Algorithms*. Boston: Birkhäuser.

Daniel, C. and Wood, F. S. (1980). *Fitting Equations to Data*, 2nd ed. New York: Wiley.

Henderson, H. V. and Velleman, P. F. (1981). "Building multiple regression models interactively," *Biometrics*, **37**, 391–411.

Hoaglin, D. C. and Welsch, R. E. (1978). "The hat matrix in regression and ANOVA," *The American Statistician*, **32**, 17–22, 146.

Hocking, R. R. (1983). "Developments in linear regression methodology: 1959–1982," *Technometrics*, **25**, 219–249 (with discussion).

Hocking, R. R. and Pendleton, O. J. (1983). "The regression dilemma," *Communications in Statistics—Theory and Methods*, **12**, 497–527.

Johnstone, I. M. and Velleman, P. F. (1985). "The resistant line and related regression methods," *Journal of the American Statistical Association*, **80**, to appear.

Schrader, R. M. and McKean, J. W. (1977). "Robust analysis of variance," *Communications in Statistics—Theory and Methods*, **A6**, 879–894.

EXERCISES

1. Section 7B asserts that we may choose to calculate b_0 once and for all, after the improvement cycle is concluded, rather than by making small adjustments at each iteration.

(a) Summarize the seven steps of the algorithm presented in Section 7B if small adjustments are made at each iteration; this method may be convenient when using a computer package for the computations, as Section 7C illustrates.

(b) Show, algebraically, why the value of b_0 must ultimately be the same, regardless of which form of the algorithm is selected.

2. (This exercise involves much computation. Release 82.1 of the Minitab statistical package is ideal for performing the calculations.) Consider the data on mental health admissions presented in Table 7-2. Suppose that, in addition to the 39 given data points, there was also a point $(X_1, X_2, X_3, X_4, X_5, y) = (202, 61, 88, 70, 15, 467)$.

(a) Give a resistant analysis of the modified data set using at least three carriers besides the constant.

(b) Does the outlier affect your selection of variables?

(c) Give the corresponding least-squares analysis and compare the results. Which analysis is more affected by the outlier, and how is it affected?

3. Suppose that you had the data for five carriers given in Table 7-2, but you did not have data on y. Instead, you are focusing on infant mortality (X_4) as a response to the four carriers X_1, X_2, X_3, and X_5.
   (a) Give a resistant exploratory analysis of this data set. How many, and which, carriers do you recommend keeping in the fit?
   (b) Compare and contrast your final fit with a least-squares fit that uses the same carriers.

4. Table 7-3 presents three different "goodness criteria" for comparing various fits to multiple-regression data.
   (a) What are the advantages of each of the three criteria? What are the disadvantages?
   (b) Which of the three criteria seems most suited for comparing exploratory fits to a data set? Can you think of other suitable criteria?
   (c) When data are known to come from a multivariate normal distribution, what mathematical considerations favor least squares as a criterion for assessing fits and for making inferences? (There may be many, and this question is somewhat open-ended.)

5. Section 7D compares a resistant analysis of the corn yield data with a least-squares analysis of covariance, and differences do emerge. Suppose now that an error in data entry on a computer changed the stand for variety A and block 1 (see Table 7-4) from 28 to 288.
   (a) Perform a resistant analysis like that of Section 7D and describe the impact of the error on the results of your resistant fit.
   (b) Do a least-squares analysis of covariance and contrast these results with those from Table 7-8.
   (c) Use parallel boxplots to compare your results from parts (a) and (b) to each other and to those shown in Figure 7-10. Discuss and interpret the outcome.

6. Reanalyze the data on corn yield in Table 7-4, taking yield per plant as the response variable. After an initial median-polish analysis, how much does including stand as a concomitant variable improve the fit?

7. Eden and Fisher (1927) describe an experiment that investigated three features of the application of nitrogenous fertilizers to a crop of oats: *timing* of the application during the growth of the crop (E = early, L = late), *form* of nitrogen (S = sulphate of ammonia, M = muriate of ammonia), and *amount* of fertilizer (1 and 2 cwt per acre of sulphate of ammonia and amounts of muriate of ammonia that yield the equiv-

alent amount of nitrogen). For the resulting eight treatments in eight randomized blocks, Table 7-9 gives the yields of grain and straw.

(a) With yield of straw as a concomitant variable, analyze the two-way table of grain yields according to the resistant strategy of Section 7D.

(b) Carry out a parallel analysis by least squares.

TABLE 7-9. Yields of grain and straw from an experiment on the application of nitrogenous fertilizers to a crop of oats.

a. Yield of grain (in eighths of a pound)

	Block							
Treatment	1	2	3	4	5	6	7	8
1*SE*	620	646	681	644	706	615	552	726
1*SL*	644	745	542	711	705	637	543	646
1*ME*	523	713	686	688	692	612	635	748
1*ML*	601	693	685	714	699	697	701	746
2*SE*	664	693	666	516	656	663	657	683
2*SL*	514	637	697	710	633	595	697	712
2*ME*	550	708	663	673	671	626	655	671
2*ML*	521	661	594	730	625	644	745	747

b. Yield of straw (in halves of a pound)

	Block							
Treatment	1	2	3	4	5	6	7	8
1*SE*	242	321	261	317	255	331	216	295
1*SL*	267	382	201	316	280	285	200	309
1*ME*	215	330	298	381	300	294	256	284
1*ML*	212	292	265	255	238	309	283	324
2*SE*	322	370	284	323	232	393	351	363
2*SL*	200	261	259	361	234	258	306	376
2*ME*	260	318	266	340	362	400	276	385
2*ML*	203	275	207	331	229	266	276	328

Source: T. Eden and R. A. Fisher (1927). "Studies in crop variation. IV. The experimental determination of the value of top dressings with cereals," *Journal of Agricultural Science*, **17**, 548–562 (data from Tables I and II, p. 556).

(c) Compare the least-squares and resistant analyses in terms of the adjusted treatment effects, the adjusted block effects, the slope against the concomitant variable, and the residuals.

8. A randomized block experiment on sugar beets involved selected combinations of superphosphate, muriate of potash, and sodium nitrate (Snedecor, 1946). Table 7-10 shows the yield and the number of beets per plot.

TABLE 7-10. Yield and number of beets per plot in a randomized block experiment on sugar beets. In each cell the upper entry is the yield (in tons per acre), and the lower entry is the number of beets. The fertilizer treatments combine superphosphate (*P*), muriate of potash (*K*), and sodium nitrate (*N*).

	Block					
Fertilizer	1	2	3	4	5	6
None	2.45 183	2.25 176	4.38 291	4.35 254	3.42 225	3.27 249
P	6.71 356	5.44 300	4.92 301	5.23 271	6.74 288	4.74 258
K	3.22 224	4.14 258	2.32 244	4.42 217	3.28 192	4.00 236
$P + K$	6.34 329	5.44 283	5.22 308	8.00 326	6.96 318	6.96 318
$P + N$	6.48 371	7.11 354	5.88 352	7.54 331	6.61 290	8.86 410
$K + N$	3.70 230	3.24 221	2.82 237	2.15 193	5.19 247	4.13 250
$P + K + N$	6.10 322	7.68 367	7.37 400	7.83 333	7.75 314	7.39 385

Source: George W. Snedecor (1946). *Statistical Methods*, 4th ed. Ames, IA: Iowa State College Press (data from Table 12.13, p. 332).

(a) Make and compare resistant and least-squares analyses of the two-way table of yields, taking the number of beets per plot as a concomitant variable.
(b) Discuss the results of analyzing yield and number of plants as two-way tables without using the number of plants as a concomitant variable.
(c) Divide yield by number of plants and analyze this response variable as a two-way table.
(d) Examine the contribution of using number of plants as a concomitant variable in the analysis of yield per plant.

CHAPTER 8

Robust Regression

Guoying Li
Institute of Systems Science
Academia Sinica
Beijing, People's Republic of China

The method of least squares has long dominated the literature and applications of regression techniques. Research on robust estimation of location, particularly during the early 1970s, naturally led statisticians to consider corresponding improvements in regression. The varied new proposals aim at providing at least one of (i) protection against distortion by anomalous data and (ii) good efficiency when the data come from the ideal Gaussian model, as well as from a range of alternative models.

Although skilled analysts have often been able to cope with ill-behaved data while using least-squares techniques, the structure of many regression data sets makes effective diagnosis and fitting a delicate matter, requiring numerous steps of interaction with data and tentative models. Part of the difficulty stems from the way that combinations of values of the explanatory variables can give some observations far greater influence than others. By limiting the impact of some types of outliers, robust regression often accelerates the process of analysis and calls attention to the unusual data. It does not, however, seek to automate regression analysis, because the diversity of structure in data prevents the careful analyst from relinquishing reasonable contact with the data. Indeed, no one robust regression technique has proved superior to all others in all situations, partly because of the challenge of handling the many forms of influential observations.

This chapter introduces some of the improvements that have brought progress toward robust regression. Section 8A establishes notation and discusses basic aspects of leverage and influence in least-squares regression. Next, Sections 8B and 8C introduce M-estimators for linear models and the closely related W-estimators, together with their computational aspects.

Some acquaintance with *M*- and *W*-estimators of location (as discussed, for example, in Chapter 11 of UREDA) will be helpful as background. In Section 8D a modest example illustrates these robust methods. Section 8E describes how some proposals for bounded-influence regression cope with the possibility of anomalous y-values in influential observations. Finally, Section 8F reviews some of the other techniques for robust regression.

Some readers may wish to postpone this chapter, which is inevitably more mathematical, until after some of those that follow.

8A. WHY ROBUST REGRESSION?

We first recall the ordinary least-squares (OLS) approach. The multiple regression model relating the response vector $\mathbf{y}$ to the explanatory variables in the matrix $\mathbf{X}$ has the form

$$\mathbf{y} = \mathbf{X}\boldsymbol{\beta} + \boldsymbol{\varepsilon} \tag{1}$$

or, in terms of the individual elements,

$$\begin{aligned} y_1 &= \beta_1 x_{11} + \beta_2 x_{12} + \cdots + \beta_p x_{1p} + \varepsilon_1, \\ y_2 &= \beta_1 x_{21} + \beta_2 x_{22} + \cdots + \beta_p x_{2p} + \varepsilon_2, \\ &\vdots \\ y_n &= \beta_1 x_{n1} + \beta_2 x_{n2} + \cdots + \beta_p x_{np} + \varepsilon_n. \end{aligned} \tag{2}$$

In the matrix notation the vector $\mathbf{y}$ consists of the n observations on the response variable. The $n \times p$ matrix $\mathbf{X}$ gives the n values of the p explanatory variables or carriers; when needed, we denote its rows by the *row vectors* $\mathbf{x}_1, \ldots, \mathbf{x}_n$. We assume that $\mathbf{X}$ has full rank. The elements of $\boldsymbol{\beta}$ are unknown parameters to be estimated. The elements of $\boldsymbol{\varepsilon}$ are unknown errors or fluctuations. For distribution theory, we customarily assume that the x_{ij} are fixed and $\varepsilon_1, \ldots, \varepsilon_n$ are independent and identically distributed with zero mean and finite variance. Usually the model has a constant term, and then all elements of the first column of $\mathbf{X}$ are 1:

$$\mathbf{X} = \begin{bmatrix} 1 & x_{12} & \cdots & x_{1p} \\ 1 & x_{22} & \cdots & x_{2p} \\ \vdots & \vdots & & \vdots \\ 1 & x_{n2} & \cdots & x_{np} \end{bmatrix}. \tag{3}$$

We continue to use β_1 for the corresponding parameter.

The method of ordinary least squares chooses the estimate $\hat{\boldsymbol{\beta}}$ as the value of $\boldsymbol{\beta}$ that minimizes the sum of squared residuals:

$$\sum_{i=1}^{n}\left(y_i - \sum_{j=1}^{p} x_{ij}\beta_j\right)^2 = \sum_{i=1}^{n}(y_i - \mathbf{x}_i\boldsymbol{\beta})^2 \tag{4}$$

$$= (\mathbf{y} - \mathbf{X}\boldsymbol{\beta})^T(\mathbf{y} - \mathbf{X}\boldsymbol{\beta}).$$

(The superscript T indicates the transpose of the matrix.) Equivalently, $\hat{\boldsymbol{\beta}}$ satisfies the "normal equations"

$$\mathbf{X}^T(\mathbf{y} - \mathbf{X}\boldsymbol{\beta}) = \mathbf{0},$$

which require the residual vector $\mathbf{y} - \mathbf{X}\hat{\boldsymbol{\beta}}$ to be perpendicular ("normal") to each column vector of $\mathbf{X}$. Solving the normal equations for $\boldsymbol{\beta}$ yields

$$\hat{\boldsymbol{\beta}} = (\mathbf{X}^T\mathbf{X})^{-1}\mathbf{X}^T\mathbf{y}.$$

The corresponding vector of fitted values is $\hat{\mathbf{y}} = \mathbf{X}\hat{\boldsymbol{\beta}}$.

The Hat Matrix: How Changes in y Change $\hat{\mathbf{y}}$

It is informative to express $\hat{\mathbf{y}}$ more directly in terms of $\mathbf{y}$:

$$\hat{\mathbf{y}} = \mathbf{X}(\mathbf{X}^T\mathbf{X})^{-1}\mathbf{X}^T\mathbf{y}.$$

Defining $\mathbf{H}$ to be the $n \times n$ matrix

$$\mathbf{H} = \mathbf{X}(\mathbf{X}^T\mathbf{X})^{-1}\mathbf{X}^T \tag{5}$$

yields the more compact notation

$$\hat{\mathbf{y}} = \mathbf{H}\mathbf{y}$$

and emphasizes that each fitted y-value, $\hat{y}_i$, is a linear combination of observed y-values $\{y_j\}$. Specifically,

$$\hat{y}_i = \mathbf{x}_i(\mathbf{X}^T\mathbf{X})^{-1}\mathbf{X}^T\mathbf{y}$$

$$= \sum_{j=1}^{n} \mathbf{x}_i(\mathbf{X}^T\mathbf{X})^{-1}\mathbf{x}_j^T y_j$$

$$= \sum h_{ij} y_j, \tag{6}$$

and we note that the coefficients h_{ij} depend only on $\mathbf{X}$. The matrix $\mathbf{H}$ is the "hat matrix" discussed by Hoaglin and Welsch (1978), Belsley, Kuh, and Welsch (1980), and Cook and Weisberg (1982), among others. The elements of $\mathbf{H}$, especially the diagonal elements h_{ii}, play an important role in the techniques of regression diagnostics, which aim at discovering whether individual observations have an unusually great influence on the fitted regression model.

To illustrate the interpretation of h_{ii}, we examine how the fitted value $\hat{y}_i$ changes when y_i varies. If we add an increment Δy_i to y_i while keeping all the other y-values fixed, then y_i becomes, from equation (6),

$$\hat{y}_i + \Delta\hat{y}_i = \sum h_{ij} y_j + h_{ii}\Delta y_i.$$

Thus

$$\Delta\hat{y}_i = h_{ii}\Delta y_i.$$

We see that the impact on $\hat{y}_i$ of a change in y_i is that change multiplied by h_{ii}. Exercise 1 examines the corresponding impact on $\hat{\boldsymbol{\beta}}$.

Deleting a Single Observation

We can also learn about the effect of the observation $(y_i, \mathbf{x}_i)$ on the estimated regression coefficients $\hat{\boldsymbol{\beta}}$ and the predicted values $\hat{\mathbf{y}}$ by (temporarily) deleting $(y_i, \mathbf{x}_i)$ from the data and recomputing without it. We denote the data after deleting $(y_i, \mathbf{x}_i)$ by

$$\mathbf{X}(i) = \begin{bmatrix} x_{11} & x_{12} & \cdots & x_{1p} \\ \vdots & \vdots & & \vdots \\ x_{i-1,1} & x_{i-1,2} & \cdots & x_{i-1,p} \\ x_{i+1,1} & x_{i+1,2} & \cdots & x_{i+1,p} \\ \vdots & \vdots & & \vdots \\ x_{n1} & x_{n2} & \cdots & x_{np} \end{bmatrix} = \begin{bmatrix} \mathbf{x}_1 \\ \vdots \\ \mathbf{x}_{i-1} \\ \mathbf{x}_{i+1} \\ \vdots \\ \mathbf{x}_n \end{bmatrix},$$

$$\mathbf{y}(i) = \begin{bmatrix} y_1 \\ \vdots \\ y_{i-1} \\ y_{i+1} \\ \vdots \\ y_n \end{bmatrix},$$

where $\mathbf{X}(i)$ and $\mathbf{y}(i)$ are often read "$\mathbf{X}$-not-i" and "$\mathbf{y}$-not-i"; and we denote the corresponding OLS estimate of $\boldsymbol{\beta}$ by $\hat{\boldsymbol{\beta}}(i)$. Then after some calculation [see, for example, Miller (1974) or Belsley, Kuh, and Welsch (1980)], we have, as the changes in $\hat{\boldsymbol{\beta}}$, $\hat{\mathbf{y}}$, and $\hat{y}_i$ caused by the omission of $(y_i, \mathbf{x}_i)$ (as long as $h_{ii} \neq 1$)

$$\hat{\boldsymbol{\beta}} - \hat{\boldsymbol{\beta}}(i) = \frac{(\mathbf{X}^T\mathbf{X})^{-1}\mathbf{x}_i^T}{1 - h_{ii}} r_i,$$

$$\hat{\mathbf{y}} - \mathbf{X}\hat{\boldsymbol{\beta}}(i) = \frac{\mathbf{X}(\mathbf{X}^T\mathbf{X})^{-1}\mathbf{x}_i^T}{1 - h_{ii}} r_i,$$

$$\hat{y}_i - \mathbf{x}_i\hat{\boldsymbol{\beta}}(i) = \frac{\mathbf{x}_i(\mathbf{X}^T\mathbf{X})^{-1}\mathbf{x}_i^T}{1 - h_{ii}} r_i = \frac{h_{ii}}{1 - h_{ii}} r_i,$$

where $r_i = y_i - \hat{y}_i = y_i - \mathbf{x}_i\hat{\boldsymbol{\beta}}$ is the residual for the OLS fit to the full data. The formula for $\hat{\boldsymbol{\beta}} - \hat{\boldsymbol{\beta}}(i)$ tells us that suppressing the observation $(y_i, \mathbf{x}_i)$ shifts the estimates $\hat{\boldsymbol{\beta}}$ by a vector proportional to $(\mathbf{X}^T\mathbf{X})^{-1}\mathbf{x}_i^T$. The constant of proportionality is $r_i/(1 - h_{ii})$. The other two formulas follow from the first upon multiplying by $\mathbf{X}$ and $\mathbf{x}_i$, respectively. Writing the third formula as $h_{ii}r_i/(1 - h_{ii})$ conveniently shows that, to determine the impact of $(y_i, \mathbf{x}_i)$ on $\hat{y}_i$, we need only know r_i and h_{ii}.

From these three formulas we conclude that the larger the residual r_i, the more change in $\hat{\boldsymbol{\beta}}$, $\hat{\mathbf{y}}$, and $\hat{y}_i$. Also, the bigger the h_{ii}, the more impact on $\hat{\boldsymbol{\beta}}$, $\hat{\mathbf{y}}$, and particularly on $\hat{y}_i$, because a large h_{ii} also produces a small denominator, $1 - h_{ii}$. Because the h_{ii} are the diagonal elements of the projection matrix $\mathbf{X}(\mathbf{X}^T\mathbf{X})^{-1}\mathbf{X}^T$, a standard result in linear algebra ensures that $0 \leq h_{ii} \leq 1$. Changes in $\hat{\boldsymbol{\beta}}$, $\hat{\mathbf{y}}$, and $\hat{y}_i$ can be very large if h_{ii} is close to 1, even though the corresponding residual r_i is small. Hence deleting even one data point whose h_{ii} is large may greatly change the OLS results. We refer to h_{ii} as the *leverage* of the data point $(y_i, \mathbf{x}_i)$. A large h_{ii} usually indicates that, in the (affine) p-dimensional space defined by the vectors whose components are the carriers or predictor variables, $\mathbf{x}_i$ stands linearly apart from the other x-vectors. In this sense $\mathbf{x}_i$ is an outlier in the carrier space.

Designating "Large" h_{ii}

When should a diagonal element h_{ii} of the hat matrix be regarded as "large"? Originally, Hoaglin and Welsch (1978) noted that the average size of h_{ii} (averaging over data points in any fixed $n \times p$ regression problem) is

p/n, and they drew upon experience to suggest that the rule of treating an h_{ii} greater than $2p/n$ as "large" would generally identify individual points of high leverage. After further experience and consideration of possible sampling models that might underlie **X**-matrices, Velleman and Welsch (1981) propose the more conservative rule of $3p/n$. (It is often convenient to consider nh_{ii}/p as a value to be compared with 2 or 3.) Huber (1981, pp. 160–162) suggests interpreting $1/h_{ii}$ as the equivalent number of independent observations entering into the determination of $\hat{y}_i$. (Thus we want $1/h_{ii}$ to be large.) For a definition of when an h_{ii} is large, he breaks the range of possible values ($0 \le h_{ii} \le 1$) into three intervals: "Values $h_{ii} \le 0.2$ appear to be safe, values between 0.2 and 0.5 are risky, and if we can control the design at all, we had better avoid values above 0.5," because they correspond to $\hat{y}_i$ based on the equivalent of at most two independent observations.

A Distance Interpretation of h_{ii}

To pursue the geometric meaning of h_{ii}, we note that, in a regression model with only one nonconstant carrier, h_{ii} primarily involves the distance of $\mathbf{x}_i$ from $\bar{\mathbf{x}}$ (Section 5D of UREDA gives the details). A similar interpretation holds in multiple regression models with a constant term. We rewrite the model (1) in terms of the centered nonconstant carriers:

$$y_i = \bar{y} + \beta_2(x_{i2} - \overline{X}_2) + \cdots + \beta_p(x_{ip} - \overline{X}_p) + \varepsilon_i$$

or

$$\mathbf{y} = \bar{y}\mathbf{1} + \tilde{\mathbf{X}}\boldsymbol{\gamma} + \boldsymbol{\varepsilon}, \tag{7}$$

where the n-vector $\mathbf{1} = (1, \ldots, 1)^T$ is the first column of $\mathbf{X}$ in (3), $\bar{y} = \Sigma y_i/n$, $\boldsymbol{\gamma} = (\beta_2, \ldots, \beta_p)^T$, and the $n \times (p - 1)$ matrix $\tilde{\mathbf{X}}$ is given by

$$\tilde{\mathbf{X}} = \begin{bmatrix} x_{12} & \cdots & x_{1p} \\ \vdots & & \vdots \\ x_{n2} & \cdots & x_{np} \end{bmatrix} - \mathbf{1}(\overline{X}_2, \ldots, \overline{X}_p)$$

with $\overline{X}_j = \Sigma_i x_{ij}/n$. For model (7) OLS gives

$$\hat{\boldsymbol{\gamma}} = (\tilde{\mathbf{X}}^T\tilde{\mathbf{X}})^{-1}\tilde{\mathbf{X}}^T\mathbf{y}.$$

Then

$$\hat{\mathbf{y}} = \bar{y}\mathbf{1} + \tilde{\mathbf{X}}\hat{\boldsymbol{\gamma}}$$

$$= \bar{y}\mathbf{1} + \tilde{\mathbf{X}}(\tilde{\mathbf{X}}^T\tilde{\mathbf{X}})^{-1}\tilde{\mathbf{X}}^T\mathbf{y}.$$

By representing $\bar{y}$ as $(1/n)\mathbf{1}^T\mathbf{y}$ we can write

$$\hat{\mathbf{y}} = \left(\frac{1}{n}\mathbf{1}\mathbf{1}^T + \tilde{\mathbf{X}}(\tilde{\mathbf{X}}^T\tilde{\mathbf{X}})^{-1}\tilde{\mathbf{X}}^T\right)\mathbf{y}. \tag{8}$$

Hence, in terms of the centered data, the hat matrix becomes

$$\mathbf{H} = \frac{1}{n}\mathbf{1}\mathbf{1}^T + \tilde{\mathbf{X}}(\tilde{\mathbf{X}}^T\tilde{\mathbf{X}})^{-1}\tilde{\mathbf{X}}^T,$$

and the diagonal elements are

$$h_{ii} = \frac{1}{n} + \tilde{\mathbf{x}}_i(\tilde{\mathbf{X}}^T\tilde{\mathbf{X}})^{-1}\tilde{\mathbf{x}}_i^T,$$

where $\tilde{\mathbf{x}}_i$ is the ith row of $\tilde{\mathbf{X}}$. Thus, whenever the model contains the constant carrier, the range of possible values for h_{ii} is $1/n \leq h_{ii} \leq 1$ rather than all of $0 \leq h_{ii} \leq 1$.

To interpret h_{ii} as a distance, we need a measure of how far the point $(x_{i2}, \ldots, x_{ip})$ is from $(\bar{X}_2, \ldots, \bar{X}_p)$ in the $(p-1)$-dimensional space of the nonconstant carriers. Because the carriers may differ widely in their units and scaling, simple Euclidean distance will not be satisfactory. The usual remedy expresses each coordinate in units of its standard deviation.

We also want to allow for possible interrelations among the carriers. If the points $(x_{i2}, \ldots, x_{ip})$ were observations from a $(p-1)$-variate Gaussian distribution with nonzero covariances, the standard approach in multivariate analysis would transform them into observations from a Gaussian distribution with unit variances and zero covariances (by factoring the inverse of the covariance matrix) and then calculate ordinary Euclidean distances from these transformed observations. When one then expresses the distance in terms of the original distribution with mean $\boldsymbol{\mu}$ (a row vector) and covariance matrix $\boldsymbol{\Sigma}$, the distance of $\mathbf{z}$ from $\boldsymbol{\mu}$ takes the form

$$(\mathbf{z} - \boldsymbol{\mu})\boldsymbol{\Sigma}^{-1}(\mathbf{z} - \boldsymbol{\mu})^T.$$

Of course, we need not assume that the carriers follow a Gaussian distribution if we want only to calculate distances. The definition of $\tilde{\mathbf{X}}$ implies that

$\tilde{\mathbf{X}}^T\tilde{\mathbf{X}}$ is $n - 1$ times the sample covariance matrix of the nonconstant carriers. Now, because $\tilde{\mathbf{x}}_i = (x_{i2}, \ldots, x_{ip}) - (\bar{X}_2, \ldots, \bar{X}_p)$, the expression $\tilde{\mathbf{x}}_i(\tilde{\mathbf{X}}^T\tilde{\mathbf{X}})^{-1}\tilde{\mathbf{x}}_i^T$ has the same form as the distance measure described above: $\tilde{\mathbf{x}}_i$ plays the role of $\mathbf{z} - \boldsymbol{\mu}$, and $\tilde{\mathbf{X}}^T\tilde{\mathbf{X}}$ plays the role of $\boldsymbol{\Sigma}$. This connection gives h_{ii} a clear distance interpretation.

In applying this instructive interpretation we may need some restraint. If we take $(\tilde{\mathbf{X}}^T\tilde{\mathbf{X}})^{-1}$ as given, we can calculate the distance between $(\bar{X}_2, \ldots, \bar{X}_p)$ and any point $(x_{02}, \ldots, x_{0p})$ in the $(p - 1)$-dimensional space of the nonconstant carriers, no matter how far from $(\bar{X}_2, \ldots, \bar{X}_p)$. Thus it might appear that we could find points for which h_{ii} exceeds 1. Because h_{ii} is defined only for rows of $\mathbf{X}$ (or $\tilde{\mathbf{X}}$), however, we would have to include any new outlying data points as additional rows of $\mathbf{X}$. These would alter $(\tilde{\mathbf{X}}^T\tilde{\mathbf{X}})^{-1}$ in a way that keeps the corresponding (new) h_{ii} from exceeding 1 (and the h_{ii} for the original rows would usually change as well).

The ratio $h_{ii}/(1 - h_{ii})$ is closely related to another, similar measure of distance—that between $\tilde{\mathbf{x}}_i$ and the mean of the remaining rows of $\tilde{\mathbf{X}}$ [see Belsley, Kuh, and Welsch (1980)].

EXAMPLE: STRENGTH OF WOOD BEAMS

Draper and Stoneman (1966) give the specific gravity, moisture content, and strength of 10 wood beams. (Strength is the response variable.) Table 8-1 lists these data along with the diagonal elements of the hat matrix for the model in which the carriers are the constant, specific gravity, and moisture content [the full hat matrix appears in Hoaglin and Welsch (1978)]. Beam 4 has the highest leverage, $h_{4,4} = .604$, $nh_{4,4}/p = 2.01$, and might seem to deserve closer scrutiny, either because $nh_{ii}/p > 2$ or, possibly, by the magnitude of h_{ii}, although 10 is not a very large number of observations. A plot of moisture content against specific gravity (Figure 8-1) confirms that beam 4 is to some extent a bivariate outlier, although it is not extreme on either carrier. The contours superimposed on the plot show distances from $(\bar{X}_2, \bar{X}_3)$ corresponding to $h_{ii} = .2$, .5, and .8.

To summarize, the OLS approach is sensitive both to high-leverage points in the carriers ($\mathbf{X}$) and to outliers in the response variable y. In the x-space, the larger the leverage h_{ii} (so that, in terms of the covariance matrix for the centered carriers, $\mathbf{x}_i$ is farther from $\bar{\mathbf{x}}$), the greater the change in the fitted value $\hat{y}_i$ when y_i varies and the stronger the impact on $\hat{\boldsymbol{\beta}}$, $\hat{\mathbf{y}}$, and particularly $\hat{y}_i$ when $(y_i, \mathbf{x}_i)$ is deleted from the data. Although statisticians developed techniques for exposing these sensitivities around 1970, we have had these problems as long as we have had least-squares fitting. They

TABLE 8-1. Data on 10 wood beams, accompanied by the diagonal elements h_{ii} of the hat matrix for the model whose carriers are 1, specific gravity, and moisture content.

Beam Number	Specific Gravity	Moisture Content	Strength	h_{ii}
1	0.499	11.1	11.14	.418
2	0.558	8.9	12.74	.242
3	0.604	8.8	13.13	.417
4	0.441	8.9	11.51	.604
5	0.550	8.8	12.38	.252
6	0.528	9.9	12.60	.148
7	0.418	10.7	11.13	.262
8	0.480	10.5	11.70	.154
9	0.406	10.5	11.02	.315
10	0.467	10.7	11.41	.187

Sources: Norman R. Draper and David M. Stoneman (1966). "Testing for the inclusion of variables in linear regression by a randomisation technique," *Technometrics*, **8**, 695–699 (data from Table 1, p. 696). David C. Hoaglin and Roy E. Welsch (1978). "The hat matrix in regression and ANOVA," *The American Statistician*, **32**, 17–22 (h_{ii} from Table 2, p. 19).

are characteristics of the method, characteristics that we must often regard as defects.

Against our background in robust and exploratory techniques, we see the problem as primarily nonresistance. An isolated unusual observation can have a disproportionate impact on the result of the OLS analysis. This effect parallels the nonresistance of the sample mean, the simple regression line, and the analysis of a two-way table by means, as discussed in UREDA. The practical difficulty, however, arises from the more complicated leverage situation. When data point i has high leverage, an anomalous value of y_i *need not* produce a large residual r_i; indeed, if h_{ii} is close enough to 1, it will be difficult for y_i to leave a large residual. At the same time, the x-vector need not look unusual when we examine the values of each carrier separately or even when we scatterplot all pairs of carriers.

Techniques of regression diagnostics have made it possible to use OLS fitting with some confidence that we will discover distortions caused by high-leverage points and outliers. [See, among others, Belsley, Kuh, and Welsch (1980), Cook and Weisberg (1982), and Henderson and Velleman (1981).] In many applications, this qualified use of OLS requires painstaking

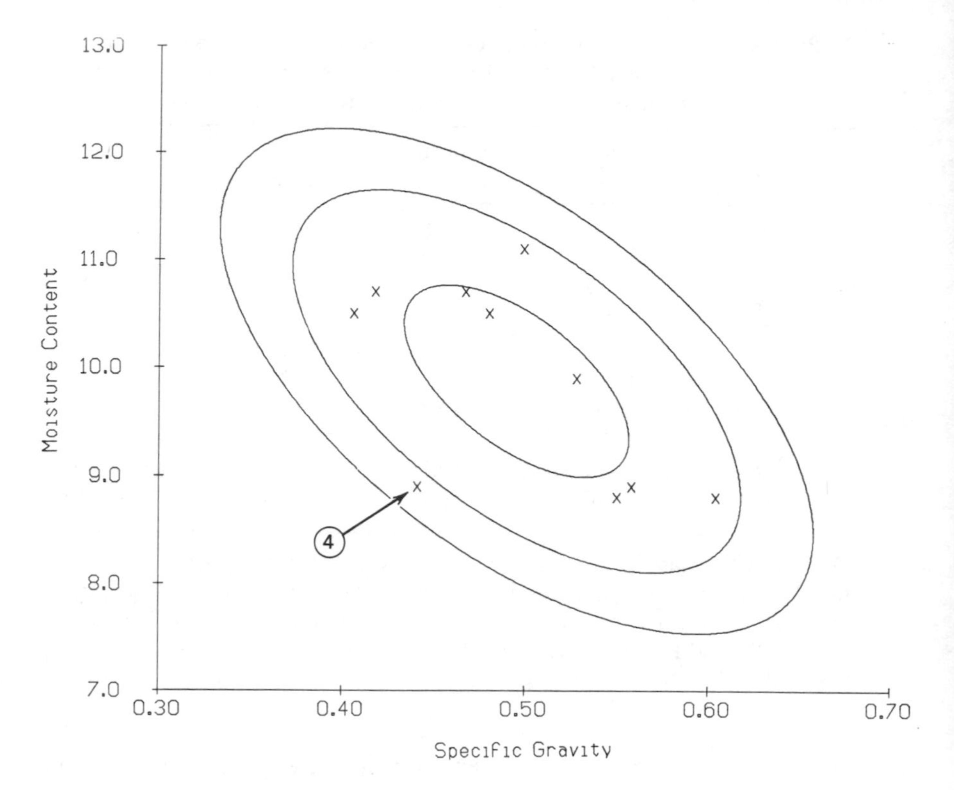

Figure 8-1. Plot of moisture content against specific gravity in the wood beams data. The contours show distances from $(\overline{X}_2, \overline{X}_3)$ that correspond to $h_{ii} = .2, .5, .8$, or, equivalently, to $nh_{ii}/p = .67, 1.67, 2.67$, respectively, from inner to outer.

diagnosis and model building. Thus we need techniques that deliver a robust/resistant fit more automatically.

In actual situations, we must recognize that asking for a *single fit* will often be a *mistake*. As we make our fitting processes more nearly automatic, we should increase the number and diversity of solutions. For the present we need to think of one OLS solution and one robust solution. Later in the chapter we will be led to ask for even more solutions.

Although still a subject of serious research, techniques for robust regression have advanced to where they can offer the routine regression user reasonable protection against both outliers and high-leverage points. We concentrate first on techniques that are resistant to outliers in the y-values and then return in Section 8E to the problems of resistance to high-leverage points.

Most popular robust methods for regression fall into three classes: M-estimators (and W-estimators), L-estimators, and R-estimators (discussed, for the location case, at the start of Chapter 11 of UREDA). M-estimation generalizes to regression models more readily than the other two approaches. Also M-estimators have several desirable properties. They offer good performance and flexibility, they are easy to compute and to handle, and they are relatively simple. Thus the next four sections focus primarily on M-estimators and the closely related W-estimators for regression. We return briefly to L-estimators and R-estimators in Section 8F.

8B. M-ESTIMATORS AND W-ESTIMATORS FOR REGRESSION

Regression models extend the location model: Instead of describing the behavior of y by a single location constant like a mean, we allow the location of y_i to depend linearly on the values of a set of carriers (using the form $\mathbf{x}_i\boldsymbol{\beta}$). Consequently, many estimators of location generalize to regression models. M- and W-estimators for regression generalize the M-estimators and W-estimators of location. Hence, parts of this section and of the next section parallel the corresponding arguments in Chapter 11 of UREDA. We generally emphasize the differences between regression and location, rather than the similarities.

Consider the regression model (1). Let us forget about scale $\sigma = \sqrt{\operatorname{var}(\varepsilon_i)}$ for a moment. That is, we assume σ is known and fixed. Without losing generality, we may assume $\sigma = 1$.

DEFINITION: The *M-estimator for* $\boldsymbol{\beta}$ in (1), based on the objective function $\rho(t)$ and the data $(y_1, \mathbf{x}_1), \ldots, (y_n, \mathbf{x}_n)$, is the value of $\boldsymbol{\beta}$, denoted by $\hat{\boldsymbol{\beta}}_M$, which minimizes $\sum\rho(y_i - \mathbf{x}_i\boldsymbol{\beta}) = \sum_i\rho(y_i - \sum_j x_{ij}\beta_j)$; that is,

$$\sum\rho(y_i - \mathbf{x}_i\hat{\boldsymbol{\beta}}_M) = \min_{\boldsymbol{\beta}} \sum\rho(y_i - x_i\boldsymbol{\beta}).$$

Equivalently, if $\rho(t)$ is a differentiable convex function, $\hat{\boldsymbol{\beta}}_M$ is determined implicitly by the set of p simultaneous equations in terms of $\psi(t) = \rho'(t)$, the derivative of ρ with respect to t,

$$\sum\psi(y_i - \mathbf{x}_i\boldsymbol{\beta})\mathbf{x}_i^T = \mathbf{0}. \tag{9}$$

W-estimators are an alternative form of M-estimators. We replace $\psi(t)$ by $tw(t)$ in equation (9); that is, we define the weight function

$$w(t) = \frac{\psi(t)}{t}.$$

If we let $w_i = w(y_i - \mathbf{x}_i\boldsymbol{\beta})$, then the p simultaneous equations become

$$\sum (y_i - \mathbf{x}_i\boldsymbol{\beta}) w_i \mathbf{x}_i^T = \mathbf{0}. \tag{10}$$

A solution (not necessarily unique) of equation (10), $\hat{\boldsymbol{\beta}}_W$, is called a *W-estimate*. The W comes from "weighted," because equation (10) can be viewed as the "normal equations" for a weighted least-squares regression problem. Rearrangement of equation (10) yields

$$\sum \mathbf{x}_i^T w_i \mathbf{x}_i \boldsymbol{\beta} = \sum \mathbf{x}_i^T w_i y_i. \tag{11}$$

In matrix terms, with

$$\mathbf{W} = \begin{bmatrix} w_1 & & & 0 \\ & w_2 & & \\ & & \ddots & \\ 0 & & & w_n \end{bmatrix},$$

equation (11) becomes

$$\mathbf{X}^T\mathbf{W}\mathbf{X}\boldsymbol{\beta} = \mathbf{X}^T\mathbf{W}\mathbf{y}.$$

These are the normal equations of a weighted least-squares regression, but now the weights w_i are neither equal nor $\mathbf{X}$-determined—they depend on the residuals $(y_i - \mathbf{x}_i\boldsymbol{\beta})$.

Theoretically speaking, M-estimators and W-estimators (when fully iterated to a solution) are the same, because equations (9) and (10) are actually two different forms of one set of simultaneous equations. In practice (as we discuss in the next section), M-estimators and W-estimators are different, because the different forms will ordinarily be solved by different computing methods. In this sense, the W-estimator is one computational procedure for solving equation (9).

In principle, we have broad latitude in choosing the function $\rho(t)$ and thus $\psi(t)$ and $w(t)$. As in Chapter 11 of UREDA, the properties of an M-estimator are essentially determined by the function $\rho(t)$ [or, equivalently, by either $\psi(t)$ or $w(t)$]. One important issue in choosing $\psi(t)$ is the balance between robustness and efficiency. Because it involves more structure than location, regression introduces more purposes and criteria, and these may shift the balance at times. For example, estimating $\boldsymbol{\beta}$ may be secondary to predicting a value of y or producing good residuals.

Five Major M-Estimators

In addition to OLS and least absolute residuals, Table 8-2 lists three proposed ρ-functions for robust regression. Figure 8-2 shows these ρ-functions and their ψ-functions. (Section 11H of UREDA discusses the corresponding location estimators and some of their properties.) Exercise 4 examines a further ψ-function.

The *OLS estimator* is well known. Taking $\rho(t) = \frac{1}{2}t^2$ yields $\psi(t) = t$ and $w(t) = 1$. Hence, equations (9) and (10) both reduce to the normal equations for ordinary least squares:

$$\sum (y_i - \mathbf{x}_i\boldsymbol{\beta})\mathbf{x}_i^T = \mathbf{0}.$$

TABLE 8-2. M-estimators for regression.

Estimator	$\rho(t)$	$\psi(t)$	$w(t)$	Range of t
OLS	$\frac{1}{2}t^2$	t	1	$\lvert t\rvert < \infty$
LAR	$\lvert t\rvert$	$\operatorname{sgn}(t)$	$\dfrac{\operatorname{sgn}(t)}{t}$	$\lvert t\rvert < \infty$
Huber[a]	$\frac{1}{2}t^2$	t	1	$\lvert t\rvert \le k$
	$k\lvert t\rvert - \frac{1}{2}k^2$	$k \operatorname{sgn}(t)$	$k/\lvert t\rvert$	$\lvert t\rvert > k$
Andrews[a]	$A^2[1 - \cos(t/A)]$	$A\sin(t/A)$	$\dfrac{A}{t}\sin(t/A)$	$\lvert t\rvert \le \pi A$
	$2A^2$	0	0	$\lvert t\rvert > \pi A$
Biweight[a] (bisquare)	$\dfrac{B^2}{6}\{1 - [1 - (t/B)^2]^3\}$	$t[1 - (t/B)^2]^2$	$[1 - (t/B)^2]^2$	$\lvert t\rvert \le B$
	$\dfrac{B^2}{6}$	0	0	$\lvert t\rvert > B$

[a] The illustrative examples of ρ-functions and ψ-functions in Figure 8-2 use $k = 1$ for the Huber, $A = 1/\pi$ for the Andrews, and $B = 1$ for the biweight.

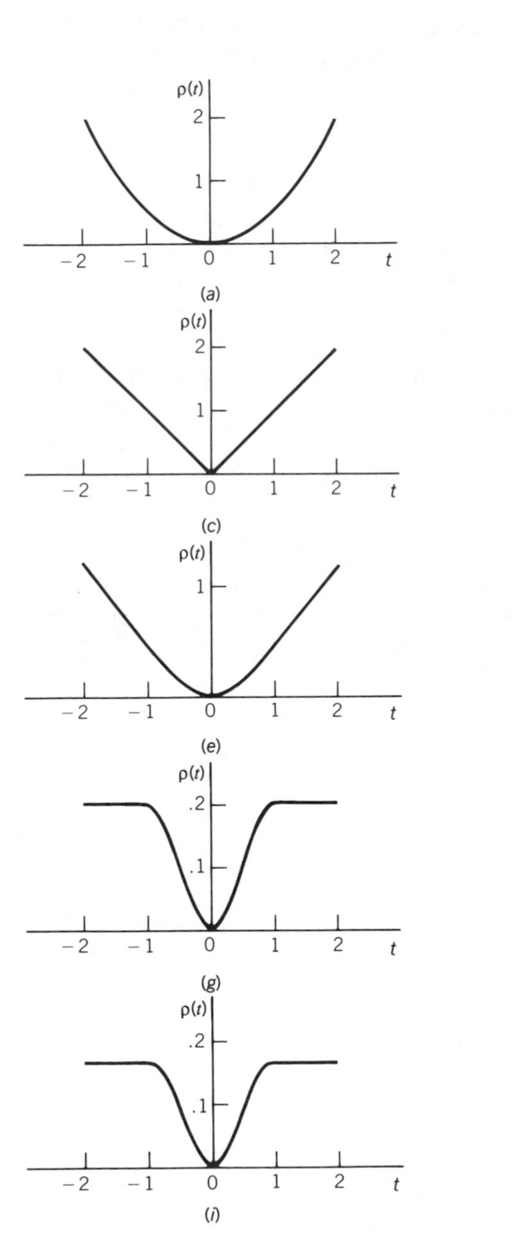

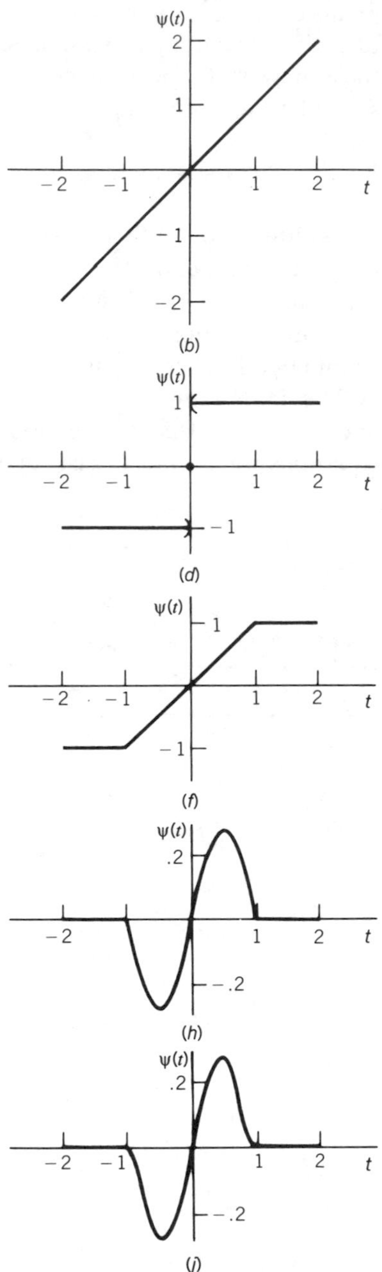

Figure 8-2. The ρ-function and ψ-function for five *M*-estimators. (*a*) OLS ρ-function. (*b*) OLS ψ-function. (*c*) LAR ρ-function. (*d*) LAR ψ-function. (*e*) Huber ρ-function. (*f*) Huber ψ-function. (*g*) Andrews ρ-function. (*h*) Andrews ψ-function. (*i*) Biweight ρ-function. (*j*) Biweight ψ-function. Table 8-2 gives scaling.

The *least-absolute-residuals* (LAR) *estimator* minimizes the sum of the absolute values of the residuals, $\Sigma|y_i - \mathbf{x}_i\boldsymbol{\beta}|$. Under a variety of names it has been studied and applied by workers in several fields. Because it allows large residuals to have less impact than does OLS, the LAR estimator may seem attractive for robust regression. (The median, its special case in estimating location, offers excellent resistance.) Unfortunately, high-leverage observations can cause the LAR estimator to break down. It may still be convenient, however, to use LAR estimates as starting values for other, more robust regression procedures (as in Sections 8C and 8F).

The *Andrews estimator* and the *biweight estimator* (also called "bisquare") belong to the class known as redescending estimators because their ψ-functions come back to 0 when the absolute value of the argument is greater than a specified positive number. By introducing the indicator function $I\{S\}$, which takes the value 1 on the set S and the value 0 elsewhere, we may combine the two cases of these simple redescending estimators. For example, the Andrews estimator has the ψ-function

$$\psi_A(t) = A\sin(t/A)I\{|t| \leq \pi A\}.$$

Substituting into equation (9), we see that the Andrews estimator $\hat{\boldsymbol{\beta}}_A$ satisfies

$$\sum \mathbf{x}_i^T \sin\left(\frac{y_i - \mathbf{x}_i\boldsymbol{\beta}}{A}\right) I\left\{\frac{|y_i - \mathbf{x}_i\boldsymbol{\beta}|}{A} \leq \pi\right\} = \mathbf{0} \tag{12}$$

and the biweight estimator $\hat{\beta}_B$ satisfies

$$\sum \mathbf{x}_i^T(y_i - \mathbf{x}_i\boldsymbol{\beta})\left[1 - \left(\frac{y_i - \mathbf{x}_i\boldsymbol{\beta}}{B}\right)^2\right]^2 I\left\{\frac{|y_i - \mathbf{x}_i\boldsymbol{\beta}|}{B} \leq 1\right\} = \mathbf{0}. \tag{13}$$

Either equation (12) or equation (13), however, may fail to have a unique solution. Even worse, when multiple solutions occur, the solutions are separated because the objective function $\Sigma\rho(y_i - \mathbf{x}_i\boldsymbol{\beta})$ of either estimator may have several local minima. This multiplicity can cause some trouble with redescending estimators, especially in multiple regression. If, as we often will, we want to get a diversity of solutions, the problem may be to find enough of the multiple solutions. In practice this makes the selection of a starting point—or of several starting points—for the iterative computation more important.

The *Huber estimator* (a family of estimators, one for each value of the parameter k) is the solution of

$$\sum_{|r_i| \leq k} r_i \mathbf{x}_i^T + \sum_{|r_i| > k} k \operatorname{sgn}(r_i) \mathbf{x}_i^T = \mathbf{0}, \tag{14}$$

where $r_i = y_i - \mathbf{x}_i \boldsymbol{\beta}$. This class of estimators includes the OLS estimator at $k = \infty$ $[\psi(t) = t]$. Also, if we replace $\psi_H(t)$ by $\psi_H(t)/k$ (a technical detail), the Huber estimators include the LAR estimator at $k = 0$ $[\psi(t) = \operatorname{sgn}(t)]$. For $0 < k < \infty$, the Huber ψ-functions combine the linearity of the OLS ψ-function at the origin and the boundedness (actually constancy) of the LAR ψ-function for large arguments. The Huber ψ-function is monotone and bounded, but it does not return to zero or toward zero. The monotonicity of the Huber ψ-function (i.e., the convexity of the Huber ρ-function) ensures that the corresponding objective function, $\Sigma\rho(y_i - \mathbf{x}_i\boldsymbol{\beta})$, has only one minimum, possibly at a single $\boldsymbol{\beta}$, possibly throughout a larger convex set. The solutions of equation (14), in case of nonuniqueness, fill a convex set. This near uniqueness can be a useful feature of the Huber estimator for regression.

The constants k, A, and B in the Huber, Andrews, and biweight estimators are usually called *tuning constants*. When the type of function ψ is fixed, the tuning constant determines the properties of the associated estimator, such as efficiency, influence function, and gross-error sensitivity. As in the location case, the tuning constant may be chosen to adjust the efficiency of the estimator at a particular underlying distribution of the fluctuations (such as the Gaussian).

Denby and Mallows (1977) have proposed systematically varying the tuning constant as a way to detect outliers. For a suitable set of values of the tuning constant k in the Huber ψ-function, extending from 0 far enough to include the OLS fit, they obtain $\hat{\boldsymbol{\beta}}_H$ and the residuals. Their plot of all the residuals against k shows which observations are receiving less weight as k decreases to 0 and hence may be possible outliers. Their plot of the individual coefficient estimates against k reveals coefficients that are unstable or heavily influenced by an anomalous y-value at a high-leverage point in carrier space.

Functionals of Distributions

In general, an M-estimator can also be viewed as a functional on empirical distributions. That is, for each empirical distribution F_n, an M-estimator, say T, determines the estimate of the parameter: $\hat{\theta} = T(F_n)$. This mathematical formulation accommodates the idea of defining the parameter as the result of applying the functional to the population distribution, F: $\theta = T(F)$.

The univariate sample $x_1, \ldots, x_n$, when ordered as $x_{(1)} \leq x_{(2)} \leq \cdots \leq x_{(n)}$, gives rise to the empirical distribution function

$$F_n(x) = \begin{cases} 0, & x < x_{(1)} \\ \dfrac{i-1}{n}, & x_{(i-1)} \leq x < x_{(i)}, \quad i = 2, \ldots, n, \\ 1, & x \geq x_{(n)}. \end{cases}$$

That is, $F_n(x)$ equals $1/n$ times the number of observations at or to the left of x.

The most familiar functional (also the most familiar M-estimator) is T_{mean}, associated with the mean. For the empirical distribution F_n, we rearrange the customary formula slightly and write

$$T_{\text{mean}}(F_n) = \sum x_i \frac{1}{n}.$$

For a continuous distribution F with density f, we resort to integration:

$$T_{\text{mean}}(F) = \int x f(x)\, dx,$$

as long as the value of the integral is finite. In a single mathematical framework these are both special cases of

$$T_{\text{mean}}(F) = \int x\, dF(x)$$

(the integral can be viewed as a Lebesgue–Stieltjes integral): $dF_n(x)$ gives probability mass $1/n$ to each of the discrete points $x_1, \ldots, x_n$, and $dF(x)$ becomes $f(x)dx$ when F is continuous.

A general M-estimator, as a functional T_M on distributions, is defined, in terms of the ψ-function, by the implicit equation

$$\int \psi[x - T_M(F)]\, dF(x) = 0.$$

Similarly, an M-estimator $\hat{\boldsymbol{\beta}}_M$ for the regression coefficients $\boldsymbol{\beta}$ in model (1) is a functional of the distribution $F(y, \mathbf{x})$ and is defined by the simultaneous equations

$$\int \psi[y - \mathbf{x}\boldsymbol{\beta}(F)]\mathbf{x}^T dF(y, \mathbf{x}) = \mathbf{0}. \tag{15}$$

In the customary setup, with fixed $\mathbf{x}_1, \ldots, \mathbf{x}_n$ in equations (2), we set

$$F(y, \mathbf{x}) = \frac{1}{n} F(y|\mathbf{x}_1)\delta_{\mathbf{x}_1}(\mathbf{x}) + \cdots + \frac{1}{n} F(y|\mathbf{x}_n)\delta_{\mathbf{x}_n}(\mathbf{x}),$$

where $F(y|\mathbf{x})$ is the (conditional) distribution of y at $\mathbf{x}$ and $\delta_{\mathbf{x}}$ is a point mass at $\mathbf{x}$. If we allow $\mathbf{x}$ to be random, $F(y, \mathbf{x})$ is the joint distribution of $(y, \mathbf{x})$.

Influence Functions

In the location problem (Section 11D of UREDA), the influence curve gives a quantitative expression of the impact on the estimate caused by perturbing the distribution F by introducing a point mass at x. For an estimator given by the functional $T(F)$, the influence curve (at F_0) is defined by

$$\begin{aligned} IC(x;\, F_0, T) &= \lim_{\varepsilon \to 0} \frac{T[(1-\varepsilon)F_0 + \varepsilon\delta_x] - T(F_0)}{\varepsilon} \\ &= \frac{d}{d\varepsilon} T[(1-\varepsilon)F_0 + \varepsilon\delta_x]\Big|_{\varepsilon=0}, \end{aligned}$$

where δ_x is the point mass at x. Thus we contaminate the distribution F_0 (underlying distribution or empirical distribution) by a point mass at x, and we examine the impact on the estimate $T(F)$ relative to the probability associated with this point mass, as this probability shrinks to zero. When the functional $T(F)$ expresses the M-estimator defined by the ψ-function $\psi(u)$, this process yields

$$IC(x;\, F_0, T) = \frac{\psi[x - T(F_0)]}{\int \psi'[x - T(F_0)]\, dF_0(x)},$$

assuming $\psi'(u)$ exists.

To derive the influence function of $\hat{\beta}_M(F)$, we follow a similar strategy. We substitute $F = (1-\varepsilon)F_0 + \varepsilon\delta_{(y,\mathbf{x})}$ into equation (15) and take the derivative with respect to ε. After a rather involved calculation we set $\varepsilon = 0$

to obtain

$$\mathbf{IC}((y,\mathbf{x});\, F_0, \hat{\boldsymbol{\beta}}_M) = \psi[y - \mathbf{x}\hat{\boldsymbol{\beta}}_M(F_0)]\mathbf{B}^{-1}\mathbf{x}^T, \tag{16}$$

where

$$\mathbf{B} = \int \psi'[y - \mathbf{x}\hat{\boldsymbol{\beta}}_M(F_0)]\mathbf{x}^T\mathbf{x}\, dF_0(y,\mathbf{x}).$$

This result illustrates the composition of the influence function without attempting to be fully general. A careful derivation would have to distinguish between the case of fixed $\mathbf{x}_1, \ldots, \mathbf{x}_n$ and the case of random $\mathbf{x}$.

The influence function of $\hat{\boldsymbol{\beta}}_M$ differs from the influence curve for the corresponding M-estimator of location in that the vector $\mathbf{B}^{-1}\mathbf{x}^T$ multiplies the value of ψ. If we impose no bounds on $\mathbf{x}$—if, in other words, we contemplate a sequence of problems (not just one) in which $\mathbf{x}$ is ever more widely spread out—then the contribution of $\mathbf{B}^{-1}\mathbf{x}^T$ is not bounded, so that the influence function is no longer bounded, even if ψ redescends.

As we consider the implications of this result, we must ask whether it is reasonable to allow a contaminating observation to have any possible $\mathbf{x}$. If, for example, it suffices to think about $\mathbf{x}$s inside the convex hull of the rows $\mathbf{x}_1, \ldots, \mathbf{x}_n$ (i.e., the smallest convex set containing $\mathbf{x}_1, \ldots, \mathbf{x}_n$), or perhaps inside some larger bounded convex set, then the influence remains bounded, although perhaps uncomfortably large.

The practical issue is how an estimator should handle points or subsets with very large influences We may have no alternative to looking *both* at results that give full credence to high-influence points *and* at results that limit their impact.

In Section 8E we briefly discuss approaches to limiting the influence of arbitrary contaminating observations.

We now turn first to the distributional behavior of regression M-estimators and then to estimation of scale. We begin with large-sample behavior because it can be described much more simply than behavior in small samples.

Asymptotic Properties

Suppose that $\psi(t)$ is monotone and $\hat{\boldsymbol{\beta}}_M$ is the corresponding M-estimator of the regression coefficients $\boldsymbol{\beta}$. Under appropriate regularity conditions, $\hat{\boldsymbol{\beta}}_M$ is consistent; that is, as n becomes large, $\hat{\boldsymbol{\beta}}_M(F_n)$ converges in probability to $\boldsymbol{\beta}$. Furthermore, the distribution of $\hat{\boldsymbol{\beta}}_M(F_n)$ approaches the p-variate Gaussian with mean $\boldsymbol{\beta}$ and covariance matrix $\mathbf{V}$ given by

$$\mathbf{V} = \frac{E\psi^2}{(E\psi')^2}(\mathbf{X}^T\mathbf{X})^{-1}. \tag{17}$$

In this straightforward expression for $\mathbf{V}$ we have had to assume that all the fluctuations ε_i in model (1) are uncorrelated observations from the same distribution, so that we can calculate the constants $E\psi^2$ and $E\psi'$ under that distribution.

Because we want only briefly to sketch the behavior of $\hat{\boldsymbol{\beta}}_M$ in large samples, equation (17) omits the dependence of $\mathbf{X}$ on n. Also, by holding p constant as n becomes large, we probably do not mimic what one would do in practice. Typically, as Mallows (1979, p. 180) points out, "the more data you have, the more structure you can find—if you look for it. We are never in the situation of having $n \to \infty$ with the model staying the same." Huber (1973) and Yohai and Maronna (1979) give the major ingredients of the theory and allow p to increase in several ways as n increases.

Estimators for the Covariance Matrix of $\hat{\boldsymbol{\beta}}_M$

In practical problems, it is useful to know the precision of an estimate of the regression coefficients $\boldsymbol{\beta}$. In one customary formulation, we construct a confidence interval for each β_i or a confidence region for $\boldsymbol{\beta}$. This requires an estimate of the covariance matrix of $\hat{\boldsymbol{\beta}}_M$.

Recall that, according to classical least-squares theory, the OLS estimator $\hat{\boldsymbol{\beta}}$ has mean $\boldsymbol{\beta}$ and covariance matrix $\sigma^2(\mathbf{X}^T\mathbf{X})^{-1}$, where $\sigma^2 = \text{var}(\varepsilon_i)$. The covariance matrix of the OLS estimator is usually estimated by

$$\left[\frac{1}{n-p}\sum r_i^2\right](\mathbf{X}^T\mathbf{X})^{-1}, \tag{18}$$

where the r_i are the OLS residuals and $\sum r_i^2/(n-p)$ is the residual mean square, an unbiased estimate of σ^2.

As we saw in equation (17), a related expression applies to other M-estimators besides OLS. Under certain conditions $\hat{\boldsymbol{\beta}}_M$ has asymptotic covariance matrix

$$\frac{E\psi^2}{(E\psi')^2}(\mathbf{X}^T\mathbf{X})^{-1}, \tag{19}$$

and this provides a starting point for approximations in finite samples. It is natural to use $\Sigma\psi(r_i)^2/n$ to estimate $E\psi^2$ and $\Sigma\psi'(r_i)/n$ to estimate $E\psi'$, where now $r_i = y_i - \mathbf{x}_i\hat{\boldsymbol{\beta}}_M$. By substituting in equation (19) we obtain an estimate of the covariance matrix of $\hat{\boldsymbol{\beta}}_M$

$$\frac{\Sigma\psi(r_i)^2/n}{\left[\Sigma\psi'(r_i)/n\right]^2}(\mathbf{X}^T\mathbf{X})^{-1}. \tag{20}$$

Because OLS is a special case of M-estimation [$\psi(t) = t$], it is straightforward to check that equation (20) reduces to $(1/n)\Sigma r_i^2(\mathbf{X}^T\mathbf{X})^{-1}$. When we compare this with the estimate in expression (18), we note that, for consistency with the usual OLS result, we should multiply expression (20) by the correction factor $n/(n-p)$. This argument thus yields

$$\frac{\Sigma\psi(r_i)^2/(n-p)}{\left[\Sigma\psi'(r_i)/n\right]^2}(\mathbf{X}^T\mathbf{X})^{-1} \tag{21}$$

as an estimator for the covariance matrix of the M-estimator $\hat{\boldsymbol{\beta}}_M$. Huber (1973) proposed this estimator and compared it with Monte Carlo results.

Unfortunately expression (21) does not capture all variations in structure that arise in finite samples. Monte Carlo studies by Welsch (1975), Hill and Holland (1977), and Gross (1977) have shown that adequacy of (21) is sensitive to outliers among the rows of $\mathbf{X}$. Hill (1979) introduced a special class of regression models known as p-point problems, for which the finite-sample covariance matrix of $\hat{\boldsymbol{\beta}}_M$ can be derived exactly, and showed that it can differ substantially from a multiple of $(\mathbf{X}^T\mathbf{X})^{-1}$. These studies also investigated some other suggestions for estimating the covariance matrix of $\hat{\boldsymbol{\beta}}_M$ in finite samples, but those alternatives were not completely satisfactory either.

Estimation of Scale

In defining $\hat{\boldsymbol{\beta}}_M$ and deriving equation (9), we assumed that the error scale parameter σ is known and fixed, and we have often avoided showing it explicitly. In practice, however, scale often must be estimated.

One reason for estimating scale is that some knowledge of it is necessary to judge the accuracy of the fitted regression model. Also, as in the location problem, an explicit scale estimate is almost always required in defining estimates and in constructing interval estimates (see Section 12F of UREDA). In regression models we may want interval estimates for the β_j or for a $\hat{y}$, or we may need to test whether a residual is an outlier.

A second reason is that, without taking scale into account, most M-estimators of $\boldsymbol{\beta}$ would not respond correctly to a change in the units of y or to a change in the scale of the errors. The only familiar exceptions are the OLS estimator and the LAR estimator.

If we take scale into account through σ, the OLS estimator satisfies

$$\sum\left(\frac{y_i - \mathbf{x}_i\boldsymbol{\beta}}{\sigma}\right)^2 = \min, \tag{22}$$

and the LAR estimator satisfies

$$\sum\left|\frac{y_i - \mathbf{x}_i\boldsymbol{\beta}}{\sigma}\right| = \min. \tag{23}$$

For any positive value of σ, both equations (22) and (23) have the same solution as before because σ^2 or σ factors out. That is, the OLS estimator and the LAR estimator do not depend on σ at all. Unfortunately, other M-estimators do depend on the scale parameter. So, even for estimating $\boldsymbol{\beta}$ alone, we must estimate some aspect of scale at the same time.

We have two basic strategies for dealing with scale in the regression problem.

1. *Estimate $\boldsymbol{\sigma}$ beforehand.* This means that, before each iterative step, we choose a scale estimator and calculate its value, $\hat{\sigma}$. Then, considering $\hat{\sigma}$ as a known and fixed constant, we proceed with M-estimation for $\boldsymbol{\beta}$; that is, we find the solution $\hat{\boldsymbol{\beta}}_M$ of

$$\sum\psi\left(\frac{y_i - \mathbf{x}_i\boldsymbol{\beta}}{\hat{\sigma}}\right)\mathbf{x}_i^T = \mathbf{0}.$$

In this setting, the most commonly used resistant scale estimator is the median absolute deviation (MAD), which yields

$$\hat{\sigma} = \frac{1}{0.6745}\operatorname*{med}_i\left\{\left|y_i - \mathbf{x}_i\hat{\boldsymbol{\beta}}^{(0)} - \operatorname*{med}_j\left\{y_j - \mathbf{x}_j\hat{\boldsymbol{\beta}}^{(0)}\right\}\right|\right\}, \tag{24}$$

where $\hat{\boldsymbol{\beta}}^{(0)}$ is a preliminary estimate of $\boldsymbol{\beta}$ and 0.6745 is the average value of the MAD for samples from the standard Gaussian distribution. Usually the LAR estimate of $\boldsymbol{\beta}$ is a convenient choice for $\hat{\boldsymbol{\beta}}^{(0)}$ (see Section 8C). An important reason for the widespread use of the MAD is its excellent resistance: its breakdown bound is nearly 50%. The MAD has also proved to be a reasonably robust estimator of scale (UREDA, Section 12C).

Sometimes we make the role of the tuning constant explicit by writing

$$\psi_0\left(\frac{y_i - \mathbf{x}_i\boldsymbol{\beta}}{c\hat{\sigma}}\right),$$

where c is a generic tuning constant and ψ_0 is the member of the particular family of ψ-functions that has tuning constant 1 (e.g., the Huber ψ-function with $k = 1$). Then, in discussing a choice of tuning constant, we must be careful to state whether $\hat{\sigma}$ involves any standardizing constants, such as the $1/0.6745$ in equation (24). Some discussions in the literature normalize MAD in this way, but others do not.

2. *Estimate coefficients* $\boldsymbol{\beta}$ *and scale* σ *simultaneously.* Again paralleling the approach for estimating location (UREDA, Sections 11A and 12D), we can set up an equation for scale that is compatible with the p simultaneous equations for $\boldsymbol{\beta}$ in equation (9). Solving the resulting system of $p + 1$ equations then yields simultaneous M-estimators of $\boldsymbol{\beta}$ and σ. Specifically, these simultaneous equations take the form

$$\sum\psi\left(\frac{y_i - \mathbf{x}_i\boldsymbol{\beta}}{\sigma}\right)\mathbf{x}_i^T = \mathbf{0}, \tag{25}$$

$$\sum\chi\left(\frac{y_i - \mathbf{x}_i\boldsymbol{\beta}}{\sigma}\right) = na, \tag{26}$$

where χ is an even function, such as $\psi(t)^2$ or $t\psi(t)$, and a is a suitable positive constant. For example, a is often chosen as $[(n - p)/n]E[\chi(Z)]$, where Z is a standard Gaussian random variable, so that the scale estimator is consistent when the fluctuations are Gaussian.

To motivate the scale equation, we note that we could make

$$\sum\rho\left(\frac{y_i - \mathbf{x}_i\boldsymbol{\beta}}{\sigma}\right)$$

as small as desired by simply allowing σ to become large enough. Equation (26) rules out such nonsensical values of σ.

As Huber (1981, p. 176) points out, if ψ and χ in equations (25) and (26) are totally unrelated, there will be trouble with existence and convergence proofs for the simultaneous M-estimators. He avoids these difficulties by relating χ to ρ and ψ through

$$\chi(t) = t\psi(t) - \rho(t).$$

To illustrate this construction, we examine the Huber estimators (parametrized by the nonnegative constant k), which have

$$\rho(t) = \begin{cases} \frac{1}{2}t^2, & \text{for } |t| \le k, \\ k|t| - \frac{1}{2}k^2, & \text{for } |t| > k. \end{cases}$$

From this objective function we obtain

$$\psi(t) = \begin{cases} t, & \text{for } |t| \le k, \\ k \operatorname{sgn}(t), & \text{for } |t| > k, \end{cases} \tag{27}$$

through the definition $\psi = \rho'$. Then substituting into $\chi(t) = t\psi(t) - \rho(t)$ yields

$$\chi(t) = \begin{cases} \frac{1}{2}t^2, & \text{for } |t| \le k, \\ \frac{1}{2}k^2, & \text{for } |t| > k. \end{cases} \tag{28}$$

We note that this formulation yields both a bounded ψ-function and a bounded χ-function, thus limiting the impact of a large residual on both the regression estimate and the scale estimate. (In terms of breakdown bound, however, the scale estimate is not as resistant as the MAD.)

8C. COMPUTATION

In order to use regression M-estimators in practice, we must arrange to calculate the solutions to sets of simultaneous equations such as those in equations (9), (10), and (25) and (26). Without becoming enmeshed in the computational details, we need to look at some of the algorithms that have proved useful. Readers with little interest in computation should still continue as far as the discussion of iteratively reweighted least squares. The software in ROBETH offers a number of alternatives for the user of robust regression (Marazzi, 1980).

Algorithms for M-estimators are generally iterative and thus require a starting value. For a regression model, however, obtaining a high-quality starting value presents a greater challenge than in the location problem, where the median is usually quite satisfactory.

Certain one-step M-estimators have the same asymptotic properties as the corresponding fully iterated estimator, and thus they sometimes offer a way to avoid much of the iteration, if we feel we are lucky enough.

Algorithms with Fixed $\hat{\sigma}$

If a scale estimate $\hat{\sigma}$ has been obtained and fixed in advance of each iterative step, the *M*-estimator for the regression coefficients is defined by equations (25):

$$\sum \psi\left(\frac{y_i - \mathbf{x}_i \boldsymbol{\beta}}{\hat{\sigma}}\right)\mathbf{x}_i^T = \mathbf{0}.$$

Generally, this system of equations is not linear, unless it corresponds to the OLS estimator. Thus iterative methods are necessary. We consider three iterative methods:

Newton's method. Applying Newton's method (often labeled "Newton–Raphson") yields the iteration formula

$$\hat{\boldsymbol{\beta}}^{(m+1)} = \hat{\boldsymbol{\beta}}^{(m)} + \hat{\sigma}\left[\sum \mathbf{x}_i^T \psi'\left(\frac{y_i - \mathbf{x}_i \hat{\boldsymbol{\beta}}^{(m)}}{\hat{\sigma}}\right)\mathbf{x}_i\right]^{-1} \mathbf{X}^T \psi\left(\frac{\mathbf{y} - \mathbf{X}\hat{\boldsymbol{\beta}}^{(m)}}{\hat{\sigma}}\right), \quad (29)$$

where

$$\psi'(t) = \frac{d}{dt}\psi(t),$$

$$\psi\left(\frac{\mathbf{y} - \mathbf{X}\hat{\boldsymbol{\beta}}^{(m)}}{\hat{\sigma}}\right) = \left[\psi\left(\frac{y_1 - \mathbf{x}_1 \hat{\boldsymbol{\beta}}^{(m)}}{\hat{\sigma}}\right), \ldots, \psi\left(\frac{y_n - \mathbf{x}_n \hat{\boldsymbol{\beta}}^{(m)}}{\hat{\sigma}}\right)\right]^T.$$

Section 11K of UREDA discusses this method for obtaining location *M*-estimates. The principle here is very similar—namely, to use a linear approximation at $\boldsymbol{\beta} = \hat{\boldsymbol{\beta}}^{(m)}$ to the function $\Sigma\psi[(y_i - \mathbf{x}_i\boldsymbol{\beta})/\hat{\sigma}]\mathbf{x}_i^T$. Now, however, the situation is more complicated because we have to use p tangent hyperplanes (p linear equations) rather than one tangent line (one linear equation).

As in the location case, Newton's method can converge to maxima (and saddle points) of $\Sigma\rho[(y_i - \mathbf{x}_i\boldsymbol{\beta})/\sigma]$, as well as to minima. Although we want to consider multiple solutions, we are unlikely to have any interest in solutions corresponding to maxima.

Huber's method. Huber (1977) suggested the iterative algorithm

$$\hat{\boldsymbol{\beta}}^{(m+1)} = \hat{\boldsymbol{\beta}}^{(m)} + \hat{\sigma}(\mathbf{X}^T\mathbf{X})^{-1}\mathbf{X}^T\psi\left(\frac{\mathbf{y} - \mathbf{X}\hat{\boldsymbol{\beta}}^{(m)}}{\hat{\sigma}}\right). \quad (30)$$

The idea of equation (30) is to replace each residual $y_i - \mathbf{x}_i\hat{\boldsymbol{\beta}}$ by the modified residual $\psi(r_i/\hat{\sigma})\hat{\sigma}$. By rewriting the OLS estimator we can obtain, because $\psi(t) = t$,

$$\hat{\boldsymbol{\beta}} = (\mathbf{X}^T\mathbf{X})^{-1}\mathbf{X}^T\mathbf{y} = (\mathbf{X}^T\mathbf{X})^{-1}\mathbf{X}^T[\hat{\mathbf{y}} + (\mathbf{y} - \hat{\mathbf{y}})]$$

$$= \hat{\boldsymbol{\beta}} + (\mathbf{X}^T\mathbf{X})^{-1}\mathbf{X}^T\mathbf{r}.$$

The iterative version is then

$$\hat{\boldsymbol{\beta}}^{(m+1)} = \hat{\boldsymbol{\beta}}^{(m)} + (\mathbf{X}^T\mathbf{X})^{-1}\mathbf{X}^T\mathbf{r}^{(m)},$$

and this becomes the more general form in equation (30) when we replace $\mathbf{r}^{(m)}$ by $\psi(\mathbf{r}^{(m)}/\hat{\sigma})\hat{\sigma}$. When ψ is the Huber ψ-function with tuning constant k, as in equation (27), this has the effect of leaving $r_i^{(m)}$ unchanged when $|r_i^{(m)}| \leq k\hat{\sigma}$ and replacing it with $k\hat{\sigma}\,\mathrm{sgn}(r_i^{(m)})$ when $|r_i^{(m)}| > k\hat{\sigma}$. Thus use of the Huber ψ-function is analogous to Winsorizing.

Reweighted least-squares method. Beaton and Tukey (1974) described another iteration for (approximately) solving equation (25), namely

$$\hat{\boldsymbol{\beta}}^{(m+1)} = \hat{\boldsymbol{\beta}}^{(m)} + (\mathbf{X}^T\mathbf{W}^{(m)}\mathbf{X})^{-1}\mathbf{X}^T\mathbf{W}^{(m)}(\mathbf{y} - \mathbf{X}\hat{\boldsymbol{\beta}}^{(m)}), \tag{31}$$

where $\mathbf{W}^{(m)}$ is a diagonal matrix with diagonal elements

$$w_i^{(m)} = \frac{\psi\left[\left(y_i - \mathbf{x}_i\hat{\boldsymbol{\beta}}^{(m)}\right)/\hat{\sigma}\right]}{\left(y_i - \mathbf{x}_i\hat{\boldsymbol{\beta}}^{(m)}\right)/\hat{\sigma}}, \qquad i = 1, 2, \ldots, n.$$

This algorithm relies on the connection between M-estimators and W-estimators, discussed in Section 8B. Like a number of other ideas in robustness, this approach has earlier roots: for example, Jeffreys (1932) sketched the main ingredients of iteratively reweighted least squares. The simple weighted least-squares estimator (with constant weight matrix $\mathbf{W}$) satisfies

$$\mathbf{X}^T\mathbf{W}\mathbf{X}\hat{\boldsymbol{\beta}}_W = \mathbf{X}^T\mathbf{W}\mathbf{y}.$$

This looks only slightly more complicated than the usual least-squares estimator, but in practice we must begin without knowing the final weights. Thus we use the preliminary estimate $\hat{\boldsymbol{\beta}}^{(0)}$ in calculating residuals $y_i - \mathbf{x}_i\hat{\boldsymbol{\beta}}^{(0)}$, from which we in turn calculate weights. Iterating these steps yields equations (31) or the equivalent form

$$\hat{\boldsymbol{\beta}}^{(m+1)} = (\mathbf{X}^T\mathbf{W}^{(m)}\mathbf{X})^{-1}\mathbf{X}^T\mathbf{W}^{(m)}\mathbf{y}$$

and gives rise to the term "iteratively reweighted least squares" (Holland and Welsch, 1977).

As in the location case, iteratively reweighted least squares will often converge only to a minimum of $\Sigma\rho[(y_i - \mathbf{x}_i\boldsymbol{\beta})/\sigma]$.

Comparing equations (29), (30), and (31), we see that Newton's method requires ψ', so that it may be more difficult to implement. Huber's method has some computational convenience, because $(\mathbf{X}^T\mathbf{X})^{-1}\mathbf{X}^T$ need be computed only once. The reweighted least-squares method permits use of an existing (unweighted) least-squares algorithm at each iteration step (after multiplying y_i and $\mathbf{x}_i$ by $\sqrt{w_i^{(m)}}$), whereas Newton's method and Huber's method do not allow this convenience.

Huber (1973, 1977), Dutter (1975, 1977a, 1977b), Birch (1980), and Dempster, Laird, and Rubin (1980) have investigated the convergence properties of these algorithms.

If ψ is monotone [that is, $\psi(u) \geq \psi(v)$ whenever $u > v$], all three methods give convergent sequences of $\hat{\boldsymbol{\beta}}^{(m)}$. Among them, Newton's method converges most rapidly, as judged by the number of iterations required. Generally, the reweighted least-squares method converges somewhat faster than Huber's method, although empirical evidence (Huber, 1981) shows only a small difference. Incidentally, the usual proofs of the convergence of the reweighted LS algorithm require symmetry of $\rho(t)$, which holds for all currently popular ρ.

When a nonmonotone ψ-function is used (for example, when ψ redescends), equation (25) may have multiple roots, including both local maxima and local minima of the objective function $\Sigma\rho[(y_i - \mathbf{x}_i\boldsymbol{\beta})/\sigma]$. Newton's method still converges to some root. Also, under mild conditions on ψ (which the usual redescending ψ-functions satisfy) Byrd and Pyne (1979) prove that the iteratively reweighted least-squares algorithm with fixed scale always converges to a root. Depending on the starting value, either method may converge to a root other than the overall minimum. Similarly, if the tuning constant is too small (usually undesirable) or the (fixed) scale estimate is too small, these algorithms may converge slowly.

Algorithms for Computing $\hat{\boldsymbol{\beta}}$ and $\hat{\sigma}$ Simultaneously

As we remarked in Section 8B, the second approach to the problem of scale is to estimate $\boldsymbol{\beta}$ and σ simultaneously by solving equations (25) and (26) with a ψ-function and χ-function related to the same ρ-function. Following the approach of Huber (1981, Section 7.8), we use

$$\sum \psi\left(\frac{y_i - \mathbf{x}_i\boldsymbol{\beta}}{\sigma}\right)\mathbf{x}_i^T = \mathbf{0},$$

$$\sum \chi\left(\frac{y_i - \mathbf{x}_i\boldsymbol{\beta}}{\sigma}\right) = na,$$

where the objective function satisfies $\rho(0) = 0$, $a > 0$,

$$\psi(t) = \rho'(t),$$

$$\chi(t) = t\psi(t) - \rho(t).$$

Then the iteration formulas for $\hat{\sigma}$ and $\hat{\boldsymbol{\beta}}$ are

$$(\hat{\sigma}^{(m+1)})^2 = \frac{1}{na}\sum \chi\left(\frac{y_i - \mathbf{x}_i\hat{\boldsymbol{\beta}}^{(m)}}{\hat{\sigma}^{(m)}}\right)(\hat{\sigma}^{(m)})^2 \tag{32}$$

$$\hat{\boldsymbol{\beta}}^{(m+1)} = \hat{\boldsymbol{\beta}}^{(m)} + \hat{\sigma}^{(m)}(\mathbf{X}^T\mathbf{X})^{-1}\mathbf{X}^T\psi\left(\frac{\mathbf{y} - \mathbf{X}\hat{\boldsymbol{\beta}}^{(m)}}{\hat{\sigma}^{(m)}}\right). \tag{33}$$

Equation (33) is essentially the same as equation (30), but equation (32) may be both unexpected and counterintuitive. Heuristically, we note that when $\hat{\sigma}^{(m)}$ is larger than the desired value, the scaled residuals $(y_i - \mathbf{x}_i\hat{\boldsymbol{\beta}}^{(m)})/\hat{\sigma}^{(m)}$ will be too small in magnitude, so that

$$\sum \chi\left(\frac{y_i - \mathbf{x}_i\hat{\boldsymbol{\beta}}^{(m)}}{\hat{\sigma}^{(m)}}\right)$$

will be smaller than na. Consequently, equation (32) will produce $\hat{\sigma}^{(m+1)} < \hat{\sigma}^{(m)}$. A similar argument applies when $\hat{\sigma}^{(m)}$ is smaller than the desired value. To explain the appearance of $\hat{\sigma}^{(m)}$ and $\hat{\sigma}^{(m+1)}$ as $(\hat{\sigma}^{(m)})^2$ and $(\hat{\sigma}^{(m+1)})^2$, we also recall that most ψ-functions are close to linear (at least near 0), so that χ is close to quadratic. Thus, χ modifies the (squared) scaled residuals, and the factor $(\hat{\sigma}^{(m)})^2$ puts them back in the scale of y.

In principle, instead of equation (33), the reweighted least-squares formula (31) can also be used. Although no general theorem yet guarantees a convergent sequence $(\hat{\boldsymbol{\beta}}^{(m)}, \hat{\sigma}^{(m)})$ when (31) is substituted for (33), there do not seem to be any known examples of unsatisfactory convergence. When the errors are distributed according to a "Gaussian-over-independent" distribution [that is, ε in equation (1) can be written as Z/V, where Z is a standard Gaussian random variable and V is independent of Z], some convergence conditions have been investigated by Dempster, Laird, and Rubin (1980).

Starting Value

With any iterative method, choosing the starting value is an important issue. From a good starting value, a procedure will converge in fewer iterations and incur less computational cost. Iterative M-estimators are particularly sensitive to the starting value when the ψ-function redescends. Given a poor starting value, an estimator with a nonmonotone ψ-function may converge to a root of equation (25) far from the overall minimum of the objective function in terms of $\boldsymbol{\beta}$.

Usually we prefer a starting value to be resistant and have high efficiency. In the location problem, the sample median is considered the best convenient choice. Its natural generalization to regression models appears to be the LAR estimator. Harvey (1977) compared the LAR estimator with two other approaches, suggested by Andrews (1974) and Hinich and Talwar (1975), respectively. The LAR estimator may be the most attractive preliminary estimator, because it has, on the whole, higher asymptotic efficiency than the other two. On the other hand, the LAR estimator may take too much computing time to be used solely as a starting value. It need not be unique, and it is even harder to calculate than most of the robust regression estimators that we intend to use.

If the ψ-function is monotone, the OLS estimator is still a conceivable choice. There is no trouble with convergence, the OLS estimator is easy to calculate, and almost every user wants to know the OLS results and compare them with the robust estimates.

For a redescending ψ-function, besides starting with the LAR estimator of $\boldsymbol{\beta}$, we can also recommend the following strategy: Use a monotone ψ-function with the OLS estimates as the starting value, iterate to convergence, and then use the nonmonotone ψ-function to iterate a few steps (perhaps only one) further.

As we have mentioned before, if $\boldsymbol{\beta}$ and σ are estimated simultaneously, only the combination of equations (32) and (33) is known to ensure convergence. For this iteration scheme, both the OLS estimates and the LAR estimates are good choices as starting values. It is usually preferable to start with the OLS estimates and then, after iterating to convergence with a Huber ψ-function and χ-function, to do a few iterations with a redescending ψ-function and either a fixed $\hat{\sigma}$ obtained from equation (32) in the Huber iterations or the MAD or fourth-spread of the residuals.

One-Step Estimators

A simple version of a regression M-estimator—a one-step Gauss–Newton approximation to equation (29)—was introduced by Bickel (1975). Under mild regularity conditions, this one-step estimator has the same asymptotic properties (consistency and normality) as the corresponding fully iterated M-estimator, provided that one uses a reasonably good preliminary estimate.

As in the location problem, the starting value for a one-step estimator is very important, because the iteration formula is applied only once. However, a convenient and broadly effective choice of preliminary estimator is not yet known. As in the location problem, the OLS estimate is not a satisfactory starting value for good one-step M-estimators. The LAR estimate may be a more satisfactory choice for regression; but it does not control leverage. Also, it is much harder to compute, and this defeats the purpose of a simple one-step procedure. Thus we suggest that the one-step method is useful mainly in location problems and some simple regression models, such as a regression line through the origin.

One natural regression analog of "one step for location" is "p steps for p regression coefficients." After some modification, this prescription can be regarded as a one-step estimator of a sort. The modification that seems likely to be adequate is

Start with OLS.

Do $2p$ iterations with Huber ψ- and χ-functions.

Take one step of a redescending estimator (with fixed scale).

8D. EXAMPLE: THE STACK LOSS DATA

This section shows how an M-estimator works in one practical problem. The data set has been discussed extensively in the books by Brownlee (1965), Draper and Smith (1981), and Daniel and Wood (1980), as well as in the papers by Andrews (1974) and Denby and Mallows (1977). Our aim here is not to search for a best model for this thoroughly studied example, but to gain more understanding of M-estimators through comparison of least squares and a more robust M-estimator. Thus we discuss only some aspects of these data. We emphasize the advantage of robust M-estimators over least squares when outliers appear. When we use least squares, we must take special care to set aside certain outliers. In contrast, a robust M-estimator often automatically reaches almost the same result as we would get from least squares with careful diagnosis.

The Stack Loss Data

The "stack loss" data (Table 8-3) represent 21 days of operation of a plant that oxidizes ammonia (NH_3) to nitric acid (HNO_3). The three explanatory

TABLE 8-3. Data from operation of a plant for the oxidation of ammonia to nitric acid.

Observation Number	Air Flow, x_1	Cooling Water Inlet Temperature, x_2	Acid Concentration, x_3	Stack Loss, y
1	80	27	89	42
2	80	27	88	37
3	75	25	90	37
4	62	24	87	28
5	62	22	87	18
6	62	23	87	18
7	62	24	93	19
8	62	24	93	20
9	58	23	87	15
10	58	18	80	14
11	58	18	89	14
12	58	17	88	13
13	58	18	82	11
14	58	19	93	12
15	50	18	89	8
16	50	18	86	7
17	50	19	72	8
18	50	19	79	8
19	50	20	80	9
20	56	20	82	15
21	70	20	91	15

Source: Cuthbert Daniel and Fred S. Wood (1980). *Fitting Equations to Data*, 2nd ed. New York: Wiley (Table 5.1 on p. 61). Their source for the data: K. A. Brownlee (1965). *Statistical Theory and Methodology in Science and Engineering*, 2nd ed. New York: Wiley (p. 454).

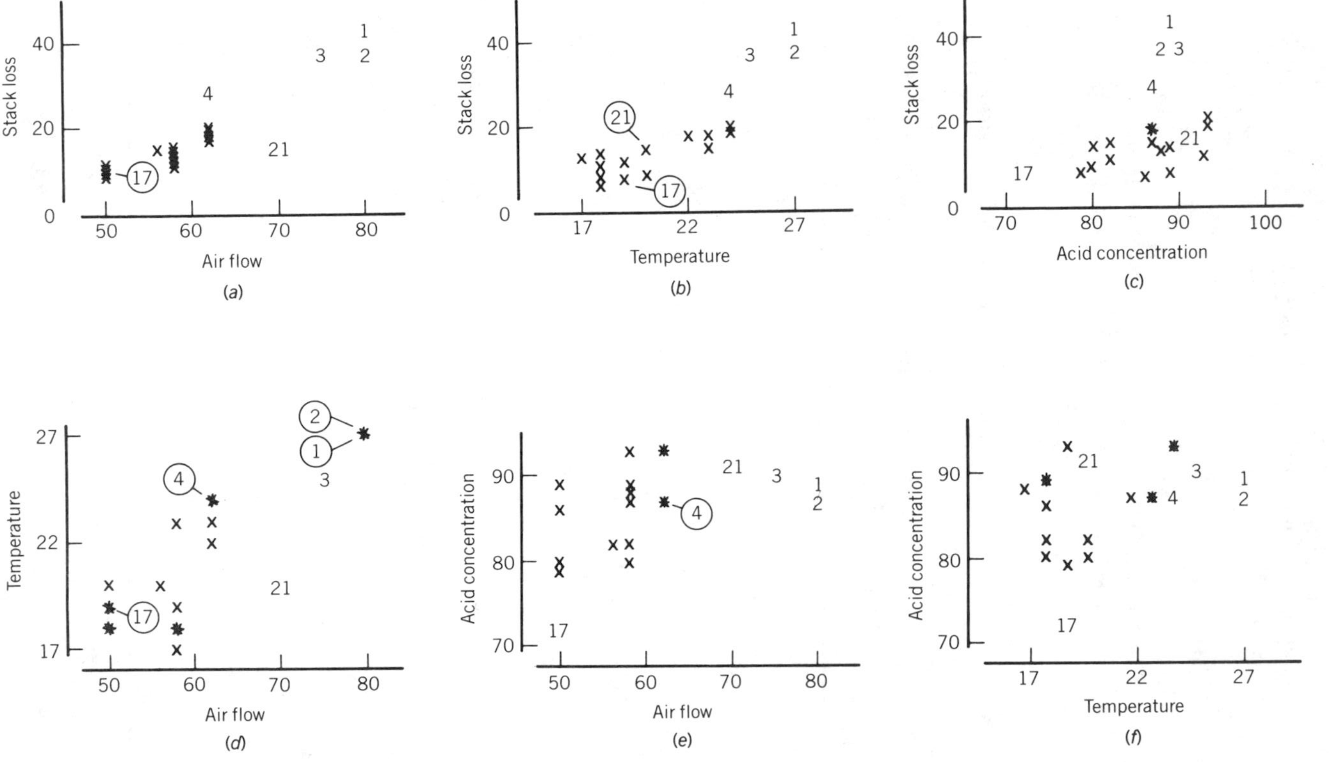

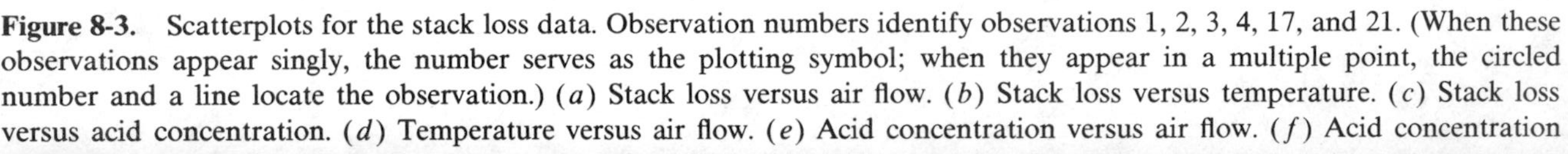

Figure 8-3. Scatterplots for the stack loss data. Observation numbers identify observations 1, 2, 3, 4, 17, and 21. (When these observations appear singly, the number serves as the plotting symbol; when they appear in a multiple point, the circled number and a line locate the observation.) (*a*) Stack loss versus air flow. (*b*) Stack loss versus temperature. (*c*) Stack loss versus acid concentration. (*d*) Temperature versus air flow. (*e*) Acid concentration versus air flow. (*f*) Acid concentration

variables x_1, x_2, and x_3 and the response variable y are as follows:

x_1 = air flow (which reflects the rate of operation of the plant),

x_2 = temperature of the cooling water in the coils of the absorbing tower for the nitric oxides,

x_3 = concentration of nitric acid in the absorbing liquid [coded by $x_3 = 10 \times (\text{concentration in percent} - 50)$],

and

y = the percent of the ingoing ammonia that is lost by escaping in the unabsorbed nitric oxides ($\times 10$)

The response variable y is the "stack loss." Figure 8-3 shows plots of y versus each of the explanatory variables, as well as plots of each pair of

TABLE 8-4. Coefficient estimates and summary statistics for three least-squares fits to the stack loss data. Standard errors of coefficient estimates are in parentheses.

Variable	All Observations	Without Observation 21	Without 1, 3, 4, and 21
Rate	0.716	0.889	0.798
	(0.135)	(0.119)	(0.0674)
Temperature	1.295	0.817	0.577
	(0.368)	(0.325)	(0.166)
Concentration	−0.152	−0.1071	−0.0671
	(0.156)	(0.125)	(0.0616)
Constant	−39.92	−43.70	−37.65
	(11.90)	(9.49)	(4.73)
Centercept[a]	17.14	17.58	16.57
s^2	10.52	6.60	1.57
R^2	0.91	0.95	0.975
n	21	20	17

[a] The value of $\hat{y}$ when the x-variables take convenient central values. Here the centercept is calculated at rate = 60, temperature = 21, and concentration = 86.

explanatory variables. Figure 8-3*d* suggests that observation 21 may have somewhat enhanced leverage, and Figure 8-3*f* gives the same impression for observation 17.

Because we want to illustrate a single fitting, we focus on the model

$$y = \beta_0 + \beta_1 x_1 + \beta_2 x_2 + \beta_3 x_3 + \varepsilon$$

and do not consider the alternative models discussed in some of the references cited above.

Least-Squares Fits

Fitting this regression equation by least squares to all 21 observations yields the estimates of the coefficients and the related information shown in Table

TABLE 8-5. Residuals from three least-squares fits.

	Fit		
Observation	All Observations	Without Observation 21	Without 1, 3, 4, and 21
1	3.23	2.06	*6.22*
2	−1.92	−3.04	1.15
3	4.56	3.25	*6.43*
4	5.70	6.30	*8.17*
5	−1.71	−2.07	−0.67
6	−3.01	−2.88	−1.25
7	−2.39	−2.06	−0.42
8	−1.39	−1.06	0.58
9	−3.14	−2.33	−1.06
10	1.27	0.01	0.36
11	2.64	0.97	0.96
12	2.78	0.68	0.47
13	−1.43	−2.78	−2.51
14	−0.05	−1.42	−1.35
15	2.36	2.09	1.34
16	0.91	0.76	0.14
17	−1.52	−0.55	−0.37
18	−0.46	0.20	0.10
19	−0.60	0.49	0.59
20	1.41	1.37	1.93
21	−7.24	*−10.13*	*−8.63*

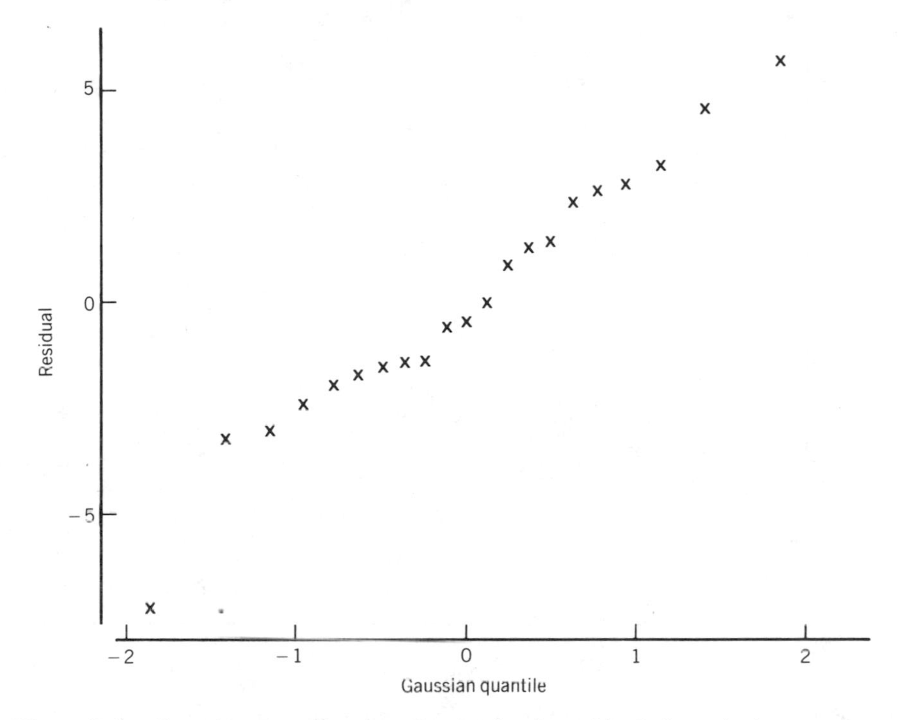

Figure 8-4. Gaussian quantile–quantile plot for the residuals from the least-squares fit to the stack loss data (21 points).

8-4. In particular, the fitted equation is

$$\hat{y} = -39.92 + 0.716x_1 + 1.30x_2 - 0.152x_3.$$

In a roughly centered form, this fit becomes

$$\hat{y} = 17.14 + 0.716(x_1 - 60) + 1.30(x_2 - 21) - 0.152(x_3 - 86),$$

where $x_1 - 60$ has mean 0.4 and values between -10 and $+20$, $x_2 - 21$ has mean 0.1 and values between -4 and $+6$, and $x_3 - 86$ has mean 0.3 and values between -14 and $+7$. The residuals from this fit appear in Table 8-5.

Perhaps the most noticeable feature of this set of residuals is the relatively large negative residual (-7.24) for observation 21. Figure 8-4, a Gaussian quantile–quantile plot (or normal probability plot) of these residuals (see Section 10C) confirms this impression by indicating that this

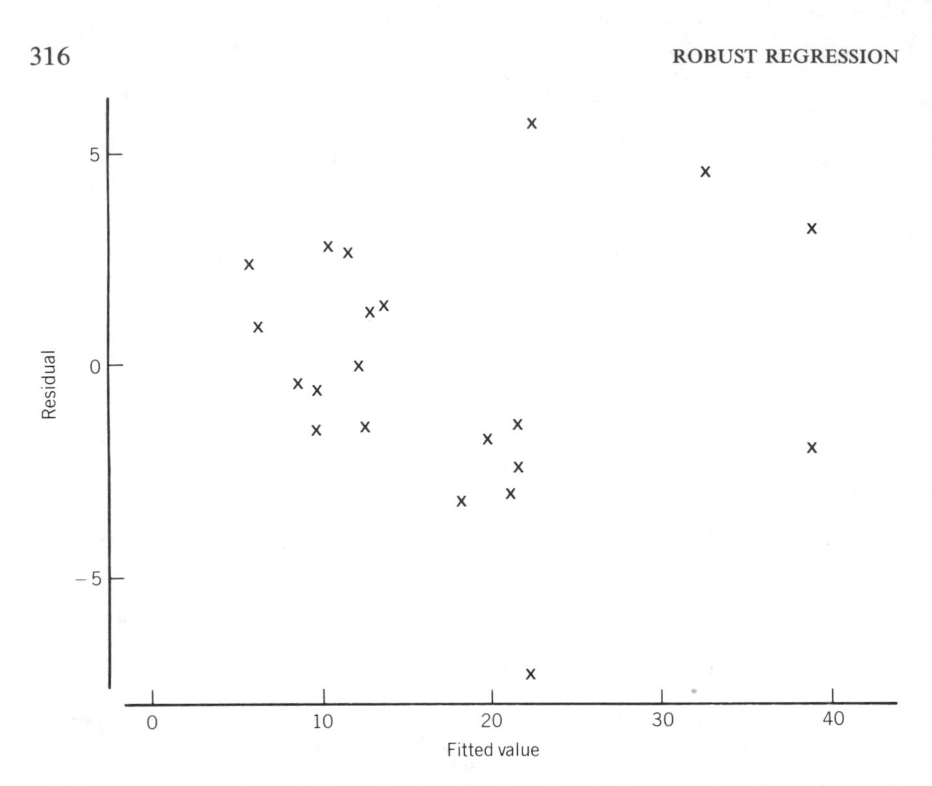

Figure 8-5. Plot of residuals versus fitted values for the least-squares fit to the stack loss data (21 points).

residual is much lower than one would expect from the other residuals. Furthermore, the plot of r versus $\hat{y}$ (in Figure 8-5) follows an unusual pattern: five residuals (observations 1, 2, 3, 4, and 21) stand apart from the rest, which, in isolation, seem to have a substantial downward slope.

Within the usual least-squares framework the messages from these displays make it imperative to set aside at least observation 21. Fitting the same model by least squares to the remaining 20 points yields

$$\hat{y} = -43.70 + 0.889x_1 + 0.817x_2 - 0.1071x_3.$$

Table 8-4 includes standard errors and summary statistics for this fit, and the residuals appear in Table 8-5. The usual measures (R^2, s^2) indicate a better fit, but a Gaussian $Q - Q$ plot and an r-versus-$\hat{y}$ plot would point again to observations 1, 2, 3, and 4 as possibly unusual. By examining the order of the observations Daniel and Wood found that, when the rate was above 60, the plant required about one day to reach equilibrium. This

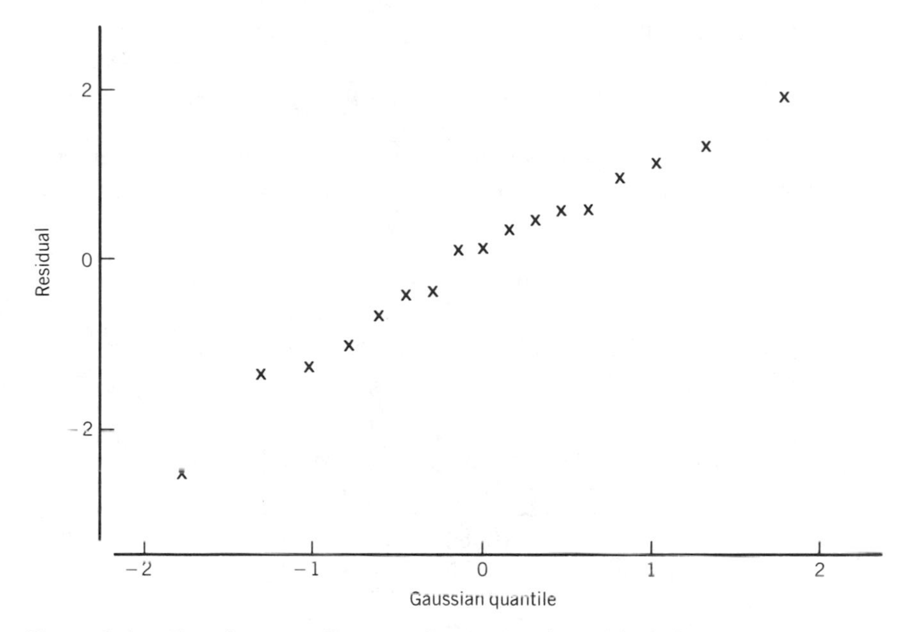

Figure 8-6. Gaussian quantile–quantile plot for the residuals from the least-squares fit to the stack loss data after removing observations 1, 3, 4, and 21.

information led them to drop observations 1, 3, and 4 as transient states. The least-squares fit to the resulting 17-point data set is

$$\hat{y} = -37.65 + 0.798x_1 + 0.577x_2 - 0.0671x_3.$$

(see also Table 8-4); the residuals appear in Table 8-5 and Figure 8-6. This fit constitutes a substantial improvement over the two previous least-squares fits, particularly as judged by s^2, the residual mean square.

Two Robust Fits

Andrews (1974) applied a robust M-estimator to the stack loss data. His ψ-function is

$$\psi(t) = \begin{cases} \sin(t/1.5), & \text{for } |t| \leq 1.5\pi, \\ 0, & \text{for } |t| > 1.5\pi, \end{cases} \tag{34}$$

with scale estimated by

$$\hat{\sigma} = \text{median}\{|y_i - \mathbf{x}_i\hat{\boldsymbol{\beta}}|\}. \tag{35}$$

Using the original 21 observations and also the 17-point data set after setting aside observations 1, 3, 4, and 21, he obtained the same robust fit (to the accuracy given):

$$\hat{y} = -37.2 + 0.82x_1 + 0.52x_2 - 0.07x_3.$$

This is quite close to the OLS fit for the 17-point data set. Table 8-6 shows

TABLE 8-6. Residuals from the robust fit and the 17-point least-squares fit.

Observation	Robust Residual	17-Point OLS Residual
1	6.11	*6.08*
2	1.04	1.15
3	6.31	*6.44*
4	8.24	*8.18*
5	−1.24	−0.67
6	−0.71	−1.25
7	−0.33	−0.42
8	0.67	0.58
9	−0.97	−1.06
10	0.14	0.35
11	0.79	0.96
12	0.24	0.47
13	−2.71	−2.51
14	−1.44	−1.34
15	1.33	1.34
16	0.11	0.14
17	−0.42	−0.37
18	0.08	0.10
19	0.63	0.59
20	1.87	1.93
21	−8.91	*−8.63*

Source: David F. Andrews (1974). "A robust method for multiple linear regression," *Technometrics*, **16**, 523–531 (columns 1, 5, and 4 of Table 5 on p. 529).

the residuals from Andrews' fit and, for comparison, repeats the residuals from the OLS fit to the 17-point data set. The differences between corresponding residuals are minor.

Discussion

Comparing the least-squares fits and the one robust fit leads us to comment on several features and then examine the reasons for them.

First, this robust fit is unaffected by the four highly unusual observations. Either with them or without them, we obtain essentially the same estimates of the coefficients. (This need not happen if their unusualness were less pronounced.) Even when the robust fitting procedure begins with all 21 observations, it manages to disregard those four points. We note, however, that the median absolute residual, used as the scale estimate in equation (35), does change from 0.97 for all 21 points to 0.71 for the 17-point subset. Thus, in other examples, we would hope that the effects of extremely unusual observations would at most be small.

Second, the robust fitting procedure is strikingly successful in identifying the four anomalous observations and in simultaneously reducing their impact on the estimated coefficients. Because it does this without an analyst's intervention, it would be safer to use in a nearly automated way in a computer package than least squares.

Third, the robust estimates of the regression coefficients are close to those in the 17-point OLS fit, which emerged only after several steps of fitting and diagnosis.

We recall that the regression M-estimator $\hat{\boldsymbol{\beta}}_M$ satisfies the implicit equations (25)

$$\sum \psi\left(\frac{y_i - \mathbf{x}_i\boldsymbol{\beta}}{\hat{\sigma}}\right)\mathbf{x}_i^T = \mathbf{0}$$

and that its ψ-function is the one given in equation (34). Thus this M-estimator of $\boldsymbol{\beta}$ is the solution of

$$\sum_I \sin\left(\frac{y_i - \mathbf{x}_i\boldsymbol{\beta}}{1.5\hat{\sigma}}\right)\mathbf{x}_i^T = \mathbf{0}, \tag{36}$$

where $I = \{i\colon |(y_i - \mathbf{x}_i\boldsymbol{\beta})/\hat{\sigma}| \le 1.5\pi\}$ and $\hat{\sigma}$ is given in equation (35). By taking into account only those observations for which $|y_i - \mathbf{x}_i\boldsymbol{\beta}| \le 1.5\pi\hat{\sigma}$, the summation does not allow points with larger residuals to affect the robust fit.

To examine the final robust fit in more detail, we start with the results for all 21 observations. The sorted absolute residuals in Table 8-7 make it easy to determine that $\hat{\sigma} = \text{median}\{|y_i - \mathbf{x}_i\hat{\boldsymbol{\beta}}_M|\} = 0.97$. Thus $1.5\pi\hat{\sigma} = 4.71$; and observations 1, 3, 4, and 21 have residuals whose absolute values exceed this cutoff. Consequently, these four points did not affect the final stages of the robust fit (except, by their presence, to increase $\hat{\sigma}$ somewhat), as Andrews confirmed by refitting after setting them aside.

We gain some further understanding of the robust fit and the 17-point OLS fit by recalling that the OLS estimator is an M-estimator, with $\psi(t) = t$. Thus, in the form of equation (25), the OLS estimator for $\boldsymbol{\beta}$ is the solution of

$$\sum\left(\frac{y_i - \mathbf{x}_i\boldsymbol{\beta}}{\hat{\sigma}}\right)\mathbf{x}_i^T = \mathbf{0}.$$

TABLE 8-7. Absolute values of three forms of residuals from the robust fit to the stack loss data: r_i = ordinary residual, $\tilde{r}_i$ = scaled residual, and r_i^* = effective residual. Observations ordered according to $|r_i|$.

Rank	Observation	$\|r_i\|$	$\|\tilde{r}_i\|$	$\|r_i^*\|$
1	21	8.91	6.12	0
2	4	8.24	5.66	0
3	3	6.31	4.34	0
4	1	6.11	4.20	0
5	13	2.71	1.863	0.958
6	20	1.87	1.285	0.960
7	14	1.44	0.990	0.836
8	15	1.33	0.914	0.792
9	5	1.24	0.852	0.753
10	2	1.04	0.715	0.655
11	9	0.97	0.667	0.618
12	11	0.79	0.543	0.517
13	6	0.71	0.488	0.469
14	8	0.67	0.460	0.444
15	19	0.63	0.433	0.420
16	17	0.42	0.289	0.285
17	7	0.33	0.227	0.225
18	12	0.24	0.165	0.164
19	10	0.14	0.096	0.096
20	16	0.11	0.076	0.076
21	18	0.08	0.055	0.055

In this *linear* equation, we can divide by 1.5 and rewrite the equation as

$$\sum \tilde{r}_i \mathbf{x}_i^T = \mathbf{0} \tag{37}$$

with

$$\tilde{r}_i = (y_i - \mathbf{x}_i \boldsymbol{\beta})/(1.5\hat{\sigma}),$$

$$\hat{\sigma} = \text{median}\{|y_i - \mathbf{x}_i \boldsymbol{\beta}|\}.$$

For convenience we refer to the $\tilde{r}_i$ as the *scaled residuals* because they have been scaled by the tuning constant (here 1.5) as well as the scale estimate $\hat{\sigma}$. If we define

$$r_i^* = \begin{cases} \sin(\tilde{r}_i), & \text{for } |\tilde{r}_i| \le \pi, \\ 0, & \text{for } |\tilde{r}_i| > \pi, \end{cases}$$

and

$$y_i^* = 1.5 r_i^* \hat{\sigma} + \mathbf{x}_i \boldsymbol{\beta}$$

(all for $i = 1, 2, \ldots, n$), these r_i^* are the *effective residuals* (or pseudoresiduals), and the y_i^* are the *pseudoobservations*. Then equation (36) becomes

$$\sum r_i^* \mathbf{x}_i^T = \sum \left(\frac{y_i^* - \mathbf{x}_i \boldsymbol{\beta}}{1.5\hat{\sigma}} \right) \mathbf{x}_i^T = \mathbf{0}.$$

This has exactly the same form as equation (37) for least squares. In this sense, we can say that any M-estimator is a least-squares estimator in which the observations y_i are replaced by pseudoobservations y_i^* or, equivalently, in which residuals are replaced by their effective counterparts.

Returning to the robust fit and the 17-point OLS fit, we have seen that the four observations (points 1, 3, 4, and 21) are actually deleted from both fits; the robust fitting procedure does it automatically, whereas with least squares the user must do it. For the remaining 17 observations, on the whole, the difference between y_i and y_i^* is small, because

$$y_i - y_i^* = 1.5\hat{\sigma}(\tilde{r}_i - r_i^*) = 1.455(\tilde{r}_i - \sin \tilde{r}_i)$$

($i = 2, 5, 6, \ldots, 20$) and $\sin x \approx x$ for small $|x|$. By including $\tilde{r}_i$ and r_i^*, Table 8-7 shows that 12 of the 21 scaled residuals $\tilde{r}_i$ are smaller in magnitude than 0.8 (actually 9 are smaller than 0.5, and 3 lie between 0.5 and 0.8) and differ little from the corresponding r_i^*; three of the $\tilde{r}_i$ are between 0.8 and 1.0 and are thus somewhat different from the corresponding r_i^*; and six $\tilde{r}_i$ are greater than 1.0 and very different from the corresponding r_i^*.

This detailed examination of residuals provides a clearer picture of why the robust fit is close to the 17-point OLS fit and how a robust M-estimator works. These M-estimators essentially work on residuals: they shrink the effects of large residuals and thus reduce the impact of outlying values of y.

The notion of shrinking residuals also allows us to characterize a basic difference between redescending and monotone M-estimators. The redescending ψ-function eliminates observations that have large residuals, whereas the monotone ψ-function only reduces these large residuals. Looking ahead to Section 8E, where we try to face up to the problems of high-leverage observations, we remark that one challenge is to turn an anomalous y-value at a high-leverage point into a large residual, so that a suitable M-estimator can reach a fit that deemphasizes that observation.

8E. BOUNDED-INFLUENCE REGRESSION

In Section 8B we saw that the influence function of a regression M-estimator (with σ known and fixed at $\sigma = 1$) takes the form

$$\mathbf{IC}((y,\mathbf{x}); F_0, \hat{\boldsymbol{\beta}}_M) = \psi[y - \mathbf{x}\hat{\boldsymbol{\beta}}_M(F_0)]\mathbf{B}^{-1}\mathbf{x}^T \tag{38}$$

with

$$\mathbf{B} = \int \psi'[y - \mathbf{x}\hat{\boldsymbol{\beta}}_M(F_0)]\mathbf{x}^T\mathbf{x}\, dF_0(y,\mathbf{x}).$$

We now face a difficulty not present in M-estimators of location: the influence of a single observation $(y, \mathbf{x})$ on $\hat{\boldsymbol{\beta}}_M$ can be arbitrarily large. This can plainly be undesirable if either the observation is "bad" or the choice of model is inadequate. It is appropriate, and necessary for an efficient fit, however, when both the observation is "good" and an adequate form is being fitted.

Looking closer at the right-hand side of equation (38), we see that $\psi[y - \mathbf{x}\hat{\boldsymbol{\beta}}_M(F_0)]$ comes from the residual, and $\mathbf{B}^{-1}\mathbf{x}^T$ comes from the position of $\mathbf{x}$. Thus, even though the ψ-functions of robust M-estimators are bounded, the contribution of an $\mathbf{x}$ that gets more and more unusual can still cause the value of the influence function to be as large as we like. Even for a redescending ψ-function which vanishes whenever the residual is large, the influence function is still unbounded *if* we consider an unbounded set of *possible* values for $\mathbf{x}$. A high-leverage point $(y, \mathbf{x})$, whose $\mathbf{x}$ is far away from the bulk of the x-data, may have a small residual r, so that the modified residual $\psi(r)$ is still positive and the influence of this point is still very large. That is, a redescending ψ-function does not suffice to bound the influence in regression, as long as we consider x-patterns arbitrarily more extreme than the one in the data.

Some readers may feel that least-squares regression diagnostics and robust M-estimators have already introduced substantial complexity into the practice of regression analysis. Because the strategies for bounding influence add to the complexity somewhat, these readers may wish to skip the details in the remainder of this section. Others, who would like to apply a bounded-influence approach in some actual analyses, should gain valuable experience from proceeding in two stages. First obtain a highly resistant starting point from least-median-of-squares regression (described in Section 8F). Then continue with the more efficient estimator of Krasker and Welsch, as sketched later in this section.

Some Basic Proposals

Several proposals offer ways to reduce the influence of outliers in x-space. Although we focus on the p simultaneous equations which these approaches substitute for equation (25), a related equation, analogous to equation (26), customarily handles the scale parameter σ.

Mallows (1973, 1975) suggested bounded-influence estimators that satisfy the p simultaneous equations

$$\sum u(\mathbf{x}_i)\psi\left(\frac{y_i - \mathbf{x}_i\boldsymbol{\beta}}{\hat{\sigma}}\right)\mathbf{x}_i^T = \mathbf{0}.$$

[In a comment on the paper by Huber (1983), Mallows discusses a related estimator in a suitable theoretical framework.] Here $u(\mathbf{x})$ offsets the contribution of an outlying $\mathbf{x}$. For example, we could take

$$u(\mathbf{x}) = \min\{1, c/d(\mathbf{x})\}$$

with a suitable constant c and

$$d^2(\mathbf{x}) = (\mathbf{x} - \mathbf{m})\mathbf{S}^{-1}(\mathbf{x} - \mathbf{m})^T,$$

where $\mathbf{m}$ and $\mathbf{S}$ are robust estimates of the location and covariance matrix of the carriers X_j, respectively. The presence of the generalized distance $d(\mathbf{x})$ in the denominator in u keeps $u(\mathbf{x})\mathbf{x}^T$ bounded. We note, however, that u will reduce the weight of any high-leverage point, regardless of whether the corresponding y-value is discrepant. Thus this strategy may sacrifice efficiency unnecessarily.

Another bounded-influence estimator, proposed by Schweppe (Handschin et al., 1975), satisfies the equations

$$\sum \sqrt{1 - h_{ii}}\,\psi\left(\frac{y_i - \mathbf{x}_i\boldsymbol{\beta}}{\hat{\sigma}\sqrt{1 - h_{ii}}}\right)\mathbf{x}_i^T = \mathbf{0},$$

where the h_{ii} are the diagonal elements of the hat matrix, $\mathbf{H}$, defined in equation (5). In effect this proposal divides each residual $y_i - \mathbf{x}_i\boldsymbol{\beta}$ by its standard deviation $\sigma\sqrt{1 - h_{ii}}$ in OLS, so that ψ can cut down larger residuals more rapidly at points of higher leverage. For example, if $h_{ii} = 0.5$, the argument of ψ is $\sqrt{2}$ times as large as it would be without the factor $1/\sqrt{1 - h_{ii}}$. At the same time the factor $\sqrt{1 - h_{ii}}$ acts as a weight to deemphasize high-leverage observations. Huber (1983) arrives at this same form of estimator in another framework. Exercise 10 verifies that the contribution of $\sqrt{1 - h_{ii}}\,\mathbf{x}_i^T$ to the influence function is bounded.

Other early proposals include those of Welsch (1977) and Hinkley (1977). The basic idea of these approaches is to put some weight on each data point, but to control it so that when $\mathbf{x}$ is far from the bulk of the data, the weight at $\mathbf{x}$ is small.

To obtain maximum resistance (essentially 50% breakdown point) Rousseeuw (1984) devised an entirely different approach, least-median-of-squares regression, discussed briefly in Section 8F.

Some Monte Carlo results have shown that the efficiency of the Mallows and Schweppe estimators (with Huber ψ-function) can be poor: under the Gaussian model, their efficiency ranges from 20% to 80% (see Hill, 1977). Thus further efforts focused on higher efficiency are called for.

Efficient Bounded-Influence Regression

A desire for bounded influence and good efficiency does not yield a single theoretical problem whose solution then settles the matter. Both influence and efficiency involve choices and leave room for alternative technical approaches. (We may need multiple choices if we are to get the multiple solutions that we require.)

What influence will we be concerned about? In regression, unlike the single-parameter location case, different definitions include the influence of a single observation $(y_i, \mathbf{x}_i)$ on the estimates of the coefficients, on its own fitted value $\hat{y}_i$, or on linear combinations of the coefficient estimates. Focusing on different influence functions will give different estimators. Some may be better, some worse.

$\mathbf{A}(F,:\hat{\boldsymbol{\beta}})$, the asymptotic covariance matrix of $\hat{\boldsymbol{\beta}}$, one could minimize the generalized variance [the determinant of $\mathbf{A}(F, \hat{\boldsymbol{\beta}})$] or the sum of the asymptotic variances of the $\hat{\beta}_j$ [the trace of $\mathbf{A}(F, \hat{\boldsymbol{\beta}})$].

In a brief overview we can neither examine all the varied theoretical approaches nor consider their technical details. The earliest approach to robust M-estimation, developed by Huber (1964), minimizes the maximum asymptotic variance of a location estimator over distributions F in some ε-neighborhood of the model, F_0. Usually F_0 is the Gaussian distribution, and the neighborhood contains all F of the form $(1 - \varepsilon)F_0 + \varepsilon G$, where G is any symmetric distribution. Huber (1983) develops the corresponding minimax theory for the one-parameter regression problem (a straight line through the origin with known scale) and derives estimators that bound influence on the parameter.

In the location problem Hampel (1968) minimizes the asymptotic variance of the estimator at the model while satisfying a bound on the influence curve. Hampel (1978), Krasker (1978, 1980), and Krasker and Welsch (1982) consider several extensions of this approach to regression. Briefly, Hampel minimizes the trace of $\mathbf{A}(F, \hat{\boldsymbol{\beta}})$. Krasker and Welsch (1982) pursue a different strategy, which we now describe in more detail as a way of illustrating the interplay among the considerations that arise in bounded-influence regression.

In various situations we may focus on linear combinations of the individual $\hat{\beta}_j$; for example, we may look at the difference between two of the coefficients, or we may plan to predict the value of y at a specified point $\mathbf{x}_0$. In general we wish to estimate $\boldsymbol{\lambda}^T\boldsymbol{\beta}$, where $\boldsymbol{\lambda}$ is a (column) vector of constants. Then the influence of a single point $(y, \mathbf{x})$ on the estimate $\boldsymbol{\lambda}^T\hat{\boldsymbol{\beta}}$ can be measured, relative to the variability of the estimate, by

$$\frac{\boldsymbol{\lambda}^T\boldsymbol{\Omega}(y, \mathbf{x})}{(\boldsymbol{\lambda}^T\mathbf{V}\boldsymbol{\lambda})^{1/2}},$$

where $\boldsymbol{\Omega}(y, \mathbf{x})$ is a briefer notation for $\mathbf{IC}((y, \mathbf{x}); F, \hat{\boldsymbol{\beta}})$ as in equation (38), and $\mathbf{V} = \mathbf{A}(F, \hat{\boldsymbol{\beta}})$ is the asymptotic covariance matrix of $\hat{\boldsymbol{\beta}}$, so that the asymptotic variance of $\boldsymbol{\lambda}^T\hat{\boldsymbol{\beta}}$ is $\boldsymbol{\lambda}^T\mathbf{V}\boldsymbol{\lambda}$. Krasker and Welsch deal with influence in terms of *sensitivity*, defined as the maximum absolute value of this ratio over all linear combinations $\boldsymbol{\lambda}$ and all points $(y, \mathbf{x})$:

$$\gamma = \sup_{(y, \mathbf{x})} \sup_{\boldsymbol{\lambda}} \frac{|\boldsymbol{\lambda}^T\boldsymbol{\Omega}(y, \mathbf{x})|}{(\boldsymbol{\lambda}^T\mathbf{V}\boldsymbol{\lambda})^{1/2}}. \tag{39}$$

What "asymptotic variance" should be minimized? The theoretical approaches work most readily with large-sample properties. In terms of

As their central model Krasker and Welsch adopt a Gaussian situation. Specifically, the $(y_i, \mathbf{x}_i)$ come from a distribution such that the conditional distribution of $y_i - \mathbf{x}_i\boldsymbol{\beta}$ given $\mathbf{x}_i$ is $\text{Gau}(0, \sigma^2)$.

They write the estimators as W-estimators and generalize the basic definition $w(t) = \psi(t)/t$ so that the weight for observation i now takes the form $w(y_i, \mathbf{x}_i, \boldsymbol{\beta})$ and the weight function is nonnegative, bounded, and continuous. Thus the estimator satisfies the equation

$$\sum w(y_i, \mathbf{x}_i, \boldsymbol{\beta})(y_i - \mathbf{x}_i\boldsymbol{\beta})\mathbf{x}_i^T = \mathbf{0}.$$

To achieve the sensitivity bound $\gamma \leq a$ (with $a > \sqrt{p}$), they use the specific weight function

$$w(y, \mathbf{x}, \boldsymbol{\beta}, \sigma, \mathbf{A}) = \min\left\{1, \frac{a}{|\varepsilon/\sigma|[\mathbf{x}\mathbf{A}^{-1}\mathbf{x}^T]^{1/2}}\right\}, \tag{40}$$

where $\varepsilon = y - \mathbf{x}\boldsymbol{\beta}$ and

$$\mathbf{A} = E\left[w^2\left(\frac{y - \mathbf{x}\boldsymbol{\beta}}{\sigma}\right)^2 \mathbf{x}^T\mathbf{x}\right].$$

The additional arguments of w in equation (40) simply make explicit the roles of the scale parameter σ and the matrix $\mathbf{A}$, both of which must be estimated. Thus, although we do not go into computational details, each iteration would seek simultaneously to improve the estimates of $\boldsymbol{\beta}$, σ, and $\mathbf{A}$.

To gain a better understanding of the behavior of w in equation (40), we note that, for a given value of the tuning constant k, the weight function corresponding to the Huber ψ-function (Table 8-2) can be written $w(t) = \min\{1, k/|t|\}$. This expression makes it easy to see that each data point receives a weight no greater than 1. When applied to the $(y_i, \mathbf{x}_i)$, equation (40) in effect assigns each data point its own value of the Huber tuning constant. Thus the Krasker–Welsch estimator is of the same type as the estimator proposed by Schweppe.

In equation (40) $\mathbf{A}$ is a sort of weighted cross-product matrix for the carriers $X_1, \ldots, X_p$, so that $\mathbf{x}_i\mathbf{A}^{-1}\mathbf{x}_i^T$ has leverage and distance interpretations analogous to $h_{ii} = \mathbf{x}_i(\mathbf{X}^T\mathbf{X})^{-1}\mathbf{x}_i^T$. Therefore, a large value of $\mathbf{x}_i\mathbf{A}^{-1}\mathbf{x}_i^T$ tends to reduce the weight given to $(y_i, \mathbf{x}_i)$. At the same time, a small residual, $(y_i - \mathbf{x}_i\hat{\boldsymbol{\beta}})/\hat{\sigma}$, tends to increase the weight. In the way it combines these two contributions the Krasker–Welsch weight function attempts to preserve efficiency by downweighting high-leverage observations with anomalous y-values while retaining high-leverage observations where the regression model fits well.

Discussion

Overall comparisons among techniques for bounded-influence regression must consider a variety of qualitative and quantitative robustness properties, including gross-error sensitivity, local-shift sensitivity, breakdown, asymptotic efficiency, and finite-sample efficiency. Our brief account in this section has emphasized one definition of gross-error sensitivity and asymptotic efficiency.

One can define the breakdown point of an estimator as the smallest amount of contamination that may force the value of the estimate off to arbitrary values. Donoho and Huber (1983) provide a valuable discussion of this concept and emphasize its application in finite samples. A high breakdown point is plainly desirable, and bounded-influence techniques in regression seek to provide this sort of protection from anomalous data. Still, the problem is difficult: essentially all the estimators that we have discussed so far have breakdown point less than $1/(p+1)$. The least-median-of-squares approach of Rousseeuw (1984) and the repeated-median estimator of Siegel (1982), although computationally more difficult, do achieve the highest possible breakdown point (essentially 50%); although they do not aim at high efficiency, they could be used to get a starting point for the iterative calculations of a more efficient estimator, such as that of Krasker and Welsch. We note also that, in estimators such as those of Schweppe and Krasker and Welsch, the generalized distance measure on the $\mathbf{x}_i$ plays an important role in determining the estimator's breakdown point. For example, Krasker and Welsch (1983) explain that the matrix $\mathbf{A}$ represents a weakness in their estimator because its breakdown point is at most $1/(p+1)$, and they discuss alternatives that may offer higher breakdown.

In a minimax framework Huber (1983) shows that the Krasker–Welsch approach corresponds to guarding against a situation in which the gross errors in y occur selectively at points with the highest leverage, and he suggests that this is an unrealistically pessimistic assumption. In the accompanying discussion Krasker and Welsch distinguish between data analysis, which involves a substantial exploratory component, and "pure estimation," which would ordinarily follow extensive data analysis and model criticism. They point out (p. 73) that "A small amount of curvature in the regression, or a failure of the model over a certain region of the x space, can generate data that behave as though nature had selectively placed gross errors on the leverage points."

To derive their weight function, Krasker and Welsch (1982) assume the existence of a weight function that satisfies both the sensitivity bound and a certain efficiency condition, and they then prove that such a weight function must have the form given in equation (40). Unfortunately, their assumption turns out to be too strong. Bickel (1984) offers a heuristic technical argument that no such weight function exists, and Ruppert (1985) has devised a counterexample. These theoretical results, however, seem to point only to small differences in performance; they do not dash all hopes for efficient bounded-influence regression. Instead they indicate that no one technique can be superior to all others in all circumstances. We can still hope that reasonable bounded-influence approaches will provide adequate protection against the adverse effects of anomalous data in a wide variety of practical situations.

8F. SOME ALTERNATIVE METHODS

The challenge of fitting regression models robustly has stimulated a rich flow of research that offers a wide variety of approaches. The present chapter has emphasized *M*- and *W*-estimation because these fruitful approaches have undergone the most vigorous development. To broaden our appreciation of robust methods in regression, we now turn to a brief selection of alternative techniques (some of which are offshoots of *M*-estimation).

In principle, any robust method for the simple linear regression model can be applied, one explanatory variable at a time, to multiple regression models, as Chapter 7 indicates. Such an approach, however, depends on the user for the order in which the variables are to enter the fit, even when the choice of model is settled.

This chapter has taken the multiple regression model as given, in the same way that least squares starts with the matrix $\mathbf{X}$. Our other multiple regression approaches continue in this vein.

The general approaches of *L*-estimation, *M*-estimation, and *R*-estimation, which have provided a framework for developing and studying a great many estimators of location (Andrews et al., 1972; Huber, 1981; and UREDA, Chapters 9, 10, and 11), may also be applied to regression. Sections 8B, 8C, and 8E have examined techniques based on *M*-estimation (or *W*-estimation). Bickel (1973) introduced a class of estimators for the linear model that generalize linear combinations of order statistics. We refer the reader to his paper for an explanation of this approach.

We begin here with two versions of trimmed least squares. One bases the trimming on regression quantiles and the other, on the preliminary residuals; both generalize the *L*-estimators of location known as trimmed means. Next we examine *R*-estimators for regression, whose motivation comes from rank tests. A fourth alternative modifies an *M*-estimator so as to adapt the ψ-function broadly to the error distribution. Outside the framework of *L*-, *M*-, and *R*-estimation, two qualitatively different approaches offer high breakdown point; one uses repeated medians, and the other minimizes the median of the squared residuals.

Trimmed Least Squares from Regression Quantiles

Among L-estimators of location, the trimmed mean is attractive because it is simple, easy to compute, and fairly efficient in a variety of situations. In the regression model, we can base a form of trimming on regression quantiles, as described by Koenker and Bassett (1978). To develop these regression quantiles, we begin with a definition of quantiles in the location model. Our account parallels that given by Ruppert and Carroll (1980).

For constant p, $0 < p < 1$, we define

$$\psi_p(u) = \begin{cases} p - 1, & \text{when } u < 0, \\ p, & \text{when } u \geq 0, \end{cases}$$

$$\rho_p(u) = u\psi_p(u) = \begin{cases} u(p-1), & \text{when } u < 0, \\ up, & \text{when } u \geq 0. \end{cases}$$

Claerbout and Muir (1973) introduced an equivalent ρ-function and gave this quantile interpretation. It is easy to check that the pth quantile of the sample $x_1, \ldots, x_n$ can be defined as a value of t that solves

$$\sum \rho_p(x_i - t) = \min. \tag{41}$$

For example, suppose $p = \frac{1}{3}$ and $n = 3$, and let $x_{(1)} \leq x_{(2)} \leq x_{(3)}$ denote the order statistics of the sample. Then

$$\sum \rho_p(x_i - t) = \frac{1}{3}\left(\sum_{x_i \geq t} (x_i - t) - 2 \sum_{x_i < t} (x_i - t) \right).$$

Straightforward checking reveals that

$$\sum \rho_p(x_i - t) = \tfrac{1}{3}[x_{(2)} - x_{(1)} + x_{(3)} - x_{(1)}]$$

for $x_{(1)} \leq t \leq x_{(2)}$ and that $\Sigma\rho_p(x_i - t)$ is larger for other values of t. Thus any value of t in the interval $[x_{(1)}, x_{(2)}]$ satisfies equation (41) and can serve, so far as this definition goes, as the $\frac{1}{3}$ quantile of the sample x_1, x_2, x_3.

In passing, we note that this quantile is actually the result of an M-estimator with $\psi_p(u)$ as the ψ-function. For instance, taking $p = \frac{1}{2}$ yields $\psi_p(u) = \frac{1}{2}\,\mathrm{sgn}(u)$, the ψ-function corresponding to the sample median (see Figure 8-2d). Thus $\rho_p(u)$ generalizes the absolute-value objective function by having different slopes for positive and negative arguments.

As with any other M-estimator, this definition easily generalizes to the regression model. The pth regression quantile, $\hat{\boldsymbol{\beta}}(p)$, is defined as any value of $\boldsymbol{\beta}$ that solves

$$\sum \rho_p(y_i - \mathbf{x}_i\boldsymbol{\beta}) = \min.$$

Koenker and Bassett define an α-trimmed least-squares estimator ($0 < \alpha \leq \frac{1}{2}$), denoted by $\hat{\boldsymbol{\beta}}_{KB}$, in this way: Remove every observation whose residual from $\hat{\boldsymbol{\beta}}(\alpha)$ is negative or whose residual from $\hat{\boldsymbol{\beta}}(1-\alpha)$ is positive, and then calculate the OLS estimator from the remaining observations.

This trimmed least-squares estimator, $\hat{\boldsymbol{\beta}}_{KB}$, has some good theoretical properties. It is asymptotically unbiased, provided that the error distribution F is symmetric; if F is asymmetric, the bias of $\hat{\boldsymbol{\beta}}_{KB}$ affects only the constant term. It is also asymptotically Gaussian and has quite good efficiency at the Gaussian (0.97 for $\alpha = 0.05$, 0.94 for $\alpha = 0.10$, and 0.84 for $\alpha = 0.25$). In the face of heavily contaminated Gaussian errors, Ruppert and Carroll found that it is relatively efficient compared with the OLS estimator, and almost as efficient as two selected M-estimators, the Huber estimator with $k = 2$ (see Table 8-2), and a Hampel estimator, whose ψ-function is piecewise linear and redescends to 0.

Trimming Based on Preliminary Residuals

Ruppert and Carroll (1980) introduce another trimmed least-squares estimator, which trims observations on the basis of their residuals calculated from a preliminary estimate, $\hat{\boldsymbol{\beta}}_0$. They denote this estimator by $\hat{\boldsymbol{\beta}}_{PE}$ or, in more detail, $\hat{\boldsymbol{\beta}}_{PE}(\alpha)$ for $0 < \alpha < \frac{1}{2}$. The observations removed are those corresponding to the $[n\alpha]$ smallest and $[n\alpha]$ largest preliminary residuals. (Here $[x]$ denotes the largest integer not exceeding x.) Then $\hat{\boldsymbol{\beta}}_{PE}(\alpha)$ is the OLS estimator determined by the remaining $n - 2[n\alpha]$ observations.

The estimator $\hat{\boldsymbol{\beta}}_{PE}$ is consistent under symmetric errors. Under an asymmetric error distribution, $\hat{\boldsymbol{\beta}}_{PE}$ is still consistent, except for the first component (the intercept). In this respect, $\hat{\boldsymbol{\beta}}_{PE}$ has the same behavior as $\hat{\boldsymbol{\beta}}_{KB}$. But the asymptotic covariance matrix of $\hat{\boldsymbol{\beta}}_{PE}$ depends heavily on the preliminary estimate $\hat{\boldsymbol{\beta}}_0$. For example, under Gaussian errors, if $\hat{\boldsymbol{\beta}}_0$ is the OLS estimate or the LAR estimate, the corresponding $\hat{\boldsymbol{\beta}}_{PE}$, denoted by $\hat{\boldsymbol{\beta}}_{PE}(\text{LS})$ and $\hat{\boldsymbol{\beta}}_{PE}(\text{LAR})$, respectively, is inefficient; but some $\hat{\boldsymbol{\beta}}_{PE}$ with other preliminary estimates can be quite efficient.

Ruppert and Carroll (1980) prove that if $\hat{\boldsymbol{\beta}}_0$ is the average of the αth and $(1-\alpha)$th regression quantiles, that is,

$$\hat{\boldsymbol{\beta}}_0 = \tfrac{1}{2}[\hat{\boldsymbol{\beta}}(\alpha) + \hat{\boldsymbol{\beta}}(1-\alpha)],$$

then for any symmetric error distribution the corresponding $\hat{\beta}_{PE}$, called $\hat{\beta}_{PE}(\alpha)$, is asymptotically equivalent to $\hat{\beta}_{KB}$. This $\hat{\beta}_{PE}(\alpha)$ has smaller asymptotic variance than $\hat{\beta}_{PE}(\text{LS})$ and $\hat{\beta}_{PE}(\text{LAR})$ under a variety of contaminated normal error distributions (Ruppert and Carroll 1980, Table 1).

R-Estimators for Regression

The least-squares estimator for regression minimizes the sum of the squared residuals and thus is very sensitive to wild y-values. To robustify the least-squares estimator, other *M*-estimators minimize the sum of a less rapidly increasing function of the residuals by using a suitable ρ-function in

$$\sum \rho(y_i - \mathbf{x}_i \boldsymbol{\beta}) = \sum \rho(r_i).$$

In a similar way, one class of *R*-estimators in the linear model, proposed by Jaeckel (1972) and discussed and implemented by Hettmansperger and McKean (1977), minimizes a sum in which each residual is weighted by a score based on its rank:

$$D(\mathbf{r}) = \sum a_n(R_i) r_i = \min,$$

where $R_i = R(y_i - \mathbf{x}_i \boldsymbol{\beta}) = R(r_i)$ is the rank of the residual r_i in $\{r_1, \ldots, r_n\}$, and $a_n(\cdot)$ is a nondecreasing score function, not identically zero, which satisfies

$$\sum_k a_n(k) = 0. \tag{42}$$

In effect, the *R*-estimates for regression coefficients minimize a kind of dispersion of the residuals. That is,

$$D(\mathbf{r}) = \sum_i a_n(R_i) r_i = \sum_k a_n(k) r_{(k)} \tag{43}$$

(where $r_{(1)} \leq r_{(2)} \leq \cdots \leq r_{(n)}$ are the ordered residuals) can be viewed as a measure of the dispersion of the residuals. To see this, we note that, because $a_n(\cdot)$ is nondecreasing and satisfies equation (42), the larger $r_{(k)}$ have positive $a_n(k)$, whereas the smaller $r_{(k)}$ have negative $a_n(k)$. Thus the sum in equation (43) usually consists of nonnegative terms. Also, replacing r_i by $A + Br_i$ merely multiplies the sum in equation (43) by B, so that it behaves like a measure of scale or dispersion. For example, the Wilcoxon scores

$$a_n(k) = \frac{k}{n+1} - \frac{1}{2}, \qquad k = 1, \ldots, n,$$

have $a_n(k) = -a_n(n+1-k)$, so that

$$\sum_k a_n(k) r_{(k)} = \sum_{k=1}^{n/2} a_n(n+1-k)\left(r_{(n+1-k)} - r_{(k)}\right).$$

Typically, $\Sigma a_n(R_i) r_i$ is a less rapidly increasing function of the residuals than Σr_i^2 but more rapidly increasing than $\Sigma |r_i|$.

For a given D, the corresponding regression estimator $\hat{\boldsymbol{\beta}}_D$ is any $\boldsymbol{\beta}$ that minimizes $D(\mathbf{y} - \mathbf{X}\boldsymbol{\beta})$. Jaeckel points out that $\hat{\boldsymbol{\beta}}_D$ may not be unique but that, for large n, the choice cannot matter much. Strictly speaking, we cannot use $D(\mathbf{y} - \mathbf{X}\boldsymbol{\beta})$ in estimating the intercept (or centercept) in the model, because D is translation-invariant; but it is easy to apply a location estimator after obtaining the other coefficients. Jaeckel proves that $D(\mathbf{y} - \mathbf{X}\boldsymbol{\beta})$ is a convex function of $\boldsymbol{\beta}$ for fixed $\mathbf{y}$ and $\mathbf{X}$, so that the minimization problem is straightforward, although possibly tedious.

In simple regression Jaeckel produces an interesting special case. Using the Wilcoxon scores

$$a_n(k) = \frac{k}{n+1} - \frac{1}{2}, \qquad k = 1, \ldots, n,$$

$\hat{\boldsymbol{\beta}}_D$ is a weighted median of the pairwise slopes $b_{ij} = (y_j - y_i)/(x_j - x_i)$ for $x_j > x_i$. (The unweighted median of these slopes is discussed in Section 5E of UREDA.)

In the general linear model Jaeckel's estimators are asymptotically equivalent to those proposed by Jurečková (1971), which have relatively good asymptotic robustness.

An Adaptive *M*-Estimator

To gain good performance over a wide range of symmetric and asymmetric error distributions, Moberg, Ramberg, and Randles (1980) devised an adaptive procedure that concludes with a one-step M-estimator. As the basic adaptive feature, it uses information from a set of preliminary residuals to select one of five ψ-functions, chosen so that they cover symmetric error distributions with light, medium, or heavy tails, as well as skewed error distributions.

Their procedure needs a resistant preliminary fit, so that outliers will have little impact. Despite its poor breakdown point, they use the LAR fit.

Using the preliminary residuals, the procedure classifies the error distribution according to its tail weight and skewness. It also forms a scale estimate.

The classification of the error distribution determines a choice of ψ-function for the M-estimator.

With the preliminary estimates as starting values, the procedure then takes one iterative step toward the M-estimator defined by ψ.

Moberg, Ramberg, and Randles undertook a substantial simulation study with $n = 20$ or 40 and $p = 2$ or 3 (each including the constant carrier). (In view of the discussion of bounded-influence procedures in Section 8E, it is worth pointing out that their **X**-matrices did not contain any rows with high leverage.) For comparison they included OLS, the LAR estimator, and a one-step M-estimator that uses Andrews' ψ-function and starts at the LAR estimate. When the error distribution is symmetric and heavy-tailed, they found that the adaptive procedure is competitive with the other two robust techniques and better than OLS. And when the error distribution is skewed or light-tailed and symmetric, the adaptive procedure does better than any of the others.

The Repeated-Median Estimator

Because a high breakdown point is often valuable in estimating regression coefficients, Siegel (1982) devised a novel algorithm based on repeated medians. For the simple regression model $y_i = A + Bx_i + \varepsilon_i$ with distinct x_i, each pair of points yields a slope $(y_j - y_i)/(x_j - x_i)$ and an intercept $(x_j y_i - x_i y_j)/(x_j - x_i)$, and the repeated-median estimates are

$$\hat{B}_{RM} = \operatorname*{med}_i \left\{ \operatorname*{med}_{j \neq i} \{(y_j - y_i)/(x_j - x_i)\} \right\},$$

$$\hat{A}_{RM} = \operatorname*{med}_i \left\{ \operatorname*{med}_{j \neq i} \{(x_j y_i - x_i y_j)/(x_j - x_i)\} \right\}.$$

In its general form the repeated-median method yields an estimate of a real (i.e., scalar) parameter θ whenever there is a positive integer k such that subsets of k data points determine an estimate $\tilde{\theta}$. The estimate from the points numbered $i_1, i_2, \ldots, i_k$ is denoted by $\tilde{\theta}(i_1, \ldots, i_k)$.

For the multiple regression model of equations (1) and (2), each nondegenerate subset of p points $(\mathbf{x}_{i_1}, y_{i_1}), \ldots, (\mathbf{x}_{i_p}, y_{i_p})$ determines a value of $\boldsymbol{\beta} = (\beta_1, \ldots, \beta_p)^T$, namely $\tilde{\boldsymbol{\beta}} = \tilde{\boldsymbol{\beta}}(i_1, \ldots, i_p) = (\tilde{\beta}_1, \ldots, \tilde{\beta}_p)$, where $\tilde{\beta}_j =$

$\tilde{\beta}_j(i_1, \ldots, i_p)$; and each β_j separately takes the role of θ. (The computations would be parallel.)

In the general setting Siegel defines the median operator M by

$$M\{\tilde{\theta}(i_1, \ldots, i_k)\} = \text{median}\{\tilde{\theta}(i_1, \ldots, i_{k-1}, j)\},$$

taking the median over all values of j for which $\tilde{\theta}(i_1, \ldots, i_{k-1}, j)$ is defined and finite. Then the repeated-median estimate of θ is

$$\hat{\theta} = M^k\{\tilde{\theta}(i_1, \ldots, i_k)\}.$$

[Each application of M leaves one fewer subscript from $\tilde{\theta}(i_1, \ldots, i_k)$.] Hence the repeated-median estimator of $\boldsymbol{\beta}$ is $\hat{\boldsymbol{\beta}}_{RM} = (\hat{\beta}_1, \ldots, \hat{\beta}_p)$ with

$$\hat{\beta}_j = M^p\{\tilde{\beta}_j(i_1, \ldots, i_p)\}.$$

Siegel proves that, asymptotically, as $n \to \infty$ with k fixed, the repeated-median procedure has a breakdown point of 50%. That is, $\hat{\theta}$ can tolerate data sets in which nearly half the observations are outliers. Even with such highly desirable resistance, however, computational expense may render the repeated-median estimator impractical for all but the most modest values of k. The amount of computation involved is essentially proportional to n^k.

Least-Median-of-Squares Regression

As another way of achieving the highest possible breakdown point, Rousseeuw (1984) introduced the idea of minimizing the median of the squared residuals, instead of the sum. That is, one determines the value of $\boldsymbol{\beta}$ that minimizes

$$\operatorname*{med}_i \left\{(y_i - \mathbf{x}_i\boldsymbol{\beta})^2\right\}.$$

In using a measure of the spread of the residuals as the explicit criterion, this approach resembles the R-estimators of Jaeckel (1972). Also, we note that when n is odd, minimizing $\text{med}\{r_i^2\}$ is equivalent to minimizing $\text{med}\{|r_i|\}$, which is essentially the scale estimator MAD.

Rousseeuw proves that, if $p > 1$ and the regression data are in general position (i.e., any p of them uniquely determine $\boldsymbol{\beta}$), then the breakdown point of the least-median-of-squares (LMS) regression estimator is $([n/2] - p + 2)/n$. As n becomes large, this approaches 50%.

It is instructive to examine two special cases. In the location problem Rousseeuw shows that the LMS estimate is the midpoint of the shortest half of the sample. Similarly, in simple regression LMS estimation corresponds to finding the narrowest strip that covers half the observations.

Applying the LMS approach to multiple regression problems involves substantial computational effort (relative to OLS). Data given by Rousseeuw (1984), however, indicate that, for $p \leq 10$ and $n \leq 200$, computation time does not become excessive.

For technical reasons the influence function of the LMS estimator is not well defined. Its high breakdown point, however, makes it an attractive starting point for various M-estimators. The computer program of Leroy and Rousseeuw (1984) should enable others to experiment with this technique.

8G. SUMMARY

Least-squares regression enjoys a number of attractive properties when the data are well behaved. Unfortunately the advantages of this popular approach quickly vanish as complications (common with practical data) set in. For example, the fluctuations may follow a heavy-tailed distribution or contain an occasional gross error. If a distinctive combination of values on the explanatory variables gives an observation high leverage, an anomalous fluctuation can be especially damaging. Thus careful examination of regression diagnostics for a variety of aspects (including leverage, impact on fit, and impact on coefficients) should be a part of any serious regression analysis.

Using accumulated insight and suitable diagnostic procedures, the experienced analyst can usually achieve satisfactory results by applying ordinary least squares to suitably edited or otherwise modified data. Automating this process, however, presents a number of challenges.

For coping with heavy-tailed fluctuations and gross errors (in the absence of points with high leverage), techniques of M-estimation minimize a function of the residuals that increases less rapidly than the sum of squares and can even be bounded. M-estimators can be implemented as iteratively reweighted least squares, in which each observation's weight depends on the corresponding residual. One reasonable strategy begins with OLS, iterates a Huber M-estimator to a moderate degree of convergence, and then finishes with a few iterations of a biweight estimator. From this process we

Learn about the OLS fit, estimated coefficients, and related diagnostics.

Obtain more robust estimates of the coefficients and a more robust fit.

Obtain residuals that are more likely to reveal anomalous observations.

Learn what weight the estimator has assigned to each observation.

Gain an opportunity to apply weighted least-squares regression diagnostics with these weights.

Learn whether any higher-leverage observations are further downweighted by the biweight.

As the example of the stack loss data illustrates, even a more streamlined strategy for robust regression can produce good results.

High-leverage observations can still cause difficulty for straightforward M-estimators. Attempts to overcome this obstacle have created a new class of robust procedures—bounded-influence regression estimators. One attractive proposal attempts to downweight an observation primarily when its leverage and its residual are both substantial. Although no single technique dominates all others in all regression models, some promising approaches have produced substantial gains. Procedures such as the Krasker–Welsch estimator deserve more extensive trial in diverse applications.

REFERENCES

Andrews, D. F. (1974). "A robust method for multiple linear regression," *Technometrics*, **16**, 523–531.

Andrews, D. F., Bickel, P. J., Hampel, F. R., Huber, P. J., Rogers, W. H., and Tukey, J. W. (1972). *Robust Estimates of Location: Survey and Advances*. Princeton, NJ: Princeton University Press.

Beaton, A. E. and Tukey, J. W. (1974). "The fitting of power series, meaning polynomials, illustrated on band-spectroscopic data," *Technometrics*, **16**, 147–192 (with discussion).

Belsley, D. A., Kuh, E., and Welsch, R. E. (1980). *Regression Diagnostics*. New York: Wiley.

Bickel, P. J. (1973). "On some analogues to linear combinations of order statistics in the linear model," *Annals of Statistics*, **1**, 597–616.

Bickel, P. J. (1975). "One-step Huber estimates in the linear model," *Journal of the American Statistical Association*, **70**, 428–434.

Bickel, P. J. (1984). "Robust regression based on infinitesimal neighbourhoods," *Annals of Statistics*, **12**, 1349–1368.

Birch, J. B. (1980). "Some convergence properties of iterated reweighted least squares in the location model," *Communications in Statistics*, **B9**, 359–369.

Brownlee, K. A. (1965). *Statistical Theory and Methodology in Science and Engineering*, 2nd ed. New York: Wiley.

Byrd, R. H. and Pyne, D. A. (1979). "Convergence of the iteratively reweighted least squares algorithm for robust regression." Technical Report No. 313, Department of Mathematical Sciences, The Johns Hopkins University.

Claerbout, J. F. and Muir, F. (1973). "Robust modeling with erratic data," *Geophysics*, **38**, 826–844.

Cook, R. D. and Weisberg, S. (1982). *Residuals and Influence in Regression*. New York: Chapman and Hall.

Daniel, C. and Wood, F. S. (1980). *Fitting Equations to Data*, 2nd ed. New York: Wiley.

Dempster, A. P., Laird, N. M., and Rubin, D. B. (1980). "Iteratively reweighted least squares for linear regression when errors are normal/independent distributed." In P. R. Krishnaiah (Ed.), *Multivariate Analysis—V*. Amsterdam: North-Holland, pp. 35–57.

Denby, L. and Larsen, W. A. (1977). "Robust regression estimators compared via Monte Carlo," *Communications in Statistics*, **A6**, 335–362.

Denby, L. and Mallows, C. L. (1977). "Two diagnostic displays for robust regression analysis," *Technometrics*, **19**, 1–13.

Donoho, D. L. and Huber, P. J. (1983). "The notion of breakdown point." In P. J. Bickel, K. A. Doksum, and J. L. Hodges, Jr. (Eds.), *A Festschrift for Erich L. Lehmann*. Belmont, CA: Wadsworth International Group, pp. 157–184.

Draper, N. R. and Smith, H. (1981). *Applied Regression Analysis*, 2nd ed. New York: Wiley.

Draper, N. R. and Stoneman, D. M. (1966). "Testing for the inclusion of variables in linear regression by a randomisation technique," *Technometrics*, **8**, 695–699.

Dutter, R. (1975). "Robust regression: different approaches to numerical solutions and algorithms." Research Report No. 6, Fachgruppe für Statistik, Eidgenössische Technische Hochschule, Zurich.

Dutter, R. (1977a). "Numerical solution of robust regression problems: computational aspects, a comparison," *Journal of Statistical Computation and Simulation*, **5**, 207–238.

Dutter, R. (1977b). "Algorithms for the Huber estimator in multiple regression," *Computing*, **18**, 167–176.

Gross, A. M. (1977). "Confidence intervals for bisquare regression estimates." *Journal of the American Statistical Association*, **72**, 341–354.

Hampel, F. R. (1968). "Contributions to the theory of robust estimation." Ph.D. thesis, University of California, Berkeley.

Hampel, F. R. (1978). "Optimally bounding the gross-error-sensitivity and the influence of position in factor space," *1978 Proceedings of the Statistical Computing Section*. Washington, DC: American Statistical Association, pp. 59–67 (with discussion).

Handschin, E., Kohlas, J., Fiechter, A., and Schweppe, F. (1975). "Bad data analysis for power system state estimation," *IEEE Transactions on Power Apparatus and Systems*, **2**, 329–337.

Harvey, A. C. (1977). "A comparison of preliminary estimators for robust regression," *Journal of the American Statistical Association*, **72**, 910–913.

Henderson, H. V. and Velleman, P. F. (1981). "Building multiple regression models interactively," *Biometrics*, **37**, 391–411.

Hettmansperger, T. P. and McKean, J. W. (1977). "A robust alternative based on ranks to least squares in analyzing linear models," *Technometrics*, **19**, 275–284.

Hill, R. W. (1977). "Robust regression when there are outliers." Ph.D. thesis, Harvard University, Cambridge, MA.

Hill, R. W. (1979). "On estimating the covariance matrix of robust regression *M*-estimators," *Communications in Statistics*, **A8**, 1183–1196.

Hill, R. W. and Holland P. W. (1977). "Two robust alternatives to least-squares regression," *Journal of the American Statistical Association*, **72**, 828–833.

Hinich, M. J. and Talwar, P. P. (1975). "A simple method for robust regression," *Journal of the American Statistical Association*, **70**, 113–119.

Hinkley, D. V. (1977). "Jackknifing in unbalanced situations," *Technometrics*, **19**, 285–292.

Hoaglin, D. C. and Welsch, R. E. (1978). "The hat matrix in regression and ANOVA," *The American Statistician*, **32**, 17–22, 146.

Holland, P. W. and Welsch, R. E. (1977). "Robust regression using iteratively reweighted least-squares," *Communications in Statistics*, **A6**, 813–827.

Huber, P. J. (1964). "Robust estimation of a location parameter," *Annals of Mathematical Statistics*, **35**, 73–101.

Huber, P. J. (1973). "Robust regression: asymptotics, conjectures and Monte Carlo," *Annals of Statistics*, **1**, 799–821.

Huber, P. J. (1977). "Robust methods of estimation of regression coefficients," *Mathematische Operationsforschung und Statistik, Series Statistics*, **8**, 41–53.

Huber, P. J. (1981). *Robust Statistics*. New York: Wiley.

Huber, P. J. (1983). "Minimax aspects of bounded-influence regression," *Journal of the American Statistical Association*, **78**, 66–80 (with discussion).

Jaeckel, L. A. (1972). "Estimating regression coefficients by minimizing the dispersion of the residuals," *Annals of Mathematical Statistics*, **43**, 1449–1458.

Jeffreys, H. (1932). "An alternative to the rejection of outliers," *Proceedings of the Royal Society, Series A*, **137**, 78–87.

Jurečková, J. (1971). "Nonparametric estimates of regression coefficients," *Annals of Mathematical Statistics*, **42**, 1328–1338.

Koenker, R. and Bassett, G., Jr. (1978). "Regression quantiles," *Econometrica*, **46**, 33–50.

Krasker, W. S. (1978). "Applications of robust estimation to econometric problems." Ph.D. thesis, Massachusetts Institute of Technology.

Krasker, W. S. (1980). "Estimation in linear regression models with disparate data points," *Econometrica*, **48**, 1336–1346.

Krasker, W. S. and Welsch, R. E. (1982). "Efficient bounded-influence regression estimation," *Journal of the American Statistical Association*, **77**, 595–604.

Krasker, W. S. and Welsch, R. E. (1983). "The use of bounded-influence regression in data analysis: theory, computation, and graphics." In K. W. Heiner, R. S. Sacher, and J. W. Wilkinson (Eds.), *Computer Science and Statistics: Proceedings of the 14th Symposium on the Interface*. New York: Springer-Verlag, pp. 45–51.

Leroy, A. and Rousseeuw, P. (1984). "PROGRES: a program for robust regression." CSOOTW/201, Centrum voor Statistiek en Operationeel Onderzoek, Vrije Universiteit Brussel.

Mallows, C. L. (1973). "Influence functions." Presented at NBER Conference on Robust Regression, Cambridge, MA.

Mallows, C. L. (1975). "On some topics in robustness." Presented at IMS Eastern Regional Meeting, Rochester, NY.

Mallows, C. L. (1979). "Robust methods—some examples of their use," *The American Statistician*, **33**, 179–184.

Marazzi, A. (1980). "Robust linear regression programs in ROBETH." Research Report No. 23, Fachgruppe für Statistik, Eidgenössische Technische Hochschule, Zurich.

Miller, R. G. (1974). "An unbalanced jackknife," *Annals of Statistics*, **2**, 880–891.

Moberg, T. F., Ramberg, J. S., and Randles, R. H. (1980). "An adaptive multiple regression procedure based on *M*-estimators," *Technometrics*, **22**, 213–224.

Ramsay, J. O. (1977). "A comparative study of several robust estimates of slope, intercept, and scale in linear regression," *Journal of the American Statistical Association*, **72**, 608–615.

Rao, C. R. (1965). *Linear Statistical Inference and Its Applications*. New York: Wiley.

Rousseeuw, P. J. (1984). "Least median of squares regression," *Journal of the American Statistical Association*, **79**, 871–880.

Ruppert, D. (1985). "On the bounded-influence regression estimator of Krasker and Welsch,"

Journal of the American Statistical Association, **80**, 205–208.

Ruppert, D. and Carroll, R. J. (1980). "Trimmed least squares estimation in the linear model," *Journal of the American Statistical Association*, **75**, 828–838.

Siegel, A. F. (1982). "Robust regression using repeated medians," *Biometrika*, **69**, 242–244.

Velleman, P. F. and Welsch, R. E. (1981). "Efficient computing of regression diagnostics," *The American Statistician*, **35**, 234–242.

Welsch, R. E. (1975). "Confidence regions for robust regression," *1975 Proceedings of the Statistical Computing Section*. Washington, DC: American Statistical Association, pp. 36–42.

Welsch, R. E. (1977). "Regression, sensitivity analysis and bounded-influence estimation." Presented at NBER Conference on Criteria for Evaluation of Econometric Models, East Lansing, MI.

Yohai, V. J. and Maronna, R. A. (1979). "Asymptotic behavior of M-estimators for the linear model," *Annals of Statistics*, **7**, 258–268.

Additional Literature

Agee, W. S. and Turner, R. H. (1979). "Applications of robust regression to trajectory data reduction." In R. L. Launer and G. N. Wilkinson (Eds.), *Robustness in Statistics*. New York: Academic, pp. 107–126.

Barnett, V. and Lewis, T. (1978). *Outliers in Statistical Data*. New York: Wiley.

Barrodale, I. (1968). "L_1 approximation and the analysis of data," *Applied Statistics*, **17**, 51–57.

Bassett, G., Jr. and Koenker, R. (1978). "Asymptotic theory of least absolute error regression," *Journal of the American Statistical Association*, **73**, 618–622.

Beckman, R. J. and Cook, R. D. (1983). "Outlier..........s," *Technometrics*, **25**, 119–163 (with discussion).

Bickel, P. J. (1976). "Another look at robustness: a review of reviews and some new developments," *Scandinavian Journal of Statistics*, **3**, 145–168 (with discussion).

Bloomfield, P. and Steiger, W. L. (1983). *Least Absolute Deviations: Theory, Applications, and Algorithms*. Boston: Birkhäuser.

Boncelet, C. G., Jr. and Dickinson, B. W. (1984). "A variant of Huber robust regression," *SIAM Journal on Scientific and Statistical Computing*, **5**, 720–734.

Brown, B. M. and Maritz, J. S. (1982). "Distribution-free methods in regression," *Australian Journal of Statistics*, **24**, 318–331.

Brown, M. L. (1982). "Robust line estimation with errors in both variables," *Journal of the American Statistical Association*, **77**, 71–79.

Clark, D. I. (1985). "The mathematical structure of Huber's M-estimator," *SIAM Journal on Scientific and Statistical Computing*, **6**, 209–219.

Cook, R. D. and Weisberg, S. (1980). "Characterizations of an empirical influence function for detecting influential cases in regression," *Technometrics*, **22**, 495–508.

Davies, R. B. and Hutton, B. (1975). "The effect of errors in the independent variables in linear regression," *Biometrika*, **62**, 383–391.

Dorsett, D., Gunst, R. F., and Gartland, E. C., Jr. (1983). "Multicollinear effects of weighted least squares regression," *Statistics and Probability Letters*, **1**, 207–211.

Eddy, W. F. and Kadane, J. B. (1982). "The cost of drilling for oil and gas: an application of constrained robust regression," *Journal of the American Statistical Association*, **77**, 262–269.

Ekblom, H. (1974). "L_p methods for robust regression," *Nordisk Tidskrift for Informations-behandling (BIT)*, **14**, 22–32.

Forsythe, A. B. (1972). "Robust estimation of straight line regression coefficients by minimizing pth power deviations," *Technometrics*, **14**, 159–166.

Gilstein, C. Z. and Leamer, E. E. (1983). "Robust sets of regression estimates," *Econometrica*, **51**, 321–333.

Green, P. J. (1984). "Iteratively reweighted least squares for maximum likelihood estimation, and some robust and resistant alternatives," *Journal of the Royal Statistical Society, Series B*, **46**, 149–192 (with discussion).

Hampel, F. R. (1983). "The robustness of some nonparametric procedures." In P. J. Bickel, K. A. Doksum, and J. L. Hodges, Jr. (Eds.), *A Festschrift for Erich L. Lehmann*. Belmont, CA: Wadsworth International Group, pp. 209–238.

Hampel, F. R., Rousseeuw, P. J., and Ronchetti, E. (1981). "The change-of-variance curve and optimal redescending *M*-estimators," *Journal of the American Statistical Association*, **76**, 643–648.

Hinkley, D. V. (1978). "Some topics in robust regression," *1978 Proceedings of the Statistical Computing Section*. Washington, D.C.: American Statistical Association, pp. 55–58.

Hocking, R. R. (1983). "Developments in linear regression methodology: 1959–1982," *Technometrics*, **25**, 219–249 (with discussion).

Hussain, S. S. and Sprent, P. (1983). "Non-parametric regression," *Journal of the Royal Statistical Society, Series A*, **146**, 182–191.

Iman, R. L. and Conover, W. J. (1979). "The use of the rank transform in regression," *Technometrics*, **21**, 499–509.

Koul, H. L. and Susarla, V. (1983). "Estimators of scale parameters in linear regression," *Statistics and Probability Letters*, **1**, 273–277.

Money, A. H., Affleck-Graves, J. F., Hart, M. L., and Barr, G. D. I. (1982). "The linear regression model: L_p norm estimation and the choice of p," *Communications in Statistics—Simulation and Computation*, **11**, 89–109.

Moussa-Hamouda, E. and Leone, F. C. (1977). "Efficiency of ordinary least squares estimators from trimmed and Winsorized samples in linear regression," *Technometrics*, **19**, 265–273.

Olshan, A. F., Siegel, A. F., and Swinder, D. R. (1982). "Robust and least-squares orthogonal mapping: methods for the study of cephalofacial form and growth," *American Journal of Physical Anthropology*, **59**, 131–137.

Parr, W. C. (1983). "A note on Hâjek projections and the influence curve," *Statistics and Probability Letters*, **1**, 177–179.

Ryan, T. A., Jr. (1976). "Robust regression—bounded leverage," *1975 Proceedings of the Statistical Computing Section*. Washington, D.C.: American Statistical Association, pp. 138–141.

Seneta, E. (1983). "The weighted median and multiple regression," *Australian Journal of Statistics*, **25**, 370–377.

Siegel, A. F. and Benson, R. H. (1982). "A robust comparison of biological shapes," *Biometrics*, **38**, 341–350.

Sposito, V. A. (1982). "On unbiased L_p regression estimators," *Journal of the American Statistical Association*, **77**, 652–653.

Wainer, H. and Thissen, D. (1976). "Three steps toward robust regression," *Psychometrika*, **41**, 9–34.

Welsch, R. E. (1982). "Influence functions and regression diagnostics." In R. L. Launer and A. F. Siegel (Eds.), *Modern Data Analysis*. New York, Academic, pp. 149–169.

Wu, L. L. (1985). "Robust *M*-estimation of location and regression." In N. B. Tuma (Ed.), *Sociological Methodology 1985*. San Francisco: Jossey-Bass, pp. 316–388.

Yale, C. and Forsythe, A. B. (1976). "Winsorized regression," *Technometrics*, **18**, 291–300.

EXERCISES

1. Show that changing y_i by the amount Δy_i in the multiple regression model (1) causes the least-squares estimate $\hat{\boldsymbol{\beta}}$ to change by $\Delta\hat{\boldsymbol{\beta}} = (\mathbf{X}^T\mathbf{X})^{-1}\mathbf{x}_i^T(\Delta y_i)$. That is, $\Delta\hat{\boldsymbol{\beta}}$ is proportional to the ith column of the $p \times n$ "catcher matrix" $\mathbf{C}^T = (\mathbf{X}^T\mathbf{X})^{-1}\mathbf{X}^T$. Comment on the interpretation of this result.

2. In the simple regression model

$$y_i = \beta_1 + \beta_2 x_i + \varepsilon_i$$

derive expressions for the general element h_{ij} of the hat matrix and for the $2 \times n$ catcher matrix $\mathbf{C}^T$ that yields the least-squares estimates via $(\hat{\beta}_1, \hat{\beta}_2)^T = \mathbf{C}^T\mathbf{y}$. Discuss interpretations of h_{ii} and the elements of $\mathbf{C}$ in terms of the x_i.

3. One might regard the variance of the fitted value $\hat{y}_i$ in the least-squares regression as equivalent to that of a mean of n_i independent observations of variance σ^2. In this setting, derive n_i as a function of h_{ii} and use the result to interpret Huber's suggested cutoff values of $h_{ii} = 0.2$ and $h_{ii} = 0.5$ in terms of stability and resistance.

4. Ramsay (1977) introduced the E_a estimators, which are M-estimators with ψ-function $\psi(t) = t\exp(-a|t|)$. The corresponding ρ-function is $\rho(t) = a^{-2}[1 - \exp(-a|t|)(1 + a|t|)]$. For $a = 0.3$, plot $\rho(t)$ and $\psi(t)$ against t, and compare their features to Figure 8-2.

5. Show that, for all values of the Huber tuning constant k greater than the maximum absolute value among the OLS residuals divided by the scale estimate, W-estimation of the regression with the Huber weight function must yield the same $\hat{\boldsymbol{\beta}}$ and residuals as OLS (Denby and Mallows, 1977).

6. From equation (16) derive the influence function of the OLS estimator.

7. For p between 1 and 2, the L_p estimator chooses a value of $\boldsymbol{\beta}$ that minimizes $\Sigma|y_i - \mathbf{x}_i\boldsymbol{\beta}|^p$.
   (a) Verify that, as happens with the OLS and LAR estimators in equations (22) and (23), the L_p estimator does not need to take scale into account.

(b) Show how one might obtain the L_p estimate by iteratively reweighted least squares.

8. In a p-point regression problem (Hill, 1979) the $n \times p$ design matrix $\mathbf{X}$ has only p distinct rows, $\mathbf{x}_1^*, \ldots, \mathbf{x}_p^*$; row $\mathbf{x}_k^*$ has n_k replications; $n_1 + \cdots + n_p = n$; and the $\mathbf{x}_k^*$ are linearly independent.

(a) Show that the regression M-estimate $\hat{\boldsymbol{\beta}}_M$ can be obtained by solving the p simultaneous equations $\mathbf{X}^*\boldsymbol{\beta} = \mathbf{y}^*$, where $\mathbf{X}^{*T} = (\mathbf{x}_1^{*T}, \ldots, \mathbf{x}_p^{*T})$ and y_k^* is the corresponding M-estimate of location for the sample of n_k y-values associated with replications of $\mathbf{x}_k^*$.

(b) If $\operatorname{var}(y_k^*) = \sigma^2 c_k / n_k$, show that the covariance matrix of $\hat{\boldsymbol{\beta}}_M$ is given by

$$\operatorname{var}(\hat{\boldsymbol{\beta}}_M) = \sigma^2 (\mathbf{X}^*)^{-1} \langle c_k / n_k \rangle (\mathbf{X}^*)^{-T},$$

where $\langle c_k / n_k \rangle$ is the diagonal matrix whose diagonal entries are $c_1/n_1, \ldots, c_p/n_p$.

(c) Show that this covariance matrix can also be written as

$$\operatorname{var}(\hat{\boldsymbol{\beta}}_M) = \sigma^2 (\mathbf{X}^T \mathbf{C} \mathbf{X})^{-1},$$

where $\mathbf{C}$ is an $n \times n$ diagonal matrix whose first n_1 entries are $1/c_1$, whose next n_2 entries are $1/c_2$, and so on. Discuss the implications of this result for the use of matrices proportional to $(\mathbf{X}^T\mathbf{X})^{-1}$ as approximations to finite-sample covariance matrices of robust M-estimators of regression.

9. (a) Verify that, for the Huber estimators, the relation $\chi(t) = t\psi(t) - \rho(t)$ yields the χ-function in equation (28).

(b) If the fluctuations come from a location-scale family whose standard density is f, derive ψ and χ in equations (25) and (26) and show that $\chi(t) = t\psi(t)$.

(c) Apply the definition $\chi(t) = t\psi(t)$ to the Huber ψ-function and compare the qualitative properties of this χ-function to those of the χ-function in part (a).

(d) Apply both these definitions of χ-function to the biweight ψ-function (Table 8-2). Discuss.

10. In the Schweppe estimator of Section 8E, show that $\sqrt{1 - h_{ii}}\,\mathbf{x}_i^T$ can make only a bounded contribution to the influence function. Hint: Use the formula

$$(\mathbf{A} + \mathbf{u}\mathbf{v}^T)^{-1} = \mathbf{A}^{-1} - \frac{(\mathbf{A}^{-1}\mathbf{u})(\mathbf{v}^T\mathbf{A}^{-1})}{1 + \mathbf{v}^T\mathbf{A}^{-1}\mathbf{u}}$$

for nonsingular **A** and column vectors **u** and **v** (Rao, 1965, p. 29) to obtain $(\mathbf{X}^T\mathbf{X})^{-1}$ from $[\mathbf{X}(i)^T\mathbf{X}(i)]^{-1}$, where $\mathbf{X}(i)$ is defined in Section 8A.

CHAPTER 9

Checking the Shape of Discrete Distributions

David C. Hoaglin
Harvard University

John W. Tukey
Princeton University and A T & T Bell Laboratories

Techniques for examining or testing the distributional behavior of a sample of continuous data have received much attention in the statistical literature. Most commonly the question is whether the data depart seriously from a Gaussian distribution, either systematically or by having a few anomalous observations. At times other families of distributions, such as the exponential or the Weibull, serve as the standard of comparison. Rather than summarize a possibly complex pattern of behavior in a single number, we learn more from graphical displays, as in a well chosen probability plot.

In a similar vein graphical techniques help us to appreciate how closely a sample of discrete data follows a member of one of the common families of discrete distributions, such as the binomial or the Poisson. The present chapter emphasizes simple graphical techniques that incorporate resistance —so as to isolate unusual portions of the data and prevent them from distorting the overall picture.

The most common types of discrete data involve only whole-number values, obtained by counting such occurrences as traffic fatalities, decayed teeth in the mouths of children, persons in low-income households, or mice that develop tumors. Especially when the individual observations are small counts, we often encounter the sample in the form of a *discrete frequency distribution* (or *table of counts of counts*), which records that n_k of the observations are equal to k ($k = 0, \ldots, L$), as in Table 9-1. To establish a convenient notation, we assume that no observation exceeds L and that we

TABLE 9-1. Frequency distribution for discrete data involving outcome values 0, ..., L. The total count is $N = n_0 + n_1 + \cdots + n_L$. Some of the n_k may be zero.

Outcome	Frequency
0	n_0
1	n_1
$\vdots$	$\vdots$
L	n_L

are interested in all integers from 0 to L, so that the total number of observations in the sample is $N = n_0 + \cdots + n_L$.

Another class of frequency distributions arises when we divide the range of a continuous variable into intervals and count the number of observations in each interval. Although the specific techniques described in the present chapter do not apply (except for the double-root residuals, Section 9G), exploratory data analysis has not overlooked such data. The approach developed by Tukey (1977, Chapter 17) often allows the data to point to a suitable frequency curve.

For all ordered frequency distributions, either discrete or continuous, a graphical technique known as the rootogram provides an effective mode of display; and the suspended rootogram focuses on deviations from a fitted comparison curve or discrete distribution (Tukey, 1971, Chapters 25 and 26; 1972; Velleman and Hoaglin, 1981, Chapter 9).

In developing diagnostic techniques for discrete distributions we try to exploit straightforward properties of the probability function $P\{X = k\}$, whose values we abbreviate $p(k)$ or p_k, and extend them to the average or expected frequencies $\eta_k = Np_k$. It is unrealistic to expect that very many families of discrete distributions will yield simple diagnostic displays that do not involve handling some parameters by either trial and error or preliminary fitting. Fortunately, however, a very straightforward approach succeeds for a number of important families of distributions, including the most common discrete families.

Four Classes of Questions

We organize our approach to discrete data and families of discrete distributions around four general classes of questions that seem natural to ask.

The first class asks whether the data are reasonably compatible with a reference distribution of a particular family.

If a family of discrete distributions involves only one parameter, we may be able to choose a monotone function $\phi(n_k)$ of n_k, the number of times k was observed, such that a plot of $\phi(\eta_k)$ against k, where η_k is the expected value of n_k according to the reference distribution, will be a straight line for each of the reference distributions in the chosen family. We then call $\phi(n_k)$ a *count metameter* (by analogy with the use, in bioassay, of "response metameter" and "dose metameter"). The slope of such a theoretical line identifies the main parameter of the theoretical distribution. (Because both $\Sigma n_k = N$ and $\Sigma \eta_k = N$, two such lines cannot be parallel.)

A plot of $\phi(n_k)$ against k can now show, by how well it follows a straight line, whether the observed n_k have a chance of being well represented by some distribution in the reference family—either for every k or for most, but not all, k. And the apparent slope suggests the value of the main parameter. We begin by deriving ϕ for the Poisson reference family in Section 9A, and later we turn to other families in Section 9D.

Sometimes the family of reference distributions involves one or more extra parameters. We recognize that we may have to explore their values by trial and error.

We can improve the plots by modifying the count metameter to $\phi(n_k^*)$ in terms of a modified count of the form $n_k^* = n_k - \delta_k$. The detailed development in Section 9B yields a simple approximation for δ_k, as well as confidence-interval half-lengths that reflect the variability of the observed counts n_k.

The second class of questions concerns the compatibility of individual n_k with some proposed line that relates $\phi(\eta_k)$ to k. Because we have counts, it may well be that each follows a binomial distribution. (Clumping or some other form of local dependence may make the n_k more variable than a binomial distribution would indicate. However, the binomial is ordinarily the natural reference. In other circumstances, heterogeneity sometimes reduces variability below that of the binomial; but, in our frequency distributions, enough heterogeneity to matter will almost surely affect the general behavior of the n_k.)

Section 9C discusses how to display the compatibility or incompatibility with binomial variability of individual n_k. Our applications all involve Poisson distributions.

Building on the improvements developed in Sections 9B and 9C, Section 9D derives and applies the count metameters for the binomial, negative binomial, geometric, and logarithmic-series families. The negative binomial distribution has an extra parameter, n, which we must assess from the data, often with the aid of trial-and-error plotting for different choices of n. Section 9E critically examines some related techniques in the literature.

The third class of questions centers on cooperative diversity: Do different parts of the data, specifically different ranges of individual counts, seem to indicate different values of the main parameter? The apparent-parameter plot in Section 9F implements this idea through local slopes in the plot of $\phi(n_k^*)$ against k. This type of analysis is especially effective in detecting heterogeneity. Another application of cooperative diversity, the indicated-parameter-change plot, lets us look at how strongly the individual n_k indicate a need to change a chosen parameter value.

The fourth class of questions focuses on the overall lack of fit of the data to the reference distribution, *after* allowing for systematic deviations. Although we must postpone discussion of them to another occasion, we believe it is highly desirable to use dissected measures, which separate systematic and irregular deviations.

Starting from a count metameter, and the idea of the n_k individually following binomial distributions and jointly following a multinomial distribution, our systematic approach to counts of counts thus asks in turn:

1. Is such a fit at all reasonable?
2. Which individual points are incompatible?
3. Do the ends (low k and high k) seem to agree with each other, and with the middle, about the value of the main parameter?
4. Allowing for the possible presence of systematic deviations, are the remaining irregular deviations larger than is compatible with a multinomial distribution?

9A. A POISSONNESS PLOT

Most often thought of as a model for the occurrence of locally rare events (such as the number of decays of a quantity of radioactive substance in a time interval of fixed length or the number of vacancies on the United States Supreme Court in a given span of months), the Poisson distribution with parameter λ has probability function

$$p_\lambda(k) = \frac{e^{-\lambda}\lambda^k}{k!}, \qquad k = 0, 1, 2, \ldots . \tag{1}$$

When one has a sample of N from what is assumed to be a Poisson distribution, it is both customary and appropriate to calculate the sample mean

$$\bar{x} = \sum kn_k/N \tag{2}$$

from the frequency distribution, and then use it to estimate λ. We should also, of course, do something to compare the observed and expected frequencies, thus checking the quality of the fit.

Because a single discrepant cell count n_k can seriously affect $\bar{x}$, we often benefit from being able to examine the agreement between the data and the Poisson distribution before finally estimating λ. Our gain will be even greater in situations where one or more values of k are unobservable, possibly because they are excluded. For example, some data on the sizes of households or the sizes of social groups may follow a Poisson distribution, with the constraint that outcomes of zero are excluded. The resulting "no-zeros Poisson distribution" has probability function

$$p_\lambda^*(k) = p_\lambda(k)/(1 - e^{-\lambda}), \qquad k = 1, 2, \ldots . \tag{3}$$

In this and other truncated Poisson distributions, numerical estimation of λ requires iterative numerical calculation, which we might be happy to avoid if such a model cannot fit well. Or we may, in other circumstances, decide to set aside an isolated unusual count (large or small) before proceeding to see how the remaining counts behave.

If possible, we would like to make a plot that follows a straight line when the observed frequency distribution is consistent with a Poisson distribution (truncated or not). A sample of N yields the expected frequencies

$$\eta_k = Np_\lambda(k) = Ne^{-\lambda}\lambda^k/k! \tag{4}$$

under the unconstrained Poisson model. Therefore, taking natural logarithms on both sides of equation (4) yields

$$\log_e(\eta_k) = \log_e(N) - \lambda + k\log_e(\lambda) - \log_e(k!). \tag{5}$$

Thus the result of plotting $\log_e(n_k) + \log_e(k!)$ against k is likely to be close to a straight line with slope equal to $\log_e(\lambda)$ and intercept (at $k = 0$) $\log_e(N) - \lambda$. (We plan to omit points for which $n_k = 0$.) This is the "Poissonness plot" derived by Hoaglin (1980). It is resistant to discrepant values of n_k because each unusual count affects only its own point in the plot. Thus, the original Poissonness plot uses $\phi(n_k) = \log_e(n_k) + \log_e(k!)$ as its count metameter.

A minor change simplifies comparisons among plots for frequency distributions with different N. We move $-\log_e(N)$ into the count metameter, so that we now plot

$$\phi(n_k) = \log_e(n_k) + \log_e(k!) - \log_e(N)$$

$$= \log_e(k!n_k/N) \tag{6}$$

against k and anticipate slope $\log_e(\lambda)$ and intercept $-\lambda$. [Kinderman (1982) has also discussed this modification.] The Poissonness plot (along with related plots for other families of distributions) takes this form throughout the rest of this chapter.

We recognize that, in the count metameter, $\log_e(n_k)$ is a fallible indicator of $\log_e(\eta_k)$. Section 9B develops a way to ease the consequences of this variability by modifying n_k.

Because $\phi(n_k)$ tends to have greater variability when η_k is small, we often use a distinctive plotting symbol (suggestively, the digit "1") for points where $n_k = 1$. This practical detail will turn out to have other benefits as well.

When the Poissonness plot adequately follows a straight line, we may use $\bar{x}$ from equation (2) as our estimate of λ. (The mean of a Poisson sample is the maximum-likelihood estimator of λ.) Otherwise, particularly if only a few isolated points are discrepant, we can fit a line to the remaining points (by eye or by some resistant technique) and use its slope, b, to estimate λ via $\hat{\lambda} = e^b$. On the other hand, systematic curvature indicates that the observed frequency distribution is not consistent with a Poisson model. We discuss some alternative models in Section 9D.

Although λ appears in both the slope $\log_e(\lambda)$ and the intercept $-\lambda$, we focus entirely on the slope, primarily for two reasons. First, the intercept becomes a more complicated function of λ in any truncated Poisson distribution, such as the no-zeros Poisson of equation (3). Second, when the points lie close to a straight line with slope $\log_e(\lambda_0)$ in the complete Poisson model, $\phi(n_0)$ must lie close to $-\lambda_0$. (If the points in the plot follow a straight line exactly, then the plot must have the correct intercept, because $\Sigma n_k = N$.)

EXAMPLE: INTERNATIONAL TERRORISM

Jenkins and Johnson (1975) compiled a chronology of incidents of international terrorism from January 1968 through April 1974. During this 76-month period they record 507 incidents throughout the world. For the 64

TABLE 9-2. Incidents of international terrorism in the United States, 1968–1974. Frequency distribution of monthly totals and calculations for Poissonness plot ($\log_e 76 = 4.33$).

Number of Incidents (k)	Number of Months (n_k)	$\log_e(n_k)$	$\log_e(k!)$	$\log_e(k!n_k/N)$
0	38	3.64	0	−0.69
1	26	3.26	0	−1.07
2	8	2.08	0.69	−1.56
3	2	0.69	1.79	−1.85
4	1	0	3.18	−1.15
12	1	0	19.99	15.66
	$N = 76$			

incidents in the United States, Table 9-2 gives the frequency distribution of the number of incidents per month, along with the calculations for the Poissonness plot.

In the plot, Figure 9-1, we cannot avoid noticing how violently the one month with 12 incidents departs from the rest of the data. Among the remaining five points, the first four lie essentially on a straight line. The point at $k = 4$, although noticeably above this line, is not seriously discrepant. On the whole, then, the number of incidents of international terrorism per month in the United States seems consistent with a Poisson model—once we set aside the month with 12 such incidents. One reasonable estimate of λ is $(64 - 12)/(76 - 1) = 0.693$.

The unusual month demands further investigation. The chronology of Jenkins and Johnson reveals that this month was July 1968. Furthermore, 11 of that month's 12 incidents were attributed to the anti-Castro group El Poder Cubano ("Cuban Power"), and the remaining incident was attributed less specifically to "anti-Castro Cuban terrorists." Thus it seems clear that, in July 1968 at least, we have a failure of the underlying assumption that individual Poisson occurrences are independent.

From the data for the United States we might reasonably ask whether a Poisson model applies to international terrorist incidents more generally. Although few other countries experienced enough incidents during the time period of the chronology to allow reasonable Poissonness plots, the evidence seems substantially negative. The plots tend to show clear upward curvature at both ends. (Interested readers can pursue this in Exercise 1.)

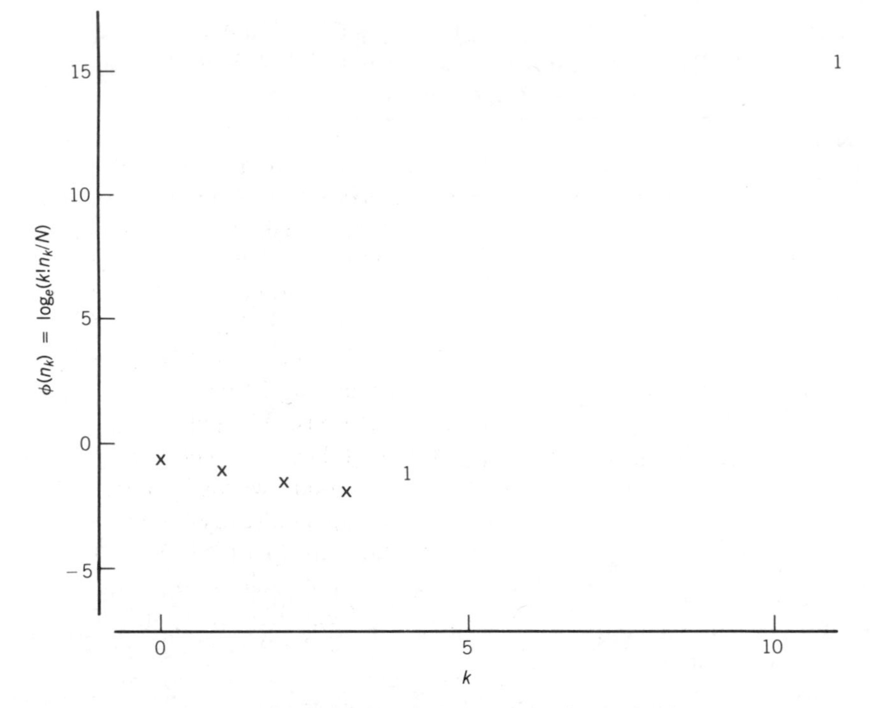

Figure 9-1. Incidents of international terrorism in the United States, 1968–1974. Poissonness plot: Poisson count metameter $\phi(n_k)$ versus k. The plotting symbol "1" indicates that $n_k = 1$.

Figure 9-1 drew our attention to the points at $k = 4$ and $k = 12$ as the most evident locations of possible departures from the Poisson model. Both n_4 and n_{12} are 1, so that no positive count can yield a lower point in the plot. Although the month with 12 incidents seems clearly discrepant because of its isolation, and the available information has provided a plausible explanation, the situation in other data sets may often be less clear. Thus we will need to take each point's variability into account in judging whether that point is discrepant.

When the data arise as occurrences over time and are available in detail rather than solely as a frequency distribution, it is possible to probe deeper by examining the counts in time order or the time intervals between occurrences. [For example, the chronology of international terrorism gives the date(s) of each incident.] In this way we can learn whether the process seems to be homogeneous over time. In the present chapter, however, we assume that we must work only with the frequency distribution.

EXAMPLE: WORD FREQUENCY

In 1787 and 1788, to persuade the people of the State of New York to ratify the United States Constitution, Alexander Hamilton, John Jay, and James Madison wrote a series of short essays. Under the general title *The Federalist*, these appeared in newspapers and were signed with the pseudonym "Publius." There is general agreement on the authorship of most of the papers. Of the 77 in the initial series, Jay wrote 5, Hamilton 43, Madison 14, and Hamilton and Madison jointly 3. Both Hamilton and Madison, however, later claimed sole authorship of the remaining 12 papers.

Using Bayesian and other statistical methods, Mosteller and Wallace (1964) concluded that Madison's claim had been correct. Part of their analysis used the frequency of certain key "marker" words in blocks of text approximately 200 words in length. For 19 papers already known to be by Madison (14 in *The Federalist* and another group of 5), Mosteller and Wallace (p. 30) give the number of occurrences of each of 10 words in each of 262 blocks of text. Table 9-3 shows the frequency distribution for the word *may*, along with the vertical coordinate for the Poissonness plot.

The curvature in the plot, Figure 9-2, suggests that a Poisson distribution cannot provide a good fit. In fact, Mosteller and Wallace obtained a much better fit with a negative binomial distribution.

TABLE 9-3. Frequency distribution of the word *may* in papers already known to be by Madison.

Number of Occurrences (k)	Number of Blocks (n_k)	$\log_e(k!n_k/N)$
0	156	−0.52
1	63	−1.43
2	29	−1.51
3	8	−1.70
4	4	−1.00
5	1	−0.78
6	1	1.01
	$N = 262$	

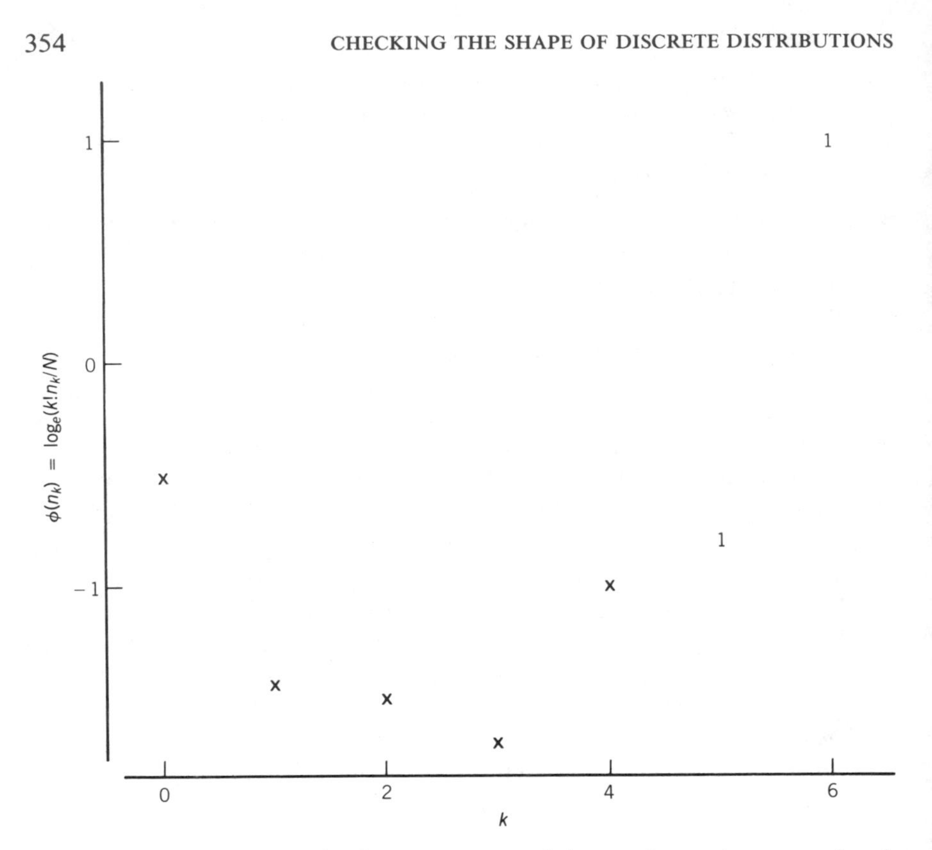

Figure 9-2. Poissonness plot for occurrences of the word *may* in papers already known to be by Madison. The plotting symbol “1” indicates that $n_k = 1$.

EXAMPLE: HOUSEHOLD SIZE

The Housing Allowance Demand Experiment (Kennedy, 1980), part of the Experimental Housing Allowance Program established by the United States Department of Housing and Urban Development, examined the effects of providing housing subsidy payments directly to low-income households as an alternative to subsidized housing. Table 9-4 gives the frequency distribution of the number of persons in the household at enrollment for the 1239 households in Allegheny County, Pennsylvania, that were still participating in the program at the end of two years. Because a household cannot consist of 0 persons, only no-zeros models [equation (3) for the Poisson case] are possibly appropriate for these data.

In the Poissonness plot, Figure 9-3, we see that the leftmost six points lie very close to a straight line, but the points for household sizes from 7 to 10

TABLE 9-4. Frequency distribution of household size at enrollment for 1239 households in the Housing Allowance Demand Experiment.

Number of Members (k)	Number of Households (n_k)	$\log_e(k!n_k/N)$
1	210	−1.77
2	315	−0.68
3	292	0.35
4	176	1.23
5	125	2.49
6	57	3.50
7	38	5.04
8	18	6.37
9	6	7.47
10	1	7.98
11	0	—
12	1	12.87
	$N = 1239$	

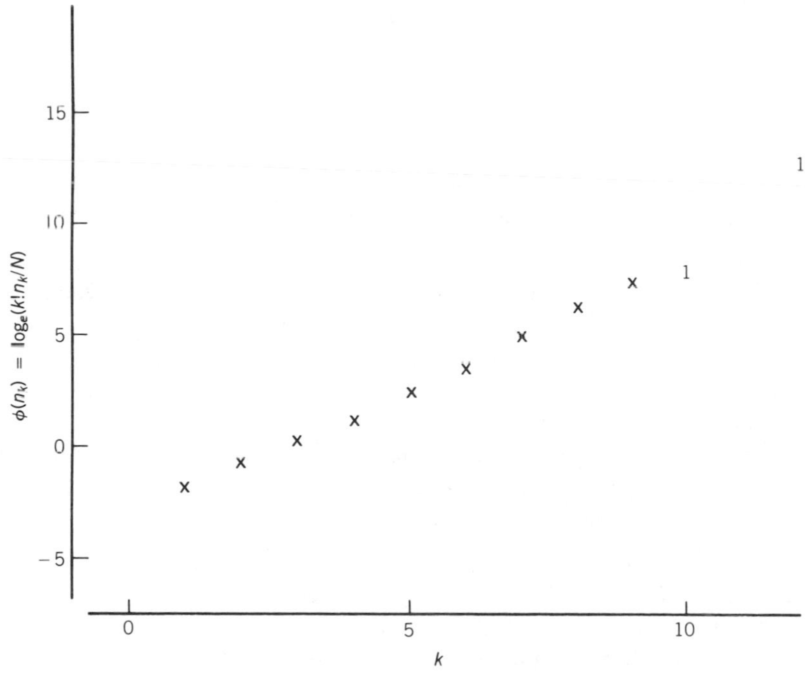

Figure 9-3. Poissonness plot for household size at enrollment of 1239 households in the Housing Allowance Demand Experiment. (Points with $n_k = 1$ are plotted as "1".)

are somewhat above that particular line, and the one household with 12 members does not fit in well with the rest of the sample. On the whole, the evidence is adequate to suggest that a no-zeros Poisson model would be a satisfactory starting point in analyzing these household size data. If one wanted to fit a truncated Poisson model to the data for $k = 1$ through $k = 6$, the estimate of λ calculated from the slope of that portion of Figure 9-3 would be an excellent starting point, either to be taken as is or to begin the maximum-likelihood calculation. Also we would like to look closer at the characteristics of the 63 households with sizes 7 through 10. Their composition might indicate that some of them were formed through a process different from that for smaller households.

If we choose the slope of the line through the points for $k = 1$ and $k = 10$, we get $\log_e(\lambda) = 1.084$ and hence $\lambda = 2.957$. The fitted numbers of households of sizes 9 through 12 are as follows:

Household	Number of Households	
Size	Observed	Fitted
9	6	3.07
10	1	0.91
11	0	0.24
12	1	0.06

confirming our impression that the single 12 is somewhat unusual but not strikingly so—at least not at a 5% level.

Although the preceding three examples illustrate only a small fraction of the diverse behavior that one can encounter in possibly Poisson data, they demonstrate that the Poissonness plot readily and resistantly reveals anomalies in observed frequency distributions and allows the user to judge the overall appropriateness of a Poisson model. One valuable ingredient in examining such plots is an appreciation of the sampling variability inherent in each point and of the fact that this variability is not constant across all points. We take up this issue in Section 9C.

Leveling the Poissonness Plot

If we want a Poissonness plot to do us as much good as possible, we will want to have our reference curve not only a straight line but even a nearly horizontal straight line. To level the plot, we find a rough value λ_0 for λ

and then plot

$$\phi(n_k) + [\lambda_0 - k \log_e(\lambda_0)] = \log_e(k!n_k/N) + [\lambda_0 - k \log_e(\lambda_0)] \quad (7)$$

against k. For a Poisson distribution with parameter λ, this new plot would have slope $\log_e(\lambda) - \log_e(\lambda_0) = \log_e(\lambda/\lambda_0)$ and intercept $\lambda_0 - \lambda$. If we have done a fair job of choosing λ_0—perhaps as $\bar{x}$, perhaps from the slope of the initial Poissonness plot—this plot will be nearly as flat as possible. An example illustrates the steps in the leveling process.

EXAMPLE: INTERNATIONAL TERRORISM

Together, Table 9-2 and Figure 9-1 suggest two easy choices for the rough value λ_0: $0.842 = 64/76$ from all months and $0.693 = 52/75$ from all months but one. The corresponding values of $\log_e(\lambda_0)$ are $-.172$ and $-.366$, respectively. For convenience in the log scale, we consider the three values $\lambda_0 = e^{-.2}$, $\lambda_0 = e^{-.3}$, and $\lambda_0 = e^{-.4}$. Table 9-5 shows the result of applying expression (7) for these three choices of λ_0. The values for $0 \leq k \leq 3$ are most nearly constant when we take $\lambda_0 = e^{-.4} = .670$, and the leveled Poissonness plot in Figure 9-4 confirms that they are flat indeed. Because of the vertical scaling in Figures 9-1 and 9-4, we learn little from Figure 9-4 that was not already evident in the rightmost column of Table 9-5. In most examples, leveling the plot produces a substantial change in scale and brings the usual gains from looking at residuals.

To look further at n_4 and n_{12}, we can ask what a Poisson distribution with $\lambda = .670$ would yield as average numbers of months with 4–12 incidents. When we calculate $76p_\lambda(k)$ for $4 \leq k \leq 7$, we get .33, .044, .0049,

TABLE 9-5. Calculations for leveled Poissonness plots of the international terrorism data, using $\lambda_0 = e^{-.2}$, $e^{-.3}$, and $e^{-.4}$.

Number of Incidents (k)	$\phi(n_k) = \log_e(k!n_k/N)$	$\phi(n_k) + [\lambda_0 - k\log_e(\lambda_0)]$		
		$\lambda_0 = e^{-.2}$	$\lambda_0 = e^{-.3}$	$\lambda_0 = e^{-.4}$
0	−0.69	.13	.05	−.02
1	−1.07	−.05	−.03	−.00
2	−1.56	−.34	−.22	−.09
3	−1.85	−.43	−.21	.02
4	−1.15	.47	.79	1.12
12	15.66	18.88	20.00	21.13

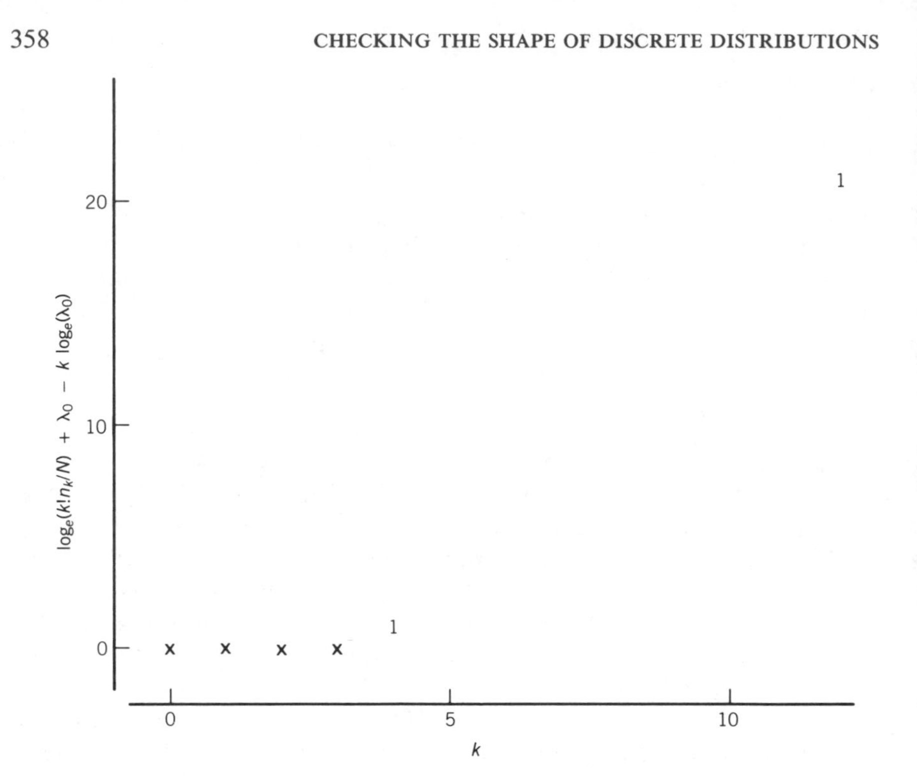

Figure 9-4. Leveled Poissonness plot for the international terrorism data, using $\lambda_0 = e^{-.4} = .670$. (Points with $n_k = 1$ are plotted as "1".)

and .00047. Thus, if $\lambda = .670$ had been the Poisson parameter, the lone occurrence of four incidents ($n_4 = 1$) should be thought of as not at all unusual, whereas the month with 12 incidents must be regarded as very exceptional.

9B. CONFIDENCE INTERVALS FOR THE COUNT METAMETER

We now begin to deal explicitly with the variability of the individual counts n_k in the observed frequency distribution, primarily by constructing approximate confidence intervals for $\log_e(\eta_k)$. The present section applies this information to develop modified counts n_k^* as the basis for an improved Poissonness plot. Section 9C goes on to determine when a point in the plot seems to be discrepant.

The customary model for an individual n_k is the binomial distribution. Specifically, under the Poisson model, each n_k follows a binomial distribution with parameters N and $p_k = p_\lambda(k)$. [This binomial behavior applies to any discrete distribution: We need only a set of underlying probabilities $p_0, p_1, \ldots, p_L$. Because we work with each value of k separately, we need not assume any particular underlying model, as far as a single value of $\log_e(\eta_k)$ is concerned. The model whose adequacy we are checking does, of course, determine the details of the count metameter $\phi(n_k)$.] When N is large, the usual first-order expansion

$$\log_e(x) \approx \log_e(x_0) + (x - x_0)/x_0$$

allows us to take $x_0 = \eta_k = Np_k$ and thus to regard $\log_e(n_k)$ as having approximate mean $\log_e(\eta_k)$ and approximate variance

$$Np_k(1 - p_k)/(Np_k)^2 = (1 - p_k)/\eta_k. \tag{8}$$

[We focus on $\log_e(n_k)$ because the other terms in the count metameter, $\log_e(k!) - \log_e(N)$ in equation (6), are constant in the present discussion.] Particularly when η_k is small, however, the distribution of $\log_e(n_k)$ will not be close to symmetric with $\log_e(\eta_k)$ as its center of symmetry. Thus we seek to improve on $\log_e(n_k)$ in $\phi(n_k)$ as the vertical coordinate of the Poissonness plot.

As we shortly explain, we accomplish this improvement by modifying n_k and using the resulting n_k^* as the basis for the count metameter $\phi(n_k^*)$. In detail,

$$n_k^* = \begin{cases} n_k - .67 - .8n_k/N, & \text{for } n_k \geq 2, \\ 1/e, & \text{for } n_k = 1, \\ \text{undefined}, & \text{for } n_k = 0. \end{cases} \tag{9}$$

Our choice of the constants that appear in the definition of n_k^* is guided by a desire to have $\log_e(n_k^*)$ as symmetrical about $\log_e(\eta_k)$ as possible. When we ask what this can mean—in an essentially unsymmetrical situation—our best choice is probably to make a single "most important" confidence interval for $\log_e(\eta_k)$ as symmetrical about $\log_e(n_k^*)$ as possible. Equivalently, we ask that the corresponding confidence interval for $\log_e(\eta_k) - \log_e(n_k^*)$ be as symmetrical about zero as possible. For the 95% interval that leaves 2.5% in each tail, we want to choose n_k^* so that $\log_e(n_k^*)$ is (approximately) the center of the interval for $\log_e(\eta_k)$. We next show that the definition in equation (9) does this with adequate accuracy.

Some readers may be more interested in the application of the n_k^* to construct improved Poissonness plots than in the detailed derivation of the formulas in equation (9). They may wish to skip to the examples starting at page 366, which try out this modification on data sets introduced in Section 9A.

For simplicity, we begin the development of n_k^* with the limiting case of a count that follows a Poisson distribution and then proceed to the binomial case. We recall that, in the limit as N becomes large and p becomes small in such a way that $\eta = Np$ remains constant, the binomial distribution with parameters N and p approaches the Poisson distribution with parameter η. Because we now deal only with a single count n_k, we omit the subscript k.

One Poisson Count

Given an observed count n from a Poisson distribution with parameter η, we want a symmetrical 95% confidence interval for η, symmetrical in the sense that it leaves out equal amounts of probability at each end. We begin with the probabilities of the three events $X < n$, $X = n$, and $X > n$ as a function of η. The confidence interval then consists of those values of η for which the outcome n is not too unlikely. Specifically, the endpoints of the interval are the values of η such that $P_\eta\{X < n\} = .975$ and $P_\eta\{X > n\} = .975$, respectively.

To illustrate, Figure 9-5 shows the situation for $n = 2$ and the relevant range of η, the latter conveniently given in the log scale. Because we need to distinguish only among $X < 2$, $X = 2$, and $X > 2$, the curves trace the boundaries of the regions corresponding to the probabilities of these three events. The endpoints of the confidence interval correspond to the intersection of the dashed line at cumulative probability .975 with the curve for $P_\eta\{X < 2\}$ and the intersection of the dashed line at cumulative probability .025 with the curve for $P_\eta\{X \leq 2\}$. The numerical values are $\log_e(\eta) = -1.418$ (or $\eta = .242$) and $\log_e(\eta) = 1.977$ (or $\eta = 7.22$); to the left of $\eta = .242$ an X of 2 or more is unlikely, and to the right of $\eta = 7.22$ an X of 2 or less is unlikely.

The count metameter involves $\log_e(n)$, so we prefer to express the confidence interval for $\log_e(\eta)$ as $[\log_e(n) - C, \log_e(n) + D]$, where the offsets $-C$ and $+D$ depend on n. In the example with $n = 2$, we have $\log_e(2) = .693$, so that

$$-C = -1.418 - .693 = -2.111,$$

$$+D = \quad 1.977 - .693 = +1.284.$$

From tables of confidence limits for the Poisson parameter, given a single observation, we could easily calculate values of $-C$ and $+D$ for a variety

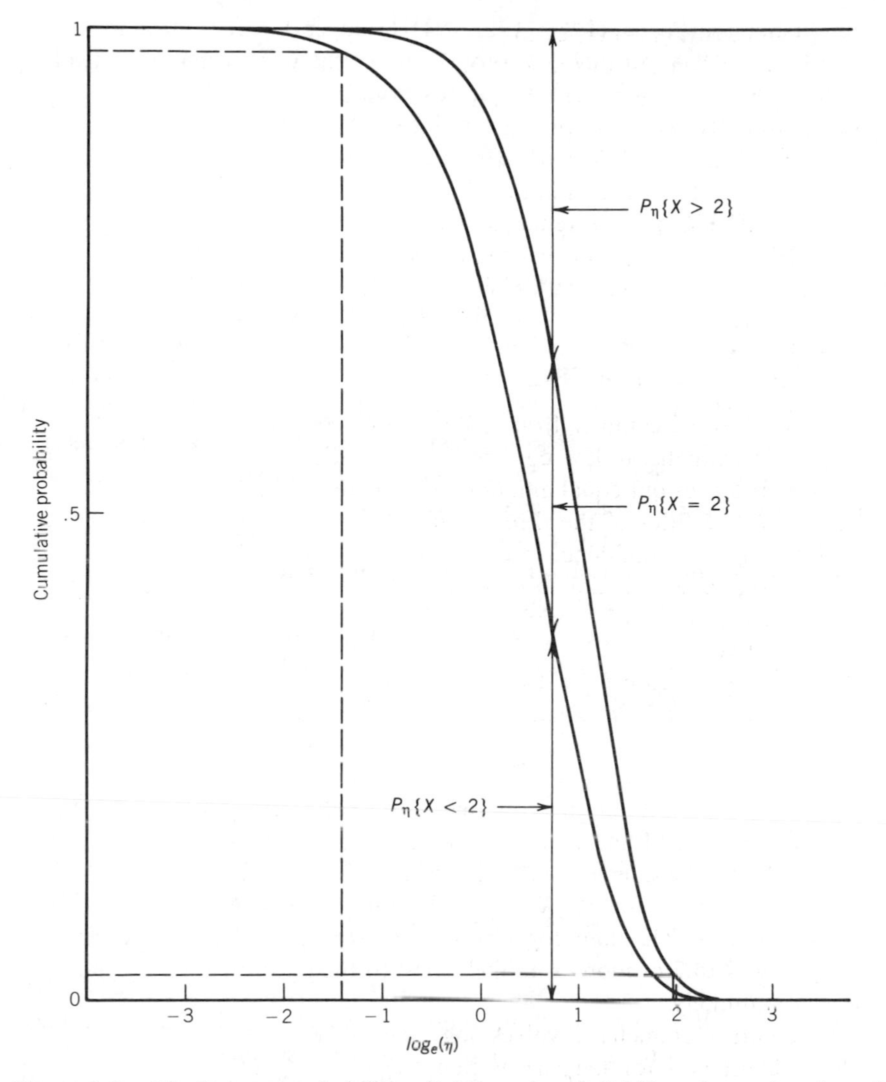

Figure 9-5. The Poisson probabilities $P_\eta\{X < n\}$ and $P_\eta\{X \leq n\}$ as a function of $\log_e(\eta)$ when $n = 2$. The vertical lines with arrows show the magnitude of $P_\eta\{X < 2\}$, $P_\eta\{X = 2\}$, and $P_\eta\{X > 2\}$ for one value of η. The dashed lines construct the 95% confidence interval $-1.418 \leq \log_e(\eta) \leq 1.977$, which yields $.242 \leq \eta \leq 7.22$.

of n. Fisher and Yates (1963, Table VIII1) cover $1 \leq n \leq 10$, and Documenta Geigy (1968, pp. 107–108) gives a more extensive version. Instead of pursuing these calculations in detail, however, we turn now to the binomial case.

One Binomial Count

If the observed count n comes from the binomial distribution with parameters N and p, and N is known, so that $\eta = Np$, the general structure of the situation is the same. We may base our calculations on confidence limits for the binomial parameter p. (These are included in the Fisher and Yates table, and their table precedes the Documenta Geigy table for the Poisson.) Because interpolation in n/N, the observed estimate of p, is almost certain to be required, it seems best to put these tables into a simple functional form, so that substitution replaces interpolation.

For a Poisson distribution with parameter η the large-sample analysis leading to equation (8) yields $1/\eta$ as the approximate variance of $\log_e(n)$, instead of $(1 - p)/\eta$. This suggests that a factor of $(1 - n/N)^{1/2}$ might take care of the difference between the Poisson and the binomial. Some numerical experimentation (not reproduced here) indicated, however, that it would be worthwhile to use

$$-C = -C_0\left(1 - c_1\frac{n}{N}\right)^{c_2}$$

$$+D = D_0\left(1 - d_1\frac{n}{N}\right)^{d_2}$$

as the form of the approximation, where C_0 and D_0 are the Poisson values, c_1 is somewhat less than 1 (substantially less at the smallest values of n), d_1 is reasonably close to 1, and c_2 and d_2 differ markedly from .5 for small values of n.

Table 9-6 gives the values of $-C_0$, c_1, c_2, D_0, d_1, and d_2 that are needed: (a) for n from 1 to 10; (b) for $n > 10$, in terms of formulas involving $g = 1/\sqrt{n}$; and (c), as illustrations of the latter, for $n = 25$, 49, 100, and 225. Given n and $\log_e(n)$, this table makes it easy to set limits on $\log_e(\eta)$ and thus to set limits on the population value of the count metameter $\phi(n_k)$. For $n > N/2$, it may pay to use $N - n$ in place of n, interchanging p and $1 - p$.

of n. Fisher and Yates (1963, Table VIII1) cover $1 \leq n \leq 10$, and Documenta Geigy (1968, pp. 107–108) gives a more extensive version. Instead of pursuing these calculations in detail, however, we turn now to the binomial case.

One Binomial Count

If the observed count n comes from the binomial distribution with parameters N and p, and N is known, so that $\eta = Np$, the general structure of the situation is the same. We may base our calculations on confidence limits for the binomial parameter p. (These are included in the Fisher and Yates table, and their table precedes the Documenta Geigy table for the Poisson.) Because interpolation in n/N, the observed estimate of p, is almost certain to be required, it seems best to put these tables into a simple functional form, so that substitution replaces interpolation.

For a Poisson distribution with parameter η the large-sample analysis leading to equation (8) yields $1/\eta$ as the approximate variance of $\log_e(n)$, instead of $(1 - p)/\eta$. This suggests that a factor of $(1 - n/N)^{1/2}$ might take care of the difference between the Poisson and the binomial. Some numerical experimentation (not reproduced here) indicated, however, that it would be worthwhile to use

$$-C = -C_0\left(1 - c_1\frac{n}{N}\right)^{c_2}$$

$$+D = D_0\left(1 - d_1\frac{n}{N}\right)^{d_2}$$

as the form of the approximation, where C_0 and D_0 are the Poisson values, c_1 is somewhat less than 1 (substantially less at the smallest values of n), d_1 is reasonably close to 1, and c_2 and d_2 differ markedly from .5 for small values of n.

Table 9-6 gives the values of $-C_0$, c_1, c_2, D_0, d_1, and d_2 that are needed: (a) for n from 1 to 10; (b) for $n > 10$, in terms of formulas involving $g = 1/\sqrt{n}$; and (c), as illustrations of the latter, for $n = 25$, 49, 100, and 225. Given n and $\log_e(n)$, this table makes it easy to set limits on $\log_e(\eta)$ and thus to set limits on the population value of the count metameter $\phi(n_k)$. For $n > N/2$, it may pay to use $N - n$ in place of n, interchanging p and $1 - p$.

Approximate Centers

When we consider using the values of $-C$ and $+D$ calculated from Table 9-6, we are struck by the substantial difference in size between C and D. Although we have known all along that these confidence intervals for $\log_e(\eta)$ are not symmetric about $\log_e(n)$, it would still be more convenient to have them in the form

$$\text{center} \pm \text{half-length}.$$

Thus we now ask about the centers and seek a simple approximation for them.

For a given value of n, the center of the confidence interval for $\log_e(\eta)$ can be written as

$$\log_e(n + \text{correction}).$$

We simply calculate "n + correction" as the geometric mean of the lower and upper confidence limits for η [or, equivalently, the antilog of the arithmetic mean of the lower and upper limits for $\log_e(\eta)$]. In the Poisson example with $n = 2$, the 95% confidence limits for η are .242 and 7.22, so that

$$2 + \text{correction} = \sqrt{(.242)(7.22)} = 1.322$$

$$\text{correction} = -.678.$$

If we begin with the 95% confidence limits for $\log_e(\eta)$, -1.418 and 1.977, we have

$$\log_e(2 + \text{correction}) = (-1.418 + 1.977)/2 = .2795$$

and again "correction" $= -.678$. For selected values of n and $\hat{p} = n/N$, Table 9-7 gives the resulting "n + correction"; the correction is negative in every case, and it goes beyond -1 only for $\hat{p} = .4$ and $\hat{p} = .5$. The empirical approximation for the correction,

$$-.67 - .8\hat{p},$$

seems adequate for $n \geq 2$. It underlies the definition of n_k^* given in equation (9).

TABLE 9-7. The value of "n + correction" for various values of n and $\hat{p} = n/N$. Each entry is the antilog of the center of the 95% confidence interval for $\log_e(\eta)$: $\log_e(n) + (D - C)/2$. The last line is a convenient approximation to the correction suggested by the tabulated values.

n	$\hat{p} = 0$	$\hat{p} = .1$	$\hat{p} = .2$	$\hat{p} = .3$	$\hat{p} = .4$	$\hat{p} = .5$
1	.375	.336	.301	—	—	—
2	1.322	1.25	1.18	—	—	1.00
3	2.330	2.25	2.16	2.09	—	1.94
4	3.341	3.26	3.17	3.08	3.00	2.91
5	4.348	4.26	4.17	4.08	3.99	3.90
7	6.366	6.27	6.18	6.08	5.98	5.90
10	9.395	9.29	9.20	9.10	9.00	8.90
36	35.45	—	—	—	—	—
144	143.45	—	—	—	—	—
$-.67 - .8\hat{p}$	$-.67$	$-.75$	.83	$-.91$	$-.99$	-1.07

Approximate Half-Lengths

For the half-lengths of the 95% confidence interval for $\log_e(\eta)$, the large-sample analysis [see equation (8)] yields

$$\pm \frac{1.96}{\sqrt{n}} \sqrt{1 - \hat{p}}\,,$$

where $\hat{p} = n/N$. The empirical approximation

$$\pm 1.96 \frac{\sqrt{1 - \hat{p}}}{\sqrt{n - (.47 + .25\hat{p})\sqrt{n}}}$$

does quite well for $n \geq 2$.

Combining the approximate center and the approximate half-length gives, as the endpoints of the confidence interval for $\log_e(\eta)$,

$$\log_e(n - .67 - .8\hat{p}) \pm 1.96 \frac{\sqrt{1 - \hat{p}}}{\sqrt{n - (.47 + .25\hat{p})\sqrt{n}}}. \qquad (10)$$

The largest error in locating the endpoints turns out to be roughly 1% of the half-length in the available cases for $n \geq 2$. This is enough accuracy for any graphical procedure.

The Smallest Observed Counts

In the preceding discussion, we claimed adequate accuracy for $n \geq 2$. We must also be prepared to handle an observed count of 1. For $n = 1$, the center is almost at $\log_e(1/e) = -1$. Straightforward calculations from the tables then give a lower confidence limit for $\log_e(\eta) - \log_e(1/e)$ equal to -2.677 (for $N = \infty, 20, 10$) and an upper confidence limit at $+2.717$ for $N = \infty$, $+2.603$ for $N = 20$, and $+2.493$ for $N = 10$. Thus we may use (for $N \geq 20$)

$$\log_e(1/e) - 2.677 \quad \text{and} \quad \log_e(1/e) + 2.717 - \frac{2.3}{N} \tag{11}$$

as our confidence limits. Their center falls at

$$\log_e(1/e) + .020 - \frac{1.15}{N},$$

which can almost always be approximated as just $\log_e(1/e)$.

Although we plan to omit from the Poissonness plot any points for which $n_k = 0$, we briefly examine this case. For $n = 0$ and $N > 1$, we can use

$$(-\infty) \quad \text{and} \quad 1.305 - \frac{1.88}{N + .4}$$

as our symmetrical 95% confidence limits for $\log_e(\eta)$, recognizing that here we have only one tail. We note also that this upper limit for $n = 0$ is higher than the interval center for $n = 1$, 2, 3, or 4, thus validating the title of Julius Bartels' paper of 1949 (in a meteorological context), which translates to "Zero (observed) *can* mean four (typically)" (emphasis and parenthetic comments added).

Redoing the Initial Examples

Now that we have convenient and adequately precise formulas for the center and half-length of the 95% confidence interval for $\log_e(\eta)$, given n and N, we can revise the Poissonness plot, as well as suggest new plots. We now use the confidence-interval centers in producing modified plots for two

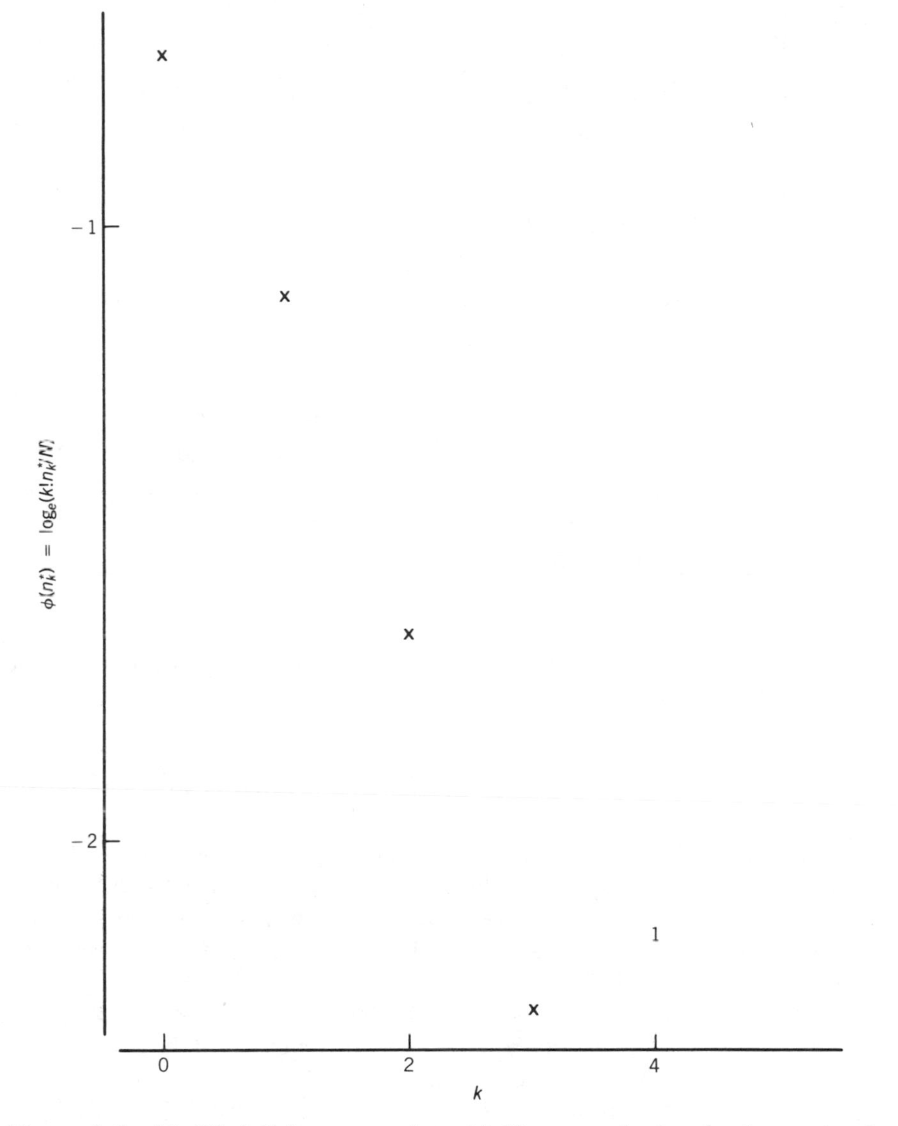

Figure 9-6. Modified Poissonness plot, $\phi(n_k^*)$ versus k, for the international terrorism data. The plotting symbol "1" identifies a point with $n_k = 1$. [The point (12, 14.66), calculated from $n_{12} = 1$, has been omitted.]

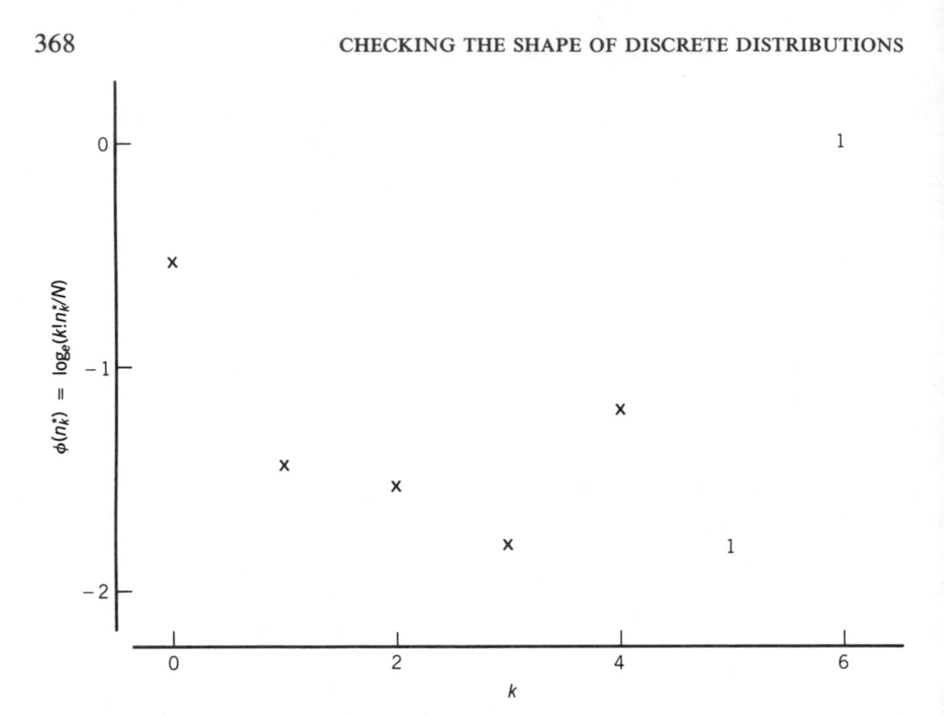

Figure 9-7. Modified Poissonness plot, $\phi(n_k^*)$ versus k, for the word-frequency data. (Points with $n_k = 1$ are plotted as "1".)

examples from Section 9A. Then Section 9C incorporates the half-lengths into some new displays.

Knowing where the centers of the confidence intervals for $\log_e(\eta)$ go allows us to use $\log_e(n_k^*)$ in place of $\log_e(n_k)$, where $n_k^* = n_k - .67 - .8n_k/N$ for $n_k \geq 2$ and $n_k^* = 1/e$ for $n_k = 1$, and thus to plot $\phi(n_k^*)$ as the vertical coordinate instead of $\phi(n_k)$. We feel that this modification often offers advantages.

Figures 9-6 and 9-7 show such modified Poissonness plots for the international terrorism data and the word-frequency data, respectively. (Tables 9-8 and 9-9 give the calculations, and Figures 9-1 and 9-2 are the original plots.) The main change is that points plotted as "1," where $n_k = 1$, no longer seem to behave differently from the other points. The clear exception (not plotted in Figure 9-6) is the month with 12 terrorist incidents, which we already agreed was quite different.

Exercise 4 invites the reader to make the modified Poissonness plot for the household size data of Table 9-4.

We can level a modified Poissonness plot in exactly the same way as an ordinary Poissonness plot. See equation (7) and Exercise 5.

TABLE 9-8. Calculations for modified Poissonness plot, using confidence-interval centers, for the international terrorism data.

Number of Incidents (k)	Number of Months (n_k)	$\hat{p}_k = n_k/N$	$n_k^* = n_k - .67 - .8\hat{p}_k$	$\phi(n_k^*) = \log_e(k!n_k^*/N)$
0	38	.500	36.93	−0.72
1	26	.342	25.06	−1.11
2	8	.105	7.25	−1.66
3	2	.026	1.31	−2.27
4	1		0.37[a]	−2.15
12	1		0.37[a]	14.66
	$N = 76$			

[a] $n_k^* = 1/e$ when $n_k = 1$.

TABLE 9-9. Calculations for modified Poissonness plot, using confidence-interval centers, for occurrences of the word *may* in papers already known to be by Madison.

Number of Occurrences (k)	Number of Blocks (n_k)	$\hat{p}_k = n_k/N$	$n_k^* = n_k - .67 - .8\hat{p}_k$	$\phi(n_k^*) = \log_e(k!n_k^*/N)$
0	156	.595[a]	154.85	−0.53
1	63	.240	62.14	−1.44
2	29	.111	28.24	−1.53
3	8	.031	7.31	−1.79
4	4	.015	3.32	−1.19
5	1		0.37[b]	−1.78
6	1		0.37[b]	+0.01
	$N = 262$			

[a] We have used the formula for n_k^* here, even though its original construction was for $\hat{p} \leq .5$.
[b] $n_k^* = 1/e$ when $n_k = 1$.

9C. WHEN IS A POINT DISCREPANT?

When an isolated point strays from an apparently linear pattern of a Poissonness plot, we may want to judge more formally whether it is unlikely to have done so by chance. The simplest reasonably accurate residual is the double-root comparison

$$\sqrt{2 + 4(\text{observed})} - \sqrt{1 + 4(\text{fitted})}\,, \tag{12}$$

which is nearly enough standard Gaussian for many purposes. (Section 9G gives some background for such residuals.) Beyond this straightforward transformation, we rely on an enhanced version of the Poissonness plot that shows the endpoints of the 95% confidence interval for each $\phi(\eta_k)$ instead of merely the center. Such a confidence-interval plot can then indicate whether any straight lines are compatible with the overall pattern of behavior. Also, after incorporating an allowance for multiplicity, it enables us to judge whether an individual point is discrepant.

Confidence-Interval Plots

As a natural next step beyond plotting $\phi(n_k^*)$ against k, we replace $\phi(n_k^*)$ with a vertical line segment whose endpoints are

$$\phi(n_k^*) \pm \text{half-length},$$

using the same basic scheme as in equations (10) and (11). We need only add in the constant $\log_e(k!/N)$ to get from those 95% confidence intervals for $\log_e(\eta_k)$ to the corresponding confidence intervals for $\phi(\eta_k)$. In practice, once we have calculated the $\phi(n_k^*)$ for the modified Poissonness plot, the straightforward formula for the half-lengths requires little further effort.

The same motivations that lead us to level a basic or modified Poissonness plot may urge us to level the confidence-interval plot, in the same way. If λ_0 seems a reasonable value for λ, we subtract $k\log_e(\lambda_0) - \lambda_0$ from both endpoints of the interval at k.

EXAMPLE: RADIOACTIVE DECAY OF POLONIUM

In a classic set of data Rutherford and Geiger (1910) gave the number of scintillations in $\frac{1}{8}$-min intervals caused by radioactive decay of a quantity of the element polonium. Table 9-10 shows their data, as well as the vertical coordinates for the modified and confidence-interval Poissonness plots. The plot in Figure 9-8a behaves almost entirely as we would expect for a large

TABLE 9-10. Scintillations from radioactive decay of polonium. Frequency distribution and calculations for modified and confidence-interval Poissonness plots.

Number of Scintillations (k)	Number of Intervals (n_k)	n_k^*	$\phi(n_k^*) = \log_e(k!n_k^*/N)$	Half-length (h)	Endpoints $\phi(n_k^*) - h$	Endpoints $\phi(n_k^*) + h$
0	57	56.31	−3.84	.27	−4.10	−3.57
1	203	202.27	−2.56	.13	−2.69	−2.42
2	383	382.21	−1.23	.09	−1.32	−1.13
3	525	524.17	0.19	.08	0.11	0.26
4	532	531.17	1.59	.08	1.51	1.66
5	408	407.20	2.93	.09	2.84	3.02
6	273	272.25	4.32	.11	4.21	4.43
7	139	138.29	5.59	.17	5.42	5.75
8	45	44.32	6.53	.30	6.23	6.83
9	27	26.32	8.21	.39	7.81	8.60
10	10	9.33	9.47	.67	8.80	10.14
11	4	3.33	10.84	1.12	9.72	11.96
12	0	—	—	—	$-\infty$	13.43[b]
13	1	0.37	13.69	−2.68[a]	11.01	16.40
14	1	0.37	16.32	+2.72[a]	13.65	19.04
	$N = 2608$					

[a] When $n = 1$, the half-lengths are −2.68 and +2.72; see equation (11).
[b] For treatment of $n = 0$, see text below equation (11).

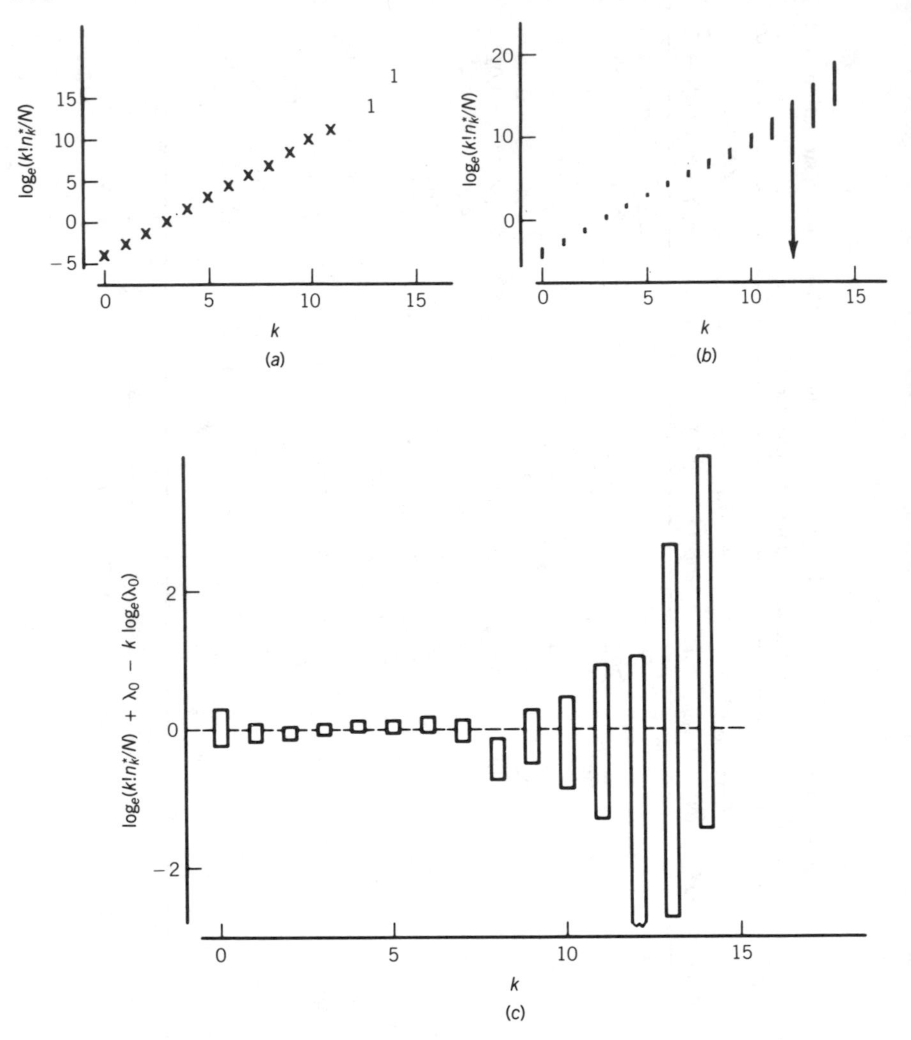

Figure 9-8. Poissonness and confidence-interval plots for scintillations from radioactive decay of polonium. (*a*) Poissonness plot. (*b*) Confidence-interval plot. (*c*) Leveled confidence-interval plot.

sample in a situation where a Poisson model should fit very well. The point for $k = 8$ is somewhat low. Because overall agreement is excellent, we can easily use $\bar{x} = 10097/2608 = 3.872$ as the λ_0 for the leveled plots.

Figure 9-8*b* shows the confidence-interval plot. Although its good behavior is still clear, the vertical scale must cover so large a range that it is hard to see what lines up with what. In the leveled confidence-interval plot,

Figure 9-8*c*, we can see that n_8 appears significantly low, but we need to recall that one miss out of 15 is very common at 5%. (The probability of *nothing* beyond 5% in 15 independent trials is only .46.)

Allowing for Multiplicity

In looking at a confidence-interval plot, we face a substantial likelihood that at least one 95% interval will miss under the best of circumstances, so that the corresponding n_k appears discrepant. We are, in effect, visually making several simultaneous significance tests, and we naturally seize upon the largest departures from a nearly linear pattern. If each of L confidence intervals has individual confidence $1 - \alpha$, the chance that all L intervals will each cover its own true value is close to $(1 - \alpha)^L$. The corresponding two-sided tail area, $1 - (1 - \alpha)^L$, indicates the *simultaneous*, rather than *individual*, significance level for the intervals.

Approaching the problem in another way, we consider sets of L simultaneous independent confidence statements, each at 95% individual confidence, and we ask how many can be violated without reaching significance at 5%. When we treat this 5% as one-sided, so that we do not declare significance when too few violations occur, straightforward binomial calculations show that we can accept 0 violations when $L = 1$, at most one violation when $2 \leq L \leq 7$, at most two when $8 \leq L \leq 16$, and at most three when $17 \leq L \leq 28$.

To make a rough visual allowance for multiplicity, we supplement our 95% intervals with broader intervals. The simplest solution is to show ± 1.5 (half-length) in addition to ± 1.0 (half-length). For two-sided intervals and a Gaussian distribution, 1.5(95% half-length) is approximately the same as a 99.67% half-length. Thus we may think of these outer limits as corresponding to quite a small individual tail area. Because we are not tailoring the allowance to the number of simultaneous intervals, we give two significance levels—one individual and the other simultaneous—for each of the two two-sided intervals. For example, with 12 values of k, the levels for ± 1.0(half-length) are 5% individual and 46% simultaneous, and those for ± 1.5(half-length) are .33% individual and 3.9% simultaneous.

To present the outer limits, we can follow a variety of display strategies. First, we may augment the vertical line segment or bar at each value of k by adding a further symbol at (center) $\pm$ 1.5(half-length). Second, the leading alternative shows the confidence interval by filling in its complement [e.g., with one vertical line segment extending downward from (center) $-$ 1.0(half-length) and another extending upward from (center) + 1.0(half-length)] and leaving the interval itself as an aperture. Again a further symbol marks the points at (center) $\pm$ 1.5(half-length). Finally, we may

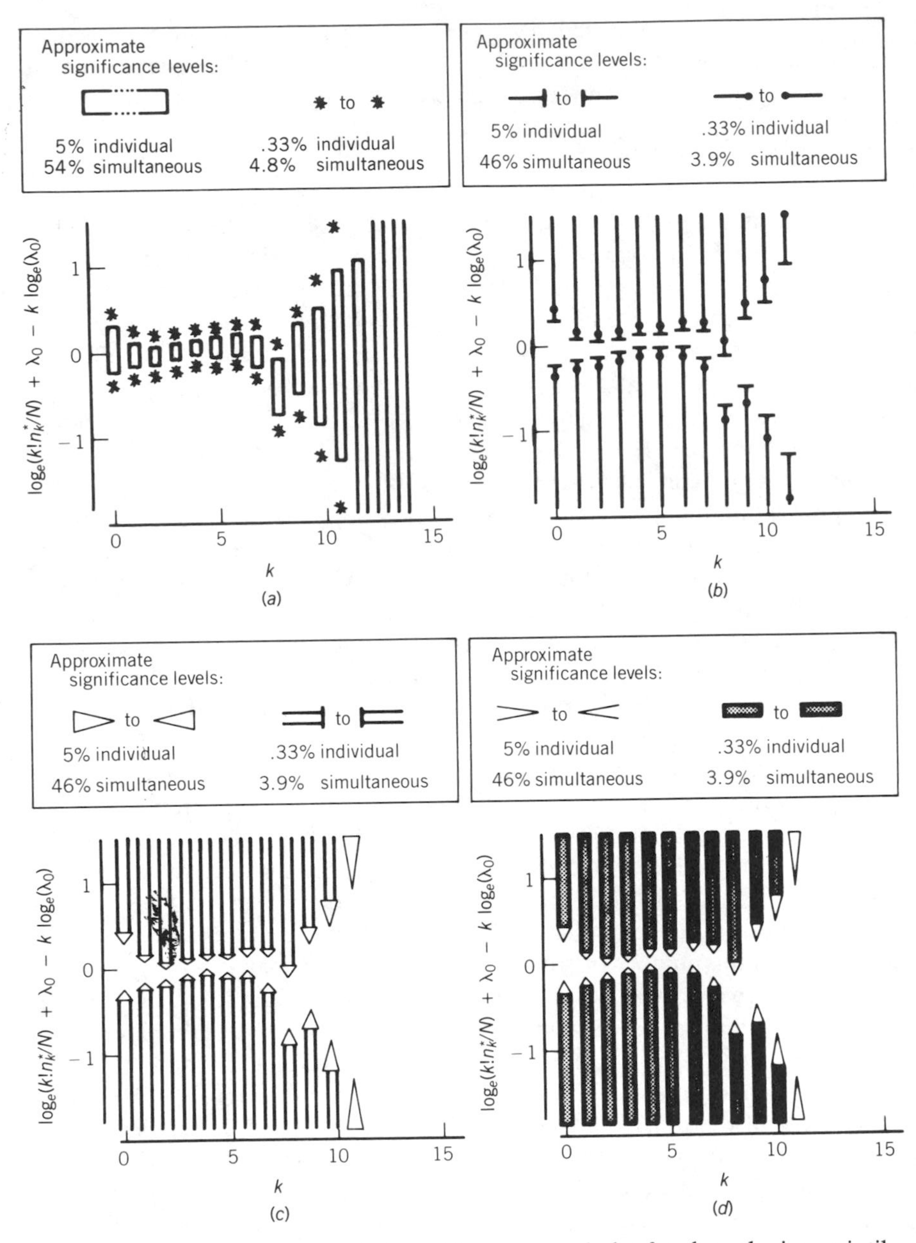

Figure 9-9. Four styles of leveled confidence-interval plot for the polonium scintillation data. (*a*) Bars and stars. (*b*) Dotted aperture. (*c*) Arrowed aperture. (*d*) Pencil-point aperture.

experiment with variations on the confidence-aperture theme. These all emphasize what is *in*compatible with the data.

EXAMPLE: POLONIUM SCINTILLATIONS (CONTINUED)

In Figure 9-9 we illustrate the addition of ± 1.5(95% half-length) to the leveled confidence-interval plot for the polonium scintillation data. Table 9-11 assembles the calculations from ingredients in Table 9-10. Because $\phi(n_8^*) + 1.5$(half-length) is positive and because the enlarged apertures would also accommodate straight lines for other values of λ besides $\lambda_0 = \bar{x}$, we conclude that n_8 is not quite discrepant.

TABLE 9-11. Values of (center) ± 1.0(half-length) and (center) ± 1.5(half-length) for the leveled confidence-interval plot of the polonium scintillation data.

	$\phi(n_k^*) + [\lambda_0 - k \log_e(\lambda_0)] + t$(half-length)			
k	$t = -1.5$	$t = -1.0$	$t = +1.0$	$t = +1.5$
0	−.36	.23	.30	.43
1	−.24	−.17	.10	.16
2	−.20	−.16	.03	.08
3	−.12	−.08	.08	.11
4	−.07	−.03	.12	.16
5	−.10	−.06	.12	.17
6	−.10	−.04	.18	.24
7	−.26	−.18	.15	.23
8	−.88	−.73	−.13	.02
9	−.70	−.50	.29	.48
10	−1.20	−.86	.48	.81
11	−1.86	−1.30	.94	1.50
12	—	−∞	1.06	—
13	−4.06	−2.72	2.68	4.03
14	−2.77	−1.43	3.96	5.32

Note: For $k = 0, \ldots, 11$, as in Figures 9-9*b* through 9-9*d*, simultaneous confidence for $t = \pm 1$ is $(.95)^{12} = .54$, so that the significance level is $1 - .54 = 46\%$, and simultaneous confidence for $t = \pm 1.5$ is $(.9967)^{12} = .961$, so that the significance level is $1 - .961 = 3.9\%$.

The four panels of Figure 9-9 exemplify four styles of showing the ± 1 and ± 1.5 limits:

Name	± 1	± 1.5
(a) Bars and stars	End of bar	Star
(b) Dotted aperture	Edge of aperture	Dot
(c) Arrowed aperture	Arrow point	Arrow base
(d) Pencil-point aperture	Pencil tip	End of solid bar

These styles offer alternatives, both as to strength of impact and as to increased plotting effort. We often prefer (c) or (d).

9D. OVERALL PLOTS FOR OTHER FAMILIES OF DISTRIBUTIONS

The simplicity and usefulness of the Poissonness plot naturally lead us to ask whether similar plots apply to other discrete distributions. In fact, for a sizable class of distributions, we can derive plots whose horizontal coordinate is k and whose standard of comparison is a straight line.

When we generalize the feature of the Poisson distribution that led to equation (5), we arrive at

$$\log(p(k)) - \log(a_k) = b(\pi) + kc(\pi) \tag{13}$$

in terms of some more general parameter, π. Here a_k depends on k but not on π, and $b(\pi)$ and $c(\pi)$ are functions of π but not k. For the Poisson distribution, for example, $a_k = 1/k!$, $b(\pi) = -\lambda$, and $c(\pi) = \log(\lambda)$.

From equation (13) it is straightforward to derive

$$p(k) = a_k \theta^k / f(\theta), \tag{14}$$

where $k = 0, 1, \ldots,$ $\theta > 0$, $a_k \geq 0$, and $f(\theta) = \Sigma a_k \theta^k$. Known as the *power-series distributions*, this class was introduced by Noack (1950). [See also Johnson and Kotz (1969, Section 2.3).] Distributions that can be written in the form of equation (14)—in which θ may not be the parameter in the usual form of the distribution—include the binomial, the negative binomial (of which the geometric distribution is a special case), the logarithmic series, and, of course, the Poisson.

Binomial Distribution

For the binomial distribution with parameters n and p, the probability function

$$p(k) = \binom{n}{k} p^k (1-p)^{n-k}, \qquad k = 0, \ldots, n,$$

may be rewritten in the form of equation (14) as

$$p(k) = \binom{n}{k} \left(\frac{p}{1-p}\right)^k \Big/ (1-p)^{-n}, \tag{15}$$

so that

$$a_k = \binom{n}{k}, \quad \theta = p/(1-p), \quad \text{and} \quad f(\theta) = (1-p)^{-n}.$$

In order to plot $\log_e(n_k) - \log_e(a_k)$ against k for the binomial distribution, we must use a value for n (which we usually know), and the observations represented in the frequency distribution must have the same value of n. This second condition is not a serious constraint, but it does force us to other ways of examining some sets of varying-n data in which we might like to check graphically on constant-p behavior.

Many frequency distributions do arise that might be binomial. For these we plot the binomial count metameter

$$\begin{aligned} \phi(n_k^*) &= \log_e(n_k^*) - \log_e\binom{n}{k} - \log_e(N) \\ &= \log_e\left[n_k^* \Big/ N\binom{n}{k}\right] \end{aligned} \tag{16}$$

against k, looking for a straight line, whose constants we interpret in terms of slope $\log_e(p/(1-p))$ and intercept $n \log_e(1-p)$. As in the modified Poissonness plot, n_k^* comes from n_k via equation (9).

EXAMPLE: WOMEN IN QUEUES

Jinkinson and Slater (1981) give the frequency of the number of females in 100 queues of length 10 observed in a London Underground station. Table 9-12 reproduces their data and shows the calculations for the plot to check whether these data are consistent with a binomial distribution.

TABLE 9-12. Number of females in 100 queues of length $n = 10$ in a London Underground station. Frequency distribution and calculations for binomial plot.

Number of Females (k)	Number of Queues (n_k)	n_k^*	$\binom{n}{k}$	$\log_e\left(n_k^*/N\binom{n}{k}\right)$
0	1	.37	1	−5.61
1	3	2.31	10	−6.07
2	4	3.30	45	−7.22
3	23	22.15	120	−6.30
4	25	24.13	210	−6.77
5	19	18.17	252	−7.23
6	18	17.19	210	−7.11
7	5	4.29	120	−7.94
8	1	.37	45	−9.41
9	1	.37	10	−7.91
10	0	—	1	—
	$N = 100$			

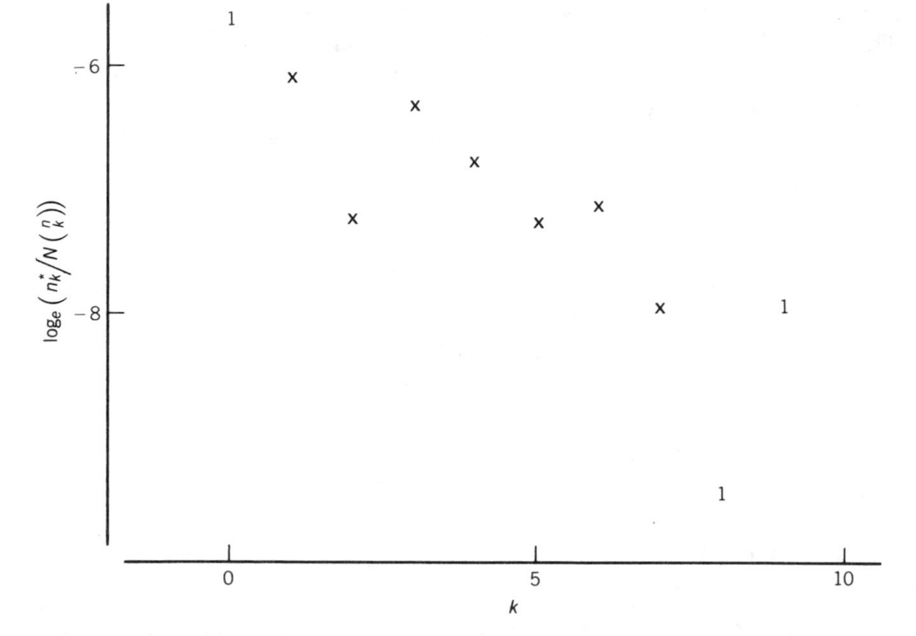

Figure 9-10. Binomial(10)ness plot, $\phi(n_k^*)$ versus k, for number of females in queues of length 10 in a London Underground station. (Points with $n_k = 1$ are plotted as "1".)

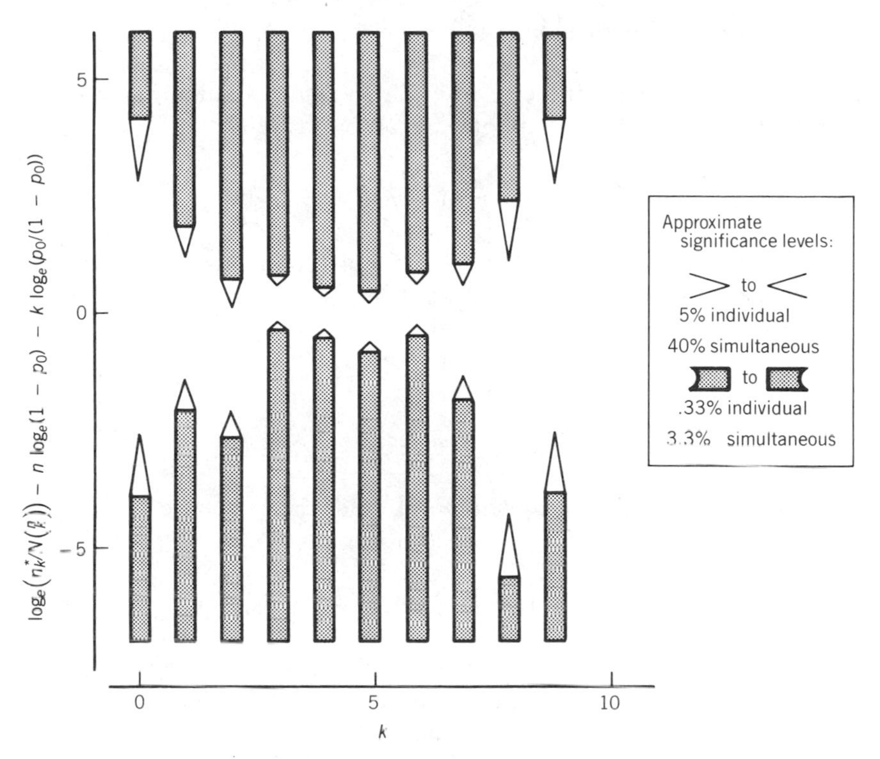

Figure 9-11. Leveled confidence-aperature plot for the data on women in queues.

Considering that N is only 100, the plot in Figure 9-10 indicates quite satisfactory agreement with a binomial model. An eye-fitted line (which can deemphasize the points at $k = 2$ and $k = 8$) yields a slope of roughly $-.3$, corresponding to an estimate of .426 for p. This value differs only slightly from the naive estimate, $\bar{x}/n = .435$.

When we use equation (10) to calculate the vertical coordinates for the confidence-interval plot and then level those coordinates by taking $p_0 = \bar{x}/n = .435$ and subtracting $n \log_e(1 - p_0) + k \log_e[p_0/(1 - p_0)]$, we have the leveled confidence-aperture plot in Figure 9-11. This plot shows good general agreement of every point with a horizontal straight line, as well as with other lines of moderate positive and negative slope. Thus, making appropriate allowance for the variability of $\log_e(n_k)$ gives us a better idea of what values of p are compatible with these data.

Negative Binomial Distribution

The probability function of the negative binomial distribution with parameters n and p (n need not be an integer) can be written

$$p(k) = \binom{n+k-1}{k} p^n (1-p)^k, \qquad k = 0, 1, \ldots \tag{17}$$

(in one of the several ways of parametrizing this distribution). Now we arrive at the form of equation (14) by putting

$$a_k = \binom{n+k-1}{k},$$

$$\theta = (1-p), \quad \text{and} \quad f(\theta) = p^{-n} = (1-\theta)^{-n}.$$

The special case $n = 1$ in equation (17) yields the geometric distribution, with probability function

$$p(k) = p(1-p)^k, \qquad k = 0, 1, \ldots . \tag{18}$$

Many processes give rise to a geometric distribution. Among these, one counts the number of failures preceding the first success in a sequence of independent trials with common success probability p. (Adding 1—for the successful trial—gives the waiting time until the first success.)

More generally, a negative binomial distribution with an integer value of n gives the distribution of the number of failures preceding the nth success in such a sequence.

Negative binomial distributions often serve as two-parameter alternatives to the Poisson. One motivation derives the negative binomial distribution as a mixture of Poisson distributions in which the values of λ are mixed according to a gamma distribution.

For plotting we use the negative binomial count metameter

$$\begin{aligned} \phi(n_k^*) &= \log_e(n_k^*) - \log_e\binom{n+k-1}{k} - \log_e(N) \\ &= \log_e\left(n_k^* \Big/ N\binom{n+k-1}{k}\right), \end{aligned} \tag{19}$$

and we interpret its slope as $\log_e(1-p)$ and its intercept as $n\log_e(p)$. Much more easily, the geometric distribution, $n = 1$, has count metameter

$$\begin{aligned} \phi(n_k^*) &= \log_e(n_k^*) - \log_e(N) \\ &= \log_e(n_k^*/N). \end{aligned} \tag{20}$$

Its slope tells us about $\log_e(1 - p)$, and its intercept has the interpretation $\log_e(p)$.

As in the case of the binomial distribution, we must adopt a value of n in order to calculate a_k and construct the negative binomial count metameter. In practice this requires either "trial and error" for several values of n or estimation by some overall calculation. As an aid to trial and error we note that, when the value of n adopted is smaller than the appropriate value, the plot of $\phi(n_k^*)$ against k tends to be concave downward, turning down more steeply as k increases. Conversely, if the adopted value of n exceeds the appropriate value, the plot tends to be convex downward, decreasing more steeply for low values of k than for higher ones.

EXAMPLE: LIBRARY CIRCULATION

Computerization of library operations has enabled large libraries to accumulate data on the circulation of books in their collections. From the records of the times at which each book is checked out, many libraries tabulate the number of books that were borrowed k times during a certain period, such as a year. Burrell and Cane (1982) give several such frequency distributions, including one year's data for the Hillman Library at the University of Pittsburgh, shown here in the first two columns of Table 9-13.

Such data often pose difficulties in determining n_0, the number of books that did not circulate during the period, because the basic records do not tell how many items in the collection were not available for circulation. We sidestep this question by setting N equal to the total number of books that were checked out at least once. This truncation does not affect the ability of a plot to indicate how closely the available data agree with a negative binomial (or geometric) distribution.

For simplicity Burrell and Cane use the geometric distribution. In the discussion accompanying the paper, at least one contributor suggests that a more general negative binomial distribution may be more appropriate and that the data from the Hillman Library seem to require n around 0.27. Our plots readily shed some light on this question.

The plot of $\phi(n_k^*)$ versus k for the geometric distribution (not shown) comes close to a straight line, but the points for $k = 1$ through about $k = 8$ indicate systematic curvature. We see this more clearly in a leveled plot, Figure 9-12a, after subtracting $\log_e(p_0) + k \log_e(1 - p_0)$ with $p_0 = \frac{1}{2}$ (chosen from the slope of an eye-fitted line). The pattern suggests a negative binomial distribution with n less than 1. To pursue this, Figures 9-12b and 9-12c show leveled plots for $n = .5$ and $n = .25$, respectively. [The leveling subtracts $n \log_e(p_0) + k \log_e(1 - p_0)$; and the values of p_0, .438 and .593, again come from eye-fitted lines.] Both of these plots show substantially less

TABLE 9-13. Circulation of books in the Hillman Library of the University of Pittsburgh in a one-year period. Frequency distribution and calculations for geometric and negative binomial plots.

				Negative Binomial ($n = .5$)		Negative Binomial ($n = .25$)	
Number of Borrowings (k)	Number of Books (n_k)	n_k^*	Geometric $\log_e(n_k^*/N)$	$\binom{k-.5}{k}$	$\log_e\left(n_k^* \Big/ N\binom{k-.5}{k}\right)$	$\binom{k-.75}{k}$	$\log_e\left(n_k^* \Big/ N\binom{k-.75}{k}\right)$
1	63526	63524.88	−.59	.500	.11	.250	.80
2	25653	25652.15	−1.49	.375	−.51	.156	.36
3	11855	11854.25	−2.26	.312	−1.10	.117	−.12
4	6055	6054.29	−2.94	.273	−1.64	.095	−.58
5	3264	3263.31	−3.55	.246	−2.15	.081	−1.04
6	1727	1726.32	−4.19	.226	−2.70	.071	−1.54
7	931	930.32	−4.81	.209	−3.25	.063	−2.05
8	497	496.33	−5.44	.196	−3.81	.057	−2.58
9	275	274.33	−6.03	.185	−4.35	.052	−3.08
10	124	123.33	−6.83	.176	−5.09	.049	−3.80
11	68	67.33	−7.43	.168	−5.65	.045	−4.34
12	28	27.33	−8.34	.161	−6.51	.042	−5.18
13	13	12.33	−9.13	.155	−7.27	.040	−5.91
14	6	5.33	−9.97	.149	−8.07	.038	−6.70
15	9	8.33	−9.52	.144	−7.59	.036	−6.20
16	4	3.33	−10.44	.140	−8.47	.034	−7.07
	$N = 114{,}035$						

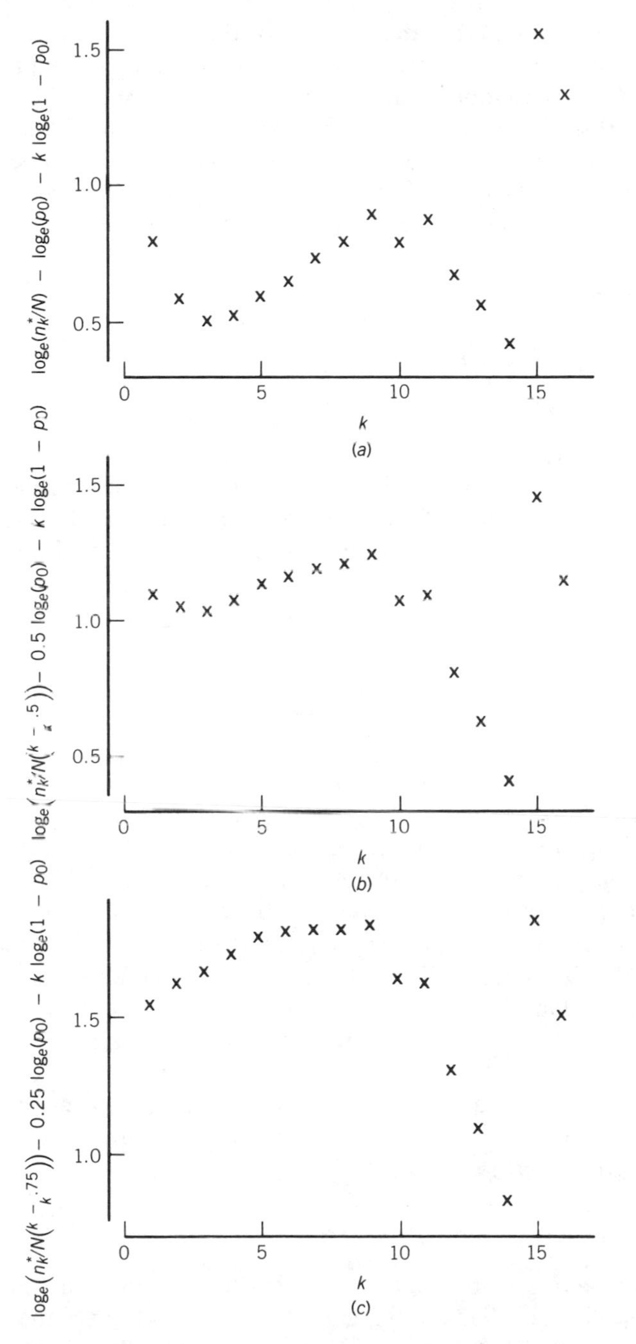

Figure 9-12. Leveled negative-binomialness plots for the data on circulation of books in the Hillman Library of the University of Pittsburgh. (a) $n = 1$ (geometric distribution), $p_0 = .5$. (b) $n = .5$, $p_0 = .438$. (c) $n = .25$, $p_0 = .593$.

curvature at the left end, but that portion of the plot for $n = .5$ seems more nearly straight. Thus a negative binomial distribution seems preferable to the geometric distribution, although $n = .27$ seems somewhat too small. Apart from this choice, another clear message of all three plots is that the observed frequencies for books borrowed nine or fewer times seem to require one value of p, whereas those for more heavily circulated books (except n_{15} and n_{16}) seem to need another, larger value of p.

Logarithmic Series Distribution

With probability function

$$p(k) = \alpha\theta^k/k, \qquad k = 1, 2, \ldots, \tag{21}$$

where $0 < \theta < 1$ and $\alpha = -1/\log_e(1 - \theta)$, the logarithmic series distribution already has the form of equation (14): $a_k = 1/k$ and $f(\theta) = 1/\alpha = -\log_e(1 - \theta)$. Thus we may plot the log-series count metameter

$$\begin{aligned} \phi(n_k^*) &= \log_e(n_k^*) + \log_e(k) - \log_e(N) \\ &= \log_e(kn_k^*/N) \end{aligned} \tag{22}$$

against k and interpret the slope of a straight-line pattern in terms of $\log_e(\theta)$ and the intercept in terms of $\log_e(\alpha)$.

Introduced into ecology by R. A. Fisher in 1943, the logarithmic series distribution has served as a model for a wide variety of processes in biology and other fields. Many of the sets of data extend to large values of k, and thus they require values of θ close to 1 in equation (21). Because θ^k/k decreases rather slowly, the corresponding distributions have many small p_k. For a plot involving individual values of k separately, these small p_k have the unfortunate consequence that the variances of the corresponding $\log(n_k)$, roughly $(1 - p_k)/Np_k$, are large. Thus even a good plot may resemble a band of scatter more than a straight line. However, because the points may come from a straight line, it is reasonable, when we are asking about the line generally, to average three or five or more adjacent points and plot the result at the central value of k.

EXAMPLE: BUTTERFLY SPECIES

One of the data sets to which the logarithmic series distribution was originally applied (Fisher, Corbet, and Williams, 1943) gives the number of individuals collected for each of 501 species of butterflies in Malaya. As a

TABLE 9-14. Number of butterfly species, n_k, for which k (≤ 24) individuals were collected in Malaya: Data and calculations for plots.

k	n_k	n_k^*	$\log_e(kn_k^*/N)$	Mean of Five[a]
1	118	117.14	−1.45	−1.45
2	74	73.21	−1.23	−1.34
3	44	43.26	−1.35	−1.40
4	24	23.29	−1.68	−1.38
5	29	28.28	−1.26	−1.39
6	22	21.29	−1.37	−1.37
7	20	19.30	−1.31	−1.25
8	19	18.30	−1.23	−1.24
9	20	19.30	−1.06	−1.25
10	15	14.31	−1.25	−1.22
11	12	11.31	−1.39	−1.37
12	14	13.31	−1.14	−1.38
13	6	5.32	−1.98	−1.50
14	12	11.31	−1.15	−1.49
15	6	5.32	−1.84	−1.51
16	9	8.32	−1.33	−1.45
17	9	8.32	−1.27	−1.42
18	6	5.32	−1.65	−1.26
19	10	9.31	−1.04	−1.16
20	10	9.31	−.99	−1.24
21	11	10.31	−.84	−1.35
22	5	4.32	−1.66	−1.58
23	3	2.33	−2.24	−2.03
24	3	2.33	−2.19	−2.19
	$N = 501$			

[a]Mean of only one at ends, and of only three next to ends.

frequency distribution (Table 9-14) these data record that k individuals were collected for each of n_k species.

We must wonder why the frequency distribution stops at $k = 24$. The answer lies in a deliberate truncation: "collecting of all individuals seen was not continued after 24 specimens had been taken." The total number of species observed was actually 620—119 had ≥ 25 individuals observed.

At first impression, the plot (Figure 9-13) hardly constitutes compelling evidence for a logarithmic series distribution. It even appears to have a slight *positive* slope instead of the required negative slope, especially if one dares to suppress some seven points that seem a little low ($k = 4, 13, 15, 18, 22, 23, 24$). The variability of the $\log(n_k)$, however, suggests that such

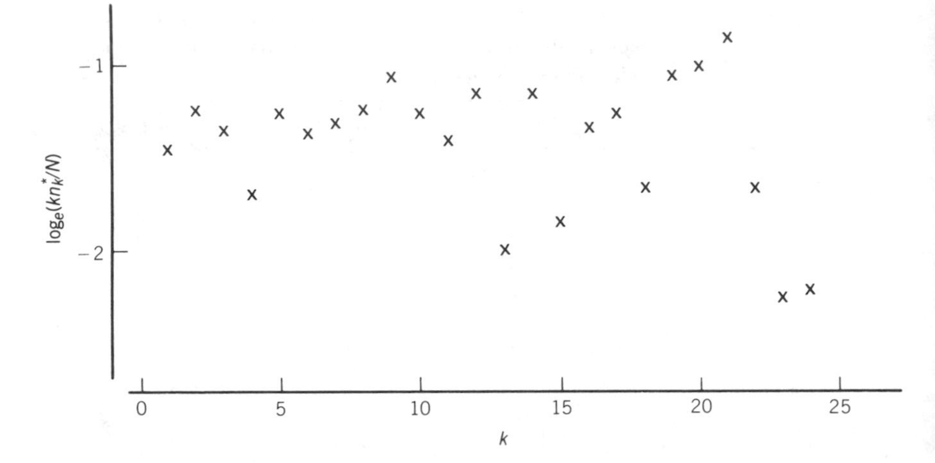

Figure 9-13. Logarithmic-series plot, $\phi(n_k^*)$ versus k, for data on butterfly species collected in Malaya.

subjective judgments can easily be incorrect. (Corbet calculated $\theta = .997$.) Fitting a 3-group resistant line (UREDA, Chapter 5) to this plot yielded a slope of $-.003593$ and hence an estimate of $\theta = .9964$. The confidence-aperture plot in Figure 9-14 shows overall behavior compatible with a straight line. Thus these plots do not reveal any systematic departure from a (truncated) logarithmic series distribution.

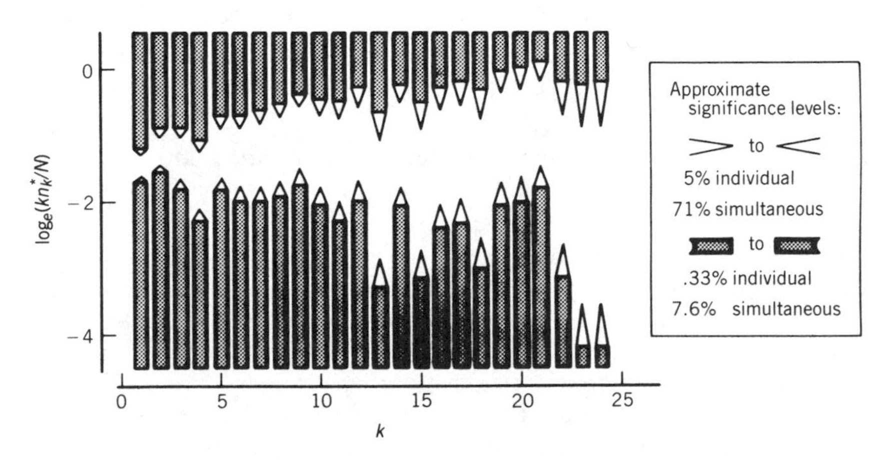

Figure 9-14. Confidence-aperture plot for the butterfly species data.

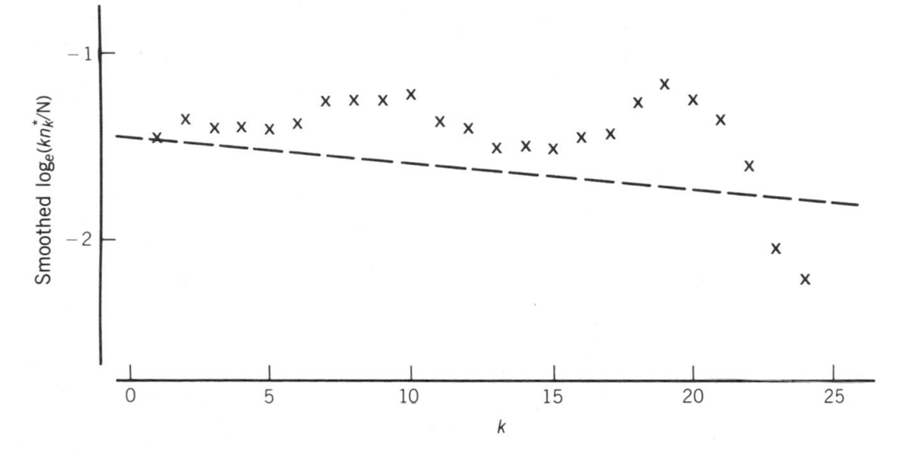

Figure 9-15. Means of five adjacent points in the plot of $\phi(n_k^*)$ versus k for the butterfly species data (Figure 9-13). The dashed line corresponds to the log-series distribution with $\theta = .98642$ suggested by the fact that 119 of 620 species had 25 or more individuals observed.

On the other hand, any fitting method will be used only if it delivers a value of θ smaller than 1. The data might have come from a logarithmic series distribution, true enough, but do they *look* as if they do?

Some local averaging gives the values of $\phi(n_k^*)$ an opportunity to contribute to a more nearly straight-line pattern. Figure 9-15 plots the means of sets of five adjacent points (fewer at the ends) from Figure 9-13. We have the impression of a small positive slope from $k = 1$ to $k = 22$ and a notable decrease for $k = 23$ and $k = 24$. What should we make of this? Because species with 25 or more individuals observed did not yield a detailed count, we wonder whether some species with 23 or 24 individuals were tallied as "25 or more" by mistake. At this distance, however, we cannot resolve such questions.

We can try to use the additional information that 119 out of 620 species had 25 or more individuals observed. A log-series distribution with $\theta =$ .98642 has $P_\theta\{X \geq 25\} = .1919 = 119/620$. This value of θ gives a slope of $\log_e(\theta) = -.0137$ and an intercept of $\log_e(\alpha) = \log_e[-1/\log_e(1-\theta)]$ $= -1.458$. The corresponding line, shown dashed in Figure 9-15, does not look like a good fit.

This example has taught us that

1. When θ is close to 1, we are likely to find it *hard* to tell from a simple plot what seems to be going on.

TABLE 9-15. Summary of plots based on $\log_e(n_k^*)$ for five discrete distributions. The vertical coordinate is the count metameter $\phi(n_k^*)$, and the horizontal coordinate is k.

Distribution	Probability Function, $p(k)$	Vertical Coordinate, $\phi(n_k^*)$	Theoretical Slope	Theoretical Intercept
Poisson	$e^{-\lambda}\lambda^k/k!$	$\log_e(k!n_k^*/N)$	$\log_e(\lambda)$	$-\lambda$
Binomial	$\binom{n}{k}p^k(1-p)^{n-k}$	$\log_e\left(n_k^*/N\binom{n}{k}\right)$	$\log_e(p/(1-p))$	$n\log_e(1-p)$
Negative binomial	$\binom{n+k-1}{k}p^n(1-p)^k$	$\log_e\left(n_k^*/N\binom{n+k-1}{k}\right)$	$\log_e(1-p)$	$n\log_e(p)$
Geometric	$p(1-p)^k$	$\log_e(n_k^*/N)$	$\log_e(1-p)$	$\log_e(p)$
Logarithmic series	$\theta^k/[-k\log_e(1-\theta)]$	$\log_e(kn_k^*/N)$	$\log_e(\theta)$	$-\log_e[-\log_e(1-\theta)]$

2. This is particularly true when we lack information about larger values of k.
3. Combining values of k(e.g., by averaging before plotting) is likely to be helpful, but we still need to be quite careful about using the evidence.

In a word, checking the logarithmic series character of a discrete frequency distribution is not easy.

Summary

We have constructed plots based on $\log(n_k^*)$ for other distributions in the class of power-series distributions besides the Poisson distribution. Table 9-15 summarizes the full set of plots by giving the probability function $p(k)$, the count metameter $\phi(n_k^*)$ for the plot (k always serves as the horizontal coordinate), and the theoretical slope and intercept.

Among these distributions the plot for the Poisson stands out as the most useful. The one for the binomial should find frequent application, the negative binomial plot perhaps less so (because of the need to assess the value of the parameter n, which is not obvious here), and the logarithmic series plot still less. In the special case of the geometric distribution, however, the simplicity of the plot may make it more attractive than that for the full negative binomial distribution. The clarity of the message in the negative binomial and logarithmic series plots is likely to be lower when large values of k are frequent (unless N is unusually large), especially if detailed information is limited to $k \leq L_0$.

9E. FREQUENCY-RATIO ALTERNATIVES

The strategy of manipulating the probability function has produced other linear plots against k besides those described in Sections 9A and 9D—for the same discrete distributions. We now review some of these related developments, again starting with the Poisson distribution.

If one wishes to calculate a sequence of Poisson probabilities for a constant value of λ, a convenient relationship is

$$p_\lambda(k) = p_\lambda(k-1)\frac{\lambda}{k}, \qquad k = 1, 2, \ldots .$$

Rewriting this equation as

$$kp_\lambda(k)/p_\lambda(k-1) = \lambda, \qquad k = 1, 2, \ldots, \tag{23}$$

almost invites us to substitute n_k for $p_\lambda(k)$, plot the resulting ratio against k, and look for a horizontal line as the ideal pattern when the frequencies are perfectly Poisson.

More generally we may ask whether the ratio on the left-hand side of equation (23) yields a straight line against k for other important discrete distributions. For the binomial distribution, Dubey (1966) derived

$$\frac{kp(k)}{p(k-1)} = \frac{(n+1)p}{(1-p)} - \frac{p}{(1-p)}k. \tag{24}$$

Ord (1967) showed that a relationship of the form

$$kp(k)/p(k-1) = c_0 + c_1 k \tag{25}$$

holds also for the negative binomial and logarithmic series distributions. We may apply equation (25) to an observed frequency distribution by plotting kn_k/n_{k-1} against k and looking for a linear pattern. It seems appropriate to refer to this and other related plots (mentioned below) as *frequency-ratio plots*. Table 9-16 brings together the four distributions and gives the corresponding values of c_0 and c_1.

TABLE 9-16. Summary of plots based on kn_k / n_{k-1} for four discrete distributions.

Distribution	Probability Function, $p(k)$	Theoretical Slope (c_1)	Theoretical Intercept (c_0)
Poisson	$e^{-\lambda}\lambda^k/k!$	0	λ
Binomial	$\binom{n}{k} p^k(1-p)^{n-k}$	$-p/(1-p)$	$(n+1)p/(1-p)$
Negative binomial	$\binom{n+k-1}{k} p^n(1-p)^k$	$1-p$	$(n-1)(1-p)$
Logarithmic series	$\theta^k/[-k\log_e(1-\theta)]$	θ	$-\theta$

Source: Adapted from Ord (1967), Table 1 on p. 233.

It is especially convenient that plotting kn_k/n_{k-1} against k yields (ideally) a straight line for four distributions that one might often consider in seeking models for discrete data—Poisson, binomial, negative binomial, and logarithmic series. This unifying feature could allow us to use the plot in choosing among alternative models. A negative slope suggests a binomial distribution, although this model would not ordinarily be a competitor to the other three. A horizontal line suggests a Poisson distribution, and a positive slope points to either negative binomial or logarithmic series—we must check the apparent intercept to decide which.

EXAMPLE: WORD FREQUENCY

For the word frequency data given in Table 9-3, Mosteller and Wallace (1964) found that they could obtain a much better fit with a negative binomial distribution than with a Poisson distribution. Table 9-17 shows the values of kn_k/n_{k-1}, and Figure 9-16 has the frequency-ratio plot. Although the plot has only six points (one of which, at $k = 6$, should receive little weight), it does convey a clear impression of an upward trend and a possibly linear pattern, suggesting a negative binomial model rather than a Poisson model. We do not try to fit a line to obtain slope and intercept for this example.

TABLE 9-17. Occurrences of the word *may* in blocks of text in papers already known to be by Madison. Frequency distribution and calculations for frequency-ratio plot.

Number of Occurrences (k)	Number of Blocks (n_k)	kn_k/n_{k-1}
0	156	—
1	63	0.40
2	29	0.92
3	8	0.83
4	4	2.00
5	1	1.25
6	1	6.00
	$N = 262$	

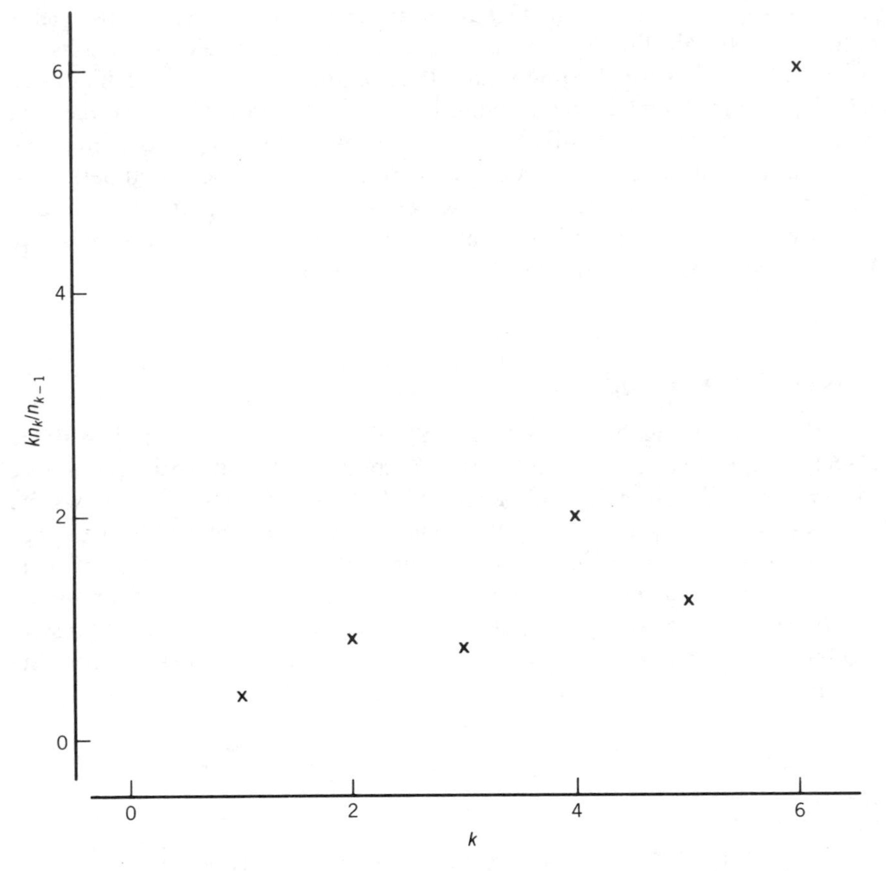

Figure 9-16. Plot based on kn_k/n_{k-1} for occurrences of the word *may* in papers already known to be by Madison.

Variants

Plotting techniques that modify the left-hand side of equation (25) have appeared in the literature. For the logarithmic series distribution Gart (1970) suggests using

$$\frac{kp(k)}{(k-1)p(k-1)} = \theta, \qquad k = 2, 3, \ldots . \tag{26}$$

Easily derived from equation (21), this relationship directly parallels the one in equation (23) for the Poisson distribution. Both lead to a plot against k

that has a horizontal line as its ideal pattern when the assumed distribution is correct.

In the same spirit Gart (1967) for the binomial distribution introduces

$$\frac{kp(k)}{kp(k) + (n - k + 1)p(k - 1)} = p, \qquad k = 1, 2, \ldots, n, \tag{27}$$

which again yields a horizontal line when plotted against k.

Gart (1970) uses the empirical versions of equations (23), (26), and (27), substituting n_k for $p(k)$, in a systematic approach to estimating the corresponding parameters and testing alternative models.

Rao (1971) works with the ratios of consecutive probabilities (or observed frequencies), plotting either $p(k)/p(k-1)$ or $p(k-1)/p(k)$ against $1/k$ or $1/(k-1)$ or $k-1$ (depending on the distribution) to get a linear pattern. For example, for the negative binomial distribution, he plots $p(k)/p(k-1)$ against $1/k$. These plots have no unity that would aid in choosing among possible models. Also their use of $1/k$ or $1/(k-1)$ as the horizontal variable makes them much less satisfactory than the plots based on equation (25).

Drawbacks

Being able to use the same plot, kn_k/n_{k-1} against k, for four common discrete distributions is advantageous, but several practical difficulties limit the effectiveness of this technique.

First, it sacrifices resistance, because one discrepant n_k affects both kn_k/n_{k-1} and $(k+1)n_{k+1}/n_k$. This weakness arises even in such well-behaved examples as the polonium scintillation data (Table 9-10 and Figure 9-8).

A further difficulty arises when $n_{k-1} = 0$ and, to a lesser extent, when $n_k = 0$. We can deal with the first of these (and, if desired, the second) by omitting the point from the plot.

Less blatant but more damaging are the problems of bias and substantially nonconstant variance. Stimulated by the plot in Figure 9-17 for the data of Table 9-12, Jinkinson and Slater (1981) analyzed the expected value and variance of kn_k/n_{k-1} (conditional on $n_{k-1} > 0$). Because n_{k-1} and n_k jointly have a trinomial distribution with parameters N, p_{k-1}, and p_k, their results hold in general; the underlying distribution enters only through p_{k-1} and p_k. (The detailed formulas are somewhat complicated.) To provide rough calibration for Figure 9-17 (for which the frequency distribution is well fitted by a binomial distribution with $n = 10$ and $\hat{p} = \bar{x}/n = .435$), Table 9-18 reproduces a numerical example from Jinkinson and Slater: the

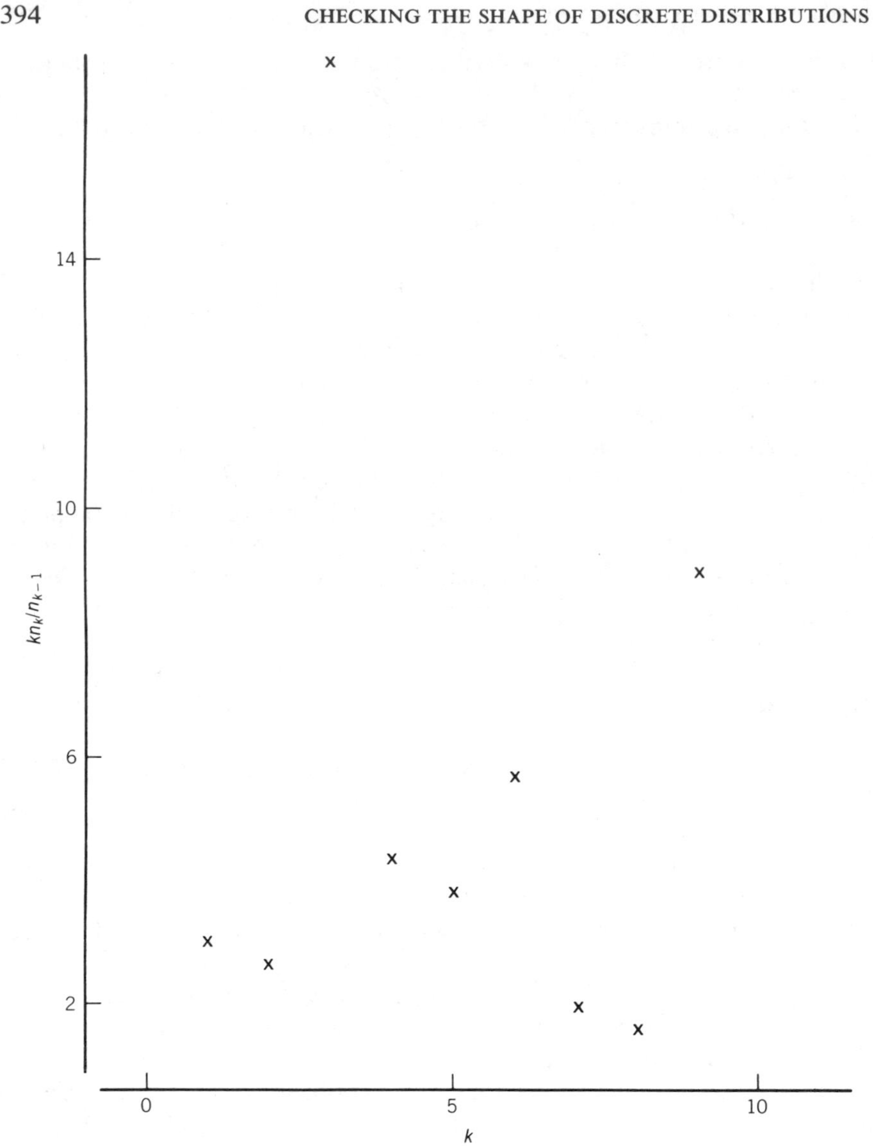

Figure 9-17. Plot of kn_k/n_{k-1} versus k for number of females in queues of length 10 in a London Underground station. *Source*: R. A. Jinkinson and M. Slater (1981). "Critical discussion of a graphical method for identifying discrete distributions," *The Statistician*, **30**, 239–248 (Figure 1 on p. 241). Used with permission.

TABLE 9-18. Conditional expectation and variance of kn_k/n_{k-1} for samples of $N = 100$ observations from the binomial distribution with $n = 10$ and $p = .5$.

k	kp_k/p_{k-1}	Conditional[a] Expectation	Conditional[a] Variance
1	10	0.944	0.935
2	9	6.772	17.445
3	8	10.503	65.149
4	7	7.729	11.767
5	6	6.322	4.131
6	5	5.219	2.694
7	4	4.215	2.668
8	3	3.313	4.370
9	2	2.626	13.458
10	1	0.752	6.652

Source: R. A. Jinkinson and M. Slater (1981). "Critical discussion of a graphical method for identifying discrete distributions," *The Statistician*, **30**, 239–248 (Table 2, p. 247). Reprinted with permission.
[a]Given $n_{k-1} > 0$.

conditional expectation and variance of kn_k/n_{k-1} for frequency distributions with $N = 100$ from the binomial distribution with $n = 10$ and $p = .5$.

The third column of Table 9-18 indicates that both the size and the direction of the bias vary. The greatest discrepancies arise at $k = 1$ through $k = 3$, and the substantial negative biases at $k = 1$ and $k = 2$ are especially troubling. Other numerical examples suggest that the biases tend to be less for $p < .5$ and even worse for $p > .5$.

The variances in the fourth column of Table 9-18 paint an even more discouraging picture, ranging from 0.935 at $k = 1$ to 65.1 at $k = 3$. Together, this violent variation and its confusing relationship to k make it difficult for a user of such a plot to attempt any mental adjustment. Worse yet, other numerical examples (still with $n = 10$) suggest that the pattern of larger and smaller variances depends in an unruly way on the binomial parameter p.

An understanding of these drawbacks of plots based on kn_k/n_{k-1} does much to explain why the plot based on $\phi(n_k^*)$ (Figure 9-10) seems much better behaved than the frequency-ratio plot (Figure 9-17). On the whole, it must leave us rather discouraged about trusting what initially appeared to be an attractive plotting technique. We may, however, be able to try the frequency-ratio plot routinely as a rough guide to the more trustworthy plots developed in earlier sections of this chapter.

9F. COOPERATIVE DIVERSITY

To look closely at data that may be well described by a few parameters, we have two natural routes in asking how well these few parameters are doing:

1. Make a plot that asks how one more parameter helps.
2. Use different parts of the data to estimate the few parameters, and compare estimates.

The basic idea underlying the second route is that of *cooperative diversity*, of asking how well diverse parts of the data cooperate with one another. In the present section we apply this notion to families of discrete distributions to develop two new plots: the apparent-parameter plot and the indicated-parameter-change plot.

Cooperative diversity also provides a convenient framework for thinking about a number of other exploratory techniques including the mid-summaries (Section 10A and UREDA Section 4C), the diagnostic plot for re-expression in a two-way table (UREDA Section 6H), the pushback technique (Section 10E), and the estimation of g in the g-and-h distributions (Sections 11A and 11C).

Apparent-Parameter Plots

To apply the idea of cooperative diversity most directly in a family of discrete distributions (such as the Poisson), we need to estimate the parameter from different parts of the data. Because we have only a single frequency distribution, we can define parts of the data only in terms of k—perhaps as low k, middle k, and high k.

We want local answers—local enough, but not too local (a single n_k seems far too little). For the Poisson, binomial, negative binomial, and logarithmic series families of distributions, whose plots are summarized in Table 9-15, the natural display of the parameter uses local slopes in the plot of the count metameter $\phi(n_k^*)$ versus k. That is, we calculate a typical slope in the vicinity of k and then plot these local slopes against k. By taking one and two steps, both forward and back, from a base point we get four slope estimates at any k not too close to either end of the frequency distribution (0 or L). The median of these four slope estimates serves as the vertical coordinate in the apparent-parameter plot, indicating what the value of the parameter appears to be near k.

More formally, for $2 \le k \le L - 2$ we calculate

$$\begin{aligned} b_1(k) &= \phi(n_{k+1}^*) - \phi(n_k^*), \\ b_{-1}(k) &= \phi(n_k^*) - \phi(n_{k-1}^*), \\ b_2(k) &= \tfrac{1}{2}\left[\phi(n_{k+2}^*) - \phi(n_k^*)\right], \\ b_{-2}(k) &= \tfrac{1}{2}\left[\phi(n_k^*) - \phi(n_{k-2}^*)\right], \end{aligned} \tag{28}$$

and then plot

$$\operatorname{med}\{b_{-2}(k), b_{-1}(k), b_1(k), b_2(k)\}$$

against k. In organizing the calculations we can take advantage of the following simple identities:

$$\begin{aligned} b_{-1}(k) &= b_1(k-1), \\ b_{-2}(k) &= b_2(k-2). \end{aligned} \tag{29}$$

TABLE 9-19. Calculations for the apparent-parameter plot for the polonium scintillation data.

		Local Slopes							
k	$\phi(n_k^*)$	One Step Apart			Two Steps Apart				Median
0	−3.84	1.28			1.30				1.29
1	−2.56	1.28	1.33			1.37			1.33
2	−1.23		1.33	1.41	1.30		1.41		1.37
3	0.19	1.40		1.41		1.37		1.37	1.39
4	1.59	1.40	1.34		1.37		1.41		1.38
5	2.93		1.34	1.39		1.33		1.37	1.36
6	4.32	1.27		1.39	1.37		1.10		1.32
7	5.59	1.27	.94			1.33		1.31	1.30
8	6.53		.94	1.68	1.47		1.10		1.29
9	8.21	1.26		1.68		1.32		1.31	1.31
10	9.47	1.26	1.37		1.47				1.37
11	10.84		1.37			1.32			1.34

A simple systematic departure from the assumed family of distributions should manifest itself as a simple deviation from constancy in the apparent-parameter plot, presumably a regular trend, up or down.

EXAMPLE: POLONIUM SCINTILLATIONS

We illustrate the apparent-parameter plot for the Poisson family, using the polonium scintillation data from Table 9-10 (but without the last three counts, $k = 12$, 13, and 14). In terms of the usual parameter λ, the theoretical slope of the plot of $\phi(n_k^*)$ against k is $\log_e(\lambda)$.

Table 9-19 lays out the detailed calculations, starting from the $\phi(n_k^*)$. The columns of local slopes form two groups, according to whether the slope comes from points one step apart (b_1 and b_{-1}) or two steps apart (b_2 and b_{-2}). Each entry in the column headed "Median" comes from the four local slopes (three or two toward each end) on the same line, whose

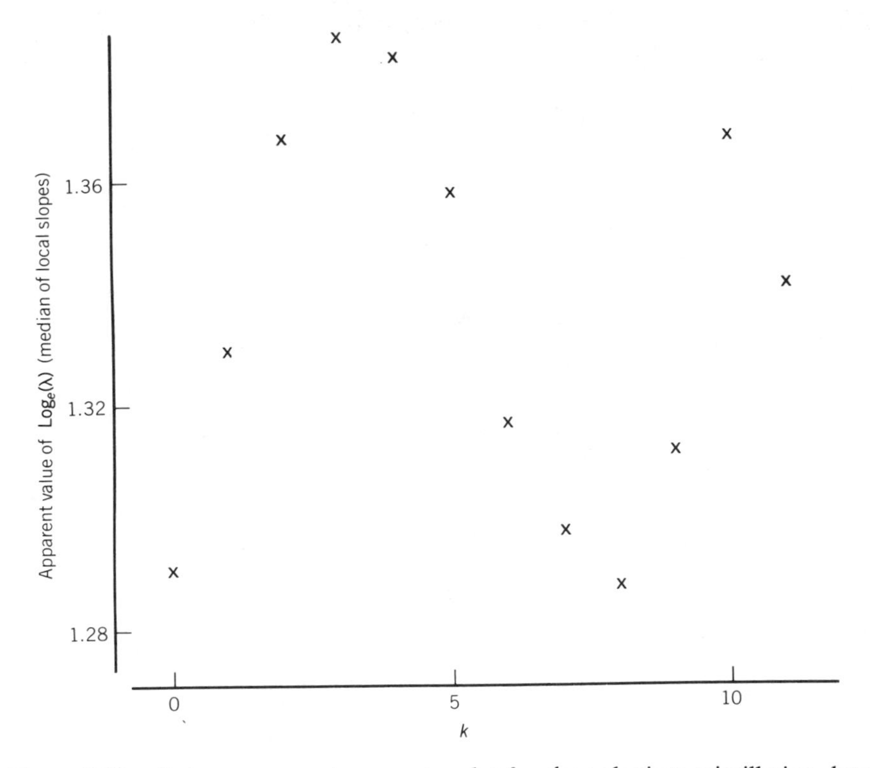

Figure 9-18. Poisson apparent-parameter plot for the polonium scintillation data.

horizontal spacing simply provides room to indicate our use of the identities (29). For example, reading from left to right at $k = 6$, we encounter $b_1(6) = 1.27 = 5.59 - 4.32$, then $b_{-1}(6) = b_1(5) = 1.39 = 4.32 - 2.93$, then $b_{-2}(6) = b_2(4) = 1.37 = \frac{1}{2}(4.32 - 1.59)$, and finally $b_2(6) = 1.10 = \frac{1}{2}(6.53 - 4.32)$. In creating the two groups of columns, we begin at the upper left and continue to shift to the right one column for each step in k until we can return to the leftmost column in the group with a blank line to separate the new entries from the preceding ones. We enter each local slope twice, according to the identities (29).

The apparent-parameter plot in Figure 9-18 shows an oscillation—up, down, up—of quite small magnitude (essentially between 1.29 and 1.38). This seems quite likely to be just noise. If we had other frequency distributions of similar measurements, we could make analogous plots and look for similarities of behavior, something not as easy to do with the other plots that we have considered.

EXAMPLE: WORD FREQUENCY

Because the original Poissonness plot (Figure 9-2) for the frequency distribution of the word *may* in papers by Madison showed definite curvature, we expect systematic behavior in the apparent-parameter plot. Table 9-20 gives the calculations, and Figure 9-19 has the plot. Here we have a steady upward trend, suggesting that one more parameter will help a lot. (In Section 9A we noted that Mosteller and Wallace had found the negative binomial a real improvement.)

TABLE 9-20. Calculations for the apparent-parameter plot for occurrences of the word *may* in papers by Madison.

		Local Slopes							
k	$\phi(n_k^*)$	One Step Apart			Two Steps Apart				Median
0	−0.53	−.91			−.50				−.71
1	−1.44	−.91	−.10			−.17			−.17
2	−1.53		−.10	−.25	−.50		.17		−.17
3	−1.79	.60		−.25		−.17		.00	−.09
4	−1.19	.60	−.59		.60		.17		.38
5	−1.78		−.59	1.79				.00	.00
6	0.01			1.79	.60				1.20

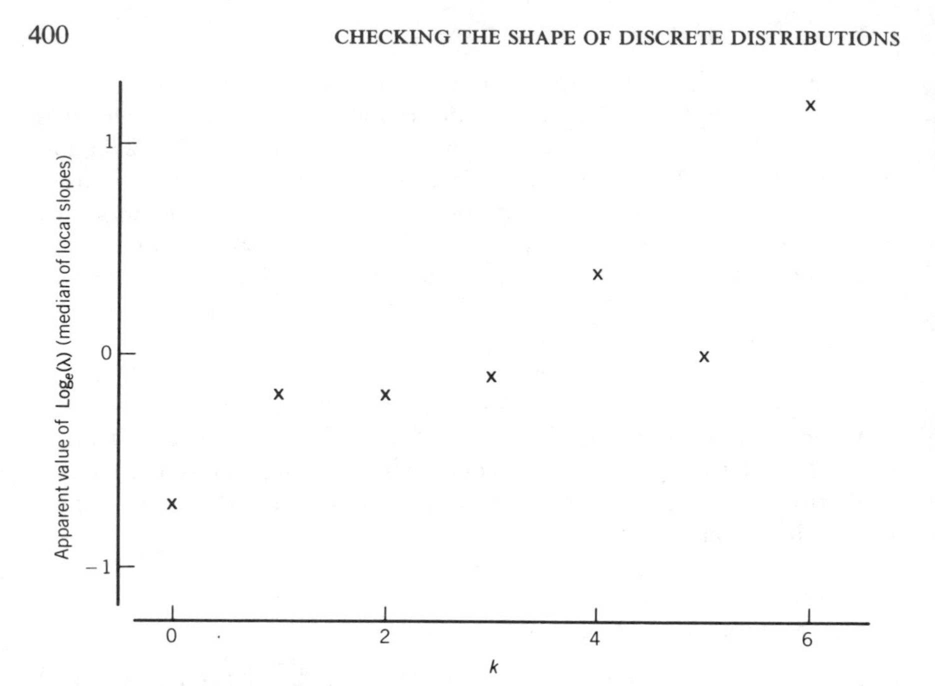

Figure 9-19. Poisson apparent-parameter plot for the word-frequency data.

Indicated-Parameter-Change Plots

Once we have chosen a parameter value (either by eye or by numerical estimation), we can ask how hard, and in which direction, each n_k is "pulling" on that parameter. When we organize this information in a plot, we would also like each point to have the same uncertainty, because this is the situation in which we are all most accustomed to looking at plots. We now show how to design and make such plots. Again we make use of cooperative diversity.

In principle, we might consider judging the pull of each n_k on an estimate of a parameter by differentiating the estimator with respect to n_k (parallel to what one can do in least-squares linear regression). The diversity of estimation procedures (graphical as well as numerical, resistant and nonresistant, some designed to handle truncated frequency distributions, most not expressible in closed form) rapidly makes such a direct approach unattractive.

Instead we pursue a less direct but much more convenient approach based on the count metameter. This choice allows us to use the understanding of variability that we developed in Sections 9B and 9C. In essence, we ask what change in the parameter would be required at each k to make the fitted count metameter $\phi(\hat{\eta}_k)$ (calculated at the chosen value of the parameter) equal the observed count metameter $\phi(n_k^*)$. For simplicity we work with $\log_e(\hat{\eta}_k)$, which differs from $\phi(\hat{\eta}_k)$ only by a constant that (with one exception) does not involve the parameter of the distribution. The Poisson family serves as the leading case.

The Poisson Family

To describe change in $\log_e(n_k^*) - \log_e(\hat{\eta}_k)$ in relation to change in the Poisson parameter λ, we use the derivative, noting that the data fix n_k^*. From

$$\eta_k = Np_k = Ne^{-\lambda}\lambda^k/k!$$

we have

$$\log_e(\eta_k) = k \log_e(\lambda) - \lambda + \log_e(N) - \log_e(k!)$$

so that

$$d \log_e(\eta_k) = \left(\frac{k}{\lambda} - 1\right) d\lambda.$$

Thus, for values of λ near the chosen value λ_0,

$$\log_e(n_k^*) - \log_e(\hat{\eta}_k(\lambda)) \approx \log_e(n_k^*) - \log_e(\hat{\eta}_k(\lambda_0)) + (\lambda - \lambda_0)\left(1 - \frac{k}{\lambda_0}\right) \quad (30)$$

(because it now matters, we explicitly show the dependence of $\hat{\eta}_k$ on λ). Asking for the value of λ that makes $\log_e(\hat{\eta}_k(\lambda))$ equal $\log_e(n_k^*)$ is (approximately) equivalent to writing

$$\log_e(n_k^*) - \log_e(\hat{\eta}_k(\lambda_0)) = (\lambda - \lambda_0)\left(\frac{k}{\lambda_0} - 1\right). \quad (31)$$

As we prepare to plot this information with $\log_e(n_k^*/\hat{\eta}_k)$ on the vertical axis and $(k/\lambda_0) - 1$ on the horizontal axis, we refer to $(k/\lambda_0) - 1$ as the *leverage* of point k. (We apply the term "leverage" by analogy with fitting a straight line through the origin to y versus x by least squares, where the leverage of each point is proportional to its x-value.) For this we write

$$\mathrm{LEV}_k = \frac{k}{\lambda_0} - 1.$$

From equation (31) the slope of the line through the origin yields the value of $\lambda - \lambda_0$ suggested by point k.

We must now deal with the minor complication that the $\log_e(n_k^*)$ for different k generally have substantially different variability. In Sections 9B and 9C we measure this variability in terms of the half-length of the 95% confidence interval (CI) for $\log_e(\eta_k)$. Thus we define

$$\mathrm{VAR}_k = [(\text{half-length of 95\% CI})/1.96]^2.$$

If we replace $\log_e(n_k^*) - \log_e(\hat{\eta}_k)$ by

$$c_k\left[\log_e(n_k^*) - \log_e(\hat{\eta}_k)\right]$$

and also replace LEV_k by

$$c_k \mathrm{LEV}_k,$$

then equation (31) will call for the same value of $\lambda - \lambda_0$ as before. As measured by VAR_k the variability of $c_k[\log_e(n_k^*) - \log_e(\hat{\eta}_k)]$ becomes

$$c_k^2[(\text{half-length of CI})/1.96]^2,$$

and this will be constant, and equal to $1/(1.96)^2$, when

$$c_k = 1/(\text{half-length of CI}).$$

All this leads us to plot the points $(\mathrm{HC}_k, \mathrm{VC}_k)$ with

$$\mathrm{HC}_k = \pm \mathrm{LEV}_k/(\text{half-length of CI})$$

$$\mathrm{VC}_k = \pm \frac{\log_e(n_k^*) - \log_e(\hat{\eta}_k)}{\text{half-length of CI}}.$$

The matching $\pm$ signs yield a further simplification, as follows: Because both (x, y) and $(-x, -y)$ lie on the same line through the origin, we may plot whichever one has a positive horizontal coordinate. Thus we choose the sign in the definition of HC to get $\mathrm{HC} > 0$, and we must take the corresponding sign in VC. Each point in the resulting indicated-parameter-change plot has the same (vertical) variability (in terms of VAR_k), and the direction of each point from (0, 0) indicates what $\lambda - \lambda_0$ that point would like (to the linear approximation).

EXAMPLE: POLONIUM SCINTILLATIONS

The polonium scintillation data from Table 9-10 provide a convenient illustration of the indicated-parameter-change plot. The calculations appear in Table 9-21, using $\lambda_0 = \bar{x} = 3.872$. The plot, Figure 9-20, uses the value of k as the plotting symbol for each point.

We look at this plot in just the same way as we look at any plot to be fitted with a line through the origin—the points have equal (vertical) variability. To aid in judging the slopes of possible lines that the whole set

TABLE 9-21. Calculations for the indicated-parameter-change plot for the polonium scintillation data ($\lambda_0 = \bar{x} = 3.872$).

k	n_k^*	$\hat{\eta}_k$	$\log_e(n_k^*/\hat{\eta}_k)$	$LEV_k = (k/\lambda_0) - 1$	Half-length of CI	HC_k	VC_k
0	56.31	54.31	.036	−1.00	.265	3.77[a]	−.14[a]
1	202.27	210.28	−.039	−.74	.134	5.52[a]	.29[a]
2	382.21	407.06	−.063	−.48	.094	5.16[a]	.67[a]
3	524.17	525.31	−.002	−.23	.077	2.91[a]	.03[a]
4	531.17	508.44	.044	.03	.077	.43	.57
5	407.20	393.69	.034	.29	.090	3.23	.37
6	272.25	254.03	.069	.55	.114	4.82	.61
7	138.29	140.50	−.016	.81	.165	4.89	−.10
8	44.32	67.99	−.428	1.07	.300	3.55	−1.42
9	26.32	29.25	−.106	1.32	.394	3.37	−.27
10	9.33	11.32	−.194	1.58	.671	2.36	−.29
11	3.33	3.99	−.192	1.84	1.120	1.64	−.16
12	—	1.29					
13	0.37	0.38	−.040	2.36	2.696[b]	.87	−.01
14	0.37	0.11	1.245	2.62	2.696[b]	.97	.46

[a] In these instances the signs of both HC and VC have been changed so that HC > 0.

[b] When $n = 1$, the half-length is half the length of the confidence interval from equation (11).

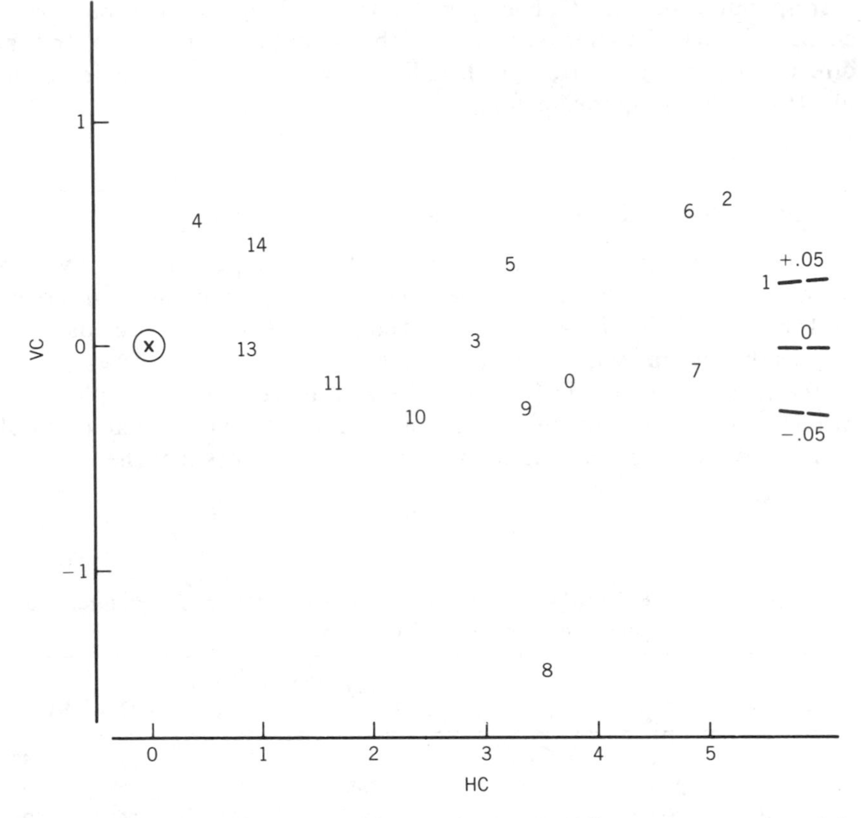

Figure 9-20. Indicated-parameter-change plot for the Poisson parameter λ and the polonium scintillation data ($\lambda_0 = 3.872$). The plotting symbol is the value of k. The scale at the right gives values of $\lambda - \lambda_0$ as slopes of lines through the origin.

of points may suggest, we show the origin prominently and include (at the right-hand edge) a brief scale of slopes.

In decreasing order of emphasis we draw these impressions:

1. Most points call for only a small change in λ away from λ_0 (a horizontal line seems quite adequate).
2. The point for $k = 8$ looks separated from the others.
3. The points for $k = 4$, 13, and 14 have noticeably less chance to affect λ (for $k = 4$ because of minute leverage, for $k = 13$ and 14 because of long confidence intervals).

TABLE 9-22. Expressions for LEV_k, the leverage of point k, in the indicated-parameter-change plots for four families of discrete distributions.

Family	Probability Function	Chosen Parameter	LEV_k
Poisson	$e^{-\lambda}\lambda^k/k!$	λ	$\frac{k}{\lambda_0} - 1$
Binomial	$\binom{n}{k} p^k(1-p)^{n-k}$	p	$\frac{k - np_0}{p_0(1-p_0)}$
Negative binomial	$\binom{n+k-1}{k} p^n(1-p)^k$	p	$\frac{n_0(1-p_0) - kp_0}{p_0(1-p_0)}$
		n	$\frac{1}{n_0+k-1} + \cdots + \frac{1}{n_0} + \log_e(p_0), \quad k \geq 1$
Logarithmic series	$\theta^k/[-k\log_e(1-\theta)]$	θ	$\frac{k}{\theta_0} + \frac{1}{(1-\theta_0)\log_e(1-\theta_0)}$

Other Families

The other families of distributions—binomial, negative binomial, and logarithmic series—yield somewhat different expressions for the derivative of $\log_e(\eta_k)$, and hence for the leverage of point k in their analogs of equation (31). For example, in the binomial family

$$\frac{d}{dp}\log_e(\eta_k) = \frac{d}{dp}\left[k\log_e(p) + (n-k)\log_e(1-p)\right]$$

$$= \frac{k}{p} - \frac{n-k}{1-p}$$

$$= \frac{k-np}{p(1-p)}.$$

Table 9-22 summarizes these results. In each case the indicated-parameter-change plot has horizontal coordinate

$$\mathrm{HC}_k = \pm \mathrm{LEV}_k/(\text{half-length of CI})$$

and vertical coordinate

$$\mathrm{VC}_k = \pm \frac{\log_e(n_k^*) - \log_e(\hat{\eta}_k)}{\text{half-length of CI}}$$

with the same linked choice of signs as discussed in the Poisson case.

We have also devised a plot for the parameter n in the negative binomial family. In Section 9D we had to adopt a value of n in order to construct the count metameter. The indicated-parameter-change plot for n offers a useful way to improve an initial adopted value of n. (Exercise 10 applies this plot to the library circulation data.)

9G. DOUBLE-ROOT RESIDUALS

When a distribution seems an appropriate model for a set of data (as indicated, for example, by the Poissonness plot or one of the other plots summarized in Table 9-15), we often fit it to the observed frequency distribution. From an exploratory point of view we should then calculate some sort of residuals and study them to learn more about quality of fit. In its use of $\log_e(n_k^*) - \log_e(\hat{\eta}_k)$ the indicated-parameter-change plot (Section 9F) organizes a special form of residual to suggest possible shifts in the value of a parameter. We now briefly discuss a more customary definition of residual.

Given the observed frequencies n_k and the fitted frequencies $\hat{\eta}_k$, we can immediately calculate the residuals $n_k - \hat{\eta}_k$ and plot them against k. Because n_k has a binomial distribution with parameters N and p_k and hence variance $Np_k(1 - p_k)$, however, the simple residuals often are not comparable.

One convenient solution transforms n_k and $\hat{\eta}_k$ to remove most of the entanglement between the typical size of n_k and its variability and then forms residuals in the transformed scale. A suitable transformation is the square root, which approximately stabilizes the variance of a Poisson random variable (see UREDA, Section 8G) or of a binomial random variable when n is large and np is small (as may happen in a frequency distribution).

TABLE 9-23. Polonium scintillation data. Double-root residuals for Poisson model with $\hat{\lambda} = 3.872$.

Number of Scintillations (k)	Number of intervals		Double-Root Residual
	Observed (n_k)	Fitted ($\hat{\eta}_k$)	$\left(\sqrt{4n_k + 2} - \sqrt{4\hat{\eta}_k + 1}\right)$
0	57	54.3	0.39
1	203	210.3	−0.49
2	383	407.1	−1.19
3	525	525.3	−0.00
4	532	508.4	1.04
5	408	393.7	0.73
6	273	254.0	1.18
7	139	140.5	−0.11
8	45	68.0	−3.03
9	27	29.2	−0.37
10	10	11.3	−0.32
11	4	4.0	0.13
12	0	1.3	−1.48[a]
13	1	0.4	0.86
14	1	0.1	1.26
	$N = 2608$		

[a] Because $n_{12} = 0$, the double-root residual is $1 - \sqrt{4\hat{\eta}_{12} + 1}$.

Freeman and Tukey (1950) suggested the modified form

$$\sqrt{x} + \sqrt{x+1}\,.$$

For a Poisson random variable X with mean η, Freeman and Tukey (1949) point out (see also Bishop, Fienberg, and Holland, 1975) that the expected value of $\sqrt{X} + \sqrt{X+1}$ is well approximated by $\sqrt{4\eta + 1}$, and its variance is close to 1. Substituting $\hat{\eta}$ for η leads to the residual.

$$\sqrt{x} + \sqrt{x+1} - \sqrt{4\hat{\eta} + 1}\,, \tag{32}$$

whose behavior is approximately that of an observation from the standard Gaussian distribution. These residuals have become known as *Freeman–Tukey deviates*.

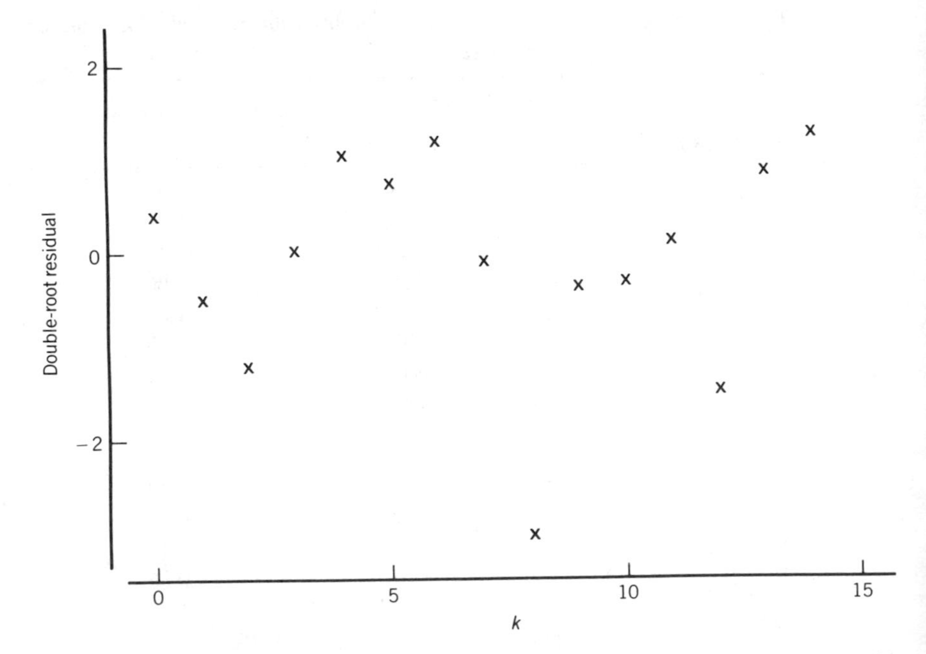

Figure 9-21. Scintillations from radioactive decay of polonium. Double-root residuals for Poisson model with $\hat{\lambda} = \bar{x} = 3.872$.

Instead of expression (32) Tukey (1971) and Mosteller and Tukey (1977, Section 11E) use the *double-root residual*

$$\mathrm{DRR} = \begin{cases} \sqrt{4x+2} - \sqrt{4\hat{\eta}+1}\,, & x \geq 1, \\ 1 - \sqrt{4\hat{\eta}+1}\,, & x = 0, \end{cases} \tag{33}$$

which is also discussed and illustrated by Velleman and Hoaglin (1981, Chapter 9). Double-root residuals and Freeman–Tukey deviates are essentially equivalent, as one can check by evaluating $\sqrt{4x+2}$ and $\sqrt{x} + \sqrt{x+1}$ at $x = 1, 2, \cdots$ (Exercise 11).

EXAMPLE: POLONIUM SCINTILLATIONS

The data on radioactive decay of polonium (Table 9-10) again provide a substantial example. Table 9-23 repeats the frequency distribution and shows the fitted counts and double-root residuals. Plotting the double-root residuals against k in Figure 9-21 shows that the fit is quite good, except for the isolated low count at $k = 8$.

9H. SUMMARY

Simple plots based on the logarithm of the cell counts in an observed frequency distribution can be instructive in checking whether each of several common families of discrete distributions is a satisfactory model for the data. The slope of a straight-line pattern in the plot identifies the main parameter of the distribution. This approach achieves resistance by working with the observed counts separately, and it readily accommodates truncated distributions.

After choosing a value of the parameter(s), one can level the plot by subtracting the corresponding straight line.

Consideration of confidence intervals for the logarithm of the expected cell count η_k leads to improved plots based on the modified cell count n_k^* and the corresponding count metameter $\phi(n_k^*)$.

The half-lengths of the confidence intervals provide a reasonable measure of variability for judging whether individual cell counts are discrepant. Expanding each half-length by the factor 1.5 copes approximately with the problem of multiplicity and forms the basis for confidence-interval plots and confidence-aperture plots. Reporting individual and simultaneous significance levels at both ± 1 and ± 1.5 half-lengths indicates the impact of the actual degree of multiplicity.

An earlier class of plots based on the ratio of adjacent cell counts appears attractive at first because a single plot applies to several common discrete distributions and thus could aid in choosing among them. Difficulties with bias and variance, however, seriously undermine the usefulness of these plots.

The idea of cooperative diversity motivates two further plots for checking agreement between a family of discrete distributions and an observed discrete frequency distribution. The apparent-parameter plot displays local information on the indicated value of the main parameter. Given a chosen value of the parameter, the indicated-parameter-change plot shows approximately how large a shift from that value would be required to fit each point exactly.

For assessing quality of fit cell by cell, double-root residuals are convenient and cope well with small fitted counts.

REFERENCES

Aitchison, J. (1955). "On the distribution of a positive random variable having a discrete probability mass at the origin," *Journal of the American Statistical Association*, **50**, 901–908.

Bartels, J. (1949). "Null kann vier bedeuten," *Annalen der Meteorologie*, **2**, 44–48.

Bishop, Y. M. M., Fienberg, S. E., and Holland, P. W. (1975). *Discrete Multivariate Analysis: Theory and Practice*. Cambridge, MA: MIT Press.

Burrell, Q. L. and Cane, V. R. (1982). "The analysis of library data," *Journal of the Royal Statistical Society, Series A*, **145**, 439–471 (with discussion).

Documenta Geigy (1968). *Wissenschaftliche Tabellen*, 7. Auflage. Basel: J. R. Geigy S. A.

Dubey, S. D. (1966). "Graphical tests for discrete distributions," *The American Statistician*, **20** (3), 23–24.

Fisher, R. A., Corbet, A. S., and Williams, C. B. (1943). "The relation between the number of species and the number of individuals in a random sample of an animal population," *Journal of Animal Ecology*, **12**, 42–58.

Fisher, R. A. and Yates, F. (1963). *Statistical Tables*, 6th ed. Darien, CT: Hafner Publishing Co. (Edinburgh: Oliver & Boyd).

Freeman, M. F. and Tukey, J. W. (1949). "Transformations related to the angular and the square root." Memorandum Report 24, Statistical Research Group, Princeton University.

Freeman, M. F. and Tukey, J. W. (1950). "Transformations related to the angular and the square root," *Annals of Mathematical Statistics*, **21**, 607–611.

Gart, J. J. (1967). "A simple nearly efficient alternative to the simple sib method in the complete ascertainment case," *Annals of Human Genetics*, **31**, 283–291.

Gart, J. J. (1970). "Some simple graphically oriented statistical methods for discrete data." In G. P. Patil (Ed.), *Random Counts in Scientific Work*, Volume 1 (*Random Counts in Models and Structures*). University Park, PA: The Pennsylvania State University Press, pp. 171–191.

Hoaglin, D. C. (1980). "A Poissonness plot," *The American Statistician*, **34**, 146–149.

Jenkins, B. M. and Johnson, J. (1975). "International terrorism: a chronology, 1968–1974."

Report R-1597-DOS/ARPA, The Rand Corporation.

Jinkinson, R. A. and Slater, M. (1981). "Critical discussion of a graphical method for identifying discrete distributions," *The Statistician*, **30**, 239–248.

Johnson, N. L. and Kotz, S. (1969). *Distributions in Statistics: Discrete Distributions*. Boston: Houghton Mifflin.

Kennedy, S. D. (1980). *Final Report of the Housing Allowance Demand Experiment*. Cambridge, MA: Abt Associates Inc.

Kinderman, A. J. (1982). "Examining a Poissonness plot." Department of Management Science, California State University, Northridge.

Meier, P. and Zabell, S. (1980). "Benjamin Peirce and the Howland will," *Journal of the American Statistical Association*, **75**, 497–506.

Mosteller, F. and Tukey, J. W. (1977). *Data Analysis and Regression*. Reading, MA: Addison-Wesley.

Mosteller, F. and Wallace, D. L. (1964). *Inference and Disputed Authorship: The Federalist*. Reading, MA: Addison-Wesley.

Noack, A. (1950). "A class of random variables with discrete distributions," *Annals of Mathematical Statistics*, **21**, 127–132.

Ord, J. K. (1967). "Graphical methods for a class of discrete distributions," *Journal of the Royal Statistical Society, Series A*, **130**, 232–238.

Rao, G. V. (1971). "A test for the fitting of some discrete distributions," *Publications de l'Institut de Statistique de l'Université de Paris*, **20**, 121–128.

Rutherford, E. and Geiger, H. (1910). "The probability variations in the distribution of α particles," *Philosophical Magazine*, Sixth Series, **20**, 698–704.

Tukey, J. W. (1971). *Exploratory Data Analysis*, Limited Preliminary Edition, Volume 3. Reading, MA: Addison-Wesley.

Tukey, J. W. (1972). "Some graphic and semigraphic displays." In T. A. Bancroft (Ed.), *Statistical Papers in Honor of George W. Snedecor*. Ames, IA: Iowa State University Press.

Tukey, J. W. (1977). *Exploratory Data Analysis*. Reading, MA: Addison-Wesley.

Velleman, P. F. and Hoaglin, D. C. (1981). *Applications, Basics, and Computing of Exploratory Data Analysis*. Boston: Duxbury Press.

Williamson, E. and Bretherton, M. H. (1964). "Tables of the logarithmic series distribution," *Annals of Mathematical Statistics*, **35**, 284–297.

Additional Literature

Bickel, P. J. and Hodges, J. L., Jr. (1983). "Shifting integer-valued random variables." In P. J. Bickel, K. A. Doksum, and J. L. Hodges, Jr. (Eds.), *A Festschrift for Erich L. Lehmann*. Belmont, CA: Wadsworth International Group, pp. 49–61.

Grimm, H. (1970). "Graphical methods for the determination of type and parameters of some discrete distributions." In G. P. Patil (Ed.), *Random Counts in Scientific Work*, Volume 1 (*Random Counts in Models and Structures*). University Park, PA: The Pennsylvania State University Press, pp. 193–206.

Holmes, D. I. (1974). "Graphical methods for the analysis of discrete data," *The Statistician*, **23**, 129–134.

Mosteller, F. and Tukey, J. W. (1949). "The uses and usefulness of binomial probability paper," *Journal of the American Statistical Association*, **44**, 174–212.

Mosteller, F. and Wallace, D. L. (1984). *Applied Bayesian and Classical Inference: The Case of The Federalist Papers* (2nd Edition of *Inference and Disputed Authorship: The Federalist*). New York: Springer-Verlag.

EXERCISES

1. The chronology of Jenkins and Johnson (1975) lists 65 incidents of international terrorism in Argentina and 28 in Turkey during the 76-month period from January 1968 through April 1974. From each frequency distribution in Table 9-24 calculate the n_k^* and make the Poissonness plots. Does a negative binomial model seem likely to give a better fit? Choose a value of n for the negative binomial distribution and examine the corresponding plots.

2. A sample of 4021 British households in 1950 yielded the following numbers of households (n_k) with exactly k children (under 14 years of age):

k:	0	1	2	3	4	5	6	7	8	9
n_k:	2303	831	565	212	67	23	15	3	1	1

Use a Poissonness plot to judge the appropriateness of a Poisson model. [Aitchison (1955) fits a no-zeros Poisson model to the data for $k \geq 1$.] About how many households with no children would you have expected?

3. Shifted Poisson distribution. As an alternative model for the household size data in Table 9-4, one could consider a Poisson distribution that

TABLE 9-24. Frequency distributions of monthly totals for incidents of international terrorism in Argentina and Turkey, 1968–1974.

Number of Incidents (k)	Number of months (n_k)	
	Argentina	Turkey
0	46	60
1	15	9
2	5	4
3	3	2
4	5	
5	1	1
6	1	

has been shifted by one unit; that is, $P\{X = k\} = p_\lambda(k - 1)$ for some λ. Calculate the probabilities for such a distribution with $\lambda = 3$, and use them in place of the n_k^* to make a Poissonness plot. How has the shift caused the plot to depart from straightness? What happens in the plot of kn_k/n_{k-1} versus k? (Option: explore further by trying shifts other than one unit and other values of λ.)

4. Calculate the n_k^* and $\phi(n_k^*)$ and make the modified Poissonness plot for the household size data in Table 9-4.

TABLE 9-25. Frequency distribution of the number of coincidences between corresponding downstrokes in comparisons of all possible pairs from 42 uncontested signatures of Sylvia Ann Howland.

Coincidences	Number of Pairs
0	0
1	0
2	15
3	97
4	131
5	147
6	143
7	99
8	88
9	55
10	34
11	17
12	15
13–30	20
	861

Source: Paul Meier and Sandy Zabell (1980). "Benjamin Peirce and the Howland will," *Journal of the American Statistical Association*, **75**, 497–506 (data from Table 1, p. 499).

5. (a) Using $\lambda_0 = e^{-.4}$ as in Figure 9-4 and the $\phi(n_k^*)$ from Table 9-8, level the modified Poissonness plot for the data on incidents of international terrorism in the United States.
   (b) Choose a value of λ_0 and make a leveled version of the modified Poissonness plot for the household size data (see Exercise 4).

6. Benjamin Peirce and the Howland will. Perhaps the earliest use of statistical evidence in an American legal case involved a contested signature, which had allegedly been traced from a genuine signature of Sylvia Ann Howland. To evaluate the likelihood that two of the woman's signatures would coincide in the length and position of all 30 downstrokes (as the contested signature and another, uncontested signature on the will did), the mathematician Benjamin Peirce matched all 861 possible pairs from 42 uncontested signatures and counted the coincidences between their downstrokes. Meier and Zabell (1980) give

TABLE 9-26. Frequency distributions of the number of borrowings of books in two libraries.

Number of Borrowings (k)	Number of Books (n_k)	
	Wishart (4B)	Sussex
1	65	9674
2	26	4351
3	12	2275
4	10	1250
5	5	663
6	3	355
7	1	154
8		72
9		37
10		14
11		6
12		2
13		0
14		1

Source: Quentin L. Burrell and Violet R. Cane (1982). "The analysis of library data," *Journal of the Royal Statistical Society, Series A*, **145**, 439–471 (data from Table 1, p. 442).

the resulting frequency distribution, shown here in Table 9-25. Make a plot to examine the agreement of these data with a binomial distribution ($n = 30$). Estimate the probability (at least on average) that the corresponding downstrokes on two of these signatures coincide.

7. From the other frequency distributions of library circulation data given by Burrell and Cane (1982), Table 9-26 presents two: a three-year period from Section 4B (advanced texts on probability and statistics) of the Wishart Library, University of Cambridge and academic year 1976–1977 from the "long-loan" collection of the Sussex University Library. Using plots for the geometric distribution and negative binomial distribution (as appropriate), compare the behavior of these data sets to that of the data from the Hillman Library (Table 9-13 and Figure 9-12).

8. Williamson and Bretherton (1964) fit a log-series model to the following frequency distribution of the number of steel blooms requested in 173 orders received by a supplier of steel:

k:	1	2	3	4	5	6	7	8	9	10	11+
n_k:	77	35	29	16	3	4	4	1	0	1	3

Calculate the log-series count metameter $\phi(n_k^*)$ and make the plot. What further impressions do you get from a confidence-aperture plot and an apparent-parameter plot? What do you think of $\hat{\theta} = .8$?

9. (a) Starting with $n = 1$ in equation (17) (i.e., the geometric distribution), calculate and plot the negative-binomial count metameter for the word-frequency data (Tables 9-3, 9-9, and 9-20). Comment on the apparent adequacy of this model.

(b) Does a different value of n seem likely to yield a better fit? What do you learn from making an indicated-parameter-change plot for this parameter?

(c) Make a leveled confidence-aperture plot for these data, using your final values of n and p.

10. The indicated-parameter-change plot for the parameter n in the negative-binomial model (Table 9-22) offers one way for the data to suggest an improved choice of n.

(a) Make such a plot for each of the three values of n considered for the circulation data from Hillman Library (Table 9-13).

(b) Taking $n = 1$ (or some other promising value), make the indicated-parameter-change plots for the two sets of library circulation data in Exercise 7.

11. Examine the agreement between double-root residuals [equation (33)] and Freeman–Tukey deviates [equation (32)] by calculating and comparing the values of $\sqrt{4x + 2}$ and $\sqrt{x} + \sqrt{x + 1}$ for $x = 1, 2, \ldots, 10$.

12. Using $p_0 = \bar{x}/n = .435$, calculate fitted counts and double-root residuals for the data on women in queues (Table 9-12). Compare the appearance of the plot of double-root residuals to that of the leveled confidence-aperture plot in Figure 9-11.

13. Investigate the behavior of the apparent-parameter plot for the Poisson distribution when the data actually come from a negative binomial distribution. Choose several values of n and, separately, several values of p. Calculate the negative binomial probabilities p_k for each (n, p) combination, and use these p_k in place of the n_k^* in calculating the Poisson count metameter $\phi(n_k^*)$ and the resulting local slopes. Characterize the appearance of the set of apparent-parameter plots.

14. Make the apparent-parameter plot for the butterfly species data (Table 9-14). Does it reveal any simple systematic deviation from the log-series pattern?

CHAPTER 10

Using Quantiles to Study Shape

David C. Hoaglin
Harvard University

In summarizing a batch of data, we almost always begin by describing location and then spread. When we have an adequate amount of data, we can profitably ask higher-order questions as well. Although it is common practice to use the third and fourth moments of the sample to characterize skewness and kurtosis, respectively, we prefer techniques based on order statistics. A moderate-sized batch, say $n = 50$ or so, offers enough information to provide a useful description of the shape of the data in terms of selected order statistics. This chapter presents techniques for using order statistics or quantiles to examine two of the simplest aspects of distribution shape: skewness and elongation.

Vague Concepts

Like location and scale (or spread), the concepts of distribution shape, skewness, and elongation are vague concepts: one begins with a general notion and can then make it precise in a variety of ways. For example, the location of a single-humped distribution or of a set of data is where its values are concentrated. The most common ways of measuring this position quantitatively are the mean and the median, although neither of these (nor any other specific measure) is necessarily *the* correct way to define location, and it is appropriate to ask what measure ought to be used under which circumstances. Mosteller and Tukey (1977, Chapter 1) discuss the role of vague concepts in data analysis and give several examples. We do not try to give a precise definition for a vague concept, but we do discuss what the concept involves.

Families of Distributions

Many theoretical distributions come in families and differ among themselves within a family only in terms of location and scale parameters. One member of the family serves as the "standard" distribution (the basis for this choice is often simplicity of notation or some historical accident), and all other members are written in terms of the standard one. Thus, if Z has the standard distribution, any other distribution in the family is the distribution of an

$$X = \mu + \sigma Z, \tag{1}$$

where μ is a location parameter and σ is a scale parameter. We think of the whole family as having—and, by extension, being—a single shape. For example, the Gaussian distributions are a shape, the standard Gaussian distribution has mean 0 and variance 1, and in equation (1) μ is the mean of X and σ is the standard deviation. Of course, μ and σ need not be the mean and standard deviation, respectively. If Z has a uniform distribution on the interval $[0, 1]$, then X has mean $\mu + \frac{1}{2}\sigma$ and standard deviation $\sigma/\sqrt{12}$.

Skewness and Elongation

Our interest in studying the distribution shapes associated with data springs, in part, from the substantial (and still growing) body of evidence that actual data rarely exhibit the ideal behavior embodied in the Gaussian distribution. We concentrate on two of the most easily visualized characteristics of non-Gaussian distributions and of data: *skewness* (departure from symmetry) and *elongation* (heavier tails). From displays that focus on these aspects we hope to learn more about the behavior of the underlying phenomenon which has produced the data.

It is easy to work with symmetric distributions because the center of symmetry serves as a convenient and readily agreed-upon definition of the distribution's location. Skewness is a departure from symmetry. Because many techniques for summarizing location work best, and are best understood, when the data come from a symmetric distribution (see Section 9B of UREDA), it can be important to check whether this assumption seems valid. Thus more attention focuses on detecting the presence of skewness than on how the departures from symmetry should be measured. This chapter emphasizes resistant techniques that leave room for some judgment. In some situations a simple re-expression can render the data much more nearly symmetric; re-expressions are the subject of Chapters 4 and 8 in UREDA. Whether or not we transform to reduce skewness, an appreciation of the skewness in the data may lead to ideas about the underlying reasons for it.

In talking about elongation, we are concerned with the stretch of the tails of a distribution. A more elongated distribution gives greater probability to outcomes that are quite notably more extreme. This aspect of distribution shape has no such natural standard as symmetry is for skewness. We must adopt a standard of reference, a distribution whose tails are neither heavy nor light, a distribution which is "neutrally elongated." As is often done, we choose the Gaussian distribution as the standard. In this chapter, we qualitatively compare the tail behavior of data and distributions with the tail behavior of the Gaussian distribution, and we deal with elongation mainly in symmetric situations. Chapter 11 describes a way of quantifying elongation, one that can be used in both symmetric and skewed situations; it also discusses the issue of removing skewness before measuring elongation.

As with skewness, an appreciation of the elongation in the data may point toward underlying explanations. Even without a tidy explanation, the information can be valuable. Occasional large outliers may require a different response from smoothly and systematically heavier tails.

Section 10A uses quantiles to diagnose skewness numerically, and Section 10B takes a similar approach to elongation. Section 10C reviews the quantile–quantile plot, a standard way to examine and compare shapes of distributions. Section 10D describes other plots for examining skewness and elongation. Section 10E discusses one way of studying the shape of the two tails separately.

10A. DIAGNOSING SKEWNESS

To probe the skewness of a distribution or a set of data, we primarily compare the upper half (above the median) and the lower half (below the median). We may use selected quantiles, selected order statistics, or all the ordered data.

Letter Values

When we can make our own choice of selected quantiles or order statistics, we usually work with the *letter values* (Tukey, 1977, Chapter 2; UREDA, Chapter 2). For a theoretical distribution the letter values begin with the median (tail area $\frac{1}{2}$) and the lower and upper quartiles or fourths (tail area $\frac{1}{4}$) and continue outward into the tails by successively halving the tail area to produce the eighths, sixteenths, and so on. Thus these selected quantiles come more from the tails than from the middle of the distribution.

In a sample of size n, the letter values stress convenience by extracting observations at certain simple depths in the ordered sample,

$$x_{(1)} \leq x_{(2)} \leq \cdots \leq x_{(n-1)} \leq x_{(n)}.$$

The *depth* of an observation is its position in the ordered sample relative to the nearer end. More technically, the depth of $x_{(i)}$ is the smaller of i and $n + 1 - i$. Thus, both $x_{(2)}$ and $x_{(n-1)}$ have depth 2.

In a fashion similar to that for distributions, the sample letter values begin with the median, at depth $(n + 1)/2$, and the extremes, at depth 1. They then work outward from the median toward the extremes, according to the sequence of depths defined by

$$\frac{[\text{previous depth}] + 1}{2},$$

where the square brackets in $[x]$ stand for the operation of finding the largest integer not exceeding x. If a depth calculated in this sequence is not a whole number, its fractional part must be $\frac{1}{2}$. We then obtain the corresponding letter value by averaging two adjacent order statistics—already a familiar operation in calculating the sample median when n is even.

We note that, except for the median, letter values come in pairs: a lower one and an upper one. If their depth is d, these are $x_{(d)}$ and $x_{(n+1-d)}$, respectively.

To refer to the letter values, we customarily use a scheme of one-letter tags: M for the median, F for the fourths, E for the eighths, and so on in reverse order through the alphabet (D, C, B, A, Z, Y, X, ...).

EXAMPLE: PRESURGICAL STRESS

In the mid 1970s researchers isolated beta-endorphin, a morphinelike chemical found in the human brain, and discovered that people have higher levels of it in their blood under conditions of emotional stress. To investigate this relationship in the absence of actual pain and drugs, Miralles et al. (1983) studied 19 patients who were scheduled to undergo elective surgical procedures. From blood samples collected 12–14 hours and again at 10 min before surgery, they determined the endorphin level in fmol/ml [a femtomole (fmol) is 10^{-15} grams times the molecular weight of the substance]. Under the presumed stress of impending surgery, most patients had a higher endorphin level. We consider the values 10 min before surgery: 6.5, 14.0, 13.5, 18.0, 14.5, 9.0, 18.0, 42.0, 7.5, 6.0, 25.0, 12.0, 52.0, 20.0, 16.0, 15.0, 11.5, 2.5, 2.0. (Exercise 1 gives the values 12–14 hours before surgery.)

The ordered observations are

i:	1	2	3	4	5	6	7	8	9	10
$x_{(i)}$:	2.0	2.5	6.0	6.5	7.5	9.0	11.5	12.0	13.5	14.0
i:	11	12	13	14	15	16	17	18	19	
$x_{(i)}$:	14.5	15.0	16.0	18.0	18.0	20.0	25.0	42.0	52.0	

and the depths and letter values are as follows:

Tag	Depth	Letter Values
M	$10 = (19 + 1)/2$	14.0
F	$5.5 = (10 + 1)/2$	8.25 and 18.0
E	$3 = (5 + 1)/2$	6.0 and 25.0
D	$2 = (3 + 1)/2$	2.5 and 42.0
	1	2.0 and 52.0

Often we arrange this information in a standard schematic format, the *letter-value display*, whose simplest form is

$n = 19$

M	10	14.0	
F	5.5	8.25	18.0
E	3	6.0	25.0
D	2	2.5	42.0
	1	2.0	52.0

Midsummaries

If a batch of data is symmetric, its median must be the point of symmetry, and each pair of letter values must be symmetrically placed about the median. That is, the lower fourth will be as far below the median as the upper fourth is above the median, and the same will be true for the eighths, the sixteenths, and so on. A simple way to check on symmetry is to define a set of midsummaries ("mids" for short), one for each pair of letter values (or other symmetric quantiles).

DEFINITION: For each pair of letter values, the corresponding *midsummary* is the average of the two letter values. The abbreviation is "mid" followed by the letter used to tag the letter value. Thus for the lower and upper fourths, F_L and F_U, the midsummary is midF, calculated as

$$\text{midF} = \tfrac{1}{2}(F_L + F_U). \tag{2}$$

Another name, midrange, has commonly been used for the midextreme.

Once we have calculated midsummaries for all pairs of letter values, we can readily examine them for evidence of systematic skewness. (If apparent skewness were due to one or two stray values, only the most extreme letter values and their midsummaries would be affected. Thus using the full set of midsummaries provides more resistance.) In a perfectly symmetric batch, all midsummaries would be equal to the median. If the data were skewed to the right,

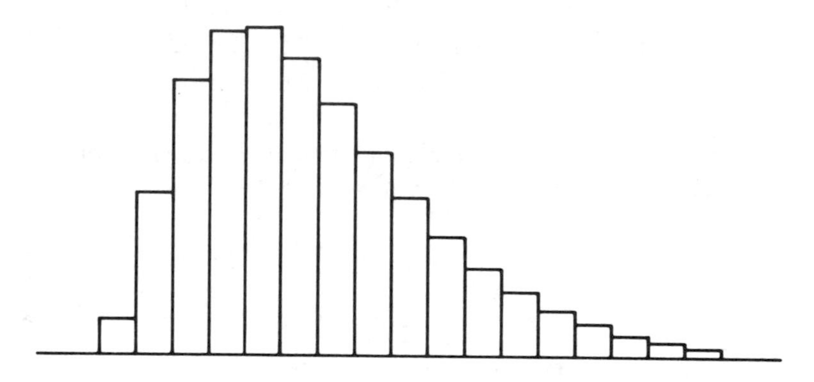

the midsummaries would increase as they came from letter values further into the tails. For data skewed to the left,

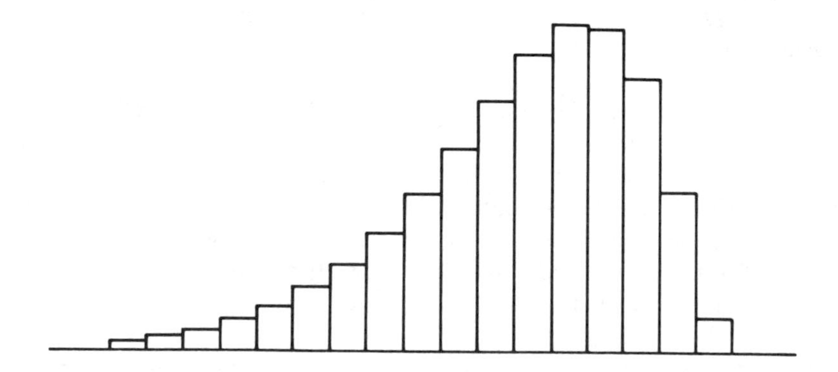

they would decrease.

TABLE 10-1. Adding midsummaries to the letter-value display for the beta-endorphin data.

$n = 19$			mid	
M	10		14.0	
F	5.5	8.25	13.125	18.0
E	3	6.0	15.5	25.0
D	2	2.5	22.25	42.0
	1	2.0	27.0	52.0

EXAMPLE:

Table 10-1 repeats the letter-value display for the beta-endorphin data. Below the median, we have added a column of midsummaries, which we calculate as follows:

$$\text{midF} = \tfrac{1}{2}(8.25 + 18.0) = 13.125,$$

$$\text{midE} = \tfrac{1}{2}(6.0 + 25.0) = 15.5,$$

$$\text{midD} = \tfrac{1}{2}(2.5 + 42.0) = 22.25,$$

$$\text{midextreme} = \tfrac{1}{2}(2.0 + 52.0) = 27.0.$$

The midsummaries suggest that, except for the high values at 42.0 and 52.0 or, alternatively, except for the impossibility of negative values, these data are reasonably symmetric. Thus, under such conditions of presurgical stress, a patient's beta-endorphin level is roughly as likely to lie a given distance below 14 fmol/ml as it is to exceed 14 by that amount.

EXAMPLE: HOUSEHOLD INCOME

The Housing Allowance Demand Experiment (Kennedy, 1980), part of the Experimental Housing Allowance Programs sponsored by the United States Department of Housing and Urban Development, studied the effects of providing direct financial assistance (in the form of housing allowances) to families in order to enable them to live in decent housing. This multiyear experiment involved four treatment groups and a control group in each of two sites: Pittsburgh, Pennsylvania and Phoenix, Arizona.

When they enrolled in the experiment, households reported their annual income. Table 10-2 gives the letter-value display (with midsummaries) for the annual incomes of 994 households in Pittsburgh. The calculations for the midsummaries go as before:

$$\text{midF} = \tfrac{1}{2}(2412 + 4944) = 3678,$$

and so on. The values of the mids increase steadily as we work down the column, and all the mids from midB on are outside the fourths. This is solid evidence of very substantial skewness. Only when we move from midY to midextreme do we fail to see an increase, and such a fluctuation at the very tail is not surprising in a batch of any size.

TABLE 10-2. Letter-value display with midsummaries for reported incomes of Housing Allowance Demand Experiment households in Pittsburgh.

$n = 994$		household income (dollars)		
			mid	
M	497.5		3480	
F	249	2412	3678	4944
E	125	1788	4115.5	6443
D	63	1517	4400.5	7284
C	32	1248	4799	8350
B	16.5	963.5	4978.75	8994
A	8.5	727.5	5241	9754.5
Z	4.5	579	5394.5	10210
Y	2.5	345	5510.25	10675.5
	1	114	5494	10874

Source: John D. Emerson and Michael A. Stoto (1983). "Transforming data." In David C. Hoaglin, Frederick Mosteller, and John W. Tukey (Eds.), *Understanding Robust and Exploratory Data Analysis*. New York: Wiley (Table 4-1, p. 106). Copyright © 1983 John Wiley & Sons, Inc. Reprinted by permission of John Wiley & Sons, Inc.

Economists often prefer to work with the logarithm of income rather than income itself. Re-expressing these data as logarithms leads to the following sequence of midsummaries, from median to midextreme: 3.54, 3.535, 3.53, 3.52, 3.51, 3.465, 3.425, 3.385, 3.265, 3.05. There is now a weaker tendency toward skewness in the opposite direction, to the left—a weak tendency we overlook for the present because we have seen how much more skewed the raw data are. The midextreme does not fit in because the smallest household income, \$114, is so low. We examine these income data in the log scale further in Sections 10B and 10C.

A regular or moderately regular pattern of overall increase or decrease from one midsummary to the next, as the two examples have shown, reveals the presence and direction of systematic skewness, and also gives a rough impression of the extent of skewness. To get a more refined idea of systematic skewness, we will gain from having an appropriate plot. We return to this in Section 10D.

10B. DIAGNOSING ELONGATION

Just as we judge skewness in terms of departure from symmetry, we must measure elongation against a standard that serves as the definition of neutral elongation. For this standard we use the Gaussian distribution, and in this section we deal only with elongation in symmetric distributions. Chapter 11 discusses a way of describing elongation in the presence of skewness.

Elongation is an aspect of distribution shape; it does not involve either location or spread. Our basic approach for studying elongation uses a sequence of measures of spread to indicate how the tails of the data behave. When compared to the Gaussian standard, these measures of spread together tell us about elongation.

Letter-Spreads

To describe the shape of the tails, we must take information from several points in each tail. The letter values, discussed in Section 10A, do just this. By their definition in terms of repeatedly halving the tail area, they pay more attention to the tails than to the middle. To carry this emphasis on the tails over into our measures of spread, we define a letter-spread for each pair of letter values.

DEFINITION: For each pair of letter values, the corresponding *letter-spread* is the difference between them, upper minus lower. The abbreviation is "-spread" preceded by the letter used to tag the letter value. Thus, for example, for the eighths, E_L and E_U, we have

$$\text{E-spread} = E_U - E_L.$$

The difference between the extremes is usually called the *range*.

The structure of the letter-value display, which places paired letter values on the same line, makes it easy to calculate these letter-spreads.

Pseudosigmas

The main step in comparing the tail behavior of a batch or a distribution against the behavior of a Gaussian distribution is to divide each of its letter-spreads by the corresponding Gaussian value. Put another way, we ask, for each pair of letter values, what standard deviation a Gaussian distribution must have if its spacing between those letter values is to equal the given spacing. In doing this, we can work with values from the standard Gaussian distribution because any other Gaussian distribution is derived from it by simple changes of location and scale. That is, if Y is a Gaussian random variable with mean μ and standard deviation σ and Z is a standard Gaussian random variable, then

$$Y = \mu + \sigma Z. \tag{3}$$

Thus, if $2z_\alpha$ is the standard Gaussian letter-spread for the pair of letter values at tail area α, then the corresponding letter-spread for the distribution of Y is $2\sigma z_\alpha$. Matching to a batch of data or to a non-Gaussian distribution is just a matter of equating its letter-spread and the corresponding Gaussian value $2\sigma z_\alpha$ and then solving for σ. For convenience we give the result a name.

DEFINITION: For each pair of letter values, the corresponding *pseudosigma* is the ratio of the letter-spread to its standard Gaussian value. To distinguish among pseudosigmas, we may use either the tag for the letter value or the percentage value of the tail area. Thus, for example, the eighths give rise to the "E-pseudosigma" or the "12.5% pseudosigma."

The tail areas, letter values, and letter-spreads for the standard Gaussian distribution are tabulated in Table 10-3.

TABLE 10-3. Standard Gaussian (upper) letter values and letter-spreads. (Each lower letter value is the negative of the upper letter value.)

Tail Area	Tag	(Upper) Letter Value	Letter-Spread
1/4	F	0.6745	1.349
1/8	E	1.1503	2.301
1/16	D	1.5341	3.068
1/32	C	1.8627	3.725
1/64	B	2.1539	4.308
1/128	A	2.4176	4.835
1/256	Z	2.6601	5.320
1/512	Y	2.8856	5.771
1/1024	X	3.0973	6.195
1/2048	W	3.2972	6.594
1/4096	V	3.4871	6.974
1/8192	U	3.6683	7.337
1/16384	T	3.8419	7.684
1/32768	S	4.0087	8.017

To judge elongation, we examine the sequence of pseudosigmas, working from the fourths out to the extremes. If the data are Gaussian (and hence, by our standard, neutrally elongated), the pseudosigmas will be nearly constant. When the pseudosigmas *increase* systematically, we say that the data or distribution from which they come is *more elongated* (has heavier tails); when they *decrease* systematically, it is *less elongated* (has lighter tails). Looking at the sequence of pseudosigmas in this way offers an advantage over using conventional measures such as kurtosis because it permits us to exercise judgment when one or two wild values break up the pattern.

The precision of the approach just discussed will ordinarily be quite adequate. If we were being extremely careful, we could take into account the slight irregularities in the behavior of letter values as the sample size n changes. Tukey (1977, pp. 632–633) gives a table of divisors that can serve as correction factors when $n \leq 100$, and we turn more systematically to the question later in this section.

To illustrate the two general patterns of elongation, we examine two theoretical distributions. We then work through a third example based on the household income data (Table 10-2).

EXAMPLE: DISTRIBUTION INDUCED BY THE FOLDED SQUARE ROOT TRANSFORMATION

It is sometimes useful to re-express percentages or counted fractions according to the folded square root transformation, which replaces the fraction u ($0 \le u \le 1$) by

$$\sqrt{2u} - \sqrt{2(1-u)} \tag{4}$$

(Tukey, 1977, Section 15B). If U is a random variable distributed uniformly on the interval $[0, 1]$, then transforming it according to expression (4) yields a new random variable whose values are symmetric about 0 and range from $-\sqrt{2}$ to $+\sqrt{2}$. The letter values of this new distribution can be calculated directly using expression (4) because the letter values at tail area α for U are just α and $1-\alpha$. Table 10-4 shows the (upper) letter values and letter-spreads (as far as W) for this distribution, as well as the resulting pseudosigmas. For example, the F-pseudosigma is $1.035/1.349 = 0.767$. As we look down the column of pseudosigmas, we see that they decrease quite steadily. This means that the letter values further into the tails of the "folded root" distribution match those of Gaussian distributions with progressively smaller standard deviation—the tails are lighter than those of

TABLE 10-4. Upper letter values, letter-spreads, and pseudosigmas for the folded square root distribution, showing that its tails are less elongated than those of the Gaussian.

		Folded Root		Gaussian	
Tag	Fraction (u)	Letter Value	Letter-Spread	Letter-Spread	Pseudosigma
F	.75	0.518	1.035	1.349	.767
E	.875	0.823	1.646	2.301	.715
D	.9375	1.016	2.032	3.068	.662
C	.96875	1.142	2.284	3.725	.613
B	.984375	1.226	2.453	4.308	.569
A	.9921875	1.284	2.567	4.835	.531
Z	.99609375	1.323	2.646	5.320	.497
Y	.998046875	1.350	2.701	5.771	.468
X	.9990234375	1.369	2.739	6.195	.442
W	.99951171875	1.383	2.765	6.594	.419

the Gaussian distribution. In the next section we make a plot for this distribution.

EXAMPLE: CAUCHY DISTRIBUTION

The Cauchy distribution often serves as an example of a violently heavy-tailed distribution. (Its tails are just heavy enough that its mean does not exist because the integral defining it does not have a finite value.) The "standard" Cauchy distribution has probability density function

$$f(x) = 1/\left[\pi(1 + x^2)\right], \qquad -\infty < x < +\infty, \tag{5}$$

and cumulative distribution function

$$F(x) = \frac{1}{2} + \frac{1}{\pi}\arctan x. \tag{6}$$

The inverse of $F(x)$ is

$$F^{-1}(u) = \tan \pi\left(u - \tfrac{1}{2}\right), \qquad 0 < u < 1, \tag{7}$$

so that it is straightforward to obtain quantiles from the corresponding fractions. The letter values, letter-spreads, and pseudosigmas appear in Table 10-5. As an illustrative calculation, the upper fourth is $\tan \pi(.75 - \frac{1}{2})$

TABLE 10-5. Letter values, letter-spreads, and pseudosigmas for the Cauchy distribution, showing its heavy-tailed behavior.

	Cauchy		Gaussian	
Tag	Letter Value	Letter-Spread	Letter-Spread	Pseudosigma
F	1.000	2.000	1.349	1.483
E	2.414	4.828	2.301	2.098
D	5.027	10.055	3.068	3.277
C	10.153	20.306	3.725	5.451
B	20.355	40.711	4.308	9.450
A	40.735	81.471	4.835	16.850
Z	81.483	162.966	5.320	30.633
Y	162.973	325.945	5.771	56.480
X	325.948	651.897	6.195	105.230

TABLE 10-6. Letter-value display with letter-spreads and pseudosigmas, showing nearly neutral elongation of the household income data in the logarithmic scale.

n = 994		mid		spread	pseudosigma
M 497.5		3.54			
F 249	3.38	3.535	3.69	.31	.23
E 125	3.25	3.53	3.81	.56	.24
D 63	3.18	3.52	3.86	.68	.22
C 32	3.10	3.51	3.92	.82	.22
B 16.5	2.98	3.465	3.95	.97	.23
A 8.5	2.86	3.425	3.99	1.13	.23
Z 4.5	2.76	3.385	4.01	1.25	.23
Y 2.5	2.50	3.265	4.03	1.53	.27
1	2.06	3.05	4.04	1.98	.31[a]

[a] The Gaussian value used for the range comes from equation (9).

= 1.000, and the F-pseudosigma is 2.000/1.349 = 1.483. The rapid and continuing increase in the pseudosigmas indicates that the Cauchy distribution is quite heavy-tailed.

EXAMPLE: HOUSEHOLD INCOMES (LOG SCALE)

When the household incomes summarized in Table 10-2 are transformed by taking logarithms, we have the letter-value display in Table 10-6. The pseudosigmas are very nearly constant; so, on the basis of this information, the transformed data appear neutrally elongated. Only the lower extreme and Y_L break up the pattern, and except for these a Gaussian distribution with standard deviation equal to .23 should fit reasonably well. Before we conclude that a lognormal distribution is an appropriate description for the original income data, however, the midsummaries remind us that we must give closer attention to the skewness that we discovered in discussing Table 10-2. We do this in the next section.

Gaussian Letter Values in Small Samples

In these three examples the calculations of the pseudosigmas have used values for the standard Gaussian population, taken from Table 10-3. This is correct when we are finding the pseudosigmas of a continuous distribution, and it provides a good approximation in large samples, but for smaller samples (say $n \le 100$) the population values are not such good typical values for the sample letter values, especially as the letter values get closer to the ends of the sample. (One example of this awkwardness appears in Table 10-6, where the population value for the range could not be used because the Gaussian distribution ranges from $-\infty$ to $+\infty$.) For small sample sizes, n, we prefer to use two times the median of the distribution of the upper letter value, assuming samples of n from the standard Gaussian distribution. In UREDA and in other chapters of this volume, we accumulated considerable experience with the median as a typical value, and the method for finding standard values is both convenient and applicable in general.

If the random variables $Z_1, \ldots, Z_n$ represent a sample of n from the standard Gaussian distribution, then we seek the median of the distribution of the ith order statistic, $Z_{(i)}$. We recall that the order statistics are in increasing order:

$$Z_{(1)} \le \cdots \le Z_{(n)}.$$

A calculation based on the inverse of the cumulative distribution function, Φ^{-1}, provides a good approximation:

$$\text{med}\{Z_{(i)}\} \approx \Phi^{-1}\left[\left(i - \tfrac{1}{3}\right)/\left(n + \tfrac{1}{3}\right)\right]. \tag{8}$$

In the present discussion, we need to use only those values of i corresponding to the letter-value depths, but opportunities to use all the order statistics arise in the next section. When d, the depth of a letter value, involves the fraction $\frac{1}{2}$, the approximation of equation (8) should still be used—we simply substitute d for i. The basis for this lies in the ideal depth that could be used for the letter values and is discussed in UREDA (Section 2E).

To summarize the discussion, we observe that in equation (8) values of i smaller than $n/2$ must yield values for med$\{Z_{(i)}\}$ to the left of zero. The symmetry of the Gaussian distribution allows us to change the sign and multiply by two to get a value for the spread. Thus

$$\text{Gaussian spread at depth } d \approx -2\Phi^{-1}\left[\left(d - \tfrac{1}{3}\right)/\left(n + \tfrac{1}{3}\right)\right]. \tag{9}$$

As an application of equation (9) we calculate the Gaussian value used for the range in Table 10-6. In this case $n = 994$ and $d = 1$, and the result is $-2\Phi^{-1}[\frac{2}{3}/(994 + \frac{1}{3})] = -2(-3.207) = 6.414$. The pseudosigma is then $1.98/6.414 = 0.309$.

Section 10G discusses numerical approximations to Φ^{-1} and gives a modification of equation (8) which is more accurate when $i = 1$.

We turn now to ways of displaying the shape of a distribution or sample graphically.

10C. QUANTILE–QUANTILE PLOTS

An important graphical technique for comparing shapes of distributions is the quantile–quantile plot (or Q-Q plot), a type of probability plot (Wilk and Gnanadesikan, 1968). In this section we review Q-Q plots and present a simplified version based on letter values.

In order to make an overall graphical comparison between two distributions or between two samples or between a sample and a distribution, we may plot quantiles of one against corresponding quantiles of the other. In the case of two distributions, with cumulative distribution functions F_1 and F_2, we vary the fraction p over a set of selected values, use the inverse functions F_1^{-1} and F_2^{-1} to obtain $F_1^{-1}(p)$ and $F_2^{-1}(p)$, and plot the points $(F_1^{-1}(p), F_2^{-1}(p))$. When the two distributions are exactly the same (that is, when $F_1 = F_2$), these points will lie on a straight line which has slope 1 and passes through the origin.

More generally, if F_1 and F_2 have the same shape, they may still differ in location and scale. Many families of distributions have this structure—like the Gaussian distributions, they are location-and-scale families. This means that, for some constant, μ, and some positive constant, σ,

$$F_2(x) = F_1((x - \mu)/\sigma) \tag{10}$$

for all x. Specifically, if F_1 has location μ_1 and scale σ_1 and F_2 has location μ_2 and scale σ_2, then standardizing corresponding random variables, X_1 and X_2, allows us to equate them:

$$\frac{X_1 - \mu_1}{\sigma_1} = \frac{X_2 - \mu_2}{\sigma_2}$$

or

$$X_1 = \mu_1 - \frac{\sigma_1}{\sigma_2}\mu_2 + \frac{\sigma_1}{\sigma_2}X_2.$$

Thus, writing $X_1 = (X_2 - \mu)/\sigma$ and equating coefficients yield

$$\mu = \mu_2 - \frac{\sigma_2}{\sigma_1}\mu_1$$

and

$$\sigma = \sigma_2/\sigma_1.$$

In terms of the quantiles, equation (10) becomes

$$F_2^{-1}(p) = \mu + \sigma F_1^{-1}(p) \qquad (11)$$

for $0 \le p \le 1$, so that the quantile–quantile plot takes the form of a straight line (with slope σ and intercept μ) when F_2 has the same shape as F_1.

Of course, when F_2 has a different shape from F_1, the plot will curve (in any one of many ways). We illustrate two important possibilities by looking again at the light-tailed and heavy-tailed distributions introduced in Tables 10-4 and 10-5, respectively.

EXAMPLE: FOLDED-ROOT DISTRIBUTION

Using the letter values as the selected quantiles, Figure 10-1 compares the folded-root distribution to the Gaussian distribution. We can see clearly what it means for a distribution to be less elongated than Gaussian. A straight line would fit this plot fairly well in the middle; but, as we go further into either tail, the curve flattens away from that line because the folded-root letter values become progressively less extreme than the Gaussian ones.

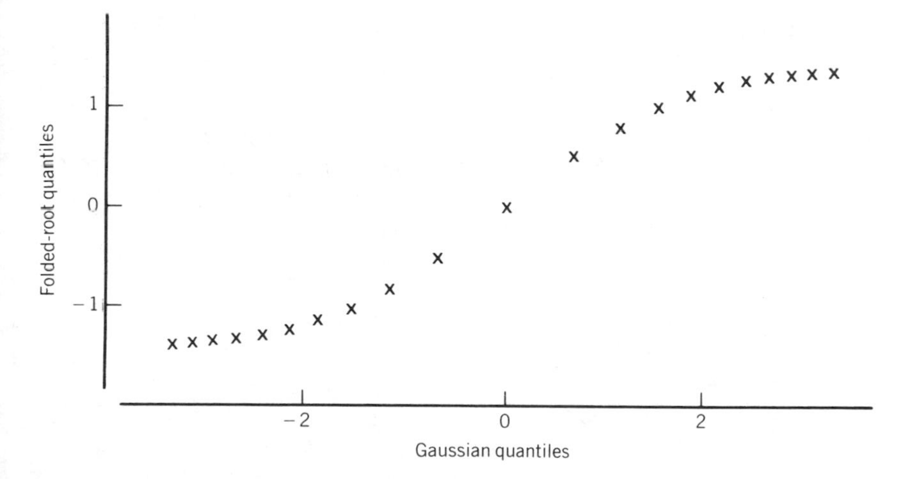

Figure 10-1. Letter values of folded-root distribution versus standard Gaussian letter values.

EXAMPLE: CAUCHY DISTRIBUTION

The Q-Q plot comparing the Cauchy distribution [see equations (5) and (6)] to the Gaussian distribution appears in Figure 10-2. The contrast between this picture and Figure 10-1 is sharp. It is still possible to visualize a straight line through the middle of the plot, but for this heavy-tailed distribution the ends of the Q-Q plot curve sharply toward the vertical. Although it will generally not be so pronounced, this qualitative pattern characterizes distributions that are more elongated than the Gaussian.

Sample versus Distribution

One common application of Q-Q plots involves comparing a sample and a distribution, generally to see whether the data depart substantially in shape from the distribution. (In this context the less specific term "probability plot" is often used.) For such a plot the quantiles of the sample play the role of $F_2^{-1}(p)$. Because we have presented the Q-Q plot in terms of varying p, we need to show how the correspondence between the sample values and the

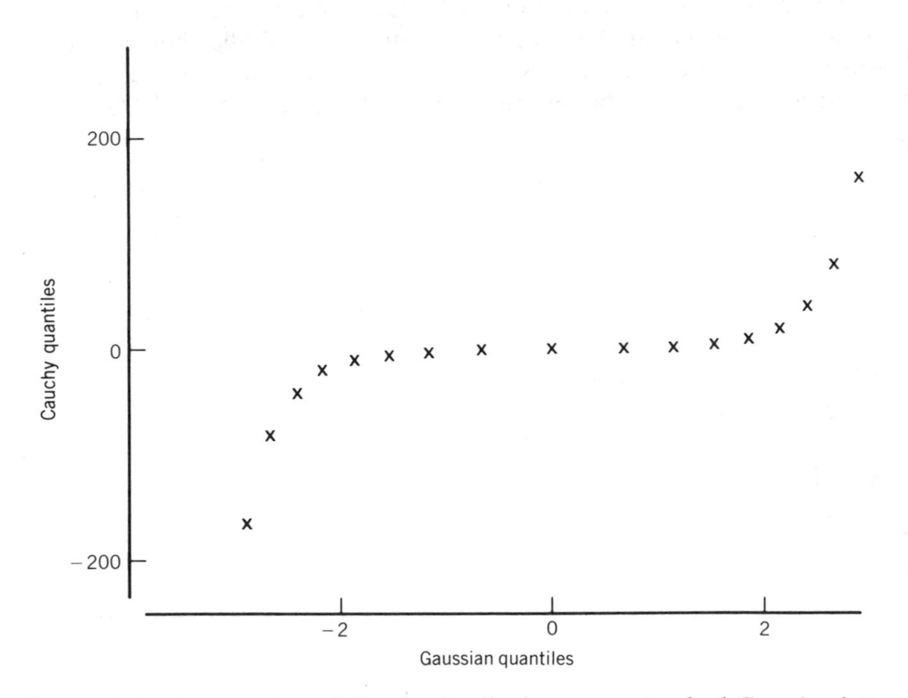

Figure 10-2. Letter values of Cauchy distribution versus standard Gaussian letter values.

distribution values, $F_1^{-1}(p)$, should be defined. We approach this through the sample order statistics, $x_{(1)} \leq \cdots \leq x_{(n)}$.

As random variables the order statistics are $X_{(1)} \leq \cdots \leq X_{(n)}$, and the generalization of equation (8) is

$$\text{med}\{X_{(i)}\} \approx F^{-1}((i - \tfrac{1}{3})/(n + \tfrac{1}{3})) \tag{12}$$

when the $X_{(i)}$ are an ordered sample of n from the distribution whose cdf is F. By regarding the observed value $x_{(i)}$ as an estimate of $X_{(i)}$, we treat $(i - \frac{1}{3})/(n + \frac{1}{3})$ as the value of p. Therefore we plot the points

$$\left(F^{-1}[(i - \tfrac{1}{3})/(n + \tfrac{1}{3})], x_{(i)}\right), \qquad i = 1, \ldots, n.$$

In such a plot, F usually belongs to the "standard" member of the particular family of distributions. The most common example is the family

TABLE 10-7. Calculations for normal probability plot of the order statistics from the beta-endorphin data.

i	$p_i = (3i - 1)/(3n + 1)$	$\Phi^{-1}(p_i)$	$x_{(i)}$
1	.034	−1.819	2.0
2	.086	−1.365	2.5
3	.138	−1.090	6.0
4	.190	−0.879	6.5
5	.241	−0.702	7.5
6	.293	−0.544	9.0
7	.345	−0.399	11.5
8	.397	−0.262	12.0
9	.448	−0.130	13.5
10	.500	0	14.0
11	.552	0.130	14.5
12	.603	0.262	15.0
13	.655	0.399	16.0
14	.707	0.544	18.0
15	.759	0.702	18.0
16	.810	0.879	20.0
17	.862	1.090	25.0
18	.914	1.365	42.0
19	.966	1.819	52.0

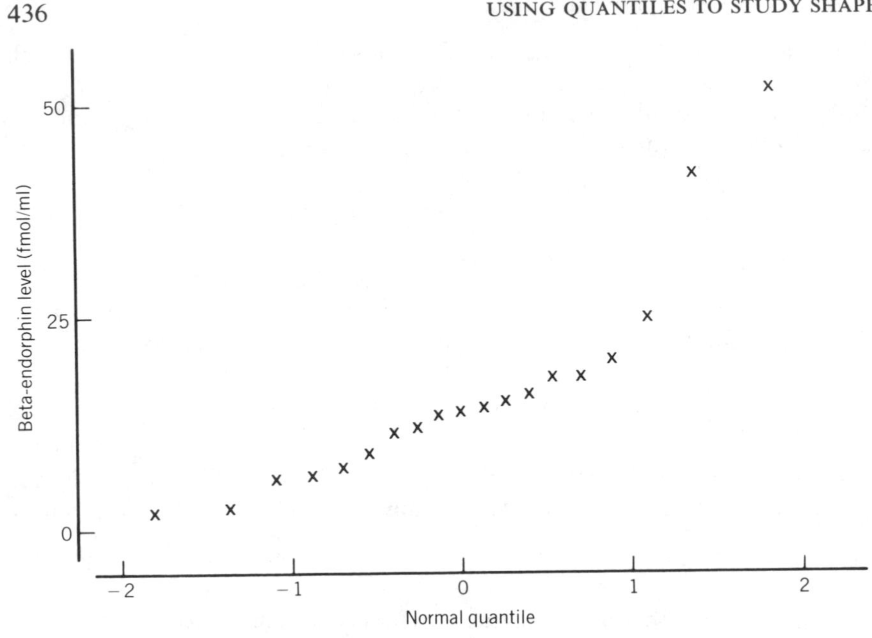

Figure 10-3. Normal probability plot of patients' blood beta-endorphin levels under conditions of presurgical stress (entire sample).

of Gaussian distributions. The "normal probability plot" uses standard Gaussian quantiles ("probability scale" graph paper has the transformation Φ^{-1} built in), and the slope and intercept yield values of σ and μ for the data.

EXAMPLE: NORMAL PROBABILITY PLOT

To illustrate the plot and its calculations, we use the presurgical stress data of Table 10-1. We also estimate σ and μ graphically. We begin by calculating the fractions, $(i - \frac{1}{3})/(n + \frac{1}{3})$, and the standard Gaussian quantiles, as shown in Table 10-7. (If we were using normal probability paper, we would need only the fractions.) We next make the plot, shown in Figure 10-3. Except for the two rightmost points (and perhaps the leftmost point and a third point at the right), the shape of the plot is close to a straight line—one has to expect some small departures because of fluctuations in data—so we find fairly good agreement with a Gaussian distribution. Using a clear plastic straightedge to fit a line to the plot yields a slope of about 7.3 and an intercept (vertical height at horizontal value of 0) of about 13.5 (for the straight-line portion). These serve as estimates of σ and μ, respectively.

In studying probability plots it is important to develop an appreciation for the routine variability that arises when the data are actually random samples from the distribution that provides the horizontal coordinate. As an aid for this sort of calibration Daniel and Wood (1980, Appendix 3A), for example, present probability plots for sets of Gaussian random samples at a range of sample sizes. Chance departures from the ideal straight-line pattern can be substantial, especially in small samples. Against this background we would view only the two rightmost points in Figure 10-3 as deserving serious attention.

Using Only the Letter Values

In the preceding example it took little effort to do the calculations and make the Q-Q plot. But the sample size was only 19. If it were several hundred, we would be reluctant to take the trouble of making the plot by hand. Even if we could turn the whole task over to a computer with good graphical output, the plot might well look cluttered. We might prefer to use only a selected set of quantiles, and we have one set readily available: the letter values. Thus we need only calculate fractions from $p = (d - \frac{1}{3})/(n + \frac{1}{3})$, where d is the depth of the letter value, find their complement $(1 - p)$ if the distribution is not symmetric, and continue as before.

TABLE 10-8. Calculations for the letter-value-based Gaussian Q-Q plot for log of household income. (We obtain upper letter values by symmetry.)

d	$p - (d - \frac{1}{3})/(n + \frac{1}{3})$	$\Phi^{-1}(p)$
1	.00067	−3.207
2.5	.00218	−2.851
4.5	.00419	−2.636
8.5	.00821	−2.339
16.5	.0163	−2.138
32	.0318	−1.854
63	.0630	−1.530
125	.125	−1.149
249	.250	−0.674
497.5	.500	0

EXAMPLE: LETTER-VALUE-BASED Q-Q PLOT

With a sample size of 994, the logs of household income in Pittsburgh offer an opportunity to save considerable effort by using only letter values. From the depths in Table 10-6 we obtain the fractions and standard Gaussian quantiles, as given in Table 10-8. Taking the letter values from Table 10-6, we plot the points $(-3.207, 2.06)$, $(-2.851, 2.50), \ldots, (0, 3.54), \ldots,$ $(3.207, 4.04)$. The result, Figure 10-4, is not as close to a straight line as we might have expected from the pseudosigmas in Table 10-6. A look at the midsummaries (given in the discussion of Table 10-2), however, reminds us that they showed a steady trend downward, corresponding to the now very evident asymmetry of the plot. The left end curves down toward the vertical, indicating that the lower tail is somewhat heavier than Gaussian. The right end has weaker curvature, and it curves toward the horizontal, giving the message that the upper tail is lighter than Gaussian. We probe this combination of skewness and elongation further in Chapter 11.

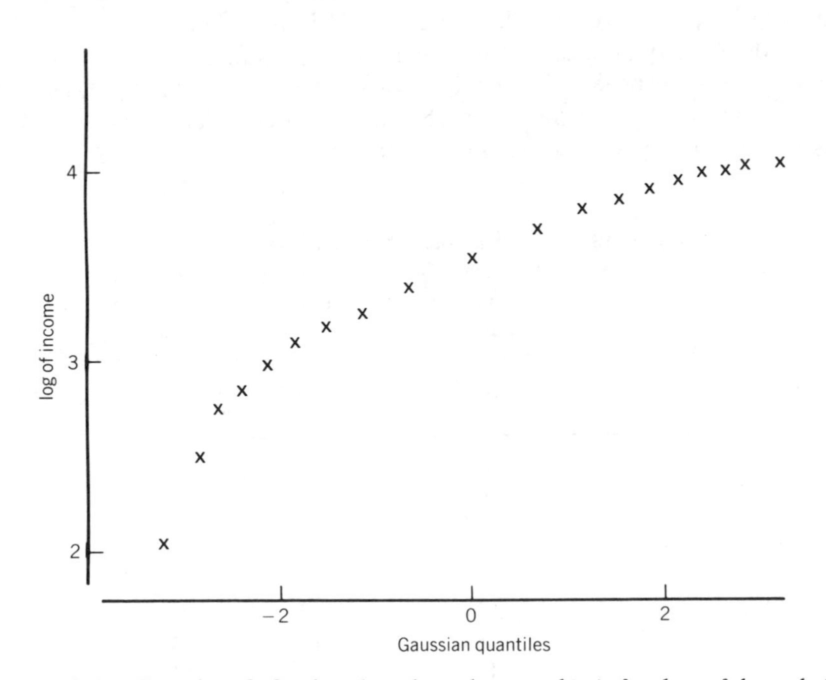

Figure 10-4. Gaussian Q-Q plot (based on letter values) for log of household income.

A Plot for Binned Data

When someone else has collected the data and recorded them as a frequency distribution, giving the count of observations in each of a set of bins, we cannot recover the full ordered sample or even the letter values for use in a Q-Q plot. We may still be able to make a satisfactory plot, however, by working with the information that is available: the edges of the bins and the corresponding cumulative counts (from the nearer edge of the ordered sample).

If the sample contains n observations and n_k of these lie in or to the left of the bin whose right edge is x_k (assuming that $n_k \leq n/2$), then we associate depth $n_k + \frac{1}{2}$ with the value x_k. (Above the median, n_k is the number of observations that lie in or to the right of the bin whose left edge is x_k.) Applying this rule at all the bin edges in the frequency distribution yields a set of depths, d_k, and a set of corresponding values, x_k, in the scale of the data. From these we produce a Q-Q plot of the sample against the appropriate theoretical distribution, as before.

To illustrate the calculation for one bin edge, we take $n = 9440$, $x_k = 11.25$, and $n_k = 62$. The depth corresponding to x_k is $d_k = 62.5$, the associated value of p is

$$\left(d_k - \tfrac{1}{3}\right)/\left(n + \tfrac{1}{3}\right) = \left(62.5 - \tfrac{1}{3}\right)/\left(9440 + \tfrac{1}{3}\right) = .006585,$$

and the corresponding standard Gaussian quantile is -2.479. Thus this bin edge would appear in the Q-Q plot as the point $(-2.479, 11.25)$. Exercise 3 extends these calculations to the full frequency distribution, which appears in Table 11-13.

Sample versus Sample

A Q-Q plot can be very effective in comparing two samples. This version is sometimes called an "empirical Q-Q plot." To allow for differences in sample size, as well as to reduce effort, we again use the letter values of the two samples as the selected quantiles. For an example of this new modification we turn to data on errors in reporting annual household income.

EXAMPLE: EMPIRICAL Q-Q PLOT

In the Housing Allowance Demand Experiment, the researchers determined how accurately households had reported their annual income. The result is a sizable sample of reporting errors in both Pittsburgh and Phoenix, and it is useful to know whether reporting errors in the two cities seem to follow the same shape of distribution. Table 10-9 gives the letter values from the two samples, and Figure 10-5 is the empirical Q-Q plot based on them. Except for three points at each end, the plot is reasonably close to a straight line with slope 1 and intercept 0. Thus the two distributions have nearly the same shape, as well as the same spread. The departure from a straight line around the center of the plot reflects a slight tendency for the two distributions to be skewed in opposite directions—Pittsburgh to the left and Phoenix to the right. On the whole the agreement in shape between these two samples is good. (Since the ns are different, the extreme points at each end are at best dubious. The remaining points fit a straight line quite well.)

TABLE 10-9. Letter values for reporting error in household income (dollars per year).

Tag	Pittsburgh	Phoenix
1	−$6,739	−$3,940
Y	−4,947	−3,938
Z	−4,197	−3,116
A	−3,070	−2,644
B	−2,232	−2,099
C	−1,669	−1,493
D	−1,099	−1,008
E	−613	−595
F	−204	−216
M	−24	−2
F	63	239
E	416	780
D	981	1,262
C	1,664	1,772
B	2,292	2,409
A	2,640	2,787
Z	3,824	2,978
Y	4,550	3,633
1	5,159	5,874
(n)	(961)	(600)

Source: David C. Hoaglin and Catherine A. Joseph (1980). *Income Reporting and Verification in the Housing Allowance Demand Experiment*. Cambridge, MA: Abt Associates Inc. (Table V-1, p. A-48).

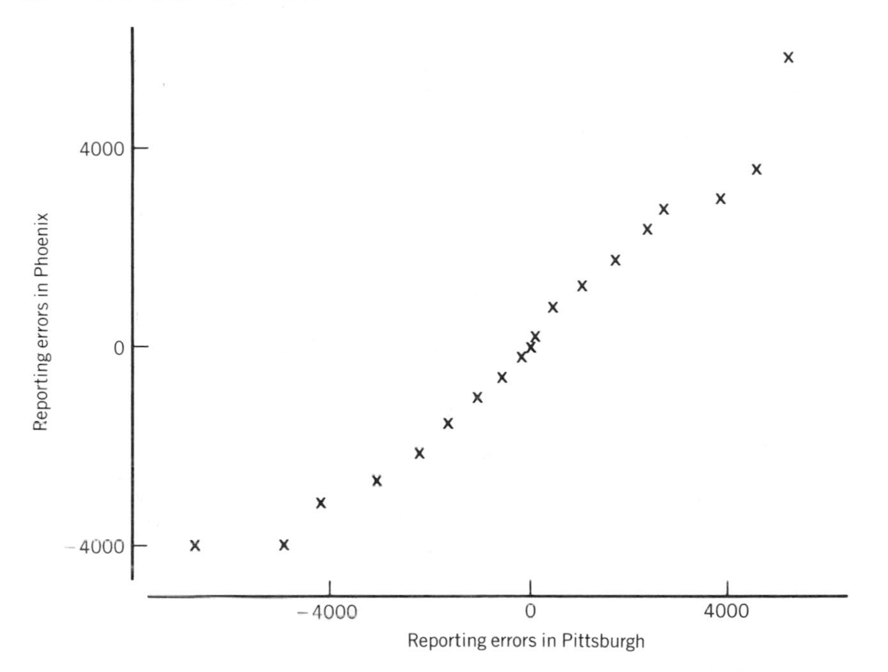

Figure 10-5. Empirical Q-Q plot for income reporting errors in Pittsburgh and Phoenix. Data (in dollars) from the Housing Allowance Demand Experiment.

Since we had no reason to expect any substantial agreement, this outcome is encouraging. We might even be so bold as to suggest this distribution shape as a standard of comparison for other data on income reporting errors observed under roughly similar circumstances. In the absence of any specific information to the contrary, this heavy-tailed shape would almost surely serve as a better starting point than the Gaussian.

Detilted Q-Q Plots

Q-Q plots readily reveal large differences from agreement, but medium and small differences can hide too easily. As always, we want to make our utopian lines horizontal. An easy way to do this is to choose a c such that cX_1 behaves rather like X_2 in the middle and then to plot

$$X_2 - cX_1 \text{ against } X_2 + cX_1.$$

Exercise 5 helps the reader to pursue this idea for the household income data in the log scale (Table 10-6 and Figure 10-4).

P-P Plots

The examples in this section have shown that Q-Q plots can reveal a great deal of information about distribution shape. For this purpose they are much more suitable than another kind of probability plot, the "P-P plot," which we mention here for completeness.

Just as we vary the fraction p and plot the points $(F_1^{-1}(p), F_2^{-1}(p))$ to obtain a Q-Q plot of F_2 versus F_1, we can vary the value x and plot the two corresponding fractions or percentages. This is the basis of the percent-versus-percent plot or P-P plot. Specifically, the plotted points are $(F_1(x), F_2(x))$ for selected values of x. When $F_2 = F_1$, the P-P plot will be a straight line connecting the point $(0, 0)$ and the point $(1, 1)$. Unlike the Q-Q plot, however, the P-P plot does not yield a straight line when F_1 and F_2 have the same shape but are not identical. This property makes the P-P plot less useful for studying shape in terms of location-scale families of distributions. Where one does not have to allow for differences in location and scale, however, the P-P plot gives greater emphasis to how the data or distributions compare in the center—much as the Q-Q plot gives more space to the tails.

Wilk and Gnanadesikan (1968) present several hybrids of the Q-Q plot and P-P plot. Fisher (1981) adapts the P-P plot to the one-sample problem of assessing symmetry about a specified point.

In the next section we discuss plots that focus more specifically on skewness and elongation than do Q-Q plots.

10D. PLOTS FOR SKEWNESS AND ELONGATION

Just as a straight-line appearance is the standard for a Q-Q plot, we like plots for skewness and elongation to yield a straight line when skewness is absent or elongation is neutral. In this section we describe four such plots.

Upper-versus-Lower Plot

In the absence of skewness the data are symmetric about the median, so a plot of upper data values versus lower data values will follow a straight line. Wilk and Gnanadesikan (1968) suggest using the two observations at the same depth for the coordinates of the plotted points; that is, we would plot $(x_{(1)}, x_{(n)}), (x_{(2)}, x_{(n-1)}), \ldots,$ continuing until we reach the median. If we assume symmetry about the median, M, we see that

$$x_{(n+1-i)} - M = M - x_{(i)} \tag{13}$$

TABLE 10-10. Selected quantiles of the chi-squared distribution on six degrees of freedom.

	Quantiles			
p	Lower	Upper	Spread	Mid
.5	5.348		0	5.348
.25	3.455	7.841	4.386	5.648
.10	2.204	10.645	8.441	6.425
.05	1.635	12.592	10.957	7.114
.025	1.237	14.449	13.212	7.843
.01	0.872	16.812	15.940	8.842
.005	0.676	18.548	17.872	9.612

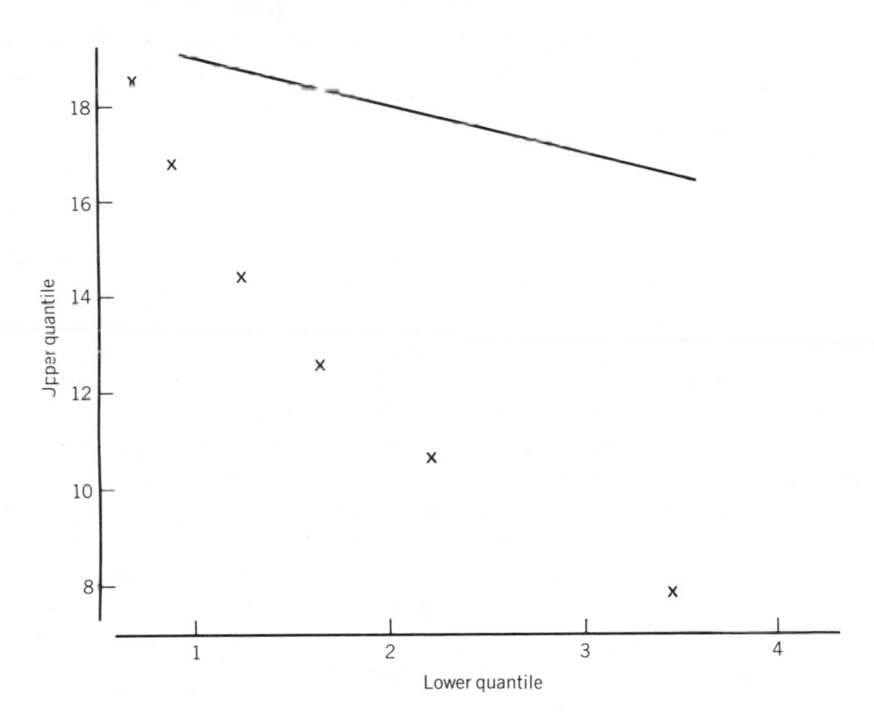

Figure 10-6. Upper-versus-lower plot for χ_6^2, showing increasing skewness to the right. (The line has slope -1.)

and hence the points lie on a line with slope -1. To simplify the plot, we can use the letter values instead of the entire sample.

EXAMPLE: CHI-SQUARED ON SIX DEGREES OF FREEDOM

To give an indication of how this plot can look when we have definite skewness, we begin with a distribution, chi-squared on six degrees of freedom (χ_6^2), which is fairly strongly skewed to the right. Table 10-10 presents selected quantiles, along with the corresponding mids and spreads, which we will use later. In this instance we have not used the letter values; instead, the quantiles are the ones readily available in a standard table. The upper-versus-lower plot is Figure 10-6. Our first step in examining this plot is to compare the pattern of the plotted points with the line having slope equal to -1. The general slope of the plot is much steeper than -1, and the curvature is clear. This agrees with the known skewness of the χ_6^2 distribution. We also note that the more extreme quantiles (or, customarily, data points) are at the left of the plot. The pattern of upward curvature from right to left, then, is what we can expect whenever a set of data is increasingly skewed to the right.

EXAMPLE: UPPER-VERSUS-LOWER PLOT FOR LETTER VALUES

The letter values of the data on income reporting errors of households in Pittsburgh were plotted on the horizontal axis in Figure 10-5. In Figure 10-7 we now see the upper-versus-lower plot. The leftmost point involves the two extremes, so it is not surprising that it seems deviant. The remaining points are not far from a straight line with slope -1. Thus the sample is reasonably symmetric. Because this sample is large (961 observations), it may be possible to deduce somewhat more about its structure without feeling that we are simply speculating on apparent patterns in noise. Other than the leftmost point, a plausible description of the remaining points is that they fall on two straight-line segments—the first three on a line whose slope is close to -1, and the other five on a line that is shifted upward and is slightly steeper. This suggests a minor departure from symmetry in the direction of skewness to the right. In the notation of equation (13), $x_{(n+1-i)} - M$ is increasing more rapidly than $M - x_{(i)}$.

A feature of the upper-versus-lower plot, as revealed by both of these examples, is the need to keep in mind a line with slope -1 and, preferably, to show it on the plot. This suggests that we can make the plot more effective by changing the coordinates so that the comparison line becomes horizontal.

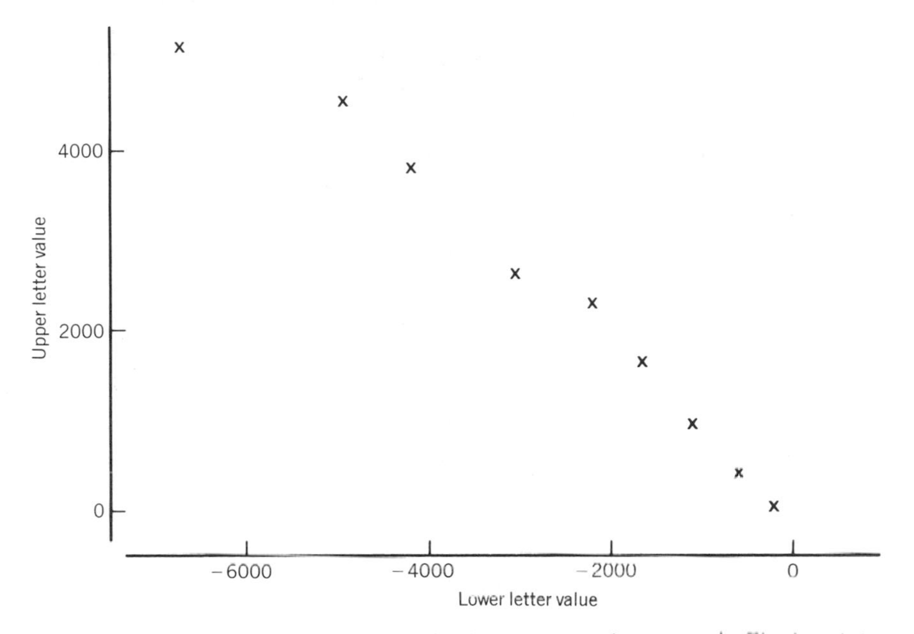

Figure 10-7. Upper-versus-lower plot for income reporting errors in Pittsburgh (in dollars) using letter values.

Doing Better: The Mid-versus-Spread Plot

An improved replacement for the upper-versus-lower plot uses the differences $x_{(n+1-i)} - x_{(i)}$ and the sums $x_{(i)} + x_{(n+1-i)}$ as the horizontal and vertical coordinates, respectively. In a perfectly symmetric set of data, simple algebra turns the relationship of equation (13),

$$x_{(n+1-i)} - M = M - x_{(i)},$$

into

$$x_{(i)} + x_{(n+1-i)} = 2M. \tag{14}$$

Because $x_{(i)} + x_{(n+1-i)}$ is the vertical coordinate on the plot and equals a constant $2M$, the standard for such a plot is a horizontal line, something that the eye can judge better than a slope of -1. When the data are nearly symmetric, the sums $x_{(i)} + x_{(n+1-i)}$ will be nearly constant. This means that we can use a large amount of vertical space in the plot to cover a small range of values and hence see small but systematic departures from symmetry more clearly.

If we work with the letter values, we can divide equation (14) through by 2 and recognize the resulting left-hand side as a midsummary. For this reason we call the plot the *mid-versus-spread plot*. In it we have the same numerical information about skewness as in the mids calculated from the letter-value display (in Section 10A), but now this information is organized so that we can see the whole situation at a glance. Usually we include the median, plotted at spread = 0.

For comparison with the upper-versus-lower plot, we use the same examples.

EXAMPLE: CHI-SQUARED

For the selected quantiles of χ^2_6, Figure 10-8 is the mid-versus-spread plot. (The calculations are in Table 10-10.) Again the skewness is clear: the points curve steadily upward away from the horizontal line through the median. This pattern is probably easier to think of as indicating skewness to the right than the pattern of Figure 10-6. If the plotted points turned downward as the spreads increased, the diagnosis would be skewness to the left.

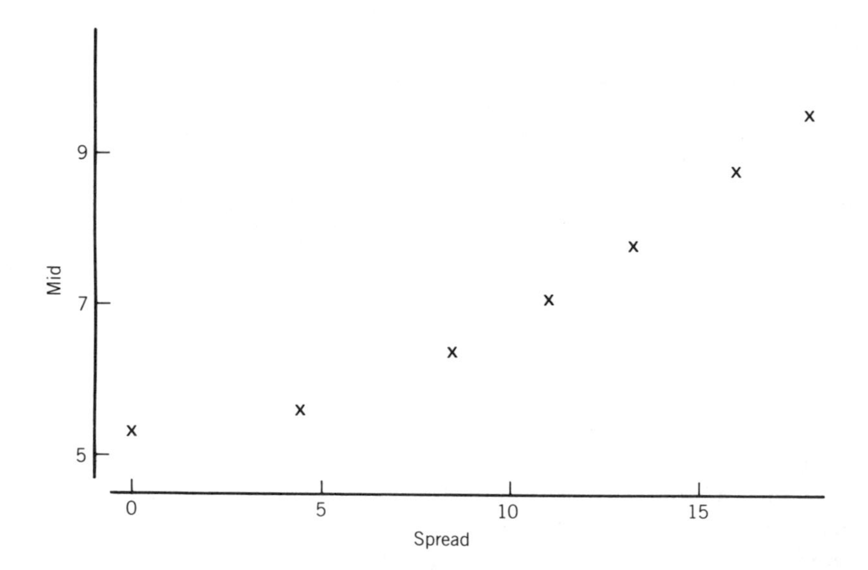

Figure 10-8. Mid-versus-spread plot for χ^2_6, showing increasing skewness to the right.

EXAMPLE: MID-VERSUS-SPREAD PLOT FOR LETTER VALUES

For the income reporting errors in Pittsburgh, Figure 10-9 shows the mid-versus-spread plot. Now the vertical scale must cover only a range of about 800 instead of about 5600, so we can see more detail. From the regular pattern we can take the impression that the middle of the sample starts to skew toward the left and then skews back toward the right as we move outward from the median. (No explanation for this behavior has yet been uncovered.) On the whole we would be willing to treat such reporting-error data as symmetric, but we would remain alert to any new sources of supporting evidence for skewness.

In the mid-versus-spread plot the primary reason for using the spreads as the horizontal plotting coordinate is convenience. They will usually be available in a letter-value display, and they increase from the median outward so that the midsummaries are laid out graphically in the proper order and in a pattern where a horizontal line corresponds to symmetry. Other choices can be useful, as we now see.

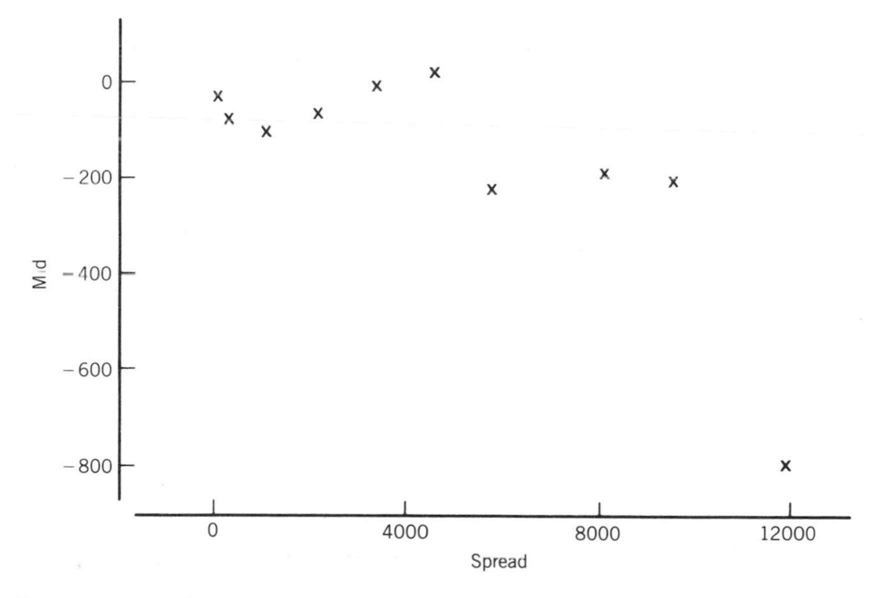

Figure 10-9. Mid-versus-spread plot for income reporting errors in Pittsburgh (in dollars).

Mid-versus-z^2 Plot

One possible drawback in the mid-versus-spread plot is the fact that the more extreme spreads can be adversely affected by wild values in the data and can thus string the plot out over a much wider range than necessary. Of course, the midsummaries will also be affected by the same wild values; but we are inspecting them for evidence of skewness, and we prefer to keep them all in the plot until we have had a chance to exercise our judgment.

An alternative horizontal coordinate is the square of the corresponding standard Gaussian quantile (often, the square of the corresponding standard Gaussian letter value). This also lays out the midsummaries in a reasonable spacing, and it is unaffected by the data values. We call this plot the *mid-versus-z^2 plot* because standard Gaussian quantiles are often denoted by z. Using z^2 as the horizontal coordinate also brings the advantage that, when the plot is close to a straight line, the skewness has a convenient numerical description in terms of the distributions introduced in Section 11A. The discussion in that section shows this theoretical basis for the mid-versus-z^2 plot.

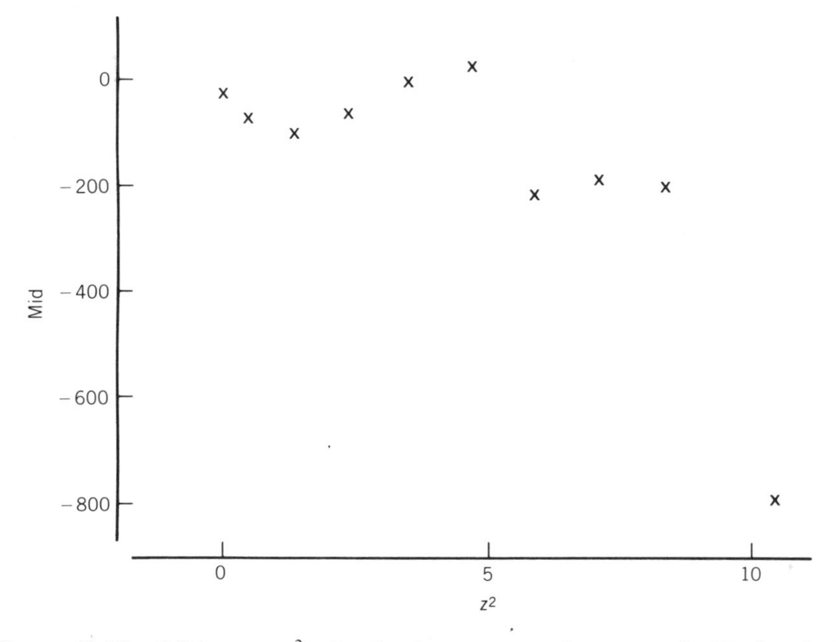

Figure 10-10. Mid-versus-z^2 plot for income reporting errors in Pittsburgh (in dollars).

EXAMPLE: INCOME REPORTING ERRORS

Figure 10-10 gives the mid-versus-z^2 plot for the income reporting errors in Pittsburgh. Its qualitative features are very much the same as in Figure 10-9, but the overall appearance is somewhat more regular. The interpretation is the same as before.

In general it is possible that using z^2 instead of the spread will straighten out parts of a mid-versus-spread plot. This did not happen in the example.

Pseudosigma-versus-z^2 Plot

We want also to be able to picture elongation. As a horizontal plotting coordinate we can again use z^2, which generally works well in spacing out the midsummaries in the mid-versus-z^2 plot. The important question now is what to use as the vertical plotting coordinate. Section 10B introduced the pseudosigmas to measure elongation, and they are what we use here. The resulting plot is called the *pseudosigma-versus-z^2 plot*. It follows a horizontal line when the data are neutrally elongated because then the pseudosigmas are constant. Plotting log(pseudosigma) versus z^2 is often more helpful, though not in the example below. Some theory related to these plots also appears in Chapter 11.

EXAMPLE: INCOME REPORTING ERRORS

We continue to examine the income reporting errors in Pittsburgh, now concentrating on elongation. Table 10-11 gives the calculations for pseudo-

TABLE 10-11. Calculations for the pseudosigma-versus-z^2 plot of the income reporting errors in Pittsburgh.

Tag	Spread	z	Pseudosigma	z^2
F	267	0.674	197.9	0.455
E	1029	1.150	447.2	1.323
D	2080	1.534	678.0	2.353
C	3333	1.863	894.8	3.470
B	4524	2.154	1050.1	4.639
A	5709	2.418	1180.8	5.845
Z	8021.5	2.660	1507.8	7.076
Y	9497.5	2.886	1645.7	8.327
1	11898	3.197	1860.8	10.224

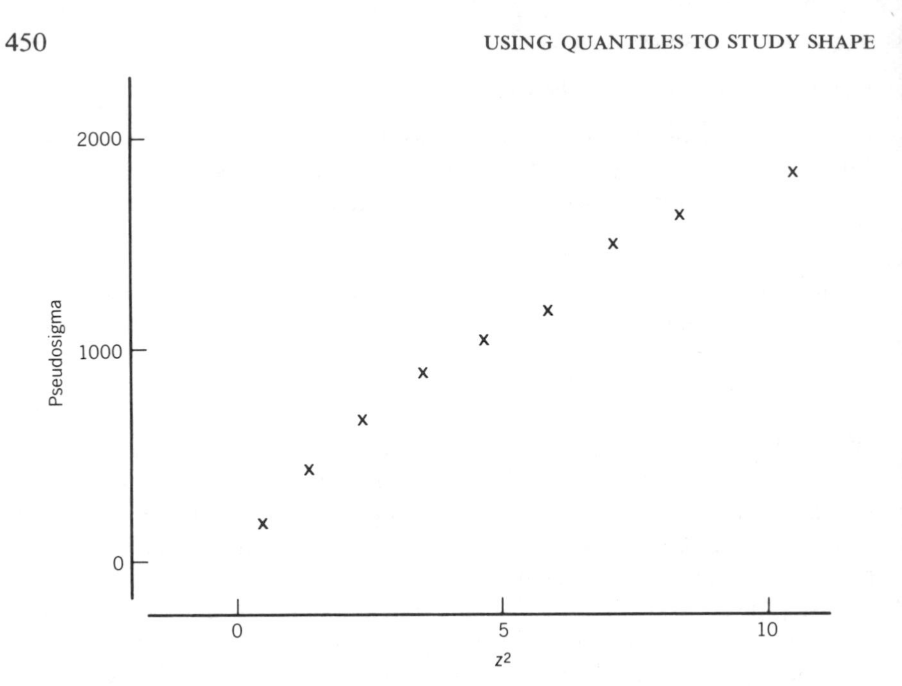

Figure 10-11. Pseudosigma-versus-z^2 plot for income reporting errors in Pittsburgh (in dollars).

sigmas and z^2 values, using standard Gaussian values from Table 10-3 (for all except the extremes). Figure 10-11 shows the plot, which is clearly not a horizontal straight line. Except for one point at each end, where pseudosigmas have large sampling variation, the remaining points are close to an inclined straight line. We must remember that this plot is designed primarily for symmetric situations, and Figures 10-9 and 10-10 give the clear message that this set of data is not symmetric. Skewness can introduce some elongation, and we discuss this in Chapter 11.

In examining elongation we have worked with spreads, taking information from both tails together. When we can assume symmetry, this is a sensible strategy. When some asymmetry is possible, a technique for looking at the two tails separately can be of benefit. The next section takes this step beyond mids and spreads.

10E. PUSHBACK ANALYSIS

Earlier in this chapter we have characterized the skewness or elongation of a distribution or a set of data by combining information from its lower tail and its upper tail. (The midsummaries and the letter-spreads work with pairs of letter values in just this way.) For skewness this is unavoidable; we must study departures from symmetry by comparing the two halves of the distribution. But elongation is another matter. We can study the elongation of the lower and upper halves of a set of data separately. We might find that the two tails have the same shape and scale (as we would expect in a symmetric situation), or they may exhibit different patterns of elongation. It is even possible that the skewness of the data reflects the fact that the two halves have the same shape but differ substantially in scale—Tukey (1977, Section 19D) gives an example of this nature.

To study individual letter values or examine the elongation of the two halves separately, we may begin by subtracting an overall Gaussian shape from the full set of letter values. This can, in particular, be done by fitting a line to the Gaussian Q-Q plot for the data. In any event, it lets us look at how the flattened letter values depart from a horizontal straight line. Specifically, we can use the fraction p (with $0 < p < \frac{1}{2}$) to index the pairs of letter values, x_p and x_{1-p} for the data (lower and upper, respectively) and z_p and z_{1-p} for the standard Gaussian values. Then we estimate a scale for the data, s, and calculate the flattened letter values $x_p - sz_p$ and $x_{1-p} - sz_{1-p}$ for each p. Because the flattened letter values have, in effect, been pushed in toward (or beyond) the middle of the data, we call this a *pushback technique*.

One quick, resistant way of obtaining s is to find the median of whatever pseudosigmas have been calculated, that is,

$$s = \text{med}\{(x_{1-p} - x_p)/(z_{1-p} - z_p)\}. \tag{15}$$

(Of course, a suitably fitted slope from the Q-Q plot might serve as well.)

If the data follow a Gaussian shape, then plotting the flattened letter values against the corresponding standard Gaussian quantiles will yield a horizontal straight line. And if the two halves of the data are Gaussian with different scales, two different straight lines will result. More complicated patterns reveal other kinds of departures from Gaussian shape.

EXAMPLE: INCOME REPORTING ERRORS (LOG SCALE)

As an illustration of this technique, we look at the income reporting errors, in a logarithmic scale, for the 174 households in the "Housing Gap, Minimum Rent" treatment group in Phoenix. Here the "error" is the logarithm (to the base 10) of the ratio of reported income to true income. Table 10-12 gives the letter values and the steps in adjusting them. The letter-spreads and pseudosigmas are in Table 10-13. Using equation (15), we pick out the median of the pseudosigmas, .086, as the overall value of s. This is the basis for the adjustment in Table 10-12.

TABLE 10-12. Flattening the letter values of the income reporting errors (logarithmic scale) of households in the "Housing Gap, Minimum Rent" treatment group in Phoenix.

Tag	Letter Value	z	$s \times z$ ($s = .086$)	Difference between Letter Value and $s \times z$
a. Both Tails				
1	−.1623	−2.667	−.229	.067
A	−.1618	−2.418	−.208	.046
B	−.1354	−2.154	−.185	.050
C	−.1204	−1.863	−.160	.040
D	−.0970	−1.534	−.132	.035
E	−.0582	−1.150	−.099	.041
F	−.0260	−0.674	−.058	.032
M	−.0018	0	0	−.002
F	.0132	0.674	.058	−.045
E	.0585	1.150	.099	−.041
D	.1042	1.534	.132	−.028
C	.1991	1.863	.160	.039[a]
B	.3327	2.154	.185	.148
A	.4100	2.418	.208	.202
1	.6218	2.667	.229	.393[a]

Tag	Letter Value	z	$s \times z$ ($s = .526$)	Difference between Letter Value and $s \times z$
b Upper Tail				
E	.0585	1.150	0.6049	−.5464
D	.1042	1.534	0.8069	−.7027
C	.1991	1.863	0.9799	−.7808
B	.3327	2.154	1.1330	−.8003
A	.4100	2.418	1.2719	−.8619
1	.6218	2.667	1.4028	−.7810

[a]Additional slope based on these two points is (.393 − .039)/(2.667 − 1.863) = .440, so that a pushback for the upper tail alone would use s = .086 + .440 = .526, as in panel *b*.

In the plot, Figure 10-12, we see that the elongation comes entirely from the upper tail. The flattened letter values in the lower tail very nearly follow a straight line with a small negative slope, whereas those in the upper tail increase sharply from C outward.

Because this behavior is not too far from linear, we can learn more about the upper tail by trying a scale constant adapted only to that tail. A line fitted to Figure 10-12 gives additional slope .440. The calculations continue in Table 10-12*b*. Only one of the last four pushed-back values lies outside the interval $-.7905 \pm .0100$. Similar eye fitting of an additional scale constant for the lower tail yields a slope of $-.013$. Thus it appears that the extreme upper tail is close to Gaussian, but with a scale (in log units) that is about $.526/.073 = 7.2$ times the scale of the lower tail.

We have been able to see the shape of this batch quite clearly. The interpretation of Figure 10-12 is that overreporting of income is longer-tailed than underreporting—it more frequently involves larger logarithmic errors. This finding runs counter to what one might have expected in a financial assistance program. Through poor recordkeeping or imperfect memory or even deliberate concealment, a household could conceivably report as little as one-tenth of its true income, but it would be unlikely to report ten times its true income. These data point to the opposite direction of skewness and a much narrower range of reporting errors.

Pushback analysis takes two useful steps beyond the classical probability plot (or sample-versus-distribution Q-Q plot, as in Section 10C). First, it

TABLE 10-13. Letter-spreads and pseudosigmas for the income reporting errors (letter values in Table 10-12).

Tag	Letter-Spread	Gaussian Value	Pseudosigma
F	.0392	1.349	.029
E	.1167	2.300	.051
D	.2012	3.070	.066
C	.3195	3.730	.086
B	.4681	4.310	.109
A	.5718	4.840	.118
1	.7841	5.334	.147
			median = .086

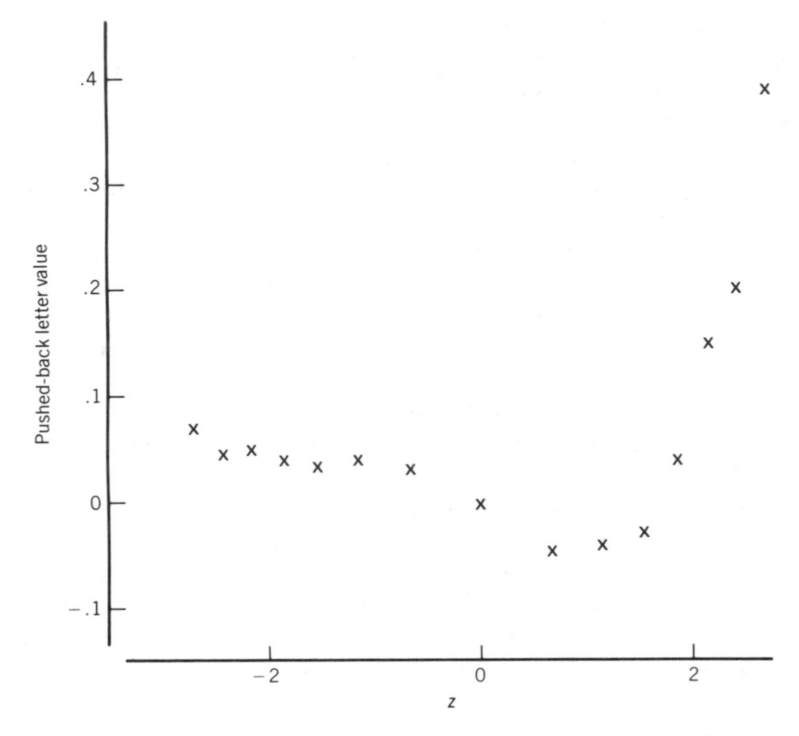

Figure 10-12. Pushed-back letter values versus z for the logarithmic income reporting errors.

removes a fitted line and thus better reveals medium and small departures. Second, it examines the upper and lower halves of the data separately.

10F. SUMMARY

This chapter has described a variety of qualitative techniques—some graphical and some numerical—for examining two key aspects of distribution shape: skewness and elongation. For both data and theoretical distributions, the quantile-based midsummaries and pseudosigmas aid in giving specific definitions to the vague concepts of skewness and elongation, respectively. Two corresponding displays, the mid-versus-spread plot and the pseudosigma-versus-z^2 plot, provide effective graphical representations of the information.

TABLE 10-14. Summary of plots based on quantiles. (The quantiles may be a selected subset, such as the letter values.)

Focus	Name	Notation	Plotting Coordinates	
			Horizontal	Vertical
Comparison of shape	Quantile–Quantile[a]			
	Distribution versus distribution	cdfs F_1 and F_2 $0 < p < 1$	$F_1^{-1}(p)$	$F_2^{-1}(p)$
	Sample versus distribution	cdf F, order statistics $x_{(1)} \le \cdots \le x_{(n)}$	$F^{-1}((i - \frac{1}{3})/(n + \frac{1}{3}))$	$x_{(i)}$
	Sample versus sample[b]	order statistics $x_{(1)} \le \cdots \le x_{(n)}$ $y_{(1)} \le \cdots \le y_{(n)}$	$x_{(i)}$	$y_{(i)}$
Skewness	Upper-versus lower[c]	order statistics $x_{(1)} \le \cdots \le x_{(n)}$	$x_{(i)}$ $1 \le i \le (n+1)/2$	$x_{(n+1-i)}$
	Mid-versus-spread[d]		$x_{(n+1-i)} - x_{(i)}$	$\frac{1}{2}[x_{(i)} + x_{(n+1-i)}]$
	Mid-versus-z^2 [d]	$z_i = \Phi^{-1}((i - \frac{1}{3})/(n + \frac{1}{3}))$	z_i^2	$\frac{1}{2}[x_{(i)} + x_{(n+1-i)}]$
Elongation	Pseudosigma-versus-z^2 [d]		z_i^2	$\dfrac{x_{(n+1-i)} - x_{(i)}}{z_{n+1-i} - z_i}$

[a] Ideal pattern is a straight line (with positive slope).
[b] Selecting or interpolating quantiles can cope with unequal sample sizes.
[c] Reference pattern is a straight line with slope −1.
[d] Reference pattern is a horizontal straight line.

The quantile–quantile plot, a standard technique of data analysis, helps in examining distribution shape, especially when one is willing to pay more attention to the tails than to the center. This versatile technique readily adapts to the use of letter values and even to the use of the bin edges from a grouped frequency distribution. All can be improved by a modification that makes the reference pattern in a Q-Q plot horizontal.

Table 10-14 summarizes the major variants of the quantile–quantile plot, as well as the plots for diagnosing the presence of skewness and elongation.

Pushback analysis allows us to study elongation in the lower tail and upper tail separately by first subtracting an overall Gaussian shape and then pursuing the pattern of departures from constancy.

10G. APPENDIX

Approximations for Φ^{-1}

In order to calculate pseudosigmas (as described in Section 10B) and make Q-Q plots (Section 10C), we must often evaluate the inverse of the standard Gaussian cumulative distribution function, Φ^{-1}. Tables of this function are widely available, but computer approximations to Φ^{-1} are quite accurate enough for our purposes. Abramowitz and Stegun (1964, Section 26.2) present the standard approximations. For small electronic calculators (including the author's hand-held one) a more convenient approximation, developed by Derenzo (1977), takes $y = -\ln(2p)$ and yields the upper-tail quantiles

$$\Phi^{-1}(1-p) \approx \sqrt{\frac{[(4y+100)y+205]y^2}{[(2y+56)y+192]y+131}} \tag{16}$$

with an error no larger than 1.3×10^{-4} when $10^{-7} < p < \frac{1}{2}$, a substantially wider range than that covered by Table 10-3.

Better Accuracy for the Extremes

The approximation

$$\text{med}\{Z_{(i)}\} \approx \Phi^{-1}\left(\left(i - \tfrac{1}{3}\right)/\left(n + \tfrac{1}{3}\right)\right) \tag{17}$$

is quite adequate for $1 < i < n$; but when $i = 1$ or $i = n$, it can be improved by a simple modification. By symmetry, it suffices to give this modification for $i = 1$:

$$\text{med}\{Z_{(i)}\} \approx \Phi^{-1}(0.695/(n + 0.390)). \tag{18}$$

This corresponds to using $\alpha = 0.305$ in equation (7) of UREDA Chapter 2.

The general approach yields approximations analogous to equations (17) and (18) for any continuous distribution: we simply use F^{-1} for that distribution instead of Φ^{-1}. For this reason we can obtain an indication of the comparative accuracy of these approximations by working with the uniform distribution on the interval [0, 1], for which $F^{-1}(p) = p$.

In this case it is straightforward to find the median of the distribution of the smallest order statistic, $U_{(1)}$, algebraically. We observe that

$$P\{U_{(1)} \leq u\} = 1 - (1 - u)^n, \tag{19}$$

set this equal to $\frac{1}{2}$, and solve for u to obtain

$$\text{med}\{U_{(1)}\} = 1 - .5^{1/n}. \tag{20}$$

Table 10-15 shows this true value and the values of $2/(3n + 1)$ and $2.085/(3n + 1.17) = 0.695/(n + 0.390)$ for selected values of n. The approximation of equation (18) is an appreciable improvement over that of equation (17), and its relative error is only about 2 parts in 1000 when $n = 100$. Thus the highly cautious analyst may want to use equation (18) or even $\Phi^{-1}(1 - .5^{1/n})$ when $i = 1$ and equation (17) otherwise.

To find out whether the improvement is worthwhile, J. W. Tukey (personal communication) has pointed out that all we have to do is to calculate the percentage point actually obtained, in percentage terms, as

TABLE 10-15. Values of two approximations for the median of the distribution of the smallest order statistic in a uniform sample.

n	$\text{med}\{U_{(1)}\}$	$2/(3n+1)$	$2.085/(3n+1.17)$
5	.12945	.12500	.12894
10	.06697	.06452	.06689
20	.03406	.03279	.03409
50	.01377	.01325	.01379
100	.006908	.006645	.006923

TABLE 10-16. Upper tail areas (50% for the exact median) corresponding to two approximations for the median of the distribution of the smallest order statistic in a uniform sample. Entries are $100(1-u)^n$.

n	$u = 2/(3n+1)$	$u = 2.085/(3n+1.17)$	Next Smallest[a] $u = 5/(3n+1)$
5	51.29	50.15	50.26
10	51.33	50.04	50.34
20	51.34	49.98	50.36
50	51.34	49.94	50.37
100	51.34	49.92	50.37
∞	51.34	49.91	50.37

[a]See text. Entries are $100[n(1-u)^{n-1} - (n-1)(1-u)^n]$.

$100[1-(1-u)^n]$, where u comes from the corresponding approximation. For slightly greater clarity Table 10-16 gives the upper tail areas, $100(1-u)^n$. These are the theoretical percentage points for the *one* extreme observation that we have. To detect the difference between 50% and 51.34% reliably would require the one extreme observation from each of something like 10,000–15,000 samples. Clearly such differences are quite undetectable in practice.

The last column in Table 10-16 shows that the deviation from 50% is only about .28 as large for the order statistic next to the extreme—and thus even more difficult to detect in practice. This rapid decrease in deviation continues as we move farther and farther from the extreme.

REFERENCES

Abramowitz, M. and Stegun, I. A., Eds. (1964). *Handbook of Mathematical Functions* (National Bureau of Standards, Applied Mathematics Series 55). Washington, D.C.: U.S. Government Printing Office.

Daniel, C. and Wood, F. S. (1980). *Fitting Equations to Data*, 2nd ed. New York: Wiley.

Derenzo, S. E. (1977). "Approximations for hand calculators using small integer coefficients," *Mathematics of Computation*, **31**, 214–225.

Fisher, N. I. (1981). "One-sample probability plots," *Australian Journal of Statistics*, **23**, 352–359.

Kennedy, S. D. (1980). *Final Report of the Housing Allowance Demand Experiment*. Cambridge, MA: Abt Associates Inc.

Miralles, F. S., Olaso, M. J., Fuentes, T., Lopez, F., Laorden, M. L., and Puig, M. M. (1983). "Presurgical stress and plasma endorphin levels," *Anesthesiology*, **59**, 366–367.

Mosteller, F. and Tukey, J. W. (1977). *Data Analysis and Regression*. Reading, MA: Addison-Wesley.

Tukey, J. W. (1977). *Exploratory Data Analysis*. Reading, MA: Addison-Wesley.

Wilk, M. B. and Gnanadesikan, R. (1968). "Probability plotting methods for the analysis of data," *Biometrika*, **55**, 1–17.

Additional Literature

Antille, A., Kersting, G., and Zucchini, W. (1982). "Testing symmetry," *Journal of the American Statistical Association*, **77**, 639–646.

Boos, D. D. (1982). "A test for asymmetry associated with the Hodges-Lehmann estimator," *Journal of the American Statistical Association*, **77**, 647–651.

Doksum, K. A. (1975). "Measures of location and asymmetry," *Scandinavian Journal of Statistics*, **2**, 11–22.

Doksum, K. A., Fenstad, G., and Aaberge, R. (1977). "Plots and tests for symmetry," *Biometrika*, **64**, 473–487.

Filliben, J. J. (1975). "The probability plot correlation coefficient test for normality," *Technometrics*, **17**, 111–117, 520 (correction).

Finch, S. J. (1977). "Robust univariate test of symmetry," *Journal of the American Statistical Association*, **72**, 387–392.

Fisher, N. I. (1983). "Graphical methods in nonparametric statistics: a review and annotated bibliography," *International Statistical Review*, **51**, 25–58.

Gnanadesikan, R. (1977). *Methods for Statistical Data Analysis of Multivariate Observations*. New York: Wiley.

Harrell, F. E. and Davis, C. E. (1982). "A new distribution-free quantile estimator," *Biometrika*, **69**, 635–640.

Harter, H. L. (1970). *Order Statistics and Their Use in Testing and Estimation, Vol 2: Estimates Based on Order Statistics of Samples from Various Populations*. Washington, D.C.: U.S. Government Printing Office.

Hoaglin, D. C. (1983). "Folded transformations." In S. Kotz and N. L. Johnson (Eds.), *Encyclopedia of Statistical Sciences, Volume 3*. New York: Wiley, pp. 161–162.

Nair, V. N. (1982). "Q-Q plots with confidence bands for comparing several populations," *Scandinavian Journal of Statistics*, **9**, 193–200.

Parzen, E. (1979). "Nonparametric statistical data modeling," *Journal of the American Statistical Association*, **74**, 105–131 (with discussion).

Rocke, D. M., Downs, G. W., and Rocke, A. J. (1982). "Are robust estimators really necessary?," *Technometrics*, **24**, 95–101.

Rogers, W. H. and Tukey, J. W. (1972). "Understanding some long-tailed symmetrical distributions," *Statistica Neerlandica*, **26**, 211–226.

Stigler, S. M. (1977). "Do robust estimators work with *real* data?," *Annals of Statistics*, **5**, 1055–1098 (with discussion).

Tierney, L. (1983). "A space-efficient recursive procedure for estimating a quantile of an unknown distribution," *SIAM Journal on Scientific and Statistical Computing*, **4**, 706–711.

EXERCISES

1. For all the same 19 patients as in the example in Section 10A, Miralles et al. (1983) give the beta-endorphin levels in blood samples taken

12–14 hours before surgery (in fmol/ml): 10.0, 6.5, 8.0, 12.0, 5.0, 11.5, 5.0, 3.5, 7.5, 5.8, 4.7, 8.0, 7.0, 17.0, 8.8, 17.0, 15.0, 4.4, 2.0. Make a normal probability plot of these data. How does its appearance compare with that in Figure 10-3?

2. The ties among the $x_{(i)}$ in Table 10-7 (and in Exercise 1) raise the possibility of averaging the ranks of tied observations and calculating the horizontal plotting position for all of them from the average rank. (For example, in Table 10-7 the two observations equal to 18.0 would share $i = 14.5$ instead of having $i = 14$ and $i = 15$.) If the data do not involve large clusters of tied observations, how much is this refinement likely to change the message in the probability plot?

3. One well-known set of data in the statistical literature gives the length (in mm) of 9440 beans. The data take the form of a frequency distribution with bins of width 0.5 mm. Using "(x, n)" to mean "the bin whose center is x has a frequency of n," the frequency distribution is (9.5, 1), (10.0, 7), (10.5, 18), (11.0, 36), (11.5, 70), (12.0, 115), (12.5, 199), (13.0, 437), (13.5, 929), (14.0, 1787), (14.5, 2294), (15.0, 2082), (15.5, 1129), (16.0, 275), (16.5, 55), (17.0, 6). Make a Gaussian Q-Q plot from the bin edges. Discuss its appearance.

4. Examine further the elongation of the distribution of errors in reporting household income. Using Figure 10-5 and the two sets of letter values in Table 10-9 to estimate the letter values of the (apparently) common underlying distribution. From these, make a Gaussian Q-Q plot, and calculate the pseudosigmas. Discuss the pattern of elongation.

5. Taking the letter values of household income in the log scale (Table 10-6) as X_2 and the corresponding standard Gaussian letter values (Table 10-8) as X_1, use $c = .23$ to make a detilted Gaussian Q-Q plot ($X_2 - cX_1$ versus $X_2 + cX_1$). Discuss. Try fitting a suitable function of $X_2 + cX_1$ to the pattern in this plot.

6. Examine the elongation of the logistic distribution, whose quantiles are given by $\text{logit}(u) = \log_e(u) - \log_e(1 - u)$ for $0 < u < 1$ [for example, the lower fourth is $\text{logit}(\frac{1}{4}) = -1.0986$]. Calculate letter values and pseudosigmas, and make the Gaussian Q-Q plot and the pseudosigma-versus-z^2 plot.

CHAPTER 11

Summarizing Shape Numerically: The *g*-and-*h* Distributions

David C. Hoaglin
Harvard University

Knowing that data are skewed left or right and that they are positively or negatively elongated still does not assign numbers to these aspects of shape. The present chapter describes how to assign numbers by using suitable monotonic functions of a standard Gaussian random variable. With this approach, we can readily explore a set of data and determine how effectively its shape can be summarized in a skewness parameter and an elongation parameter. In this way, we can describe a considerable range of data, as well as approximate many non-Gaussian theoretical distributions. As a further advantage, once we have arrived at a fitted distribution, we can easily use the monotonic functions to calculate quantiles from the standard Gaussian quantiles.

Thus this chapter continues, more quantitatively, the study of distribution shape that we began in Chapter 10. After appreciating the skewness and elongation that we find in a body of data, we proceed to summarize them numerically, using an approach introduced by Tukey (1977). Our goals may be practical or theoretical. In practice it takes a lot of data to assess either skewness or elongation with precision. So we need to be able to combine evidence across samples—by writing down a number (or numbers) for each sample. In theoretical work, especially when we use synthetic data to study the behavior of techniques in controlled situations, we often need to generate samples from distributions of specified shape. The techniques of this chapter allow us to do so easily. Furthermore, they can readily be expanded to provide more detailed descriptions of distributions, in both practical and theoretical settings.

Section 11A deals with skewness, and Section 11B shows how to describe positive, symmetric elongation. Section 11C combines the two. Section 11D shows how these techniques can be extended to describe more complicated patterns of skewness and elongation. When the data take the form of an observed frequency distribution instead of a theoretical distribution, we can still apply all the techniques, as Section 11E shows. A discussion of a new family of distributions is seldom complete without information on its moments; Section 11F fills this gap. Finally, Section 11G examines the relation between the new family of shapes and two well-known systems, the Pearson curves and the Johnson curves.

11A. SKEWNESS

Our method of describing a skewed random variable, X, is based on expressing it as a monotonic function of a standard Gaussian random variable, Z. Because we are concerned with shape, we must first make allowance for the location and scale of X. We do this by writing

$$X = A + B \times Y \tag{1}$$

and thinking of Y as a "standard" random variable with the same shape as X. In this standardization we fix the median of Y at zero, so that A is the median of X. The precise definition of the scale parameter, B, will become clearer after we express Y in terms of Z.

We write Y as a function of Z, $Y = Y(Z)$. We think of skewness as produced by a reshaping function, G, which affects positive values of Z differently from negative ones. Specifically,

$$Y(Z) = G(Z) \times Z. \tag{2}$$

Aside from the special case of $G(z) \equiv 1$, we customarily consider G such that $G(-z) \neq G(z)$ for all nonzero z. Because the appearance of skewness generally becomes stronger as one looks further away from the median, we want the reshaping effect of G to be slight near zero. This can be had by asking that $Y(z) \approx z$ near 0; the constant B in equation (1) can take care of any other scaling.

Furthermore, in order to mimic the behavior of many data sets and theoretical distributions, Y must be unbounded in at least one direction.

g-Distributions

If the function $Y(z)$ is to satisfy the conditions $Y(0) = 0$ and $Y(z) \approx z$ for z near 0, its series expansion must look like $Y(z) = z + \cdots$. A convenient one-parameter family of functions that have these properties is

$$Y_g(z) = (e^{gz} - 1)/g, \tag{3}$$

where the constant g controls the skewness. We can check the leading term by recalling that

$$e^{gz} = 1 + gz + \frac{(gz)^2}{2!} + \frac{(gz)^3}{3!} + \cdots .$$

Thus

$$Y_g(z) = (e^{gz} - 1)/g = z + g\frac{z^2}{2!} + g^2\frac{z^3}{3!} + \cdots . \tag{4}$$

This form of $Y_g(z)$ corresponds to taking

$$G_g(z) = (e^{gz} - 1)/(gz) \tag{5}$$

and using $G_g(Z)$ as the multiplier of Z on the right-hand side of equation (2). We call this family of skewed distributions the "g-distributions."

DEFINITION: If Z is a standard Gaussian random variable and g is a real constant, the random variable $Y_g(Z)$ given by

$$Y_g(Z) = G_g(Z)Z = \frac{e^{gZ} - 1}{gZ} \times Z, \tag{6a}$$

or, equivalently, by

$$Y_g(Z) = (e^{gZ} - 1)/g \tag{6b}$$

is said to have the *g-distribution* with the given value of g. The parameter g controls the amount and direction of skewness.

From equation (6b) it is easy to see that changing the sign of g changes the direction, but not the amount of skewness; that is,

$$Y_{-g}(z) = -Y_g(-z). \tag{7}$$

Figure 11-1 plots $Y_g(z)$ and $Y_{-g}(z)$ against z for one value of g. When

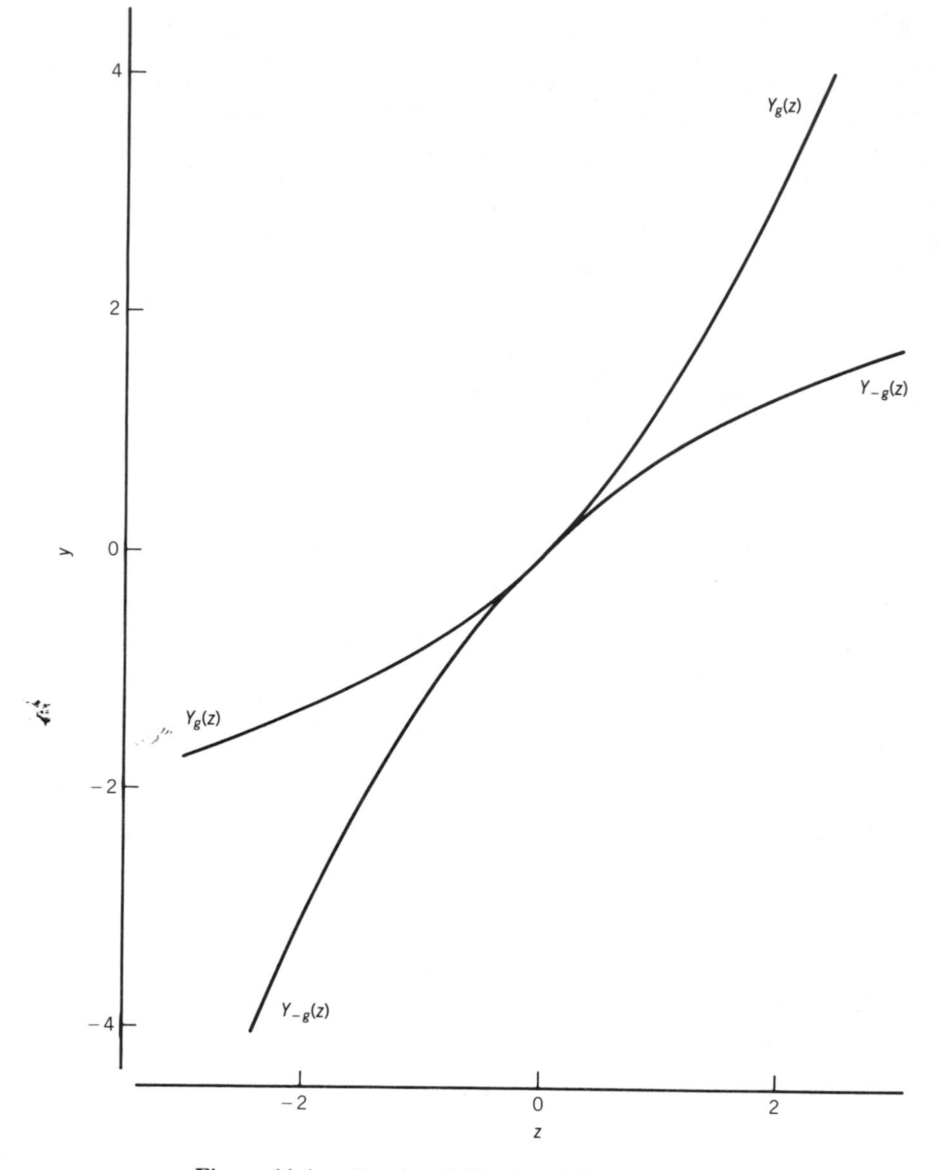

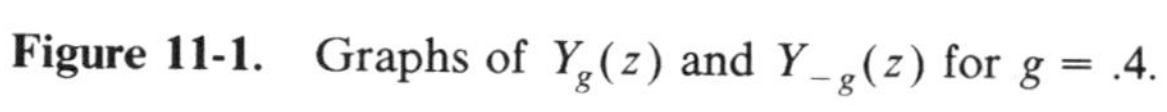

Figure 11-1. Graphs of $Y_g(z)$ and $Y_{-g}(z)$ for $g = .4$.

TABLE 11-1. Letter values of $Y_g(Z)$ for five values of g.

Tag	Tail Area[a]	z	$Y_g(z)$ $g = .2$	$g = .4$	$g = .6$	$g = .8$	$g = 1.0$
A	1/128	−2.418	−1.917	−1.550	−1.276	−1.069	−0.911
B	1/64	−2.154	−1.750	−1.444	−1.209	−1.027	−0.884
C	1/32	−1.863	−1.555	−1.313	−1.122	−0.968	−0.845
D	1/16	−1.534	−1.321	−1.147	−1.003	−0.884	−0.784
E	1/8	−1.150	−1.028	−0.922	−0.831	−0.752	−0.684
F	1/4	−0.674	−0.631	−0.591	−0.555	−0.521	−0.491
M	1/2	0	0	0	0	0	0
F	1/4	0.674	0.722	0.774	0.831	0.894	0.963
E	1/8	1.150	1.293	1.461	1.657	1.887	2.159
D	1/16	1.534	1.796	2.118	2.517	3.015	3.637
C	1/32	1.863	2.257	2.766	3.429	4.297	5.441
B	1/64	2.154	2.692	3.417	4.402	5.752	7.618
A	1/128	2.418	3.109	4.075	5.442	7.397	10.219

[a] Tail area refers to the upper tail for upper letter values (here positive) and to the lower tail for lower letter values (here negative).

$g = 0$, it follows from equation (4) that $Y_g(Z) = Z$; thus $g = 0$ corresponds to no skewness and, in fact, to the Gaussian distribution.

Because $Y_g(z)$ is strictly increasing with z, it is easy to obtain quantiles of the distribution of $Y_g(Z)$ as follows. If z_p is the pth quantile of the standard Gaussian distribution (that is, $P\{Z \leq z_p\} = p$), then $Y_g(z_p)$ is the pth quantile for $Y_g(Z)$. We now use this relationship to examine the skewness corresponding to some selected values of g.

EXAMPLE: SKEWNESS FOR SEVERAL VALUES OF g

For an indication of how varying g affects skewness, Table 11-1 shows the value of $Y_g(z)$ at the letter values (through A) for $g = .2, .4, .6, .8,$ and 1.0. To provide a graphical feel for these distributions, Figure 11-2 shows their Gaussian probability plots, and Figure 11-3 sketches their probability density functions. The principal plot for diagnosing skewness is the mid-versus-spread plot (described in Section 10D), and Figure 11-4 combines the mid-versus-spread plots derived from the letter values for the five values of g in Table 11-1. In this plot, symmetry corresponds to a horizontal line at mid $= 0$. The plotted points lie above this line and show increasing upward

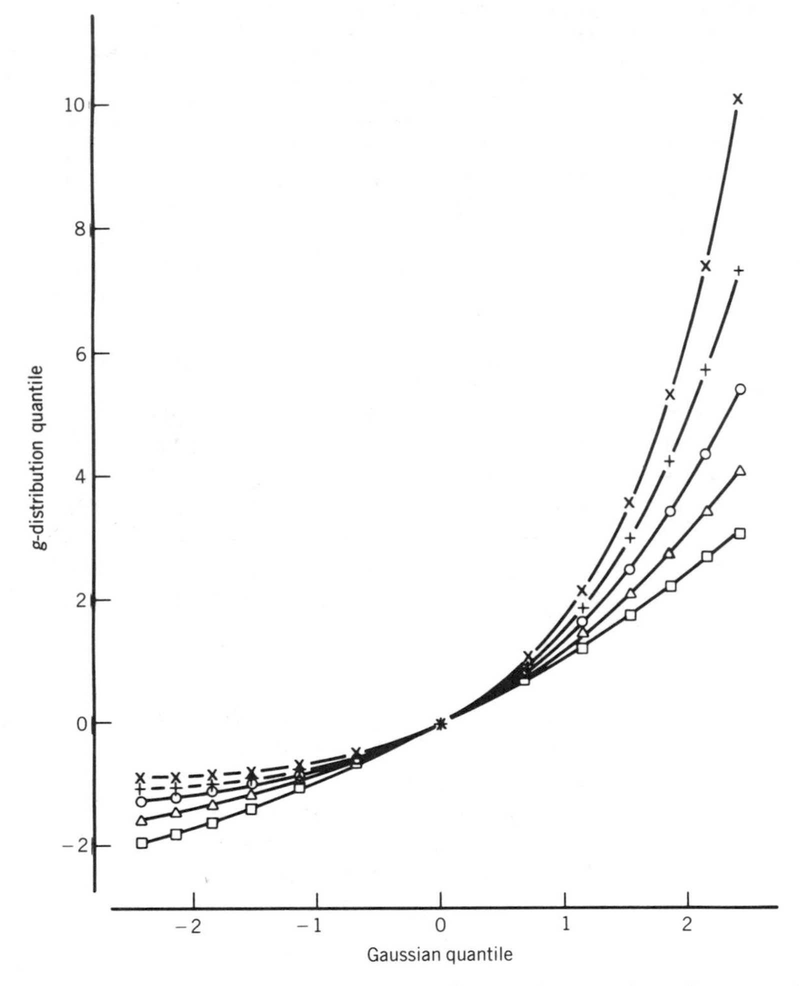

Figure 11-2. Gaussian probability plots, showing letter values, for five selected values of g ($\square$, $g = .2$; $\triangle$, $g = .4$; $\bigcirc$, $g = .6$; $+$, $g = .8$; $\times$, $g = 1.0$).

curvature as g increases. This means that the g-distributions are more skewed to the right for larger values of g. (One confusing aspect of Figure 11-4 is the way the letter-spread grows as g increases. This is an indication that skewness brings with it some stretching, relative to the Gaussian distribution. We discuss elongation in the next section and the combination of skewness and elongation in Section 11C. Later in this section we present another plot, Figure 11-7, which does not involve this confusion.)

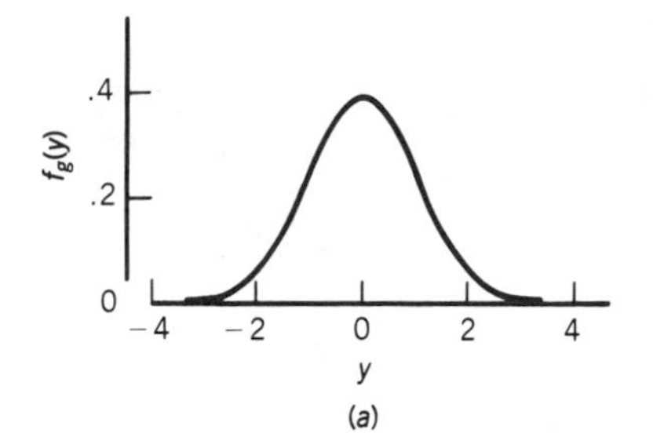

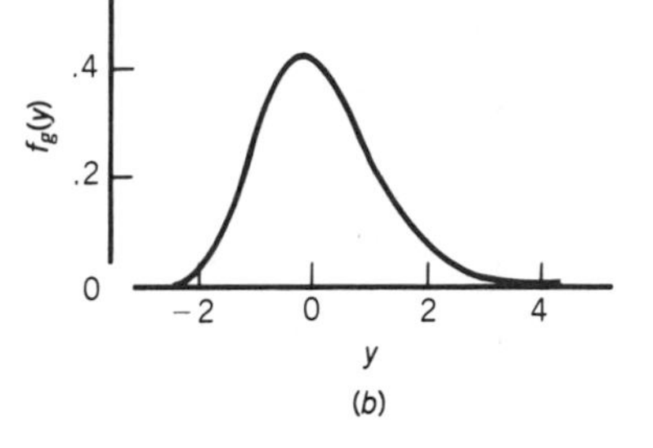

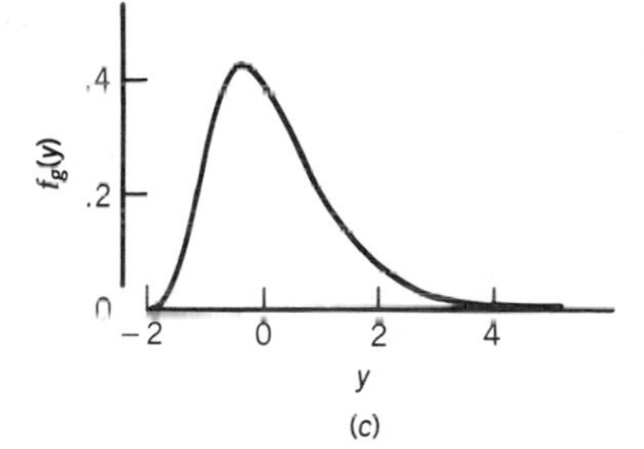

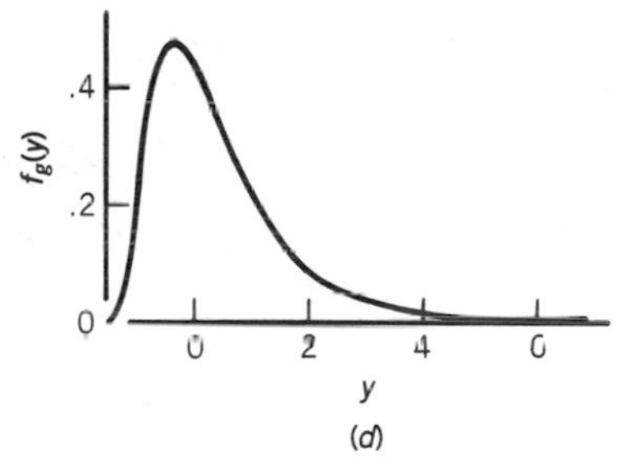

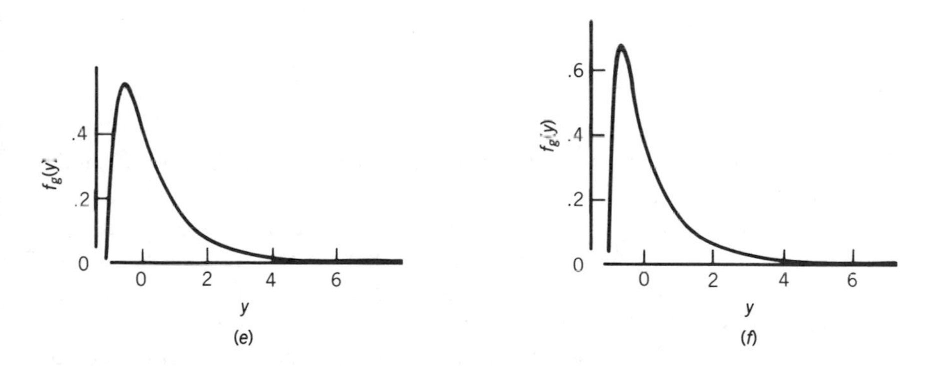

Figure 11-3. Graphs of the probability density function $f_g(y)$ for six values of g. (a) $g = 0$. (b) $g = .2$. (c) $g = .4$. (d) $g = .6$. (e) $g = .8$. (f) $g = 1.0$.

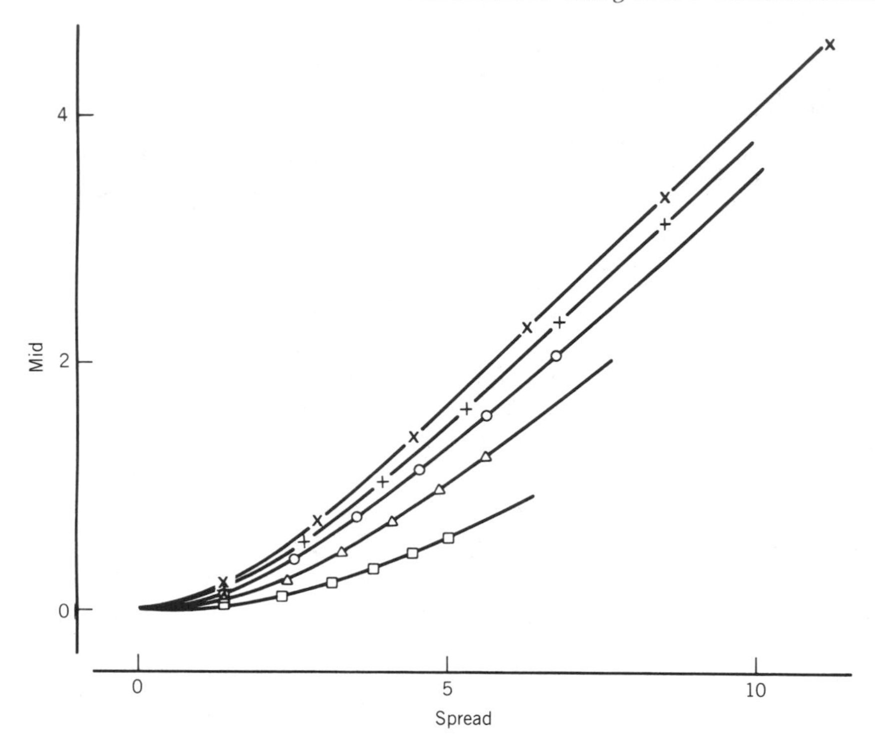

Figure 11-4. Mid-versus-spread plot showing letter values for five selected values of g ($\square$, $g = .2$; $\triangle$, $g = .4$; $\bigcirc$, $g = .6$; $+$, $g = .8$; $\times$, $g = 1.0$).

Estimating g

In working with data, we need to estimate g. The simple form of Y_g makes it possible to do this directly from quantiles. The basic idea is to find the value of g which exactly fits the pth and $(1 - p)$th quantiles of the data (as well as the median, which we have set at zero). Then, because these values of g may not be the same for different values of p, we study the relationship between g and p. In these calculations it is clearer to refer to g_p.

Determining a value of a parameter so that it exactly fits part of the data often reveals much; we use this device in defining "pseudosigma" in Section 10B. (Section 9F discusses the more general idea of cooperative diversity.) The simplest patterns of skewness are those for which g_p is almost constant across all values of p; we delay going beyond those until Section 11D.

To recover a value of g_p, we begin with the data x, related to y through $x = A + By$. Then, for $p < 0.5$,

$$x_p = A + By_p = A + B\left(\frac{e^{gz_p} - 1}{g}\right) \tag{8a}$$

$$x_{1-p} = A + By_{1-p} = A + B\left(\frac{e^{-gz_p} - 1}{g}\right) \tag{8b}$$

(because $z_{1-p} = -z_p$), and

$$x_{.5} = A + By_{.5} = A \tag{8c}$$

(because $y_{.5}$ was chosen to be 0). Using these results, we derive

$$e^{-gz_p} = \frac{x_{1-p} - x_{.5}}{x_{.5} - x_p} \tag{9}$$

and hence (remembering that $z_p < 0$), taking logarithms and solving for g (or, as we now label it, g_p), we have

$$g_p = -\frac{1}{z_p}\log_e \frac{x_{1-p} - x_{.5}}{x_{.5} - x_p}. \tag{10}$$

Thus g_p measures skewness in terms of the logarithm of the relative distances of the $(1-p)$th and pth quantiles from the median. For convenience in later discussion and examples, we refer to the (positive) distance between a quantile and the median as a *half-spread*. The pth lower half-spread and upper half-spread are $\mathrm{LHS}_p = x_{.5} - x_p$ and $\mathrm{UHS}_p = x_{1-p} - x_{.5}$, respectively.

EXAMPLE: CHI-SQUARED ON SIX DEGREES OF FREEDOM

To illustrate the procedure for estimating g, we take as "data" a set of quantiles for the chi-squared distribution on six degrees of freedom, which is fairly strongly skewed to the right but is not J-shaped. The second and third columns of Table 11-2 give these quantiles in a format similar to a letter-value display, and the columns to the right show the results of successive stages of the calculation. For $p = .25$ the half-spreads are

$$5.348 - 3.455 = 1.893 \quad \text{and} \quad 7.841 - 5.348 = 2.493,$$

TABLE 11-2. Calculations in finding g_p for the chi-squared distribution on six degrees of freedom.

	Quantiles		Half-spreads		$\log_e$ of	$z_{1-p} =$	
p	Lower	Upper	Lower	Upper	Ratio	$-z_p$	g_p
.5	5.348						
.25	3.455	7.841	1.893	2.493	0.2753	0.6745	0.4082
.10	2.204	10.645	3.144	5.297	0.5216	1.2816	0.4070
.05	1.635	12.592	3.713	7.244	0.6683	1.6448	0.4063
.025	1.237	14.449	4.111	9.101	0.7947	1.9600	0.4055
.01	0.872	16.812	4.476	11.464	0.9405	2.3264	0.4043
.005	0.676	18.548	4.672	13.200	1.0386	2.5758	0.4032

their ratio [as in equation (9)] is 2.493/1.893 = 1.317, and the natural logarithm of this is 0.2753. From Gaussian tables, $z_{.25} = -0.6745$, so that equation (10) yields $g_{.25} = 0.2753/0.6745 = 0.4082$.

Although the values of g_p (.4082, .4070, .4063, .4055, .4043, .4032) decrease steadily as the values of p become more extreme, this pattern affects only the third decimal place of the g_p. Thus, a constant value of g is not at all a bad approximation in summarizing the pattern of skewness for χ_6^2; and a suitable value is 0.406, the median of the six g_p values in Table 11-2. We will shortly (in Table 11-3) examine the fitted quantiles.

In settling on a value of g, we have established the shape corresponding to the skewness of χ_6^2, but we have not fixed the location or the scale. We can determine these and at the same time roughly examine the adequacy of the chosen constant value of g by using a quantile–quantile plot (Section 10C); we put the available quantiles of χ_6^2 on the vertical axis and the corresponding quantiles of Y_g on the horizontal axis, as in Figure 11-5. Although the result is reasonably straight, the slight curvature toward the horizontal at each end suggests that some negative elongation may be involved. (We return to this in Section 11C.) An eye-fitted line has slope 3.096 and intercept 5.348, so our approximation to χ_6^2 is

$$5.348 + 3.096\left(\frac{e^{.406Z} - 1}{0.406}\right) \tag{11a}$$

TABLE 11-3. Horizontal coordinate for Figure 11-5 and fitted quantiles for χ_6^2.

p	z_p	y_p ($g = .406$)	$5.348 + 3.096y_p$	χ_6^2 Quantile	Residual
.005	−2.576	−1.598	0.401	0.676	0.275
.01	−2.326	−1.505	0.688	0.872	0.184
.025	−1.960	−1.352	1.162	1.237	0.075
.05	−1.645	−1.200	1.633	1.635	0.002
.1	−1.282	−0.999	2.255	2.204	−0.051
.25	−0.674	−0.590	3.521	3.455	−0.066
.5	0	0	5.348	5.348	0
.75	0.674	0.776	7.750	7.841	0.091
.9	1.282	1.682	10.556	10.645	0.089
.95	1.645	2.340	12.593	12.592	−0.001
.975	1.960	2.995	14.620	14.449	−0.171
.99	2.326	3.870	17.330	16.812	−0.518
.995	2.576	4.546	19.422	18.548	−0.874

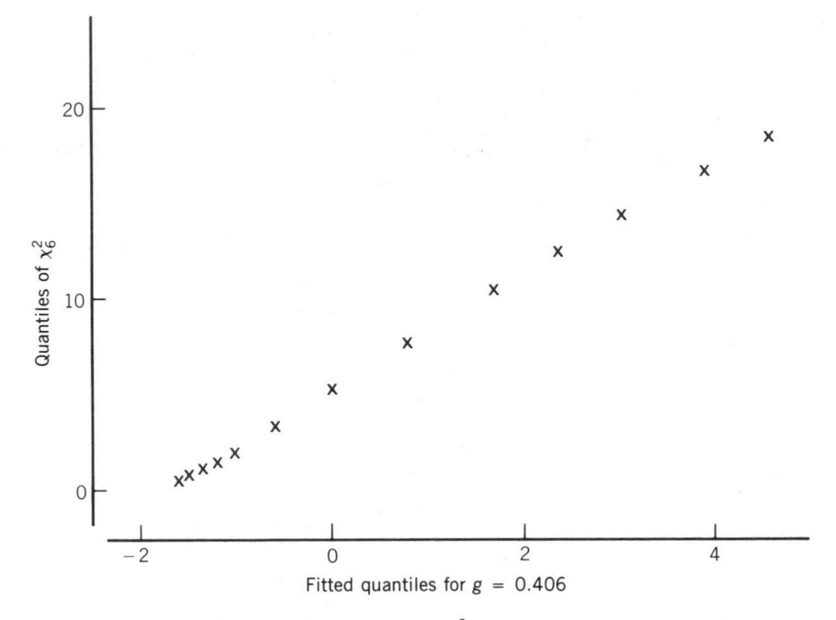

Figure 11-5. Quantile–quantile plot for χ_6^2 and the fitted g-distribution (g = 0.406).

or

$$7.626e^{0.406Z} - 2.278. \tag{11b}$$

Table 11-3 traces the major steps in the calculations from z_p through y_p to the fitted quantiles of equation (11a). For comparison it also includes the quantiles of χ_6^2. To be surer of where we stand, we could now plot the residuals of this fit; we omit this step.

We now turn to an example based on actual data instead of a theoretical distribution.

EXAMPLE: TOBACCO CONSUMPTION

For 200 male patients suffering from cancer of the lung, David and Pearson (1961, p. 4) give the amount of tobacco (in ounces per month) smoked before the onset of the illness. Table 11-4 presents a stem-and-leaf display

TABLE 11-4. Tobacco consumption of 200 male patients.

Stem-and-Leaf Display
(Leaf Unit = 1 ounce/month)

3	0*	001
8	t	22223
16	f	44455555
32	s	6666666677777777
46	0·	88888899999999
64	1*	000000000000111111
85	t	222222222222333333333
100	f	444444555555555
100	s	6666777777777
87	1·	8888888889999999
71	2*	00000001111111
57	t	22233333
49	f	444444444555555
34	s	66677
29	2·	88889999
21	3*	1
20	t	222
17	f	55
15	s	677
12	3·	89
10	4*	01
8	t	23
6	hi	44.1, 46.0, 51.9, 52.4, 54.9, 57.0,

Letter-Value Display

$n = 200$

	Depth	Lower	Mid	Upper
M	100.5		15.9	
F	50.5	10.2		23.5
E	25.5	7.15		28.8
D	13	5.2		37.8
C	7	2.8		43.2
B	4	2.1		51.9
A	2.5	1.05		53.65
	1	0		57.0

TABLE 11-5. Calculations in finding g_p for the tobacco consumption data.

		Half-Spreads				
Tag	Depth	Lower	Upper	Ratio	$-z_p$	g_p
F	50.5	5.7	7.6	1.33	0.673	.427
E	25.5	8.75	12.9	1.47	1.147	.338
D	13	10.7	21.9	2.05	1.528	.469
C	7	13.1	27.3	2.08	1.835	.400
B	4	13.8	36.0	2.61	2.090	.459
A	2.5	14.85	37.75	2.54	2.297	.406
	1	15.9	41.1	2.58	2.700	.352

and a letter-value display for these data (which are given to tenths). Some skewness to the right is evident in the stem-and-leaf display.

Table 11-5 goes on to determine the value of g_p for each pair of letter values. The values of g_p vary from .338 to .469, but they do not show any regular dependence on p. Thus we take their median, .406, as a constant-g description of the skewness. (That this agrees to three decimal places with the value of g found from Table 11-2 seems to be entirely a coincidence!)

To see how well this approximation fits the letter values of the data, we turn to Figure 11-6, a quantile–quantile plot with the quantiles of the data on the vertical axis and those of the fitted distribution on the horizontal axis. Some of the points depart from a straight line, especially at the upper end, but there is no systematic curvature. Thus a g-distribution with $g = .406$ seems to be an adequate description of these tobacco consumption data.

We can keep this value of g in mind to see how it compares with those for any other similar batches of data. Also it provides a starting point if we need to assess the adequacy for these data of inferences based on the assumption that the data follow a Gaussian distribution. We can state that the data depart from this assumption and can offer a non-Gaussian distribution that seems to be a more accurate description. If necessary, we could determine (perhaps by simulation) just what behavior we should expect of the usual significance tests or confidence intervals when the data are drawn from this particular g-distribution. Or if results were already available for a variety of g-distributions, as in Land (1972), we would know which ones to call upon.

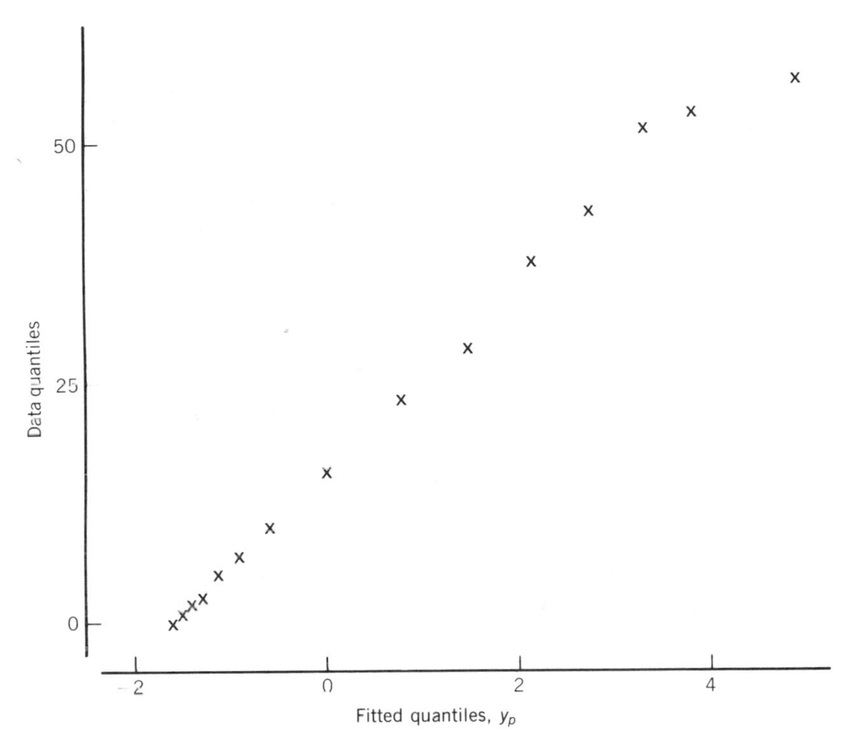

Figure 11-6. Quantile–quantile plot for the tobacco consumption data and the fitted g-distribution ($g = 0.406$).

Lognormal Distributions

A closer look at the definition

$$Y_g = \frac{e^{gZ} - 1}{g} \tag{12}$$

in equation (6b) reveals that the "g-distributions" with constant and positive g are lognormal distributions (so named because, after allowing for location and scale, the logarithm of such a random variable follows a Gaussian distribution). One common way of writing a random variable of the lognormal family (Johnson and Kotz, 1970, Chapter 14) is

$$X = \theta + e^{(Z-\gamma)/\delta}. \tag{13}$$

If we write $X = A + BY$ (with $B > 0$) and equate the two expressions for X, we arrive at the relations

$$\theta = A - \frac{B}{g}, \tag{14a}$$

$$\delta = \frac{1}{g}, \tag{14b}$$

$$\gamma = -\frac{1}{g}\log_e\left(\frac{B}{g}\right). \tag{14c}$$

In the last of these relations, we see the reason for the condition that g be positive: the argument of the logarithm must be positive. Thus, in contrast to the g-distributions, the family of lognormal distributions offers only positive skewness (i.e., skewness to the right), as in the sketch. Of course, we can easily obtain negative skewness from a lognormal distribution by using the mirror image of the distribution, just as the g-distributions for negative g are the mirror images of those for positive g.

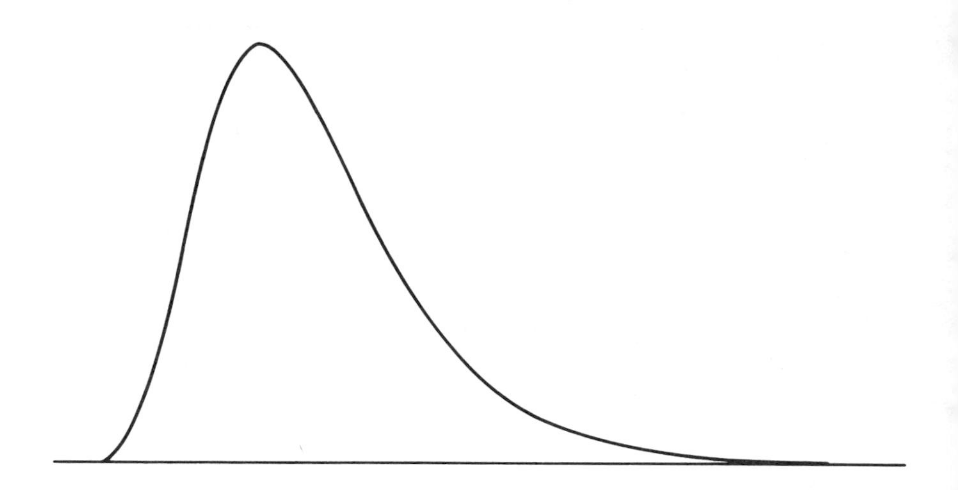

Subject to the condition on g, the relationship between (θ, δ, γ) and (A, B, g) shows how to convert one set of parameters into the other: any lognormal distribution can be written as a g-distribution with positive g and vice versa. The lognormal parameter γ plays the role of a scale parameter (through $e^{-\gamma/\delta}$); the definition of the g-distributions leaves the scaling to the parameter B. By using B for scaling and putting g in the denominator of the right-hand side of equation (12), we have gained some additional generality; we can handle negative skewness just as easily as positive skewness.

Mid-versus-z^2 Plot

Chapter 10 introduces the mid-versus-spread plot as one way of examining skewness. The primary reason for using the spreads as the horizontal plotting coordinate is a desire to watch the mid drift as we move from center to tails. We can respond to the same purposes by using the square of the standard Gaussian quantile (z_p) as the horizontal plotting coordinate, and we refer to the result as a "mid-versus-z^2 plot." Such a plot has the advantage that, although fluctuations or wild values in the data can affect the midsummaries (because a midsummary is just the average of a pair of letter values), they do not affect the horizontal plotting positions (because these are not calculated from the data). Also, when we compare samples or distributions, the mid-versus-z^2 plot will not be affected (as was the mid-versus-spread plot in Figure 11-4) by the tendency for a given letter-spread to be larger for more skewed situations.

If the mid-versus-z^2 plot follows a straight line, the data may be well approximated by a distribution with a constant value of g. To see this, we calculate a midsummary for $X = A + BY_g(Z)$. As in the calculation of half-spreads, equations (8a) and (8b) give

$$x_p = A + B\frac{e^{gz_p} - 1}{g}$$

and

$$x_{1-p} = A + B\frac{e^{-gz_p} - 1}{g},$$

so that

$$\tfrac{1}{2}(x_p + x_{1-p}) = A + \frac{1}{2}\frac{B}{g}(e^{gz_p} - 2 + e^{-gz_p}). \tag{15}$$

Now, using the first few terms of the power series

$$e^x = 1 + x + \frac{x^2}{2!} + \cdots,$$

we approximate the quantity in parentheses on the right-hand side of equation (15) by $(gz_p)^2$, omitting terms in $(gz_p)^4$ and higher even powers. Thus

$$\tfrac{1}{2}(x_p + x_{1-p}) \approx A + Bgz_p^2/2 \tag{16}$$

as long as g is not too large.

To see how the mid-versus-z^2 plot goes for a few values of g, we make this plot for the five distributions ($g = .2, .4, .6, .8, 1.0$) represented in Table 11-1. For the letter values as far as A, Figure 11-7 shows the result. In this plot, some slight curvature is evident at $g = .6$, and at $g = 1.0$ it has become very noticeable. Thus, for letter values no more extreme than letter value A ($p = 1/128$), the approximation of equation (16) should be adequate for $|g| \leq .5$. For comparing the skewness of a variety of g-distributions, the mid-versus-z^2 plot (Figure 11-7) is more effective than the mid-versus-spread plot (Figure 11-4) because variations in elongation do not enter in.

Because the slope in a straight-line mid-versus-z^2 plot involves both g and the scale factor of the data, B, further analysis will be necessary to

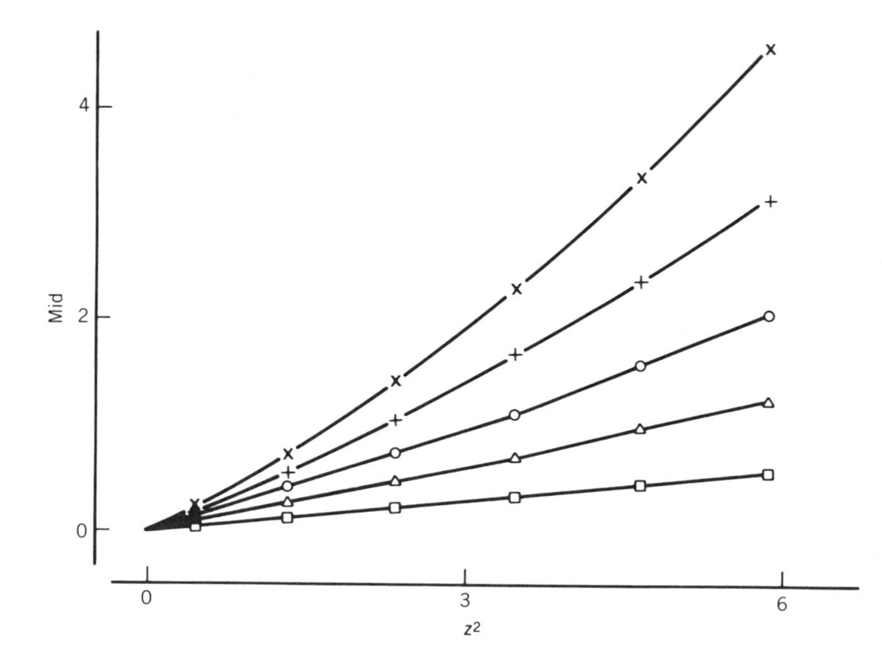

Figure 11-7. Mid-versus-z^2 plot based on letter values for five selected values of g (□, $g = .2$; △, $g = .4$; ○, $g = .6$; +, $g = .8$; ×, $g = 1.0$).

determine g. As one method, we can find the half-spreads and use the technique developed earlier in this section. We may need to be prepared for the presence of some elongation; and so we turn next to modeling elongation.

11B. ELONGATION

To model symmetric distributions whose tails are heavier than those of the Gaussian distribution, we can reshape a standard Gaussian random variable by stretching its tails. The basic idea is to write

$$Y = H(Z) \times Z \tag{17}$$

and choose the function H so as to preserve symmetry while stretching the tails. This means that H must be a positive, even function; that is, $H(-z) = H(z)$, and $0 < H(z) < +\infty$ for all finite z. And, to accomplish the stretching, H must be increasing for $z \geq 0$; this ensures that Y is an acceleratedly increasing function of Z, as desired.

h-Distributions

A simple family with the desired behavior consists of the functions

$$H_h(z) = e^{hz^2/2}, \tag{18}$$

so that

$$Y = Ze^{hZ^2/2}. \tag{19}$$

We call this family of distributions the "h-distributions."

DEFINITION: If Z is a standard Gaussian random variable and h is a real constant, the random variable $Y_h(Z)$ given by

$$Y_h(Z) = Ze^{hZ^2/2} \tag{20}$$

is said to have the *h-distribution* with the given value of h. The parameter h controls the amount of elongation.

The particular functional form given in equation (20) is suitable because it can provide a wide variety of tail-stretchings, from none at all ($h = 0$) to those associated with quite heavy-tailed distributions. Positive values of h produce positive elongation; and the larger the value of h, the more elongation. [A negative value of h is not impossible, but special treatment may be required because the monotonicity of $Y_h(z)$ fails for $z^2 > -1/h$.]

In theoretical work on characterizing the tail behavior of heavy-tailed distributions [that is, the relationship between x and $1 - F(x)$ as x becomes large], the Pareto distributions play an important role. These have the property that if F is the cumulative distribution function, then

$$1 - F(x) \approx \left(\frac{x}{k}\right)^{-\alpha} \quad \text{as } x \to \infty \tag{21}$$

for some fixed $\alpha > 0$. (The constant k is a scale parameter.) Equation (21) is concerned only with the upper tail, but it is possible for a symmetric distribution to exhibit this tail behavior, and the h-distributions do so. We can calculate their tail behavior from (20) and from the approximation that, for the standard Gaussian distribution, the upper-tail probability p and the quantile z_p are related by

$$z_p^2 \approx C - 2\log_e p, \tag{22}$$

where C is a constant whose value need not concern us here. There are more accurate approximations relating p to z_p [see, for example, Feller (1968, p. 175)], but (22) is adequate for the present discussion. From equation (20) we write y_p in terms of z_p (for simplicity, we omit h as a subscript):

$$y_p = z_p e^{hz_p^2/2}. \tag{23}$$

Then substituting for z_p^2 according to equation (22) yields

$$y_p \approx \sqrt{C - 2\log_e p}\; e^{(hC/2) - h\log_e p} \tag{24a}$$

$$= e^{hC/2}\sqrt{C - 2\log_e p}\; p^{-h}. \tag{24b}$$

Now, as p goes to zero, the behavior of (24b) is dominated by the factor p^{-h}, and this is equivalent to the Paretian tail behavior of equation (21), with $\alpha = 1/h$. Thus we see that the factor $\frac{1}{2}$ in the exponent in equation (20) produces $\alpha = 1/h$ instead of the less simple $1/2h$. The Cauchy distribution, often used as an example of a heavy-tailed distribution (see Sections 10B and 10C), has Paretian tail behavior and is well approximated

TABLE 11-6. Letter values of $Y_h(Z)$ for five values of h.

Tag	Tail Area	z	$Y_h(z)$ $h = .1$	$h = .2$	$h = .3$	$h = .4$	$h = .5$
F	1/4	0.674	0.690	0.706	0.722	0.739	0.756
E	1/8	1.150	1.229	1.313	1.403	1.499	1.601
D	1/16	1.534	1.726	1.941	2.184	2.456	2.763
C	1/32	1.863	2.216	2.635	3.134	3.728	4.435
B	1/64	2.154	2.716	3.425	4.320	5.447	6.870
A	1/128	2.418	3.238	4.337	5.810	7.781	10.423

by an h-distribution with h close to 1, as we shall see shortly. We now look at the elongation of some h-distributions with somewhat less extreme values of h.

EXAMPLE: ELONGATION FOR SEVERAL VALUES OF h

For $h = .1, .2, .3, .4, .5$, Table 11-6 presents the upper letter values as far as letter value A, calculated according to equation (20). Figure 11-8 sketches

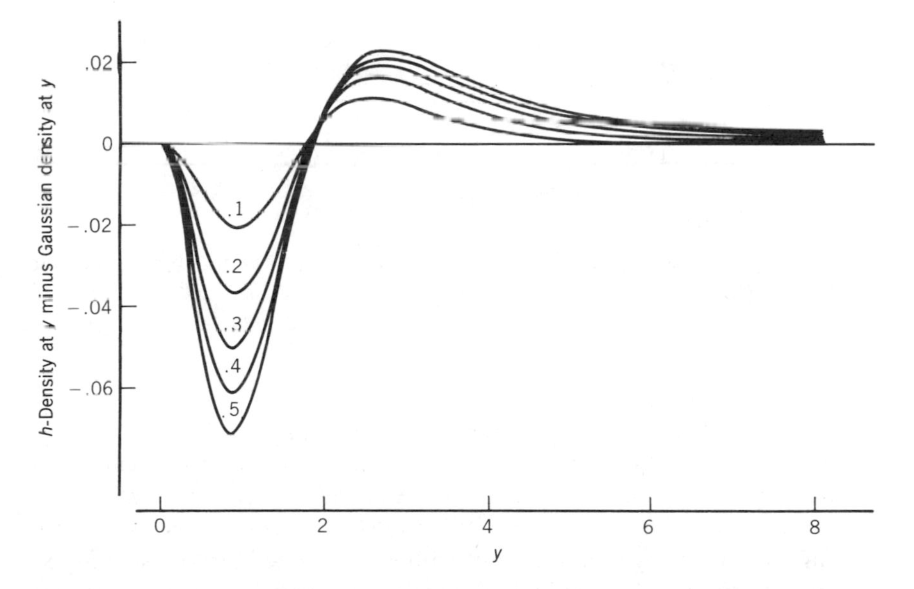

Figure 11-8. Difference between the probability density function for each of five h-distributions ($h = .1, .2, .3, .4, .5$) and the standard Gaussian density (positive arguments only). The Gaussian density itself equals .40, .24, .054, .0044, and .0001 when $y = 0, 1, 2, 3$, and 4, respectively.

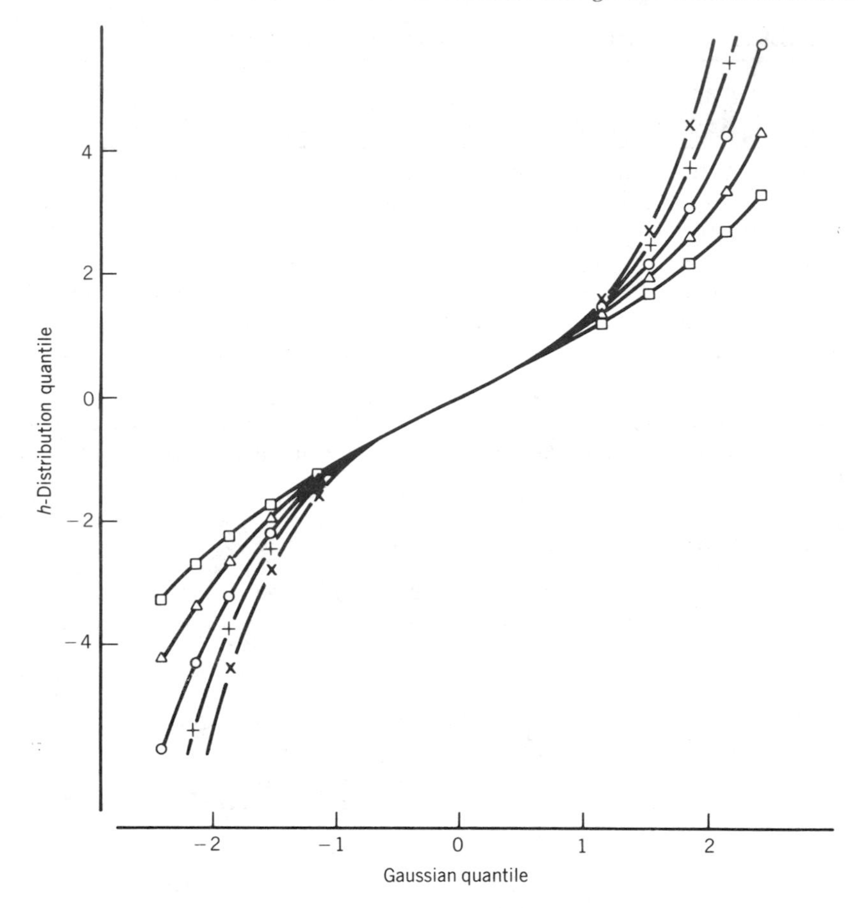

Figure 11-9. Gaussian probability plots, showing letter values, for five selected values of h ($\square, h = .1$; $\triangle, h = .2$; $\bigcirc, h = .3$; $+, h = .4$; $\times, h = .5$).

the difference between the probability density function for each of these h-distributions and the standard Gaussian density (for positive values only). Comparing the column of Table 11-6 for each value of h to the column of z values, we can see the extent of the tail stretching. Looking across the rows, we can see how elongation increases with h. To show both of these aspects graphically, Figure 11-9 combines the five Gaussian quantile–quantile plots. (The point at zero is common to all five distributions, and the vertical scale of the plot does not separate the fourths.) It is evident that $h = .5$ corresponds to quite substantial elongation.

Estimating h

In working with data, we must allow for changes in location and scale, so we write

$$X = A + BY,$$

just as we did with the g-distributions. If we know A and B and can, therefore, work with Y directly, we can find the value of h which exactly fits the pth quantile. We refer to this value as h_p. For $0 < p < 0.5$ or $0.5 < p < 1$, equation (23) readily yields

$$h_p = \left[2 \log_e(y_p/z_p)\right]/z_p^2. \tag{25}$$

Note that when we deal with a symmetric distribution, $h_{1-p} = h_p$. This equality need not hold for a set of data. In general, h_p may not be constant across all values of p. For the present, however, we work only with constant h and leave more complicated patterns of behavior until Section 11D.

When A and B are unknown, symmetry still allows us to eliminate A by using the letter-spreads (or other differences of symmetric quantiles) of the data. For $0 < p < 0.5$ we have

$$x_p = A + Bz_p e^{hz_p^2/2} \tag{26a}$$

and (remembering that $z_p < 0$ and $z_{1-p} = -z_p$)

$$x_{1-p} = A - Bz_p e^{hz_p^2/2}, \tag{26b}$$

so that the corresponding quantile spread is

$$x_{1-p} - x_p = -2Bz_p e^{hz_p^2/2}. \tag{27}$$

From Chapter 10, we may recall that a pseudosigma is defined as a quantile spread of the data divided by the corresponding quantile spread for the standard Gaussian distribution. This allows us to rearrange (27):

$$(\text{pseudosigma})_p = \frac{x_{1-p} - x_p}{-2z_p} = Be^{hz_p^2/2}. \tag{28}$$

Thus a plot of $\log_e(\text{pseudosigma})_p$ against $z_p^2/2$ for a set of values of p (such as those for the letter values) would be linear with intercept $\log_e B$ and slope h.

The plot of $\log_e$(pseudosigma) against $z^2/2$ is a modification of the pseudosigma-versus-z^2 plot, which can be used to diagnose the presence of elongation (Section 10D). Now that we are using the h-distributions to measure elongation quantitatively, we have a clear reason for a logarithmic scale on the vertical axis. For an illustration of this plot and the other calculations in estimating h, we turn to an example based on the Cauchy distribution.

EXAMPLE: CAUCHY DISTRIBUTION

The standard Cauchy distribution (considered also in Sections 10B and 10C) is symmetric about 0, so $A = 0$ in equations (26), but we still need the appropriate value of B. Table 11-7 gives the letter values and pseudosigmas and includes columns of $\log_e$(pseudosigma) and $z^2/2$. For example, using the fourths, we have F-pseudosigma $= 1.000/0.6745 = 1.483$, and its natural logarithm is 0.394. Figure 11-10 shows the plot of $\log_e$(pseudosigma) against $z^2/2$. The pattern is very close to a straight line, with the two leftmost points lying slightly above it. A good eye-fitted slope is 0.975, and the intercept is quite close to 0—too close to estimate reliably from the plot. Thus, for these letter values, the constant value $h = 0.975$ summarizes the elongation of the Cauchy distribution quite well. (The scale constant, B, is not of any great interest in this example because it is not involved in describing shape. Indeed, the value of B is entirely a consequence of which member of the Cauchy family is used as the standard member. See UREDA, Section 12A.) In general we can again use a quantile–quantile plot to determine values of A and B, as well as to check the adequacy of the

TABLE 11-7. Calculation of $\log_e$(pseudosigma) for the Cauchy distribution.

Tag	Tail Area	Cauchy Letter Value	z	Pseudosigma	$\log_e$(pseudo-sigma)	$z^2/2$
F	1/4	1.000	0.6745	1.483	0.394	0.23
E	1/8	2.414	1.1503	2.098	0.741	0.66
D	1/16	5.027	1.5341	3.277	1.187	1.18
C	1/32	10.153	1.8627	5.451	1.696	1.73
B	1/64	20.355	2.1539	9.450	2.246	2.32
A	1/128	40.735	2.4176	16.850	2.824	2.92
Z	1/256	81.483	2.6601	30.633	3.422	3.54
Y	1/512	162.973	2.8856	56.480	4.034	4.16
X	1/1024	325.948	3.0973	105.230	4.646	4.80

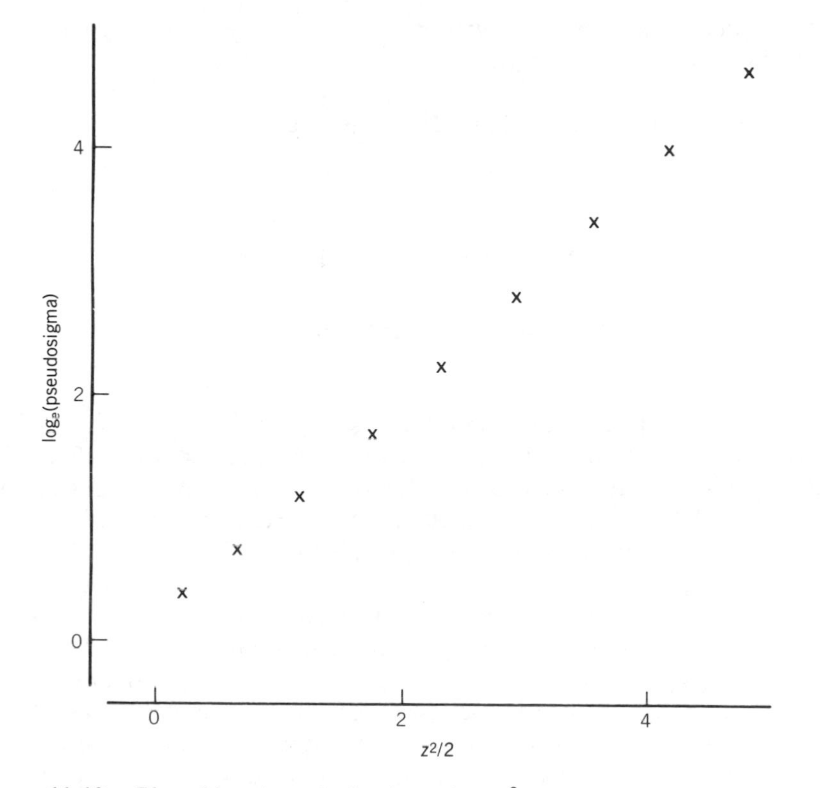

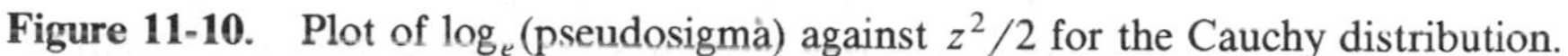

Figure 11-10. Plot of $\log_e$(pseudosigma) against $z^2/2$ for the Cauchy distribution.

approximation. We can also refine either or both of these graphical approaches by plotting residuals.

11C. COMBINING SKEWNESS AND ELONGATION

Often a distribution or a set of data is both skewed and elongated, so we will need to be able to handle these two aspects of shape simultaneously. Our approaches to skewness and elongation in the previous two sections both involved reshaping the unit Gaussian random variable, Z. We multiplied it by $G(Z)$ for skewness and $H(Z)$ for elongation; now, to combine the two, we again use multiplication, yielding

$$Y = G(Z)H(Z)Z. \tag{29}$$

For the particular choice of functions G and H in equations (5) and (18), this gives

$$Y_{g,h}(Z) = Z\left(\frac{e^{gZ} - 1}{gZ}\right)e^{hZ^2/2} \tag{30a}$$

or, equivalently,

$$Y_{g,h}(Z) = \left(\frac{e^{gZ} - 1}{g}\right)e^{hZ^2/2}; \tag{30b}$$

and, as before, we allow for location and scale by taking

$$X = A + BY,$$

so that A is the median of X and B is a scale constant.

Combining skewness and elongation in this way extends our definition of "neutral elongation." For $h = 0$ in Section 11B the shape is Gaussian, the customary definition of "neutral elongation" where only symmetric distributions are concerned. We continue to define neutral elongation in skewed situations as $h = 0$, but now we have symmetry (and the Gaussian shape) only when $g = 0$ as well. For $g \neq 0$ we have the lognormal distributions, as shown in Section 11A, which we now take as neutrally elongated.

If we had approached elongation by simply comparing data and distributions to the Gaussian distribution, we could have found that skewed data and distributions usually appear to be elongated. For example, in terms of the pseudosigmas (Section 10B) and the pseudosigma-versus-z^2 plot (Section 10D), we would see that the pseudosigmas from more extreme quantiles tend to become larger. It is convenient to avoid this entanglement between skewness and elongation by first summarizing the skewness and allowing for it and then treating as elongation any tail heaviness not associated with skewness. In a sense this approach extends the customary requirements that measures of scale be free of location and that measures of shape be free of location and scale. We return to this point in Section 11G.

One advantage of handling g and h in separate, multiplicative reshapings becomes apparent when we try to recover the value of g. If, as in equation (10) (in Section 11A), we calculate g_p, the value of g that exactly fits x_p and x_{1-p}, we find (with $0 < p < 0.5$) that

$$\frac{x_{1-p} - x_{.5}}{x_{.5} - x_p} = \frac{A + B\dfrac{e^{-gz_p} - 1}{g}e^{hz_p^2/2} - A}{A - \left(B\dfrac{e^{gz_p} - 1}{g}e^{hz_p^2/2} + A\right)}$$

$$= e^{-gz_p}$$

because $e^{hz_p^2/2}$ conveniently cancels. Thus we can determine values of g without having to do anything about h. Just as in Section 11A, we have

$$g_p = -\frac{1}{z_p}\log_e\frac{x_{1-p} - x_{.5}}{x_{.5} - x_p}; \tag{31}$$

and, if appropriate, we try to summarize skewness with a constant value of g.

Once we have chosen a value of g, we can turn to h. We begin with the upper half-spread, $\text{UHS}_p = x_{1-p} - x_{.5}$, which we are describing as

$$x_{1-p} - x_{.5} = \frac{B}{g}(e^{-gz_p} - 1)e^{hz_p^2/2}. \tag{32}$$

Dividing through by $(e^{-gz_p} - 1)/g$ adjusts for the skewness (at least as long as the g_p are constant) and leaves only elongation. Thus we can use

$$\text{UHS}_p^* = \frac{g(x_{1-p} - x_{.5})}{e^{-gz_p} - 1} = Be^{hz_p^2/2} \tag{33}$$

in much the same way as we used $x_{1-p} - x_p$ in Section 11B: A plot of

$$\log_e(\text{UHS}_p^*) \text{ against } z_p^2/2$$

would have intercept $\log_e(B)$ and slope h.

In working with data, and thus necessarily with approximate values of g (which may not be nearly enough constant in the data), it may help to use both half-spreads, either by averaging the logarithmic results or by dividing the full spread by the appropriate denominator before taking the logarithm.

EXAMPLE: CHI-SQUARED ON SIX DEGREES OF FREEDOM

To illustrate the process of adjusting for skewness and estimating elongation, we continue the χ_6^2 example begun in Table 11-2. Table 11-8 shows the calculations, and Figure 11-11 is the plot. The points lie close to a straight line with an eye-fitted slope of $-.033$, so we can take $h = -.033$. We have been discussing elongation and emphasizing positive values of h, but we warned in Section 11B that negative values are not impossible. The message of a negative h here is that the χ_6^2 distribution has slight negative elongation. The approximation based on $g = 0.406$ and $h = -.033$ will be fairly good, but it cannot be fully precise because h is negative—as we mentioned in Section 11B, the monotonicity of $Y_h(z)$ fails for $z^2 > -1/h$. In the

TABLE 11-8. Adjusting for a constant g ($g = 0.406$) in the χ_6^2 example.

p	UHS_p	$-z_p$	$(e^{-gz_p} - 1)/g$	$\log_e(\text{UHS}_p^*)$	$z_p^2/2$
.25	2.493	0.6745	0.7759	1.1672	0.227
.10	5.297	1.2816	1.6812	1.1476	0.821
.05	7.244	1.6448	2.3397	1.1301	1.353
.025	9.101	1.9600	2.9954	1.1113	1.921
.01	11.464	2.3264	3.8709	1.0857	2.706
.005	13.200	2.5758	4.5458	1.0660	3.317

present situation this yields $z^2 > 30.30$ or $|z| > 5.505$, not a particularly serious limitation for most purposes (the tail probabilities that are folded back amount to roughly 1.9×10^{-8} each).

We must still find the intercept, $\log_e(B)$, from Figure 11-11. Subtracting $-.0165z_p^2$ from the vertical coordinate of each point and taking the median of the resulting partial residuals yields 1.1748. From this value of $\log_e(B)$ we could recover B as 3.237, but this is not necessary to write out the

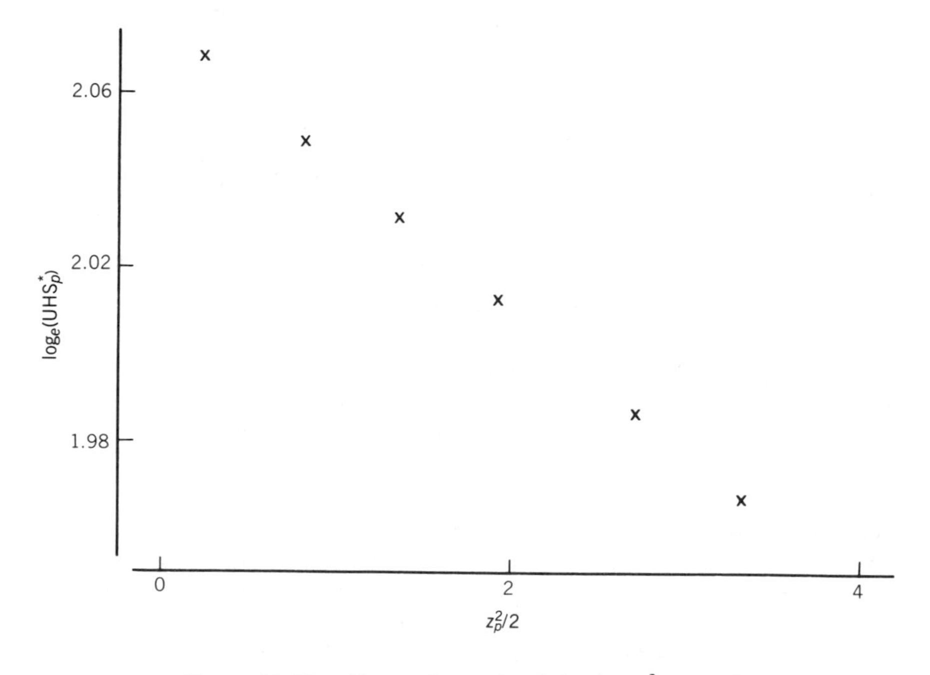

Figure 11-11. Plot to determine h in the χ_6^2 example.

TABLE 11-9. Accuracy of the approximation $\chi_6^2 \approx 5.348 + 7.974(e^{.406Z} - 1)e^{-.0165Z^2}$

p	z_p	χ_6^2 quantile	Approximation	Residual
.005	−2.576	0.676	0.712	−0.036
.01	−2.326	0.872	0.891	−0.019
.025	−1.960	1.237	1.241	−0.004
.05	−1.645	1.635	1.633	0.002
.1	−1.282	2.204	2.199	0.005
.25	−0.6745	3.455	3.452	0.003
.5	0	5.348	5.348	0
.75	0.6745	7.841	7.841	0
.9	1.282	10.645	10.647	−0.002
.95	1.645	12.592	12.593	−0.001
.975	1.960	14.449	14.450	0.001
.99	2.326	16.812	16.806	0.006
.995	2.576	18.548	18.540	0.008

approximation we have found. Putting together all the pieces, we get

$$\chi_6^2 \approx 5.348 + e^{2.0762}(e^{.406Z} - 1)e^{-.033Z^2/2} \tag{34a}$$

or

$$\chi_6^2 \approx 5.348 + 7.974(e^{.406Z} - 1)e^{-.0165Z^2} \tag{34b}$$

To see how well this fits all the quantiles in Table 11-2, we turn now to Table 11-9. Except for $p = .005$ and $p = .01$ the residuals are all smaller in magnitude than .01, so the approximation is a reasonably accurate way of calculating quantiles of χ_6^2 from standard Gaussian quantiles. Because the values of g_p are not quite constant and because we determined h from the upper half of the distribution by using a single value of g, we should not be surprised that the approximation is somewhat less accurate at the lower end of the distribution.

The problem of approximating quantiles of a chi-squared distribution in terms of the corresponding standard Gaussian quantiles has received considerable attention in the statistical literature [see, for example, Johnson and Kotz (1970, Section 17.5)]. One of the most accurate approximations, devised by Wilson and Hilferty (1931), applies to all chi-squareds: If χ_ν^2 is a random variable whose distribution is chi-squared on ν degrees of freedom, then the distribution of

$$\sqrt[3]{\frac{\chi_\nu^2}{\nu}}$$

is approximately Gaussian with mean $1 - 2/9\nu$ and variance $2/9\nu$. Thus, $\chi_{\nu,p}^2$, the pth quantile of χ_ν^2, is well approximated by

$$\nu\left(z_p\sqrt{\frac{2}{9\nu}} + 1 - \frac{2}{9\nu}\right)^3.$$

However, as Exercise 7 shows, the g-and-h approximation is substantially closer.

Pursuing a different approach, Hoaglin (1977) discovered that, for $\nu \geq 5$ and $.005 \leq 1 - p \leq .1$, quantiles of χ_ν^2 in the upper tail can be reasonably well approximated directly (that is, in terms of ν and p, rather than ν and z_p). The approximation

$$\chi_{\nu,p}^2 \approx \left[1.00991\sqrt{\nu} + 1.95188\sqrt{-\log_{10}(1-p)} - 1.14485\right]^2$$

generally has relative error less than 1%. UREDA (Section 6H) describes the initial steps of the data-analytic approach underlying such a direct approximation.

11D. MORE GENERAL PATTERNS OF SKEWNESS AND ELONGATION

As the preceding sections show, when g and h are constant, we can readily work with

$$Y = \frac{e^{gZ} - 1}{g} e^{hZ^2/2}. \tag{35}$$

In the χ_6^2 example, however, the steady decrease of g_p as p decreases (Table 11-2) indicates that a more complicated description might sometimes be required. We can have this greater generality by allowing g and h to be functions of z. One way to do this is to treat them as polynomials in z^2. Thus we can begin with

$$g(z) = g_0 + g_2 z^2 \tag{36}$$

and

$$h(z) = h_0 + h_2 z^2 \tag{37}$$

and later include higher powers of z^2 if the situation calls for them. Expressing g and h as functions of z^2 ensures that H will continue to be an even function of z and G will continue not to be an even function.

To determine g_0 and g_2 in $g(z)$, we begin, in just the same way as in Sections 11A and 11C, by calculating g_p. We continue by plotting g_p against z_p^2 and choosing suitable values for g_2 and g_0. (The residuals, that is, $g_p - g_0 - g_2 z_p^2$, may reveal a need for higher-order terms.)

Before we can summarize the pattern of elongation, we must adjust for skewness. The idea is the same as in Section 11C, where we found $g(x_{1-p} - x_{.5})/(e^{-gz_p} - 1)$ and expected a plot of

$$\log_e \frac{g(x_{1-p} - x_{.5})}{e^{-gz_p} - 1} \quad \text{against} \quad \frac{z_p^2}{2}$$

to have slope h and intercept $\log_e(B)$. We must allow for the new feature that $g(z)$ may not be constant. Thus the divisor will be $(e^{-g(z_p)z_p} - 1)/g(z_p)$, and the logarithm of the adjusted upper half-spread,

$$\log_e(\text{UHS}_p^*) = \log_e \frac{(x_{1-p} - x_{.5})g(z_p)}{e^{-g(z_p)z_p} - 1}, \tag{38}$$

should now be fitted by

$$\log_e(B) + \frac{h_0}{2} z_p^2 + \frac{h_2}{2} z_p^4. \tag{39}$$

Again, plotting is the first step. An example shows how these steps fit together.

EXAMPLE: HOUSEHOLD INCOMES

Table 11-10 begins with the half-spreads for the annual incomes reported by the 994 Pittsburgh households that enrolled in the Housing Allowance Demand Experiment (described briefly in Section 10A; a letter-value display for these data appears in Table 10-2). The next two columns give the calculations for describing skewness: the ratio of the half-spreads and the value of g_p. Because the values of g_p decrease steadily and substantially, a constant value of g will not be a very effective description of the pattern of skewness. Plotting g_p against z_p^2 (Figure 11-12) reveals that a straight line comes much closer, and fitting by eye yields

$$g_p \approx 0.493 - 0.025 z_p^2.$$

TABLE 11-10. Calculations for describing skewness in the household income data.

Tag	Half-Spreads[a] Lower	Upper	Ratio = (Upper)/(Lower)	$g_p = -(1/z_p)\log_e(\text{ratio})$	z_p^2
F	1068	1464	1.371	0.468	0.455
E	1692	2963	1.751	0.487	1.323
D	1963	3804	1.938	0.431	2.353
C	2232	4870	2.182	0.419	3.470
B	2516.5	5514	2.191	0.364	4.639
A	2752.5	6274.5	2.280	0.341	5.845
Z	2901	6730	2.320	0.316	7.076
Y	3135	7195.5	2.295	0.288	8.327
X	3366	7394	2.197	0.254	9.593

[a] Median = 3480.

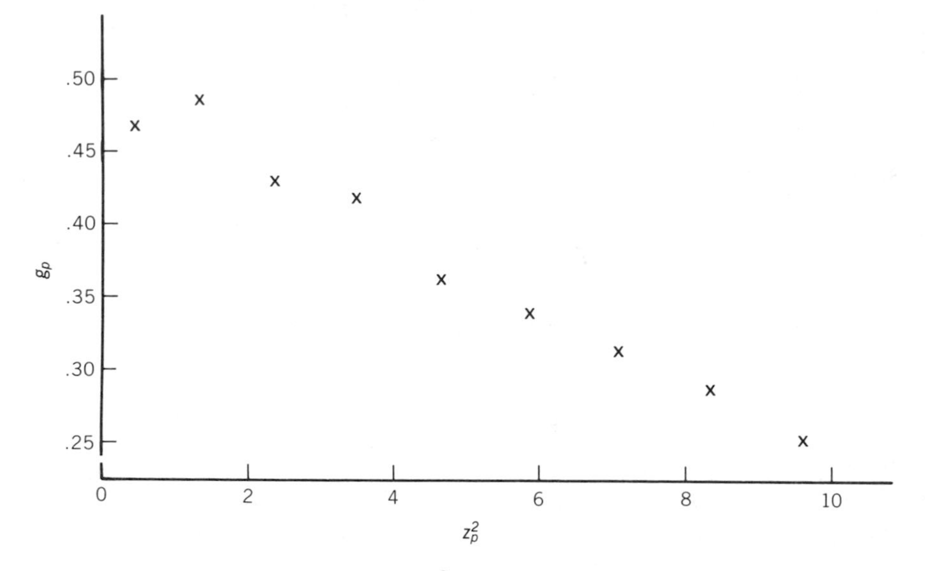

Figure 11-12. g_p versus z_p^2 for the household income data.

The point based on the eighths departs noticeably from the line, but no explanation for this suggests itself. Overall, this set of data is less skewed in the tails than a lognormal distribution with the same skewness in the shoulders. Correspondingly, these data are more skewed in the middle than such a lognormal distribution. (Recall that a lognormal distribution has a constant value of g.) Because these are low-income households, there is

TABLE 11-11. Adjusting the upper half-spreads of the household income data for the fitted pattern of skewness, $g(z) = 0.493 - 0.025z^2$.

Tag	UHS_p	$-z_p$	$g(z_p)$	$G^*(z_p)^a$	$\log_e\left(\frac{\text{UHS}_p}{G^*(z_p)}\right)$	z_p^2
F	1464	0.6745	0.482	0.797	7.516	0.455
E	2963	1.1503	0.460	1.516	7.578	1.323
D	3804	1.5341	0.434	2.180	7.465	2.353
C	4870	1.8627	0.406	2.784	7.467	3.470
B	5514	2.1539	0.377	3.322	7.414	4.639
A	6274.5	2.4176	0.347	3.786	7.413	5.845
Z	6730	2.6601	0.316	4.170	7.386	7.076
Y	7195.5	2.8856	0.285	4.477	7.382	8.327
X	7394	3.0973	0.253	4.701	7.361	9.593

[a] $G^*(z) = (e^{-g(z)z} - 1)/g(z)$.

some constraint on how far to the right the data can extend, but more detailed information would be required to discuss the nature of this constraint, which is not a simple cutoff. In fact, the eligibility limit on income depended on both the treatment group and the number of persons in the household.

To see whether the household income data also involve elongation, we must first adjust for the fitted pattern of skewness. Table 11-11 shows the steps in the calculations, and Figure 11-13 plots

$$\log_e(\text{UHS}_p^*) \quad \text{versus} \quad z_p^2.$$

There may be a suggestion of curvature, but most of this comes from the point for the eighths. That point continues to be discrepant because it does not fit into the overall pattern of skewness. Fitting a straight line (by the resistant technique described in Section 7A and in UREDA, Chapter 5) yields

$$7.52 - 0.0168z_p^2,$$

and the corresponding set of residuals does not curve enough to warrant trying an additional term in z_p^4. Interpreting the coefficients of the line as

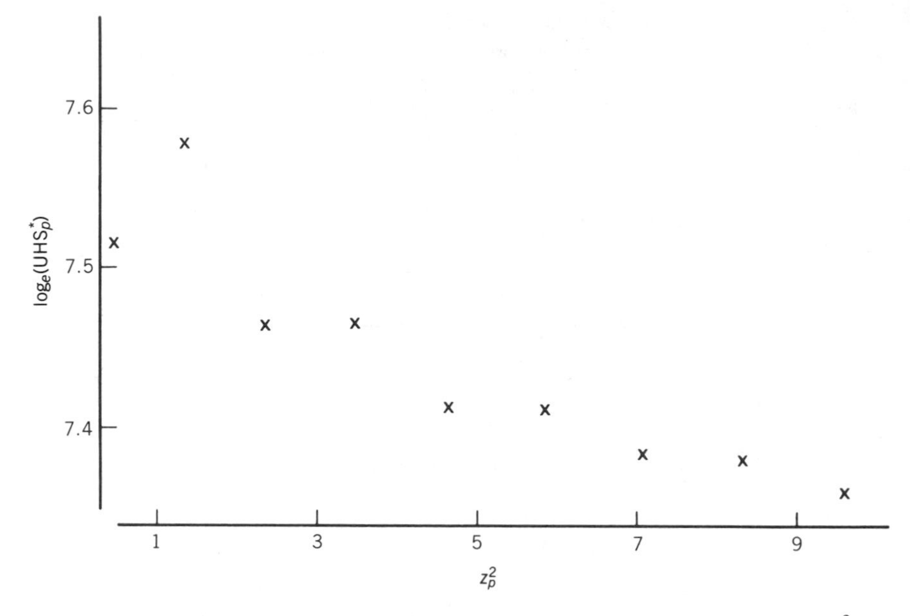

Figure 11-13. Plot of natural logarithm of adjusted upper half-spread versus z_p^2 for the household income data.

$\log_e(B) = 7.52$ and $h_0/2 = -0.0168$ brings us to $h_0 = -0.0336$ and $B = 1845$. Thus this set of data has a slight negative elongation.

The fact that the median is 3480 gives $A = 3480$ and completes the description in terms of the generalized (g, h) family of distributions. The result is

$$3480 + 1845 \frac{e^{g(Z)Z} - 1}{g(Z)} e^{-0.0168Z^2}, \tag{40a}$$

$$g(Z) = 0.493 - 0.025Z^2. \tag{40b}$$

Instead of adjusting the upper half-spreads for the fitted pattern of skewness (as we did in Table 11-11), we could use the lower half-spreads and apply a similar adjustment. Although we do not show the calculations and plot (corresponding to Figure 11-13), we did make such a preliminary examination of the lower tail as a partial check on the fit. The result was essentially the same.

For an overall examination of how the fitted distribution agrees with the data, we can compare the observed letter values with the fitted ones calculated from equations (40). Table 11-12 shows these and the corresponding residuals. Except for three letter values in the upper tail, the fit is satisfactory, and even these residuals are not large relative to the data values.

By allowing g to be a simple function of z^2, we have managed (working with the letter values) to reduce a sample of 994 incomes to a description of their whole distribution in terms of five constants, as recorded in equations (40). From this fit we should be able to recover rather accurately other quantiles of the distribution, especially if these do not require extrapolation beyond the observed extremes.

TABLE 11-12. Observed and fitted letter values for the household income data.

	Letter Value		
Tag	Observed	Fitted	Residual
X	114	109	5
Y	345	324	21
Z	579	533	46
A	727.5	743	−15
B	963.5	963	1
C	1248	1206	42
D	1517	1494	23
E	1788	1868	−80
F	2412	2426	−14
M	3480	3480	0
F	4944	4939	5
E	6443	6216	227
D	7284	7346	−62
C	8350	8327	23
B	8994	9150	−156
A	9754.5	9811	−57
Z	10210	10312	−102
Y	10675.5	10660	16
X	10874	10865	9

11E. WORKING FROM FREQUENCY DISTRIBUTIONS

When the data come in the form of a frequency distribution, which gives the edges of a set of bins and the number of observations in each bin, we generally cannot obtain exactly complementary quantiles, x_p and x_{1-p}. Often it will be adequate to approximate the letter values of the sample by interpolation and then proceed as before. Once we have chosen values of g and h, we can make a Q-Q plot to check the fit by working with the bin edges and the corresponding cumulative counts (as mentioned in Section 10C).

Letter Values by Interpolation

Usually a frequency distribution has only enough detail to indicate which bin contains each of the letter values. To come closer to the actual letter values of the sample, we must use some form of interpolation.

The simplest procedure treats the observations in a bin as if they were spread uniformly across the bin. Specifically, we denote the bin boundaries by

$$x_0, x_1, \ldots, x_k$$

and the bin counts by

$$n_0, n_1, \ldots, n_k, n_{k+1}$$

so that n_i is the count in the bin whose right-hand boundary is x_i. The last bin, with count n_{k+1}, is open to the right; the first bin, with count n_0, is open to the left. (Often both n_0 and n_{k+1} are zero.) If the lower letter value at depth d lies in the bin whose boundaries are x_{L-1} and x_L, then we treat this bin as n_L subintervals of width $(x_L - x_{L-1})/n_L$, each with an observation at its center. Thus we act as if the leftmost observation in the bin falls at $x_{L-1} + 0.5(x_L - x_{L-1})/n_L$, the next observation at $x_{L-1} + 1.5(x_L - x_{L-1})/n_L$, and so on. We place the interpolated lower letter value at

$$x_{L-1} + \frac{d - (n_0 + \cdots + n_{L-1}) - 0.5}{n_L}(x_L - x_{L-1}). \tag{41}$$

Similarly, if the upper letter value at depth d lies between x_{U-1} and x_U, we place the interpolated upper letter value at

$$x_U - \frac{d - (n_{U+1} + \cdots + n_{k+1}) - 0.5}{n_U}(x_U - x_{U-1}). \tag{42}$$

Of course, when a letter value falls in either of the open-ended bins, we will be unable to interpolate.

EXAMPLE: LENGTHS OF BEANS

One well-known frequency distribution in the statistical literature is based on the lengths (in mm) of 9440 beans (Pretorius, 1930; Kendall and Stuart, 1963). Table 11-13 shows the bin boundaries and bin counts along with cumulative counts from each end. (For later use the last two columns give the tail fraction based on the depth at the bin boundary and the corresponding Gaussian quantile.) Table 11-14 gives the interpolated letter values and the values of g_p, found according to equation (10). As an example of the calculation for interpolated letter values, the lower X, at depth 10, is

$$10.25 + \frac{10 - 8 - 0.5}{18}(10.75 - 10.25) = 10.292.$$

Although the values of g_p range from $-.251$ to $-.131$, they do not do so in a way that is linearly related to z_p^2. Thus it is simplest to begin by taking their median, $-.205$, as the basis for an initial fit with constant g.

TABLE 11-13. Frequency distribution for the lengths of 9440 beans.

Bin Center	Bin Count	Cumulative Count (d)	Fraction $(d + \frac{1}{6})/(n + \frac{1}{3})$	Gaussian Quantile
9.5	1	1	.0001236	−3.665
10	7	8	.0008651	−3.133
10.5	18	26	.002772	−2.774
11	36	62	.006585	−2.479
11.5	70	132	.01400	−2.197
12	115	247	.02618	−1.940
12.5	199	446	.04726	−1.672
13	437	883	.09355	−1.319
13.5	929	1812	.1920	−0.871
14	1787	3599	.3813	−0.302
14.5	2294	3547	.3757	0.317
15	2082	1465	.1552	1.014
15.5	1129	336	.03561	1.804
16	275	61	.006479	2.485
16.5	55	6	.0006532	3.215
17	6			
	(n = 9440)			

TABLE 11-14. Interpolated letter values and values of g_p for the bean data ($n = 9440$).

Tag	Depth	Letter Value Lower		Letter Value Upper	$-z_p$	g_p
M	4720.5		14.494			
F	2360.5	13.903		15.035	.6745	−.131
E	1180.5	13.410		15.376	1.150	−.179
D	590.5	12.915		15.638	1.534	−.210
C	295.5	12.371		15.825	1.863	−.251
B	148	11.817		16.093	2.154	−.239
A	74.5	11.336		16.226	2.418	−.248
Z	37.5	10.903		16.468	2.660	−.225
Y	19	10.542		16.636	2.886	−.212
X	10	10.292		16.718	3.097	−.205
W	5.5	10.036		16.833	3.297	−.196
V	3	9.857		17.042	3.487	−.172
U	2	9.786		17.125	3.668	−.159
	1	9.500		17.208	3.842	−.159

Note: Except at the extremes [which correspond to tail area $2/(3n + 1)$], the values of $-z_p$ are population values (see Table 10-3 and Section 10G).

TABLE 11-15. Adjusting the upper half of the bean data for constant g ($g = -.205$).

Tag	UHS_p	$-z_p$	$(e^{-gz_p} - 1)/g$	$\log_e(\text{UHS}_p^*)$	$z_p^2/2$
F	0.541	.6745	0.630	−.1522	0.227
E	0.882	1.150	1.024	−.1498	0.661
D	1.144	1.534	1.316	−.1402	1.177
C	1.331	1.863	1.548	−.1514	1.735
B	1.599	2.154	1.741	−.0853	2.320
A	1.732	2.418	1.907	−.0960	2.923
Z	1.974	2.660	2.050	−.0380	3.538
Y	2.142	2.886	2.178	−.0169	4.164
X	2.224	3.097	2.293	−.0304	4.796
W	2.339	3.297	2.397	−.0243	5.435
V	2.548	3.487	2.491	.0225	6.080
U	2.631	3.668	2.578	.0202	6.727
1	2.714	3.842	2.659	.0205	7.380

To learn what values of h these data may suggest, we adjust for the constant value of g that we have just chosen, using the upper half-spreads for convenience, as in equation (33) and Table 11-8. The results of the calculation appear in Table 11-15, and Figure 11-14 plots $\log_e(\text{UHS}_p^*)$ against $z_p^2/2$. Although they do not all lie close to a single straight line, the points in this plot indicate a positive value of h. An eye-fitted line with slope .040 and intercept $-.205$ seems a reasonable compromise. From it we obtain $h = .040$ and $B = .815$. Thus our initial g-and-h fit for the bean data is

$$X \approx 14.494 - 3.974(e^{-.205Z} - 1)e^{.020Z^2}. \tag{43}$$

We could examine the adequacy of this fit by calculating residuals between the interpolated letter values of Table 11-14 and the corresponding fitted values from equation (43). At this stage for a frequency distribution, however, we should return to the bin boundaries of the data, quantiles that do not involve interpolation. The corresponding Gaussian quantile in the right-hand column of Table 11-13 becomes Z in equation (43), and we obtain the values of X and the residuals shown in Table 11-16. Both these

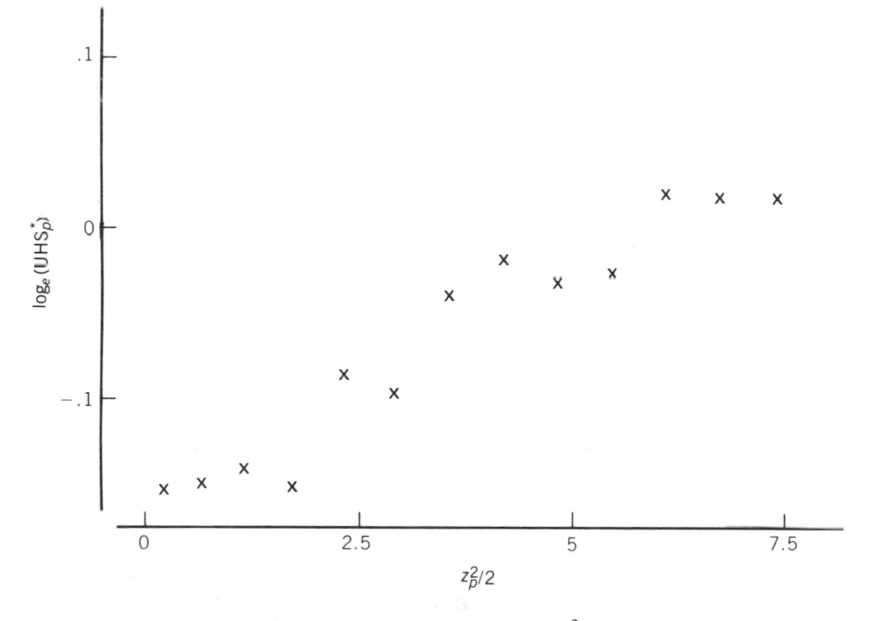

Figure 11-14. Plot of $\log_e(\text{UHS}_p^*)$ against $z_p^2/2$ for the bean data.

TABLE 11-16. Checking the g-and-h fit to the bean data at the bin boundaries. $X = 14.494 - 3.974\,(e^{-.205Z} - 1)e^{.02Z^2}$.

Bin Boundary	Z	X	Residual
9.75	−3.665	8.67	1.08
10.25	−3.133	10.14	0.11
10.75	−2.774	10.94	−0.19
11.25	−2.479	11.52	−0.27
11.75	−2.197	12.00	−0.25
12.25	−1.940	12.40	−0.15
12.75	−1.672	12.78	−0.03
13.25	−1.319	13.22	0.03
13.75	−0.871	13.71	0.04
14.25	−0.302	14.24	0.01
14.75	0.317	14.74	0.01
15.25	1.014	15.26	−0.01
15.75	1.804	15.81	−0.06
16.25	2.485	16.29	−0.04
16.75	3.215	16.85	−0.10

residuals and the Q-Q plot, Figure 11-15, indicate that the fit in the lower tail could be improved (although no residual is as large as twice its standard error).

The most noticeable discrepancy in Figure 11-15 is the leftmost point, which indicates that the smallest observations are larger than one would expect from the bulk of the data. Otherwise, the deviations from straightness are consistent with the differences between $-.205$ and the g_p in Table 11-14 and with basing the choice of the value of h on the upper tail.

Although we could apparently do better with nonconstant g, the fit of equation (43), involving constant g, constant h, a scale constant, and a location constant, provides a satisfactory first approximation. Slifker and Shapiro (1980) obtain a somewhat closer fit by using an S_U distribution (see Section 11G) and determining the parameters that match the data quantiles corresponding to $z = -3, -1, +1,$ and $+3$. In terms of the usual chi-squared goodness-of-fit criterion for frequency distributions, both of these fits (and others in the literature) are less adequate than one would expect by chance if the distribution being fitted were correct. If we were directly involved with the original investigators in such a situation, we should try to inquire more closely into the details of the data. Negative skewness seems

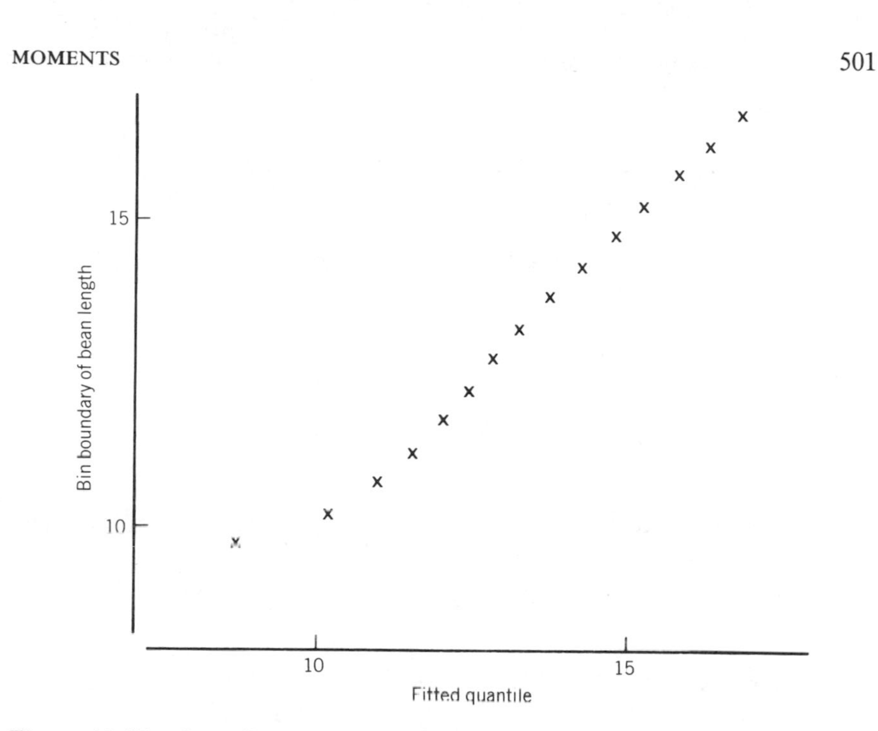

Figure 11-15. Quantile–quantile plot for the bean data: bin boundaries versus corresponding quantiles of the fitted g-and-h distribution.

much rarer in actual data than positive skewness, and so we might even wonder about the homogeneity of the sample of beans.

Whatever the outcome, this example has demonstrated how to apply the g-and-h distributions to data in the form of a frequency distribution and how the g-distributions accommodate negative skewness as easily as positive.

11F. MOMENTS

Discussions of new distributions customarily include their moments, when these exist and can be calculated. For the g-and-h distributions, Arthur (1979) has shown that the calculations are straightforward, although they rapidly become tedious as the order of the moment increases. We begin with the simpler special cases, the g-distributions and the h-distributions, and then proceed to the more general case involving both nonzero g and nonzero h. Throughout this discussion we work only with constant values of g and h. Also, we calculate the moments for the "standard" random variable Y defined in equations (6b), (20), and (30b).

g-Distributions

In this subfamily, $Y = (e^{gZ} - 1)/g$. The usual first four moments are

$$E(Y) = \left(e^{g^2/2} - 1\right)/g, \tag{44a}$$

$$\operatorname{var}(Y) = e^{g^2}\left(e^{g^2} - 1\right)/g^2, \tag{44b}$$

$$E\{[Y - E(Y)]^3\} = e^{(3/2)g^2}\left(e^{3g^2} - 3e^{g^2} + 2\right)/g^3, \tag{44c}$$

$$E\{[Y - E(Y)]^4\} = e^{2g^2}\left(e^{6g^2} - 4e^{3g^2} + 6e^{g^2} - 3\right)/g^4. \tag{44d}$$

h-Distributions

In this subfamily, $Y = Ze^{hZ^2/2}$, and the symmetry of the distributions about zero implies that the odd-order moments of an h-distribution are zero (when they are finite). The second moment and fourth moment are

$$E(Y^2) = \frac{1}{(1 - 2h)^{3/2}}, \qquad 0 \le h < \tfrac{1}{2}, \tag{45a}$$

$$E(Y^4) = \frac{3}{(1 - 4h)^{5/2}}, \qquad 0 \le h < \tfrac{1}{4}. \tag{45b}$$

Thus the variance of Y is finite only when $h < \frac{1}{2}$, and the fourth moment is finite only when $h < \frac{1}{4}$.

g-and-h Distributions

The general distribution with constant g and constant h has

$$Y = \frac{1}{g}(e^{gZ} - 1)e^{hZ^2/2}.$$

Its mean and variance are

$$E(Y) = \frac{1}{g\sqrt{1-h}}\left(e^{g^2/2(1-h)} - 1\right), \qquad 0 \le h < 1, \tag{46a}$$

$$\operatorname{var}(Y) = \frac{1}{g^2\sqrt{1-2h}}\left[e^{2g^2/(1-2h)} - 2e^{g^2/2(1-2h)} + 1\right]$$

$$- \frac{1}{g^2(1-h)}\left[e^{g^2/2(1-h)} - 1\right]^2, \qquad 0 \le h < \tfrac{1}{2}. \tag{46b}$$

More generally, Martinez (1981) derives the nth moment about the origin, when $g \neq 0$ and $0 \le h < 1/n$:

$$E(Y^n) = \frac{1}{g^n\sqrt{1-nh}} \sum_{i=0}^{n} (-1)^i \binom{n}{i} e^{[(n-i)g]^2/2(1-nh)}. \tag{47}$$

Higher-order moments about the mean are less attractive because of the tedious algebra involved. However, a computer system for algebraic manipulation could handle them with ease. [Moses (1971) gives an overview of such systems.] Also, one could obtain them numerically for particular values of g and h by using equation (47) and the identities linking moments about the mean to moments about the origin.

Skewness and Kurtosis

Quantitative discussions of distribution shape and routine summaries of data often focus on the third and fourth moments about the mean. Specifically, in terms of $\mu = E[X]$ and $\mu_k = E[(X-\mu)^k]$, the population skewness is measured by

$$\sqrt{\beta_1} = \frac{\mu_3}{\mu_2^{3/2}} = \gamma_1,$$

and the population kurtosis is defined as

$$\beta_2 = \frac{\mu_4}{\mu_2^2} = 3 + \gamma_2.$$

The corresponding sample measures are defined in an analogous way in terms of the sample moments. We recall that for a Gaussian distribution $\sqrt{\beta_1} = 0$ and $\beta_2 = 3$.

For completeness, we examine these measures of skewness and kurtosis in the simpler special cases: the g-distributions with constant g and h-distributions with constant h. Even with constant g and h, the general g-and-h distribution leads to such a thicket of algebra that formulas for its skewness and kurtosis would not be instructive.

From equations (44b), (44c), and (44d), the g-distributions have

$$\sqrt{\beta_1(g)} = (e^{3g^2} - 3e^{g^2} + 2)/(e^{g^2} - 1)^{3/2}, \tag{48a}$$

$$\beta_2(g) = (e^{6g^2} - 4e^{3g^2} + 6e^{g^2} - 3)/(e^{g^2} - 1)^2. \tag{48b}$$

After straightforward algebra these reduce to

$$\beta_1(g) = e^{3g^2} + 3e^{2g^2} - 4, \tag{49a}$$

$$\beta_2(g) = e^{4g^2} + 2e^{3g^2} + 3e^{2g^2} - 3. \tag{49b}$$

The h-distributions have

$$\beta_1(h) = 0, \qquad 0 \le h < \tfrac{1}{3}, \tag{50a}$$

$$\beta_2(h) = 3(1-2h)^3/(1-4h)^{5/2}, \qquad 0 \le h < \tfrac{1}{4}, \tag{50b}$$

by equations (45a) and (45b).

11G. OTHER APPROACHES TO SHAPE

To provide a link with the extensive existing literature related to distribution shape, we briefly touch on two important systems of distributions—the Pearson curves and the Johnson curves. Both systems approach distribution shape primarily through moments, so that each possible pair of values, $\sqrt{\beta_1}$ and β_2, corresponds to exactly one distribution in the system. (Not all combinations of $\sqrt{\beta_1}$ and β_2 are possible because $\beta_2 \ge \beta_1 + 1$ for any frequency distribution.) For comparisons with the g-and-h distributions this reliance on moments has two main consequences.

First, because values of h outside a relatively small interval produce infinite third and fourth moments, the g-and-h distributions offer a much wider variety of tail behavior. In practice, of course, what matters is how much of this greater flexibility we need in fitting actual data. (Earlier sections of this chapter have demonstrated that working with quantiles provides resistance and convenience unavailable with sample moments.)

Second, the notion of "neutral elongation" receives a different interpretation. In terms of kurtosis, the neutral value is that for the Gaussian distribution: $\beta_2 = 3$. Within the family of g-and-h distributions, $h = 0$ corresponds to neutral elongation; that is, the g-distributions (or lognormal distributions) define the neutral amount of elongation that accompanies a given degree of skewness. (Neutral skewness corresponds to symmetry in both approaches.) Thus, from equation (49b),

$$\beta_2(g) = e^{4g^2} + 2e^{3g^2} + 3e^{2g^2} - 3,$$

which equals 3 only when $g = 0$.

Pearson Curves

The frequency curves, $f(x)$, in the Pearson system [see, for example, Johnson and Kotz (1970, Section 12.4.1)] can be approached in a unified way because they all arise as solutions to the differential equation

$$\frac{1}{f}\frac{df}{dx} = -\frac{a + x}{c_0 + c_1 x + c_2 x^2} \tag{51}$$

with real coefficients a, c_0, c_1, and c_2. The principal resulting functional forms of f, listed in Table 11-17, fall into seven types (when special cases

TABLE 11-17. Simplest functional forms and ranges of values for the seven types of Pearson curves.

Type	Simplest Functional Form	Range of x
I	$x^{m_1}(1 - x)^{m_2}$	$0 < x < 1$
II	$(1 - x^2)^m$	$-1 < x < 1$
III	$x^m e^{-x}$	$0 < x < \infty$
IV	$(1 + x^2)^{-m} e^{-b \tan^{-1} x}$	$-\infty < x < \infty$
V	$x^{-m} e^{-1/x}$	$0 < x < \infty$
VI	$x^{m_2}(1 + x)^{-m_1}$	$0 < x < \infty$
VII	$(1 + x^2)^{-m}$	$-\infty < x < \infty$

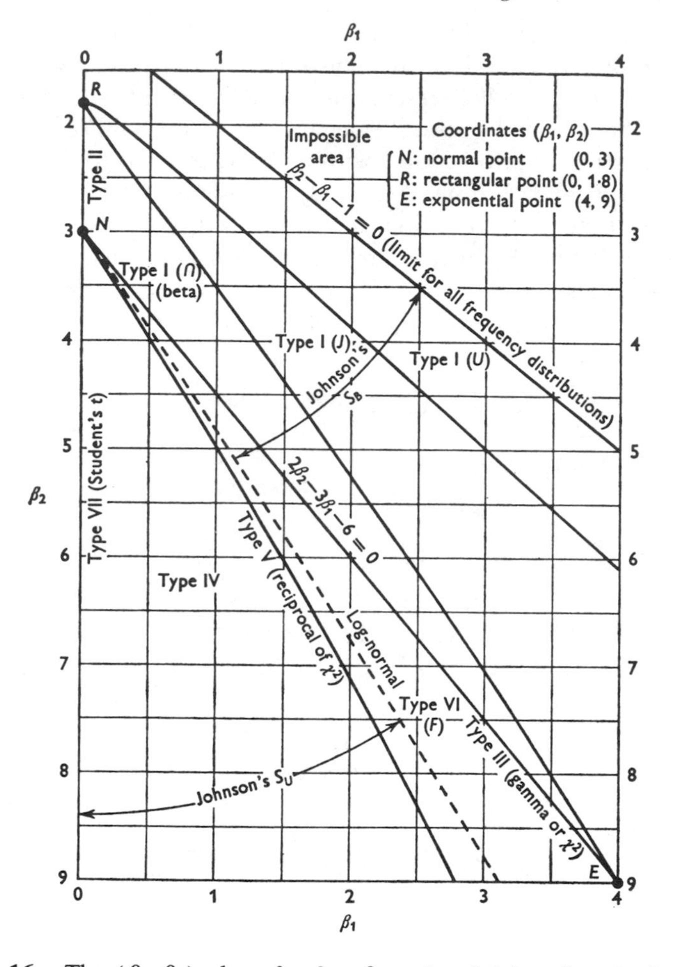

Figure 11-16. The (β_1, β_2) plane for $0 \le \beta_1 \le 4$ and $1.5 \le \beta_2 \le 9$, showing regions that correspond to the types of Pearson curves and Johnson curves. *Source*: E. S. Pearson and H. O. Hartley (Eds.) (1972). *Biometrika Tables for Statisticians*, Volume II. Cambridge: Cambridge University Press (Figure 7, p. 78). Printed by permission of the Biometrika Trustees.

are omitted). In giving these functional forms, we have retained only the parameters related to shape. If desired, we can introduce location and scale constants as in equation (1). Figure 11-16 shows how the seven types of Pearson curves partition the (β_1, β_2) plane and also includes the distributions in the Johnson system (to which we will turn shortly). Note that the β_2 axis increases from the top to the bottom of the figure.

The ranges of values of x for the g-and-h distributions lead us to consider primarily Types III through VII. The Type III distributions are the gamma distributions, which include the χ^2 distributions as special cases. Because a Type III distribution has smaller β_2 than the lognormal distribution (i.e., g-distribution) with the same β_1, we now see that it would have been reasonable to expect a small negative value of h for the χ_6^2 example in Section 11C.

Type VI are F distributions, and Type VII are Student's t distributions (including the Cauchy distribution). Both are Pareto distributions, as in equation (21).

Pearson and Hartley (1972) give extensive tables of the standardized percentage points of Pearson distributions, according to the values of $\sqrt{\beta_1}$ and β_2. These quantiles make it easy to investigate how closely Pearson distributions can be fitted by members of the g-and-h family. Unpublished research along these lines by James M. Landwehr and John W. Tukey indicates that, on the whole, Pearson distributions of Types III through VII can be approximated quite closely by g-and-h distributions, but not always with constant g and h. In some instances they find it useful to include g_2 in equation (36) and h_2 in equation (37).

Pearson distributions of Types I and II, whose ranges have both endpoints finite (and, to some extent, other Pearson distributions with a single finite endpoint), generally produce reasonable g-and-h shapes after transformation along the lines of "first aid," as discussed by Mosteller and Tukey (1977, Section 5H). In working with data that are subject to finite endpoints, however, the adequacy of the fit can depend heavily on how accurately one knows the endpoint(s).

Johnson Curves

The distributions in the system introduced by Johnson (1949) come about by taking a standard Gaussian random variable Z, translating it to $(Z - \gamma)/\delta$, and then applying a suitable transformation T to get

$$Y = T\left(\frac{Z - \gamma}{\delta}\right). \tag{52}$$

The three families in this system are (1) a family of bounded distributions, denoted by S_B and based on

$$T(x) = \frac{e^x}{1 + e^x}; \tag{53}$$

(2) the family of lognormal distributions, for which

$$T(x) = e^x; \tag{54}$$

and (3) a family of unbounded distributions, denoted by S_U and based on

$$T(x) = \sinh(x) = \tfrac{1}{2}(e^x - e^{-x}). \tag{55}$$

In the S_B and S_U families, the parameters γ and δ govern the shape of the distribution. For example, in S_U, because sinh is an odd function—that is, $\sinh(-x) = -\sinh(x)$—taking $\gamma = 0$ yields a symmetric distribution, whose elongation is determined by δ. On the other hand, a nonzero value of γ introduces skewness by translating or shifting the basic Gaussian distribution away from the origin.

In the lognormal family, δ governs the shape while γ influences only the scale, through the relationship $e^{(Z-\gamma)/\delta} = e^{-\gamma/\delta}e^{Z/\delta}$. We examined this aspect in Section 11A in discussing the connection between lognormal distributions and g-distributions [see equations (13) and (14b)].

Figure 11-16 shows the portions of the (β_1, β_2) plane covered by the three families of distributions in the Johnson system.

Aside from the lognormal distributions, which are g-distributions, the fitting of Johnson curves by g-and-h distributions has received little systematic attention. For given values of γ and δ, however, equations (52), (53), and (55) make it easy to calculate quantiles of an S_B or S_U distribution from the corresponding standard Gaussian quantiles. Thus a systematic investigation would be straightforward. It seems reasonable to expect results qualitatively similar to those for the Pearson system: close approximations to S_U distributions (by going as far as g_2 and h_2) and satisfactory approximations to S_B distributions after application of "first aid" transformations.

11H. SUMMARY

For exploring the distribution shape of data and summarizing the results quantitatively, the family of g-and-h distributions offers resistance and flexibility. The basic constants, g for skewness and h for elongation, can be used separately or together, and each generalizes to a function of z^2 by adding powers higher than the constant.

Although the descriptive techniques associated with the g-and-h distributions are designed for use with suitably chosen sets of quantiles, they easily accommodate interpolated quantiles from a frequency distribution. In this setting, information from the bin boundaries provides a framework for checking adequacy of fit.

Within the g-and-h family, calculations of moments reveal the difference between neutral elongation ($h = 0$) and the moment-based notion of neutral kurtosis ($\beta_2 = 3$). Neutral skewness still corresponds to symmetry.

Comparisons with the Pearson system of frequency curves indicate that g-and-h distributions (in their more general form) closely approximate Pearson curves over a substantial part of the (β_1, β_2) plane.

REFERENCES

Arthur, E. P. (1979). "Skew/stretched distributions and the t-statistic," Ph.D. dissertation, Princeton University.

David, F. N. and Pearson, E. S. (1961). *Elementary Statistical Exercises*. Cambridge: The University Press.

Fama, E. F. and Roll, R. (1968). "Some properties of symmetric stable distributions," *Journal of the American Statistical Association*, **63**, 817–836.

Feller, W. (1968). *An Introduction to Probability Theory and Its Applications*, Vol. 1, 3rd ed. New York: Wiley.

Hoaglin, D. C. (1977). "Direct approximations for chi-squared percentage points," *Journal of the American Statistical Association*, **72**, 508–515.

Johnson, N. L. (1949). "Systems of frequency curves generated by methods of translation," *Biometrika*, **36**, 149–176.

Johnson, N. L. and Kotz, S. (1970). *Distributions in Statistics: Continuous Univariate Distributions-1*. Boston, MA: Houghton Mifflin Company.

Kendall, M. G. and Stuart, A. (1963). *The Advanced Theory of Statistics*, Vol. 1 (*Distribution Theory*), 2nd ed. London: Charles Griffin and Company Limited.

Land, C. E. (1972). "An evaluation of approximate confidence interval estimation methods for lognormal means," *Technometrics*, **14**, 145–158.

Martinez, J. (1981). "Some properties of robust scale estimators," Ph.D. dissertation, Temple University, Philadelphia, PA.

Moses, J. (1971). "Algebraic simplification: a guide for the perplexed," *Communications of the ACM*, **14**, 527–537.

Mosteller, F. and Tukey, J. W. (1977). *Data Analysis and Regression*. Reading, MA: Addison-Wesley.

Pearson, E. S. and Hartley, H. O., Eds. (1972). *Biometrika Tables for Statisticians*, Vol. II. Cambridge: Cambridge University Press.

Pretorius, S. J. (1930). "Skew bivariate frequency surfaces, examined in the light of numerical illustrations," *Biometrika*, **22**, 109–223.

Slifker, J. F. and Shapiro, S. S. (1980). "The Johnson system: selection and parameter estimation," *Technometrics*, **22**, 239–246.

Tukey, J. W. (1977). Modern techniques in data analysis. NSF-sponsored regional research conference at Southeastern Massachusetts University, North Dartmouth, MA.

Wilson, E. B. and Hilferty, M. M. (1931). "The distribution of chi-square," *Proceedings of the National Academy of Sciences*, **17**, 684–688.

Additional Literature

Aitchison, J. and Brown, J. A. C. (1963). *The Lognormal Distribution*. Cambridge: Cambridge University Press.

Bickel, P. J. and Lehmann, E. L. (1975). "Descriptive statistics for nonparametric models. I. Introduction," *Annals of Statistics*, **3**, 1038–1044.

Bickel, P. J. and Lehmann, E. L. (1979). "Descriptive statistics for nonparametric models. IV. Spread." In J. Jurečková (Ed.), *Contributions to Statistics, Jaroslav Hájek Memorial Volume*. Prague: Academia, pp. 33–40.

Burr, I. W. and Cislak, P. J. (1968). "On a general system of distributions. I. Its curve-shape characteristics. II. The sample median," *Journal of the American Statistical Association*, **63**, 627–635.

Chissom, B. S. (1970). "Interpretation of the kurtosis statistic," *The American Statistician*, **24** (4), 19–22.

Churchill, E. (1946). "Information given by odd moments," *Annals of Mathematical Statistics*, **17**, 244–246.

Darlington, R. B. (1970). "Is kurtosis really 'peakedness?'," *The American Statistician*, **24** (2), 19–22.

Davis, C. S. and Stephens, M. A. (1983). "Algorithm AS192: approximate percentage points using Pearson curves," *Applied Statistics*, **32**, 322–327.

DuMouchel, W. H. (1983). "Estimating the stable index α in order to measure tail thickness: a critique," *Annals of Statistics*, **11**, 1019–1031.

Efron, B. (1982). "Transformation theory: how normal is a family of distributions?" *Annals of Statistics*, **10**, 323–339.

Elderton, W. P. and Johnson, N. L. (1969). *Systems of Frequency Curves*. Cambridge: Cambridge University Press.

Finucan, H. M. (1964). "A note on kurtosis," *Journal of the Royal Statistical Society, Series B*, **26**, 111–112.

Gaver, D. P. (1983). "Stochastic modeling: ideas and techniques. 3. Additional modeling topics." In G. Louchard and G. Latouche (Eds.), *Probability Theory and Computer Science*. London: Academic, pp. 36–49.

Hall, D. L. and Joiner, B. L. (1982). "Representations of the space of distributions useful in robust estimation of location," *Biometrika*, **69**, 55–59.

Hettmansperger, T. P. and Keenan, M. A. (1975). "Tailweight, statistical inference and families of distributions—a brief survey." In G. P. Patil, S. Kotz, and J. K. Ord (Eds.), *Statistical Distributions in Scientific Work*, Vol. 1 (Models and Structures). Dordrecht, Holland: D. Reidel, pp. 161–172.

Heyde, C. C. (1975). "Kurtosis and departure from normality." In G. P. Patil, S. Kotz, and J. K. Ord (Eds.), *Statistical Distributions in Scientific Work*, Vol. 1 (Models and Structures). Dordrecht, Holland: D. Reidel, pp. 193–201.

Hildebrand, D. K. (1971). "Kurtosis measures biomodality?," *The American Statistician*, **25** (1), 42–43.

Hoaglin, D. C. (1983). "g-and-h distributions." In S. Kotz and N. L. Johnson (Eds.), *Encyclopedia of Statistical Sciences*, Vol. 3. New York: Wiley, pp. 298–301.

Hogg, R. V. (1972). "More light on the kurtosis and related statistics," *Journal of the American Statistical Association*, **67**, 422–424.

Martinez, J. and Iglewicz, B. (1984). "Some properties of the Tukey g and h family of distributions," *Communications in Statistics—Theory and Methods*, **13**, 353–369.

Nichols, W. G. and Gibbons, J. D. (1979). "Parameter measures of skewness," *Communications in Statistics—Simulation and Computation*, **B8**, 161–167.

Oja, H. (1981). "On location, scale, skewness, and kurtosis of univariate distributions," *Scandinavian Journal of Statistics*, **8**, 154–168.

Parzen, E. (1979). "Nonparametric statistical data modeling," *Journal of the American Statistical Association*, **74**, 105–131 (with discussion).

Pearson, E. S. (1963). "Some problems arising in approximating to probability distributions, using moments," *Biometrika*, **50**, 95–111.

Pearson, E. S., Johnson, N. L., and Burr, I. W. (1979). "Comparisons of the percentage points of distributions with the same first four moments, chosen from eight different systems of frequency curves," *Communications in Statistics—Simulation and Computation*, **B8**, 191–229.

Shenton, L. R. and Bowman, K. O. (1975). "Johnson's S_U and the skewness and kurtosis statistics," *Journal of the American Statistical Association*, **70**, 220–228.

Tadikamalla, P. R. (1980). "On simulating non-normal distributions," *Psychometrika*, **45**, 273–279.

Tadikamalla, P. R. and Johnson, N. L. (1982). "Systems of frequency curves generated by transformations of logistic variables," *Biometrika*, **69**, 461–465.

Wheeler, R. E. (1980). "Quantile estimators of Johnson curve parameters," *Biometrika*, **67**, 725–728.

EXERCISES

1. For $g = .2, .4, .6, .8$, and 1.0, calculate the value of $\sqrt{\beta_1(g)}$, using equations (49a) and (48a). Reexamine Figures 11-2 and 11-4 in light of this numerical assessment of skewness.

2. By completing the square in the exponent of e, derive the first four moments of the g-distributions in equations (44) and the second and fourth moments of the h-distributions in equations (45).

3. Equation (19) gives the random variable for the standard h-distribution as $Y = Ze^{hZ^2/2}$, where Z is a standard Gaussian random variable and h is a positive real constant. Show that the corresponding probability density function can be written implicitly, in terms of z, as

$$f_h(y) = \frac{1}{\sqrt{2\pi}} \cdot \frac{e^{-(h+1)z^2/2}}{hz^2 + 1},$$

where $y = ze^{hz^2/2}$.

4. For the lower half-spread LHS_p, work through the derivation parallel to equations (32) and (33) to adjust for a nonzero constant value of g.

5. In the χ_6^2 example (Tables 11-3, 11-8, and 11-9) investigate elongation by adjusting the *lower* half-spreads for the fitted value of g, .406.

Choose values of h and B to produce an approximation in the form of equations (34) and examine the accuracy of this approximation.

6. In the household income example (Tables 11-10 through 11-12 and Figures 11-12 and 11-13), adjust the lower half-spreads for the fitted pattern of skewness and compare the resulting information on elongation to that obtained from the upper tail.

7. Use the Wilson–Hilferty approximation for the pth quantile of the chi-squared distribution on ν degrees of freedom,

$$\chi^2_{\nu,p} \approx \nu\left[z_p\sqrt{\frac{2}{9\nu}} + 1 - \frac{2}{9\nu}\right]^3,$$

to obtain approximate quantiles of χ^2_6. How do the absolute accuracy and relative accuracy of this approximation compare with the g-and-h approximation given in Table 11-9?

8. For the chi-squared distributions on 3, 4, 8, 12, 24, and 100 degrees of freedom, Table 11-18 gives the same selected quantiles as Table 11-2

TABLE 11-18. Selected quantiles for six chi-squared distributions.

Lower	Degrees of Freedom					
Tail Area	3	4	8	12	24	100
.005	0.072	0.207	1.344	3.074	9.886	67.328
.01	0.115	0.297	1.646	3.571	10.856	70.065
.025	0.216	0.484	2.180	4.404	12.401	74.222
.05	0.352	0.711	2.733	5.226	13.848	77.930
.1	0.584	1.064	3.490	6.304	15.659	82.358
.25	1.213	1.923	5.071	8.438	19.037	90.133
.5	2.366	3.357	7.344	11.340	23.337	99.334
.75	4.108	5.385	10.219	14.845	28.241	109.141
.9	6.251	7.779	13.362	18.549	33.196	118.498
.95	7.815	9.488	15.507	21.026	36.415	124.342
.975	9.348	11.143	17.535	23.337	39.364	129.561
.99	11.345	13.277	20.090	26.217	42.980	135.807
.995	12.838	14.860	21.955	28.300	45.558	140.169

Source: Milton Abramowitz and Irene A. Stegun (Eds.) (1964). *Handbook of Mathematical Functions* (National Bureau of Standards, Applied Mathematics Series 55). Washington D.C.: U.S. Government Printing Office (selected entries from Table 26.8, pp. 984–985).

gives for chi-squared on six degrees of freedom.

(a) Determine a fitted value of g for each of these distributions.

(b) Together with the value $\hat{g}_6 = .406$ obtained in Section 11A, use these values to examine $0.95/\sqrt{\nu - 0.5}$ as a summary of the relation between $\hat{g}_\nu$ and the degrees of freedom, ν.

(c) Adjust the quantiles for the skewness associated with $\hat{g}_\nu$ and determine a fitted value of h for each chi-squared distribution. (It may be desirable to do this for both the upper tail and the lower tail and choose a compromise value of h.)

(d) Together with the value $\hat{h}_6 = -.033$ obtained in Section 11C, use these values to assess $-0.19/(\nu - 0.1)$ as a summary of the relation between $\hat{h}_\nu$ and ν.

9. Using one choice of scaling, Fama and Roll (1968, Table 2) give selected quantiles of 12 symmetric stable distributions. For the stable distribution with index 1.5 and tail areas near those for the letter values, the points (upper tail area, stable quantile, standard Gaussian quantile) are as follows: (.38, 0.427, 0.3055), (.26, 0.921, 0.6434), (.12, 1.837, 1.1750), (.06, 2.763, 1.5548), (.03, 4.049, 1.8808), (.015, 6.043, 2.1701), (.01, 7.737, 2.3264), (.005, 11.983, 2.5758), (.0005, 54.337, 3.2905). Fit an h-distribution to the quantiles of this stable distribution and discuss the quality of the approximation. (Hint: h need not be constant.)

10. From Table 11-14 (the bean data) plot g_p against z_p^2. What functional forms, $g(z)$, seem likely to be more effective than the constant value of g, used in the example?

11. To handle negative elongation more conveniently than one can by using h-distributions with negative values of h, Tukey (1977) has suggested squeezing the tails according to $Y = Z/\sqrt{1 + jZ^2}$ with positive real j.

(a) Verify that this transformation of z is monotonic and converges to $\pm 1/\sqrt{j}$ as z goes to $\pm\infty$.

(b) Show how to estimate j from symmetric data.

(c) Apply this approach to the folded-root distribution, whose letter-spreads are given in Table 10-4.

(d) Use j together with g to approximate the quantiles of the chi-squared distribution on six degrees of freedom.

Index

Applied Probability and Statistics (Continued)

HOLLANDER and WOLFE • Nonparametric Statistical Methods
IMAN and CONOVER • Modern Business Statistics
JAGERS • Branching Processes with Biological Applications
JESSEN • Statistical Survey Techniques
JOHNSON and KOTZ • Distributions in Statistics
Discrete Distributions
Continuous Univariate Distributions—1
Continuous Univariate Distributions—2
Continuous Multivariate Distributions
JOHNSON and KOTZ • Urn Models and Their Application: An Approach to Modern Discrete Probability Theory
JOHNSON and LEONE • Statistics and Experimental Design in Engineering and the Physical Sciences, Volumes I and II, *Second Edition*
JUDGE, HILL, GRIFFITHS, LÜTKEPOHL and LEE • Introduction to the Theory and Practice of Econometrics
JUDGE, GRIFFITHS, HILL, LÜTKEPOHL and LEE • The Theory and Practice of Econometrics, *Second Edition*
KALBFLEISCH and PRENTICE • The Statistical Analysis of Failure Time Data
KISH • Survey Sampling
KUH, NEESE, and HOLLINGER • Structural Sensitivity in Econometric Models
KEENEY and RAIFFA • Decisions with Multiple Objectives
LAWLESS • Statistical Models and Methods for Lifetime Data
LEAMER • Specification Searches: Ad Hoc Inference with Nonexperimental Data
LEBART, MORINEAU, and WARWICK • Multivariate Descriptive Statistical Analysis: Correspondence Analysis and Related Techniques for Large Matrices
McNEIL • Interactive Data Analysis
MAINDONALD • Statistical Computation
MANN, SCHAFER and SINGPURWALLA • Methods for Statistical Analysis of Reliability and Life Data
MARTZ and WALLER • Bayesian Reliability Analysis
MIKÉ and STANLEY • Statistics in Medical Research: Methods and Issues with Applications in Cancer Research
MILLER • Beyond ANOVA, Basics of Applied Statistics
MILLER • Survival Analysis
MILLER, EFRON, BROWN, and MOSES • Biostatistics Casebook
MONTGOMERY and PECK • Introduction to Linear Regression Analysis
NELSON • Applied Life Data Analysis
OSBORNE • Finite Algorithms in Optimization and Data Analysis
OTNES and ENOCHSON • Applied Time Series Analysis: Volume I, Basic Techniques
OTNES and ENOCHSON • Digital Time Series Analysis
PANKRATZ • Forecasting with Univariate Box-Jenkins Models: Concepts and Cases
PIELOU • Interpretation of Ecological Data: A Primer on Classification and Ordination
POLLOCK • The Algebra of Econometrics
PRENTER • Splines and Variational Methods
RAO and MITRA • Generalized Inverse of Matrices and Its Applications
RIPLEY • Spatial Statistics
SCHUSS • Theory and Applications of Stochastic Differential Equations
SEAL • Survival Probabilities: The Goal of Risk Theory
SEARLE • Linear Models
SEARLE • Matrix Algebra Useful for Statistics
SPRINGER • The Algebra of Random Variables

(*continued from front*)

THE MYTH OF THE GOOD CORPORATE CITIZEN

DEMOCRACY UNDER THE RULE OF BIG BUSINESS

Murray Dobbin

Published in 1998 by Stoddart Publishing Co. Limited
34 Lesmill Road, Toronto, Canada M3B 2T6
180 Varick Street, 9th Floor, New York, New York, USA 10014

ORDERING INFORMATION

Distributed in Canada by General Distribution Services Inc.
34 Lesmill Road, Toronto, Canada M3B 2T6
Toll-free tel. for Ontario and Quebec 1-800-387-0141
Toll-free tel. for all other provinces and territories 1-800-287-0172
Fax (416) 445-5967
Email Customer.Service@ccmailgw.genpub.com

Distributed in the U.S. by General Distribution Services Inc.
85 River Rock Drive, Suite 202, Buffalo, New York 14207
Toll-free tel. 1-800-805-1083 Toll-free fax 1-800-481-6207
Email gdsinc@genpub.com

02 01 00 99 98 1 2 3 4 5

Cataloguing in Publication data available from the National Library of Canada.

ISBN 0-7737-3087-7

Research Associate: Ellen Gould

Cover design: Bill Douglas @ The Bang
Text design: Tannice Goddard

Printed and bound in Canada

CONTENTS

Acknowledgements

As with other books I have written, this is work of political intervention, as much a part of my involvement in social justice politics as of my work as an activist. In this sense, there are too many people to acknowledge for their contributions, intended or otherwise. However, I will mention a few.

The conclusion that this book needed to be written was the result of many years fighting the initiatives of provincial and federal governments, fights involving many thousands of Canadians. As we struggled against free trade and NAFTA, the GST, cuts to social programs, and privatization, a number of activists began to focus predominately on the fact that while government was implementing these policies, it was corporations that initiated them. Working this realization through to an evolving analysis with Maude Barlow, Ed Finn, Tony Clarke, David Robinson, David Langille, Jim Turk, and others has led to this book and, as well, I hope, to the beginnings of a new democratic politics. I thank these friends and colleagues for their insights and their commitment to anti-corporate work.

I am also indebted to others who helped in a variety of ways, from reading chapters, providing information, to helping with particular points of analysis. These include Jim Grieschaber-Otto, Jack Warnock, John Dillon, Duncan Cameron, Seth Klein, and Gil Yaron. I have leaned heavily on the work of others in the writing of this book. I am deeply indebted, for their contributions in exposing corporate power, to David Korten, Richard Barnet, John Cavanagh, William Greider, and Allan Engler.

Living with someone writing a book is like caring for someone with a chronic illness. I thank Ellen Gould, my partner, for her excellent advice, her assistance on the book — and her forbearance.

I dedicate this book to the memory of my parents: to my father, whose "downsizing" by Eaton's after twenty-five years as a dedicated employee was the seed of my activism, and to my mother, whose unfailing sense of right and wrong have helped to guide it.

MURRAY DOBBIN
Vancouver, B.C.

Introduction: How Did Things Get This Bad?

As Canadians approach the millennium we find ourselves in a world dominated at every turn by large corporations — banks, trust companies, transnationals, corporations involved in currency speculation, corporations that buy and sell government bonds, corporate agencies like bond rating companies that pronounce on government policies. Virtually every aspect of our lives is dominated by these huge corporations: what we buy and who we work for (or whether we work at all), the information we receive, the entertainment we "consume," and the physical environment that surrounds us. And as corporate control increases, the control we used to exercise over those areas of our lives, collectively or individually, diminishes. Corporate domination, brazen, ruthless, ever more powerful, is evident everywhere as civilization seems to devolve in its path.

The values of what we normally understand as civilization are swept aside by the new corporate imperative and its accompanying ideology. Thousands of years of human development and progress are reduced to the pursuit of "efficiency," our collective will is declared meaningless compared to the values of the marketplace, and communitarian values are rejected in favour of the survival of the fittest. A thinly disguised barbarism now passes for, is in fact promoted as, a global human objective. It is not only the best we can expect; it is the goal we should be seeking.

We are increasingly dependent, directly and indirectly, on large corporations for our jobs, for the kinds of jobs we have, and for how well they pay even if we do not work for them. Governments are (or say they are) prevented from passing progressive public policy because of corporate power; we watch, seemingly helpless, as corporations downsize, move jobs off-shore, and in general do whatever they please. And do it with virtual impunity because, well, business is their business. But more particularly because corporations have the status, rights, and protection of super-citizens.

Increasingly, we experience corporate power through the actions of the so-called global corporate citizen. With its awesome rights and powers, this artificial citizen can ignore national borders, national citizenship, and the rights and obligations of real flesh-and-blood citizens. This is a true ruling elite, a few hundred corporate citizens with more power than most nations, who rule the world and accumulate riches and power like the wealthiest of families run amok.

Real citizens are asked to sacrifice everything because corporations say we must. We must sit by as mere observers, wringing our hands while our most precious communitarian creations — medicare, public education, a protected environment — are eroded because collecting sufficient taxes to pay for these things will make us (in the lexicon, we become indistinguishable from corporations) "uncompetitive." These erosions are inevitable, we are told, and the destruction of the nation is somehow akin to an act of God: terrible in its consequences but irresistible. Indeed, we are told by the pundits that to resist is not only irresponsible, not only futile, but just as irrational as trying to command the skies to produce rain.

Not only do corporations want more — more cuts to their taxes, more cuts to UI and pension premiums, ever greater cuts to social programs, more repeals of environmental laws and protections for workers' health and safety, and more and better ways to squeeze more from their employees. They revel in their power and reward themselves obscenely for their success in bullying their employees, citizens, and governments into letting them have their way with our country or any other they choose. It is as if this abuse of power was a reflection of ability, even genius. Indeed, inside the corporate culture the more ruthless a CEO is in dealing with expendable employees, the higher the esteem in which he is held and the more he is paid.

Corporate spokesmen are the loudest voices for cutting deficits by cutting social spending, a brazen demand given that billions in revenue that would help pay for social programs are lost every year through corporate tax evasion, avoidance, outright fraud, and tax breaks. CEOs of 63,000 corporations making $13 billion in profits and paying no taxes clamour for ever greater cuts to welfare and medicare.

These corporate chiefs can afford to be brazen because through their media conglomerates, the free-market political parties, and think-tanks they fund, they have come to dominate the political debate in Canada and even the language with which we conduct that debate. So complicit is the media in popularizing this dominant ideology that anyone who describes a different take on reality is dismissed as a crank, out of touch, or, even worse, a "special interest." Only corporations whose lobbying is for one thing and one thing only — more and more money for shareholders, CEOs, and assorted courtiers — are not a vested interest.

Much of the increased power of corporations in Canada has developed with the advent of the Free Trade Agreement and NAFTA, deals with the United States and Mexico that had little to do with trade but are in reality corporate bills of rights. These deals give rights to corporate "citizens," rights to profit from investment with ever decreasing obligations to employees, community, environment, and consumers. These deals give corporations political rights by making it possible for them to stop governments from acting in ways that might affect their profits or growth. In vast areas of public policy, conflict between the interests of citizens and corporations is decided in favour of corporations by laws that are unchangeable by Parliament, unchallengeable in court. And there is more on the way.

We watch as many of our governments (those institutions that supposedly speak for us) grovel at the feet of corporations, passing and repealing laws to suit their new masters, or begging them to come to this or that province, gleefully offering up as sacrifices the communities they represent: low-waged workers, low or no taxes, interest-free loans, free land, cheap water and energy. Our governments have not yet offered our first-born but it might take such an offer to surprise anyone, given how subservient many of our elected representatives have become and how passive most citizens seem to be.

Thus we saw then New Brunswick premier Frank McKenna giddily grinning out at us from the pages of the *Globe and Mail* like some second-string used-car salesman desperate for a sale: "Act now! Give me a call." Why invest in New Brunswick? "Reduced workers' compensation costs: one more reason New Brunswick is Canada's best place to do business . . . and the best place for your bottom line." Phone 1-800-McKenna and, the ad implied, Brian would answer. "Open for business. Always." As if this wasn't enough, McKenna established a special zone where corporations have even fewer obligations. A kind of northern *maquiladoras*.

Who is responsible for this sorry state of affairs? As this book documents, the huge transnational corporations, the corporate super-citizens, have redesigned the planet and its economic system. The political elites of the developed world and the institutions they control have been complicit in this transformation. They have implemented an enormous transfer of power from government — that is, from citizens — to corporations. The media, corporate think-tanks, political parties, universities, and other institutions that we created to serve and even define civil society have instead betrayed it.

Ultimately, in a democracy the people who are responsible are its citizens. If their institutions fail them it is the responsibility of citizens to reform those institutions or create new ones that will do the job. Ironically, in the past fifty years we have collectively gone through a process of first creating such institutions and then watching them fail us. The other side of the seizure of power by corporate citizens is the ceding of that power by flesh-and-blood citizens. Comfortable in the thought that we had made progress, and believing or hoping that progress was assured, too many Canadians became observers instead of citizens. We forgot our history.

We are so accustomed to corporations and their power that we forget that they are our creation. Governments passed and continue to pass laws that give corporations their legal right to exist and behave the way they do. They are not aliens amongst us, created by forces beyond our control. To be sure, they behave as virtual aliens, demanding the rights of citizenship while refusing the responsibilities. But it is citizens, through governments, who created them and allowed them, bit by bit, to become what they are. And it is citizens who can, if they choose, change the laws to remake corporations and remove the citizen rights they enjoy.

We have also forgotten how we first achieved the things we are now losing. None of these community institutions was simply given to us by benign governments and good corporate citizens. The struggles of hundreds of thousands of men and women — citizens — in labour movements, farm movements, and social organizations made democracy real at a time when many of them did not know where their next meal was coming from. For them democracy was a process to be engaged in, not an institution to revere in the abstract and then blame for its failures.

Because the practice of democracy in Canada has become so truncated, and the language of participation so impoverished, the notion that the things we cherish are the result of citizen participation has largely been lost. The social and psychological distance between citizen and government has become so great that the notion of government as an expression of community is weaker now than at any time in the post-war period.

For too many people political citizenship is an afterthought or, at best, is reduced to brief and alienated participation in electoral rituals. Maybe we got into this habit because we believed medicare, public education, a full-time job, and a future for our children would always be secure. These things are our right. But as corporations prove every day, rights, like democracy, are a process, not an institution. As a community, as citizens, we haven't caught up with the new reality.

At its root democracy is a constant struggle for rule by the majority. Just saying those words implies what we instinctively know: most of the time it is a minority that rules. The increasing domination of huge corporations, the advent of so-called globalization, is just the current expression of the historic contest between social classes over the distribution of wealth and power.

We can't know where we are going if we don't know where we

came from. That is the thought that motivated the writing of this book. As citizens we are faced with a crisis greater than any since the Great Depression, and there are good reasons to believe this one is far greater. The consequences of not accepting the challenge of this crisis will be catastrophic. We have no choice. Either we change the world by becoming active citizens or we watch our world descend into a barbarism that is inherent in the amoral nature of the transnational corporations and their manic drive for wealth at any cost.

In the first part of this book I look at what has happened and what is likely to happen to the world as transnational corporations gain ever more power over our lives. In the second part of the book I examine just how the corporations, and those who benefit from their power and wealth, accomplished the neo-conservative counter-revolution that is now evident all around us.

The assault on democracy and equality by corporations and those who benefit from their power is couched in such terms as *globalization* for good reason. It creates an atmosphere of inevitability. We can, by this definition of reality, do nothing but attempt to adapt to the predetermined course of history. Globalization is like the Borg. We will be assimilated. Resistance is futile. But of course history teaches us that the only thing that is inevitable is change. What kind of change depends on the kind of people making it.

The social madness inherent in the drive for ever greater growth and profits, and its parallel creation of crushing poverty worldwide, will have a fearsome backlash. It could be chaotic, violent, and self-destructive or it could be enormously creative, rebuilding towards a truly egalitarian and ecologically sustainable world. The former is a virtual certainty if real citizens don't consciously direct the response to the corporate destruction of civil society. For this to happen there will have to be a revolution in citizen consciousness and a massive increase in participation in democratic politics.

This self-conscious citizenship has to include an explicit rejection of the logic of modern mass consumerism, the exchange of our rights and obligations as citizens for rights as consumers of global products. Democracy in the new millennium will be either the democracy of the customer — one dollar, one vote — or the democracy of the conscious, deliberate, self-aware citizen.

1

Globalization and the Rise of the Transnational Corporation

Since trade ignores national boundaries and the manufacturers insist on having the world as a market . . . the doors of the nations which are closed against him must be battered down. Concessions obtained by financiers must be safeguarded by ministers of state, even if the sovereignty of unwilling nations be outraged in the process.

— Woodrow Wilson, 1907

Globalization is one those words (like *deficit*) so loaded with ambiguous meaning, so packed with propaganda value, and so symbolic of modern-day angst that it carries with it the whole story of our current social and economic life. So much of what causes our anxiety about the economy, social programs, and the future of our children is encompassed by it. The dictates of the "global economy" imply that we simply have to accept all the misery, as if there was some uncontrollable source of the decline of the civil society in which we are all a part of, regardless of social class.

Globalization has a will-of-God dimension to it. It is presented to us as at once all-powerful, incomprehensible, impossibly complex, seemingly unchallengeable, and, on top of that, unprecedented. It is as if globalization is a new phenomenon and not just the latest mutation of capitalism. Nothing like this has ever happened before. It is unique. None of what we knew before, none of the solutions, none of the understanding, is of any use to us. Before this juggernaut, even our values are swept aside as impractical, unachievable, naive, archaic. Those who resist this vision of the future are dismissed as children. All of this conveniently lends itself to a portrayal of the new world order as inevitable.

Globalization is also conveniently class-neutral and effectively obscures the fact that what we are really talking about is the power of capital that is increasingly concentrated in the hands of transnational corporations. That globalization happens to benefit the top 15 to 20 percent of the world's population, that it has produced thousands of millionaires and hundreds of billionaires is rarely referred to. Like the deficit, which is now being paid back by those who can least afford it rather than by those who created it, globalization as an idea serves the most powerful extremely well. With virtually the entire political elite saying we simply must adapt, those who benefit from globalization hope they can enjoy the spoils of unrestrained capitalism without worrying about a revolt of the dispossessed.

But of course nothing is inevitable in the rich and complex universe of human relations, except change itself. Globalization is no more a "natural" phenomenon than electronic banking. Dissect it and you find that it has a history consisting of thousands of decisions, some unconsciously contributing to the whole and others explicitly creating the conditions for the advance of the global economy and the transnational corporations that direct it.

The deceptive neutrality of globalization and its effective use as an ideological tool mask the powerful reality of the domination of the world by a few hundred enormously powerful transnational corporations, or TNCs. It is these corporations, many American and backed by American might, which are global in character and reach, and the part of the world economy they dominate is likewise enormous. But it is also true that billions of people are completely excluded from this TNC economy even though, precisely because they are excluded, their lives are profoundly affected by it.

While transnational corporations have benefited enormously from the ideology of inevitability that the notion of globalization provides, governments, too, have profited. Far from being helpless and outmoded entities facing a threat from without, the nation-states of the developed world are directly responsible for creating the agencies and institutions that have permitted TNCs to accumulate their unprecedented global power. To be sure, governments often find themselves working at cross purposes, responding to the needs of citizens (to the extent that they must to get re-elected) and of corporations. But the "global economy" could not have developed and TNCs could not have accumulated their power without the cooperation and complicity of governments. The so-called downsizing of governments is restricted to ordinary citizens. For the largest corporations in the world, the most powerful corporate citizens, governments are becoming more active, more interventionist, not smaller but much larger.

TRANSNATIONAL CORPORATIONS: SEEKING "FREEDOM"

The imperative of globalization and its principal agents, the TNCs, is the gradual elimination of all restrictions on capital investment and claims on the profits from that investment. Corporations have been seeking such freedom from the time they were conceived. Restrictions on the investment of capital from environmental regulation and labour law to taxation and requirements to create jobs are for corporations the costs of doing business and as such are to be eliminated. The more restrictions you eliminate, the bigger your profits.

Although the popularization of the idea of the global corporation is recent, corporations and their political servants have been fantasizing about total corporate freedom for decades. George Ball, the U.S. undersecretary of state for economic affairs, stated in 1967: "The political boundaries of nation-states are too narrow and constricted to define the scope and activities of the modern business . . . By and large, those companies that have achieved a global vision of their operations tend to opt for a world in which not only goods but all factors of production can shift with maximum freedom."[1]

Thirty years later, that global "vision," the dream of corporate domination, is close to being a reality for the largest corporations. The sheer power of the TNC today is breathtaking. American researchers

Sarah Anderson and John Cavanagh in their book *The Top 200: The Rise of Global Corporate Power* give graphic evidence of the economic power of TNCs. Of the one hundred largest economies in the world, fifty-one are now corporations. Wal-Mart, number twelve on the list, is larger than 161 countries; in other words, its gross revenue is greater than the total wealth, or gross domestic product, of any of these countries. General Motors is larger than Denmark, Ford is bigger than South Africa, and Toyota surpasses Norway. Canada ranks number eight. The largest ten corporations had revenues in 1991 exceeding the combined GDPs of the hundred smallest countries. Put another way, the two hundred largest corporations have more economic clout than the poorest four-fifths of humanity.[2]

Although we have been accustomed over the years to identify multinational and transnational corporations with the U.S., it turns out that the largest TNCs are no longer U.S.-based. Only three of the top ten corporations are American, while six are Japanese. All but fourteen of the Top 200 are headquartered in just six countries: the U.S., Japan, Germany, France, the U.K., the Netherlands. Of the Global Fortune 500 corporations, the U.S. is home to 153, Japan to 141. Canada counts six amongst the 500.[3]

And while the Top 200 have sales that account for 28.3 percent of the world's GDP, the percentage of the jobs they account for is minuscule in comparison. Out of a worldwide paid workforce of approximately 2.6 billion, the Top 200 TNCs employ 18.8 million, less than three-quarters of 1 percent of the world's workforce.[4]

When just five firms control more than half of the global market, economists consider that market to be highly monopolistic. The *Economist* recently listed twelve industrial sectors that demonstrate this highly monopolistic pattern. In consumer durables, the top five corporations control 70 percent of the global market; in the automotive, airline, aerospace, electronic components, electrical/electronics, and steel sector, the top five control more than 50 percent. Other sectors show equally strong monopolistic tendencies, including oil, personal computers, and media, where control of more than 40 percent of the respective world markets rests with five or fewer corporations.[5]

Monopolistic markets are nothing new; they have existed in developed countries since the 1970s. In the U.S. in that decade, one to four firms controlled 75 percent or more of the market in more than a dozen product areas. In Canada in 1991, according to Allan Engler in *Apostles of Greed*, "ten non-financial corporations accounted for more than one-fifth of the GDP. These were Bell Canada, G.M., Ford Canada, Canadian Pacific, Imperial Oil, Alcan, Chrysler, George Weston, Noranda and Thomson Corporation."[6]

In food retailing, the independent operator is a relic. The corner grocer has been replaced by 7-Elevens and Mac's and the local drug store by huge chains. George Weston owns Loblaws, Super Valu, and Real Canadian Superstores and the largest bakery chain in the country. The Weston empire shares the domination of the retail food business with U.S.-owned Safeway Corporation.

The monopolization of retail business is not always so obvious. How many shoppers have heard of Dylex? Not many unless they are also investors. Yet any given mall in Canada and other countries might have as many as ten outlets owned by this transnational giant, shops such as Tip Top, Harry Rosen, Big Steel, Fairweather, Suzy Shier, Town and Country, Club Monaco, Alfred Sung, Biway and Thriftys.[7]

This illusion of choice and competition is repeated in supermarkets. Although the aisles are full of hundreds of brand names, the truth is that a mere handful of companies produce the vast majority. In Canada these are Weston, Labatt, and McCain, and internationally, Nestlé and Unilever. And it is going to get much worse. Corporations are still merging at an incredible rate, and it was predicted in 1990 that the largest six hundred would become the world's largest three hundred by 2000, the same year that the international banks will merge into ten banking giants to form an elite cabal that will control the banking industry worldwide.[8]

THE MYTH OF THE FREE MARKET

Surveying this incredible concentration of economic power, and control of markets nationally and globally, exposes the sheer nonsense of any talk about competitive markets, the free market, the free enterprise system, and the critical economic role of the entrepreneurial spirit. The notion that the attraction of real investors and businesspeople to market ideology leads them to enjoy the excitement of competition flies in the face of everyday business practice. For flesh-and-blood owners, their worst enemy is uncertainty. And competition embodies uncertainty.

Corporations, particularly large ones, expend enormous effort to eliminate competition and avoid risk and uncertainty. David Korten writes in *When Corporations Rule the World*: "In a globalizing market, the widespread image is one of the corporate titans of Japan, North America, and Europe battling it out toe-to-toe in international markets. This image is increasingly a fiction that obscures the extent to which a few core corporations are strengthening their collective monopoly market power through joint ventures and strategic alliances with their major rivals."[9]

These TNCs, Korten says, share "access to special expertise, technology, production facilities, and markets; spread the costs and risks of research and new product development; and manage the competitive relationships with their major rivals."[10] The auto industry is a prime example, with Chrysler owning parts of Mitsubishi, Maserati, and Fiat; Ford owning 25 percent of Mazda; and GM having a 37.5 percent stake in Isuzu. The same alliances, joint ventures, and cooperation prevail in computer hardware and software, aerospace, pharmaceuticals, telecommunications, defence, electronics, and many others. The 1997 collaboration between Apple Computers and Bill Gates's Microsoft didn't surprise investment gurus at all.

Risk avoidance at the level of the global corporation absorbs a huge amount of effort because it pays off. And only huge corporations can successfully engage in such activity. When we hear corporate CEOs talk about the need to be ever larger in order to be globally competitive, what they really mean is they need to be gigantic in order to ensure monopoly control of their markets and gain enough clout to engage in the modern-day version of collusion.

What most TNCs are now doing is not new. It was pioneered at the turn of the century by what Allan Engler calls the technological monopolies. Westinghouse, GE, and Bell realized that they could not take full advantage of their innovations so long as they were competing in court over hundreds of patent suits. As a result, Engler writes, GE and Westinghouse "decided to pool their patents and divide the growing electrical business among themselves: 62.5 percent was assigned to GE and 37.5 percent to Westinghouse. In 1920 Bell joined the cartel."[11]

The widespread operation of free markets received its death notice with industrialization. With mechanized production came socialized labour. Since that time capital has been in the process of increasingly collectivizing itself in giant national and transnational corporations. The allocation of capital, labour, technology, and raw materials, theoretically accomplished by the invisible hand of the market, is ideology at its grandest. The market has long since been replaced by the administrative coordination of corporate managers. The transnational corporation is simply taking that administrative process to its logical conclusion. As Korten describes it:

> Central management buys, sells, dismantles, or closes component units as it chooses, moves production units around the world at will, decides what revenues will be given up by subordinate units to the parent corporation, appoints and fires managers of subsidiaries, sets transfer prices and other terms governing transactions among the firm's component organizations, and decides whether individual units can make purchases on the open market or must do business with other units of the firm.[12]

The propensity of corporate spokesmen, business columnists, and free-market think-tanks to engage in ideological hyperbole in favour of free markets and rail against state planning is amusing given the extent to which country-size corporations plan their "economies." Indeed, no country, short of the former Soviet Union with its command economy, comes even close to the command structures and decision-making power of the modern TNC.

Management guru Peter Drucker argues: "There is a remarkably close parallel between General Motors' scheme of organization and those of the two institutions most renowned for administrative efficiency: that of the Catholic Church and that of the modern army." Drucker says that the industrial corporations of the nineteenth century adopted the structure of the U.S. Armory at Springfield, Massachusetts. "CEOs played the role of general; below them senior managers commanded corporate divisions; junior managers commanded departments.

Foremen and supervisors were non-commissioned officers."[13] The top-down hierarchies that resulted from the collectivization of capital are as bureaucratic as any government department. The main difference with large corporations is that they are not accountable to the millions of people whose lives they affect. They are equally hierarchical and far more authoritarian.

The modern transnational corporation has taken the historic efforts to control markets, technology, and capital to heights never before imagined. One of the strategies of the super-corporation is to rid itself of as much risk as possible by contracting it out to smaller firms. Instead of internalizing the costs of doing business, the sign of a genuinely free-market economy, the global corporation does its best to externalize them.

U.S. agribusiness is a good example. Rather than have huge production facilities themselves, like the plantations of the past, the large corporations download the risk to smaller producers. The large firms have almost complete control of the market and therefore can dictate conditions to individual producers, no matter how large. Even as early as 1980 the percentage of U.S. farm production controlled by marketing contracts to huge buyers was 95 percent in milk, 89 percent of chickens, 85 percent of processed vegetables, and 80 percent of all seed crops.[14]

Under true competition thousands of buyers and sellers determined the price of each good. Today, as a result of corporately administered markets, the producer gets less and the consumer pays more than the free market would have dictated. This is true as well in developing countries, where the huge banana plantations of the past have been largely replaced by dozens of "independent" producers who are in fact dependent on one large buyer that has externalized the risks of production and restricts its activity to processing and marketing.

The *Economist* magazine, one of the most prominent cheerleaders for globalization, has actually suggested that the TNCs, already creating semifeudal relationships worldwide, go the last step in risk avoidance: don't produce anything, just rent out the "right" to produce. Those who have exclusive control over patented technologies would set a price for renting the patent to producers at a price equal to what they would have made if they had done the work themselves.[15]

But the ultimate transnational will look quite different, and the connections already developing between the corporate giants are the likely harbinger of the future. According to management consultant Cyrus Friedheim the world economy will become dominated by what he calls the relationship-enterprise. Rather than a single corporate entity the relationship-enterprise would be a continuously adapting network of strategic and tactical alliances. These alliances could be geographic, technological, or within sectors.

Friedheim gives as an example talks between the largest aerospace companies — Boeing, McDonnell Douglas, and Airbus, and Mitsubishi, Kawasaki, and Fuji — about a joint project to build a new super-jumbo jet. Such relationship-enterprises would dwarf anything seen today and could easily have revenues exceeding a trillion dollars. The Apple/Microsoft alliance fits this relationship-enterprise model as well. This is the model for the largest corporation in history.

MULTINATIONALS VERSUS TRANSNATIONALS

Today's TNCs are described as being separate from nation-states, in conflict with them and operating independently of state actions. As will be shown throughout this book, this is a convenient fiction. The TNCs are largely the creations of dominant nation-states and the agencies these dominant states establish. Not only is there no contradiction between the administrators of the nation-states and those who run the TNCs, but there is a coincidence of interests in the overlapping of memberships in the national economic and political elites, an interest reinforced by an almost universal acceptance of neo-liberal ideology.

It wasn't always thus. The phenomenon of the multinational corporation, the precursor of the transnational, was born out of competition between Europe and the United States. In the late 1960s the domination of U.S. multinationals was beginning to attract the attention of European policy makers. The alarm was raised by Jean-Jacques Shreiber in his book *The American Challenge*. He warned that American-based corporations were becoming the third-greatest economic power on Earth after the U.S. itself and the Soviet Union and would soon dominate Europe if not stopped.

The European counter-attack was the catalyst for the growth of the global-reach corporation. Led by Britain and France and followed by Germany, Austria, and Italy, state governments promoted the mergers of their own large corporations to counter the power of the U.S. giants. They also passed laws preventing further consolidation of American corporations. "France concentrated its steel, electronic, chemical and computer industries as preparation for the counter-offensive of the 1970s . . . In each [country] the ties between government and big corporations grew closer."[16] As government-sponsored mergers changed the economic landscape of Europe, Japan was already merging state and corporate interests with its integration of public and private institutions.

It became commonplace to talk about the growth of the multinational corporations and their influence in the world economy. While European states were promoting their corporate giants, U.S. corporations spent a decade dumping many ill-conceived subsidiaries and concentrating investment in those corporations best positioned to plan on a global scale and most able to integrate local operations into a worldwide enterprise. One lasting result was the merging of corporate and state interests and the increasing emergence of a single political and economic elite in all the developed countries.

The evolution from multinational corporation to transnational corporation, however, is a change of the past twenty years, and has its roots in complex changes in geopolitics (specifically the decline in American power) and in long-term corporate growth strategies. The old multinational corporation was characterized by its multiple national identities. It established complete operations in the countries it chose to do business in; that is, the operations were relatively autonomous and included production, marketing, financing, distribution, and sales, all developed largely with local labour and locally recruited management. Such practices meant that the corporation inevitably established roots in the communities where it chose to set up. While much of the profits were funnelled out of the country to the head office, some financed expansion and development.

Thus the multinational corporation was portrayed as a good local "citizen" because its local incarnation was genuinely integrated into the community. Even though owned and controlled beyond the borders, the managers and even the CEO were a part of the community. This reality was particularly significant in Canada, where so many sectors of the economy were dominated by American corporations. That domination, however, was not as visible precisely because the companies operated in ways that were not obviously different from those of Canadian-owned corporations.

Much of the economic and social integration of multinationals was determined by government policy. The more conditional investment regulations were, the more integrated the corporation had to be. When investment was contingent on providing jobs, transferring technology, hiring locally, doing research, and paying the statutory tax rate, corporations were not volunteering for good corporate citizenship. They were obligated by the rules of civil society. The more protected individual national markets were, the more obligated a corporation was to behave in the multinational mode. The initially gradual and now rapid removal of barriers to local markets freed up the corporate giants to take advantage of varying conditions in the countries they did business in.

The transnational corporation, according to David Korten, exists when a corporation's operations are integrated "around vertically integrated supplier networks." Korten gives the example of Otis Elevator when it set about to create an advanced elevator system. "It contracted out the design of the motor drives to Japan, the door systems to France, the electronics to Germany and small geared components to Spain. System integration was handled from the United States."[17]

Whereas the multinational corporation made efforts to create a local identity, the transnational is motivated to do just the opposite, to eliminate as much as possible any consideration of national identity or local corporate "citizenship." As the TNC is continuously calculating the advantages of producing here, financing there, and coordinating somewhere else, flexibility of movement is a priority and the location of any aspect of its operation is seen as contingent.

With such a motivation, the last thing a TNC wants is any kind of real integration into the local community or economy. The more ties there are, the less flexible it is; the more commitments it has made — everything from where it gets its supplies to who it has on its local board of directors — the messier it is to pack up and leave. The fewer the organic connections with the nation and the local community the better. Put another way, the fewer rules and regulations there are to cement and formalize those connections, the better it is for the TNC. As we will see, the free-trade deal with the U.S. created the conditions for multinationals to become transnationals virtually overnight.

The trend towards TNCs is so pronounced that their CEOs brag about their lack of any connection with a nation-state, as if such a disconnection indicates success in itself, as indeed it often does. Charles Exley, of National Cash Register, boasted to the *New York Times*, "National Cash Register is not a U.S. corporation. It is a world corporation that happens to be headquartered in the United States." Another example is IBM, once the quintessentially American corporation, which now employs eighteen thousand Japanese workers as one of that country's largest computer exporters. General Motors has an agreement with Toyota whereby GM produces twenty thousand Toyota cars in the U.S. for export to Japan.[18]

The TNC plots its global operations by identifying where it is cheapest (labour costs, taxes, and environmental regulation, for example) to produce its goods, where it is cheapest to arrange financing, in which jurisdiction it is most convenient to declare profits and expenses, where research subsidies and tax breaks are most generous, and what markets are the most lucrative and the easiest to penetrate. And it is not difficult to understand why. As the international trade and investment rules by which corporations operate are swept away, the need to establish full-blown operations in each country where there may be a viable market simply vanishes. The purpose of the corporation has always been to enhance the bottom line. Liberalization of corporate regulations has been both a cause and an effect of the power of TNCs and has made it easy to split corporate operations into many divisions, each contributing to the maximization of profits on the basis of geographic advantage, with each discrete aspect of the operation acting almost as an autonomous unit.

That describes the mechanics of the transnational corporation, but how would one describe the ethos of the development of TNCs? David Korten sees globalization as the most sweeping transformation in human history: "It is a conscious and intentional transformation in search of a new world economic order in which business has no nationality and knows no border. It is driven by global dreams of vast corporate empires, compliant governments, a globalized consumer monoculture, and a universal ideological commitment to corporate libertarianism."[19]

He goes on to list the key objectives of those carrying out the global corporate agenda:

- The world's money, technology, and markets are controlled and managed by gigantic global corporations;
- A common consumer culture unifies all people in a shared quest for material gratification;
- There is a perfect global competition among workers and localities to offer their services to investors at the most advantageous terms;
- Corporations are free to act solely on the basis of profitability without regard to national or local consequences;
- There are no loyalties to place and community.[20]

I suggested earlier that governments have been instrumental in establishing the conditions for the rise of transnationals. But it is not just individual governments that play this role. The International Monetary Fund, the World Bank, the World Trade Organization (formerly the General Agreement on Tariffs and Trade), Asia Pacific Economic Co-operation (APEC), and the secretive Multilateral Agreement on Investment (MAI) being brokered by the OECD are run by, funded by, and responsible to governments. And every one of them has, at different times and in different ways, aided and abetted the growth of transnationals and the internationalization of capital. More than that, they have been absolutely key in the design and development of transnationals. Without these institutions and the role they have played, it is fair to say that the current shape of the world's economy and the nature and role of corporations would be totally different.

TRANSNATIONALS IN CANADA

Canada features just six of the top 500 global corporations, and none of the world's two hundred largest transnational corporations are based here, though of course many operate here as if they are Canadian, accorded, under free-trade agreements and other international accords, virtually the same rights as Canadian corporations. But there are homegrown corporations that typify the TNC in their behaviour and lack of any commitment to their home country, its communities, or the workforce that built their company.

One of the most prominent Canadian transnationals is Northern Telecom, the crown jewel of the high-tech sector in Canada. It has demonstrated at one time or another in its recent history all of the characteristics of the TNC: the transfer of domestic jobs to low-wage countries, its efforts to undermine unions, the concerted attempts to arrange its activities to take maximum advantage of tax, environmental, labour, and other laws in the various countries in which it operates.

Nortel began in 1882 as a Canadian subsidiary of AT&T, and its later relations with Bell Canada gave it a large and secure base that facilitated the early development of its technological capabilities. It benefited considerably from defence department spending in the Second World War and from research conducted by the National Research Council. By the late seventies it was selling into the American market; by 1986 half its market was there, as were half its jobs. By 1991 Nortel had operations in eight countries.[21]

As early as 1975 Northern Telecom had identified the importance of low-wage facilities in the U.S. to its manufacturing strategy. The firm shut down a unionized plant it had purchased in Michigan, and shifted the production to a non-union facility in Nashville. Throughout the latter stages of its U.S. expansion Nortel increasingly chose Southern right-to-work states for its facilities, simultaneously closing down those in the northern U.S. where its employees were unionized. Between 1984 and 1989, Nortel campaigns decertified five of the seven unionized plants.[22]

If the company's North American workers were denied the benefits of Nortel's tremendous success, its workers in the cheap labour locations fared even worse. In Turkey, Northern Telecom fired sixteen hundred union members after they won a three-month strike. In

1992 the company decertified a union in Britain months after it acquired the company the union bargained with. Over an eighteen-year period, Nortel worked hand-in-hand with the Malaysian government to thwart five attempts by its employees to unionize.

Nortel's labour relations improved somewhat by 1993, but its restructuring did not stop. In the early 1990s serious restructuring in the new global mode saw the loss of three thousand jobs in Canada. The company, which is itself controlled by huge conglomerate BCE Inc., now has 68,000 employees in fifteen countries; only 22,000 of those are in Canada. As is the pattern worldwide, it is the shareholder who benefits and the employees and communities who pay the price.[23] In 1993 when Nortel closed or sold plants in Montreal, London, Winnipeg, and Saskatoon, it increased its dividend; it increased it again the next year after announcing the closing or selling of four more plants. During those same two years, Nortel opened plants in two of the most notoriously low-wage, politically repressive, and environmentally unregulated countries in the world: China and Mexico.[24]

In 1997 the company announced that it was adding six thousand jobs in Canada as part of a boom riding the electronics revolution and massive deregulation of telecommunications.[25] But consistent with another Nortel pattern, these jobs were in marketing and development, not the most valuable production jobs. Partly as a result of Nortel, Canada has a large electronics trade deficit, and it has grown by over 300 percent in the past decade, reaching $12 billion in 1994.[26]

Yet the principal objective of government R&D tax incentives is to build Canada's high-tech productive capacity. Nortel has been a huge beneficiary of those credits, yet uses the research to export jobs abroad. Nortel has received at least $880 million in federal research and development tax credits. The Canadian taxpayer has subsidized each job remaining at Nortel's Canadian operations (as of 1995) to the tune of $140,000.[27]

Lest it be assumed that Nortel pays back the Canadian taxpayer for the largesse it receives, its worldwide income tax bill in 1994 totalled $35 million, just 0.4 percent of total revenue.[28] As of 1996, the company owes $213 million in deferred taxes.[29] Even its expansion plans in Montreal in 1997 saw it get a two-year property tax holiday and reduced taxes for three more years.

Without laws and regulations, Canadians might well be working sixty-hour weeks and be subject to extremely dangerous work conditions in order to make a living. The proof seems close at hand. The same chief executives and board members who run our large corporations in Canada also run corporations in Mexico, the Philippines, Indonesia, Thailand, and a host of other underdeveloped countries.

Canadian-based Placer Dome consists of four regional business units, in Vancouver; Santiago, Chile; Sydney, Australia; and San Francisco. It also has an international exploration office in Miami and a joint-venture office in London, England.[30] The company is infamous for an environmental disaster at the Marcopper Mine in the Philippines. On March 24, 1996, a tunnel from a containment pit collapsed and sent four million tonnes of tailings into the Boac River. A U.N. report cited negligence on the part of the mine's management. The spill was so serious and the culpability of Placer Dome so obvious that the Philippine government charged Marcopper president John Loney and mine manager Steve Reid with criminal gross negligence.

It now appears that all charges will be dropped, however, as Placer Dome has succeeded in persuading the government to quash the prosecution. Charges were "indefinitely" postponed and President Fidel Ramos pointedly "welcomed" Placer Dome to "continue its investment in the Philippine mineral industry."[31]

It wasn't the first time that Placer Dome had the president directly intervene to save its skin. Since it began its operations, Marcopper has intentionally dumped the overburden from the mine directly into the waters off Santa Cruz on the island of Marinduque. By 1992 there was a 4.7-kilometre causeway into Calancan Bay, and the overburden covered 71 square kilometres of the coral-rich bottom of the bay. This kind of dumping is outlawed in all developed countries.

As early as 1974, residents had tried to stop the practice. In 1981 and 1988, they succeeded in having Philippine government agencies order Marcopper to cease its dumping, and twice Marcopper executives intervened to have the Office of the President overturn both orders. Placer has refused to compensate villagers for the loss of their livelihood, denying that the mine has had any environmental impact.[32]

The same year as the Boac River disaster, the company was being condemned for its environmental record at its joint-venture Porgera mine in Papua New Guinea. That mine discharges 40,000 cubic metres of tailings a day directly into the Maiapam-Strickland River. The waste rock includes heavy metal sulphides, hydroxides, and cyanide compounds up to three thousand times the country's legal limits. The Mineral Policy Institute of Australia has documented 133 unusual deaths of villagers along the river between 1991 and 1993.[33]

Placer Dome's operations elsewhere provide little promise of any better treatment of local populations or the environment. In a Mineral Policy Institute news release, Placer Dome is cited for its operations at the Misima gold mine at Milne Bay in Papua New Guinea, whose processing plant dumps tonnes of softrock into the oceanic trench next to the island. The institute warns that "the effects of unrecovered cyanide in the Misima's tailings are potentially disastrous." At the proposed Placer Dome mine at Namosi in Fiji, one of the largest open-pit mines in the world, the intention is to engage in "submarine dumping of 98,000 tonnes of tailings daily."[34]

A gold mine planned for Costa Rica is being opposed by environmentalists and residents. According to a report in the newspaper *La Nacion*, the area where the mining exploitation would take place runs along a border area with Nicaragua of enormous ecological value.

In Venezuela, Placer Dome managed to get the currency control laws changed, allowing it to repatriate all of its profits, a condition for its developing its Las Christinas gold property. Placer Dome's plan to mine this property is now the subject of a lawsuit brought by several environmental organizations. Las Christinas is smack in the middle of the Imataca Reserve of the Amazon rain forest, a spread as big as Holland. Placer Dome's plan is for a 2-square-kilometre open-pit mine, 300 metres deep. In the judgement of the World Conservation Union, "The potential destruction here is much larger than in other parts of the world" because of the reserve's ecological richness.[35]

Placer Dome's conduct and its public image building inhabit different worlds. In its 1994 corporate profile, the company goes on at length about its international record with such phrases as "living the ethic of environmental responsibility in our business practices"; "working towards high standards of environmental protection and safety at all our operations worldwide"; "sensitive to cultural differences . . . [striving] for harmonious relationships in different cultural environments."[36]

What the examples of Northern Telecom and Placer Dome demonstrate is not just that corporations flout ethics and morality when they operate in less developed countries. Their behaviour extends to participating and cooperating with local authorities to undermine democracy and suppress political dissent. Placer Dome's intervention with President Fidel Ramos, and Northern Telecom's enlisting the Malaysian government to thwart its workers' legal attempts to unionize, reveals a willingness of Canadian corporations to align themselves with reactionary and repressive governments in order to enhance their bottom line.

As corporations become ever larger and the regulations governing them increasingly weaker, we get to see the raw power of capital in ways not observed since the last century. It is an instructive, if frightening, revelation. As transnational corporations use their unprecedented power and global reach to expand their freedom, all large corporations take advantage of that power and that freedom.

As we observe the steady erosion of communities, social programs, the power of workers, and environmental and other regulations we are observing the true nature of the corporation. As it strips away the features of civil society, the corporate citizen reveals itself as a dangerous fraud. Without the state's strict enforcement of the obligations of citizenship, the corporation is exposed for what it is. It is not a citizen, or anything even vaguely resembling such a multifaceted and complex entity. The corporation is strictly one-dimensional, demonstrating only the narrowest of characteristics: greed.

2

DEMOCRACY, THE STATE, AND THE CORPORATE CITIZEN

We are calling upon [those who wield corporate] power and property, as mankind once called upon the kings of their day, to be good and kind, wise and sweet, and we are calling in vain. We are asking them not to be what we have made them to be. We have put power into their hands and ask them not to use it as power."

— HENRY DEMAREST LLOYD, AMERICAN WRITER AND POPULIST ORGANIZER AGAINST CORPORATIONS, 1894

When approaching these issues [of financial regulation] our obligation is to our constituents but also our obligations [are] to you . . . . You recognize that as members of Parliament our first duty is to our electors . . . but you are also our electors because you provide jobs. We will go about that difficult task in working with you in the future and if we don't get it right this time we'll work with you to make sure we will get it right.

— LIBERAL MP JIM PETERSON, CHAIR OF THE COMMONS STANDING COMMITTEE ON FINANCE, TO GORDON FEENEY, CHAIR OF THE CANADIAN BANKERS' ASSOCIATION, 1996

Corporations are to all intents and purposes — and certainly their own — legal citizens. They communicate their views directly to other real citizens through advertising, they assume the status of citizen in dozens of public forums, they have the standing of citizens in court, are sought out for their opinions on public policy, take part in elections by contributing to political parties. Indeed, corporations enjoy many of the same "freedoms" that flesh-and-blood citizens enjoy, although the power and influence of the corporate citizen far outweighs that of the flesh-and-blood variety.

An obviously artificial person, the corporate citizen in the era of the TNC is now a kind of super-citizen, with overwhelming power on the one hand and ever decreasing responsibilities on the other. The TNC is a special case, with the status of global corporate citizen whose "rights" are transnational, in effect superseding the cultural and social basis of the citizenship enjoyed by you and me. Of course, it is precisely these features — the excessive power, the refusal to accept responsibility, the distinct international status — that should disqualify corporations from citizenship. What is so astonishing about their continued status is that almost no other entity on Earth is less qualified to be a citizen.

THE EVOLUTION OF THE CORPORATION IN CANADA

Somewhere along the line, we seem to have simply accepted the notion that corporations have a far superior status to a real citizen, more power, more influence — access to power that no ordinary citizen, or group of citizens, could possibly imagine. If this super-citizen had super-abilities and superior ethics and yet was benign, it might be argued that giving corporations this status made sense. But in fact corporations have gained this status by stealth and accident over centuries of organizational and political evolution.

When the Charter of Rights and Freedoms was passed in 1982 many Canadians felt that their rights as citizens had just taken a major leap forward. Having watched American crime dramas on TV in which people were read their rights, we felt that we too would now be protected from the abuses of the state. Like the Americans, we would now have constitutional rights. Democracy, the right of ordinary people to determine their own future, would be enhanced. Our fundamental freedoms — of expression, of association, of religion, against false arrest — would be protected by the power of a constitution, not just by the legislature, the common law, or by our demanding those rights in the streets.

Thirteen years later, in the fall of 1995, Canadians received a powerful signal that their initial assessments of the Charter may have been mistaken. The Supreme Court of Canada, building on a number of lower-profile, precedent-setting cases over the years, ruled that the right of citizens not to become addicted as children to cigarette smoking was subordinate to the right of the powerful tobacco companies to freedom of expression. Specifically, the highest court of the land said: "Freedom of expression, even commercial expression, is an important and fundamental tenet of a free and democratic society. If Parliament wishes to infringe this freedom, it must be prepared to offer good and sufficient justification for the infringement."[1] (Apparently, the deaths of forty thousand Canadians annually and the targeting of children with sophisticated marketing did not provide sufficient justification.)

The ruling was just the latest instance of the rights of corporate citizens being enhanced and written in stone in Canadian law. Sometimes the rulings bordered on the farcical, such as the one declaring the right of a margarine company to "express" itself by colouring its margarine to look like butter. And other times they were so far-reaching that they changed the course of the country's history.

In 1984, for example, an Alberta judge threw out a law limiting third-party spending in elections. Had this law been in place, it would have prevented the most powerful corporations in the country from engaging in an unprecedented $19-million propaganda campaign to subvert the 1988 election, in which free trade was the central issue. Corporations, said Mr. Justice Donald Medhurst, had to be free to express themselves as citizens. He dismissed the argument that corporations' overwhelming financial resources represented a threat to democracy; he said there had to be "actual demonstration of harm or a real likelihood of harm to a society value" before any government could impose limits on the freedom of expression guarantee in the Charter.[2] The challenge to the law was brought by the National Citizens' Coalition, a right-wing lobby group funded in large part by corporations.

The Charter of Rights and Freedoms, though it has narrowly advanced the rights of some groups in Canada, has, according to Osgoode Hall law professor Michael Mandel, done far more to damage democracy than it has to enhance it. Mandel believes all such charters have been designed precisely to counter the effects of majoritarian democracy. He argues that the Charter is "an antidote to democracy that appears throughout the world when you have universal suffrage, or when representative institutions start to cause problems. Then the powerful people — the property owners, the wealthy — in society start to get scared. They look for means to control them . . . judicial review, charters of rights, as an antidote to really representative government."[3]

The property Mandel refers to is not personal property but capital. Under representative democracy the rights of capital — that is, corporations — to do as they please in their pursuit of profits, such as targeting children in tobacco advertising, or clear-cutting forests, or building a factory on a riverbank, can be challenged by citizens. Charters, argues Mandel, give corporations a whole new arsenal in their fight to be free of restrictions.

Explicit property rights are not included in the Charter, and so Canadian courts have looked south for guidance in giving citizens' rights to corporations. The U.S. has long experience with efforts to protect property from the sovereign people. According to Mandel, "the American constitution and its Bill of Rights put interference with property beyond state and congressional reach." That is, in his view, precisely what the Canadian Charter does in placing the unelected court above the authority of parliamentary sovereignty. Not only can corporations challenge governments on whether they are obeying the laws but they are given "the unlimited right to veto those laws, indeed the right to final approval of the legitimacy of all government action."[4]

While the Charter gives new, substantive citizen rights to corporations, it simultaneously puts explicit restrictions on governments. In effect, it puts forward the notion that we have more to fear from our governments, our elected representatives, than we do from corporations. Allan Blakeney had grave reservations about the Charter, which came into being while he was premier of Saskatchewan. He has said: "Charters of rights are predicated on the belief that . . . the only people who interfere with rights are governments, and therefore the charter will restrict governments and no one else; and that government action always curtails freedom and . . . never enhances freedom. I think that many government actions, including minimum wage laws . . . expand freedom."[5] (It all depends on what you mean by freedom.

For corporate libertarians it is freedom to do what you want; for Blakeney, governments provide freedom from want, insecurity, freedom based on the protection of citizens from powerful interests.)

And Canada ended up with a very different form of government with the Charter of Rights and Freedoms, one that greatly weakened the principle of representative democracy through parliamentary sovereignty. In so doing it handed over some of that sovereignty to that fictitious citizen, the corporation. It is powerfully reminiscent of the time when Canada was a sovereign corporation.

CANADA AS A CORPORATION

Canada's intimate relationship with the corporation may well be unique. The country's character and beginnings were established within the bounds of a single corporation, the Hudson's Bay Company. In addition to organizing the fur trade, "the Company" carried out the activities of government, including making laws and regulations governing the lives of its employees and the aboriginal population within its jurisdiction. It maintained its own policing and arrested, tried, and jailed people. It was, of course, completely undemocratic, with no accountability to those it governed.

Similar to trends we can recognize with the transnationals of today, the Hudson's Bay Company (HBC) was an amalgam of state and corporation. The charter for the HBC and similar companies did establish obligations as well as money-making privileges for them. The Royal Proclamation of 1763, one of Canada's founding documents, granted to the governor the power to incorporate but cautioned: "It is our express will that you do not, upon pain of Our highest displeasure, give your assent to any Law or Laws setting up any Manufactures and carrying on any Trades, which are hurtful and prejudicial to this Kingdom."[6]

But by the end of the eighteenth century, a new colonial economic elite wanted freedom from any obligation and insisted that governments serve them. According to historian Donald Creighton, the colonial business elite demanded changes to the way the colony would be governed: "The state should make the way smooth for the commercial concerns, the banks and the land companies. The state should provide harbours, ship channels and canals. The state, in fact, must become the super-corporation of the new economy, dominated like any other corporation by commercial interests and used to its full capacity as a credit instrument in a grand programme of public works."[7]

According to legal researcher Gil Yaron, by the mid-1880s the act of incorporation was changed to the simple one of registration rather than royal charter, and from then on incorporation was a right, not a privilege. In contrast to the limited life of the chartered corporation, the registered version opened the door to "perpetual succession" or, in effect, a corporation that never went out of existence.[8]

Corporations so dominated colonial life in Canada that our nation's key constitutional document, the British North America Act, is completely void of any mention of human happiness or other lofty goal. Unlike its American counterpart it was strictly a business proposition, reflecting the business objective of creating a separate market for the Canadian business class. According to Gerry Van Houten, writing on the history of corporate Canada, "Canada's ruling class was able to determine the main content and characteristics of the new constitution . . . The constitution had nothing to say about the rights of Canada's farmers and workers. It also had nothing to say about the First Nations or Quebec's unique position in the Canadian state."[9]

THE AMERICAN EXPERIENCE WITH CORPORATIONS

The United States is now home to most of the world's most powerful transnational corporations. The U.S., which we now see as almost synonymous with the transnational corporation, developed that character only after a dynamic and popular struggle against corporate domination. While Canada was established almost as a corporate project, the U.S. began as a revolutionary democracy, founded on the principles of the sovereignty of the citizen and the subordinate status of all other entities, including corporations. Corporations existed only with the explicit permission of citizens. Early citizen-imposed constraints on what corporations could do help remind us just how artificial a being the corporation is and may help point the way to re-establishing sovereign authority over corporations in the future.

Richard Grossman in his booklet "Taking Care of Business: Citizenship and Charter of Incorporation" tells many stories of how corporations were held to account. In an 1809 judgment, the Supreme Court of Virginia decreed that if an applicant's purpose in seeking a corporate charter "is merely private or selfish; if it is detrimental to, or not promotive of, the public good, they have no adequate claim upon the legislature for the privileges."[10] Corporate charters were routinely denied because the corporation failed to prove it would serve a clear public purpose.

The principle for challenging a corporation's charter was *quo warranto* — by what authority? — underlining the fact that in most states the corporation was allowed to engage only in those activities explicitly allowed in the charter legislation. The revocation of corporate charters was commonplace. President Andrew Jackson, to great popular acclaim, in 1832 vetoed a law that would have extended the charter of the Second Bank of the United States. Pennsylvania revoked the charters of ten banks in that same year and in 1857 passed a constitutional amendment instructing its legislators to "alter, annul or revoke any charter of a corporation . . . whenever in their opinion it may be injurious to citizens of the community."[11]

The tide began to turn in favour of American corporations in the latter part of the nineteenth century, when they began to successfully use the courts to challenge or get around the chartering powers of the states. The Civil War had made many corporations fabulously rich. According to Grossman, "Following the Civil War and well into the twentieth century, appointed judges gave privilege after privilege to corporations. They freely reinterpreted the U.S. constitution and transformed common law."[12]

The biggest victory of the corporations came in 1886, when the Supreme Court ruled that a private corporation was in fact a "natural person" under the U.S. constitution. In a fundamental reversal of practice, and reflecting their new "person" status, corporations were given the right to operate in any fashion not explicitly prohibited by law. By the turn of the century the judiciary had led the way in positioning the corporation to become, in the words of a Ford executive, "America's representative social institution, an institutional expression of our way of life."[13]

By the early part of the twentieth century the battles were over. In 1941 a congressional committee concluded, "The principal instrument of the concentration of power and wealth has been the corporate charter with unlimited power." Indeed, that power was used in the Supreme Court to block the New Deal of President Roosevelt. Although Roosevelt eventually forced the court to back down, the legal power of the corporation was not changed. Says Grossman, "Many U.S. corporations are transnational. No matter how piratical or where they roam, the corrupted charter remains the legal basis of their existence."[14]

THE ORIGINS OF THE WELFARE STATE

The Ford executive's proud declaration that the corporation had become the institutional expression of the American way of life was an accurate description of the reality for ordinary Americans and Canadians. Once corporations were accorded the right to exist forever and operate with limited liability, under legislation that made no mention of any responsibility to society, they were by far the most important organizing institution in society. With government so conspicuous today, it is hard to imagine just how dominant corporations were in the early part of the twentieth century.

Hardly any area of life wasn't affected by the power of corporations. The romantic view of western grain farmers challenging the elements in a heroic struggle to improve their lot masks the monopolistic stranglehold that banks, grain companies, and railways had on their lives. The same was true for workers. Factories and mines employed only the most minimal safety standards, and there were virtually no legal mechanisms to protect workers' rights during industrial disputes.

Almost every important decision affecting ordinary citizens was either made by corporations or made in their interests. There was no social safety net, no medicare, no unemployment insurance. Economic life was completely in the hands of private capital. The corporate way of life was one in which the corporate leaders, at least when it came to all important governance questions, were indistinguishable from the political leaders.

This was particularly the case in Canada. The extreme suffering experienced during the depression of the 1930s turned Canadian politics into a cauldron of conflict, which shook the confidence of corporate elites. The respective roles of the state and corporations had to be reconceived. With the goals of social control and infrastructure support for business, the Canadian state undertook a broad series of social reforms.

Programs such as government mortgage lending, unemployment insurance, a federal manpower agency, and a variety of marketing boards were proposed, and some were implemented by the Conservative government of R. B. Bennett. Bennett himself was a multimillionaire, and his cabinet was dominated by men with ties to large corporations, and the senior civil service was made up of former businessmen. Their apparent change of heart was explained by A. E. Grauer, writing for the Royal Commission on Dominion-Provincial Relations in 1939: "The note sounded has not been so much the ideal of social justice as political and economic expediency."[15]

In fact many businessmen supported a workers' compensation scheme because it was ultimately cheaper than paying private insurance companies for protection against workers' lawsuits; a shorter work week was popular because it "spread work around"; unemployment insurance was supported, said Grauer, because of "the appalling cost of relief" and the hope that UI would "give some protection to public treasuries in future depressions."[16]

The Bennett government was the same one that was utterly ruthless when it came to suppressing workers' demands for rights of any kind and union rights in particular. The state directly assisted corporations' efforts to stop organizing drives. It jailed communists, deported non-British immigrants who were seen as the source of communist ideas, forced the unemployed into camps, and ended a trek to Ottawa by the unemployed with a violent assault in Regina. Reforms were one thing; rights, another.

Liberal prime minister William Lyon Mackenzie King was not convinced of the necessity of social reforms and spent his early years in power doing everything possible to block them, claiming that governments could do nothing about unemployment and should not run up debts to deal with crises. It was an argument that the war years rendered obsolete. Canadians daily had the evidence before them that if a situation was considered serious enough (like the threat of Nazism that affected more than just the poor and the unemployed), their government was willing to undertake an extraordinary amount of intervention in the market, creating full employment, controlling trade, taxing wealth, providing incentives for industry — and going into massive debt.

In addition, during the war years the union movement doubled in size and increased in militancy. Still King resisted, until he felt he was in a position to run on a platform of social reform without alienating the more reactionary wing of the business elite. Both the Progressive Conservatives and the Co-operative Commonwealth Federation, or CCF, which was now calling for reforms rather than fundamental change, had policy platforms calling for state intervention, and were thus serious threats to Liberal dominance. King's program included a family allowance scheme and spending to help in converting the war economy back to its domestic version. It also included historic legislation that legalized unions and collective bargaining. The debate about social legislation continued to focus on tactical questions about what was necessary for preserving the relations between business and labour.

Alvin Finkel documents how intimately the corporate chieftains of the day were involved in determining what combination of repression and social reform was best to preserve labour peace. Fully integrated into both the Liberal and Conservative parties, corporate executives embodied corporate rule in its quintessential Canadian mode, with the political and corporate elite deciding what was best for corporations.

The historic state-mediated compromise between corporations and working people was also based on a major weakness of corporations at that time. Corporate investment was relatively hard to move right through the 1960s, which meant that foreign and domestic corporations that established in Canada had to respond to conditions as they were, including a militant labour movement. It wasn't until the 1970s, when investment became more mobile, that the practical basis for the social contract and corporations' willingness to compromise would begin to disappear.

The rising expectations of workers and farmers following the war became a part of the political culture. The notion of citizenship was greatly enhanced as a result of the victories Canadians had achieved. In Alvin Finkel's view, "Canadian citizenship came to mean more than simply having a formal set of 'negative' constitutional rights such as 'life, liberty and the security of the person,' it meant you were entitled to an assortment of more 'positive' social welfare rights that 'you are entitled to simply because you are a Canadian.'"[17]

Canadian citizenship in a sense came of age. The notion of a collectivity based on the values of mutual responsibility and social solidarity became a theme in our political culture. The individualism of the early part of the century, enforced as a cultural value by the power of corporations, was fundamentally altered in the post-war period. The corporate elite had allowed the development of a political culture hostile to its interests, one that it would have to confront eventually. The ideological attacks on the notions of rights and entitlement, and the explicit promotion of the values of individualism, that we are witnessing today are no accident but a calculated effort to reinvent the notion of citizenship.

Another key factor in the development of the welfare state in Canada was the crucial role of the civil service. Mackenzie King had become extremely indecisive as a leader in the post-war period. As Robert Chodos argues in his book *The Unmaking of Canada*, "It was into this vacuum that the civil service stepped, overflowing with policy ideas acquired from [John Maynard] Keynes in Cambridge and brought to Canada by way of the New Deal Washington . . . The fledgling Canadian mandarinate was made up almost entirely of disciples of John Maynard Keynes."[18]

By the 1960s, the Canadian state was pursuing policy associated with the work of Keynes, objectives such as full employment, high economic growth, price stability, and an equitable distribution of income. New measures were aimed at establishing equity between regions through regional development programs and equalization payments. It also introduced nationwide medicare in 1966 and a greatly enhanced UI program in 1971. Payments to the provinces to share the costs of post-secondary education and social assistance programs completed the main features of the second wave of the egalitarian welfare state.

Two decades of virtually uninterrupted growth, and robust profitability, fostered in part by the Keynesian programs of Liberal governments, resulted in a de facto policy of benign political neglect on the part of corporations. The state had delivered to the corporations what it had promised: stability, the conditions for mass consumption, programs that smoothed out the economic cycles, and absorption of the costs of maintaining a healthy, educated, and contented workforce. And all of this was done with a fairly regressive tax system, one which taxed the better-paid workers and the middle class to pay for social programs for everybody. While corporations paid a much greater portion of the tax bill than they do today, that was partly a consequence of high profits, not high tax rates.

LIBERAL DEMOCRACY — AN OXYMORON?

The approach of Canadian governments both before and after the war was the classic expression of liberal democracy: an uneasy marriage between free-market individualism and democratic government policies that provided for a modicum of economic and social equality. But the question would be raised in the 1960s and early 1970s about just how democratic liberal democracy was. With the growth of the economy there were rising expectations not only with respect to the delivery of government programs and regulation of the economy but of substantive democracy. In short, there began to develop a strong sentiment for a more democratic process, not just state-sponsored equality.

Not only was the voice of corporations muted, there were New Democratic governments in place in three western provinces, the Parti Québécois was strongly social democratic, and even the national Conservative Party was dominated by Red Tories. Social movements had captured the imagination of thousands of young people across the country. The anti-Vietnam War movement, the radical student movement, the rebirth and rapid growth of feminism, anti-poverty groups, and a sometimes militant aboriginal movement grabbed headlines. Common to all of them was the demand for a more substantive democracy.

This was a demand for democracy of a different sort, a radical democracy in the sense of returning to the roots of the idea, based as it was in effective rule by the majority. Aboriginal people demanded self-determination, women demanded equality, and the anti-war movement focussed on the role of corporations in the Vietnam War. The student movement insisted on an end to privilege, demanding that universities serve the community, and publicly denounced professors for their failure to teach anything relevant about the world around them.

A powerful moral and ethical spirit was at the centre of much of the new politics, and young people targeted both corporations and governments. The latter were seen as complacent and ossified, the former as self-serving and immoral. Alongside high levels of poverty, estimated at 27 percent, were corporations that received all sorts of tax breaks. NDP leader David Lewis effectively played to this audience with his "corporate welfare bums" campaign in 1972. Even Prime Minister Pierre Trudeau acknowledged, in the debate over tax reform in 1971, "It's likely that we heard more from the vested interests than we did from the little taxpayer who didn't . . . have high-paid lawyers to speak for him."[19]

In addition, the labour movement, awakening after years of complacency, was increasingly aggressive at the bargaining table and much of it was adopting social unionism, a broader involvement of all the social issues of the day, as part of its revival.

It was the first time since the conflicts of the 1930s and 1940s that liberal democracy was seriously exposed to a radical critique. And liberal democracy was very vulnerable to those who wanted the real thing. The renewed passion for democracy, and the critique of the existing model, was perhaps best captured by C. B. Macpherson in a 1965 CBC lecture series called *The Real World of Democracy*, to this day one of the most compelling essays on democracy written in Canada.

Until the latter part of the last century, wrote Macpherson, democracy was considered a "bad thing . . . feared and rejected by all men of learning, men of substance, men who valued civilization."[20] It was a "levelling" doctrine, and giving the franchise to those who were "dependent" on others would level it downward. In Western societies the democratic franchise was established only after the liberal capitalist society was firmly established.

In liberal democracies, liberalism, understood as the freedom of the individual, came first. Macpherson explained that before democracy "there came the society and politics of choice, the society and politics of competition, the society and politics of the market. This was the liberal society and state . . . Both society as a whole and the system of government were organized on a principle of freedom of choice."[21] The impact of the market on society had been profound, changing relationships between people into market relations. Democracy, said Macpherson, was an "add-on."

It was inevitable that the principle of freedom of choice would be taken up by those without property who were denied any say in the governing of the state. But the very nature of the liberal state and society, governed by market relationships, was such that the demand for democracy was not a radical demand. "By the time democracy came . . . it was no longer opposed to the liberal society and state. It was, by then, not an attempt by the lower class to overthrow the liberal state or the competitive market economy; it was an attempt by the lower class to take their fully and fairly competitive place within these societies. Democracy had been transformed. From a threat to the liberal state it had become a fulfilment of the liberal state."[22]

With democracy transformed by, and absorbed into, the competitive political culture of market society, the substance of democracy was reduced to the welfare state and the regulatory state. In fact, argued Macpherson, these two elements would have been introduced even without the extension of the franchise because the capitalist economy could not have flourished without them. It was often the most conservative, pro-business governments, like R. B. Bennett's, that introduced the welfare state. Not only had democracy not fundamentally altered market society but for the most part people's understanding and expectations of democracy were as much transformed as the actual fruits of democracy were.

The flowering of democratic movements in the 1960s and early 1970s was briefly accommodated and even promoted by the government of Pierre Trudeau. Perhaps uniquely amongst Western developed countries, the Trudeau Liberals institutionalized the political movements of previously marginalized groups through a program of participatory democracy. At least part of the motivation of this funding was a genuine attempt to give access to policy making to those who had so far been excluded. But, unfortunately, the era of participatory democracy turned out be extremely short-lived.

Just when social movements seemed to have hit their peak, corporations faced their biggest crisis since the Great Depression. Both profitability and growth were in decline. The crisis was brought to a head by the 1973 Arab oil embargo and the enormous price rises that followed. Suddenly everything changed in the world of the corporation and capital investment. The oil crisis led to a severe decline in profitability (except, of course, for the oil companies), a flight of capital out of North America looking for cheaper manufacturing locations, and a resulting deindustrialization. In addition, the drive to reduce production costs and increase productivity prompted a technological revolution that between 1971 and 1979 would see the loss of 630,000 jobs in Canada.[23]

Classic liberalism, with its emphasis on individual (and, by extension, corporate) freedom, was back; majoritarian democracy was once again seen as a threat to profitability, at least in the corporate circles that ultimately determined what the liberal democratic state did. In the mid-seventies the state was scrambling to fix the system, but nothing seemed to work. A new phenomenon, "stagflation," caused by the sudden and rapid jump in oil price, put Keynesians inside the government on the defensive. The country was facing growing unemployment and rising inflation, something that wasn't supposed to happen. In a desperate effort to appease business and stimulate growth, a new finance minister, John Turner, introduced dozens of tax breaks aimed at creating a quarter-million jobs by 1979. Instead, the number of jobs decreased. As a result of all of these factors, the federal government began accumulating the debt that would become the issue of the decade fifteen years later.

Corporations were eager to get out from under the regulatory welfare state, not cooperate in its expansion. The relationship had soured as the state had failed to maintain the conditions for continued profitability. According to Duncan Cameron, chair of the Canadian Centre for Policy Alternatives, corporations rejected the other option, more regulation of investment. The solution to inflation would be the crude instrument of monetarism, the control of the money supply. Keynesian theory on investment allowed for creating the conditions for private investment. The Keynesian theory on inflation required wage and price controls. Cameron says: "It really required the socialization of the investment process. It meant that investors had to agree that if we're going to, as a society, give you lower wages through wage controls you are going to make more money and in return we want you to invest that money in more jobs and more income."

But private property was sacrosanct. The government went instead to monetarism. Curb the growth of the money supply; if interest rates go sky-high, so be it; if the economy crashes, so be it. "The business community preferred to see the economy crash, as it did in 1981–82 and again in 1990–91, rather than go into a tripartite negotiating process, which the logic of Keynesian economics would have led to."[24]

It was the answer that corporations gave in virtually every Western country. They were strongly assisted by the Organization for Economic Co-operation and Development, made up of all the industrialized nations, which was hugely influential in promoting tight control of the money supply as the solution to the crisis, arguing against both corporate cooperation with labour and government and wage and price controls. The larger corporations in particular were receptive to the idea of breaking from the welfare state and the social contract implied in the historic compromise with labour. The state had provided stability and subsidized the "reproduction" of labour through provision of a healthy, educated, and reliable workforce. But it had also shifted the balance of power in favour of labour, and in a time of continuing falling profits, the welfare state was seen as a liability.

When the corporate elite in Canada and elsewhere decided to break its historic compromise with labour and the welfare state, it freed itself from the restrictions of national boundaries, already severely weakened between 1974 and 1979 by the increased mobility of capital. That political decision changed everything, and effectively accelerated the globalization of the economy. Growth based on healthy and stable domestic economies in the developed world was abandoned, in large part out of necessity. Preferred now was growth through global expansion, and the drive for "efficiency" that international competitiveness demanded.

The abandonment of the social contract, and with it Keynesian economics, was explicitly an abandonment of a compromise between social classes and a tacit recognition that class tensions would increase and political debates would polarize along class lines. But in Canada, situated as it is next door to the American superpower, it meant much more. No other nation in the developed world has relied so much and for so long on the state for its existence. As Steven McBride and John Shields argue in *Dismantling a Nation*, in Canada "an active role for the state has been an instrument of nation-building and national identity . . . The attack on the state's role and the promotion of unrestricted market relations involve a challenge to the definition of the country rather than merely to established social relations within it."[25]

The Liberal government's efforts at regional development, from its unique UI program to equalization payments and regional development strategies, were strategic devices that helped resist the powerful north-south pull created by the huge American economy. That north-south pull was already powerful because of the continentalist policies of previous Liberal governments. The freeing up of market forces inevitably meant a return to that continentalist strategy. It also inevitably jettisoned the moral imperative that was implied in Keynesianism. Nation building necessarily contained the moral element of equality and fairness as well as the principle of democracy, which were much more difficult to impose on international capital.

The new imperatives of international trade and competitiveness were the perfect context for ideas about freeing the market from interference by government. Arguments in support of the new global economy, and those who stood to benefit, poured through the cracks in the consensus underpinning the welfare state. They soon amounted to the equivalent of an ideological holy war against the old forces of Keynesian economics, state social intervention, collective rights, and egalitarian values.

The sweeping changes that have taken place in Canada since 1980, accelerating since the election of the Mulroney Tories in 1984, are referred to as neo-liberal because they represent a new version of the old liberal free-market values of the nineteenth century. Neo-liberal prescriptions have a profound effect on building up the power of corporations at the expense of citizens. These prescriptions, which include eliminating public-sector jobs, privatizing government corporations, and cutting the social safety net, all make citizens more dependent on corporations for their livelihoods.

Dependence on corporations reflects the new balance of power between capital and labour or, put another way, the erosion of the enhanced citizenship that arose out of the struggles of the 1930s and 1940s. It is, then, a decline of the more substantive democracy that citizens in those struggles took from the corporate state. Indeed, the new economic order, with its roots in the crisis of the 1970s, pits a weakened real citizen against a new breed of corporate citizen goliath whose power and capacity to influence political events dwarfs that of the corporations of the past.

The abandonment of Keynesianism and as a consequence domestic economies as the strategy for growth unleashed the forces that would create the transnational corporation. The strategy of international investment and trade was a catalyst to the growth in size and reach of the corporation. At the same time, it freed the corporation from the state-enforced "good citizenship" that seemed to characterize developed nations for the two decades after the war. However meagre the benefits of that level of corporate responsibility, it had always been expedient for both the state and the corporations. After the mid-1970s that expedient was gone.

This did not mean that the corporations wanted the state to leave the playing field. Far from it. As we will see, corporations have since used their power to enlist the state in implementing their new global strategy. The power they now wield, individually and collectively, as global corporate citizens is formidable. Yet their transnational character does not free them from acting at the national level where state power is exercised.

In the history of transnational corporations, some governments are obviously more important than others. The U.S. is now the only superpower and what it does both domestically and internationally has enormous influence on how the world works. This superpower status makes it important to know how American-based TNCs operate to influence the U.S. government and, by extension, the world. The gradual merging of corporate and state governance is easily illustrated in the U.S. The most "free market" of any of the developed countries, it still has the power to set trends if not dictate the behaviour of other nation-states. And how corporations dictate to the U.S. government is important to our understanding of the development of global corporate rule.

CITIZEN GE

Some corporations take their political role more seriously than others, but it is fair to say that none works harder at governing than General Electric, rated the fifty-fifth largest "economy" in the world in 1995 (ahead of the Philippines, Iran, Venezuela, Pakistan, and New Zealand). William Greider in his book *Who Will Tell the People?* talks at length about "Citizen GE."

Greider begins his tale with an account of how the corporation built up its team of political operatives by hiring people who were among the smartest political figures in government, often those whose job it had been to develop policies regulating GE and other corporations. Benjamin Heineman Jr. (Harvard, Yale, Oxford) had been assistant secretary for planning at the department of health, education and welfare under President Carter. He went to GE because the corporation offered him more opportunities to make public policy than any government job could have — "from the tax code to defense spending, from broadcasting to environmental regulation, from banking law to international trade, from Head Start to Star Wars."[26]

William Lacovara, a former Watergate prosecutor, went to GE because "GE recognizes that as a major economic entity it has the stature and responsibility to form opinions."[27] One of his first efforts was to derail new corporate-sentencing guidelines being drawn up for the federal courts. Stephen D. Ramsey had been assistant attorney general for environmental enforcement under Reagan. GE hired him to produce a "play book" on how to "confound the government's efforts to collect billions owed by polluters." GE has one of the worst pollution records in the U.S., with the highest number (47) of priority clean-up sites of any U.S. company.

GE hired a former legislative counsel for the treasury department to head its tax-planning policy and a former counsel for the energy department as its chief lawyer for its appliance division. The list goes on, and while GE is not typical because it is so large, its behaviour demonstrates a growing reality: for corporations like GE the politics of governing is as natural as producing light bulbs. Indeed, it is an integral part of all that it does. GE's breadth of activity has been almost as broad as the government's. The products it makes and the corporate interests they reflect have converged with nearly every aspect of Washington's decision making.

Its very size and the breadth of its economic activities, from making hydrogen bombs to owning media conglomerates, from banking and stock brokerage to jet engines, determined that GE would begin to act, in Greider's words, "like a mediating institution." In other words, like an instrument of government. It begins to behave as if it actually speaks for whole groups of people, but people to whom it has no accountability and with whom it never consults: workers, consumers, shareholders, communities, other corporations, and the country at large.

GE has the resources to develop and promote new political ideas and to organize public opinion around its political agenda. "It has the capacity to advise and intervene and sometimes to veto. It has the power to punish political opponents." And its political opponents are usually Democrats, because like most large U.S. corporations GE has consistently supported the party of business. It sponsored Ronald Reagan's TV career in the 1950s and "launched him on the lecture circuit as a crusader against big government."[28]

The job of governance is important enough at GE that it is directed by a senior vice-president. He is backed by a sophisticated program and a staff to match, a whole infrastructure of different elements including twenty-four full-time lobbyists in Washington, and strategic financial investments in charities, politics, propaganda, and education.

GE nurtures its good corporate citizen image with an $18.8-million fund for its foundations (less than half of 1 percent of its net earnings). But even here much of the money is directly self-serving, going to business think-tanks, business associations, and coalitions fighting government regulation. So the Institute for International Economics lobbies for the corporate line on trade, and the Americans for Generational Equity campaigns for cuts to entitlement programs like social security. GE was a key funder of the Committee on the Present Danger, which propagandized for the massive military buildup of the 1980s, and of the Center for Economic Progress and Employment, which, despite its name, is a front group of industry giants determined to gut product-liability laws. GE is also prominent on the Business Roundtable, the prototype corporate governance lobby group copied in Canada by the Business Council on National Issues.

Like most mega-corporations in the U.S., GE combines a soft image created by tens of millions of dollars in yearly image advertising with a ruthless hardball politics on Capitol Hill. As the congressional vote on a corporate tax cut approached in 1981, Congress members were flooded with telegrams and lobbied personally by CEOs (organized by the Carlton Group of corporate tax lobbyists, headed by a GE executive) warning them that lack of support would mean a loss of jobs in their districts.

GE was one of the biggest winners in the tax cut bill, getting a $283 million tax rebate for the period 1981–83 despite making profits of $6.5 billion. The tax windfall allowed GE to go on a buying spree of other companies, including RCA and NBC. This was a pattern for all the corporate giants as capital investment increased dramatically, with almost all of it going to other countries. In effect, the billions in tax dollars sacrificed by the American people were used to further deindustrialize their own heartland and boost the growth of super-corporations.

GE is the quintessential corporate citizen of the global era — with virtually unlimited resources, which it uses to intervene at every level of government from the White House down. Indeed, its political branch operates almost as a parallel government, with policy-making capacity and its own propaganda arm capable of selling its policies to the public at the same time that it persuades or blackmails elected representatives into carrying out those policies. Whenever the nominal government veers off the desired course, GE intervenes to bring it back.

CORPORATE DEMOCRACY IN CANADA

There are no parallels to "Citizen GE" in Canada, largely because Canada does not host any of the truly huge transnational corporations with the breadth and depth of a GE. As well, the stakes in changing Canadian policy are not normally as high as they are for American TNCs. There was one instance, though, in which the stakes were extremely high and when Canadian corporations, in partnership with their American branch plants, intervened as forcefully as a U.S. corporation: the corporate campaign to buy the 1988 federal election and ensure that the free-trade deal would become law.

Corporations intervene politically everywhere in the world wherever they have interests. It is part of doing business. In Canada, every industry has had its methods and key issues. But what if they could all intervene just once, in order to secure a law that could render them super-citizens, that would make interventions on hundreds of separate issues virtually unnecessary?

That is essentially what the Canada-U.S. Free Trade Agreement (FTA) promised. It has been called a bill of rights for corporations, but even this description underestimates its eventual impact. In effect, the FTA allowed corporations to begin the final stage of opting out of the social contract altogether. It established the principle that corporations had no inherent obligations to the nation-states in which they did business. They had all the legal rights of citizens but had the obligations waived. It was a kind of unilateral declaration of transnational corporate citizenship.

The Mulroney government's reversal on free trade was the direct result of the intervention of the Business Council on National Issues (BCNI), the voice of the 150 largest corporations in Canada. The BCNI, which will be examined in detail in chapter 7, had already developed a consensus amongst the dominant corporations and simply delivered that message to Brian Mulroney.

According to writer Nick Fillmore, analysing the corporate role in the election for *This Magazine*, in the two years leading up to the 1988 "free trade" election, the Tory government spent an estimated $32 million promoting free trade, most of it through the International Trade Communications Group in external affairs.[29] But when the time came to put on a serious push, the group, headed by Philippe Beaubien, CEO of the giant Telemedia Corp. and a BCNI executive member, decided that a public advocacy campaign was needed to sell the deal. That campaign would be headed up by the Canadian Alliance for Trade and Job Opportunities.

The alliance, a creature of BCNI behind-the-scenes organizing, was announced on March 19, 1987. It consisted of thirty-five business organizations, including the big four: the BCNI, the Canadian Manufacturers' Association (CMA), the Canadian Federation of Independent Business (most CFIB members would eventually suffer under free trade), and the Canadian Chamber of Commerce. It had as its figurehead co-chairs Alberta's former Tory premier Peter Lougheed and Liberal Donald Macdonald, who had chaired the royal commission on Canada's future economic prospects and had called for a "leap of faith" in free trade. But it was David Culver, Alcan's CEO and the chair of the BCNI, who ran the alliance.

The alliance would end up spending $6 million in the months leading up to the election. Although that was the single largest amount spent by a single organization, big business altogether would spend more than $19 million promoting their deal of the century.[30] In addition to the alliance, other member organizations spent heavily, with the BCNI putting up $900,000, the Chamber of Commerce $400,000, and the CFIB $750,000. Seventy-five of the largest corporations came up with $3.75 million, with Alcan, the Royal Bank, and Noranda each pledging $400,000.[31]

The alliance spent $1.5 million on a four-page insert that went into thirty-five English-language daily newspapers across the country and $75,000 on a five-page insert in *Maclean's* magazine that arrived at the homes of 600,000 subscribers the week before the election. The Alberta government, eager to permanently end federal "interference" in the oil patch, spent $600,000 in advertising.[32]

In addition, corporations made a major propaganda assault on their own employees. Stelco and Loblaws, among others, put pro-FTA leaflets in pay packets; others, such as Crown Life Insurance, gathered their employees together for dire warnings about plant closures or layoffs if they didn't vote for free trade; Enfield Corp., a finance and management company, sent out ten thousand pro-FTA pamphlets to employees and clients.

Last but not least, corporate Canada united beyond one of the probusiness parties in a way that had not been witnessed at any time in the post-war period. Making its new consensus very clear, and in an explicit expression of displeasure with the Turner Liberals, big business filled the Tory coffers to overflowing. The Conservatives were able to spend to the maximum in the 1988 election, in excess of $14 million. The Liberals went into the election $5 million in debt.[33]

The free-trade initiative was the most concerted and massive corporate intervention in an election in the history of the country and probably unprecedented in any other developed country except the

U.S. It made a mockery of election rules that restricted what political parties could spend but said nothing about powerful corporations. The grassroots opposition to free trade, organizations representing several million citizens, raised an extraordinary budget to fight the deal but were still outspent by almost fifteen-to-one by a relative handful of corporate "citizens."

There is more than a little irony in the growth and development of the corporate super-citizen. For while corporations have spent millions of dollars and enormous effort to increase their influence, their strategies, and their effectiveness as citizens, they and the think-tanks they fund promote the notion that government and nations are virtually obsolete. In fact, the power of the state has never been more important to the success and power of the corporation. And because of this, the role of the corporate citizen is increasingly important.

The flip side of the enhanced status and activity of the corporate citizen is the reduced power, influence, and effectiveness of the real citizen. As governments withdraw from regulating the economy and providing programs and services, the ordinary citizen faces a world with less and less public space, less collective expression. In a world dominated by corporations selling goods and services, people are more and more defined by what they buy and less by their participation in the community. We are being reduced to little more than customers in the world marketplace, our citizenship increasingly reduced to a set of negative constitutional freedoms.

3

From Citizen to Customer

I shop, therefore I am.
— Bumper sticker

The advent of the transnational corporation is just one way in which the world economic order has changed. The TNC is both the agent of that change and a product of it. But the driving force of that change and the resulting strategy of the transnationals, the governments that serve them, and the agencies designed to facilitate their growth have as great an influence on daily life and democracy as the structure of the global economy.

That strategy focusses on the world's middle-class consumers, the "global shoppers," those individuals with discretionary income, a (manufactured) desire to immerse themselves in mass consumerism, who increasingly have no separate cultural identity or loyalty and who, willingly or otherwise, begin seeing themselves less as citizens and more as consumers. Even Prime Minister Jean Chrétien has absorbed this ideology of the global economy. In a speech to the National Press Club in Washington, explaining his conversion to free trade, Chrétien expressed the confidence in the transnational corporate future: "I'm not pessimistic about the twenty-first century because you have 1,200,000,000 people in China and they will develop a middle class; they will need to buy all sorts of products and both you and I will be there selling."

It is this consumerism that produces the social base for the globalization revolution. Cause and effect begin to blend into an indecipherable amalgam as citizens, under a barrage of business propaganda, expect less from their governments and identify less with their countries, in exchange for increased access to more and more consumer goods. As communities fracture and disintegrate under the withering assault of corporate power, the things that once held people together are giving way to a new paradigm of common identity and connectedness: the products we consume. The transnational consumer, jettisoning citizenship and local culture, is the key to the growth of the transnational corporation.

The consumer culture is hardly new. The whole counter-culture of the 1960s was preoccupied with opposing it, ridiculing it, and opting out of it. But the counter-culture was no match for the consumer culture's capacity for adapting and co-opting its icons and the corporate/government alliance promoting it. The hippies lost the battles and the war. And new factors have strengthened consumer culture (and weakened communitarian, citizen-based culture) beyond anything imagined by the flower children or the targets of their movement of thirty years ago.

Given that so much of Canadian history has unfolded within a developing consumer culture, it is easy to forget just how artificial this culture really is. Living next door to the country where the alliance between mass production and mass consumption was pioneered, we could be forgiven for thinking it is a natural phenomenon. But William Leach, in *Land of Desire: Merchants, Power and the Rise of a New American Culture*, reminds us just how artificial and manipulative consumerism is. He argues that "the culture of consumer capitalism may have been among the most nonconsensual public cultures ever created."[1] It is manipulative in two ways, first because it was created not by ordinary people but by organized commercial interests, and second, because it was promoted so forcefully it took on the characteristic of being the only path to personal fulfilment.

The current push to create a global consumer culture is simply the final stage in the transformation of citizen into consumer, and of what remains of civil society into a society based almost exclusively on the consumption of private goods. The overwhelming message of the last ten years of corporate assault on the egalitarian state is that we shall have more choices in the marketplace (if we have the money) but less choice about what kind of society we have. By weakening our environmental laws, we might get cheaper goods but we cannot "buy" breathable air, and only the privileged few will be able to buy drinkable water. Few consumers are aware that the price of an automobile is subsidized by billions of dollars in infrastructure, health care, and more expensive food as arable land is used up for highway construction. In comparison, the cost of a mass-transit system is an obvious public expenditure, made more obvious by those who attack such spending.

The most recent phase in the promotion of the consumer culture includes every individual, worldwide, with enough income to take part. It began in the early 1980s with the realization by American (and other) banks that the possibilities for profits to be made from lending to underdeveloped countries was virtually exhausted. That recognition coincided with the presidency of Ronald Reagan, arguably the most unabashedly pro-business president in fifty years.

With banks desperate to find new sources of profitability, Reagan was only too willing to comply. That new source was vastly increased consumer spending at home, and the door to that new wealth was easy credit. And it was just a matter of time before that easy credit in the U.S. market was extended to the wealthy and middle classes of every country in the world. That combination of stateless production and a stateless (and increasingly cultureless) class of consumers to purchase global products represents the essential character of the global economy. The application of easy credit opened the door to the growth of transnational corporations.

The role of credit was a key factor in the formative stages of the global economy because it created the conditions for the rapid expansion of the consumer economy. The stimulation of mass consumerism by mass credit is well documented by Richard Barnet and John Cavanagh in their book *Global Dreams*. The authors feature the story of Citicorp, the American bank that in effect pioneered credit as just another consumer product. Citi figured out that credit could be sold like any other product using the tried and tested tools of smart packaging and aggressive promotion. Millions of people who rarely used a bank could be enticed into buying on credit.

Citicorp sent out tens of millions of application forms and, after only cursory checks to identify current bankrupts, sent out tens of thousands of new credit cards without further credit checks. Since the business depended on volume, defaults were factored in as a normal business cost. Based on Citi's criteria for creditworthiness, all but 20 to 30 million of the 150 million American adults would qualify for credit.

The newly stimulated consumer culture kept American banking happy and profitable through most of the 1980s, but declining real incomes rang alarm bells. In 1991 Citi lost $1 billion. Yet the power of consumer credit was borderless, and Citi had anticipated the decline. Even as it took that large loss at home, its European operations were already more important than its U.S. base, and it was spending millions expanding its credit reach. It already had 13.8 million cardholders in thirty-seven countries.

Citi believed it had discovered the secret to its continued success: "People's attitudes about their finances are a function of how they're raised, their education, and their values, not their nationalities. What works in New York also works in Brussels, Hong Kong, and Tokyo," claimed Citi chairman John Reed.[2] The notion that financial services could be marketed with little or no reference to local cultural differences helped create the notion of the global middle class. The bank's strategy — and the world's transnational providers of goods and services would not be far behind — was to focus exclusively on the wealthiest 10 percent of the developing world's population.

Examining the Indian market, Pei-yuan Chia, Citi's head of global consumer operations, said: "Forget about 90 percent of the people and focus on the top 10 percent. That's 80 million people, larger than West Germany, and if you look at their standard of living, it's higher than the average German's."[3] How do you identify this top 10 percent? Simple. Look them up in the phone book. If they could afford a telephone, Citi could afford to offer them credit.

The idea of the global customer with identical desires unmediated by local culture had tremendous appeal for all global corporations. It was the perfect concept and marketing strategy for corporations wishing to end their multinational character, which required complete operations in each country. Marketing to the global customer further eliminated the need for a local, corporate citizen, presence. Providing the top 10 percent of the world's population with credit and then bombarding them with advertising for truly global products and services was enough to make any corporate strategic planner giddy with fantasies of unlimited growth.

No corporation presents a stronger image of the global corporation and global product than Disney. Its success at peddling its relentless cheer, as Barnet and Cavanagh call it, is well known. As if to deliberately take on the most formidable cultural challenge in the world, Disney, after a huge success in Japan, decided to set up a Disney World outside Paris. The initial visitor count was disappointing and it was universally attacked by the country's cultural and political elites as a cultural Chernobyl.

But millions now pay high ticket prices to see Disney's version of the American Dream. Disney remains confident that its "universal themes of love and adventure as seen through unmistakably American eyes . . . is exactly the experience large numbers of non-Americans want." As Barnet and Cavanagh explain, Disney World and other theme parks have much in common with early American expositions, which, with government participation, were used to impart explicit political and ideological messages — "views of empire, views of science, views of racial superiority and views of what constituted progress."[4]

As well as selling consumerism itself, Disney is marketing something quite different. Disney theme parks are, say Barnet and Cavanagh, "mood enhancers" designed to give people a few days of diversion, offering "illusions of connectedness while providing protected private space and canned dreams."[5]

Of course Disney World is just one example amongst hundreds of heavily marketed American pop culture products. The U.S. movie, music, TV, and video industries have for decades been penetrating the markets of virtually every country. And it wasn't just relentless cheer that the American culture exporters were selling. It was the American brand of personal freedom, reflected in every cultural product offered, which was so compelling that even efforts to slow it down seemed to have the opposite effect. It was as if by trying to suppress American culture, governments were trying to suppress the very freedom promoted by the songs, films, and videos. It was a marketing result that could not have been more effective if it had been planned.

Of all the products that have been most successful at becoming global, none surpasses Coca-Cola, which now sells 560 million cans and bottles every day in 160 countries. The marketing of Coke reflects the view of global marketing guru Theodore Levitt, who argued that "the products and methods of the industrial world play a single tune for all the world and all the world eagerly dances to it." The global marketing strategy was the advertising world's contribution to the shift from multinational corporations to transnational. Multinationals, according to Levitt, paid too much attention to local customs, tastes, quirks, and religions, adjusting their practices accordingly. The global corporation, on the other hand, "sells the same things the same way everywhere."[6]

THE THREAT OF CULTURE TO CORPORATE DOMINATION

The sophisticated psychological manipulation by the world's advertising agencies has had as much to do with the evolution of the transnational corporation as the changes in communications technology. Armed with the proposition that diverse cultures were not the obstacle they were thought to be, the advertising industry itself globalized and grew exponentially. It portrayed the notion that culture was a barrier as a lie. "The lie is that people are different. Yes, there are differences among cultures, but a headache is a headache," claimed Norman Vale, international area director for Grey Advertising Agency.[7]

While the idea of the global product was a boon to international sales, creating and marketing its appeal didn't come cheap. Corporations in 1989 were putting $240 billion into advertising and $380 billion into creative design and packaging, an amount exceeding $120 for every man, woman, and child on the planet.[8] The need to spend huge sums to get people to engage in mass consumption underlines the fact that the actual cost of producing a product is becoming a smaller part of the end price.

In other words, manufacturing the desire for the product is more costly than manufacturing the product itself, and indeed drives the efforts to reduce production costs. Japanese industrialist Kenichi Ohmae states, "Production costs are typically only about 25 percent of the end-user price; the major contribution to a product's price comes increasingly from marketing and support functions such as financing, systems integration and distribution."[9] Ohmae fails to mention the costs that we all pay in increased pollution, social costs, crime, and stress and other health problems that result from creating "cheap" products.

While not every product has the potential to be easily transformed into a global item pitched the same in every country, the efforts are still being made. And the advertising for Western products, global or otherwise, has a reach and depth that is the cultural equivalent of saturation bombing. Television now penetrates nearly every village in the world, and where it doesn't other forms of advertising, including door-to-door cosmetics sales in African villages, almost certainly do.

Amway, the American-based direct seller of shampoo, detergent, and other household products, now has a worldwide sales force of 2.5 million in forty-three countries. The Philippines is especially vulnerable to American pop culture and products. At just one of dozens of such Amway events, fifteen hundred eager salespeople snapped up starter kits for $77. The company trained expatriate Filipinos from the U.S., Australia, and New Zealand to return to their country to recruit thousands of new salespeople.

In the developed world, dozens of TV channels bombard consumers, increasing the number of commercial "hits" on each American from sixteen hundred a day in 1980 to three thousand a day ten years later. Increasingly, no matter where you turn, the efforts to commercialize daily life and human consciousness are unavoidable. Ads appear everywhere there is a flat surface to put them on — above urinals, on parking meters, on vehicles of all kinds, and even projected onto clouds. In New Zealand cuts to medicare have resulted in huge billboards being placed atop hospitals and on the sides of ambulances as a new source of revenue.

Advertising executive Norman Vale's declaration that culture is not a barrier is, however, more a declaration of cultural war than a description of current reality. It is a reality nonetheless that Vale and other advertising giants are trying their best to bring about. It is the inherent barrier to global products presented by "culture" that explains the ferociousness of the TNCs' assault on national cultures. The attack on Canada's remaining cultural protections by the giant American entertainment companies is not simply a demand that they have a level playing field for investment. It is the Canadian front of a general corporate war against individual national cultures that act as a barrier to mass consumption of global products.

The identification of culture as a barrier to trade was stated clearly by Sony chairman Akio Morita in an influential article, "Toward a New World Economic Order," in the *Atlantic Monthly*. Calling for the complete elimination of every barrier to trade and investment worldwide, and lauding NAFTA as a good start, Morita stated: "A level playing field cannot be declared into existence . . . The history of each country and region generates political, social, economic, and cultural factors that result in a business environment unique to that nation or region. This in turn gives rise to a variety of competitive factors . . . which can be seen as non-tariff barriers."[10]

The comic bumper sticker "I shop, therefore I am" has become a serious description of modern life. And the message to those who cannot "shop" is just as real: if you don't consume, you don't exist. For the new transnational corporations that is a brutally exact description of how they view humanity. The focus of all their attention on people able to afford consumer goods, like those in India defined by telephone ownership, excludes most of humanity. At first glance that might seem a blessing in disguise, but being excluded from the global economy does not mean you are not affected by it. For all but a few who have managed to create community economic models, the alternative is dangerously close to no economy at all.

YOUTH: THE FIRST GENERATION OF GLOBAL CUSTOMERS

While the world's middle-class adult population is still by far the largest consumer market, the American marketing juggernaut is aimed carefully and deliberately at teenagers everywhere. The long-term advantages of hooking a whole generation into the consumer culture are difficult to exaggerate. As civil society, and the connectedness it provides, steadily erodes, this generation and the one to follow will identify by default with whatever does provide that feeling of connectedness. And about the only thing left, given the erosion of citizenship and the decline of religion in most countries, is the global product.

The global teenager already exists in the developed world. Says William Roedy, director of MTV Europe: "Eighteen-year-olds in Paris have more in common with eighteen-year-olds in New York than they do with their parents . . . They go to the same movies, they buy the same music, sip the same colas. Global advertising merely works on that premise."[11] More important, they have more connection with other Coke drinkers than they have with their own communities and their own history.

The enormous industry picking apart the entrails of the consumer leaves no subculture unexamined. There are companies that compete with each other to tell producers what nine-year-olds like. Others specialize in age groups from twelve to fifteen, or from six to fifteen. Teenage Research Unlimited claims that 13.4 million Americans aged twelve to fifteen spend $10.5 billion a year of their own money and much more when they shop with their parents' money. Children start to become the focus of corporate attention when they reach the age of nine because that is when they start to take on part-time jobs.[12]

Business Week magazine ran a major feature article in 1997 titled "Hey, Kid, Buy This!" and subtitled "Is Madison Avenue Taking 'Get 'Em While They're Young' Too Far?" It also carried an editorial calling for a return to the days when marketers pitched to parents, who acted as gate-keepers. The enlightened scepticism of the editors aside, there is no sign that the trend to hooking kids will end any time soon.

Youth are now subjected to a dizzying array of advertising formats, from logos on underwear to dozens of magazines and Web sites aimed at them. Advertising targeted to kids in the U.S. increased by 50 percent from 1993 to 1996, to $1.5 billion. It starts, writes *Business Week*'s David Leonhardt, about three days after the newborn gets home. "As an infant, Alyssa may wear Sesame Street diapers and miniature pro-basketball jerseys. By . . . 20 months she will start to recognize some of the thousands of brands flashed in front of her every day."[13] By seven she will be seeing twenty thousand TV commercials a year.

The systematic targeting of ever younger children reveals the desperation of mass consumerism to pump up the North American market where the adult market is largely flat. The effect on these children as future citizens is potentially enormous. Says Leonhardt: "As kids drink in the world around them their cultural encounters . . . have become little more than sales pitches, devoid of any moral beyond a plea for a purchase. Even their classrooms are filled with corporate logos . . . Instead of transmitting a sense of who we are and what we hold important, today's market-driven culture is instilling in them a sense that little exists without a sales pitch attached and that self-worth is something you buy at a shopping mall."[14]

The consumer-profile industry is not only huge, it operates largely without the knowledge of the people being manipulated. How many citizens know that they and their children are regularly divided up into dozens of categories and subcategories of "belongers, outer directeds, inner directeds, need drivens (poor people), emulators and achievers," to name just a few, by psychologists, market experts, and product designers?[15]

Choices: Public versus Private

That the sinister manipulation of children, teenagers, and adults is not even considered an important social and political issue is itself a sign of the times. Ordinary citizens are being subjected to the most sophisticated and well-financed manipulation in human history. And the impact is not trivial. Having these forces manoeuvring and micromanaging what people buy and how much, not to mention what they watch and listen to, goes a long way towards determining what kind of society we end up with. Imagine if the world's societies spent even one-tenth of the world's advertising budget on providing information designed to nurture self-aware citizens.

Money spent on artificially induced needs and desires in the consumer world is money not spent on other things: books, magazines, travel to heritage sites, cultural pursuits. Immersing oneself in the cornucopia of consumer products also takes time, time otherwise spent (in the past, at least) with family, community, friends. And the more money people feel they need to engage in this consumer world, the less they are able and willing to "spend" in taxes for public services.

The industry argues that people still have the choice to buy or not buy, yet this is true only at the most abstract level. It is the same argument used by the tobacco industry when it claims that its advertising of sports and cultural events has no effect on whether young people start smoking. If that were true, the billions they spend would be completely wasted. That they keep spending is proof enough that they believe the money well spent.

The impact of the global consumer revolution is felt everywhere. Countries now define their national purpose almost exclusively in terms of how well they will fare in the global competition for investment. History, culture, equality, civil society are all swept aside as nonconsumables and therefore incapable of being factored into the new order of things. Some countries have little choice in the matter; other states do so willingly as their elites eagerly surrender sovereignty to the TNCs.

The irony is that global competition is not focussed as much on consumption at home in developed countries where real incomes are steadily dropping, but on developing world consumption. And as governments deregulate investment and labour so we can compete with the lowest-paid, lowest-taxed, and least regulated developing countries, corporations depend more and more on the spending of other nations' elites just as they depend on the labour of other nations' workers. The result is a global levelling involving two classes, those who can earn enough to consume the global products and those who can't, including working people in the developed world.

The new world of the devolving citizen/evolving customer is not a parallel world to civil society, it is a replacement world. Freedom is no longer freedom from want, freedom from insecurity. It, too, has devolved. Now freedom means consumer choice. The success of the TNC in creating global consumerism is at the same time the gradual destruction of civil society. The cause becomes the effect; as more people acquiesce to the lure of the marketplace, they abandon their status as self-conscious citizens and with it the ability to resist the slow destruction of their community.

The key to the global marketplace is the global customer whose identity is found in the stuff they buy. This is precisely why Sony's Akio Morita declared that not only culture but all the particularities of civil society are a barrier to commerce. What he was talking about was connectedness. The more connected we are to community, neighbourhood, tradition, family, and history, the more likely we are to resist the lure of consumerism.

The gulf between the evolving and increasingly powerful corporate citizen and the devolving citizen/customer has been growing greater. Real citizens have abandoned the field, voluntarily or because they can no longer find ways of engaging civil society; corporate citizens rule the world, and liberated from even the minimal social responsibility once imposed by civil society, reveal themselves for what they are.

4

The Transnational Corporate Citizen

The foundation of Chile's strong economic position today . . . was courageously laid by [former Chilean military dictator Augusto] Pinochet. [The coup was justified] because it brought wealth to an enormous number of people, I mean in my terms. If you ask somebody in jail, he'll say no. But that's the wonderful thing about our world: we can have the freedom to disagree.
— Peter Munk, CEO of Barrick Gold, whose company is active in Chile

[The large corporations] are latter-day versions of colonial empires. They defend their positions not with machine guns or cruise missiles but with trademarks, copyrights and huge marketing budgets that dwarf the resources of many governments.
— Angus Reid, *Shakedown: How the New Economy Is Changing Our Lives*

If, for the sake of argument, we accept the notion that corporations can be citizens, then these particular citizens exhibit the behaviour of textbook sociopaths. Incapable of telling right from wrong, demonstrating severe antisocial behaviour, they are often a threat to the community, even a deadly threat. The men (they're almost exclusively men) who are the collective mind, heart, and soul of the corporate citizen are somehow capable of deciding, as Ford executives did in the 1970s, that it was not cost effective to fix the gas tank of its Pinto model, although they knew it could explode on rear impact. Ford calculated that the cost of fixing the defect in existing and future cars, $11 per car, was greater than the cost of court settlements arising from the predictable number of fatal accidents, estimated at 180 deaths per year at $200,000 each.[1]

We have to assume that the men who made this decision were not evil monsters capable of murdering people and then going home to their families as if nothing had happened. Indeed, what makes this story so horrendous is that these men, as individual citizens, were not evil monsters at all, and were very likely no less ethical and decent than average Americans. Walking through the doors of Ford Motor Company had transformed them into corporate citizens, mutated by corporate imperatives. This contrast between corporate and personal behaviour was identified by former General Motors vice-president John Delorean in the insider's profile he provided of his company: "It seemed to me . . . that the system of American business often produces wrong, immoral and irresponsible decisions, even though the personal morality of the people running the businesses is often above reproach. The system has a different morality as a group than the people do as individuals."[2]

The Pinto example is just one of many where corporations demonstrate that they do not operate as good citizens. Corporate Crime, a study, done in 1975–76, reported that 60 percent of large corporations in the U.S. had been charged with at least one form of criminal behaviour. And the largest corporations accounted for over 72 percent of the most serious crimes.[3]

In recent years, North Americans have witnessed the spectacle of tobacco executives declaring solemnly before government committees that they do not believe smoking is addictive. As it was soon revealed, these men knew that in the files of their corporations were stacks of documents disclosing the lie of what they were claiming. Not only did corporate records show nicotine is addictive, they also detailed efforts over two decades to increase a cigarette's effectiveness as a "nicotine delivery system" by adding ammonia.

Today, clearly, corporate culture does not allow executives to make moral choices. Publicly traded corporations that undermine their profitability by behaving as good corporate citizens would soon find their shareholders demanding a change in management.

There is a very short list of corporations in North America that try to behave like good corporate citizens. But these companies are so rare that they become newsworthy because they are such aberrations. The *Globe and Mail*'s *Report on Business* has featured a couple of Canadian companies whose treatment of its employees is exemplary. Husky Injection Molding Systems is a small transnational (seventeen hundred employees in sixteen countries) and, though not unionized, has a record for treating its employees with respect in a management structure that is almost democratic. Management consultant Jim Collins calls Husky one of the ten greatest firms he has ever seen. "They do not exist to maximize profits; they practice a set of unwavering core values; they embrace constant change; they set audacious goals; they instill unwavering devotion in their employees." Husky is also very profitable and growing, from sales of $72 million in 1985 to $609 million in 1995.[4]

A number of features of today's corporate environment make Husky the rare exception rather than the rule. In fact, in the ruthless world of corporate raiding it is precisely the features of a good corporate citizen that make a company a prime target for a takeover. A company that has large cash reserves, a well-funded pension fund, a large resource base, or a highly paid workforce is a plum waiting to be picked by those who make their money in the raiding business. The ferociousness of speculative finance to find ever greater levels of profitability and ever higher rates of return for investors does not permit many examples like Husky. Indeed, it is the speculative markets that are at the root of much of corporate pathology in the nineties.

MONEY FOR NOTHING

Every day the amount of money that changes hands in the world's foreign exchange markets amounts to nearly $2 trillion. Over just three days the trade in currencies equals the dollar value of the gross domestic product of the U.S. More important, 97.5 percent of that currency trading is devoted purely to speculation. Just 2.5 percent involves the financing of real economic activity — trading, buying, and selling goods, services, and capital assets. In the 1970s, only about 10 percent of the currency trading was speculative.[5]

Much of what we see in the daily business press that seems inexplicable and irrational — stock prices plummeting with news of decreasing unemployment rates, years of high interest rates despite near zero inflation — is explained by the financial markets, including currency speculation. The apparent abandonment of integrated domestic markets by corporations and government policy makers, corporate strategies of downsizing and decentralization, the almost maniacal focus on short-term corporate profits can all be traced in large measure to the power of the financial markets.

The influence of the international foreign exchange system has been nothing short of revolutionary. It has severely eroded the links between the productive economy — the real economy of creating and selling goods and services — and the money system. It is now more profitable to speculate in the money system than it is to invest in the production of goods and services.

In a healthy money system, creating and using money is directly linked to real things, to exchanges of goods and services that we use or productive capital such as machinery, land, technology, and labour. But as David Korten argues, ours is a sick money system in which money and productive activity become delinked and a class of "money manipulators," whose specialty is extracting money that is no longer linked to real wealth, develops. Yet while delinked from real wealth, this money still provides those who hold it with a demand on the real wealth that is being produced.

According Bernard Lietaer, a former currency trader, "This new [currency] asset class has some tremendous advantages from an investor viewpoint. The first is the extraordinarily low transaction costs. Placing a few billion dollars in foreign exchange costs very little, for major currencies just a fraction of a percent. Compared to stock transactions this is 10 to 20 times cheaper."[6] It costs only eighteen cents and takes just a few seconds to complete a multimillion-dollar currency transaction across the ocean.

But perhaps the most critical aspect of the foreign exchange market is its dependence on volatility for its profitability. The more volatile a currency, the greater the potential profits from buying and selling. The worst thing that could happen to a currency trader is to buy a million French francs and then nothing happens; it just sits there. The longer you wait for a change in its value, the more money you lose.

As David Korten puts it, "The global financial system has become a parasitic predator that lives off the flesh of its host — the productive economy."[7] The domination of the financial sector has had profound effects on investment patterns. The wealth to be extracted from currency speculation has directly spurred other areas of speculation as well as indirectly affected other phenomena of the 1980s and 1990s such as the mergers and acquisitions mania that continues to this day.

The financial markets present a number of opportunities to make money, to extract wealth, without engaging in any productive activity. Besides currency speculation, there is the international bond market. Just as in the currency market, very small fluctuations in bond prices can yield large profits for institutional investors. A change of just three to ten cents per $100 is enough to persuade such investors to make trades, to get out of one country's bonds and into another's.

All of this speculation has huge repercussions for the global economy and for the ability of governments to create public policy in the interest of their citizens. It also affects what business does with its money. For example, Bernard Lietaer suggests, "for real international businesses that do mining, make cars, or electronics, the foreign exchange risk is the biggest single risk they face. If a chemical company wanted to set up a plant in India the business risk is minimal compared to the foreign exchange risk. Real investments are not being made because the foreign exchange risk is not manageable."[8]

Because of this new risk TNCs are obliged to develop strategies for dealing with the volatile currency markets, and it is here that the behaviour of large corporations is partly determined by speculative activity. According to Lietaer, corporations cope with currency markets in two ways: decentralization of production and marketing, and recentralization of financial and treasury functions. Both are aimed at increasing profitability to the level attained in the speculative financial markets. The decentralization strategy allows for externalizing the costs of production. For example, instead of costing in pollution control as part of the picture, a particularly polluting industry will set up a plant in Mexico or the Philippines, where it can pollute with impunity.

The centralization of the finance and treasury functions gives maximum control over exchange markets — if you can't beat 'em, join 'em. Most of the largest transnational corporations now have huge financial divisions, and many of them engage in speculation themselves. John Dillon reports that a survey of 530 U.S. nonfinancial corporations found that 75 percent used derivatives to hedge their financial exposure, and fully 40 percent were actively speculating.[9]

These are not minor parts of these businesses. British Petroleum's director of money transactions has stated, "Our currency trading is as important as our oil trading." This pattern, says Dillon, prevails for some of the world's largest corporations. "General Motors and General Electric both made more profits from their internal financial subsidiaries than they did manufacturing real automobiles or electronic products."[10] In fact, the largest financial institution in the U.S. today is General Electric.

The enormous amounts of money to be made in these sophisticated forms of gambling has created a new class of investor. Geared to the expectation of instant profits and fast turnarounds, they are impatient with the old world of buying shares in a company and waiting for the company to grow. As Magna Corporation owner Frank Stronach put it, "Why would you pour a foundation, buy machines, hire employees if you could make as much money buying bonds?"[11]

A NEW CLASS OF INVESTORS

The speculative markets not only create a whole new class of investors. Their impact — setting the standard for profit taking — has an enormous influence on the behaviour of nonfinancial corporations. It means such corporations are in a constant race to make profits as high as those in speculative markets and as fast. The behaviour of large corporations is driven by shareholders, and shareholders today have changed. The owners of corporate stocks are now dominated by huge institutional investors who "speak" for the millions of ordinary citizens in Canada and elsewhere who have money in pension funds or mutual funds.

In the U.S. one-third of all corporate shares, $4 trillion in assets, are held in workers' pension funds managed by the trust departments of the giant banks. And in 1996 mutual funds overtook pensions as the most important players on stock exchanges. In Canada, pension fund administrators manage $360 billion. In addition, more than $260 billion is invested in mutual funds. And although these funds are managed for ordinary citizens, including union members who often suffer from the lean and mean behaviour of their corporate employers, they are the most influential force in the market today.

The fund managers feel under tremendous pressure to perform well, and, says David Korten, in the hot-money atmosphere of the 1990s that means providing "nearly instant financial gains. The time frames involved are far too short for a productive investment to mature, the amount of money to be 'invested' far exceeds the number of productive investment opportunities available, and the returns the market has come to expect exceed what most productive investments are able to yield even over a period of years."[12]

If nonfinancial corporations are constrained by the heady atmosphere of the speculative markets, governments are even more affected. The power of money speculators to transfer massive amounts of money instantly between markets has given this handful of traders a weapon by which they hold public policy hostage to their own narrow pursuits. And corporate advocates aren't shy about using their power to threaten even the most powerful governments. The Cato Institute, a leading free-market think-tank, issued such a warning to President Bill Clinton and Congress in the pages of the *New York Times*, reminding them that "equity investors have developed a global perspective and they prefer markets where government is downsizing and the prospects for economic growth are good." The author suggested that the government get rid of "hundreds of ill-considered laws that benefit special interests" and "thousands of counter-productive rules in the Code of Federal Regulations."[13]

Walter Wriston, the former chair of Citicorp, made this boast about the government of François Mitterand of France: "The market took one look at his policies and within six months the capital flight had made him change course." The "decision," said Wriston, was made by those sitting at "200,000 monitors in trading rooms all around the world [who conduct] a kind of global plebiscite on the monetary and fiscal policies of governments issuing currencies."[14]

The most dramatic example of how a large-scale speculator can affect a country's currency involved the British pound. Currency speculator George Soros actually caused the British pound to fall 41% against the Japanese Yen, and in effect sabotaged the system of fixed exchange rates planned for the European Union, eliminating a major threat to speculators who depend on floating rates.

Speculators also wield enormous influence over government policy in the bond market. Political science professor James Laxer refers to the bond market as "the most powerful force in contemporary capitalism."[15] It is the floating-rate bonds that allow bondholders to sell government financial instruments, such as Treasury bills, in such quantity that buyers can successfully demand a discount to purchase them. Because financial markets are all linked together, the effect is an upward pressure on all that country's interest rates.

The collapse of the speculation house of cards that many had predicted finally happened in late 1997. The madness of unfettered currency speculation came home to roost — and right where global capital had been pinning most of its hopes, among the so-called Asian Tigers. First Thailand was obliged to devalue its currency, then Indonesia, followed by the Philippines, all second-string Asian economies. But then came Hong Kong and South Korea, the eleventh-largest economy in the world, humiliated by being forced to sign a bail-out from the IMF of $57 billion. Estimates of the total for all countries was well over $100 billion. It did not stop even there as the "Asian flu" spread to Japan, where the country's fourth-largest brokerage house collapsed.

The stock market followed the downward plunge as the ridiculous overvaluation of real world assets at last collapsed. The fully integrated, speed-of-light, twenty-four-hour-a-day world markets virtually guaranteed that a currency crisis in Thailand, a bit-player in the global scheme, would bring global chaos. The crisis spread to Latin America, where Brazil's stock market plunged in sympathy. In Southeast Asia, it was estimated that purchasing power dropped by $300 billion in just a few weeks.

While shareholders around the world lost big, middle-class Asians as well as workers and peasants suddenly found their money worth 30 to 50 percent less than it had been. Tens of billions were drained from central banks, much of it headed for numbered bank accounts. The crisis for individuals was paralleled by the crisis for nations as debts owed in dollars were suddenly 30 percent higher, tax revenues started dropping, and money for economic activity disappeared.

The response of the elite to this global catastrophe was in perfect synch with free-market ideology. Avoiding at all costs any call for bringing money traders and the speculative economy to heel, everyone in authority from President Clinton to the market pundits began with the mantras of "sound fundamentals" and "market corrections." They then tried to blame the "structurally weak economies" of Southeast Asia, the very same economies that just weeks before retained their status as Tigers. Just two months before the collapse, the director general of the IMF praised the economy of Malaysia as a model economy for the rest of the world to follow.

The crisis underlined the fatal contradiction of the global system. It must have stability of function, otherwise every mini-crisis has the potential to become a catastrophe. Yet the system itself is driven by and is inextricably meshed with a subsystem of uncontrolled speculation that depends on instability.

None of this broke through the armour of the true believers among the world "leaders." Only Malaysia's eccentric and largely discredited Prime Minister Mahathir Mohamed broke ranks. During the 1997 APEC summit, Mahathir, guilty in his own right of economic mismanagement, attacked the free market as the source of the problem. Demanding the imposition of strict regulations on currency traders, he stated, "Currencies don't fall on their own . . . We want to know how much money [currency traders] made and are they being taxed and who should tax them."[16] For his efforts at declaring that the emperor had no clothes, he was accused by Canada's trade minister, Sergio Marchi, of uttering conspiracy theories.

Instead of dealing with the issue of money speculation and re-regulation of international trade, Canada backed the U.S. in its call for even greater powers for the IMF to force countries to further liberalize their economies. The deal finally worked out between the IMF and South Korea will take a staggering human toll. The IMF demanded that the country cut its growth rate from 6 percent to 3 percent and its deficit by $10 billion. The result will be a rise in unemployment from 2.5 percent to 7 percent, throwing 1.5 million people out of work.[17]

VULTURES THAT TARGET THE HEALTHY

The search for super-profits and increased share prices has unleashed a ruthlessness in the business world that destroys not only individual lives but companies themselves. In his book *When Corporations Rule the World*, David Korten provides a number of compelling examples of corporations whose social responsibility made them vulnerable. One was the Pacific Lumber Company (PLC), acquired by corporate raider Charles Hurwitz. Before the acquisition, PLC was known as "one of the most economically and environmentally sound timber companies in the United States. It was exemplary in its . . . use of sustainable logging practices on its substantial holdings of redwood timber stands, was generous in the benefits it provided to its employees, overfunded its pension fund to ensure it could meet its commitments, and maintained a no-layoffs policy even during down-turns in the timber market. These practices made it a prime takeover target."[18]

After taking control Hurwitz doubled the cutting rate and described the mile-and-a-half-wide clear-cut into thousand-year-old forest as "our wildlife biologist study trail." His contempt for ecology was matched by his contempt for his employees as he raided the $93-million pension fund to the tune of $55 million and invested it in the insurance company that had financed the junk bonds he had used to underwrite the takeover. This insurance company eventually failed.

The attitude demonstrated by such corporate raiders and CEOs in general is vastly different from that of recent generations of corporate executives, although each decade has had its share of robber barons. Revelling in the mass destruction of thousand-year-old redwood trees and gleefully dismissing tens of thousands of employees seems to be a new form of ostentatiously antisocial behaviour. Indeed, these corporate takeover executives almost go out of their way to humiliate their new employees. According to Korten, "To justify the mass firings and wage cuts that followed the takeover of the Safeway supermarket chain, investor George Roberts told the *Wall Street Journal* that the supermarket chain's employees 'are now being held accountable . . . They have to produce up to plan, if they are going to be competitive with the rest of the world.'"[19]

At the top of this list of celebrity CEOs is "Chainsaw" Al Dunlap, who gained notoriety for his turnaround of Scott Paper. Self-cast as a saviour of badly run companies, Dunlap brags about being the superstar in the field of downsizing and restructuring, claiming to be "much like Michael Jordan in basketball and Bruce Springsteen in rock 'n' roll."[20] By laying off a third of Scott's employees, selling hundreds of millions in assets, slashing research and development, closing plants, and eliminating all charitable donations, Dunlap increased the value of the company's shares from $2.9 billion to $9.4 billion in less than two years.

Dunlap's book *Mean Business: How I Save Bad Companies and Make Good Companies Great* describes his business philosophy. "Executives who run their businesses to support social causes . . . would never get my investment dollars. They funnel money into things like saving the whales . . . That's not the essence of business. If you want to support social causes . . . join the Rotary International." His pleasure at bullying people is legendary: "If people balk at something and say 'We've tried that before,' that's the kiss of death. They're a disease; cut them out."[21]

But Dunlap's bragging conceals as much as it reveals. Scott Paper was not in nearly as bad shape as Dunlap claims (it was number one in the world in sales) and many of the reforms he implemented, other than the asset stripping and layoffs, had been in the planning stages for years. When he bragged about creating 107 new products, it turned out that 44 of them were the same two products packaged in twenty-two countries. His macho slashing of R&D has resulted in Scott losing market share in all three of its core businesses, paper towels, bathroom tissue, and facial tissue.[22] Following his devastation of Scott Paper (which Dunlap delivered into the merger arms of its long-time competitor, Kimberly-Clark), Dunlap went on to "save" Sunbeam by cutting its workforce of twelve thousand in half and closing 18 of its 26 factories and 37 of its 61 warehouses.

Corporate raiders spend much of their time poring over annual reports to identify companies that have not already downsized and stripped their assets in the name of quick returns. There are even investment funds that specialize in looking for opportunities in labour-intensive industries that have resisted the trend of moving to low-wage countries.

American and Mexican investors set up the AmeriMex Maquiladora Fund to target U.S. companies. AmeriMex's prospectus reads: "The fund will purchase established domestic United States companies suitable for maquiladora acquisitions, wherein a part or all of the manufacturing operations will be relocated to Mexico to take advantage of the cost of labor."[23]

AmeriMex goes on to describe how much money could be saved by companies that currently pay wages in the $7–$10-an-hour range by moving to Mexico, where they would pay $1.15 to $1.50 an hour: "It is estimated that this could translate into annual savings of $10,000–$17,000 per employee . . . It is anticipated that most investments will be retained for three to eight years." At maximum savings, a company with a thousand production employees could see the market value of its stock increase by $170 million over a very short period.

It is exactly these kind of super-returns that the rogue financial system demands of any publicly traded company. The problem, beyond the obvious moral and ethical questions, is that once the savings have been realized the company has to come up with another trick to keep its growth rate at such a peak. Having already externalized its greatest cost, it is unlikely to be able to do so, which accounts for the short period that ownership of the company is held by AmeriMex. The raiders are not builders and leave their targets crippled, decimated by asset stripping, or bankrupt.

As Korten points out, examples such as AmeriMex underline the ultimate futility of demanding that individual corporations act as good corporate citizens. The global financial system effectively renders such management a recipe for corporate suicide through raiding and asset stripping or shifting production off-shore. Says Korten, "Those who call on corporate managers to exercise greater social responsibility miss the basic point. Corporate managers live and work in a system that is virtually feeding on the socially responsible."[24] Simply to head off being a victim of a corporate raider, even responsible managers are forced to mimic those who would take them over. In the end, the result is much the same, often differing only in the degree of restructuring or the speed at which it takes place.

The irony is that in most cases downsizing doesn't accomplish what its promoters said it would. Michael Hammer, co-author of *Re-engineering the Corporation*, essentially a handbook on cutting and slashing, is now scrambling to recast his advice. His examples of successful companies now include a joint venture between GE and Fanuc of Japan, "which boosted revenue by 18 percent during the past two years, while the number of employees has risen by just 3 percent."[25]

A 1993 study of Canadian firms that went through restructuring supports Hammer's about-face. The study, by Watson Wyatt Worldwide, concluded that "relatively few companies accomplished their goals." Of the 148 firms studied, 62 percent said they reduced costs but only 37 percent increased profits; just 17 percent improved their competitive advantage, and only 40 percent increased their productivity.[26]

Another study, by University of Saskatchewan commerce professor Marc Mentzer, found similar results. Mentzer studied 250 large Canadian companies over eight years and concluded: "There is no relationship between downsizing and profit," he found. In fact, there is no clear relationship even to share price. The trend is for a share to increase in the short run and then go into a long-term decline. Repeated waves of layoffs damage companies severely. Mentzer argues, "This water-torture style of downsizing kills morale and forces everyone to think about job hunting instead of doing their current job well."[27]

Instead of using layoffs as a last resort as companies should, they do it almost as a fad. Mentzer concluded that unsure managers downsize because big companies like GM and IBM do it. Downsizing becomes "a badge of honour because it shows other executives and the investment community how tough you are. It should be a badge of shame because it's a sign of bad management. Any idiot can go through a payroll and fire every fifth person."[28]

The downsizing mania has an enormous effect on economies where it is practised. A comparison of G7 countries by economist

Frank Lichtenberg showed that countries in which companies lay off employees in response to a downturn suffered the most prolonged recessions. The U.S. and Canada were the two least stable countries; Japan and Italy the most stable. In the U.S., "when output falls by $1 the income of workers tends to decline 48 cents and profits 52 cents. In Japan, by contrast, virtually 100 percent of output fluctuations was borne by shareholders." By laying off workers, companies take demand out of the economy in a way that doesn't happen when shareholders' dividends go down.[29]

Despite the evidence that downsizing doesn't accomplish what its practitioners say it does, and despite the decamping of some of its gurus, in late 1997 downsizing showed signs of increasing again. The return to the trend was signalled by Kodak, which announced a layoff of 10 percent of its workforce over two years — ten thousand employees worldwide. Five of the ten largest cuts in the U.S. in 1997 occurred in the fall. The reasons for the continued trend, according to the U.S. Conference Board, was "intense import competition . . . Businesses have to drive earnings up."[30] Sometimes it appears as though the layoffs are carried out in order to pay for outrageous CEO compensation packages. Such was the case at Levi Strauss. In 1996 it paid its retiring president $126 million; in 1997 it fired 6,395 employees and closed eleven plants.[31]

Robert Hare of the University of British Columbia is Canada's leading researcher in psychopathic behaviour. He describes the corporate downsizers like Al Dunlap as "sub-criminal psychopaths." These are the people who work amongst us but are incapable of forming genuine allegiances or loyalties even in their personal lives. They are equally incapable of feeling guilt or remorse and package all of this antisocial behaviour in a veneer of charm. In the end they are extremely destructive to the companies that hire them and, in Hare's words, "they manoeuvre to have detractors fired and ruin other people's careers without a hint of remorse."[32]

According to Canadian psychologist Paul Babiak, it is precisely in the atmosphere of restructuring that corporate psychopaths flourish. They do not do well in stable corporate bureaucracies, where there are well-established controls and operating systems. In companies undergoing massive change, Babiak says, "there is chaos, or breakdowns of norms and values in the culture, and in that chaotic milieu, the psychopath can move in and do very well."[33]

MILLION-DOLLAR CEOS: REWARDING GREED

The breakdown of societal constraints on corporate behaviour manifests itself in increasingly obscene CEO compensation packages, especially in Canada and the U.S., while employee incomes are held at the levels of the early 1980s. In 1996, eighty Canadian CEOs took home more than a million dollars in compensation — usually a combination of salary, bonuses, and the exercising of stock options.

Including stock options in compensation goes a long way to explain why asset stripping and massive layoffs happen: the rapid driving up of share prices means that the stock options of the CEO and other executives will be worth that much more. The CEO's incentive to drive up the stock price in the short term works against any ambition to build the company over the medium term or plan for its long-term future.

The constant hype around the market and its supremacy tells us that money is everything; in Al Dunlap's words, "Money is a CEO's report card." Apparently not in Canada. According to a *Globe and Mail* report of the one hundred top-paid executives in Canada in 1996, CEOs did very nicely whether or not their report cards were good. Twenty-two of the companies managed by these men saw their profits decline from the year before. But of the twenty-two CEOs responsible for those declining results, only four had their compensation reduced. The rest had increases.[34]

Some of the increases were spectacular, starting with number one on the list, Laurent Beaudoin at Bombardier, whose compensation increased by 1,335 percent (to $19,100,317) while profits dipped by 36 percent. Michael Brown of Thomson Corporation got a 166 percent increase for his accomplishment — decreasing profits by 28 percent. Nova's Edward Newall got a whopping 283 percent increase (up to $6,350,750) while presiding over a decline in profits of 39 percent. Gulf Resources chief James Bryan achieved a decline in profits of 232 percent but got a nice surprise in his pay packet with an increase of 65 percent to $1.75 million. Peter Munk, one of the most aggressively outspoken advocates of the free market, saw his pay increase for the two companies he runs when both had decreased profits, and in each case part of those increases were in fatter bonuses. Sixteen of the twenty-two CEOs presiding over declining profits earned increased bonuses, presumably a measure of their gall as well as a sign of the gross irresponsibility of their boards of directors respecting shareholders' interests.

How much can one man (all of the top 100 CEOs in 1996 were men) be worth? The notion that a CEO is worth millions a year suggests that there is real genius involved, that the increase in share price is the result of an extraordinary business talent. But in fact corporations don't work this way. No one person is that smart, nor can their presence be so pervasive in a huge organization that they can affect all its decisions and planning.

In his study of General Motors, Peter Drucker dismissed the idea that corporate organizations naturally bring executives with exceptional ability to the top. "No institution can possibly survive if it needs geniuses or supermen to manage it. It must be organized in such a way as to be able to get along under a leadership composed of average human beings. No institution has solved the problem of leadership . . . unless it gives the leader a sense of duty and a sense of mutual loyalty between him and his associates; for these enable the average human being . . . to function effectively in a position of trust and leadership."[35]

By this measure, the current crop of CEOs, whose expertise is increasingly in the area of finance and cost cutting, should be paid less, not more. Loyalty and duty have been so devalued in the current corporate culture that they have virtually disappeared. The drive for short-term share price increases in effect invites the selection of CEOs whose logic is disturbingly similar to the U.S. general in Vietnam who declared in all seriousness that he had to destroy a village to save it.

The notion that professional managers must be paid astronomical sums to motivate them to do their jobs seems absurd on the face of it, not least because these same men support government policies that take money away from the poor to motivate them to work. Former Harvard University president Derek Bok believes that CEOs are paid a bonus precisely because they are being asked to work against their better judgement as managers and human beings. The obsession with short-term returns means that managers must ignore the interests not only of their employees and the community but also of the long-term interests of the company they are working for. In short, their bonuses are paid because their jobs have become so distasteful.

There are no doubt some managers who are uncomfortable over their destructive roles, yet they are clearly in the minority of those who are receiving gargantuan pay packages. There may be a simpler explanation of why CEOs get paid so much. It has more to do with social class than with psychology or motivation. They pay themselves obscene amounts quite simply because they can get away with it. No matter how ridiculous the amounts, the other shoe never drops.

The comparison of what corporate executives pay themselves and what they pay their average employee reveals just how out of proportion their compensation is. In 1996, Bombardier's Beaudoin took home 627 times what his average Canadian employee earns; Francesco Bellini of BioChem Pharma got 328 times.[36] The CEOs of all the banks were paid at least a hundred times more than their average employee. It is worth noting that these ratios are unique to North American and other English-speaking developed countries. In Japan, executives by informal consensus take home on average about seventeen times what their average employee earns. There is no evidence that Japanese CEOs lack motivation or success.

CORPORATIONS AND DEMOCRACY, LIKE OIL AND WATER

When the political and economic elites of developed countries promote democracy in Third World countries, clearly it is that brand of democracy advocated by libertarians and not the sort fought for by the people themselves. Stability and the protection of property rights are what corporations are seeking from governments. Substantive democracy, in its original sense of rule by the majority, is a threat to the corporate bottom line. Corporations have a long history of ridding the world of that kind of democracy.

One of the favourite investment locations for Canadian companies is Chile, and for good reason. Following the military coup in 1973, the entire country was restructured in the exclusive interests of corporations and those with property. Every deregulation fantasy of corporate CEOs and finance capital was put in place. Labour has been largely silenced, taxes are low, environmental protection is minimal, nearly every public service has been privatized. And the corporations that operate there have their fellow corporate citizen, International Telephone and Telegraph (ITT), to thank for it all.

The military takeover of Chile was engineered by the U.S. government in close cooperation with three of the major multinational corporations with investments in that country, ITT and mining giants Anaconda and Kennecott. The 1970 election victory of Salvador Allende, a socialist who had campaigned on a platform of nationalization of Chilean resources, had taken the U.S. and the American corporations by surprise. As soon as the election results were known, an ITT official contacted Henry Kissinger's senior adviser on Latin America to indicate how determined ITT was to prevent Allende from becoming president and how much it was willing to spend in the effort: "I told [him] to tell Mr. Kissinger [ITT chairman] Mr. Geneen is willing to come to Washington to discuss ITT's interest and that we are prepared to assist financially in sums up to seven figures."[37]

In the wake of the Allende victory ITT established close contacts with the State Department, the National Security Council, the U.S. Information Agency, the CIA, the Inter-American Development Bank, the Senate Foreign Relations Committee, and other agencies with the purpose of pushing the U.S. to intervene covertly in Chile. Efforts were made to prevent Allende from taking power, which he could do only with the help of the centrist Christian Democrats. The CIA immediately implemented a program of economic destabilization to demonstrate to the Christian Democrats the folly of supporting the inauguration of Allende as president. When that failed, tactics changed. An ITT memo from field operatives in Chile read: "A more realistic hope . . . is that a swiftly deteriorating economy (bank runs, plant bankruptcies, etc.) will touch off a wave of violence resulting in a military coup."[38]

Once Allende was assured of the presidency, the CIA moved quickly to foment a military coup, and ITT with its many contacts and operatives in Chile was close at the CIA's side. ITT executive William Merriam gave the following description of a meeting at CIA headquarters: "Approaches continue to be made to select members of the Armed Forces in an attempt to have them lead some sort of uprising."[39] In fact, ITT was ahead of many transnationals in its capacity to respond to the actions of national governments. It was, as ITT chronicler Anthony Samson stated, effectively a sovereign state. In Chile it had a foreign policy, a foreign service, an information service, a clandestine service, and other operations more typical of governments than of corporations. The coup in Chile cost the lives of thirty thousand civilians and ensconced in power one of the most ruthless dictatorships in all of Latin America.

Equally important was the role that the dictatorship would play in the application of new-right economic theory after the coup. An entire package of Milton Friedman–inspired policies was implemented — deregulation, suppression of labour rights, and massive privatization of public services.

It is no wonder that corporate planners love Chile. What was done there is exactly what they would like to do in every country — and would were democratic freedoms not in place. The coup in Chile wiped clean the social, political, and cultural landscape of a nation and prepared the ground for a massive experiment in new-right social engineering. It was carried out at the behest of transnational corporations in their interests, and the subsequent "new society" was designed by the ideologues of the counter-revolution. Economists from Friedman's Chicago School of Economics applied their neoliberal tools to the defenceless people of Chile, creating social divisions and economic disparities unprecedented in the country's history.

That Chile is touted as a proud accomplishment of the forces of globalization reveals a great deal about where corporate rule would take us. Among the Canadian corporations eagerly taking advantage of the fruits of military-enforced restructuring of Chile are the Bank of Nova Scotia, the National Bank of Canada, Bata, Canadian Airlines, CanWest Global Communications, Shell Canada, and mining giants Falconbridge, Inco, Noranda, Placer Dome, Rio Algom, and Munk's Barrick Gold.

GLOBAL GREED AND PRIVATE ARMIES

The counter-revolution in Chile took place in the global context of the cold war. The United States spent hundreds of billions of dollars keeping the Soviet Union in check and suppressing movements for social and economic justice wherever they occurred and regardless of the size of the threat. As the invasion of Grenada demonstrated, no opposition was too small to crush.

In the post–cold war years everything has changed and everything has stayed the same. Authoritarian governments, dedicated primarily to protecting private property, are in place in most of the countries that the U.S. considers important sites for corporate investment.

Where governments are not powerful enough to enforce property rights on behalf of corporations, another strategy is adopted. The extent to which corporations now feel little compunction in challenging the sovereignty of governments and assuming their role was dramatically revealed in early 1997. Executive Outcomes, a private army for hire, was established by two Britons, a businessman and a former army officer. Antony Buckingham was the CEO of Heritage Oil and Gas, which had close links to a Canadian company, Ranger Oil. In 1993 the companies teamed up with Eeben Barlow, a former South African Defence Force officer who had served with some of the most ruthless units of the apartheid military, including its assassination network, which was organized along corporate lines.

Ranger Oil recruited Executive Outcomes to drive the rebel Unita army out of Soyo, a centre of the oil industry in Angola. For $30 million, Ranger, in cooperation with the Angolan government, purchased a force of five hundred former South African soldiers as well as sophisticated weaponry (including fuel-air bombs and helicopter gun ships). The intervention of this force virtually changed the course of the civil war. From Angola, Executive Outcomes moved on to Sierra Leone, where it shored up the corrupt regime of Valentine Stroesser against an armed rebellion that was on the verge of seizing the capital city.

Executive Outcomes is a complex network of some eighteen companies, the advance guard for major business interests engaged in a latter-day scramble for the mineral wealth of Africa. The corporation has worked with more than thirty countries, mostly African, and also has links with South Korea and Malaysia. And according to a British Intelligence report, it is growing rapidly. The *Guardian* predicts that "Executive Outcomes will become ever richer and more potent, capable of exercising real power even to the extent of keeping military regimes in being. If it continues to expand at the present rate, its influence in sub-Saharan Africa could become crucial."[40]

It seems more than a coincidence that Executive Outcomes, first identified in 1991, has developed into such a potent force at the same time that the cold war has come to an end. Until now, transnational corporations could count on the U.S. to either intervene or use its threat of military force to ensure compliant, pro-business governments no matter where they were. With the Soviet threat gone, such intervention is politically harder to sell to the American public. The major powers see Africa as a geopolitical morass and are happy to hand off its minor conflicts to corporations with their limited liability. In the absence of any efforts from major powers or the U.N. to stop them, corporations are now free to hire armies to enforce their interests. Corporate use of private armies represents the expropriation of one of the most fundamental roles of government, the sanctioned use of violence. The agreement by citizens not to resort to violence is rooted in this aspect of the state. We each agree to give up any "right" to use violence in defence of our individual interests on the understanding that everyone else will also give up that right or face the democratically sanctioned coercion of the state.

The very existence of Executive Outcomes raises the question of who will be next to decide that it is legitimate to create their own armed force. As we will see in chapter 6, that answer is not hard to come by. With private security forces outnumbering public police by three to one in the U.S. and Canada, wealthy individuals and corporations are blurring the lines between state-sanctioned violence and the private use of force.

SLAPPing Democracy

The use of private armies and the growth of private police services are not the only examples of the assault on democracy by corporations. They are also intervening through the courts to intimidate individuals and community organizations that speak up in the public interest. Corporate efforts to strategically target longstanding democratic rights have become known as SLAPP suits, for Strategic Lawsuits Against Public Participation.

SLAPP suits explicitly target the exercise of democratic rights, from citizens' submissions to government and published articles to public speaking or boycotts. In virtually every case the defendant in SLAPP cases has been successful in bringing public attention to the issue. SLAPP suits target individuals or community groups because they are the most vulnerable to intimidation and cannot afford the court costs.

American corporations are far ahead of their Canadian counterparts, although these suits are being used more frequently here. In the U.S., the strategy of tying up advocacy groups in the courts started in the 1970s. "Tens of thousands of Americans have been silenced by threats," say George Pring and Penelope Canan in "Slapps: Getting Sued for Speaking Out."[41] Nine states have passed legislation to stop SLAPP suits, seeing them as almost always frivolous and aimed only at silencing critics.

Several such corporate-sponsored suits in B.C. have targeted members of municipal councils who have spoken out against irresponsible practices. But perhaps the most notorious is the case of Daishowa, a huge Japanese conglomerate, against the Lubicon Indian Band of northern Alberta.

The Lubicon case pits one of the most desperately impoverished aboriginal bands in the country and their supporters against an enormously powerful transnational corporation. The suit is against Kevin Thomas and the Friends of the Lubicon, an Ontario-based group that organized an effective nationwide boycott of Daishowa paper products. The boycott was prompted by Daishowa's decision to log an area that is part of the Lubicon land claim. It was so effective that it was credited with shutting down Daishowa's operations in the area. Daishowa responded by going to court, arguing that the boycott had cost it $5 million. The court found in favour of Daishowa, granting it a temporary injunction against the boycott.[42]

It needn't be a huge corporation that brings a SLAPP suit. In Nova Scotia the Jacques Whitford Group, a consulting firm, sued a citizen in that province's Supreme Court for speaking at a Halifax City Council meeting as a member of one of council's committees. He had criticized a waste-incineration proposal put forward by the Whitford Group, referring to two previous projects in which the firm had been involved, both of which had gone over budget.[43] Clearly, the threat of a SLAPP suit delivers a strong message to anyone thinking of becoming an active citizen.

The use of SLAPP suits is not the only modern addition ot the corporate bag of tricks to subvert democracy. Corporations spend millions to support their interests. This practice is documented in *Masks of Deception: Corporate Front Groups in America*,[44] which identifies thirty-six industry-sponsored groups whose public interventions are regularly portrayed as those of citizens' groups. Thus the green-sounding National Wetlands Coalition is actually sponsored by industries fighting to ease government restrictions on the commercial conversion of wetlands. When the U.S. Senate was debating the new clean-air legislation in 1990, car companies hired Bonner and Associates, a sophisticated peddler of grassroots democracy. It convinced groups as diverse as Big Brothers, the Easter Seal Campaign, and the Paralyzed Veterans' Association to lobby against the legislation. They were persuaded by Bonner that the new rules would make it impossible to produce any cars larger than a Honda Civic.[45]

The critical development that makes corporations in the late twentieth century use such a threat is that their power is now applied on a global scale, increasingly immune to the authority of nation-states and citizens. The power of governments to moderate the behaviour of market forces, in place for almost three generations, has been ceded to a social organization second only to nation-states in power and without their moral or democratic imperative. The response of corporations has been anything but benign. They have not simply taken advantage of government deregulation and withdrawal from the field. Sensing their power and assuming that governments will not intervene, they have now become aggressive in further challenging democratic institutions and citizen rights.

The popular notion of the nation-state becoming obsolete in the face of the power of transnational corporations is more accurately described as the amalgamation of corporate and state rule. As consensus has developed between economic and political elites and resistance to corporate domination is undercut, the lines between corporate governance and national governance have blurred.

There is, of course, a new twist to the old theme of corporate efforts to end government meddling in their affairs. What transnational corporations bring to this new era of corporate control is precisely their global reach and aspirations. They don't want the state to disappear because they could not operate for a week without it. But they do want to put in place appropriate state institutions that serve their particular global agenda. That agenda is driven by fierce competition for trade and investment opportunities that can provide the greatest return on investment. And the key to success in this global struggle is total flexibility to move, change, merge, and otherwise manipulate every aspect of corporate operations on a worldwide scale.

The demands of transnational corporations have outgrown the services that the traditional nation-state can provide. In order to assist corporate profitability, it was necessary for those industrial states to create transnational institutions that would meet the requirements of the transnational corporation. These institutions, mechanisms, and agencies would superimpose on nation-states a regime of rules, regulations, governing bodies, and effective sanctions that would reflect the new era and its new paradigm. Such institutional mechanisms are now rapidly being put in place, and Canada and Canadians have played a special role in testing them out. That role was one of guinea pig in global corporate governance, and the tests were the free-trade agreements.

5

How Corporations Rule the World

We are writing the constitution of a single world economy.
— Renato Ruggiero, director-general, World Trade Organization, describing the Multilateral Investment Agreement

The strategy [on the Free Trade Agreement] should rely less on educating the general public than in getting across the message that the trade initiative is a good idea. In other words a selling job. It is likely that the higher the profile the issue attains, the lower the degree of public approval will be. Benign neglect from a majority of Canadians may be the realistic outcome of a well-executed communications program.
— From a memo leaked from the office of Prime Minister Brian Mulroney, 1985

How do corporations "rule the world"? In many respects they do so in the same ways that they always have. First, they rule the world because they make many of the important decisions that affect the lives of ordinary people and their communities, decisions about where to establish a plant, what to produce, who will work and who won't, how much pollution will go into the environment, and which countries will get the next factory. But they also rule politically.

They rule country by country; they influence politicians and political parties with huge amounts of cash; they threaten governments that dare to consider any laws that lessen their power or privilege; they withdraw capital to punish governments that don't take them seriously; they spend millions on lobbyists to persuade legislators; they spend hundreds of millions on public relations experts and advertising firms to convince the public that what's good for corporations is good for them; they have phalanxes of lawyers to get around regulations and more lawyers and tax accountants to avoid and evade taxes.

And they own newspapers, TV networks, and radio stations whose editorial policies and programming they control, directly or indirectly, and thus can publicly punish politicians who defy their agenda, and determine the information and analysis given to citizens to help them decide who to vote for and, more generally, how to view the world.

They use their virtually unlimited resources in a relentless, continuous, and self-interested assault at every level of government, from presidents and prime ministers down, from senior policy makers to schoolboard members, town councillors and the lowest-level functionaries of municipal governments. They have the resources to hire the cream of the crop in any area of expertise, and they have the ability to be everywhere at once promoting their bottom line.

Most large corporations have at one time or another intervened in the political process to advance their own interests. Such activity is a normal part of doing business. But in the era of the transnational corporation the sheer size of some corporations means that there has been not just a quantitative change in the way corporations "do" the politics of business. There has been a qualitative change. That change involves not only how and how much TNCs intervene in the governing process but the scope of the demands they now make on governments. They have enormous capacity to intervene, and they now do so with global objectives in mind. They don't simply want the local, provincial or national rules changed. They want international rules established so that they don't have to be bothered by local or national rules or take the time to intervene. The future that the corporations are seeking has no surprises.

Corporations now exercise the power not only to force governments to back off environmental legislation but to create international treaties and agencies and regimes of punishment for countries that dare to enact environmental policies that might restrict their activities. The Canada-U.S. free-trade deal, the North American Free Trade Agreement (NAFTA), APEC (Asia Pacific Economic Co-operation forum), the World Trade Organization (replacing the General Agreement on Tariffs and Trade), and the OECD's Multilateral Agreement on Investment (MAI) are all examples of the international agreements and laws aimed at permanently changing the rules by which corporations, and by default people, are governed.

The old-fashioned interventions into government, the traditional ways of thwarting citizens' efforts to build humane communities, are being superseded by a level of corporate political intervention appropriate to the new era of TNCs. We have a transnational regime of world governance to match the transnational nature of corporations, and the "governing" institutions are as remote and unaccountable as the corporations that designed them. This is world government with only a few thousand "citizens" — the largest corporations in the world — and with laws that exclusively serve their interests.

Corporate rule in the 1990s does not entail destroying democratic institutions, halting elections, or jailing opposition politicians. What we are witnessing in Canada and elsewhere is not the classic political coup, not an illegal seizure of power, but what John Ralston Saul calls "a coup d'état in slow motion." It is as if the FTA and NAFTA deals, and other international agreements, are slowly dissolving the substance of democracy while leaving the institutional facade intact.

Long before the Free Trade Agreement, NAFTA, the WTO, MAI, and other such agreements were in place, the world's largest companies had extremely powerful agencies working on their behalf. The International Monetary Fund, the World Bank, and GATT were all institutions created as a result of the famous Bretton Woods meetings in July 1944. Forty-four nations gathered in the U.S. to create the foundation for a world economic system. The new multilateral institutions promised general prosperity and economic interdependence as a bulwark against further armed conflict.

John Maynard Keynes, the intellectual father of the modern egalitarian state, had prepared a set of principles for the design of a new financial system in preparation for the Bretton Woods conference. The basic features of the plan included a world central bank, which he intended as a "clearing union" to extend credit to countries suffering from balance-of-payments problems and which would hold deposits from nations with a surplus. There would also be an international currency, which would be used for international transactions. An international trade organization would work to stabilize commodity prices by establishing buffer stocks of commodities and setting up binding commodity agreements between states. Lastly, and directly through the U.N., there would be an aid program to provide unconditional below-market-rate loans and grants to less developed countries.[1]

Keynes had expressed serious doubts that freer trade would lead to a more equitable international division of labour. "Ideas, knowledge, science, hospitality, travel — these are the things which should of their nature be international. But let goods be homespun wherever . . . possible and above all let finance be primarily national."[2]

Yet it was power, not the strength of ideas promoting social and economic justice, that would prevail, and the U.S., with its large holdings of Britain's debt, had that power. In the discussions leading up to Bretton Woods the two key figures, Keynes of Britain and Harry Dexter White of the U.S., had agreed that it was essential to control the movement of international capital and to allow national governments to pursue full employment policies. White, in line with Keynes, had suggested in an early draft that deposits or investments from another country not be permitted without the approval of that country's government.

Keynes's vision would have both creditor and debtor nations share the burden of international adjustment, something White did not disagree with personally. A crucial issue at Bretton Woods centred around whether nations would have automatic access to the IMF. Keynes's position was that automatic access was crucial to the development of poorer nations. Whatever Harry White's personal conviction, U.S. corporations had their own thoughts. The deliberations, like those that created the United Nations, were dominated by the U.S., and there was another agenda at Bretton Woods — in the words of David Korten, "to create an open world economy unified under U.S. leadership that would ensure unchallenged U.S. access to the world's markets and raw materials."[3]

When New York bankers heard of White's proposal for currency controls, they intervened with the U.S. treasurer, and the country's position changed. During the negotiations White pressed for conditions to be attached to access to the new IMF. The U.S. position won out. The idea of an international currency was dropped and the new world currency system, the basis for world trade, would be the American dollar, with the U.S. given the right to print and spend the world's principal currency. The new international agencies would assist the world's largest corporations, at the time mostly American, in their efforts to accumulate capital and power.

THE WORLD BANK AND THE IMF

The World Bank was intended to assist European countries and Japan to rebuild in the post-war period. But within a couple of years of Bretton Woods, European countries did not need World Bank loans because they were getting billions from the American Marshall Plan. Third World countries were disinterested for another reason. Their elected officials and policy analysts were divided about lining up behind World Bank objectives. Economic nationalists wanted to follow a policy of protecting local markets and avoiding international financial alliances and dependencies.

Amongst the elites the debate turned around whether a country should follow an industrial policy of import-substitution or an export-led strategy of producing goods locally for export to the developed world. The import-substitution strategy ran counter to the World Bank's policy because it required fewer imports and therefore reduced the need for foreign exchange. The whole point of the World Bank was to provide financing for the purchase of corporate products from advanced industrial countries.

Like the corporations whose goods it was intended to help sell, the World Bank soon set out to create new customers in the Third World. In doing so it began circumventing and manipulating elected governments and accountable officials, actions that continue today with trade and investment treaties. The World Bank, as David Korten points out, "gave priority to 'institution-building' projects aimed at creating [in the developing nations] autonomous governmental agencies that would be regular World Bank customers . . . staffed primarily by transnational technocrats with strong ties to the bank."[4] As a result elected officials were increasingly subject to advice from their own bureaucrats that was crafted by the bank in the interests of transnational corporations eager to sell into their countries' markets.

The less developed countries descended into the vortex of World Bank loans and the contradictions they produced. The bank itself was so blinded by its free-market, comparative advantage ideology that each failure to help a country increase its wealth and pay down its debt was met with irresistible offers of more loans and a repeat of the cycle.

While the economic nationalists of the developing countries were prevented from effectively pursuing policies of self-sufficiency, their debt situation, already difficult, was about to take a turn for the worse. Basing the world economy on the U.S. dollar worked for about fifteen years. But by the late 1950s the U.S. was experiencing a continuing balance-of-payments problem, and in 1971 it arbitrarily ended the fixed exchange rate. What followed was a wave of currency speculation involving what were called Eurocurrencies, currencies deposited in other countries, such as French francs held in Germany or yen in Britain. The new floating exchange rate led to an enormous growth of Eurocurrencies, supplemented by OPEC dollars, from $150 billion in 1971 to $5.4 trillion by 1988.[5] By mid-1997 it was up to $13 trillion. That money had to go somewhere. Starting in the 1980s much of it went to the least developed countries.

These nations went on a borrowing spree promoted by the World Bank's extremely low interest rates and aggressive "selling" of loans by multinational banks. Larger loans, of course, meant even more loans were needed to pay the interest and the ever increasing principal. From 1970 to 1980 the long-term external debt of low-income countries increased from $21 billion to $110 billion. That of middle-income countries rose from $40 billion to $317 billion.

The crushing debt burden was a further boon for the international lenders and the transnationals, as the debt crisis opened the door to what would become known as structural adjustment. This began the period of the World Bank's and the IMF's policy of conditionality, a euphemism for the elimination of the sovereignty of indebted nations. The two agencies of corporate libertarianism dictated macro-economic and social policies to scores of countries. To get more loans to pay the interest on the old ones countries were obliged to alter their policies to channel more and more of their resources and productive capacity towards foreign trade and debt repayment.

The IMF was the debt collector for the system, while the World Bank was literally an institution of governance, as Jonathon Cahn described it in the *Harvard Human Rights Journal*, "exercising power . . . to legislate entire legal regimes and even to alter the constitutional structure of borrowing nations. Bank-approved consultants often rewrite a country's trade policy, fiscal policies, civil service requirements, labor laws, health-care arrangements, environmental regulations, energy policy, resettlement requirements, procurement rules, and budgetary policy."[6]

FREER TRADE: THE MISSING THIRD LEG

When the major capitalist powers established the IMF and World Bank there was a third agency that the U.S. administration had wanted but didn't get. In place of the planned International Trade Organization, twenty-three countries, including Canada, signed in 1947 the General Agreement on Tariffs and Trade (GATT). Its operating principle was that of "most-favoured nation," under which any member country reducing the tariff on a commodity for one of its trading partners would automatically reduce it to all other members.

By the late 1970s tariffs were no longer the major barrier to trade and certainly not the greatest threat to the world capitalist system. The crisis facing the industrialized nations was much deeper. In Europe corporate profitability fell from 11 percent to 5 percent between 1970 and 1975 and the American balance of trade worsened. In the early 1980s there was nearly a trade war between the U.S. and Europe over the latter's agricultural price support system, designed to maintain farmers' incomes. The GATT did not address this issue, and Europe was in no mood for free trade.[7]

At the same time, mass consumption in the U.S. seemed to have reached a saturation point. The new phenomenon of stagflation revealed both high inflation and stubbornly high unemployment at the same time. The golden age of capitalism was well and truly dead, and with the 1980–82 recession and its aftermath came the threat of cutthroat international competition and, for Canada, increased fears of American protectionism.

As for the Americans, barriers to trade, and more particularly to investment, were not coming down fast enough. The U.S. was prepared to pursue economic integration, the means to knock down investment barriers, one country at a time. So when free-trade negotiations began Canada wanted some firm guarantees against U.S. protectionist legislation, and the U.S. wanted greater access to Canadian resources and energy, and new investment rights. But it was also clear that for the U.S. the negotiations were, as Duncan Cameron argues in *Canada Under Free Trade*, integral to a long-term strategy designed to set precedents for future multilateral agreements.[8]

CORPORATIONS: REAPING THE BENEFITS OF FREE TRADE

The intractable problem of stagflation and the intervention of Milton Friedman's free-market ideology provided the basis of an assault on the historic compromise between capital and labour. The crisis of the welfare state allowed its opponents to attack the whole notion of using Keynesian economic policies to moderate the excesses of the marketplace. That foundering social contract demanded a new state/corporate equilibrium, one that would be dominated by corporations and their ideological allies. Its first major manifestation for Canada was the Free Trade Agreement, a constitutional charter of rights for transnational corporations that embodied the principles of corporate libertarianism. Its designation as a trade deal obscured its essential purpose: to economically integrate Canada with the U.S.

It was clear to the largest corporations in both countries what this deal meant, not just for what was in it but for what it presaged for multilateral agreements. If they won, it would be a huge step forward in their desire for a permanent counter-revolution against the egalitarian state. The deal reconstructed democracy and the state so that corporate interventions in the political process would require little more than minor adjustments.

Much of the attention on the FTA (and later NAFTA) focussed on how Canada "lost" to the Americans. But in effect what happened was that all workers and most citizens, Canadian, American, and Mexican, lost to corporations. In a myriad of ways average citizens lost human rights, lost citizen power, lost control over their communities, lost the ability to determine what kind of society they and their children would have. In one sense Canadians lost more because we had more to lose: the state was more active in redistributing wealth, ensuring public services, financing culture, and promoting regional equality. Americans never had the egalitarian state in the first place.

To the extent that it was nominally "nations" negotiating with each other, Canada clearly lost and the U.S. won. Yet even here it was U.S. trade laws working in the interests of U.S. corporations that defined the American victory. The FTA and NAFTA were not trade "deals" between sovereign nations so much as they were declarations by national governments of their willingness to give up sovereignty and political power to corporations.

The collaboration of Canadian and American corporations to get these deals passed proves the point. It did not matter to the very largest corporations which country they were based in because they were seeking a new charter of transnational corporate rights that would permit them to treat the two countries as one jurisdiction.

Between the end of 1988 and 1994, as many as 334,000 manufacturing jobs were lost in Canada, 17 percent of the pre–free trade total. Most of these were shed by the largest corporations, members of the Business Council on National Issues (BCNI) who spent millions promoting the deal. Between 1988 and 1996, thirty-three BCNI member corporations laid off 216,004 employees (eleven others increased employees but by just 28,073). The big losses were in labour-intensive industries. By the end of 1994 the clothing industry had lost 32.8 percent of its jobs, electronics 24 percent, paper 22.5 percent, food and primary metals both 16.5 percent. At the same time, those thirty-three corporations increased their combined revenues by more than $40 billion, to $158.2 billion.[9] During the free-trade debate the deal's proponents promised "more and better jobs" — 350,000 of them. The jobs that have been created have been in low-wage personal and commercial services.

A major feature of life under free trade is the development of a dual economy, one global and the other domestic. According to the

Canadian Centre for Policy Alternatives (CCPA), "We have a growing export-oriented manufacturing and resource sector, alongside a depressed domestic market. The export boom has failed to generate any jobs. The domestic economy is held down by high unemployment and stagnant demand for locally produced goods and services."[10] This is the "jobless" economy.

FTA opponents pointed out a hidden agenda: corporations' desire for a level playing field in social programs. It was always understood that this would mean lowering the field to the American level. The denials by the Mulroney government and its corporate allies came fast and heavy during the lengthy debate over free trade.

Corporate leaders were not so foolish as to make their intentions known during the debate but they weren't always so reticent. In 1980, Laurent Thibault, a corporate executive who later headed the Canadian Manufacturers' Association, told a Senate committee: "It is a simple fact that, as we ask our industries to compete toe to toe with American industry . . . we in Canada are obviously forced to create the same conditions in Canada that exist in the U.S. whether it is unemployment insurance, workmen's compensation, the cost of government, the level of taxation or whatever."[11]

As the CCPA analysis of the FTA shows, within a few weeks of the federal election, the Canadian Manufacturers' Association, the Canadian Chamber of Commerce, and the BCNI began campaigning for spending cuts, especially to UI, arguing that it was crucial to Canada's competitiveness — a self-fulfilling argument after the deal was signed. The chair of the powerful Canadian Manufacturers' Association demanded that every federal and provincial program be re-examined: "All Canadian governments must test all their policies to determine whether or not they reinforce or impede competitiveness. If a policy is anti-competitive, dump it."[12]

The Mulroney government's first free-trade budget, five months later, began the series of unprecedented cuts to UI, old age security, and transfers to the provinces for health and education that continued unabated for eight years. These cuts have taken us back to the level of community spending achieved in 1949. Of course these cuts were made without any reference to free trade. As we will see in chapter 9, the level playing field was accomplished by debt hysteria. Canada's high debt, created almost exclusively by policies promoted by corporations, was a convenient weapon against our social programs, which U.S. corporations saw as unfair subsidies to Canadian business.

One of the most oft-repeated assurances of corporate Canada and the Mulroney government was that medicare would not be touched. Yet the impact of the agreement was clear: medicare would be changed to accommodate the increased involvement of private corporations. According to the CCPA, "Chapter 14 of the FTA on Services defined health care management as a 'commercial service' rather than a 'social' or 'public' service . . . It opened the door for private foreign firms to enter into an array of health services, including hospitals, nursing homes, homes for the disabled, ambulance services, rehabilitation clinics, home care services, laboratories and blood banks."[13]

Under NAFTA rules, medicare was further threatened. A study prepared for the Canadian Health Coalition and the Canadian Union of Public Employees by Manitoba law professor Bryan Schwartz concluded that parts of the agreement that were supposed to protect medicare were so vague that they were open to all sorts of challenges.

While corporations wanted the area of health care opened up, they wanted environmental protection closed down. A number of trends in environmental deregulation have been identified in the aftermath of FTA/NAFTA. There has been a shift, promoted by TNCs, towards establishing continental and international regulations, a trend that makes national initiatives more difficult and restricts citizens' access to the process. The language of "sustainability" is gradually giving way to that of "competitiveness," reflecting the fact that policy makers are adopting the corporate perspective.

Perhaps most alarming is the trend away from regulatory regimes to ones of "voluntary compliance." A comprehensive 1994 survey by the forecasting firm KPMG of Canadian corporations and public institutions revealed that whereas 69 percent had an environmental management plan, only 2.5 percent could claim to have incorporated all the vital components. Ninety-five percent said compliance with existing state regulations was their strongest motivation. Determined lobbying from corporations on both sides of the border convinced the Chrétien government to introduce the Regulatory Efficiency Act, which would have allowed corporations to seek waivers from environmental laws. The federal government is devolving its authority over the environment to the provinces, which are in the process of "harmonizing" their regimes. That means harmonizing them with the province whose level of protection is the lowest and whose enforcement is weakest.

The FTA provisions on energy and natural resources, repeated in NAFTA, make conservation next to impossible. Under the "proportionality clause," Canada is obliged to export the same proportion of energy and other natural resources to American corporations as it does currently, even in times of shortage. This not only undermines conservation goals. In the event of an energy emergency it subordinates the interests of Canada to the abstract principles of economic efficiency and free trade. We could face a situation where we would have to shut down Canadian industries to be able to meet our natural gas commitments to U.S. energy corporations.

It is difficult to find a more compelling example of how the free-trade agreements put commercial contracts between corporations ahead of the interests of a nation and its citizens. Had Canada been militarily conquered, its humiliation might have been understandable. This humiliation, however, was gleefully imposed by the Canadian government at the urging of TNCs and domestic corporations.

Of all the areas of government policy that have been affected by free trade, the protection of Canadian culture has been one to which the federal government has been forced to pay lip service. As we saw, the long-term strategy of the TNCs is built on a foundation of marketing the mass consumer culture to every corner of the planet. As Sony's Akio Morita proclaimed, culture is a trade barrier to building a global mass consumer society. The federal government's schizophrenic record on cultural issues reveals its basic support for the corporate position and its countervailing need to appear protective.

The widespread attacks compelled Mulroney to negotiate a cultural "exemption" in both the FTA and NAFTA. But the exemption included what amounted to a poison pill. It allowed American corporations, through the U.S. government, to punish Canada for taking measures to protect Canadian culture. The U.S. can implement measures of "equivalent commercial effect" against Canada.

The result of the poison pill has been a whole series of decisions by Canada to either back off new support for culture or directly favour U.S. conglomerates. Legislation to restrict foreign takeovers of Canadian publishers was dropped; soon after that the educational publisher Ginn was bought by Paramount. When the CRTC forced the American Country Music Channel off the air in favour of the Canadian New Country Network, the U.S. threatened all sorts of retaliatory action. Part of the resulting compromise was a promise by Canada that it would not "de-list" any other cable service.

In 1997, Canada put a brave defensive face on a World Trade Organization ruling against protecting Canadian magazines — the so-called split-run issue involving *Sports Illustrated* — a decision reflecting the principles behind free trade. But promised legislation to help Canadian books and sound recordings to increase their share of the market has been dropped, and continued huge cuts to the CBC mean that private broadcasters have less competition and a weakened standard of genuinely Canadian programming to live up to. Cuts to the National Film Board and Telefilm Canada, like those to the CBC, were far out of proportion to cuts to other areas and saved relatively small amounts of money, indicating that Ottawa is essentially committed to bringing down the cultural barriers to investment so detested by global corporations.

Perhaps the most important overarching principle enshrined in the FTA and NAFTA is that of national treatment. National treatment means that the rights of citizenship, already accorded to the fictitious corporate person in Canada, are now extended to American and Mexican corporations. What this means is that Canadian governments must treat American and Mexican corporations no differently than they treat Canadian ones.

Clearly these deals represent a serious erosion of democracy and the democratic rights of citizens to determine the future of their country. They so fundamentally shift power in favour of corporations that they are the equivalent to constitutional changes. Here are some other ways these deals undermine democratic governance.

- The dispute settlement mechanism of NAFTA allows corporations to challenge government actions at international arbitration panels. Just the threat of such actions will put a chill on policies that encroach upon corporate interests. A 1997 case was seen as a harbinger of things to come. The American firm Ethyl Corporation demanded $350 million in compensation from Ottawa, claiming that new legislation banning one of its products, the fuel additive MMT, because of its health threat, amounts to "expropriation without compensation."

- Not only do the FTA and NAFTA restrict how crown corporations can be used, they make the establishment of new ones difficult if not impossible. The best example of this was the decision by the Ontario government of Bob Rae to back off its election promise to establish public auto insurance. Rae, to his lasting shame, declined to challenge the FTA by proceeding with such insurance and even declined to make it clear to the public that the FTA was the reason for his decision. Yet the American insurance industry had made it abundantly clear that it was poised to demand billions in compensation if the government made such a move.
- The rationale corporations and the Mulroney government put forward for pursuing free trade was that the agreements would protect Canada from U.S. "trade remedy" laws — in other words, punitive retaliation (countervailing duties and anti-dumping duties). Canada got no such thing, of course, and is just as subject to these laws as it was before the deals. All the dispute settlement mechanism now in place provides is a method of ensuring that U.S. laws are applied fairly.

Many critics saw the trade remedy argument as simply a pretext for the free-trade negotiations from the beginning, as trade experts knew all along that GATT negotiations would have provided Canada with the same trade remedy measures within a few years. Mel Clark, formerly a senior Canadian trade negotiator, did a comprehensive cost-benefit analysis comparing the FTA to GATT in those areas he calculated would "especially impinge on Canada's interest." His conclusion: "The FTA ceded to the U.S. and the private sector the right to shape many of Canada's economic, social, environmental and cultural policies . . . If one assumes the Mulroney government was guided by even a normal regard for the national interest, the FTA defies understanding . . . while vital national powers are protected under GATT, the same cannot be said of the FTA."[14]

Analysis shows that Canada would have been better off with GATT. But even more important is that the free-trade deals were the models for changes to GATT that now threaten to apply to every country in the world the corporate libertarian rules at the core of FTA/NAFTA. This reinforces the claim of many critics that the free-trade deals were intended as stalking horses for global trade and investment regimes, and Canada as a guinea pig in an experiment in corporate rule.

THE WORLD TRADE ORGANIZATION

On January 1, 1995, the multilateral agreement known as GATT became the World Trade Organization, heralding a frightening era of world corporate domination. In the words of the Common Front on the WTO, a Canadian coalition opposed to its initiatives, "The establishment of the WTO represents a watershed in the process of establishing a truly global economic order and it is likely to exert a more profound influence over the course of human affairs than any other institution in history."[15] Yet despite the implications of this development, it received almost no publicity in Canada. It was slipped through Parliament in late 1994 as legislation ratifying the latest GATT agreement, with virtually no public attention let alone any debate. It was scarcely even announced.

While it might be argued that this is true of other multilateral accords, there is something terribly wrong when an agreement with profound implications for national sovereignty is designed in such a way that it can become law with no effort to inform the public, literally unbeknownst to the millions whose lives will be affected. By allowing trade to be defined narrowly as something of interest only to corporations that engage in it, nation-states implicitly surrender their sovereignty as if it were a trifle.

With the advent of the WTO, says David Korten, "a trade body with an independent legal identity and staff similar to that of the World

Bank and the IMF is now in place, with a mandate to press forward and eliminate barriers to the free movement of goods and capital. The needs of the world's largest corporations are now represented by a global body with legislative and judicial powers that is committed to ensuring their right against the intrusions of democratic governments."[16]

For transnational corporations the WTO is the long-awaited third leg in their decades-long campaign for complete liberation from government interference. The IMF and the World Bank successfully established corporate rights throughout the less developed countries of the world. With the WTO it is now the turn of the industrialized countries to turn over their decision-making authority to global corporations.

In transforming itself into the World Trade Organization, the GATT evolved by a huge leap. Its importance, according to the Common Front, is threefold. First is simply the increasingly important role of trade in the world economy: international trade grew by 9.5 percent in 1994 alone. The TNCs control more than a third of the world's productive assets and they organize them — or try to — without reference to national borders. A growing portion of world trade, now at 40 percent, takes place between corporations owned by the same TNC conglomerate.

Second, trade is no longer trade as we used to understand it, that is, the buying and selling of goods across borders. Trade and investment are now indistinguishable and virtually inseparable. With so much trade taking place within corporations, any rule affecting investment — local sourcing, performance requirements, technology transfers — has consequences for "trade" between the companies of the affected TNC.

Perhaps most important, the WTO has extremely powerful enforcement rules that even the largest nations cannot easily ignore. Under GATT strong sanctions existed but were imposed only if there was consensus amongst its 120 members, including the offending country. The rulings of the WTO bureaucracy are now automatically implemented unless a consensus of members (including the member bringing the complaint) votes against the sanction. These are not puny penalties. The first WTO trade complaint was a challenge to the U.S. Clean Air Act. Its regulations were found to be in violation of

WTO rules and the U.S. would have faced a penalty of $150 million each year had it not withdrawn the offending measures.

The World Trade Organization is the institution of choice for transnational corporations quite simply because it works. The ramifications of its decisions far exceed those of the General Assembly of the U.N. and even those of the Security Council. The lives of the vast majority of people in the world are affected by investment decisions, and those are increasingly made by TNCs. The WTO, essentially designed for and dominated by the transnational corporate agenda, is effectively a United Nations for TNCs, providing corporations with the powers of an international state.

Yet, as we will see in the discussion of one WTO initiative, the Multilateral Investment Agreement, there is a crucial difference between the WTO and its two sister organizations, the IMF and World Bank. Decision making in the WTO is based on one country, one vote and not weighted by financial contributions as in the IMF and World Bank. Decisions are made with the unanimous agreement of all member countries. It is a key contradiction, and reflects the world forum in which corporations face down what remains of the sovereignty of nation-states.

As early as 1972, the U.S., on behalf of domestically based TNCs, began shifting its attention away from the United Nations and towards other forums in order to achieve open investment rules. That year the less developed countries, looking for some system to curb the abuses of TNCs, used their voting majority in the General Assembly to set up the Group of Eminent Persons. Its mandate was to examine transnational corporations and their impact on host countries. With the release of their report in 1973 it was clear that U.N. action on a code governing the behaviour of TNCs was near completion.

From that point on it was clear to the U.S. and its corporate clients that a regime to liberate corporate investment from the performance requirements, technology transfers, and controls over foreign exchange established by national governments was not going to be achieved through the U.N. At first that meant pursuing those goals through other means. But in recent years the political leadership of the U.S., media pundits, and corporate spokespeople have conducted an all-out assault on the United Nations.

With their relentless campaign to cast the U.N. as inept, unrepresentative, dominated by an unaccountable, self-serving, and bloated bureaucracy, the political representatives of TNCs have tried to denigrate the world body while creating another one that will speak unflinchingly for their interests. Just as the Free Trade Agreement's promoters tried to put it in place by stealth and negotiated it behind closed doors, TNCs at the international level circumvented the only democratic forum for the world's nations.

One of the enormous advantages of the WTO is that negotiations take place in almost total secrecy, away from the prying eyes of the media and citizens. When a corporation challenges a law under the WTO (or NAFTA), both parties present their arguments to a panel of three trade experts, usually trade lawyers with experience representing corporations. This hearing is held in secret, and the documents presented are secret. Although the ruling itself is made public, the WTO forbids revealing how the panelists voted or what they said.

In short, citizens have absolutely no way of knowing why — or by whom — their democratically determined laws have been declared null and void. As for those defending the said law, they are obliged under WTO rules to prove that the law is not a restriction of trade — that is, countries are guilty until they prove their innocence. Corporations, on whose behalf complaints are made, cannot be charged or found "guilty" of anything or held accountable in any way in any international forum.

It is not just ordinary citizens who are kept in the dark. According to Martin Khor, director of the Malaysia-based Third World Network, the leaders of dozens of the poorest countries are excluded from negotiations and often do not have the sophisticated trade expertise to effectively represent their people when they are presented with agreements signed by the major industrial countries.

The negotiations involve corporations and nation-states to the virtual exclusion of all other interests, reflecting the historical pattern of trade being the sole purview of those doing the trading. Government negotiating teams, particularly the American and Canadian delegations, are heavily dominated by corporate advisory committees and representatives.

One of the most important of these corporate groups was the

Advisory Committees for Trade Policy and Negotiations on GATT in the U.S. Its members, all members of the Business Roundtable (the American equivalent of the Business Council on National Issues), included IBM, AT&T, Bethlehem Steel, Time Warner, 3M, Corning, BankAmerica, American Express, Dow Chemical, Boeing, Eastman Kodak, Mobil, Pfizer, Hewlett Packard, and General Motors. Again, corporations have a legitimate interest in trade issues, but so do ordinary citizens. Workers, environmentalists, the churches, women, consumers, educators, health-care providers, and many more are affected by such agreements. Yet of the 111 members of the three main U.S. advisory committees only two represented unions; the single environmental position was not filled, and there was no consumer representative. All meetings were closed to the public.[17]

In contrast to the U.S. government, which makes public the names of its corporate advisory committees, the Canadian government refuses to do so. Brian MacKay, the deputy director of the Advisory Secretariat for the department of foreign affairs, pointedly refused to reveal the names of any corporations involved in the sixteen sectoral committees advising the government on the MAI. When asked why the names were not public information, MacKay replied, "The corporations asked that their names not be made public." Asked if the corporations individually requested that their names be kept secret, he replied, "Yes." The BCNI in Canada and the European Round Table of Industrialists have been in the forefront of pushing for these deals and the advantages they provide. Japanese corporations have historically worked closely with their governments on all trade and investment policies. It is inconceivable that transnational corporations from these countries would not dominate the process that works out the details of the deals they lobbied for. The whole process is immersed in the corporate culture and reflective of corporate objectives. The bureaucrats who negotiate with each other are trade department officials, many of whom came from the corporate sector and see no difference between the corporate and public interests. While the Canadian government keeps the names of its corporate advisers secret, there is no question that corporations completely dominate the shape of the MAI, here and elsewhere.

Of all the areas on which TNCs have set their libertarian sights, that of investment is by far the most important, both to their interests and to nations and communities. For communities the attraction or loss of capital investment are life and death events, and unregulated investment can be catastrophic. Corporations are preoccupied with investment because they are driven by an imperative of unrestrained growth. And countries, dozens of them impoverished from years of debt strangulation, are competing desperately for that capital.

THE MULTILATERAL AGREEMENT ON INVESTMENT

The initiative for a global treaty on investment was begun as early as 1986, when the U.S. challenged Canada's Foreign Investment Review Agency (FIRA) under GATT rules. American corporations did not get everything they wanted but they got enough that their foot was in the door. By 1987, GATT articles were being reviewed and discussions were taking place about which government investment measures could be considered trade related.

From the beginning the struggle was between the capital-exporting industrial countries and the less developed countries, which opposed restrictions on their laws regulating foreign investment. Efforts to get a multilateral investment agreement through GATT continually ran up against resistance from less developed nations, often led by India. TNCs made slow progress under GATT, and in 1992 a number of international business groups approached the OECD through its business advisory committee, pressing it to take the initiative.[18]

The corporate advisory committee to the U.S commerce department was also pushing for an OECD initiative, and the U.S. made it clear that it wanted "to develop a full treaty it could fast track through Congress and would bind states."[19] There were two advantages to going through the OECD. First, it would prevent backsliding within OECD countries. Second, it would establish a precedent; these standards could then be promoted outside the OECD. The OECD as a consensus-building organization for the dominant market economies thus played the role of missionary for neo-liberal policies.

The U.S. also supported the OECD track because it was confident that it could get a much tougher investment regime negotiated amongst industrialized countries in the absence of non-OECD countries. It was clear from U.S. actions that American-based transnationals believed an OECD treaty would have the advantage of leveraging poorer countries into line with a promise of access to markets in return for increased liberalization. In the end, both sides of the debate amongst developed countries agreed to a two-track approach, pressing for an investment agreement in the World Trade Organization while getting one in place through the OECD, with each initiative reinforcing the other.

In 1995, an OECD report on a multilateral agreement on investment launched negotiations between the twenty-nine developed countries. The target date for a deal was just two years away. The MAI was designed to set a high standard for the treatment of investors and provide transparent rules on liberalization, dispute settlement, and investor protection. More important, the OECD's 1995 *Multilateral Agreement on Investment Report* stated: "The MAI would provide a benchmark against which potential investors would assess the openness and legal security offered by countries as investment locations. This would in turn act as a spur to further liberalization."[20]

A key meeting for the WTO initiative on the multilateral investment agreement was held in Singapore in 1996. By the previous June opposition amongst the less developed countries was growing. In September, India hosted a meeting of fourteen countries and raised the possibility of shifting the discussion to the U.N.'s responsible body, UNCTAD, the U.N. Commission on Trade and Development.

There was no consensus by the time the Singapore meeting was held, and the ministerial declaration outlining the meeting's agenda barely mentioned investment as a major consideration. Opposition from the less developed countries prevented the agreement from getting off the ground, in large part because WTO decisions are all made by consensus. Technically, a single no vote can scuttle any agreement. But the way the industrialized countries attempted to get the agreement through revealed just how the poorest countries are treated by their more powerful neighbours to the north.

The agenda at Singapore was to have focussed on GATT-related problems for the less developed countries. Instead, the developed countries hijacked the conference. The official conference featured ministers from each country giving their speeches in a great hall holding three thousand people. But there was, according to Martin Khor of the Third World Network, no discussion, no debate. According to Khor, "After the first morning no one attended anymore. So, that hall for three thousand people had five people inside. By the second afternoon, the chairman and general secretary of the WTO called an informal meeting of thirty countries which they selected. They did not inform the hall of this meeting; did not tell who the thirty were or how they were selected. They met for five days . . . [and] decided the whole agenda."[21]

The agenda focussed on how the WTO was going to expand its mandate beyond trade in the years to come. That expansion would include not only investment but two other major areas of economic policy, both part of the transnational corporate agenda. The first was competition, by which the agenda setters meant government "monopolies" — state-owned enterprises and state-sanctioned private monopolies granted their status for public policy purposes. The second mandated expansion would be the elimination of "discrimination" in favour of national and local suppliers in the area of government procurement.

The transparency of the manipulation at the WTO was in large measure responsible for a core of about a dozen developing countries firmly rejecting the investment agreement. Despite this initial victory, Martin Khor is not optimistic. In the high-powered world of WTO politics, the poorest countries are most often ill equipped to see the implications of what they are agreeing to. Says Khor, "The WTO rep visits the country and says you have to change your laws in this way. '*Why*?' Because you signed this agreement. '*But the agreement goes against our constitution*.' Yes, we know — it goes against it in five ways and we have some constitutional amendments for you to take account of that."[22]

Nonetheless, the WTO is still a democratic organization, and any agreement eventually will have to be voted on. Khor believes that the fight against agreements like the WTO's investment agreement will be determined by how quickly and thoroughly developing countries can learn the rules of the game. "If [developing countries] are saying 'Liberalization has gone too far' but the U.S. wants to push it forward they will, because if [less developed countries] are not prepared they will not even know how to articulate that you want to push it backward."[23]

The failure of the WTO track for a global investment treaty put the ball firmly back in the OECD court, where the U.S. government and American-based TNCs had wanted it all along. Given the secrecy surrounding the affairs of global corporations, few people in any of the OECD countries had ever heard of the agreement. In Canada word didn't get out until a couple of months before the 1997 federal election, just two months before the deal was supposed to have been signed.

CORPORATIONS = NATIONS

The MAI confers on transnational corporations comprehensive political powers that they have not had since the early nineteenth century. This transfer of power is a deliberate act of nation-states; a collaboration between states and TNCs that in effect sees a willing surrender of authority from one institutional structure to another, within a single elite class that increasingly sees itself as international.

As Tony Clarke pointed out in his analysis of the MAI for the Canadian Centre for Policy Alternatives, the MAI throughout its text refers to investors when it means corporations. The agreement confers upon corporations exactly the same political status as nation-states. The precise expression of this status is still being debated, but some of the delegations "go so far as to propose investors (i.e. corporations) and the 'contracting parties' (i.e. governments) be given the same definition in the MAI." Provisions dealing with how corporations entering Canada will be treated serve "to establish corporations as having a superior class of citizenship rights."[24]

Canadian corporations already have extensive citizenship rights in Canada, but under the MAI foreign corporations operating in Canada could be afforded more favourable treatment than even domestic corporations. This is because the agreement provides for treatment of such corporations that is "no less favourable" than that given to Canadian companies. This treatment must also provide foreign corporations "equality of competitive opportunity." This means that governments cannot provide Canadian corporations with any unfair advantages. But in practice Canada may treat foreign corporations *more favourably* than domestic corporations to induce them to invest here.

The final piece to this revolutionary set of rules is that Canada will not be able to impose performance requirements on foreign corporations — things like job content, use of local suppliers, transfer of technology — even if it does so with domestic companies. In other words, if the Canadian government wants to use performance requirements to pursue social, employment, or environmental objectives it will have to discriminate *against* Canadian corporations to do so, holding them to a higher standard of corporate responsibility than it can foreign corporations.

The majority of OECD members, including Japan and the U.S., want the MAI to apply to all levels of government so that provincial and municipal governments would face the same dilemma. This has enormous implications for provincial constitutional powers, which, if there is a conflict between these and the provisions of the MAI, will be superseded by the MAI.

One of the most striking aspects of the MAI is the degree to which it explicitly shifts power away from communities and into the hands of corporations. Historically, one way that governments have attempted to ensure that citizens benefit from the investment of capital is the myriad of regulations that are a prime target of the MAI and of NAFTA before it. But the other, more direct method has been the use of direct capital investment by governments.

Crown corporations in Canada have played a vital role in building communities and defining the nature of the country. Power and natural gas utilities, publicly owned telephone companies, the post office, joint ventures with private corporations, ferry services, airlines, bus companies, the CBC and other cultural agencies — there are hundreds of examples of crown enterprises that provided services that corporations either refused to provide or would have done so at much higher cost.

In Saskatchewan, telephone service, electricity, and natural gas reached the remote areas many years earlier than if a private corporation had been in charge. All of these people's corporations had policies of purchasing and hiring locally where possible. As well, no private bus service would have provided transportation to the hundreds of small towns that were a feature of settlement in the west; the government bus service has provided subsidized service to small communities for decades. When the potash industry brought lengthy court challenges to government increases in taxes and royalties in the 1970s, the Saskatchewan government nationalized the industry and subsequently made hundreds of millions of dollars by developing a resource that belonged to the citizens of the province. (The Tory government of Grant Devine privatized it, selling it for half its assessed value.)

The MAI will make such community investment next to impossible. The WTO has identified crown corporations, pejoratively called state monopolies, as its next target for a multilateral deal, but the MAI begins the process. Severe restraints are placed on what governments can do with their own enterprises and assets. Governments could no longer treat these enterprises any differently than they do foreign corporations. They could not instruct them to buy local goods, nor could they establish noncompetitive prices for their goods or services. They would be required to act "solely in accordance with commercial considerations" and not with a public purpose.[25] Had this rule been in place in Saskatchewan in the 1940s, none of the public service enterprises just described could have been established. In effect, the MAI wipes from the map any competition that nations or communities might present to TNCs.

If a government privatizes a state enterprise, the MAI requires that the "national treatment" principle be applied so that any foreign corporation be allowed to bid on the assets. Some OECD members are pushing for a rule that would also prevent any "special share arrangements" that allocate a certain percentage of shares to the general public or that allows employees or the community to buy the company. At every turn, the MAI demonstrates the ruthless determination of transnational corporations to eliminate any measure that would guarantee either democratic control or a community benefit from an investment decision.

The corporate authors of the MAI have left nothing to chance. Other rules give corporations unprecedented legal protection in terms that would have pleased the merchant colonial states of two centuries ago: no government is permitted to "impair . . . the operation, management, maintenance, use, enjoyment or disposal of investments in its territory of investors of another Contracting Party."[26]

The most powerful aspect of the MAI is its mechanism for enforcing all the corporate rights it establishes. What is called the investor-state dispute mechanism, like NAFTA, provides an almost unlimited scope for corporations to sue a government over any breach of the provisions "which causes (or is likely to cause) loss or damage to the investor or his investment." Even "a lost opportunity to profit from a planned investment" would be just cause for a corporation to bring suit against a government.[27]

A sovereign government would have no choice but to appear before the tribunal set up to hear the corporate complaint, regardless of how frivolous or vexatious that complaint might be. The tribunal judges the case exclusively on the basis of the MAI provisions, with no reference to the laws of the sovereign state. The decisions of the unelected and politically unaccountable panel are binding and are enforced "as if it were a final judgment of [the country's] courts."[28] A footnote to this provision explains that it is intended to ensure that no government declines to accept a judgment on the basis that it would be "contrary to its public policy."

Not satisfied with having countries sign a deal giving up huge chunks of their sovereignty, those drawing up the MAI have built in "political security" provisions, which ensure that even without amendments or additions to the MAI it will continue to liberalize investment beyond the terms already signed by the governments. The MAI contains two provisions, one called rollback and the other a "standstill" provision, that together produce a ratchet effect, clawing away at government regulations and democratic authority long after the ink has dried on the deal.

The rollback provision states that "any regulatory measures of nation-states which do not conform with the principles and conditions of the MAI are to be reduced and eventually eliminated."[29] Even current levels of corporate taxation could be challenged as "creeping expropriation," in other words the illegal seizure of a company's "property" in the form of its rightful return on investment.

The standstill provision commits governments to "agree not to introduce any new non-conforming laws, policies or programs." For example, if a newly elected social democratic government wanted to regain some control over its economy through even a modest amount of public ownership, it could not do so. If it wished to reintroduce environmental regulations to address a growing crisis, any corporation affected by the move could prevent it.

Of all the ground-breaking aspects of the MAI, the most extraordinary is the provision providing for cancellation, or, as the agreement says, "withdrawal." Once the deal is signed, "Contracting Parties" (even the language suggests countries are now just parties to a commercial contract) will have to wait five years before they can withdraw. In addition, all investments made during those five years would are governed by the agreement for a further fifteen years. In short, if this agreement is ratified by the twenty-nine countries of the OECD, the democratic rights of their citizens will be suspended for an entire generation. At least with the FTA and NAFTA, the deals could be cancelled with six months' notice.

What do Canadians get in return for this unprecedented loss of sovereignty? According to its promoters, more of the same benefits we were supposed to achieve through NAFTA — that is, increased foreign investment. But the record of the FTA and NAFTA demonstrates that the promise of these agreements has not been fulfilled. According to Simon Fraser University political economist Marjorie Cohen, the trade deals have led to a huge increase in Canadian investment abroad. The outflow of direct and portfolio (stocks and bonds) Canadian investment was around 5 percent of total business investment in 1993; by 1997 it was 14 percent. There was a net outflow of capital from Canada in 1996 of $3 billion in direct investment and $8 billion in portfolio investment. Says Cohen, "Even when direct investment occurs, it does not necessarily improve the productive capacity or employment levels in Canada . . . Direct foreign investment most often takes the form of acquisition of existing corporations."[30]

THE CORPORATE BLUEPRINT FOR THE MAI

The MAI is so transparently a corporate rights document that it scarcely matters who determined its main features. Yet the key document that determined those features, obtained by the Council of

Canadians in late 1997, revealed just how integrated the developed nations and the TNCs have become. Entitled "Multilateral Rules for Investment," it was produced by the International Chamber of Commerce (ICC).[31]

The document was, in effect, the first draft of the MAI. All of the key aspects of the MAI described above are present in the ICC document. Amongst the critical elements was the description of a global marketplace in which "trade and investment become indistinguishable parts of single strategy except to the extent that national regulation demands distinction."

Under "Requirements for an International Investment Regime," the now famous phrase "high standard" agreement is front and centre. The ICC calls for a "broad definition of investment"; national treatment; an extremely broad definition of what corporations can "repatriate" without interference; and a dispute settlement procedure in which the corporation gets to choose the type of arbitration rules.

As is the case in the MAI text, the ICC insists on very broad definitions of the terms of the agreement, including "investor," "investment," "returns" on investment ("profit, capital gains, interest, dividends, royalties, and fees"), and the term "national," which should include not just persons but any "corporation, trust, partnership, joint venture" of the party in question. The ICC document also calls for the MAI to apply to all subnational levels of government. It provides many of the terms found in the 1997 drafts of the deal — including "standstill" and "rollback" — proposes tax policy as a barrier to investment, and contains the principle that "the MAI should prohibit screening for economic policy purposes."[32]

Corporations have driven the MAI from the beginning but were able to do so only because there is now almost no difference between global corporations and those aspects of state dealing with trade, investment, and foreign affairs. Merging these elements of the state and the business organizations representing TNCs results in an almost seamless institutional regime. The corporatized market state systematically excludes any other elements of civil society and therefore excises any internal dissent or debate. Thus, when business speaks there is no civil-society filter for the message to go through; it is in effect received wisdom. This truncated version of the state, effectively reduced to bureaucrats from finance, trade, and foreign affairs, has the same objectives as the TNCs it deals with.

The MAI is arguably the most important "law" Canada will be faced with in its history. Yet Canadians have had absolutely no say in the negotiation of this treaty. A bureaucrat, Bill Dymond, of the International Trade Branch of the foreign affairs department, speaks for all of us when he sits at the OECD table for Canada. He never consults with ordinary Canadians and speaks only to a very select group of elected officials. He gets his direction mostly from other unelected bureaucrats. To be sure, it is the minister in charge and the cabinet that will officially make any decision. But even here it is trade bureaucrats, committed to liberalization, who advise the ministers in charge. These men and women are politicians who, with little or no knowledge of trade matters, are totally dependent on the expert advice they receive. These advisers are preoccupied with trade and investment issues; ministers and officials of other departments who might be able to warn of the social and environmental consequences of the MAI are not involved. For example, when the Council of Canadians contacted Heritage Minister Sheila Copps's office about the MAI in the spring of 1997, she and her staff had not heard of the agreement.

The perception that the MAI is just another trade deal, of interest only to narrow interests, permeates the department of foreign affairs and international trade. Nicole Bourget, an assistant to then trade minister Art Eggleton, revealed just how disconnected the government is from any notion of accountability. Bourget declared that "there's nothing secret" about the talks, which were in fact going on behind closed doors. Casually confirming exactly what the MAI's critics were saying, Bourget told the *Globe and Mail* that negotiations for other trade deals "proceeded in a similar fashion. Ottawa regularly consults the provinces and has met and sought advice of industry groups, which will also be affected by any agreement."[33]

The apparent innocence of this response is more frightening than one with a public relations gloss. It demonstrates that the lack of awareness regarding who will be "affected" by such deals and the resulting absence of any understanding of public accountability is now firmly entrenched in the public service.

Indeed, while Canadians were kept in the dark about the agreement, the government was busy briefing Canadian business organizations throughout the entire process. One of those organizations was the Canadian Chamber of Commerce. On the occasion of Art Eggleton's appointment as minister for international trade, the chamber's chairman, Gary Campbell, and president, Timothy Reid, wrote to congratulate him. They also took the opportunity to lobby him with the chamber's position paper on the MAI and thanked a trade department official, Phil Somerville, "for the thorough briefing on the MAI that he provided to a select group of our corporate members."[34] Over a year later Eggleton would publicly deny that any such agreement was even being considered. In April 1997, just weeks before the deal was to have been signed, he told *Maclean's* magazine, "It's too early" to debate the agreement.[35]

Eggleton played down Canada's role in the MAI, but his protests were undermined by others closer to the scene. Reinforcing the claim that the politicians are just cheerleaders in the process, Eggleton denied any special role, saying, "As far as I know — I am sure somebody told me this not too long ago — we did not initiate this matter." Not so, according to Alan Rugman, a University of Toronto business professor who prepared the 1995 OECD background study. Rugman claims that Chrétien himself gave the MAI his okay at the G7 meeting in 1995. Donald Johnston, a former Trudeau cabinet minister, now heads the OECD, arguably the most aggressive promoter of trade and investment liberalization of any international agency. Said Rugman, "There are Canadian fingerprints all over the MAI. The untold story is that we're the real heroes getting it going."[36]

ASIA PACIFIC ECONOMIC CO-OPERATION

The multilateral investment initiatives are only the most grandiose schemes being undertaken by transnational corporations. If they cannot push their way through one door they will try another, and another, until their global regime of corporate libertarianism is in place. The most comprehensive efforts outside the formal jurisdiction of either the OECD or WTO is the Asia Pacific Economic Co-operation (APEC) forum, a free-trade initiative encompassing eighteen countries, including Canada and the U.S.

APEC is the free-trade equivalent of a floating crap game: it does not involve binding agreements or enforcement measures and does not have a large bureaucracy like the OECD and WTO. It exists only as an annual series of ministerial meetings prepared by a small secretariat in Singapore. But that is where the dissimilarities end. With its secrecy, its unabashed adherence to the imperatives of corporate rule, the source of its policy advice, and the nature of its ideology, APEC is very much a cousin of the MAI.

◆

APEC is made up of Malaysia, Singapore, Indonesia, Thailand, the Philippines, Brunei, Canada, Australia, New Zealand, Chile, Mexico, China, Taiwan, Hong Kong, Japan, Papua New Guinea, South Korea, and the U.S. According to New Zealand political analyst Jane Kelsey, "APEC has always been market driven and is heavily influenced by big business and the private sector free marketeers. It mainly relies for its research on the tripartite think tank of business representatives, academics and [state] officials . . . known as Pacific Economic Advisory Council."[37]

Its members have committed themselves to trade and investment liberalization. Its agenda includes a long list of "freedoms" for corporations, from deregulation, privatization, market-driven services, minimal controls on resource exploitation, and completely unrestricted foreign investment. Indeed, APEC has added a new twist to the new-right manipulation of language by referring to its members as economies rather than countries, a convenient fiction that allows them to deal with each other as if none of them actually had any responsibility for flesh-and-blood citizens. As TNCs become increasingly like states in their reach and power, and states devolve their role to being corporate facilitators, the distinctions between them begin to blur.

APEC did not start out as a free-trade group. It was formed in the late 1980s as a counterweight to the emerging trade blocs in Europe and North America. It was intended as a regional forum for economic cooperation. The U.S. was not initially invited (neither was Canada) but forced itself in and in 1993 used its influence to see APEC develop its current investment liberalization direction. Canada has been a major booster of this agenda and, with the other English-speaking developed nations, has been pressing for a more legalistic, binding approach. For this reason the leading powers in APEC have insisted that the IMF and World Bank be a part of the deliberations.

Canada's commitment to corporate libertarian principles is contained in a document entitled "Canada's Individual Action Plan," which details Ottawa's efforts at compliance with the goals of APEC.[38] Among the commitments under the heading "Deregulation" is the Regulatory Efficiency Act (REA), which the Chrétien government withdrew in 1996 after considerable public pressure. The REA is designed, says the document, "to encourage sensible regulation conducive to business growth thereby enhancing Canada's global competitiveness."

In its application, the REA allows for the waiving of environmental impact hearings, compliance with existing pollution limits, and other environmental regulations. This same approach, says the "Action Plan," will be applied in other areas. "When regulating, authorities must ensure that government intervention is justified, that Canadians are consulted, that benefits outweigh costs, that adverse economic impacts are minimized."[39] Just which Canadians will be consulted is not made clear, but the trend suggests a narrow definition of those "affected" by trade and investment matters, that is, corporations.

The "Action Plan" reveals a whole variety of changes to the way governments work, changes that Canada is committing itself to in the total absence of any public consultation or even knowledge. For example, it reveals that Canada is working on a series of bilateral deals that will protect the interests of investors, implicitly protecting them against government action. Eighteen of these Foreign Investment Protection Agreements have already been signed and others are being negotiated.

Throughout the "Action Plan" the government makes clear that APEC negotiations and principles are being coordinated with the WTO and OECD initiatives on multilateral investment agreements. These coordinated and reinforcing multilateral deals paint a picture of a government hell bent on surrendering the most important aspects of Canadian sovereignty as fast as it can. Canada now applies the high

NAFTA investment-review thresholds (the level of investment at which review kicks in) to all WTO members. Ottawa is also busy signing deals and making unannounced commitments to liberalize government procurement regulations — in short, reducing government's ability to purchase goods and services from Canadian suppliers. The authority and power of governments to enhance their communities are being rapidly eroded through agreements that are, as far as the public is concerned, clandestine. Only a handful of Canadians, excluding most parliamentarians and premiers, have heard of the Bogor Declaration, the Ministers for Cartagena, or the Osaka Action Agenda.

As with all of these multilateral agreements, the negotiations nominally take place between governments, masking the fact that the exclusive beneficiaries are corporations. Evidence does occasionally emerge, however, of just how crucial these deals are to large corporations. The merging of the market state and corporations was revealed in the days leading up to the APEC summit in Vancouver in 1997. Sixty-seven corporations contributed $9.1 million towards the $57.4-million cost of the summit. Seven corporations contributed half a million each: Canadian Airlines (infamous for squeezing its employees for wage concessions), TD Bank, Federal Express, BC Tel, Nortel, General Motors of Canada, and the Export Development Corporation. Other big contributors included the Royal Bank ($250,000), Panasonic ($250,000), and SunLife ($320,000). These contributions, raised by former TD Bank chair Richard Thompson at Prime Minister Chrétien's request, are a powerful reminder of who APEC is intended to benefit. Equally important, they strengthen the hand of these huge corporations in their efforts to control the outcome of negotiations. Whoever pays the APEC piper calls the APEC tune.[40]

There are other initiatives in corporate libertarianism in various regions around the world. Some are just multilateral agreements like the recent one deregulating telephone services (expected to dramatically increase local-use charges). Others are corporate-state initiatives meant to drive the free-trade initiative faster. This is the case with the Enterprise for the Americas Initiative (EAI), announced by George Bush in 1990 and aimed at creating a hemispheric free-trade zone and liberalized investment and debt restructuring. The U.S has already signed more than thirty framework agreements with Caribbean and Latin American governments. If they wish to get on the NAFTA train, each country must commit to "opening markets in goods and services; removing barriers to investment, and safeguarding intellectual property."[41]

The U.S. is prepared to sign "mini-agreements" on investment and intellectual property rights with Latin American states while it continues its efforts to get a hemispheric free-trade deal. Chile, one of the most aggressively market-oriented countries in the Americas, has already joined a free-trade deal with Canada. These bilateral deals represent the second of a two-track strategy and became more important in late 1997 when President Clinton failed to get approval for so-called fast-track authority for deals like the EAI and MAI.

Common to all of these deals and potential deals is the complete absence of any reference to social, environmental, or labour standards or human rights. In the developed countries, governments have responded to this criticism with what are now commonly referred to as "side deals." These usually concern labour and the environment, the two areas where the protest has been the most effective, and a social charter. So far, under the FTA and NAFTA, the side deals have proven ineffective, though supporters suggest the jury is still out.

There is a fatal flaw in the notion of social charters being attached to liberalizing multinational agreements. Pressing for improved labour, social and environmental standards fundamentally contradicts the whole point of these deals, which is to lower costs and investment barriers to transnational corporations. Free trade and investment demand deregulatory requirements that make social protection ideologically and politically unacceptable. It would be difficult to make them effective precisely because, to the extent that they are effective, they undermine the purpose of the agreement and its corporate libertarian philosophy. You might just as well try to add rabbit-like features to a fox to make it behave more like a rabbit.

The same is true of international environmental protection agreements. Unlike trade and investment regimes, they have no teeth. The Rio Summit of 1992 is a good example. The record of governments "promising" to meet environmental goals established in Rio de Janeiro is appalling because there are no consequences for breaking the promises. And that's how corporations wish to keep it. It is for this reason that they insist on having such issues dealt with, if at all, in side deals rather than integrated into the main agreement.

As for the willingness of corporations to entertain even these toothless "protections," the heavy hitters behind the MAI have drawn a line in the sand. The U.S. Council for International Business, the most powerful lobby speaking for U.S. TNCs, wrote to senior U.S. officials in 1997. "The MAI is an agreement by governments to protect international investors and their investments and to liberalize investment regimes. We will oppose any and all measures to create or even imply binding obligations for governments or business related to environment or labor."[42]

Viable or not, side agreements on labour and the environment have not been pursued by Canadian negotiators in MAI negotiations. In late 1997 it was revealed that Canada was not among those nations speaking up on the issue during the most recent negotiations. Instead, the government accepted weak and nonbinding language in the preamble of the agreement. This is not the only area in which the Canadian government is failing to protect the country's interests. In response to charges that the MAI threatens medicare, education, and social services, the government declared that key areas of the economy and social infrastructure will be listed as Canadian "reservations," allegedly protecting them from the terms of the agreement.

But according to a study by trade lawyer Barry Appleton, commissioned by the Council of Canadians, the way the specific Canadian reservations are worded makes them virtually meaningless. Appleton, who is acting for the Ethyl Corporation against Canada on the MMT case, pointed out that reservations in the context of international treaties are always interpreted restrictively. "Reservations are a form of exception to international treaty commitments and are interpreted narrowly by international courts and tribunals."[43] As well, the definition of "measures" to be reserved in the MAI is itself very narrow compared to the definition in NAFTA, which explicitly lists "laws, regulations, procedures, requirements or practices."[44]

Appleton stated unequivocally that, judged on this basis, Canada's reservations intended to protect social services, health care, and public education are, because they protect only the federal government's programs, "inadequate to permit provincial governments to provide these services without compensating affected foreign investors."[45]

As worded, the reservations would protect only public law enforcement and correctional services.

The environment too is left exposed. According to Appleton, Canada has "chosen to voluntarily bind itself, its provinces and its municipalities to obligations which protect investments over the environment."[46] And culture is left even more vulnerable, since the government made no reservations at all in this area. Appleton's report concluded: "Any policy or program that advantaged Canadian culture or content directly or indirectly, would run afoul of the national treatment or performance requirements obligations."[47] Six areas, including Canadian content in TV, Canadian film distribution policy, and postal subsidies to Canadian publications, require specific reservations; otherwise, they are subject to demands for compensation from foreign corporations.

Reservations to treaties are problematic at the best of times. With the MAI, of course, they also face the rollback clause. In effect, the reservations that are made become a corporate hit-list — conveniently targeted areas of public policy that the government is committed to gradually eliminating.

Transnational corporations, in the process of becoming the most powerful organized force on the planet, have made huge steps towards effectively ruling the world. Corporate influence in public life has always been a reality in capitalist economies, and that influence has always been characterized by a lack of democracy. The last quarter of the twentieth century, however, has witnessed a profound change in the exercise of corporate power and influence.

Going beyond the exercise of influence on individual national politics, corporations are now circumventing the institutions of democracy and creating their own supra-national institutions of domination, institutions that are inaccessible to ordinary citizens, whose laws supersede those passed by democratic legislatures, and whose express purpose is to choke off the ability of communities to control their own destiny. Most alarming of all is the willing collusion of our democratically elected governments in this global counter-revolution.

The Multilateral Agreement on Investment and APEC, like the FTA and NAFTA before them, will make enemies of citizens and their government. By aligning themselves so exclusively with transnational corporations, governments virtually by definition declare a kind of social war on their own citizens, a declaration of acceptance of permanent class divisions as a feature of modern society. Most of what government does, for ordinary citizens, involves moderating the unequal effects of the marketplace. If we see democracy as principally a process of and for equality, then the fruits of democracy can be achieved only by exercising significant control over the market economy.

Transnational corporations have transcended the national governments that gave them identity and corporate status, but only with the aid of those governments. If nations can sign multilateral agreements ceding their individual sovereignty to the so-called larger purpose of international trade, they could create international agencies of democratic governance. Yet no major nation-state has proposed an international tax scheme, an enforceable labour code, an enforceable environmental code, or international standards for health and education assured by international financing. What explains the fact that nation-states are willing to surrender sovereignty to transnational corporations but not to a world government? Only that the economic and political elites that currently rule the developed nations are complicit in the development of corporate rule over their own citizens.

The speed with which global corporate rule is being implemented is breathtaking; the eagerness of democratically elected governments to dismantle what they have constructed over three generations is just as stunning. The headlong rush into a totally deregulated world is a kind of social madness or, as John McMurtry says in his book *The Global Market as an Ethical System*, a cancer, eating away at a body politic whose immune system no longer recognizes the mortal danger. We are creating ever greater poverty at home and in less developed countries, contrasted by ever greater obscene wealth concentrated in fewer and fewer hands. The rulers and the ruled have not been divided by such an economic, social, and moral gulf since the Great Depression.

6

RULERS AND RULED IN THE NEW WORLD ORDER

The total cost of the 700 social events laid on for delegates during the single week was estimated at $10 million . . . A single formal dinner catered by Ridgewells cost $200 per person. Guests began with crab cakes, caviar and crème fraiche, smoked salmon and beef Wellington. The fish course was lobster . . . the entrée was duck with lime sauce, served with artichoke bottoms filled with baby carrots. Dessert was a German chocolate turnip sauced with raspberry coulis, ice cream bonbons and flaming coffee royale.
— JOURNALIST GRAHAM HANCOCK, DESCRIBING A JOINT IMF–WORLD BANK MEETING IN WASHINGTON, IN *LORDS OF POVERTY*

Potential troublemakers were threatened with arrest and violence. The Philippine government put out a blacklist of undesirable anti-APEC visitors, and turned away, among others, the east Timorese activist and Nobel Peace Prize winner Jose Ramos Horta, for fear of upsetting Indonesian President Suharto. The shanties of over 400,000 families were demolished in an attempt to create an "eyesore free zone" for visiting dignitaries . . . Their land, which has been confiscated . . . is slated for the largest super-mall in the world, which will house the first Disneyworld of the region.
— MAUDE BARLOW, COUNCIL OF CANADIANS DELEGATE TO PEOPLE'S SUMMIT ON APEC, 1996

Ruling elites and demonstrations of their power are as old as human organization. itself, and in modern times those elites have been easily identified, whether in communist, capitalist, democratic, or authoritarian regimes. Yet the elites in the age of transnational corporations are playing a particular role. Increasingly we are witnessing an international ruling elite developing apace with international capital and the institutions of global governance such as the WTO, the OECD, and the multilateral deals they design and implement.

As we will see in the next chapter, underlying this global ruling elite is an unprecedented political and ideological consensus. The speed with which the new order is being put in place owes a great deal to this consensus. But for the flesh-and-blood individuals who constitute that elite it also has to do with how they make their money, the things they buy, and who they identify with and have contact with. Their wealth depends on giant corporations and currency speculation, their work has been internationalized, and they take comfort in the decline of civil society by agreeing, with a hint of regret, that the great social experiment in egalitarianism simply didn't work.

The rich are different from us. Of course, they have always had more money and have always gained it at the expense of the huge majority who don't have as much, or any. But today they are even more different from us. The rulers of the world are not only are getting obscenely rich. They are getting rich through increasingly unchallenged corporate rule and through the exponential growth of the power of transnational corporations, both financial and political.

There are now so many billionaires in the world that *Fortune* magazine's list no longer names them all. So many are added each year that only the "Top" 200 are now rich enough to warrant being listed. The 1997 list nearly surpassed 500; in 1996 it was 447. In 1995 there were "only" 357 billionaires. Their net worth was $760 billion, more wealth than the bottom 45 percent of humanity. That is, 357 people in the world owned more combined wealth than 2.7 billion other people. And this, we are told, is the result of the "market," the invisible hand by which each individual's pursuit of wealth is alleged to benefit the whole of humanity.[1]

And the billionaires are just at the very top of the human pyramid. Below them are tens of thousands of multimillionaires and millions who have assets of more than a million dollars. The inequality is staggering. It is, nonetheless, portrayed in the popular culture as natural.

The egalitarian state, on the other hand, has always been denounced by free marketeers as social engineering.

Yet what the world has experienced over the past twenty years is a massive and accelerating transfer of wealth from the poor to the rich, a social engineering project matched only during colonial times. When wealth is created it is divided between those who labour and those who invest; between owners and workers. And everywhere that division is becoming more unequal.

This is just as true with nations as it is with individuals, with the less developed world every year pouring billions of dollars into the pockets of the elites in the developed world. It is also true within developed countries. A 1993 U.N. Human Development Report revealed that in 1960 the richest fifth of the world's population was thirty times better off than the poorest fifth. The latest estimate is that they are now 150 times better off.[2]

In Canada the situation parallels the international scene. *Financial Post* magazine reports that the wealthiest fifty Canadians now have assets of $39 billion, equivalent to two and a half times the total income of more than five million Canadians who earn less than $10,000 a year.[3] In 1993 alone, according to Statistics Canada, the top 30 percent of Canadian families "took an extra $14.3 billion over and above the amount they would have received had their share of the income pie not increased since 1973. And virtually all of their income was siphoned off the bottom 50 percent."[4]

The wealth transfer has been accelerating, with the share going to the wealthy growing two to three times faster from 1987 to 1993 than from 1973 to 1987. A November 1997 Ernst and Young survey revealed that Canada was home to 220,000 millionaires, an incredible threefold increase since 1989. The number is expected to triple again by 2005. (The millionaire status was determined by liquid assets and did not include home equity or pension plans.)[5]

The transfer of wealth has a variety of sources. Much of it, according to Ernst and Young, comes from inheritances, which in Canada are not taxed. (Canada is the only OECD country, other than Australia, without such a tax.) In addition, the speculation-driven stock market has made millionaires out of many who used to consider themselves middle class. Ironically, a large part of the new wealth comes from interest on government debt, paid by average taxpayers to the wealthy who hold government debt. The incomes of managers and professionals have gone up much faster than average earnings. The wealthy have also benefited from the years of high interest rates on their assets. Tax breaks and lower marginal tax rates have helped, too. After two decades of tax reform Canada has a virtually flat tax system, where everyone pays the same rate. A Carleton University study revealed that, taking all taxes into account — income, sales, payroll, property, and others — those earning between \$100,000 and \$150,000 pay 32.6 percent of their total in taxes; those earning between \$40,000 and \$50,000 pay 34.1 percent, and those earning less than \$10,000 pay 30.1 percent.[6]

The ruling elite consists of those who make the most important decisions that affect our daily lives. What aspects of life do we normally consider the most important? At the top of the list for most people is the ability to make a decent living. Living in a community and neighbourhood in which you feel not only safe but a part of that community is a key aspect of daily life. So too is access to safe food, clean air, drinkable water, and an environment that is generally free of threats of any kind. Housing, transportation, and access to services such as medicare and education are also high on the list.

When we talk about who "rules," we necessarily contrast how many of these aspects of daily life are determined by government, over which ordinary people have some nominal influence, and how many are determined by corporations and, more specifically, those who run them. Even a cursory examination of what government does reveals its efforts to moderate the "normal" behaviour of corporations in the economy either by regulating them or by providing public services that corporations won't provide equitably.

For most people the ability to make a living depends on whether private capital invests in some activity. Yet how capital behaves in providing that living is not entirely in its hands. Labour laws have something to say about child labour, working hours, minimum wages, employer deductions for pensions and EI, protection of employees' health and safety, and last but not least their right to form a union to speak for themselves collectively. A similar list could be drawn up for other areas of daily life.

But when government withdraws from these areas, or cuts back its involvement, it does not mean that no one is left governing. When government abandons its responsibility to create jobs or loosens regulations about working conditions or passes "regulatory efficiency" laws that bypass environmental protection, then the economic elites simply increase the social territory over which they rule.

Countries maintain the trappings of democratic, popular governance while CEOs of large corporations, hidden from public view, seize more and more of the ground of governance and make more and more of the decisions that affect our daily lives. This shift from democratic governance to corporate rule takes place incrementally, each step carefully rationalized by the purveyors of market ideology. The state still governs, but corporations rule.

It is not just the decisions these individuals make in their capacity as corporate executives that constitute corporate rule; neither is it just those at the pinnacle of the corporate order who are part of the ruling elite. This corporate upper echelon has always made decisions that affect the lives of millions of people. Corporate rule's "governing" elite involves two other dimensions. First, the role that the truly powerful play goes beyond the decisions they make for their corporations. Their role in lobbying government and political parties, their support for and involvement in think-tanks, their networking with other elite members through company directorships, their personal alliances with national and international corporate leaders, their private and occasional public intervention on policy issues, all of these activities extend their ruling capacity beyond the confines of the corporate suite.

The second dimension of the ruling corporate elite is its wider base, today including those who share the power, influence, and financial benefits with the top 1 percent of the truly powerful. The large numbers who populate the managerial class of corporations and the law firms, accounting firms, brokerages, and financial houses that make up the modern corporate world are all part of that elite, sharing in the privileges and adopting the perspectives and depending on the dominance of those at the top. Add to that the political class of senior university administrators and academics, newspaper publishers and other guardians of information, senior bureaucrats and senior cabinet ministers and you have a conglomeration of people who constitute the ruling elite.

This class has always been integral to the functioning of a liberal democratic society. But two things have changed in the past twenty years. First, there is stronger consensus about the direction of the state than there has been for decades, and it is radically different from the one that arose in the post-war period. That consensus amounts to a virtual counter-revolution against the egalitarian state, a determination to turn back the clock on social programs, regulatory regimes, development aid, human rights, and the commitment to social equality that has been the definition, or at least the operating myth, of the Canadian nation-state for almost three generations.

The second change, and indeed part of what drives the radical new consensus, is the international perspective of this reconstituted ruling elite. The attachment to market ideology is at least in part a rationalization for the source of the new elite's vast new economic wealth and its future well-being. With the advent of free-trade deals, the power of transnational corporations to manipulate the world economy, and the wealth to be made in stock and speculative markets, the rulers have lost interest in their local domain. Their dominion, like the mercantile class of the eighteenth century, is now the world.

THE SECESSION OF THE SUCCESSFUL

One of the keys to the development of this neo-mercantilist elite is how they make their money. Those who work directly in the highest ranks of the private sector not only engineer profits for their companies through the international markets in goods, services, and currencies. They are in many ways more linked to that international world than they are to their own communities or country. Robert Reich, a Harvard economist who became Bill Clinton's labour secretary, describes the new elite this way: "The highest earners now occupy a different economy from other Americans. The new elite is connected by jet, fax, modem, satellite and fiber-optic to the great commercial and recreational centers of the world but not particularly to the rest of the nation."[7]

Not all countries have devolved their civil societies to the extent that the U.S. has, but it is instructive to examine the American experience as a window on the future everywhere if current trends continue. In a piece for the *New York Times Magazine* called "Secession of the Successful," Reich describes the work of the American equivalent of this elite as symbolic analyzers. "Most of their work consists of analyzing and manipulating symbols — words, numbers or visual images. Among the most prominent of these 'symbolic analyzers' are management consultants, lawyers, software and design engineers, research scientists, corporate executives, financial advisers, strategic planners, advertising executives, television and movie producers, and other workers whose job titles include terms like 'strategy,' 'planning,' 'consultant,' 'policy,' 'resources,' or 'engineer.'"[8]

It is a long time since Henry Ford realized that his new mass production methods were so efficient that he was making more cars than there were Americans who could afford to buy them. His solution, to double the wages of his workers so they *could* afford the sticker price, was dubbed Fordism. Reich's anecdotal version is the rich industrialist on the hill who worries about his workers down below — because they produce his wealth through their work. No work, no wealth. That scenario is pretty much history. In the post-Fordist world, the ruler on top of the hill doesn't care if the ruled at the bottom works; he doesn't ever see him. His wealth, in any case, is more likely produced by the labour of a Mexican or Filipino or Thai worker whom he doesn't see either. He doesn't even care if that worker buys his product so long as some middle-class consumer somewhere buys it.

Reich argues in his article and in his book *The Work of Nations* that the "fortunate fifth," the wealthiest 20 percent of Americans, were "quietly seceding from the rest of the nation." That trend is now highly advanced in the U.S. Reich cites many examples, beginning with the fact that the wealthy now pay only a fraction of the taxes they used to pay, decimating public finances and services. They can thus justify shelling out even greater amounts of money for private services provided in what have become known as gated cities. Everything from libraries, street repair, fire service, and police are now privately funded. Much of America is becoming characterized by a kind of social apartheid. The elite "feel increasingly justified in paying only what is necessary to insure that everyone in their community is sufficiently well educated and has access to the public services they need to succeed."[9] The social solidarity that characterized the recent past is being replaced by class solidarity at the top.

Large U.S. urban centres are being "splintered into two separate cities. One is composed of those whose symbolic and analytic services are linked to the world economy" while the other is the whole panoply of service workers who depend on the "other" city for their employment. Most blue-collar jobs have disappeared from big cities. Between 1953 and 1984, New York lost 600,000 manufacturing jobs while gaining 700,000 "symbolic analyst" and service jobs. Wealth-creating jobs, which used to link the social classes, have gone elsewhere.

The symbolic analysts no longer even live in the cores of the big cities. Many, connected from home to the centre of trade and finance around the world, live in pastoral settings far from the eroding cities and the lives of their marginalized fellow citizens. Those still living in the urban centres create their own cities within cities. "One such New York district, between 38th and 48th streets and Second and Fifth Avenues," writes Reich, "raised $4.7 million from its residents in 1989, of which $1 million underwrote a private force of uniformed guards and plainclothes investigators."[10]

The trend to private security is now so advanced that there are three times as many private cops as there are public police. And rather than deal with the poverty of the "other cities" the U.S. is putting people in jail at a record-setting pace. It now jails 2 percent of its working-age population, at a rate of 529 inmates for every 100,000 population.[11] This is nothing more than warehousing of the unemployed, as the record shows that being tougher on crime in this manner has no effect on crime rates.

The use of private cops suggests more than just a "secession" of the elite. It also suggests that they believe they can protect themselves from threats from the underclass they are helping create. Though the private police do not yet have the same powers of coercion as their public counterparts, the lines between them are blurring. The wealthy are gradually seizing the "right" to use violence now exclusively in the hands of the democratic state.

THE ELITE SECESSION IN CANADA

The international orientation of the ruling elite is a generally accurate description of Canada's economy, too; however, because so many TNCs are centred in the U.S., the role of symbolic analyzers is necessarily larger in that country. But what Canada lacks in that global field it makes up for in the export economy. Since the first free-trade deal was passed, the government and the business elite, as represented by the Business Council on National Issues, have focussed almost exclusively on policies promoting international trade, to the detriment of the domestic economy.

The great puzzle of the "jobless recovery" is no puzzle at all when one examines the extent to which the domestic economy has been savaged by government spending cuts, the erosion of income redistribution, the flattening of the tax system, and the loss of more than 200,000 industrial jobs. These are all deliberate government policies, not some globally dictated inevitability. It is not just that the domestic economy is being neglected. It is, as we will see later in this chapter, being subjected to a "controlled economic stagnation" whose purpose is to discipline labour and check inflation to ensure the international competitiveness of the corporations oriented to international trade.

The trade economy, while constantly touted in government and business propaganda as key to Canada's future, is dominated by large corporations, which are the big job destroyers. Despite our loss of jobs, there has been an enormous increase in overall trade: 33.2 percent of GDP went to exports in 1994, compared with 26.3 percent in 1988, while imports increased from 25.8 percent of GDP to 32.5 percent. The majority of that increase has been with the U.S. as the result of free trade. And 90 percent of the trade with the U.S. is accounted for by the two hundred largest exporters.[12]

Robert Reich's description of the secession of the successful does not fit what's happening in Canada, at least not yet. But like so many trends connected to economic development, especially given the pressures for "harmonization" from the FTA and NAFTA, there are clear signs that Canada is headed in the same direction. We can simply look at the U.S. to see where we will be ten or fifteen years from now.

As Canadian budgets for social programs and infrastructure steadily erode it is hard to imagine Canada not following the same path as the

U.S. One of the key developments leading to the "two separate cities" phenomenon in the U.S. is diminishing federal funding to the cities, from 25 percent of their budgets in the 1970s to 17 percent in the mid-1990s. The same trend is developing here as provincial governments, facing cuts in federal transfers, download costs to municipalities. Canada is also following the U.S. trend of putting more of the poor and marginalized in jail.

The evidence that the Canadian ruling elite is seceding is mounting. The anecdotal evidence is everywhere, of course; editorials and columnists in almost every newspaper in the country rail against "big government," and the vast majority of academics on radio and TV panels express the same consensus. Polls show that the elite in this country are gradually withdrawing their commitment to social equality. The social contract "signed" by their predecessors is being unilaterally broken.

A 1994 project launched by the Ekos research and polling group, called "Rethinking Government," interviewed 1,000 of the key decision makers in the country — top civil servants, elected officials, and corporate executives — and 2,500 members of the general public. The results were dramatic. "The comparison between the elites and the general public suggests that . . . a profound gap exists between the public and decision-makers in the area of preferred government values."[13] Given twenty-two value choices for government action, the two groups' attitudes are almost totally reversed. "Competitiveness" and "minimal government" ranked first and third for the elite but at the bottom of the list, numbers 20 and 22, for the general public. Virtually all of the policy values related to equality, social justice, collective rights, full employment and regulation of business, and even personal privacy were low on the elite's list of priorities and high on the general public's.

The overall conclusion of this study was that the elites look at government through the other end of the public's telescope. Everything is attenuated. Not only is everything related to government reduced, it is also purged of its moral content. One finding of the study was the degree to which the elite was "homogeneous in their values and attitudes . . . due to both their shared social class and internal cohesion . . . A chasm exists between those charged with governing and those being governed."[14]

The Ekos poll is far from the only sign of the elite's abandonment of nation building and social justice. One of the results of the elite's resentment of enforced social solidarity is an unprecedented increase in tax evasion. *Taipan*, one of many "insider" financial newsletters, shamelessly promoted illegal tax evasion: "You'll be able to conduct your financial life in absolute privacy and security . . . The government won't know whom you're sending money to or getting money from. The whole transaction will disappear from the face of the earth . . . Those who amass wealth and profits will be able to keep it. The tax burden will shift towards those who make money in the traditional way."[15] A book of advice on tax avoidance, *Take Your Money and Run*, sold eighty thousand copies in Canada.

As of 1995, Canadian wealth sitting in off-shore tax havens already amounted to tens of billions of dollars. By diverting their wealth off-shore, the elites don't have to bother demanding lower tax rates, already slashed during the 1980s; they just don't pay taxes, period.

Secession is well under way. The *Financial Post*, in an article entitled "A System for the Rich," described how millionaires bypass the public medicare system to get the very best treatment immediately from specialist clinics in the U.S.[16] As the elite avail themselves of private services they increasingly resent paying taxes for public services they no longer use. And as public services decline in the face of lost revenue, the wealthy abandon them even more, widening the gap between public and private, rich and poor. And with the principle of universality now breached with respect to pensions, the wealthy feel even more that they pay for programs that they won't benefit from.

The restructured ruling elite has managed to create a broad social base for its continuing counter-revolution. As we will see later in this chapter, the traditional middle class in Canada and elsewhere is fracturing. Once the very expression of social solidarity, it is now a class with no clear social role. Those who have managed to attach themselves to the new economy now identify with the elite and have done extremely well. The result is a portion of society — the most politically literate and class conscious — that now accounts for nearly 40 percent of the population.

The willingness of the elite, and by default their followers, to abandon the historic compromise is rationalized, according to the Ekos study, by a stated belief that the whole post-war project failed to deliver the goods. But the ease with which many in this upper social class savage their fellow human beings goes well beyond any such practical conviction. The ruthlessness of the corporate world and the esteem with which downsizing CEOs are held in that class cannot help but infect the elite as a whole. The moral detachment of money traders and speculators adds to that sense of separateness and secession.

One of these traders, Dennis Levine, eventually imprisoned for insider trading, described how the whole Wall Street culture was like a big game. "When a company was identified as an acquisition target, we declared that it was 'in play.' We designated the playing pieces and strategies in whimsical terms: white knight, target, shark repellent, the Pac-Man defense, poison pill . . . Keeping a scorecard was easy — the winner was the one who finalized the most deals and took home the most money . . . It was easy to forget the . . . material impact upon the jobs, and thus the lives, of millions of Americans."[17]

In the U.S., the twenty-year exodus from the cities, now accelerated by changes in communications, is creating two class solitudes, which increasingly never see or experience each other. It isn't just the elite that gets isolated. The poor do as well, as the high-poverty ghettoes grow at an alarming rate, from four million ten years ago to eight million today. There have always been geographic class divisions, yet the current trend to both extreme poverty and extreme wealth is changing the character of U.S. cities. American sociologist William Wilson, author of *When Work Disappears*, argues that the poor are becoming so hostile that widespread social unrest is increasingly likely within the next ten years.

Yet the elites seem complacent about the possibility of such unrest, confident that increased private security will be enough. A combination of blaming the poor, which prevents a moral response, and a blind confidence in their continued success absolves them from taking any action to address the issues. And why would we expect anything else? The downsizing champions carry on with impunity. They trash the lives of thousands of their employees and the other shoe never drops. With the exception of a mere handful of dissenting voices, mostly private sector, this sociopathic behaviour is endorsed by default.

Canada has not yet created the terrible divisions we see south of the border. But all the ingredients for that class division and potential class violence are now in place in Canada. There is nothing to suggest that the new global-oriented elite here is any more conscious of its history, more aware of its amorality, or any less arrogant in its triumphalism than its American counterpart. Giddy with its success, content to blame the poor and count on the police, the rulers in Canada have left themselves no way out.

THE RULED

The historic efforts at democratic self-determination of anti-colonial governments, socialist Third World states, and the unique one-party democracies of Africa have long since disappeared. Subjected to the geopolitics of the cold war and the pro-corporate machinations of the IMF and the World Bank, dozens of less developed countries were never given the chance to develop civil societies. And now that communism is dead the U.S. has lost interest in democracy and human rights, one of its tools in its cold war battle with the Soviet Union. Many of these nations are, as a result, now governed by the opportunists and thugs recruited by the U.S. to fight communism.

Or they now have peace with no justice, as in the states of Central America where the same economic elites and the same political gangs who spoke for them in the past still run their countries. The poverty, unemployment, and exploitation are exactly the same, and the leaders of the developed world expect the exploited to be thankful that they are not being dragged from their beds in the middle of the night and murdered.

The best the ruled can hope for is that the transnational corporations' expectation that countries play by accepted rules, will mean no surprises. That is the substance of the democracy being promoted by TNCs and their state allies. Perhaps the only benefit of NAFTA for the people of Mexico is that pressure from the U.S. and the TNCs to end rampant corruption and political dictatorship has helped bring a greater semblance of parliamentary democracy to that country. Jack Warnock, author of *The Other Mexico*, says, "The free trade move undermined the old way of doing business in Mexico. TNCs want transparency in dealing with government. They want . . . assurances that contracts will be honoured." Also, Mexico has become a more open society. "The foreign press, business interests, and some governments have felt free to criticize the Mexican government for its corruption, ties with the drug industry, and human rights abuses."[18]

Transparency involves rules, not the conditions under which wealth is created and labour exploited. And that is the irony in this moderate democratic opening. For while it arrives in part because of NAFTA, the very fact of NAFTA and economic liberalization renders the democratic state virtually powerless to improve the lives of its citizens. The corporate and political power brokers of the world are delivering on a decades-old promise of institutional democracy only when it is bereft of any meaning, stripped of its egalitarian essence, and reduced to little more than a service to transnational corporations. Indeed, judging just by economic and social conditions, most Mexicans were far better off before the democratic opening.

Before the signing of NAFTA, Mexico went through a radical restructuring based on the same neo-liberal principles underlying free trade. That restructuring, which followed Mexico's debt repayment crisis in 1982, involved replacing its longstanding import-substitution strategy with an export-driven strategy. In effect, NAFTA was just another step along the road to the neo-liberal game plan. The results for ordinary Mexicans have been devastating.

Despite its much-touted low-wage advantage, Mexico has experienced the same de-industrialization that Canada and the U.S. have experienced since the early 1980s. In 1994, about 60 percent of the membership of the chambers of commerce in Mexico reported that they were in "dire need of assistance." The competition with the U.S. (indirectly, competition with extremely low-wage-based imports from Asia) saw the number of sporting goods manufacturers drop from 150 to 38; 114 drug companies were reduced to 40. In 1993–94, fully 40 percent of Mexican clothing manufacturers went bankrupt. The shift to exports was supposed to help Mexico's debt situation, but in 1992 the country had a balance-of-payments deficit of $18.8 billion. In 1994, the first year of NAFTA, that deficit rose to $28 billion.[19]

The fall-out for the average Mexican worker has been a cruel ratcheting down of living standards. The share of GDP going to wages dropped from 37 percent in 1980 to 24 percent in 1991, and real wages (money wages corrected for inflation) of Mexican workers dropped by 40 percent. A major factor in this catastrophic drop was the decline in real minimum wage from $7.49 a day in 1981 to $2.65 in 1994.[20] The jobs created by NAFTA have actually tended to bring down the average. The overall average industrial wage was $2.61 an hour in 1994, while in the *maquiladoras* (where the number of plants has increased by six hundred since NAFTA was signed in 1993) the average is $1.75 an hour.[21]

The enormous disparities in wealth in Mexico, historically very high, have become even worse with NAFTA and neo-liberal policies. The selling of billions of dollars in state assets has created over two dozen new billionaires; 54 percent of the country's total wealth is now held by its richest thirty-six families. Far from helping to bring Mexico into the "first world" as promised, NAFTA and corporate libertarianism have driven Mexico closer to the bottom of the Latin American ladder.

The race to the bottom that has been unleashed on the peoples of the world has profound consequences for hundreds of millions of workers and their families. The drive to externalize the costs of production is most ruthless in the very countries that are supposed to benefit from off-shore production and investment. The competition for the lowest wages, taxes, and environmental and human rights standards produces predictable results.

According to the International Labour Organization, the most reliable figure on the level of child labour for 1997 was 250 million. This is not just children working for their families' businesses or assisting at home. This is also factory labour, much of it extremely hazardous. It is indentured labour, in which the children's pay is never enough to pay back the employer's "costs" of keeping them alive; it is children making the things Canadians purchase from transnational corporations like Disney, Nike, and dozens of other corporations who contract out their production to the contractor with the lowest bid.

The race to the bottom involves everyone, everywhere. If democratic reforms and workers' fights for better wages improve conditions in one developing country, corporations move to one where labour is cheaper. An exposé of conditions in Thailand followed a fire at a toy factory in which 188 workers, mostly thirteen-year-old girls, lost their lives because the owner violated safety laws. As a result of improved law enforcement, companies moved toy production from Thailand, Indonesia, and the Philippines to plants in free-trade zones in southern China. There the daily wages averaged 80 cents compared with $5.52 in Thailand.[22]

South Korea, emerging from decades of military dictatorship, implemented democratic reforms that resulted in unions gaining a tripling in wages for Nike shoemakers and workers in general. Nike led the way in reacting to these changes. The company contracted factories in China, Indonesia, and Vietnam, where hourly wages are as low as seventeen cents an hour.

Nike, the target of citizen campaigns around the world, is a quintessential TNC. Its products are "global," providing middle-class kids everywhere with the kind of product identification that is the envy of other corporations. It does not have factories of its own but hires thousands of subcontractors who employ half a million workers. For these employees, who labour sixty to seventy-five hours a week, corporate rule directly determines working conditions, which for many are appalling

A 1997 report by Hong Kong human rights groups revealed conditions at a Wellco subcontractor in Dongguan, China, employing children as young as thirteen "that have resulted in workers losing fingers and hands; beatings by security guards; . . . fines levied for workers who talk to each other on the job; 72 hour work weeks; and pay less than the . . . minimum wage of US$.24 an hour." In Vietnam workers for another Nike subcontractor had their mouths taped shut for talking on the job.[23] Meanwhile, Nike pays its star advertiser, Michael Jordan, $20 million a year, just about the same as it pays in one year to its twelve thousand Indonesian workers, most of whom are young girls who work fifteen hours a day.[24]

In country after country competing for TNC investment, unions are smashed, their organizers beaten or killed, and human rights violated. In 1996 workers and others in El Salvador signed a letter asking the GAP clothing company to allow independent investigation of conditions in its Mandarin International plant. In response, the Salvadorean minister of the interior threatened the death penalty for the signers and deportation for foreigners.

The family of Ken Saro-Wiwa has filed suit in the New York District Court against Royal/Dutch Shell for its collusion with the Nigerian military dictatorship in the detention, trial, and subsequent judicial murder of the poet-activist and eight colleagues. The suit will include evidence that Shell called in the military to suppress demonstrations, bribed witnesses at the trial, and had direct involvement in human rights violations against the Ogoni people.

THE RACE TO THE BOTTOM

The impact of transnational corporate power on social life and democracy is not limited to developing countries or military dictatorships. As dropping real wages combine with the steady shredding of the safety net, more and more working people in Canada and other developed countries fall into a social class characterized by almost constant insecurity. The benefits to corporations driving this agenda are both economic and political.

The economic benefits are obvious. The political benefits are more subtle. But the decline in democratic participation by working people is every bit as important to corporations as the decline in wages and social programs. The historic battle for who gets the biggest share of new wealth produced by corporations — employees or shareholders — turned in favour of workers in the 1960s and early 1970s. Full employment gave working people the upper hand in a sellers' labour market.

And workers achieved success not only at the bargaining table. They achieved equally important gains at the polling booth. A revitalized and militant labour movement combined with dynamic and forceful social movements of women, youth, aboriginal people, and the poor wrenched away some of the formally exclusive influence of corporations on government. It was this brief flowering of democratic participation that drove the social policy agenda in that period.

The changes affecting the security of working people and their communities are easy to identify. Restructuring the world economy and the TNCs that dominate it has created dual economies in two quite different ways. Within TNCs, according to David Korten, the corporation's internal operations are reduced to its core competencies, "generally the finance, marketing, and proprietary technology functions that represent the firm's primary source of economic power." These core functions are centralized at company headquarters and the employees performing these functions "are well compensated, with full benefits and attractive working conditions."

The second tier of employees are in what is now called peripheral functions, which, ironically for firms known for their products, includes manufacturing jobs. These are the functions that are most often farmed out to independent contractors or foreign-based units of the company. The pattern here is extremely low paid work, often by part-time employees who have few if any benefits.[25]

Not only are these functions of the TNC seen as peripheral and therefore given no commitment but they are also the areas in which competition is the most fierce. These production units, even the ones remaining in the firm, function as independent small contractors forced to compete with each other for the firm's business.

With the de-industrialization of North America, and the major investments in labour-displacing technologies, the job market in Canada and the U.S. is characterized by a good job/bad job phenomenon. The good jobs are just like the good jobs of old: well paid, generous benefits, a promise of career advancement, and a measure of security. The bad jobs — and these are the ones the economy is mostly creating — offer almost no hope of advancement. These jobs are designed to be dead-end, for the just-in-time contingent workers who must be on call twenty-four hours a day if they want work. These workers are effectively competing with those in the developing countries.

The trend to part-time work is developing with astonishing speed and throughout the whole economy. A 1996 study done for the *Globe and Mail Report on Business* revealed that "part-timers now make up 29 percent of the average firm's total workforce, more than triple the 1989 level." More than half of Canadian companies now employ part-timers, compared with a little over a third in 1989. In construction the increase between 1989 and 1996 was from 8.7 percent to 34.7 percent; in finance, from 2.4 percent to 16.3 percent. The trend is most dramatic amongst the largest firms, with 69 percent employing over 250 employees using part-timers.[26]

The combination of technological change, downsizing, two-tier wage agreements, and the proliferation of part-time jobs has resulted in a dramatic worsening of the income levels for the bottom rung of wage earners. A study by McMaster University economists Peter Kuhn and Leslie Robb revealed that the poorest 10 percent of men saw their income drop by a full 28 percent between 1977 and 1991. The trend to contracting out, a major issue in the 1996 confrontation between the Canadian Auto Workers union and General Motors, is slowly eroding high-wage jobs, too.

A 1996 Statistics Canada report shows another result of the increase in the number of contingent, part-time workers. In 1995 just over half of Canadian workers worked a traditional thirty-five-hour week, down from 67 percent in the mid-1970s. This change was triggered by the recession of the early 1980s but, said the report, "seems to have entrenched itself as a permanent feature in the Canadian labour market framework." While unskilled workers find themselves in the part-time job ghetto, there is a growing use of overtime with skilled workers and university graduates. Corporations can treat unskilled workers who require little training "as roughly interchangeable." This, said the report, gives employers "flexibility and saves money on fringe benefits."[27]

Unskilled workers in Canada are rapidly becoming a permanent underclass. As of October 1997, only 43.5 percent of Canadians with only some high school education were employed. Between October 1996 and October 1997, fully 319,000 new jobs were added to the economy. But Canadians with only some high school education lost 67,000 jobs. Canada now has the second highest percentage of low-paid jobs of any industrial country, with 25 percent of full-time jobs falling into this category. It helps produce the fifth-highest poverty rate in the OECD.[28]

Overlaid on all of these disparate trends is the increasing insecurity of all but the highest-paid workers. Few now express a desire for a shorter work week, fearing that they may not have a work week at all in the near future. Statistics Canada reported in 1997 that only 6 percent of paid workers preferred a shorter work week even if it meant less pay. When the Conference Board asked the same question in 1987, fully 17 percent gave that answer. Now almost a third of workers want more hours.

The work world thus reflects another duality, the overworked and the underemployed. This pattern has become a key feature of the race to the bottom. Says labour market analyst Armine Yalnizyan: "The approach taken must reinforce one's primary link to the world through providing one's labour services. This is to be achieved by handcuffing people ever more tightly to the labour market, which increasingly demands a twenty-four-hour-a-day, seven-days-a-week, on-call-as-needed commitment. The propelling force behind getting people to accept this model is a pervasive sense of insecurity and rapid change."[29]

WOMEN AND YOUTH: NOWHERE TO GO BUT DOWN

The new forces at work have hit women by far the hardest of any group. According to researcher Jane Jenson, throughout developed countries, "while women are not confined to this non-standard labor force they are disproportionately located in it rather than in the core sectors."[30] Part-time work remains "a female ghetto." The loss of thousands of full-time private-sector jobs in the clerical, banking, and receptionist fields, traditional employers of women, accounts for much of this shift. A 1996 International Labour Organization report confirmed this view: "While more and more women are working, the vast majority of them are simply swelling the ranks of the working poor."[31] This is not just a truth in the industrialized nations. In the developing countries, women provide 80 percent of the cheap labour for export industries.

Huge cuts to social programs, medicare, and education in Canada and elsewhere have a double impact on women. Two-thirds of union jobs in the public sector, such as health, education, social services, and the postal service, historically have been held by women, and cuts there hurt them the most. When they lose those jobs because of cutbacks, they also suffer because of the loss of the services. As well, women involuntarily take up the slack of the eroded social services. An enormous part of the newly "externalized" costs of production, once covered by state programs, is now increasingly provided as unpaid work by women. Thousands of young women who had been living independently, once able to rely on UI or social assistance when unemployed, now end up back at home. The cutbacks in health care have meant that elder care falls increasingly on families; that is, on women.

The other group that has suffered disproportionately under restructuring is young people. Canadians are aware of the high rates of joblessness of young people but few realize that jobs for young people are declining. Since the recession of 1992, the situation has been getting worse. Between January 1996 and July 1997, there were 330,000 jobs created for adults, while jobs for young people fell by 74,000.[32]

The high rate of youth joblessness has remained extremely high (at 16 percent almost double the average) even though the number of young people actively in the workforce has dropped dramatically. In 1989, 71 percent of people between the ages of fifteen and twenty-four were working or looking for work. That figure hit 61 percent in December 1996.[33] There are many unskilled workers in this age group, yet there are many others who are overeducated and to get an entry-level job have to "dummy down" their résumés, putting in question the current elite infatuation with training as a solution to unemployment.

The overall pattern of low-paying part-time jobs is worsened by the steady twenty-year erosion of the minimum wage. In 1976 the minimum wage in all provinces except Ontario provided an income above the poverty line, and the federal government provided an income equivalent to 106 percent of the poverty line. By 1994 the federal minimum was at 53 percent of the poverty line, and in the provinces it ranged from 67 percent to 89 percent.[34]

Driving this pattern of low-wage and part-time jobs is the persistently high level of unemployment. If those who have stopped job hunting are included, the rate exceeds 14 percent, and if those who want full-time work are added it shoots up to at least 17 percent. These figures would be alarming enough if the social programs of the early 1980s were still in place. But in every case the safety net has been shredded. Welfare rates have been slashed so severely that many provinces now have rates lower than most U.S. states. Out of sixty-two jurisdictions, B.C. placed highest in Canada but just sixteenth best in North America; Nova Scotia was thirtieth, Saskatchewan fifty-third.[35] Only 46 percent of the unemployed now qualify for the EI they paid into. Several provincial governments are introducing workfare schemes, making welfare recipients work or enroll in training as a condition to receiving benefits.

THE PERMANENT RECESSION

Of all the government strategies to reverse the political and economic gains of labour, the use of monetary and fiscal policy has been the most powerful and effective. Jim Stanford, of the Canadian Centre for Policy Alternatives, argues that since 1981 Canada's monetarist, anti-inflationary strategy has kept the country in a state of permanent recession. Even periods of "recovery," such as that experienced in 1997, hardly seem worthy of the term.

The reason, according to Stanford, is that "elected and unelected economic institutions have since [1981] embarked on a policy that deliberately fostered a state of controlled economic stagnation. Permanent unemployment was reestablished in order to repair the 'damage' done to the profit system by 30 years of full employment, rising wages, growing unions and an increasingly interventionist public sector."[36]

It may be hard to credit that democratically elected governments would deliberately create permanent high levels of joblessness. But in following this path the Canadian government and the Bank of Canada were simply falling in line with the IMF, the OECD, and other agencies devising ways out of the capitalist crisis of falling profits and increasing inflation.

The IMF criticized countries for their policies aimed at "the achievement and maintenance of unduly low unemployment rates." The IMF went on to recommend wage limits and tight monetary policies to "crush" inflation. The Bank for International Settlements, the central bankers' central bank, concluded with satisfaction in 1981: "Many if not most governments now appear to believe that restrictive demand management offers the main — and perhaps only — hope of a gradual return to more satisfactory levels of unemployment."[37]

The permanent recession is referred to in the language of the downsized state as the non-accelerating inflationary rate of unemployment, the so-called NAIRU rate. It is that rate that must be maintained to keep inflation, the scourge of the bondholders, in check. Governments used to set the maximum socially acceptable level of unemployment based on workers' interests; it now sets a minimum level based on the interests of corporate investors.

While unions in Canada have maintained an almost 35-percent level of unionization, their ability to defend their members' interests is being steadily eroded. In nonunion workplaces the situation is so desperate that Statscan reported that almost 20 percent of Canadians worked overtime in the first quarter of 1997 and 60 percent of them earned no extra pay. In addition, thousands of employees put up with blatant violations of the labour codes, from sexual harassment to unsafe working conditions to unannounced cuts in pay, because they are desperate to keep their jobs.

The poverty rate in Canada has been increasing since the early 1980s and has accelerated to 18 percent in 1996. A European study comparing the ratio of the highest and lowest income earners of seventeen developed nations showed that Canada was fourth-worst in terms of income inequality, behind only the U.S., Ireland, and Italy.[38] The 1994 World Competitiveness report compared thirty countries, developed and less developed, examining the amount of household income going to the lowest quintile. This study revealed that Canada, with just 5.7 percent of income going to the bottom 20 percent of the population, placed twenty-second, behind Indonesia, the Philippines, Thailand, and South Korea.[39] This study was done in 1994, before the huge cuts to social transfers and the devastating loss of full-time jobs. A Statscan study of the wage gap, comparing 1990 with 1996, revealed that 5.4 million Canadians reported income of less than $10,000 a year; over 300,000 reported income of more than $100,000.[40]

The attack on the living standards of millions of Canadian workers has two distinct but closely connected objectives. One is to simply reduce the costs of production through lower wages and greater flexibility in the use of labour. But equally important is the effect of this chronic insecurity on the functioning of democracy. During the period of rising expectations in the 1960s and early 1970s, democratic participation in Canada reached a height not achieved since the years immediately before and after the Second World War. The economy was at near full employment and the expression of dissent both in the workplace, regarding working conditions, and outside it, regarding the role of government, was driven by a sense of security and a vision of a more equal society.

Ten years of high unemployment, the loss of hundreds of thousands of high-paying jobs, the advent of part-time and temporary job ghettoes, and the discovery that education is no longer a ticket to economic security have had the effect of shutting down democratic participation for hundreds of thousands of Canadians. Two and half million workers are now either unemployed or face the very real possibility of becoming unemployed.

Poverty, near poverty, and the economic insecurity generated by cutbacks and corporate downsizing increase social isolation and exclusion. The poor are disenfranchised because to take part in any meaningful way in the political life of the community you need, at a minimum, to be free of fear of where your or your child's next meal is coming from. This insecurity can cripple an individual's ability and motivation to engage with the community, to take part in the political process.

Overrepresented in low-waged work and in the ranks of the un- and underemployed, women and youth, two of the groups most active in the sixties movements for social justice and equality, are now the ones most likely to be effectively disenfranchised by their economic situation. Many women, who are consistently to the left of men when it comes to social policy and the role of government, are being slowly silenced by poverty; nearly half of young people, with arguably the greatest interest in maintaining government as a force for equality, are socially and culturally isolated and excluded from the political life of the country.

For working people the only effective political voice most ever have is through their unions. Trade and investment agreements, however, are undermining this collective voice. The trend in the U.S. is alarming. An analysis of the NAFTA Labor Commission's study "Final Report: The Effects of Plant Closing or Threat of Plant Closing on the Right of Workers to Organize" showed that "plant-closing threats and actual plant closings are extremely pervasive and effective components of U.S. employer anti-union strategies. From 1993 to 1995, employers threatened to close the plant in 50 percent of all union certification elections and in 52 percent of all instances where the union withdrew from its organizing drive." After a certification victory, 15 percent of employers followed through on threats made during the organizing campaign, triple the rate before NAFTA went into effect.[41]

The use of threats to close plants is symbolic of what is happening to democracy in the new global economic order. The fundamental right of workers — the right of association — which is protected in human rights covenants around the world, is gradually being eroded or made irrelevant by the increasing "rights" of so-called corporate citizens. This is not only a direct assault on a fundamental human right, it is also an attack on democracy in that it prevents workers from speaking with a collective voice.

Another corporate tactic to undermine the unions' collective voice turned up in the aftermath of an extremely bitter two-year strike at the Irving oil refinery in St. John's. Returning strikers were obliged to go through what union leaders called a "brainwashing" and political cleansing program. "Former strikers must socialize with replacement workers who crossed the picket line . . .; they must be co-operative; they must be appreciative of work and they must accept that the union was wrong." The shop-floor "test" lasted four weeks and workers were assessed every day. They had to pass the test before they received full pay.[42]

Nearly every aspect of these developments is either designed by government policies or it is allowed to flourish because governments refuse to act to change it. It is governments that have refused to maintain the minimum wage above the poverty level; it is governments that have savaged the unemployment insurance system; it is governments that have cut back on enforcing labour standards and workplace health and safety, and it is governments that have established draconian and punitive welfare policies.

THE MIDDLE CLASS: CAUGHT IN THE MIDDLE

When it comes to nation building it has historically been the middle class that has provided much of the commitment, the imagination, and the defining characteristics of the national project. No nation can be built without the cooperation of the ruling economic elite, but it was the broad middle class, whose numbers were large enough and whose political motivation was strong enough to command the attention of the elites, who were critical to the process.

Within the middle class were writers, academics, journalists, broadcasters, those occupying all the areas of the arts and culture and also those who chose to put their time and money into supporting the arts and cultural activities. They had the background, the resources, the education, the motivation, and the self-confidence to take on the task of shaping their communities and, as a result, the nation. In Canada this group included the thousands of cultural workers and planners in the dozens of publicly funded institutions such as the CBC, the National Film Board, the Canada Council, and the National Arts Board. Included in this class as well were the key public policy thinkers at all levels of government, from those imagining a national health-care system to those designing provincial public parks or municipal recreation programs.

This is not to say that other social classes had nothing to do with nation building. The historic compromise of the post-war period obliged the Canadian elite to support state policies that were nation building in their consequences if not their design. Indeed, Canadians today identify the broad range of social programs and the social equality they bring as the defining characteristic of their country. It was the working class's struggles for social justice in the 1930s that created the conditions for that compromise and for the subsequent egalitarian state. But once the compromise was reached it was the broad middle class that worked out the details and coloured the national landscape.

Today the middle class might well be called the muddle class. No longer imbued with a clear view of what the nation means, they have lost that cultural and ideological homogeneity that was key to their success and their influence. The unity that characterized the middle class now belongs to the ruling elite. The political power of the middle class is waning because it is divided between old notions of social solidarity and acceptance of the new global view promoted by the elites. No longer a single coherent force in society, the middle class has lost its mediating role.

A large portion of the middle class has seen its dream of steadily increasing living standards dashed with a brutal finality and breathtaking speed. The victims are left in stunned disbelief as they go through the doors of the welfare office for the first time. Another portion has hit the global gravy train. They are among the "symbolic analyzers" that Reich talks about.

For them the adaptation is not to a shocking new insecurity and even poverty but to membership in a new elite, one that requires a major ideological shift, from a conviction that government can be a force for good to a belief in the essential virtue of the market. With the spectre of their middle-class friends living steps away from poverty, they don't want to look back and they don't dare do anything that might send them there, too.

A study by Clarence Lochhead and Vivian Shalla revealed this split reality of the Canadian middle class. To make comparisons, researchers in income distribution typically divide the population into five quintiles, or fifths of the population. Between 1984 and 1993, the top 20 percent saw their income increase 5 percent (to $102,792). The next quintile, that segment of the middle class benefiting from the new corporate order, saw their incomes rise by 3.6 percent (to $61,333). But the next two quintiles, the middle- and lower-middle-income groups, both saw declines, of 2.68 percent (to $43,103) and 10.2 percent (to $26,291) respectively. The lowest 20 percent of the population was by far the hardest hit, with a 31.9 percent drop (to $5,325).

The decline of market income of the three bottom groups was not insignificant. Had they maintained their 1984 levels they would have had another $5.2 billion a year to spend. The overall average income for a family with children was at a virtual standstill between 1984 and 1993, rising from $47,663 to $47,777.[43] The beleaguered middle class is getting poorer, paying more taxes, and getting fewer services. That is a recipe for a decline in the support of government, except that the insecure middle class still relies heavily on transfers and other programs for their standard of living.

There are other pressures, too. Many of those who have lost full-time jobs are now "self-employed." Too often that is a euphemism for having to administer your own downward spiral in living standards. The self-employed fall disproportionately into two large categories, those earning $100,000 or more and those earning $20,000 or less.

The trend to self-employment has been portrayed as a return to the free-market era of small independent producers. In fact, this self-employment, in the trucking industry, for example, reflects the efforts of large corporations to externalize their costs and the associated risks. The self-employed, whether truck drivers or those in the consulting business, are now just as dependent on large corporations — and often the same corporation — as when they received a salary.

Much of the middle class are maintaining their status by going ever deeper into debt, using credit as a private-sector safety net. In the twelve months ending March 1997, consumer spending increased by 5.1 percent. The only problem was that this amount was seven times greater than the increase in income. Personal debt as of mid-1997 out-paced Ottawa's debt by $75 billion, a level very close to 100 percent of personal disposable income, the highest level in Canada's history. The personal savings rate is plummeting as personal debt rises. According to Statscan, the savings rate was 17.8 percent in 1982. It dropped to 10 percent by 1993 and to 5.4 percent by June of 1996.[44] In the first three months of 1997 it sank almost out of sight, to 1.7 percent.[45]

The fracturing of the middle class into "self-employment" means that tens of thousands of Canadians have now entered a much more individualistic economic and social subculture. Instead of relying on stable employment, it is now every man and woman for themselves. Many of the self-employed left voluntarily to escape the pressures of the downsized workplace, where layoffs have meant a sometimes doubled workload for those remaining. Whatever their reasons, the exodus of the middle class from the traditional work world disengages them from the workplace community and leaves them more isolated. It has also meant a greater reliance on investment income for those who have any money to invest.

There is, of course, an enormous irony in those affected by the downsizing mania turning to investments to survive or prosper. The billions of dollars pouring into mutual funds and employee pension funds are being used to finance the very corporations that are doing the downsizing. Indeed, the reason that a mutual fund buys a stock is quite likely to involve the company's dedication to cutting costs. The laid-off employee of Bell Canada who puts part of her severance money into a mutual fund reinforces Bell's antisocial behaviour. Hundreds of thousands of Canadians are unwitting accomplices in their own declining economic security at the same time that they reinforce the logic of the market and corporate domination.

Yet as Canadian attitudes become increasingly characterized by possessive individualism there is little practical choice for many. Whatever a person's values may be, however much one would like the world to be different, the world people face when they get up in the morning is no less real because they had no say in making it. A large portion of the middle class who were the foundation of the communitarian character of Canada for two generations now find themselves thrown into the dog-eat-dog marketplace. They don't have the option of refusing to go along.

Not only does this force them to adopt market values that aren't theirs, or at least behaviour that reflects those values, but it leaves many unable to find work that reflects their old values. The hundreds of thousands of civil servants, CBC broadcasters, artists, writers, librarians, people in the publishing business, in public recreation and theatre, and in public education have been thrust not just out of jobs but out of the work that is the basis for a communitarian culture. The work available in the private sector is simply not the same.

The change is in both the content of the work and the attitude required in a high-pressure private sector focussed on profits and obsessed with efficiency. Former journalists working for public relations firms, ex-public policy analysts working for corporations, former public teachers working in private schools, independent university researchers going begging for corporate research money, former tax auditors working for corporate tax law firms — all of these changes augur profound long-term consequences for the political culture of the country.

The decline of much of the middle class has an immediate effect on government legitimacy. The propaganda about the inefficient public sector, failed social programs, bloated bureaucracies, and wasted taxpayer money is far more effective if the middle class has lost faith in government. As we will see in the next chapter, this loss of legitimacy in the eyes of citizens, instead of creating a crisis for the state, has so far allowed it to dramatically reduce Canadians' expectations of what government can or will do.

The apparent ease with which the radical restructuring of democracy has proceeded is explained not in the change in Canadians' values but in the simple fact that the majority of Canadians now have to cope with a situation for which the politics of the country has never prepared them. So long as the shifts in policy implemented by the governing elite took place within certain parameters, proscribed by the forty-year-old historic compromise, we could cope. These were, after all, incremental changes, and the consensus, though often challenged and always vulnerable, had never been rejected outright. Until now. What Canada is experiencing is nothing short of a counter-revolution in the role of government, imposed from the top, designed by and carried out openly in the interests of transnational corporations.

The corporate domination of the world was no more an accident of history than was the struggle of ordinary people for progressive government. Corporations have waged a twenty-year campaign to regain the power they lost in the post-war period. They have carried out a deliberate propaganda campaign, built around the deficit and debt and aimed at lowering people's expectations of government. It has been carried out with the ruthless determination of rich and powerful people protecting their hold on that wealth and power. And the state, "our" government, has been enlisted to serve the corporate agenda despite the opposition to it by the vast majority of citizens.

But guiding this broad campaign to change the political culture of Canada and the developed world in general was a process of building a consensus amongst the world's elites and within the elites of the most important developed countries. The world's most powerful corporate executives, its most influential political leaders, together with supportive opinion makers, senior government officials, and academics meet regularly in a number of high-profile but largely secret forums. They plan the future of the world, and their power and influence are such that their conclusions are directly reflected in the decisions our governments make. That most citizens have no idea of this powerful influence on their democratic governments is no accident.

7

CREATING THE ELITE CONSENSUS

Some of the problems of governance in the United States today stem from an excess of democracy.
— *THE CRISIS OF DEMOCRACY*, TRILATERAL COMMISSION, 1975

Revolution is never brought about by mass armies. Every revolution whether you look at Lenin's or Castro's or any other . . . [is] led by a small group of people with powerful ideas.
— TOM D'AQUINO, PRESIDENT AND CEO OF THE BUSINESS COUNCIL ON NATIONAL ISSUES

If we were to create an image of the advent of global corporate rule we would first have to picture dozens of democratic institutions, parliaments, legislatures, congresses, supreme courts, the offices of senior government planners and bureaucrats, the United Nations, all as facades, with familiar and reassuring fronts, like the main street set in a western movie. Empty vessels, these physical trappings of what used to be governing bodies and agencies maintain the fiction that societies and communities are engaged in making choices about what kind of society they will have, about the prospects for their children.

It is here, according to the old script, that elected representatives, duly chosen by the people, debate and decide these questions. And of course the representatives are still there and elections do still happen.

Important laws are passed and repealed. The democratic authority to make decisions still resides in these institutions. But in reality the script has changed. The pomp and ceremony of Parliament's opening, the acrimonious debates, and the occasional scandals are becoming little more than ritual. There is a new script, one that ordinary citizens don't even get to see.

For almost twenty years the most important policy directions taken by Western governments have been discussed, refined, and agreed upon not in the established democratic forums but in closed, exclusive clubs established for this purpose. Such consensus-building organizations and forums as the Trilateral Commission, the World Economic Forum, and the Bilderberg forum have not been established by or even acknowledged by governments. Yet they, and their nationally based counterparts like the Business Council on National Issues, are among the most important policy forums in the world. Officially recognized or not, they have an enormous influence on the lives of ordinary citizens in every country, in every village, in the world.

Elite consensus, then, is no longer determined in democratic forums where, ostensibly at least, people of all classes, and parties of various world-views, contribute to some semblance of pluralist compromise. Abandoning the social contract in the mid-1970s entailed abandoning the democratic traditions that both created the historic compromise and became part of it. Just as there is no longer an elite commitment to the welfare state, there is no longer a commitment to democracy.

The proper place for developing a consensus on corporate libertarianism is away from the compromising "art" of politics, from the prying eyes of citizens and the media. Indeed, corporate libertarianism is by its very nature totalitarian, and it follows that its planning would be carried out in private, where the interests of others need not be taken into account.

The organizations that have fostered the development of an elite consensus operate at both the national and international levels. At the international level some have been operating for almost as long as the United Nations, which was supposed to be the international forum for consensus building. Today, virtually none of the important decisions affecting the world's nations are taken there. Drained of its prestige, moral authority, and power, it is of all the democratic institutions in the world the most thoroughly reduced to ritualistic pretence.

There are numerous international forums in which political leaders and corporate executives get to mingle informally, but two go back several decades and have proven to be extremely influential in directing the world's economic order. One of these is so informal that it has no name other than that of the hotel where its first meeting was held in Oosterbreek, the Netherlands. Prince Bernhard of the Netherlands founded the Bilderberg forum in 1954 to address the broad issue of managing world capitalism. The U.S. was concerned about the Soviet threat but even more about a repeat of the national rivalries within Europe that led to the Second World War. In the 1920s, it had pushed for a "United States of Europe" that would have provided a stable European entity to participate in managing the world economy. Bilderberg was, in part, a response to the failure of these pre-war efforts. It was a forum for informal deliberations by the most important political and corporate players in the European movement for unification.

Joseph Retinger, a key figure in the Polish government in exile during the war, headed the organization and saw it as a way of addressing the historic weakness of Europe. The solution was to move towards a federation of European countries in which states would "relinquish part of their sovereignty."[1] The Bilderberg meetings are credited more than any other forum with building the foundation for European unity and the European Union itself.

The Bilderberg has no permanent membership. It includes some of the most influential people on the planet — heads of state, other leading political figures, key industrialists and financiers. It also includes academics, diplomats, sympathetic media figures, and a smattering of trade unionists to round out the broader definition of the ruling elite. Prince Bernhard (who resigned in 1976 after being implicated in the Lockheed bribery scandal), at the head of a core of fewer than half a dozen officials, appointed all members of the steering committee and decided who would be invited. No invitations were ever sent out to representatives of developing countries.

Bilderberg is exceptionally secretive; the confidential report drafted after each meeting contains no names, just the speaker's nationality.

The Bilderberg has no institutional powers but, according to researcher Peter Thompson, "Bilderbergers are in positions of such considerable executive power that if a consensus is reached and acted upon, the advanced capitalist West is likely to act more or less as a unit."[2] A German participant claimed that through the Bilderberg elite network a dozen people got the world's monetary system working again after the OPEC crisis.

As early as 1968 Bilderberg was to have taken up the issue of the internationalization of the economy but was prevented by the Vietnam War and student revolts in France. Nonetheless, the issue of managing global capitalism was the forum's constant preoccupation. It played a key role in creating out of the OECD a club of rich nations and global corporations. Its members were also instrumental in the establishment of the Trilateral Commission, the pre-eminent establishment consensus body.

The Bilderberg was not a decision-making body. The consensus it develops is much more general and gradual. In the words of Joseph Retinger:

> Even if a participant is a member of a government, a leader of a political party, an official of an international organization or of a commercial concern, he does not commit his government, his party or his organization by anything he may say . . . Bilderberg does not make policy. Its aim is to reduce differences of opinion and resolve conflicting trends and to further understanding, if not agreement, by hearing and considering various points of view and trying to find a common approach to major problems. [3]

As an informal forum for developing and refining an international elite consensus, the Bilderberg is challenged only by the World Economic Forum (WEF). The forum was established in 1970 by its president, Professor Klaus Schwab. Its Web page openly describes its role: "The Annual Meeting in Davos, Switzerland, sets the world agenda for the year to come. Members of the foundation, represented at the CEO level, converge on the Swiss ski resort of Davos at the end of January for six days of intensive discussion on those issues of greatest relevance to the future of world business."[4] In 1997, fully 1,000

CEOs of the world's largest corporations, 200 senior government officials, 40 heads of state, and 300 "experts" gathered to take in 236 seminars, plenaries, social events, and so-called brainstorming sessions.

The WEF, like Bilderberg, provides the perfect format for corporate governance. It is not an organization, has no administrative staff, and is very protective of its exclusivity. In his gushing praise of the forum, *Globe and Mail* publisher William Thorsell described it as "a set of circumstances rather than a place, a network rather than an organization. It integrates the most powerful, wealthiest and most capable sector of global society — corporations — with the most important governments . . . It lowers the stakes for the main political players precisely because it involves so many others with significant power."[5]

Apparently sanguine about the world being run by a secret club of fifteen hundred men, Thorsell declares, "The World Economic Forum is very much a precursor to 21st century institutions . . . [It is] one of the informal, unofficial . . . agents of communication and decision-making now emerging as the operative instruments in human affairs."[6]

Some of the corporate delegates at the 1997 meeting were Matthew Barrett, Bank of Montreal CEO; Thomas Bata, the right-wing chairman of Bata Shoes; Jacques Bougie, CEO of Alcan; Jacques Lamarre, president of SNC-Lavalin; and Charles Sirois, chairman of Teleglobe. All were there hoping to be wooed by government ministers from the hot economies of Asia, all offering up their workers, their environment, and their police protection in return for investment dollars.[7]

A recent initiative of the WEF is its computer-networking service, intended to keep the world's elite in constant touch with each other, sharing ideas and helping each other deal with "crises." The forum's Web page service, called WELCOM, promises "crisis-management support and direct access to experts and media leaders. Direct face-to-face counselling with experts at leading business and knowledge centres around the globe, such as the John F. Kennedy School of Government, Harvard University, the Brookings Institution, the MIT Laboratory for Computer Science and the Hong Kong University of Science and Technology."

THE OECD

The Organization for Economic Co-operation and Development is not usually seen as a consensus builder, and in most public references to the organization it is portrayed as little more than a research institute for the developed Western countries. But the OECD goes far beyond being a source of the statistics we see in newspapers. Its key role in pushing the MAI has in effect revealed the true power and influence of this elite-nation organization.

The OECD in the 1970s comprised almost exclusively advanced, industrial nations (it now includes Mexico, Poland, the Czech Republic, Hungary, and Korea). Among its key founding goals were "efforts to reduce and abolish obstacles to the exchange of goods and services and current payments and maintain and extend the liberalization of capital movements."[8]

During the 1970s efforts to establish favourable rights for transnational corporations occurred within the OECD, for obvious reasons. This was an organization whose membership was restricted to those nations that were the home base for all the world's transnationals and provided a forum for developing global policy; unlike the U.N., then, it excluded developing countries. The organization has been used extensively by developed countries to prepare consensus positions that were then taken to broader and more democratic forums, such as the U.N. or GATT. This was particularly true when developing countries began to challenge the role of multinational and transnational corporations.

THE TRILATERAL COMMISSION

Of all the consensus-building forums of the world's economic and political elites, the Trilateral Commission stands out as the most important and the best known. Less secretive than either Bilderberg or the World Economic Forum, the Trilateral Commission has permanent offices in the three regions it was established to represent: Japan, North America, and Europe. It commissions and publishes studies that, along with the names of its members, are available to the public. It meets yearly and produces a report on its proceedings.

The Trilateral Commission was formed in 1973 in response to a series of crises and worsening chronic problems facing the world capitalist system. First, the consensus so carefully built up over the previous twenty-five years was suddenly violated by President Richard Nixon, head of the very country that had worked so hard to establish it. Second, the oil-producing countries of the Middle East created a crisis for the world economy by dramatically increasing the price of oil — a move that would result in the transfer of huge numbers of American dollars to countries that previously had Third World status. And within the rest of the Third World, many countries were vigorously pursuing a policy of anti-colonialism through self-sufficiency, which, if successful, could shut the door on foreign investment or place restrictions on how it operated.

Perhaps more important, because it affected the way the developed countries could deal with these problems, there was a surge of participatory democratic activity throughout the developed world, particularly at its imperial centre, the United States. This increased involvement in democratic politics was threatening the elite governance that had prevailed to the end of the 1950s. And it was threatening at the very time that concerted elite action was crucial for the stability and even the survival of the world economic system.

But the immediate incentive for the commission was the crisis precipitated by Nixon, who was determined to reverse the gradual decline of the U.S. as an economic superpower. As part of his New Economic Policy, Nixon unilaterally delinked the U.S. dollar from the gold standard, violating an IMF agreement to which the U.S. was a signatory. He also violated the General Agreement on Tariffs and Trade by slapping a 10 percent tariff on most goods entering the U.S. With these two moves Nixon declared the U.S. intention to reassert its economic dominance over Japan and Europe. But he also effectively declared war on the most powerful corporate groups in the world: the increasingly dominant transnational corporations and international financiers whose interests lay in even freer trade, free investment, and a completely interdependent capitalist economy.

The result of the "Nixon shocks" in the transnational camp was almost immediate. Several key administration officials resigned and went to work for groups promoting the interests of finance capital, groups like the Brookings Institution and the Council on Foreign

Relations. Some of the biggest names in finance capital, including David Rockefeller, as well as Zbigniew Brzezinski, George Ball, and Cyrus Vance, and European and Japanese representatives began to blitz the Nixon administration through a series of high-profile articles, conferences, and exchanges of opinion.[9]

These actions led in early 1973 to the formation of the Trilateral Commission, an organization that has been described by author Holly Sklar as "the executive advisory committee to transnational finance capital."[10] According to Richard Falk, writing in the *Yale Law Review*, "The vistas of the Trilateral Commission can be understood as the ideological perspective representing the transnational outlook of the multinational corporation [which] seeks to subordinate territorial politics to non-territorial economic goals."[11]

The Trilateral Commission was also going to be assigned another difficult and growing problem. Brzezinski (who would become Jimmy Carter's secretary of state), credited along with Rockefeller with founding the commission, described the problem of the developing world as follows: "Today we find the international scene dominated . . . by conflict between the advanced world and the developing world than by conflict between trilateral democracies . . . The new aspirations of the Third and Fourth worlds united together seems to me to pose a very major threat to the nature of the international system and ultimately to our own societies."[12]

For the trilateralists, whose deliberations and global prescriptions came to be known as the Washington consensus, the weapon of choice, argues Sklar, would be debt dependency, a neo-colonial "leash around a Third World country's neck. The leash is let out to allow Western-directed development projects to gallop ahead . . . Or the debt leash can be pulled in tight — as part of an economic and political destabilization campaign — to strangle a rebellious nation into submission."[13]

The Trilateral Commission was initially designed to operate for just three years, enough time, it was thought, to create a foreign and domestic policy framework for the trilateral areas. But it kept renewing its three-year mandates and has become the permanent forum for global trade and investment policy making. Its initial funding came from some of the elite from the American corporate world — General

Motors, Sears Roebuck, Caterpillar Tractor, Exxon, Texas Instruments, Coke, Time, CBS, Wells Fargo Bank — as well as numerous foundations, with the Ford Foundation putting up $500,000 and the Rockefellers' Fund $150,000.

The commission is more than just a consensus-building organization amongst policy intellectuals. It is explicitly an action-oriented body that develops consensus through studies and consultation and then makes concerted efforts to implement that consensus. The whole membership of the Trilateral Commission (it started with 180 and is now about 300) meets once a year in one of the three regions. Keynote speakers address key issues of the day and task forces present reports. These plenary meetings, as well as executive meetings, are highlighted by meetings with heads of state and high government officials.

The commission's task forces are its key policy development tool. Each task force is coordinated by three "rapporteurs," one each from the three regions. But it is the direct lobbying of decision makers that gives the task force reports (and their subsequent refinement) their clout. The commission holds "impact meetings" in order to "sharpen the impact of its work among decision-makers." The meetings, often facilitated by former political leaders in its membership, take place with the media, legislators, "executive and congressional staffers; embassy officials, leaders of international organizations and political bodies and top government officials . . . In this fashion the circle of 'influential persons' directly in contact with the Commission is widened considerably."[14]

The original Canadian membership was dominated by Liberal politicians, government officials, and academics. The current Canadian membership is equally high profile but completely dominated by business. Since 1991 it has been chaired by Allan Gotlieb, former ambassador to the U.S., a key promoter of free trade and NAFTA, and an influential member of the Canadian establishment. Gotlieb joined the commission in 1989 just as the Free Trade Agreement came into force. At the same time, he was appointed senior trade adviser to Burson-Marsteller, the world's largest public relations and lobbying firm, which that same year bought a 49-percent interest in Executive Consultants, one of the oldest government relations firms in Ottawa.

Burson-Marsteller was eager to have more lobbying clout with the Canadian government. Its list of clients includes many of the trilaterally represented corporations: AT&T, DuPont, GE, Shell, Westinghouse, Coca-Cola, and Bank of America. It is not surprising, then, that Burson-Marsteller and Gotlieb were among the most aggressive promoters of NAFTA. Reinforcing the original goal of the Trilateral Commission, Gotlieb told a meeting of the Americas Society in 1991 that NAFTA was critical to the completion of "the new world order [in which] the withering of the nation-state . . . is the dominant feature."[15]

Among the thirteen Canadian commissioners are Hollinger's Conrad Black; Jacques Bougie, CEO of Alcan; Mickey Cohen, CEO of Molson's and former finance department architect of many of Ottawa's corporate tax breaks; Yves Fortier, former ambassador, and Paul Desmarais, head of Power Corporation, a man who has helped finance the careers of three prime ministers.

The membership is rounded out by three powerful western millionaires: Jimmy Pattison, of Vancouver car dealership fame; Ron Southern, of Calgary-based ATCO, and Michael Phelps, CEO of Westcoast Energy. Of the thirteen Canadian members in 1996, eight were CEOs; one was a current MP; three were in the diplomatic or civil service corps, and one was an academic, Marie-Josée Drouin, a member of the powerful Washington-based Council on Foreign Relations and former executive director of the conservative Hudson Institute. All the companies represented are also on the Business Council on National Issues, except for Black's Hollinger (privately owned companies are not invited into the BCNI).[16]

For the first three years the commission was run by Brzezinski, reflecting the fact that American finance capital took the initiative in redirecting international policy. And although it was the Nixon shocks that ignited this effort, the commission's founders were also concerned with chronic political problems with American democracy.

The issue was quickly addressed in one of the Trilateral Commission's first commissioned studies, entitled "The Governability of Democracies." Eventually published as *The Crisis of Democracy*, the study detailed what the most vigorous proponents of the global economy and transnational corporations believed was the gravest threat to the future of capitalism: the growing demands of citizens on Western democratic governments. The major conclusion of the study was summed up in the section on the U.S. authored by Samuel P. Huntington. "Al Smith once remarked that 'the only cure for the ills of democracy is more democracy.' Our analysis suggests . . . [that] instead, some of the problems of governance in the United States today stem from an excess of democracy . . . Needed, instead, is a greater degree of moderation in democracy."[17]

The study looked fondly back at the days when American democracy was run quietly by an elite of political figures, industrialists, academics, and media owners. "Truman had been able to govern the country with the co-operation of a relatively small number of Wall Street lawyers and bankers," states Huntington.[18] By the early 1970s that era was long past and government authority weakened. The public now questioned "the legitimacy of hierarchy, coercion, discipline, secrecy, and deception — all of which are in some measure inescapable attributes of the process of government"[19] and "no longer felt the same compulsion to obey those whom they had previously considered superior to themselves in age, rank, status, expertise, characters or talents."[20]

Stating that a "governable" democracy requires "apathy and non-involvement" on the part of marginalized groups, Huntington describes the current crisis as it developed in the sixties and seventies: "Previously passive and unorganized groups in the population, blacks, Indians, Chicanos, white ethnic groups, students, and women now embarked on concerted efforts to establish their claims to opportunities, positions, rewards and privileges, which they had not considered themselves entitled to before."[21]

The principle of democracy was infecting all sorts of American (and other Western nations') institutions, which were better run as autocracies loyal to the capitalistic ethic. The study identified universities and the media as needing major reform, for their role in turning the American public against the war in Vietnam was particularly galling to trilateralists. The study called for action to restore a balance between government and the media through more rigorous self-censorship.

The anti-war upheaval on U.S. campuses was seen as a deadly threat to the functioning of American democracy. "This development constitutes a challenge to democratic government which is potentially at least as serious as those posed in the past by the aristocratic cliques, fascist movements and communist parties."[22]

The trilateralists' solution was that rather than have "value oriented intellectuals," there was a need for "technocratic policy-oriented intellectuals." As for students, the study suggested that more teenagers from the previously marginalized groups should be steered towards vocational training. In addition, "an attempt should be made to lower the expectations of 'surplus' people with college degrees."

The trilateralists' fears of the threat of democracy extended well beyond its effects at home. Indeed, it was on foreign policy and the freedom of transnational corporations that much of the discussion focussed. In Huntington's words: "If American citizens challenge the authority of American government, why shouldn't unfriendly governments? . . . A decline in the governability of democracy at home means a decline in the influence of democracy abroad."[23]

The kind of democracy the trilateralists wanted reflected the future strategy of TNCs, discussed in chapter 3. It was the democracy of consumption. The success of American democracy had been, according to Huntington, the adoption by millions of Americans of middle-class values and "consumption patterns." An article by Daniel Boorstin in *Fortune* described that democracy as the "Consumption Community." "Consumption Community is democratic. This is the great American democracy of cash which has so exasperated the aristocrats of all older worlds. Consumption Communities generally welcome peoples of all races, ancestry, occupations and income levels, provided they have the price of admission."[24]

Applying this principle of consumption democracy to all those, worldwide, with the price of admission is now the global strategy of the TNCs. This not just an economic strategy. One of the most advantageous aspects of the consumption community is social control, for this community is by its nature politically apathetic.

At the time that *The Crisis of Democracy* was published the elite consensus in Canada was not in sync with that of the Americans. The reaction of Canadian commission members and others invited to a Montreal forum to discuss the study was disagreement verging on alarm. The political elite, at least, was going in the opposite direction — supporting Pierre Trudeau's participatory democracy initiatives, which actually nurtured the kind of interest in politics that Huntington equated with the threat of fascism.[25]

Canadian business, however, like its counterpart in the U.S., was alarmed at political developments. It is interesting to note that of all the participants in that forum on the governability of Canadian democracy there was only one business representative. That man was Simon Reisman, who would twenty-two years later negotiate the Canada-U.S. Free Trade Agreement, which did more to make Canadian democracy "governable" than any other single government initiative. That business was not invited to the forum was in marked contrast to how the Trilateral Commission functioned in the U.S. and it may have reflected the fact that in Canada the political elite and the business elite were more separate than at any time in the country's post-war history.

The Canadian business elite was acutely aware of something the Trudeau government, from the business perspective, seemed oblivious to. The so-called golden era of capitalism was over. Though Canada did have a modest public sector, mostly in utilities, for three decades the private sector had taken care of job creation and investment more or less by itself. An almost unprecedented profit boom, reinforced by the multiplier effect of higher wages and mass consumerism, had given the federal government the resources to expand social programs, increase the public sector, and introduce measures to reduce poverty.

But the end of that economic expansion suddenly left the corporate world with major problems. As economist Jim Stanford notes, "The bottom-line consequence for employers . . . was a long historic rise in the wage share of output, mirrored by a decline in the profit share . . . Full employment and expanded unionization enhanced labour's bargaining power over both wages and work practices, and expanded social programs provided workers with a degree of social and economic security," and thus independence from employer power.[26]

International competition and new tax measures, combined with low productivity, pressured CEOs and boards of directors to find solutions. In a few short years, from 1974 to 1976, powerful business figures would take major initiatives to reverse the trend of participatory democracy, labour power, and a social safety net that tipped the balance of power towards workers.

THE BUSINESS COUNCIL ON NATIONAL ISSUES

Few people took much notice when three groups formed to pursue this objective in their own ways. The Business Council on National Issues (BCNI), the Fraser Institute, and the National Citizens' Coalition would each play a key role in fashioning a new elite consensus and changing the political culture to make it amenable to that consensus. The Fraser Institute and the National Citizens' Coalition focussed their attention on public opinion and will be examined in the next chapter. The BCNI was the Canadian branch of trilateral elite consensus.

One of the reasons that the business elite in Canada was seemingly absent from policy making had to do with the nature of the Trudeau government and the makeup of Parliament itself. Trudeau did not come out of the business community and was not easily accessible to it; in fact he publicly criticized the role of corporations. The leader of the Conservative Party, Robert Stanfield, was a Red Tory and generally supportive of a policy of state intervention. And David Lewis of the NDP gained the balance of power in the House of Commons on the strength of a campaign that branded big business as "corporate welfare bums" because of the tax breaks and subsidies they received. Not only was the business elite disconnected from its political counterpart but corporations had hit a new low in public opinion. It, more than government, was blamed for the economic ills of the country.

But more important, the political voice of business had become weak. The imposition of wage and price controls in 1975 politicized the economic sphere, and the Trudeau government was eager to consult more closely with business. The trouble was, there was no one to effectively consult. Michael Pitfield, the powerful secretary to the cabinet, claimed that "the existing business organizations had atrophied so badly they had become part of the problem."[27] No organization addressed the issues and implications of an increasingly internationalist capitalist economy.

That changed in 1976 when two powerful corporate executives created that voice in the form of the Business Council on National

Issues. W. O. Twaits, retiring chairman of Imperial Oil, and Alfred Powis, president of Noranda, were the first co-chairs of the new organization. This would be an organization that would go beyond the "carping and special pleading" with which business groups had become identified in the public mind. It would, in Twaits's words, "strengthen the voice of business on issues of national importance and put forward constructive courses of action for the country."[28]

The BCNI modelled itself after the Business Roundtable in the U.S., an organization whose members were the CEOs of the 192 largest corporations in the country. Only active CEOs could be BCNI members, a strategy aimed at solving one of the problems of the traditional business lobby groups, whose spokespeople and staff could not make decisions quickly; because they did not speak for their corporations it was impossible to develop a consensus. The BCNI began as an organization of the CEOs of the 150 largest corporations in the country. Careful not to offend existing business groups, the council included senior officers of the Canadian Chamber of Commerce, the Canadian Manufacturers' Association, and the Conseil du Patronat du Québec as associate members and ex-officio members of the Policy Committee.

The BCNI also copied the structure and general approach of its American counterpart. It was the perfect national complement to the Trilateral Commission in the sense that it tried to speak for the system as a whole. Just as the Trilateral Commission was a conglomerate of all sectors and interests addressing issues of global capital, the BCNI addressed those same issues in the Canadian context.

It was clear from the beginning that the council would be fundamentally different from any other business organization in Canadian history. According to York University political scientist David Langille, it would explicitly reflect the interests of multinational and transnational corporations under the leadership of finance capital, led by the chartered banks. Long before the term *global economy* had become a common phrase, the BCNI was organized to reinvent the Canadian state in order to facilitate globalization.

The coalition of sectors in the council was consistent with Canadian capitalist history. The alliance between finance capital and companies involved in resource extraction and primary manufacturing dominated the council and its thirty-member Policy Committee just as they did the economy as a whole. Nearly a third of the BCNI members are foreign-owned corporations.

The most important purpose for which the BCNI was established was to transform public policy. Here, too, its approach was critically different from the crude lobbying of the past. The BCNI's efforts were cast in the mould of a newly enlightened business class. Its public relations spin claimed that "members were sought for their public spiritedness and commitment to the betterment of public policies, for their leadership abilities." The council's approach was to address carefully chosen and broad areas of public policy on which to influence the government. According to former council president William Archibald, the council aimed at taking a pre-emptive approach "at emerging public policy areas" and would "develop approaches and solutions . . . that they can put forward early in the process so that they have a chance of being considered."[29]

The council thus presented to senior civil servants and cabinet ministers business-oriented policies that also addressed the potential political problems of those policies. In effect, with its high quality of research the BCNI would become virtually a parallel government, combining the skills of the senior policy analyst with the sensitivity of a politician. In many instances, BCNI policy papers proposed the wording of legislation that would reflect their recommendations.

Indeed, the BCNI's structure of task forces made it a virtual shadow cabinet. The task forces were established and dissolved according to the political priorities of the day. Among the task forces were those addressing national finance (that is, taxation), international economy and trade, social policy and regional development, labour relations and manpower, government organization and regulation, foreign policy and defence, competition policy, education, and corporate governance. The BCNI has a small staff compared to other business organizations but its task forces draw on the staff of its member organizations and the resources of the C. D. Howe Institute, another consensus-building agency funded by many of the same corporations.

Until the early 1980s the BCNI had such a low profile that few Canadians ever heard of it. But that changed when Tom d'Aquino was chosen president and CEO of the council. A former Trudeau trade adviser, d'Aquino would give the council a greater public profile and influence just at a time when the Trudeau government, in trouble with its traditional business supporters over its National Energy Program, felt obliged to make overtures to corporate Canada. It was also at a time when key corporate leaders and d'Aquino had decided to push within the corporate elite for a consensus on a free-trade agreement with the U.S.

With d'Aquino at its helm, the BCNI effectively seized control of national policy making. It began with the Trudeau Liberals as they faced the new reality of a hostile and aggressive corporate class, moved effortlessly into the regime of the Mulroney Tories, who despite a landslide victory in 1984 had no clear agenda for the country, and picked up again with the Paul Martin and Jean Chrétien Liberals after 1993. Governments and prime ministers come and go, but the BCNI, the voice and organizational embodiment of corporate rule, is a permanent presence.

The rejection of the historic compromise by business cannot be traced neatly to any single expression or event. But perhaps the clearest and most brutal chapter in the Liberals' awakening to the new corporate agenda was the response of the corporate elite, all BCNI charter members, to the government's efforts at tax reform in 1981.

When Allan MacEachen was appointed finance minister in 1980 big business requested that government examine the tax system with a view to making changes. But MacEachen's senior advisers soon focussed his attention on how billions of dollars were being lost yearly to scores of dubious corporate tax breaks. Finance officials put together a tax reform package designed, among other things, to eliminate 165 of the most costly and counter-productive tax expenditure measures and in the process increase revenue by close to $3 billion. When he introduced the legislation it caused a firestorm of protest from the corporate elite.

Neil Brooks, now professor of tax law at Osgoode Hall Law School, was working for the finance department on the tax reform package and has recalled the tactics of the large corporations. "It's almost a classic example of what's called a capital strike. I mean, business simply said to the government that if you go ahead with these measures we will stop investing in Canada." The development industry reacted instantly. "Literally the next day they were closing jobs down and . . . pulling cranes off construction jobs." Measures designed to prevent developers from deducting up to 30 percent of the cost of construction as an immediate expense were "reversed on very quickly, within a week and a half."[30]

Life insurance companies had their own strategy. The industry, which for years had paid income tax rates of close to zero, wrote to every one of its policyholders, telling them the new measures to tax investment revenue would greatly increase their premiums. "The government," says Brooks, "at one point was receiving thousands of letters a day from people across the country."

The ferocious assault by the corporate and wealthy elite proved to be the wakeup call that moved the Liberals off their social agenda and back to business. The government's signal that it was serious about an improved relationship with business was Trudeau's dumping of MacEachen as finance minister. His replacement was Montreal lawyer Marc Lalonde. Formerly in the social liberal camp, Lalonde changed colours quickly, meeting with and smoothing the feathers of a stream of representatives from various sectors in the economy. But, in a move pregnant with the symbolism of BCNI power, Lalonde went to the home of Tom d'Aquino in the tony Rockcliffe area of Ottawa to meet with council leaders. There, according to David Langille, "he is alleged to have signed a peace pact with the business leaders and to have promised them the government's support."[31]

That support wasn't long in coming and it set a precedent for how corporate-government public policy making would take place in the new era. The BCNI's role in remaking Canadian energy policy illustrates its role in elite consensus building, a process that included reconnecting the previously divided political and business elites. The council arranged a series of secret meetings involving oil executives, the government of Alberta, industrial representatives, and the Ontario and federal governments, and managed to create a temporary peace.

But the BCNI was always anticipating problems. Its Task Force on Energy Policy, headed by the CEO of Imperial Oil, sent d'Aquino and Union Gas president Darcy McKeough on a mission to Alberta premier Peter Lougheed, Ontario premier Bill Davis, and energy minister Jean Chrétien. Out of that junket came a two-day high-level secret meeting held at the Niagara Institute. Eight months later, in June 1984, a deal was hammered out at another high-level meeting. The new legislation was a virtual carbon copy of most of the BCNI's key proposals, including deregulating Canadian oil and gas prices, lowering federal taxes and provincial royalties (thus contributing to the federal debt), and shifting government incentive grants to the tax expenditure system, where they would be less visible.

A good example of the BCNI's approach to rewriting government policy involved its intervention on anti-combines legislation. A year before the Trudeau government introduced its own discussion paper, the BCNI put twenty-five corporate legal counsels to work and literally wrote a new act. The 236-page report, released in 1981, was comprehensive, with precise recommendations and detailed statutory amendments that it stated were intended to "ensure that the public interest in having Canadian business compete vigorously in domestic and international markets can be served." Included in the amendments were several that addressed interests "outside the business community" so as to "reach a 'reasonable consensus' that would be 'practically workable.'"[32] The new Competition Act was virtually the same as the BCNI's recommended package of changes. It generated almost no comment let alone opposition, and business got just what it wanted, legislation that did almost nothing to promote competition.

By far the most important initiative taken by the BCNI was its early decision to pursue a free-trade deal with the U.S. More than any other policy area, the council's position on trade represented the interests of transnational corporations in Canada. And more than any other political victory, its initiative and lobbying on free trade fundamentally altered Canadian democracy and public policy.

Until the early 1980s the capitalist class in Canada was divided over freer trade versus protectionism. But that changed by 1983, according to David Langille. "As oil prices dropped and the National Energy Program came apart, so did any hopes of repatriating the Canadian economy. For their efforts, Canadian government and business leaders were so severely chastised by the Reagan administration that they [were] unwilling to attempt further experiments in economic nationalism which might jeopardize their 'special relationship' with the Americans."[33]

The push for free trade was in the interests of the U.S. simply by virtue of its status as the home base for so many of the world's transnational corporations. But by the mid-eighties Canadian capital was onside, with the manufacturing sector ready to join the finance and resource sectors in a continentalist strategy vis-à-vis the U.S. Growing international competition and the trend towards ever larger corporations were forcing a consensus amongst Canadian business. The BCNI was able to bring all sectors of capital together to support a comprehensive deal with the U.S.

It was left to d'Aquino to state the broad objectives of a free trade deal and other policy initiatives of big business. He gave a speech to Toronto's Empire Club, responding to a highly publicized statement by Canadian bishops calling for less corporate greed. D'Aquino said that he and the BCNI were working on a plan to "reconstruct" Canada. "And by reconstruction we mean fundamental change in some of the attitudes, some of the structures and some of the laws that shape our lives."[34]

Having achieved consensus amongst the business and corporate elite in Canada, the BCNI turned to the other players who needed to be onside: the government of Brian Mulroney (who had opposed free trade during the 1984 election) and the top echelons of U.S. corporate and political leadership. To help accomplish the first task, the BCNI initiated a forty-five-member task force to present a united front to the government.

The BCNI's initiative on free trade was taking place at the same time that the Royal Commission on the Economic Union and Development Prospects for Canada, the Macdonald Commission, was deliberating. Rather than mount a frontal assault on the government itself, d'Aquino launched a coordinated corporate campaign to determine the outcome of the commission. The council orchestrated a host of presentations to the commission by some of its key member companies, all calling for free trade. Combined with a strategically timed media blitz, the council virtually determined the commission's report. The commission, in an unprecedented move, came out in favour of free trade, calling for a "leap of faith" before its own researchers had even completed their work.

The spring of 1985 saw the council and its allies promoting a trade deal on both sides of the border. In March the council dispatched seventeen representatives on a three-day blitz of Washington, meeting with key corporate leaders and some of the top representatives of the Republican political elite: Secretary of State George Shultz, Defense Secretary Caspar Weinberger, Senate Majority Leader Robert Dole, and the chair of the Senate's International Trade Subcommittee, John Danforth.

By April, after the "successful" Shamrock Summit between Mulroney and Ronald Reagan, d'Aquino felt the time was right to ask the Conservative prime minister to seek early discussions with the U.S. about a comprehensive deal on trade liberalization. That task turned out to be little more than a formality. So complete was the business consensus that the story is told of d'Aquino convincing Mulroney of free trade in a fifteen-minute conversation on an Ottawa street. More than any other victory, the BCNI's consensus building on free trade would turn out to be the key to achieving d'Aquino's goal of "reconstructing" Canada.

But it by no means stopped there. Having set the framework by persuading the Tories to move on free trade, the council continued to add other elements to its corporate libertarian agenda. Perhaps the most important of these was its relentless campaign to set an inflation target of 2 percent or lower. As we will see in chapter 9, the war on inflation accomplished several corporate objectives: it protected corporate assets (the alleged motivation for low inflation) and at the same time accomplished the crucial political objective of lowering expectations among working and middle-class Canadians.

One of the other BCNI initiatives worthy of mention was its determined support for the goods and services tax. So powerful was the corporate consensus on this issue that the Conservative government of Brian Mulroney virtually committed political suicide to get it through. Dedicating themselves to a corporate initiative (the GST transferred $18 billion in taxes from corporations to individuals), the Tories showed just how much the political elite and its agenda are now virtually indistinguishable from the wishes of corporate Canada.

❖

The BCNI's success in writing public policy during the Mulroney years is breathtaking. The council's defence paper, for example, was so closely mimicked by the government's own White Paper that the magnanimous d'Aquino agreed to delay its release for several months so as not to embarrass the government.

In 1981, the BCNI had established a task force on foreign policy and defence, and "discovered, in painful detail, how manifestly incapable this country is of defending itself from external aggression and effectively protecting and asserting Canadian sovereignty." D'Aquino declared it was "scandalous" that Canada did not have modern submarines.[35]

In 1984, the BCNI was saying universality in social programs should be abandoned and was recommending that $5 billion be cut from federal spending. Yet it said military spending should be increased by 6 percent after inflation each year for ten years. Fortunately, this is one instance where the council did not prevail. Had it done so, national defence spending in 1995 would have been $25 billion rather than the $11 billion budgeted. Yet the BCNI campaign for a military buildup did not exactly fall on deaf ears. The federal government increased military spending by 40 percent after inflation through the 1980s and the military's budget has never returned to where it was before the buildup rationalized by the cold war.

The BCNI is responsible for two long-lasting changes to the very nature of how the federal government works. First, in its pro-business victories with free trade, taxation, competition policy, and regional development (the latter killed any sort of industrial policy at the national level, the only place it counts), the BCNI put into place a policy environment that is a permanent feature of federal governments, regardless who is in power.

Second, and related to that policy environment, the BCNI's approach to corporate intervention has permanently changed the way that public policy is made in Canada. It's almost as if the traditional forums for debating and determining a public consensus are now little more than bread and circuses, rituals tossed out to the public as a sop to democratic legitimacy.

The BCNI's control of both the policy process and the federal agenda is so effective that prime ministers are reduced almost to the status of figureheads. Within months of taking power the Chrétien government, elected on a promised return to social liberalism, was receiving its marching orders from the BCNI. Jean Chrétien was, in any case, long accustomed to listening to Tom d'Aquino. During a brief stint as finance minister in the early 1980s he had declared that he never prepared a budget without seeking the opinion of the BCNI. In fact, long before the 1993 election, d'Aquino had begun getting close to the Liberal Party, partly through the vehicle of corporate funding. According to Duncan Cameron of the Canadian Centre for Policy Alternatives, "While other people were debating with the Tories about free trade and other issues, d'Aquino was quietly briefing the Liberals in opposition on the deficit issue and on how to deal with the NAFTA issue. He was back-rooming it with the Liberals because he knew the Tories were going to lose. By the time of the election the transition was almost seamless."[36]

Most of the policies of the Mulroney government dealing with the economy had been written by the BCNI and delivered to Brian Mulroney for the formality of implementation. The first term of the Chrétien government delivered on most of the BCNI's social policy agenda. And in 1994, the BCNI in effect delivered to the Liberal government instructions for the next phase of corporate governance. Its plan, "A Ten-Point Growth and Employment Strategy," covered almost every area of government policy.[37]

In social policy the Liberals were urged to make the huge cuts that the Tories had failed to make, to end universality, and to revamp programs like UI to remove alleged disincentives to work. The 150 member CEOs insisted that, though the battle with inflation had been largely won, the government must continue with a policy of "non-inflationary growth." The ten-point plan recommended that the government eliminate the deficit entirely by 1998–99.

And on trade it pushed for more of the same along the lines of NAFTA, calling for the elimination of provincial trade barriers and telling the government to continue, at the international level, to pursue aggressive international trade development and diversification. Finally, signalling that it really would like to see the federal government decamp from national governance altogether, the strategy document called for a more decentralized federation.

The BCNI's plan is well on its way to completion as the Liberals enter their second term. The Red Book of Liberal promises is now little more than a historical curiosity, a testament to political duplicity and government subservience. Yet in a sense the BCNI had just provided its particular preferences for the further liberalization of the economy. One wing of the Liberal Party had been onside all along. While the social liberals provided the deception needed for an election victory, Chrétien and Paul Martin and the party's business Liberals had been primed to implement the BCNI's wish list.

At the most important policy conference the party had held in years, the Liberals gathered at Aylmer, Quebec, in the fall of 1991 to hammer out their direction for the new millennium. Here, consolidation of the elite consensus took a huge step, with the Liberal Party decisively shifting in favour of its business faction. According to Maude Barlow and Bruce Campbell, "Chrétien declared that the old left-right split in the party was obsolete; from now on there was only the inevitability of the global economy, and Canada would have to adapt. 'Globalization is not right wing or left wing. It is simply a fact of life.'"[38]

At the Aylmer gathering, the presentation that heralded the transformation of the Liberal Party was given by Peter Nicholson, a Nova Scotia Liberal and free marketeer. In his speech "Nowhere to Hide" he endorsed the economic rationalism that now guided nearly all of the English-speaking developed world. "What seems beyond question is that the world has entered an era where the objectives of economic efficiency . . . will hold sway virtually everywhere. Societies which fail to respond effectively to the market test can, at best, look forward to a life of genteel decline, and at worst a descent into social chaos."[39]

The most dramatic demonstration that business was back and firmly in control, and that the elite consensus was solidified, was the cabinet choices Chrétien made upon achieving power. Chrétien himself had impeccable corporate credentials and his previous cabinet posts — finance; national revenue; industry, trade, and commerce; energy and mines; and Treasury Board — had brought him into contact with the most powerful business sectors in the country. After his defeat in 1988 he worked for corporate lawyers Lang Michener and sat on the boards of Toronto Dominion Bank, Stone Consolidated, and Viceroy Resources.

He was also close to Quebec power broker Paul Desmarais, who agreed to be his chief fundraiser in his bid for the Liberal leadership.

Chrétien's cabinet choices reinforced business Liberals' control over the cabinet and the national agenda. Finance Minister Paul Martin owns Canada Steamship Lines, one of the most notorious tax-dodging companies in the country. The other economic ministries went to people like businessman John Manley at industry, the hard-right Doug Young at transport, former Toronto mayor Art Eggleton at the Treasury Board, and the right-wing friend of the oil industry, Alberta's Anne McLellan, in natural resources. Marcel Massé, head of the Privy Council, was former Canadian director of the IMF and once president of CIDA, and is a hard liner on structural adjustment who had already helped apply the program in the Third World. "It isn't just the Third World that needs structural adjustment," Massé once opined. "We all do, in one form or another. We should avoid the temptation to let our desires for justice in the world obscure the view of reality."[40]

Perhaps most important, next to Martin himself, was Chrétien's choice for minister of international trade. Roy MacLaren had been the point man in setting up formal contacts between the Trudeau cabinet and the BCNI. MacLaren, as Chrétien's trade critic in opposition, worked closely with Tom d'Aquino in developing a position on NAFTA intended to appease the social liberals while it assured the corporate elite that its agenda was still in place.

Interviewed for *Canadian Forum* magazine in 1992, Tom d'Aquino scoffed at the idea that he was the de facto prime minister. It was false modesty. Yet it is important to note that d'Aquino, able as he is, speaks on behalf of his corporate cabinet. Its members, who make up the all-important Policy Committee of the council, are the cream of the corporate CEO crop.

Amongst the thirty members of this group, which effectively determines social and economic policy for the country, are the CEOs of three of the big banks, including the Bank of Montreal's Matthew Barrett and A. L. Flood, CEO of CIBC and chair of the BCNI; Brian Levitt of Imasco; David Kerr of Noranda; David O'Brien of Canadian Pacific; Paul Tellier of CN; Guy St. Pierre of SNC-Lavalin; Alfred Powis, the retired founding chairman; John Mayberry of Dofasco;

Loram Corporation's Ronald Mannix, and Jean Monty of Northern Telecom (now at its parent company, BCE).[41] The members reflect the dominant sectors of the Canadian economy, heavily weighted in favour of financial institutions and the resource industry. There is also strong representation from American transnationals (which make up one-third of the council's membership), including Imperial Oil, Cargill, GM, DuPont, and Hewlett-Packard.

In the summer of 1997 the Policy Committee, having gained almost everything it could think of in the way of social, fiscal, and economic policy from three successive governments going back sixteen years, took a political initiative intended to reduce the federal government's role to an absolute minimum. A memo sent to the prime minister reinforced the BCNI's position that the federal government should give up its leadership role to the provinces. With an arrogance befitting a de facto prime minister, d'Aquino states, "Prime Minister, we acknowledge that you too feel passionately about the future of the country." He then goes on to urge Chrétien to surrender leadership on the Quebec issue and support "any worthwhile initiative" that might come from the provinces.[42]

After sending this memo the BCNI conveniently provided the provinces with just such a "worthwhile initiative." In one of its rare high-profile public interventions, it issued a position paper on national unity and ostentatiously gave it to Alberta premier Ralph Klein.[43] Klein, a favourite son of the Western resource companies, was to be the BCNI's errand boy in delivering the national unity word to the annual premiers' conference.

Choosing Klein was no coincidence, as he is onside with the BCNI's call for radically decentralized constitutional powers. While the BCNI and its members are concerned about Quebec separation, the use of the unity issue as a vehicle for the further devolution of the federal government is a theme common to Klein, Ontario's Premier Mike Harris, and the federal Reform Party. And gutting the federal government's authority is consistent with other BCNI initiatives.

Ostensibly about Quebec, the memo to the premiers focusses almost exclusively on shifting powers from Ottawa to the provinces. It states that "all provinces have their right to equal treatment as partners in the Canadian federation."[44] It then lists seven principles that should guide the "evolution" of the federation, almost all of which imply the gutting of the federal government, including the Reform Party's favourite propaganda line that services should "be provided by the level of government that can do so most effectively and efficiently within the framework of the constitution."[45]

The Business Council on National Issues is the quintessential agency of corporate rule, perhaps unique in the developed world. There are more powerful organizations, such as the Business Roundtable in the U.S., but none have likely dominated political life to the degree that the BCNI has. It has forged an elite consensus so powerful that, as Duncan Cameron put it, some corporations whose interests have been harmed by free trade, for example, "have fallen on their swords" as a sacrifice to the general good of corporate rule. And the council has conducted its affairs with a strategic intelligence that has, with few exceptions, allowed the state to implement a comprehensive corporate agenda while appearing to act in the public interest. In classic role-of-the-state terms it has facilitated an enormous increase in the government's "accumulation" role for capital while maintaining its legitimacy.

OTHER PLAYERS, DEEPENING CONSENSUS

The BCNI is so dominant in its role of creating elite consensus that its supporting organizations are often overlooked. But it does have help from many other players, and the consensus is as powerful as it is because it has now permeated many other institutions and agencies that once produced broad policy alternatives. Among its most powerful allies (and ex-officio members) is the Alliance of Manufacturers and Exporters (formerly the Canadian Manufacturers' Association). In addition, the council and some of its key members have brought onside organizations whose membership is not at all united on the issues at hand.

Two such organizations are the Canadian Chamber of Commerce and the Canadian Federation of Independent Business (CFIB). Both consist for the most part of small and medium-size businesses whose interests depend not on the so-called global economy or even on trade but on the health of the domestic economy. The tens of thousands of bankruptcies over the past ten years have been almost exclusively among smaller firms, and many of those failures can be traced to the permanent recession created at the behest of the country's largest corporations. While only one-third of CFIB members said they would benefit from free trade, its chief at the time, John Bullock, claimed otherwise, and supported it from the start.

There are fundamental conflicts of interest between small and medium businesses and large corporations in Canada, which normally would be reflected in conflicts in public policy needs. Yet the organizations that supposedly speak for small and medium companies are almost always onside with the BCNI. Here, too, the dominant BCNI members have paid attention to strategy and work hard to ensure that there are no cracks in the business consensus. Duncan Cameron gained some insight into how they "manage" their lesser cousins:

> I was invited to speak at the Canadian Chamber of Commerce on a panel to discuss the budget. I arrived there and the banks had funded all the events. The lunch was provided by Scotiabank, the dinner was the Royal Bank and so on down the line. The banks were all over it. They were on the program and on the program committee. I think the chamber is a legitimate representative of small businesses which join at the local level but once it gets to the senior level it's a different story. Which members can afford to fly to an annual meeting and who pays for the tickets? Who's actually running the Chamber of Commerce? When Jean Chrétien says he checks with the Chamber who is he actually checking with? He might as well just check directly with the banks.[46]

The BCNI also has a strong ally in the C. D. Howe Institute, the principal corporate think-tank in the country. The C. D. Howe plays a dual role of helping to create the elite consensus and also helping to change the political culture to reflect it. It has played a major role in bringing key elements of the elite onside, including senior public policy makers at all levels.

As with all such think-tanks, the C. D. Howe provides an air of neutrality and intellectual legitimacy to an ideology that does little more than serve the interests of its large corporate members. The institute's history reflects precisely the shifting consensus in the corporate and political elite. Until the late 1970s it promoted policies of full employment, tax reform, and social program enhancement. By the 1980s it was pressing Ottawa to pursue free trade with the U.S. and urging all governments to attack deficits through massive cuts to social spending. One of its most prominent policy papers, *Social Policy in the 1990s*, by Tom Courchene, provided the ideological foundation for the Liberals' gutting of the UI program, one of the most important elements in the corporate agenda.

Its membership is dominated by the same Canadian corporate heavyweights that dominate the BCNI. There are 280 members and sponsors, including Alcan, Canadian Pacific, GM, MacMillan Bloedel, all the big five banks and the bankers' association, Quebec's Power Corporation, and Great-West Lifeco, as well as such key individuals as Thomas d'Aquino and Peter Bronfman. Its board of directors, like the corporate sector it promotes, is dominated by the financial sector.[47]

Once the ideas of the right were reflected in public policy — ideas of "smaller" government, market solutions to public problems, the debt as the principal issue of society — they began to permeate every public institution that dealt in ideas. Just as Keynesian economic policy had enjoyed this status as "common sense" until the mid-1970s, new-right neo-liberal prescriptions achieved that status within government by the late 1980s. And the aggressive corporate promotion of these ideas, whose claim was to have simple answers for everything, inevitably infected institutions that were once forums for genuine debate. Business and its ideology are now everywhere. According to Duncan Cameron:

> Any successful operation eventually needs more money. Once you let the business groups use their network to invite people to the Canadian Club, for example, pretty soon they're going to be invited to be on the executive, and the next thing you know they're going to be inviting businesspeople to come and talk.
>
> For example, once the Canadian Institute of International Affairs got a journal going it was sort of natural to reach beyond its immediate membership. And the corporations, once they

> became international, got interested in international affairs, so they got involved and started providing funding. The department of external affairs got interested and soon the institute is dominated by a certain way of thinking.[48]

Universities, too, are rapidly changing, partly because of the dominant ideology but more because of huge funding cuts. The battle between public space and private space is determined by the withdrawal of public monies. Says Cameron: "The entire university community in Canada has become hostage to the new type of university president, who is in large part a fundraiser. The university president is no longer someone who takes a leadership role and speaks out on matters concerning society. They lunch, they raise money, and the terms of the relationship are quite clear. Money for services rendered."[49]

Notwithstanding the economic power of corporations and the political power of the state, decision making must still appear to be democratic. And indeed it is still politicians who vote in Parliament. The crude application of elite rule, especially when contrasted with a recent history of more open democracy, would have been unacceptable. Government must still balance its two key roles in a liberal democratic capitalist society: accumulation and legitimation. Without coercion the first cannot succeed in the long term without the second.

We grudgingly accept the new world order being created by transnational corporations only because of an unprecedented ideological assault on Canada's communitarian political culture by corporate-funded think-tanks, lobby groups, right-wing populist organizations, Preston Manning's Reformers, conservative foundations, and the corporate media. This relentless attack on social programs, public services, the idea of government, and the ideas of the Left and small-l liberals has smoothed the way for the corporate domination of the country. In an effort to foment a counter-revolution of falling expectations, the free-market propaganda machinery has tried to bully Canadians out of their expectations of government and their own vision of the country. It is machinery largely designed and paid for by the corporations that stand to gain from the changes.

8

Changing the Ideological Fabric of Canada

If you really want to change the world you have to change the ideological fabric of the world.

— Michael Walker of the Fraser Institute

I understand: the Fraser Institute is the wholesaler and the NCC is the retailer!

— Milton Friedman

Alfred Powis was a busy man in 1975. Chairman of the powerful Noranda Corporation, he was also a founder and the first chair of the Business Council on National Issues, a lobby group that would redefine the political role of corporations in Canada. He was motivated by the waning influence of corporations and believed fervently in the role they should play. It angered him that corporations had hit rock-bottom in the polls. He complained in an address to the Canadian Club that "the private sector is increasingly subject to uninformed, but strident and highly publicized attacks which seem to have a pervasive impact on government policies."[1]

Even in times of relative public disfavour, CEOs can depend on two key factors to tilt the political scales towards corporations. The media's newsrooms are, in theory, independent of their advertising departments, but in practice their dependence on corporate advertising exerts a powerful pull towards corporate views. Corporate funds also play a significant role in the finances of most Canadian political parties, making it unlikely that politicians will bite the hand that feeds them.

In the mid-1970s, though, Powis and other key executives of transnational corporations felt that these pro-business filters were letting too many critical voices through. While corporations needed a powerful voice through an organization like the BCNI, they needed something else, too. They needed an organization that could counter the ideas of the Left.

This was clear to others in the corporate world, especially in B.C., where the NDP government of Dave Barrett was not just expressing ideas of social justice, human rights, and public ownership. It was implementing them, and by doing so was giving broad public legitimacy to ideas that until then were more or less confined to marginal groups with no claims on power. What was needed was a think-tank that would re-establish the dominance of free enterprise ideas, the value of the market, and property rights. To fill this need, Alfred Powis joined other corporate heavyweights and created the Fraser Institute, the now famous right-wing think-tank headed by the ubiquitous Michael Walker.

At about the same time, another businessman, not nearly so well connected but certainly well heeled, was getting just as angry at what he viewed as the dominance of unions, a growing and alarming presence of the government in the economy, a plethora of universal social programs like medicare, and a general loss of economic freedom. For Colin M. Brown Sr., "big government" was a threat to everything he held dear. He had been acting on his beliefs for several years but in 1975, on the advice of his friend retired Alberta premier Ernest Manning, he formally established the National Citizens' Coalition. It would become the pit-bull terrier in the corporate assault on government and communitarian values.

Of all the institutes and organizations working to promote minimalist government and market solutions, the NCC and the Fraser Institute are among the most important. This chapter deals with them more extensively than other agents of right-wing change because they were established to undermine public support for public programs. Their strategy and tactics, their reliance on corporate funding, and their questionable claims to legitimacy all need to be exposed. We need to understand how corporations, by changing the political culture of Canada, have made the country more amenable to corporate rule.

The Fraser Institute has worked systematically to change the country's "ideological fabric," focussing on media coverage, and on grooming a right-wing intellectual elite. The NCC, as its long-time president (until 1998) David Somerville says, is "very directly engaged with the political process" and has more leeway than the Fraser Institute because it is set up as a nonprofit organization rather than as a charitable organization. Through its political activism, the NCC has provided a model for a variety of pro-business groups such as the Canadian Taxpayers' Federation and the Canadian Progressive Group for Independent Business. The NCC's successful court challenges to limitations on third-party spending in elections have opened the door for money to play an even greater role in Canadian politics.

The role these organizations play can scarcely be exaggerated. They and other agencies such as the media, public relations firms, and the Reform Party, also examined in this chapter, have been key players in the corporate domination of Canada. The development and growth of corporate rule in Canada and elsewhere could not have taken place without an ideological assault on the values and expectations of ordinary citizens. Ideology has been called meaning in the service of power, that is, the creation of rationalizing myths, ideas, and, in today's lexicon, "common sense," that pave the way for people to accept conditions they would otherwise protest against. Had Canadian values and expectations remained untouched from the mid-1970s, the assault on medicare and public education, the casual savaging of people in poverty, and the criminally high levels of unemployment would never have been tolerated. Certainly governments espousing these policies would not have been elected.

It is the job of ideological agencies to scorch the cultural earth to prepare it for the assault on government services, employees, and the environment. The terminology used to describe this ideology is often confusing, as the terms *neo-liberal*, *neo-conservative*, and *new right* often seem to be used interchangeably. Desmond King has described the relationship between neo-liberalism and neo-conservatism this way: "Liberalism is the source of New Right economic and political beliefs and policy objectives; conservatism provides a set of residual claims to cover the consequences of pursuing liberal policies."[2] He gives the example of neo-liberal-inspired cutbacks that force women to take on the burden of social problems. This consequence for women is justified by neo-conservative appeals to "traditional family values" and the traditional role of women. The NCC and the Fraser Institute have been disseminators of both neo-liberal and neo-conservative ideology.

The Fraser Institute's policy prescriptions are the answer to a CEO's prayers. Its likely candidates for membership are, says its literature, "owners of property or those who seek to own property." You can see why by reading any of its annual reports. Seeking legislation that will let your company hop jurisdictions and invest anywhere you get the most favourable terms? The institute campaigns for free trade, removal of interprovincial trade restrictions, and the decentralization of government programs. For people interested in cannibalizing public services for private profit, the Fraser Institute's papers proclaim the failures of the public health and education systems and assure the reader that privatization will offer more "choice." For executives enraged with organized labour, the institute organizes right-to-work conferences promoting legislation that makes it harder for workers to unionize. Are you challenged by women to improve fairness in hiring and promotion? The Fraser Institute's senior analysts deny there is any significant inequality in the workforce or that what does exist is just a result of women's lifestyle choice to bear children. Are you embarrassed by skyrocketing profits as poverty worsens? An institute analysis claims that poverty statistics are overstated. And if you are worried about being sued for faulty products or polluting, the institute undertakes a Law and Markets Project advocating limits on citizens' ability to take you to court.

But the Fraser Institute's work goes beyond the strategic publishing of papers and studies. It is a centre of extreme neo-liberal ideology, an ideology it resolutely promotes. And by implication this ideology is supported by all the executives and senior officials of the corporations that fund the institute.

Does the Bank of Nova Scotia's senior vice-president, Warren Jestin, think employers should be allowed to ask women if they plan to have children and discriminate against them if they do? Does Brian Levitt, president of Imasco, want Canada to model itself on Singapore, where prisoners are beaten so severely they are permanently scarred, citizens can be arrested without warrant, and opposition politicians have been locked up? Does Richard Currie, who received $2.4 million in 1996 as president of Loblaws and George Weston Companies, agree with Michael Walker that "poverty is simply a reflection of the fact that the sufferers were dealt an unlucky intellectual or physical allocation from the roulette wheel of genetic inheritance"?[3]

All of these positions have emerged from the Fraser Institute, self-described as Canada's "largest, privately-funded, public policy research organization."[4] This organization backed by Canada's blue-chip corporations has prospered and become increasingly influential in stamping a right-wing agenda on Canadian public life. More than half of the top one hundred most profitable corporations in Canada have contributed to the Fraser Institute. That is according to a list of corporate donors compiled by the institute in 1989, a practice it has not repeated.[5]

The institute declares it is "an independent Canadian economic and social research and education organization." Its "diversity of revenue" is supposed to guarantee its independence. However, while corporations contribute 31 percent and individual members 11 percent, fully 57 percent of its money comes from business-oriented charitable foundations, such as the John Dobson Foundation, whose declared purpose is to "educate the public with respect to the free enterprise system," or the fabulously wealthy right-wing Donner Foundation, whose stated purpose is to promote market solutions to public policy issues. (The Donner Foundation gave it $450,000 in 1994 to make government debt a dominant public concern.) Foreign foundation donations appear to be increasing rapidly, from $223,000 in 1994 to $342,000 in 1995, so that 17 percent of the institute's funding is now from foreign sources. The Fraser Institute has a small membership of two thousand and is able to raise $2.7 million from a very small pool of donors.[6] The Royal Bank handed over $20,100 to the institute, according to the bank's charitable contributions report for 1996.

Because it can count on such significant donations, the Fraser Institute's fundraising costs are minimal and more than 90 percent of its budget is spent on pursuing its anti-government goals. In 1996, the institute produced eight books, six studies, a monthly issue of *Fraser Forum* magazine, its youth newsletter, *Canadian Student Review*, sixty-five op-ed articles, forty-nine speeches, and four parliamentary submissions, hosted twenty-five luncheons and conferences, and funded six student seminars.

The institute says it takes no government funding. That should be qualified: it takes no direct government funding, but governments in Canada and the U.S. give up tax revenue worth an estimated $1 million to institute donors by allowing them to deduct charitable contributions to the Fraser Institute or its supporting foundations.

The claim is also made in its annual reports that it does not "undertake lobbying activities." However, it does claim credit for "changing the conventional wisdom about many areas of public policy." According to the institute's own documents, the organization offers free dinners to members of Parliament who attend its presentations, packages its products specifically for politicians, holds policy conferences in provinces where it thinks those governments are likely to adopt its ideas, and targets those it approaches for funding on the basis of specific kinds of projects.

Yet, to retain its status as a charity, the institute theoretically is not supposed to be "political." Ron Davis, director of Revenue Canada's charities division, has said that his department defines an organization as political "if its purpose is to affect government directly or indirectly, or to sway public opinion." It is hard to think of a single aspect of the institute, either its self-proclaimed goals or the activities it reports, that would not qualify as political under this definition.[7]

The Fraser Institute is best known for its prescriptions that put the highest value on "freeing" society so that people can pursue wealth. Democratically determined government policies are presented as gross impositions on freedom, as is conveyed in such institute titles as "Breaking the Shackles: Deregulating Canadian Industry."

Executive director Michael Walker is forthright in explaining what he and the institute are up to. In the 1990 annual report, Walker states: "The Institute is in the ideas business. In a way which is not possible for those in business who are perceived as having a vested interest, the Institute forcefully argues the case for the competitive enterprise system at every opportunity and in every forum."

The advantage of concealing vested interests becomes clear in the context of specific institute "products." Executives at major food companies must have been delighted when the institute produced reports slamming marketing boards. Likewise, Bramalea, Equitable Real Estate Investment, the Vancouver Board of Trade, and Vanac Development must have been very pleased with the institute's series of studies attacking rent control and public restrictions on land development in British Columbia.

Although the institute is usually circumspect about its service to the country's largest corporations, someone at head office leaked its five-year plan to the *Edmonton Journal* in 1996. According to the plan, the Fraser will "enlist the help of no less than 25 multinational companies in supporting the development of the [Economic Freedom] index." A person will be hired exclusively "to work with multinational firms and foundations for the purpose of getting more resources to support the Economic Freedom Project."

THE IDEOLOGICAL ENTERPRISE

In the fall of 1973, Michael Walker was working for the federal finance department when he got a call from an old college friend, Csaba Hajdu. Hajdu's boss, MacMillan Bloedel's T. Patrick Boyle, and other business executives in B.C. were greatly agitated by the NDP government of Dave Barrett and wanted advice on how to bring about its demise. In the spring, Walker met with Boyle, who twenty-three years later is still a Fraser Institute trustee. While a think-tank was not an ideal way to deal with the immediate problem of getting rid of the NDP government, Boyle and his mining-executive friends were apparently willing to take the long view. Walker's pitch was good enough to persuade fifteen of them to hand over a total of $200,000 to get the project started.[8] It was the seed money for the Fraser Institute.

Institute publications now brag that the organization's success is reflected in the shift in policies of all federal and provincial governments, regardless of who is in power. The Fraser Institute, in fact, was virtually a Chicago School Trojan horse sent into Canadian political life. Not only was Michael Walker a Milton Friedman follower but over the years they became close friends and racket-ball partners. Friedman does not seek elimination of the state; he wants the state dedicated to the interests of corporations and to social control. This relationship between state and business underlines the international aspect of monetarist, free-market ideology. In some ways, it is a movement, though one with a small and select membership. Throughout Europe, North America, New Zealand, and Australia, Friedmanite "franchises" play a similar role to that of the Fraser Institute.

Inevitably, the Fraser Institute's free-market advocacy conflicts with notions of democratic majority rule. In 1984 that conflict was made explicit with the launching of the Economic Freedom of the World project.

This project is essentially a political effort by the Fraser Institute, and other right-wing institutes collaborating with it, to redefine freedom in the public debate about the direction of governments. The project originated in a 1984 paper Michael Walker presented to the international club of free marketeers who belong to the Mont Pelerin Society. Rather than majority votes, opposition parties, freedom of association and press, or other such criteria commonly used to define the degree of freedom in a country, the freedom for individuals to do whatever they wanted with their wealth was to be the new standard. By this standard, Singapore, a virtual one-party state that administers public beatings to its prisoners, locks up citizens for their religious beliefs, and drives opposition figures into exile, is rated as far more free than Sweden.

Fraser Institute chair R. J. Addington explained the ideological goal underlying the Economic Freedom Index project: "The Institute's ambition in producing the Economic Freedom Index is nothing short of changing the nature of public discourse about the role of government in society. It is our ambition, by creating an international measurement movement, to ensure that adequate attention is paid to the implications of government actions for the level of economic freedom."[9] Under the Fraser Institute's Freedom Index, countries that focus on ensuring the basic needs of all citizens by promoting social equality are given demerits because such policies infringe on the freedom of investors. The Fraser Institute publication "Economic Freedom: Toward a Theory of Measurement" argues against "value-laden rating systems which indicate that democracy is the best way to advance economic freedom."

The Fraser Institute is forcefully promoting its ideological attack on democracy. In 1996, it hosted a conference in San Francisco that provided ideological training in the use of the index for participants from thirty-seven countries. The American foundation Liberty Fund, with its annual budget of $115 million, is paying for the institute's work on the index.

PURSUING ANTI-DEMOCRATIC ENDS

At a Fraser Institute symposium, "Freedom, Democracy, and Economic Welfare," held in 1986, Milton Friedman went after a speaker who dared to say "democracy is an ultimate value, given protection of minority rights and basic fundamental rights." Friedman said flatly, "You can't say that majority voting is a basic right . . . That's a proposition I object to very strenuously."[10] He argued that the ability to freely pursue the acquisition of wealth should be considered the ultimate social value, whereas the pursuit of social justice would "ruin the world." In *Fraser Forum*, Friedman has said: "One of the things that troubles me very much is that I believe a relatively free economy is a necessary condition for a democratic society. But I also believe there is evidence that a democratic society, once established, destroys a free economy."[11]

How far would the Fraser Institute roll back democracy? Here are the views of Walter Block, who co-authored the report "Economic Freedom of the World — 1975–1995," worked as the Fraser Institute's senior economist from 1979 to 1991, and whose opinion pieces are still published in *Fraser Forum*. At the 1986 Fraser Institute symposium on democracy, Block said: "Why does it follow that we should have an equal right to vote in the political process? Voting in a political process is not a negative freedom, it is a positive freedom, and it is an aspect of wealth. We don't say that everyone has an equal right to vote in IBM . . . It depends upon how many IBM shares they bought. If we look upon the polity as a voluntary organization, we must recognize the legitimacy for unequal votes."[12]

Block disputes the idea that freedom of assembly and freedom to form unions should be considered positive. He has called unions "bands of criminals," and says, "Unions are just institutions that engage in prohibition of entry into labour markets. They are anti-free labour markets, and I'll be damned if I can see why they get a plus. And the same goes for political demonstrations, which are often organized violations of private property rights."[13] Block also makes comments reminiscent of Nazi philosophers about how human rights are a sign of the decline of the strength of a people. "The first settlers in the land meet harsh conditions and this resolve and strong character carries over until the third or the fourth generation. But eventually later generations get weaker. They become involved in pornography and rights for homosexuals and things like that."[14]

Women's rights are also expendable in Block's world-view, sacrificed on the altar of private property rights: "Consider the sexual harassment which continually occurs between a secretary and a boss . . . While objectionable to many women [it] is not a coercive action. It is rather part of a package deal in which the secretary agrees to all aspects of the job when she agrees to accept the job, and especially when she agrees to keep the job. The office is, after all, private property."[15]

Michael Walker summarized the conference on democracy as finding that "majority rule of itself has no particular virtues."[16] In 1991, when he appeared before the Standing Committee on Finance to lobby for legislated limits on government spending, he argued against the "tyranny of the majority" and said an amendment to the Constitution was needed to place "a limitation on the ability of Parliament to legislate with regard to the extraction of a person's income . . . The Fraser Institute, through its National Tax Limitation Committee, has been investigating ways in which the self-destructive economic forces unleashed by democratic political choice might be restrained."[17]

He made similar arguments in 1993 to Finance Minister Paul Martin, who had invited the institute to an all-day meeting shortly after the Liberals won the 1993 federal election. Walker insisted in a paper entitled "The Political Problem" that just as employers did not downsize on the basis of the votes of their employees, nor should the government base its budgetary decisions on majority opinion.[18]

CHANGING CANADA'S IDEOLOGICAL FABRIC

In the booklet for the Fraser Institute Endowment Fund, Walker promises potential funders that "the organized involvement of informed and concerned individuals can greatly affect Canadians in their choice between two ways of life." Given its ideological project, it is easy to predict how the Fraser Institute would operate. Throughout its twenty-three-year history, it has targeted all the major institutions in the country that influence how Canadians think: the media, churches, political parties, nongovernmental organizations, and universities.

According to the institute's leaked five-year plan, "a central focus of our program . . . during the next five years will be the expansion of our penetration of the national media." The Fraser Institute has so penetrated the Canadian media that it is hard to imagine where it could possibly expand. Its 1996 annual report claims 3,108 references to the institute in the media that year, 51 percent more than the previous year. The institute holds news conferences and issues news releases, and makes sure it always has staff available to be interviewed. Its fax news-broadcasting operation sends a two-page news sheet to 450 radio stations every week. It provides packaged editorials for newspapers and radio designed "to explain the merits of the free-market system, issue by issue." As well, its seminars are extensively covered by cable stations, which gave it 105 hours of coverage in 1996.

The receptivity of the media is not surprising. Over the years, key media institutions have funded the Fraser Institute, including Sterling Newspapers, Southam, Thomson Newspapers, and Standard Broadcasting. David Radler, president of Hollinger, is an institute trustee, and Barbara Amiel Black, another Hollinger executive as well as a columnist for Southam papers, was a trustee until 1996.

Media-friendly gimmicks help the mass media get Fraser Institute views across. The institute's "Tax Freedom Day" — the day in the year when the average family has earned enough to pay its total tax bill to all levels of government — receives wide coverage. Fraser Institute report cards predictably flunk governments that, in the view of institute staffers, have not slashed or privatized enough.

The institute also claims to provide an objective evaluation of media bias through its National Media Archive. The authors of this department's publication, *On Balance*, review major news programs on

CBC and CTV for bias and claim their analyses are "completely objective." However, *On Balance* often demonstrates that these analysts are viewing programs through the coloured glasses of the institute's motto, "Public Problems . . . Private Solutions." For example, they criticized the CBC for emphasizing negative economic news such as youth unemployment rather than broadcasting good-news stories about increases in corporate and bank profits. CBC and CTV were applauded for substituting the term *scabs* with the management-friendly term *replacement workers*.

An independent group, NewsWatch Canada, evaluated seventeen issues of *On Balance* published in 1995 and 1996. It found generally that "while purporting to promote 'objectivity' and expose a lack of balance in journalism, the National Media Archive itself manifests a consistent pattern of innuendo, decontextualized results, and selective interpretation."[19] Yet despite these failings, Fraser Institute staff are sought for comment as experts on the media.

Young people are a particular target of Fraser Institute efforts. According to its 1996 annual report: "Over the years, particular attention has been paid to the development of the student program as the Institute and its supporters recognize the importance of bringing free-market economic ideas to university, college, and high school students in their prime learning years." Through sympathetic professors and students on campuses, the institute distributes twenty thousand free copies annually of its newsletter, *Canadian Student Review*. The institute encourages like-minded professors to place its publications on course lists and to assign its essay contests as part of class work. Typical of the themes for these contests is "Free Market Solutions to Environmental Problems."

Paul Havemann, writing about how right-wing ideas have been marketed in Canada, has described the Fraser Institute as "a finishing school for the salesmen of the new Establishment Ideology."[20] Each year, the institute runs six student seminars in cities across Canada attended by 1,000 students at a cost of $200 per student, and plans to increase the number of seminars to fourteen. Every year, it spends $1,250 per student on an in-depth leadership training colloquium for twenty participants; by 2001, three more will be added. Fourteen graduates from the student seminars have been deemed worthy enough to receive a week's worth of intensive training in right-wing ideology at the Institute for Humane Studies in Fairfax, Virginia. With funds from the Donner Foundation and the Hunter Family Foundation, the Fraser Institute hires seven students as interns.

"We have already had the satisfaction of seeing some of the earlier participants in our programs rise to positions of influence within political parties, or within the policy-making apparatus of government."[21] Ezra Levant, a former columnist for the *Edmonton Sun* and the *Calgary Sun*, and now legislative assistant to Reform Party leader Preston Manning, is perhaps the best-known institute acolyte. Levant worked for the institute in 1995 as an intern. Recently, the institute has published *Youthquake*, his polemic against government social spending. In this book Levant pits young people against baby boomers, accusing the latter of living high off the government hog and leaving the next generation to pick up the tab.

The Fraser Institute brags that "the current public policy agenda reads like an index of past Institute publications." Five programs specifically target politicians and government officials: seminars for MPs; a "hot line" that MPs can phone to get "direct personal assistance"; free copies of all institute publications, including the *Fraser Forum*; a new series of publications, *Public Policy Sources*, which supply short position papers on topical issues, and the National Media Archive, which provides transcripts of all CTV and CBC news programming.

During the 1993–97 Parliament, it often sounded as if the Fraser Institute was the research wing of the Reform Party. Twenty-two of the fifty-one Reform MPs drew on institute materials for their speeches. The party helps promote Fraser Institute gimmicks like Tax Freedom Day, and in return gets advance copies of institute studies; Michael Walker's study on cutting social programs was first presented to a Reform Party policy session. Reform MPs Rob Anders and Jason Kenney are also Fraser Institute members. Support for the institute is not confined to Reform, however. Finance Minister Paul Martin signalled his own comfort in being associated with the institute by presenting the Fraser Institute's $20,000 Prize for Economy in Government in 1994.

The institute is invited to make submissions to parliamentary standing committees and is consulted by civil servants. Some who have made the pilgrimage to the institute include the federal finance department's assistant deputy minister Don Drummond, members of the B.C. premier's round table on social program renewal, and foreign affairs staff accompanied by foreign guests.

INSTITUTE CLAIMS OF LEGITIMACY

Fraser Institute studies are replete with charts, graphs, and numbers that appear to demonstrate the seriousness of a problem, often in terms of what the cost of a program is to the average family. It makes the self-serving claim that it has a "well-deserved reputation for the quality of its work, which earns its recommendations the attention of policy makers around the globe."[22] But the methods used in these studies often would not pass muster in an undergraduate social sciences class.

A much-quoted hospital waiting list released in 1997 provides a glaring example: the survey consists entirely of the "impressions" of medical specialists about how long patients have had to wait for an operation; no random sample is used, and the results compiled are only from those motivated enough to send in their "data" to the institute. It has no more credibility than a TV phone-in survey asking a deliberately provocative question. Steven Lewis, who was head of the National Forum on Health's committee on private versus public medicine, described the survey as "methodological garbage."[23] Obviously, specialists have a vested interest; promoting the notion that there are lengthy waits allows specialists to argue for increased spending on their services.

As if the biases in that survey were not enough, the institute plans to carry out another one on patient satisfaction in conjunction with a vociferous U.S. opponent of public-sector health care, the National Center for Policy Analysis. The survey will double as a media campaign since "patients will be asked to indicate their willingness to give interviews to the press about their experience so that the results of the survey can be translated into terms that are more directly understandable by the general public."[24]

Michael Walker and Walter Block in their 1985 study denouncing employment equity programs reported in a footnote that they were engaged in "an informal competition" with a researcher at the American

Enterprise Institute on who could come up with the better numbers indicating women were faring better in the workforce than men.[25] They also used Helen Gurley Brown's 1964 pop-psychology book, *Sex and the Single Girl*, as reference for their claim that "other things equal, they [women] will accept lower pay for a job which puts them in contact with large numbers of eligible bachelors."[26]

Another example of the quality of the institute's work was reported in the *Vancouver Sun*. The story, headlined "Beloved Regulations Squeeze Us Badly," featured an institute study attacking government regulations.[27] Researcher Fazil Mihlar had used U.S. data compiled by an adviser to President Reagan to claim that government regulation is harmful to the public. His solution: hand regulation over to corporations. Journalists reporting on the proposal seemed blithely unconcerned about its implications for Canadian democracy. Nor did they comment on the fact that not a shred of Canadian data was used.

With the sheer volume of Fraser Institute activity (eleven conferences in 1996 alone), it would take a separate institute to evaluate how well its research stands up to academic standards. However, an international team of economists and labour lawyers did evaluate the presentations made at one of the institute's 1996 conferences. In a report entitled "Bad Work: A Review of Papers from a Fraser Institute Conference on 'Right-to-Work' Laws," they documented a multitude of errors that could only be seen as the result of an anti-union political agenda.[28]

One speaker gave a talk entitled "Closed Shop Provisions Violate Canadian and Provincial Charters of Rights and Freedoms." But he somehow neglected to mention a key Supreme Court ruling that closed-shop provisions do not violate the Charter. Another speaker argued for reform of a B.C. law that he said gave too much power to unions. But the law had been repealed a decade earlier.

The Institute repeatedly demonstrates a willingness to make claims that are unsubtantiated or even contradicted by available data. That the media do not reject them may reflect a bias in favour of the same ideas or may simply be because reporters, stretched to the limit by downsizing, lack the ability or resources to critique the institute's work.

With medicare on the ropes, governments dealing with debt by cutting spending, the CBC in peril, NAFTA in place, and the MAI looming, the Fraser Institute might be expected to be pacified. Not so. Its five-year plan calls for doubling its budget, to $5 million, and aggressively pursuing its radical vision of corporate libertarianism. Public schools, universities, hospitals, and public land all will be on the auction block if the details of this plan are carried out. Unions and environmental organizations could be crippled if government spending is lowered to a minimalist 30 percent of GDP from its current 33 percent. Fazil Mihlar has said that governments should reduce their legislation until all that remains are "framework laws," the ones that protect property and ensure contracts.[29]

THE NATIONAL CITIZENS' COALITION

An insight into the division of labour between the Fraser Institute and the National Citizens' Coalition was provided by NCC president David Somerville at a 1996 gathering of right-wing libertarians. Asked about the merits of joining organizations like the NCC, Somerville replied: "If you want red meat for breakfast then you want to get involved in something like the National Citizens' Coalition. We are very directly engaged with the political process. I just want to offer one caution and that is to do the kind of job that I'm doing you not only have to be strongly intellectually engaged but strongly emotionally engaged."[30]

The NCC's "red meat for breakfast" style of operating was established by its founder, millionaire businessman Colin M. Brown. In 1967 this London Life insurance agent set the pattern for NCC campaigns by placing a full-page ad in the *Globe and Mail* attacking medicare and asking for donations to spread the word. According to an NCC description of Brown's legacy, he waged a "ceaseless battle against big government and big unions."[31] Brown himself described the NCC as "a hobby that went berserk."[32]

The NCC has tried to make secrecy a democratic virtue. The organization has repeatedly refused to reveal who backs it, and does not even provide a breakdown of corporate versus individual supporters. A May 15, 1992, NCC news release criticized a proposal for electoral reform because those who contributed more than $250 to a third-party group would have to reveal their names. Somerville said, "This is an affront to freedom of speech. No Canadian should be forced to divulge his political beliefs," and he compared funding disclosure to the right to a secret ballot.

The NCC campaigns to allow the wealthy to exert a disproportionate influence on Canadian political life. Secret ballots enable all citizens to exercise a right, regardless of wealth. In contrast, keeping secret the names of corporations and individuals who have bankrolled particular campaigns conceals key information about these campaigns. For example, farmers might want to know if Cargill Grain is covering the costs of the NCC's current support for challenges to the Canadian Wheat Board.

In 1985, Canadian Labour Congress president Dennis McDermott accused the National Citizens' Coalition of being "nothing more than a front for some of the wealthiest and most powerful individuals and corporations in the country." Somerville responded, "There's nothing wrong with private associations . . . They're throughout society."[33]

One source of information that lists individuals and corporations willing to be associated with the NCC is the program given out at the NCC's annual Colin M. Brown Freedom Medal awards dinner. At the 1996 dinner for Ontario premier Mike Harris, some of those paying for advertising in the program or named in it as an NCC "Patron of Freedom" include John D. Leitch, Edward Bronfman, Jack Pirie, and Thomas Bata; Upper Lakes Shipping, Magna International, and Rogers Cable; and the John Deere Foundation. The money raised from this dinner-program advertising alone was at least $15,000.[34]

A list of the members of the NCC's advisory council and board of directors is another source of information about its backers. Very powerful members of Canada's corporate establishment have lent the NCC credibility and assisted its fundraising.

John Leitch acted as chair in 1987 and has been a member of the advisory council for more than two decades.[35] Leitch is past-president of Upper Lakes Shipping International and has been a director of ten major corporations, including the Bank of Commerce, Massey-Ferguson, Canada Life, Dofasco, and American Airlines. One-time advisory council member John Clyne was the chair of MacMillan Bloedel for sixteen years. Like Leitch, Clyne was a director of the Bank of Commerce and he has held directorships in five major corporations. Ernest Manning, yet another Bank of Commerce director, was involved with the NCC. After being the Social Credit premier of Alberta from 1943 to 1968, Manning went on to sit as a director of eleven major corporations. Gerald Hobbs brought his connections as director of the Bank of Nova Scotia to the NCC advisory council. Hobbs was head of Cominco, and was also a director of B.C. Tel, North America Life Assurance, MacMillan Bloedel, and Pacific Press.[36]

Nick Fillmore's analysis of key NCC advisory board members in 1986 revealed ties to thirty-nine major Canadian corporations, including, in addition to the above, Canadian Pacific, Brascan, Bank of Montreal, Royal Trust, Power Corporation, and Bell Canada. There were also links to "eight major insurance companies, seven advertising agencies and more than fifty lesser corporations."[37]

Reinforcing its connections with the country's business elite at the national level, the NCC got Keith Rapsey, the former president of the Canadian Manufacturers' Association, to serve on its advisory council during the 1980s. As well as these nationally well-connected businessmen, the NCC has drawn on the prestige of members of regional business elites in appointments to its advisory council, including Harold P. Connor, the former head of National Sea Products, and Jack Pirie, president of Pirie Resource Management.

In a response to an article critical of the NCC, communications director Gerry Nicholls claimed: "We do not receive a single cent in government handouts, and contributions to the NCC are not tax-deductible."[38] The NCC is registered as a nonprofit society, rather than a charity, so it cannot issue charitable receipts. However, Revenue Canada allows businesses to deduct contributions to such organizations. Businesses can pay to become NCC members and get its newsletters, then write off this support as business expenses. As well, businesses that advertise in the programs distributed at NCC events qualify for tax deductions.

Noted Canadian tax lawyer Arthur Drache, writing in the *Financial Post*, has said that the deductibility of contributions is how business lobby groups like the NCC "are able to operate on handsome budgets." Drache argues that "the end result is that . . . businesses are much more able to make their views heard with deductible tax dollars than are individuals."[39]

Every year, the NCC publishes "Tales from the Tax Trough," which ridicules government spending. Grants to organizations like the National Action Committee on the Status of Women come in for particular attack; feminists are portrayed as fat pigs with hairs sprouting from their chins. But government grants to organizations are at least subject to review by democratically elected politicians, whereas the public has no say at all over how the NCC's budget, $2.6 million in 1996, is spent. Yet if businesses are allowed tax deductions for their donations to the NCC, all Canadian taxpayers are forced to support the organization through the revenue the government gives up. This support includes helping to disseminate arguments about why it is not in the nature of the French to be democratic;[40] why Vietnamese refugees should not have been allowed into Canada (because they would bring in thousands of their relatives and would not "fit in" as well as Europeans with the same "blood lines" as Canadians),[41] or how there is almost no real poverty in the U.S.[42]

THE NCC'S GOALS

Colin Brown campaigned against medicare in the 1960s and 1970s and battled in the early 1980s against the Canada Health Act, claiming in an NCC fundraising appeal that people "*would die*" if the act was passed. He asked prospective donors to contemplate "how you would like your open-heart surgery done by a civil servant?"

Brown summarized his anti-government philosophy by claiming, "The less a government does, the more the man-on-the-street has to do. And he enjoys doing it."[43] Articles in NCC publications decry the public school system as a failure, advocate privatization of crown corporations including the post office and the CBC, ridicule government grants for research and the arts, and praise the private health-care system in the U.S. as superior to Canadian medicare. The NCC has organized Canadian speaking tours of British consultant Madsen Pirie, who advocates "the privatization of everything." The NCC sent copies of Pirie's guides on how to privatize to 1,465 federal, provincial, and municipal politicians, as well as to senior civil servants. Alberta premier Ralph Klein made it required reading for every member of his caucus.[44]

Consistent with the views of Milton Friedman, the NCC calls for the elimination of government, except in a policing and military role. Although the NCC's motto is "More Freedom Through Less Government," the organization stands for "a strong defence." Brown and Somerville made a special trip to Washington in 1985 to oppose what they called the "strident anti-Americanism of Canada's peace movement."[45] NCC ads placed in American newspapers said it was time Canadians acknowledged our "debt" to the U.S. While they were in Washington, Brown and Somerville met with Republican senators and members of Congress who opposed arms control; the Canadians promised to maintain contact with them.

The kind of political advocacy the NCC engages in is expensive. One NCC project alone, Ontarians for Responsible Government (O.R.G.), budgeted $560,000 in 1995 to wage what it termed an "all out electoral war" against the provincial NDP.[46] O.R.G. was created in 1991 "to bring down the government of Bob Rae."[47] Billboards with anti-government messages appeared first in downtown Toronto and then right across the province, with pictures of Rae and slogans like "How Do You Like Socialism So Far?" The Ontario news media gave extensive coverage to this campaign.

As community and labour organizations mounted opposition to the cuts introduced after the Conservatives defeated the NDP government, O.R.G. began a counter-campaign, using its substantial media-buying budget to support Premier Mike Harris. O.R.G. spent $20,000 in the week preceding the 1995 general strike in London, Ontario, calling for a "fight back" against the unions. It ran a similar campaign to combat the protest in Hamilton in 1996.

NCC radio and TV commercials in the 1996 B.C. election could hardly have been more direct: "Don't vote for Glen Clark's NDP." Because B.C. law prohibits third parties from buying ads that tell people how to vote, the NCC spent $44,000 to have its message broadcast on American television and radio border stations. Somerville issued a news release saying the law limiting third-party spending should be opposed because it was "dangerous and oppressive" and "citizens who value freedom must resist tyranny."[48] When the NDP had won, Somerville announced the NCC would be funding Kelowna businessman David Stockell's attempt to overturn the election results.

How did the NCC get to play such a major role in Canadian politics? According to writer Nick Fillmore, until 1984 "the coalition was very much an unimportant right-wing fringe group, paid little attention by most politicians, the media and even shunned by other right-wing lobby groups . . . The first breakthrough came in July, 1984, when the NCC successfully used the Alberta Supreme Court to overturn the federal government's bill C-169, a . . . law aimed at preventing third parties . . . from advertising a political position during an election campaign."[49]

Bill C-169, a bill with all-party support, was designed to block spending in elections unless it was approved and accounted for by the party that stood to gain from the spending. Judge Donald Medhurst in striking down the law said there had to be proof that such spending undermined democracy before any government could impose limits on the freedom of expression guarantee in the Charter of Rights and Freedoms.

The NCC's court victory opened the door to virtually unlimited corporate spending in the 1988 federal election, arguably the most important election in Canada in decades. Advocates of free trade were able to far outspend opponents; the Canadian Alliance for Trade and Job Opportunities spent $1.5 million on one booklet alone, double the total amount spent by the Pro-Canada Network to fight the deal.[50]

In 1993 Justice D. I. MacLeod of the Alberta Court of Queen's Bench declared that the renewed federal efforts to prevent third-party spending through Bill C-114 were also unconstitutional. Bill C-114 was partly the result of the Royal Commission on Electoral Reform and Party Financing, which described the 1988 free-trade election as "the most striking intrusion of third-party advertising in a national campaign in over 40 years."[51] The federal government appealed MacLeod's decision, but in 1996 the Alberta Court of Appeal ruled against the government and blocked electoral reform once again.

Along with its campaigns to eliminate funding for social advocacy groups, the NCC has established its reputation in attacking the ability of unions to support social causes. Perhaps inspired by similar and successful action by corporate interests in the U.S., in the 1980s the NCC funded community college teacher Merv Lavigne's court case, which challenged the social-action objectives of Canadian unions.

The argument Lavigne's NCC-paid lawyers made was that his right to freedom of expression as defined in Canada's Charter of Rights was infringed upon by having to pay dues to his union that were used for purposes he personally did not support.

Had the case succeeded, it would have meant that unions would have had to go back to each individual member every time they were deciding whether to spend even a couple of pennies. The Canadian Labour Congress and other unions pointed out that Lavigne had every right to engage in the democratic process within his union to determine what political causes were funded.

The NCC won the case at the Ontario Supreme Court, but the Ontario Court of Appeal overturned the decision in 1989, saying in part that "it is not the courts' job to decide what is collective bargaining and what is not." Lavigne, again financed by the NCC, took the case to the Supreme Court of Canada, where in 1991 the court ruled again in favour of the unions.

In her judgment, Chief Justice Bertha Wilson observed how limits to union social-action spending had worked in the U.S.: "When [American] unions speak out on political matters, for example, they must (upon request) refund to dissenting members the prorated cost of such activity. Corporations do not have this problem; corporations may speak out on political subjects in spite of shareholder dissent. Corporations also speak with a far louder voice, heavily outspending labour on dissemination of their views."

The Lavigne case cost the NCC close to a million dollars, demonstrating just how much its backers were willing to spend to cripple unions. Having lost at the Supreme Court, the NCC kept up its anti-union efforts. In 1995 it launched a project called "Canadians Against Forced Unionism" that was dedicated to the introduction of right-to-work legislation. The project's spokesperson, Robert Anders (now a Reform MP), declared, "The time has come to free Alberta's workers." Ralph Klein's Conservative government decided against right-to-work legislation, partly on the basis that, with the existing laws, the rate of unionization was already decreasing in the province.

NCC LEGITIMACY AND LOBBYING

In the early 1970s many of the social movements the NCC criticized seemed to be reaping the benefits of years of activism: social programs appeared secure, and labour, human rights, and women's groups were having a significant influence on public policy. An NCC newsletter recounts how at one point in this period Ernest Manning "urged our founder to transform his one-man crusade into a citizens' action group. Shortly thereafter, Colin [Brown] incorporated the National Citizens' Coalition." Brown got $100,000 in seed money to do this from wealthy Canadians.[52]

The National Citizens' Coalition may have taken on a grassroots name for Brown's crusade, but it took on nothing of the character of a membership-controlled organization when it was incorporated in 1975. It is not national, its citizen members can't vote, and it is not a coalition. In its bylaws the NCC distinguishes between two categories of membership — public and voting. According to Bylaw 27: "Public members shall not be entitled to receive notice of or to attend any meeting of the members of the Corporation and shall not be entitled to vote at any such meeting."

Voting members, on the other hand, are entitled to receive notification of NCC meetings, attend them, and vote. Voting members choose the board of directors, the president, and the vice-president. The NCC does not reveal who its voting members are (this is a highly select group) and, according to Bylaw 28, *only two* voting members are required to make up a quorum at meetings.

A voting member also qualifies to become one of four NCC directors, and only three of these have to be present at any meeting to conduct NCC business. These three or four people decide who is allowed to become an NCC member, both voting and public, and the fees. Directors can also force members to resign. This structure makes the NCC seem more like a private lobbying firm than a non-profit citizens' organization.

Every NCC ad campaign includes a reference to the number of supporters (virtually all of whom are in Ontario, Alberta, and B.C.) the organization claims to have. However, there is inconsistency in the figures the NCC reports — both 40,000 and 45,000 have been published in different NCC documents for 1996.[53] It reported a membership of exactly the same number from 1979 to 1986, a constancy that is hard to credit given the variation in the popularity of its campaigns. Efforts to find out what membership entails have repeatedly failed, but anecdotal evidence suggests that once you join you are on the NCC's membership list for many years.

The NCC claims, "We do not lobby politicians or bureaucrats — we speak directly to our fellow citizens." This statement, from "Who We Are and What We Do 1996," the most recent organizational pamphlet, is another attempt to cultivate a grassroots image. In fact, the people that run the NCC rarely speak with fellow citizens or even with their own supporters. What the NCC does do (a typical campaign is its right-to-work crusade in Alberta) is commission opinion polls, try to generate public pressure on politicians through opinion pieces and mass-media advertising, cultivate political friends who will push its policies to the forefront, and make submissions to government. These are all the traditional activities of lobbyists.

The NCC has always had privileged backroom access to politicians, starting with its founder Colin Brown, who took Ontario Conservative premiers Bill Davis and John Robarts along with him on annual chartered flights to the Masters Golf Tournament in Georgia. And over the years, well-connected right-wing politicians have been on its advisory board: Sarah Band, who ran for the federal Tory leadership; Eric Kipping, a former Tory MLA from New Brunswick who led a delegation of former Conservatives from the Maritimes to join the Reform Party; Ernest Manning, and the late Robert Thompson, former national Social Credit leader.

The NCC gives $10,000 prizes to politicians who have contributed to "freedom" — as the NCC defines it. Politicians who have made the grade have been Tory cabinet minister John Crosbie, Reform senator Stan Waters, and Tory premiers Ralph Klein and Mike Harris.

The NCC has hired former Reform MP Stephen Harper to take over from David Somerville as president. Harper worked with Stan Waters on the Reform Party's original policy book. Somerville, commenting on the founding convention of the Reform Party in 1987, said, "If NCC supporters notice a remarkable similarity between the political agendas of the RPC [the Reform Party of Canada] and the NCC, it may be because an estimated one third to one half of the delegates were NCC supporters."[54] As guest speaker to the NCC's Colin M. Brown Memorial Dinner at the Hamilton Golf and Country Club in Ancaster, Ontario, in 1994, Harper cited the achievements of the Reform Party and the NCC since the last time he had addressed the same gathering in 1989:

> What has happened in the past five years? Let me start with the positive side . . . Universality has been severely reduced: it is virtually dead as a concept in most areas of public policy. The family allowance programme has been eliminated and unemployment insurance has been seriously cut back . . . These achievements are due in part to the Reform Party of Canada and . . . the National Citizens' Coalition.[55]

In 1994, NCC president David Somerville was invited to speak to the Ontario Conservative Party's Policy Advisory Council. He recommended the government "come out strongly in favour of privatization, contracting-out, repeal of pro-union labour laws and immediate action to eliminate the deficit and reduce the province's debt."[56] All of these recommendations coincide with the current policies of the Conservative government.

THE NCC AND THE MEDIA

By 1987, the NCC could report to its donors: "The NCC has attained a high profile in the major media. Hundreds of articles about the Coalition appear weekly in newspapers and magazines across the country. Television and radio stations follow the activities of the NCC very closely. In 1986, over 1,000 letters to the editor from Colin Brown and David Somerville were printed in newspapers across Canada."[57]

Starting in 1986, the NCC was supplying 160 newspapers with a weekly column. A comparison of NCC news releases on a campaign such as "Tales from the Tax Trough" with the articles major dailies printed reveals how uncritically some Canadian editors deal with NCC material. On issues like election spending limits and MP pensions, Canada's major dailies have adopted NCC terminology — for example, election spending limits are called gag laws, and MPs receive "gold-plated pensions." In its publication *Consensus* the NCC reports on numerous success stories of its influence in the media from prominent columnists and journalists. And many open-line shows use NCC spokesmen on its "Pigs-at-the-Trough Day."[58]

The NCC rewards journalists who have the "right" perspective. Three of the eight Colin M. Brown Freedom Medals and $10,000-cash awards the NCC has given out have gone to journalists: Barbara Amiel Black in 1987, Lubor Zink in 1989, and Diane Francis in 1995.

OTHER PLAYERS

The Fraser Institute and the National Citizens' Coalition are, of course, not the only corporate-funded institutions working overtime to change Canada's communitarian political culture. In 1994, an East Coast clone of the Vancouver-based Fraser Institute was established with $450,000 from the Donner Foundation. The Atlantic Institute for Market Studies was set up as an "economic and social policy think tank . . . with a view to determining whether and to what extent market-based solutions can be successfully applied to the myriad social and economic problems facing Atlantic Canada."[59]

And the C. D. Howe Institute has played a dual role of building elite consensus and of changing public opinion. Linda McQuaig has documented in her book *Shooting the Hippo* the influence the C. D. Howe Institute wields within government circles. One institute study, touting the elimination of inflation as an overarching goal for government policy, was "used by the Bank [of Canada] and by the government to justify John Crow's highly experimental zero inflation policy."[60]

Widely distributed C. D. Howe Institute studies foster panic about government debt, recommend radical cuts to social programs, and undermine confidence in the Canada Pension Plan. While it does not claim to speak for business in the way the BCNI does, the C. D. Howe Institute and the BCNI share corporate sponsors, as detailed in chapter 7. In 1996, the Royal Bank contributed $52,000 of the institute's $2-million budget. A December 1995 *Globe and Mail* survey of Canadian think-tanks quoted McQuaig's concern: "You see them referred to in the press as an independent think tank and that leaves out the fact that they're almost exclusively funded by Bay Street."

The C. D. Howe Institute seems to try to position itself vis-à-vis its larger rival, the Fraser Institute, by emphasizing that it is less committed ideologically and more objective, saying this "means refraining from polemics, keeping an open mind about solutions to difficult problems, encouraging support and input from a broad private sector membership base, and engaging in regular, substantive discussions with federal and provincial government policy makers." How much the institute avoids polemics is open to question. By calling the Canada Pension Plan a "Ponzi" scheme in its promotional material, the institute boasts that it has been able to "electrify" the public debate.

Right-wing foundations in Canada used to be a minor force relative to their counterparts in the U.S. The main player in recent years has been the Donner Canada Foundation, the tenth largest in the country with an endowment of close to $100 million. Established in 1950 by American steel magnate William Henry Donner, who had taken an interest in Canada, the foundation was until 1993 "the epitome of middle-of-the-road Canadian liberalism."[61] But since it was taken over by conservative Donner family members, it has become one of the principal funders of right-wing propaganda in the country. Under the direction (until late 1996) of American Devon Cross, former editor of the libertarian journal *The Idler*, the foundation gave out $2 million a year, almost exclusively to right-wing causes. By giving $400,000 to the Fraser Institute, bestowing large sums to fund the charter school lobby, bankrolling the conservative journals *Next City* and *Gravitas*, funding the market-oriented Energy Probe and its various spin-offs, and coming up with the cash to launch the Atlantic Institute for Market Studies, the Donner is a major actor in the campaign to "change the ideological fabric" of Canadian society. The Donner Foundation has adopted the strategy of its American cousins with a very focussed, deliberate promotion of vehicles for the dissemination of neo-liberal ideas.

MUTUAL SEDUCTION: RIGHT-WING ORGANIZATIONS AND THE MEDIA

The mere fact that the Fraser Institute, the NCC, and other right-wing organizations work so hard to attract media coverage should not guarantee that they will get it. Ideally, the amount of coverage would reflect the representativeness of these organizations and the reliability of their statements. Minimally, the media should be alerting the public to their political agendas, such as when the NCC announced it intended to wipe the Ontario NDP government off the political map.

Instead, as the Fraser Institute's annual reports highlight, there is a cosy relationship between prominent Canadian journalists and the Fraser. The *Financial Post* co-sponsors the institute's Economy in Government prize, which has rewarded such ideas as creating publicly funded private schools. Diane Francis, editor of the *Financial Post*, is pictured in an annual report photo addressing a Fraser Institute fundraising luncheon. *Financial Post* editor-at-large Neville Nankivell and *Globe and Mail* columnist "Terry" Corcoran are shown "sharing a joke" and "taking part in a discussion" with staff at the institute's offices.

The Fraser Institute seems to have a particularly cosy relationship with CTV news. In 1994, chief anchor and senior news editor Lloyd Robertson lent his support to the institute by serving as guest speaker at one of its fundraising luncheons. Mike Duffy, host of CTV's public-affairs program *Sunday Edition*, also helped the institute fundraise by being a guest speaker in 1995.

Some reporters do provide balanced coverage of the Fraser Institute and the NCC. But as right-wing organizations increased their media-influencing capability, forces within the corporate media were making it more open to influence. In 1970, Keith Davey's senate committee on mass media sounded a warning about the increasing concentration of ownership. Eleven years later, with the disappearance of even more newspapers, another federal investigation, this one headed up by Tom Kent, raised the alarm again. Not only were independent newspapers being bought out by such major chains as Southam and Thomson but chains were now swallowing up other chains. Government remained complicit in this steady erosion of democracy by declining to act on the key recommendations coming out of these reviews, a press ownership review board, and a Canadian newspaper act.

In the 1990s, Conrad Black's Hollinger Inc. has gobbled up the Sifton newspaper chain in Saskatchewan, bought into the company that publishes the *Toronto Star*, taken over seven Thomson dailies in Atlantic Canada and six in Ontario, vied to take over the *Financial*

Post, and attempted to buy out all of the remaining shares that Black does not already own in Southam. Hollinger spent half a billion dollars in 1996 gorging itself on Canadian newspapers. It now owns 60 of Canada's 105 daily newspapers, has monopolies in three provinces, and controls the Canadian Press wire service.

Hollinger's voracious appetite makes statistics on media concentration in Canada continually outdated. The ultimate Black conquest of all of Canadian newspapers, including the last significant holdouts in the Toronto market, seems inevitable. As it is, Black influences all but four Canadian daily newspapers. Through Canadian Press's Broadcast News, Black's influence extends to 425 radio stations, 76 TV outlets, and 142 cable stations — at a time when at least six countries have taken steps to encourage more diversity in newspaper ownership. Hollinger now controls 43 percent of Canadian newspaper circulation. In contrast, Gannett, the largest American chain, controls only 10 percent of that country's circulation.

With successive takeovers, more and more Canadian newspaper staff lost their jobs — 1,550 over three years in the Southam chain after Hollinger took over. Hollinger president David Radler, a.k.a. "The Human Chainsaw," radically cuts staff at small-circulation papers to create cash flow for new acquisitions. With fewer journalists on staff, news editors increasingly turn to the copy provided by organizations like the Fraser Institute to fill the "news holes" between advertisements in their papers.

The preference for right-wing copy starts at the top of Hollinger, with CEO Conrad Black and vice-president of editorial Barbara Amiel, whose neo-conservative views are documented in Maude Barlow and James Winter's *The Big Black Book: The Essential Views of Conrad and Barbara Amiel Black*. As well as running Amiel's weekly column, Black hired his cousin Andrew Coyne and Amiel's ex-husband, George Jonas, to flog their conservative views in Southam papers. David Radler, who has said it is important to have his employees fear him, states flatly that Hollinger papers, on principle, will endorse only free-enterprise parties, explicitly ruling out any paper's support for the NDP.[62]

Even with all the firings at Southam, journalists have been willing to admit that Black's influence is a factor in the newsroom. Karen

Sherlock, national editor at the *Edmonton Journal*, has said, "His shadow is a big shadow and we're feeling it."[63]

The transformation of *Saturday Night* magazine after Black bought it has also been a factor in the prevalence of right-wing opinion in the Canadian print media. With former *Alberta Report* staffer Kenneth Whyte as the magazine's editor, *Saturday Night* has been serving up a steady diet of Whyte's "advice for the right" columns, mean-spirited critiques of such Canadian heroes as anti–child labour activist Craig Kielburger and Farley Mowat, and articles on why women should be in the home rather than the workforce. *Saturday Night* gives yet another platform for Southam columnists Andrew Coyne and George Jonas to air their views, as well as to neo-conservative journalists from the Sun newspaper chain, such as David Frum, Michael Coren, and Peter Worthington.

In his biography of Conrad Black, *The Establishment Man*, published in 1982, Peter C. Newman provided an insight into the fate that would inevitably befall *Saturday Night* once Black took it over. Newman's book contains the following excerpt from a letter Black wrote to American arch-conservative William F. Buckley on how to change a magazine the way Buckley had transformed *National Review*:

> I take the liberty of writing to you on behalf of many members of the journalistic, academic and business communities of this country who wish to convert an existing Canadian magazine into a conveyance for views at some variance with the tired porridge of ideological normalcy in vogue here as in the U.S.A. [during the 1970s]. We are aware of the lack in Canada of serious editorial talent of an appropriate political coloration . . . We are, however, people of some means as well as of some conviction, and unless faced by an insuperable economic barrier, intend to persevere with our plans, to execution.[64]

As though the rightward turn of Canada's self-described "most influential magazine" was not enough, the Donner Foundation financed two new right-wing magazines. *Next City*, established in 1994 with a $1.4-million commitment from the foundation, seems to specialize in eroding compassion for the poor; writers since 1995 have celebrated Latin American shantytowns, portrayed beggars as scam artists, and declared that poverty is a matter of personal choice.

Editor Lawrence Solomon wrote in the summer 1996 issue that "for most people lifestyle choices dictate income levels, not the other way around." *Next City* has carried articles by Andrew Coyne, Fraser Institute senior fellow Filip Palda, and Fraser Institute writer Karen Selick, and has been distributed nationwide as an insert in the *Globe and Mail.*

The Donner Foundation gave a $390,000 grant in 1995 to neo-conservative Ian Garrick Mason to transform his newsletter, "Gravitas," into a quarterly magazine. *Gravitas* serves as a vehicle to distribute output from other Donner-funded organizations, such as the Fraser Institute, the Atlantic Institute for Market Studies, and the Society for Academic Freedom and Scholarship, whose members attempt to use research on brain size to attack anti-discrimination policies at universities.

Beyond the direct control that some of these owners exert over "their" media outlets (a *Vancouver Sun* reporter who asked to remain anonymous claims that David Radler calls the paper at least once a day to monitor content), there is the process of what Noam Chomsky calls "manufactured consent." Reporters, too, are subject to the tidal wave of neo-liberal and social conservative ideology sweeping over Canada from the dozens of sources outside the media. As in academia, those with left-wing or even moderately small-l liberal views feel pressured even by their colleagues to suppress their opinions, which in so many other venues are dismissed as archaic, naive, or flying in the face of the new "common sense." Public broadcasting is no less subject to these pressures, especially when right-wing commentators repeatedly single out the CBC for criticism.

The role of the public relations industry in selling corporate rule deserves a book of its own. The application of extremely sophisticated manipulative techniques in polling and the advertising that results from it are a critical element in the ideological assault on democracy. Some of the corporations engaged in this activity, like the giant Burson-Marsteller, are large transnationals in their own right. The largest, Burson-Marsteller, played a major role in selling NAFTA in the U.S.

Another public relations giant, Hill and Knowlton, gained notoriety for its "sales pitch" for the U.S. war on Iraq. President George Bush faced a disinterested American public not keen to save an obscenely wealthy and authoritarian elite in one country from the ruthless dictator of another. Hill and Knowlton, hired by the Kuwaiti government, put their people to work to come up with an image that would shake Americans out of their peaceful complacency. They came up with the story of Iraqi soldiers "ripping" premature babies from incubators in Kuwaiti hospitals, tossing them on the floor and stealing the incubators. The story was fabricated in a brainstorming session, as Hill and Knowlton later admitted, but it had the desired effect. In a famous documentary by the CBC's *fifth estate*, the young executive who "handled" the Kuwaiti account seemed completely unaffected by the fact that, as a result of his work (the incubator story was widely credited with shifting American public opinion), hundreds of thousands of people would die.

The political public relations industry's lack of morality is legendary; indeed, it almost defines the industry. It is characterized by the placing of powerful weapons of psychological manipulation in the hands of people who, in order to do their jobs, must cleanse themselves of ethical considerations. The most successful firms boast about how many corporate disasters they have managed to smooth over or spin into obscurity; how many deaths have been explained away. Typical services the spin industry provides to corporations are revealed in the brochure for Vancouver-based Verus Group International, which boasts in its list of accomplishments: "450 negative stories prevented, nine chemical spills explained, three mergers supported, six industrial deaths explained, 23 environmental protests handled."[65]

The application of public relations techniques to the political process is now a science. A particular type of polling, which reveals under what conditions people will accept change that their values would otherwise cause them to reject, is now commonplace. First used in the early 1980s in Canada, one of its original practitioners was Decima Research and its chief, Allan Gregg. One of Social Credit leader Bill Bennett's advisers described the service Gregg provided: "Decima believes that you can take [poll results] . . . and we can change your mind. We can move you to do something that you may not have agreed is the logical thing to do . . . We can move you to the other side of the ledger."[66]

Recent federal governments have taken the final step in merging policy making and public relations. They contract out much of the important policy development work to public relations firms that do polling, provide the research, and set the agenda. According to the CCPA's Duncan Cameron, "Even under the [Chrétien] Liberals most of the policy work in finance is done by the Earnscliffe [Research and Communications] Group, run by a Tory and ideologically preoccupied by the deficit."[67]

THE REFORM PARTY AND FRIENDS

The ideological activities of organizations like the Fraser Institute, the National Citizens' Coalition, and the C. D. Howe Institute all combine to undermine communitarian values and Canadians' faith in government. But these organizations focus primarily on neo-liberal economic, social, and fiscal policies. The assault on political culture comes from another source as well, and that is neo-conservatism, the expression of what are often called traditional values, particularly so-called family values.

Conservative ideology is expressed in the formal political arena by the Reform Party and its leader, Preston Manning. Manning's political roots are found in extremely conservative evangelical Christianity and are a significant part of the ideological fight against government. Manning's strategy since founding the Reform Party has been to promote the neo-liberal state by harnessing populist anger. The debates about abortion, gun control, immigration, the death penalty, youth crime, and human rights legislation such as affirmative action and so-called special rights for Quebec, aboriginals, and gays and lesbians have a dual impact on the political culture.

First, if one group is set against the other, any efforts to organize people around economic and social equality issues are made that much more difficult. Second, and perhaps more important, focussing on these volatile issues diverts attention from broader social and economic issues. In part, the appeal of such issues is rooted in the sense that here, at least, there are simple answers. People often feel ill equipped to cope with the larger questions of economic policy, the deficit, interest rates, "jobless" recoveries, and corporate power because all of these issues are put to them as inextricably linked to globalization. But youth crime seems open to obvious solutions — lock up young offenders at younger ages, for longer periods, and punish them with harsher treatment.

Central to this populist message is anger at or contempt for politicians and government, a sentiment that Manning, like his allies in the National Citizens' Coalition, carefully nurtures. This relentless attack serves to legitimize people's cynicism about politics (legitimate for other reasons as well), a feature that showed up dramatically in the 1997 federal election, which had one of the lowest voter turn-outs in sixty years.

In addition to undermining respect for the idea of government, Reform promotes the conservative values of the family, of women once again taking on the responsibility for social ills, of individual self-reliance, and charity. As the neo-liberal state cuts back and eliminates social programs, these "traditional" values are of great utility because they provide the rationale for shifting community and government responsibilities back onto women.

The influence of Reform has helped spawn supportive conservative social movements, which it then draws on for its electoral purposes. The Canadian Taxpayers' Federation has signed up tens of thousands of members at $50 a head in support of its campaign against the egalitarian state. Rather than attack social programs directly, the CTF typically goes after the "high taxes" Canadians pay. Originally organized as a virtual pyramid selling scheme, with most of the money going to its officers, the CTF and its provincial wings are, like the NCC, totally undemocratic. Members have no say in who runs the organizations, who is hired, how money is spent, or what the policies will be. There is considerable overlap between the Taxpayers' Federation and the Reform Party, as there is in the anti-gun control lobby, the anti-abortion movement, and the Ontario-based Alliance for the Preservation of English in Canada. All of these groups and movements have in common a visceral contempt for government, a theme they promote in virtually everything they do.

While Reform carefully nurtures its populist image, it is a party founded with corporate money from the Alberta oil patch. From the time he started the party, Preston Manning has made great efforts to attract corporate money, and in the past two years has been increasingly successful. Conrad Black's Hollinger and Canadian Pacific are two notable contributors in the past few years.

❖

Starting in the mid-1970s, the largest corporations in Canada launched a series of independent projects aimed at establishing a solid elite consensus regarding the long-term strategy of capital and the appropriate political direction of the country. The elite consensus was the job of groups like the Business Council on National Issues, other business organizations, and the C. D. Howe Institute. But the project to fundamentally change the political culture of the country, away from its communitarian tradition and towards a more individualistic bent, would be the job of a different set of agencies: arm's-length groups, funded and directed by corporations or corporate executives, but capable of acting as independent voices. They would be purveyors of ideology, not lobby groups for particular corporate interests.

Yet the massive output of the Fraser Institute, the NCC, and the C. D. Howe, popularized by the Reform Party and its allies, is only as good as the efforts to implement it into public policy. It is the strategic use of the issues these institutes and political organizations raise that determines the outcome for ordinary citizens. The goal of all of these efforts is, after all, the dismantling of the egalitarian state. Exactly how that was accomplished was played out in the broader political world in which governments used the scorched earth provided by the corporate propaganda agencies to implement the corporate counter-revolution.

9

Propaganda Wars: The Revolution of Falling Expectations

[A] large part of my message as a politician is to say: we have to put an end to rising expectations. We have to explain to people that we may even have to put an end to our love for our parents or old people in society, even our desire to give more for education or medical research.

— Pierre Trudeau, January 1977

A campaign to lower our collective and individual expectations as citizens and as income earners has been under way for more than twenty years. That campaign has been waged by the two most powerful organizations in history, transnational corporations and the state. The range of institutions, organizations, and resources dedicated to selling this counter-revolution against civil society is staggering. And the targets are obvious. The notions of equality and fairness, of democracy, of governments acting for the public good, of a public service dedicated to the ideals of community, all of these ideas and values have come under an unprecedented assault by corporations and the institutions of ideological warfare labouring at their behest.

The reason is brutally simple. Corporations and the wealthy elite who run them and benefit from their power are determined to experience democracy never again. The post-war period, particularly the 1960s to early 1970s, was as close as liberal democracy has ever come to achieving equality and security in Canada. That period was an aberration, and the ruling elite intends to keep it that way. To do that, the expectations that citizens here and in other industrialized countries developed in that period must be wiped from the collective memory. Rid the world of those high expectations, pacify those who would otherwise challenge you, and you make it safe for capitalism.

This campaign is reminiscent of the U.S.-backed Contra war against the Nicaraguan revolution. The strategy was not to win a military victory; it was to destroy the dream of social justice. Thus the Contras attacked and destroyed schools, day-care centres, health clinics, workers' co-ops, indeed any government agency or service that was working to make life easier for peasants and workers. The message was unmistakable: forget about things getting any better. If you try to make them better, we will kill you.

In Canada the corporate agenda is the same even if the tactics are considerably less brutal. The goal is to ratchet down the expectations of the majority of Canadians regarding their standard of living and the quality of life of their communities. The campaign, started in earnest in the mid-1980s, has been multifaceted. The strategy is to attack government on a whole series of fronts with the intention of lowering expectations of what the state can or will do. The Fraser Institute, the NCC, the C. D. Howe Institute, the Reform Party, and the corporate media have spent their money with great strategic care on the propaganda machinery that the corporations finance directly.

First came debt terror. Then came the attack on public services; if public services were no good, then the more "efficient" private sector could do it better. When the free market failed to deliver on its promises and people worried about unemployment, the campaign declared regretfully that governments can't create jobs. And now we are entering the final campaign: tax cuts. Governments having been proven no good for anything, we will give customers their money back.

There will be, if these campaigns are successful, nowhere to run. Corporations and governments will have established a permanent state of insecurity whose impact on democracy will be profound.

The campaign to lower expectations was paralleled by the very real decrease in what government and the private sector delivered. First, the huge cuts to government spending reduced the sphere of action in which governments exercise their sovereignty. Second, the multilateral trade and investment deals further restrict what governments can do. In the private sector, the strategy is simple: in the context of a manufactured recession, demand more and more of workers and provide less and less; drive up the amount of work to be done, provide fewer full-time jobs, demand rollbacks or freeze wages. Put the fear of the depression years back into the hearts of workers.

THE DEFICIT SCARE

In the world of government "downsizing" propaganda, no other single idea, no other single aspect of public policy has been so effectively manipulated to turn back the egalitarian state as the debt/deficit issue. Indeed, the debt is so effective precisely because every other issue leads back to it. Try to force a government to defend medicare and the conversation turns to the deficit. If citizens demand that the government do something about unconscionably high unemployment rates, the government shrugs apologetically and says, as a title to a Canadian Centre for Policy Alternatives booklet put it, "The Deficit Made Me Do It." You simply cannot have a discussion about public policy without reference to the debt and deficit. Even with the deficit disappearing, the huge debt looms in the background like some monster the government can set loose whenever it chooses.

The deficit hysteria campaign began in earnest the day after the free-trade election on November 22, 1988. The noon-hour news on CBC Radio announced that the country faced a debt "crisis." It must have developed overnight: there had been no mention of it during the election. And, in what became a consistent pattern, the reporter did not question this declaration, its source, or its suddenness. The love affair between the media and the deficit story had begun.

For the Tories, the debt crisis campaign could not be launched too early. Deficit talk had begun some years earlier and was simply eclipsed by the free-trade debate that raged months before the election was called. But cranking up the deficit volume became a critical strategy precisely because of the free-trade fight. Opponents of the deal had zeroed in on medicare and social programs as Canadian institutions threatened by the deal. The government and its allies had been obliged so many times to deny this claim that their denial became imprinted in the public memory. Any hint that medicare or education were going to be sacrificed to free trade, and the government would have been pilloried.

The Free Trade Agreement was, in any case, just one piece in the restructuring of the country. The free-trade imperative worked inexorably to lower all Canadian government social programs and regulatory regimes to U.S. levels. The use of the debt to justify cuts to social programs was a useful back-door route to implement the implied objectives of deregulated trade and investment. But it had a much wider application. The persistent claim that social programs caused the debt had a dual purpose. By justifying reduced spending on social programs, the state increased the insecurity of most Canadians. And by blaming the debt on social programs, government reduced expectations by implying that we could not afford them and by making people feel greedy for wanting them.

I will show decisively, later in this chapter, that social programs have not contributed significantly to the country's debt, nor to the provincial debts. The neo-conservative ideologues know this; the corporate think-tanks know it, too. But only on rare occasions do any of them admit it publicly. As the Canadian Centre for Policy Alternatives noted, "In the United States . . . David Stockwell and other officials in the Reagan administration now openly admit that, at the behest of their corporate friends, they deliberately increased the deficit so that it would justify later cuts in social programs funding."[1]

In Saskatchewan the Devine government dramatically decreased the royalties on oil and gas and thus added more than $4.5 billion to the provincial debt inherited by the NDP.[2] Next door in Alberta the debt story was almost identical. The Klein government railed on about public spending "skyrocketing" out of control. The truth lay elsewhere. Former civil servant Kevin Taft, himself a Tory, revealed in his exposé *Shredding the Public Interest* that government spending had been declining since the 1980s — long before Klein came to office in 1992. The debt was due exclusively to massive subsidies to corporations and to falling resource revenue.

It was important in the neo-conservative plan that the issue of the debt be presented as a crisis and not just a problem. The word *debt* appeared only attached to the word *crisis*. If it was just a problem, it would be commonplace; if it was a crisis, people would be prepared to make sacrifices. This political wisdom was put forward in 1993 at a colloquium put on by the International Institute for Economics, a Washington-based think-tank, to examine how governments in various countries could sell the policies of structural adjustment. The goal was to produce a "manual for technocrats and technopols involved in implementing structural adjustment programmes."[3]

A key hypothesis tested at the event was formulated by its convenor, economist John Williamson. According to New Zealand analyst Jane Kelsey, Williamson argued that societies tend to become "sclerotic and their flexibility declines. When a major crisis occurs within the existing system, it creates new opportunities for actors who until then have been prevented from taking the initiative. Where a crisis does not occur 'naturally,' it might make sense to provoke one to induce reform."[4] Williamson could have used Canada as a case study.

The campaign to sell the idea of a debt crisis involved dozens of players from nearly every sector of the ruling elite: politicians, corporate think-tanks, media owners, key columnists and broadcasters, academic economists, financial analysts, the Bank of Canada, and even, on occasion, foreign commentators as well as foreign bond-rating agencies. Even the IMF gave the debt warriors a hand. One might be forgiven for suspecting a conspiracy. But this just underlines the strength of the elite when it has reached a consensus, particularly one rooted in an ideology that all can adopt on faith rather than reason.

Corporate think-tanks and alliances took the lead in creating deficit hysteria. The BCNI's key anti-deficit document was entitled "Canada's Looming Debt Crisis." According to Seth Klein, coordinator of the B.C. branch of the Canadian Centre for Policy Alternatives, "During January and February, 1993, a team from the BCNI took the report on the road, visiting the premiers of New Brunswick, Quebec, Saskatchewan, British Columbia, Nova Scotia, and Manitoba, the finance minister of Ontario, Kim Campbell, Jean Charest, and the Liberal shadow cabinet."[5] Sam Boutziouvis, the BCNI's senior economist at the time, was happy with the council's overall efforts. "Did we have an effect there? I believe we did."[6] The C. D. Howe Institute was also pleased with its propaganda efforts, including its major piece, "The Courage to Act: Fixing Canada's Budget and Social Policy Deficits." According to the co-author Tom Kierans, "This has been a time of high visibility . . . In recent months our studies have reverberated in newspaper editorials across the country, on television news, and in parliamentary debates."[7]

The campaign was waged with particular ferocity by columnists such as Peter Cook, Terence Corcoran, and Diane Francis and media commentators like the C. D. Howe's William Robson and right-wing darling David Frum. Andrew Coyne, then writing for the *Globe and Mail*, told Peter Gzowski that "denying the debt problem is akin to denying the holocaust" and implied that writer Linda McQuaig was "out of her mind" for criticizing the Bank of Canada's high interest rate policy. Popular broadcasters also took up the call, with Mike Duffy, Michael Campbell, and Eric Malling all hammering on the issue. Malling produced an inflammatory and extremely influential W5 documentary on New Zealand's alleged "debt crisis" asking the rhetorical question about Canada's chances of hitting the "debt wall."

The debt warriors even managed to get some mercenaries from the U.S. onside. One prominent commentator generously called Canada a "banana republic" because of its debt. The more vigorous of Canada's debt warriors were hard on the mercenaries when they didn't perform up to snuff. In her book *Shooting the Hippo*, Linda McQuaig tells the story of interviewing Vincent Truglia, one of the most feared bond raters in New York, specializing in Canada for Moody's Investors Service. McQuaig expected a lecture about Canada's high debt but instead got a tirade "against members of the Canadian investment community for overblowing Canada's debt problems."[8]

The fact that Canada's debt situation never resulted in a credit downgrade was almost never commented upon. At the height of the ideological assault on the deficit, Moody's Investors Service stated: "We see no significantly negative trends in Canada's debt burden that could justify a change in the (triple-A) ratings of the nation's Canadian dollar and foreign currency debt. Canada's an extraordinarily low risk; that's the message."[9] While such expressions of confidence in Canada's debt management were quietly reported in the business pages, they were studiously avoided in the popular media.

Canada's bonds were always snapped up immediately, and in 1997, when the federal government issued no new bonds for the first time in twenty-five years, investors panicked about a "bond shortage." As economist Jim Stanford pointed out, if Canada's $600-billion debt suddenly vanished, "brokers would be leaping from their skyscrapers over the sudden loss of such a lucrative and risk free investment outlet."[10]

One of the most effective pieces in the debt arsenal was the argument that government was just like a household and no household could survive for long spending more money than it took in. "Putting our fiscal house in order" appeals to people's sense of personal and social responsibility. Just as it would be irresponsible to burden your family with debt by spending beyond your means, so too is it irresponsible to create a "legacy of debt" for future generations.

This argument is simply false. Governments, unlike families, have enormous authority and capacity to determine their revenues. They can increase taxes, lower interest on their own debt, stimulate the economy to create jobs and revenue. As well, no indebted family would slash everything in its budget; it would cut out luxuries first and only then cut back on essentials. Canadian governments, however, have slashed such essentials as health care, education, and social services. There is little talk among the elite of the legacy to our children of gutted social programs and a crumbling education system.

Seth Klein documented the propaganda techniques in his study on the debt, "Good Sense versus Common Sense: Canada's Debt Debate and Competing Hegemonic Projects." The arguments were designed to overwhelm any sceptics, with so many negative consequences of government deficits that no one in the opposition ranks is left standing. "The country will soon hit a 'debt wall,' the debt prevents Canada from lowering interest rates, and the foreign debt threatens Canadian sovereignty; the debt discourages and 'crowds-out' private sector investment; . . . the government has no room to raise taxes; the country cannot escape its debt crisis through increased economic growth; and the private sector, left free of government interference, will undertake productive investment," if the debt is brought down.[11]

The source of the debt was rarely spoken of by neo-conservative players. And in the end it didn't matter. The debt was so serious that only by huge cuts could we solve the problem. The debt took on the character of an infestation that had to be destroyed at any cost.

THE REAL DEBT PROBLEM

Let's be clear here, the debt was and is a problem. Any time we are spending thirty cents out of every tax dollar in interest payments we have a problem. But the debt is not *the* problem. The crisis is not the debt itself but in how and why it was accumulated and the way it is being addressed by current governments and their corporate benefactors. Canada and other Western countries have had very large debts before, but they were never viewed as a "crisis" in the same way the debt is portrayed today in Canada.

Debts were viewed as a problem, and governments set out to solve the problem at the same time as they set out to build a growing and stable economy. In short, the solution was part of a broad economic policy. The debt issue was resolved as the result of economic and industrial policies designed to accomplish other objectives, economic and social.

Jim Stanford, an economist with the Canadian Auto Workers, suggests a number of reasons why the debt is a problem. First, the accumulation of the huge debt means that it must be paid off, and that represents a massive transfer of wealth from poor, working, and middle-class Canadians to the wealthiest Canadians and the largest corporations. Tax money collected from the vast majority of Canadians goes to wealthy and corporate bondholders. Stanford argues that this transfer of tens of billions of dollars is "more regressive than other income redistribution issues (such as the negative distribution effects of consumption taxes)."[12]

The fall-out of debt terrorism on the public can hardly be over-estimated. Even in 1997, when the federal deficit and several provincial deficits are under control and even disappearing, the effect of ten years of bullying and fear-mongering is still ingrained in the public psyche. It is no longer as prominent in the propaganda of the neo-conservative forces, yet it underlies other campaigns and haunts any talk of increased public spending.

It is important to know where the debt came from, so we can

understand where it fits in the whole restructuring package and in the propaganda used to sell it. Deliberately created or not, the debt was the result of a series of interrelated neo-conservative policies, all of which were designed to enhance the corporate bottom line.

The federal debt was accumulated over a period of twenty years starting in 1975, the result of three government actions. The first was a shift to monetarism, which handed the money-creation powers of the Bank of Canada to the private banks. The second was a number of policies, including the deindexing of income tax and a huge increase in the number of tax expenditures (grants provided through the tax system) for the wealthy and Canada's largest corporations, which contributed to a dramatic decline in revenue beginning in the mid-1970s. The third and arguably the most important government policy was the crusade against inflation, a high interest rate/high unemployment policy aimed at bringing inflation to zero, which caused interest payments on the debt to skyrocket.

In 1975, when the Bank of Canada adopted monetarism as its broad policy approach, it abandoned its post-war practice of financing a good part of federal and provincial deficits through what amounted to interest-free or low-interest loans. In 1974 the Bank of Canada held 20.7 percent of the federal debt; in 1994 it held just 5.8 percent.

As the government adopted monetarism, it launched an unprecedented number of tax incentives for business and wealthy investors. The measures failed to create jobs, and the resulting loss of revenues has amounted to tens of billions of dollars over the past two decades. Researcher Kirk Falconer looked at corporate tax breaks and found that untaxed profits had steadily increased from $9.9 billion (from 62,619 corporations) in 1980 to $27 billion (from 93,405 corporations) in 1987. The total exceeded $125 billion for the eight years. This was just for corporations that managed to reduce their taxes to zero; many more billions were lost in reduced taxes.

The majority of this largesse went to the largest corporations, many of them foreign-owned. In 1987, fully 84 percent of those untaxed profits were earned by corporations earning $1 million or more in profits; 57 percent were accounted for by those earning $25 million or more. For this latter group, 145 of the very largest corporations in Canada, the average untaxed profit was $106.4 million.[13]

By 1980 the accumulating debt was getting serious; it was about to go through the roof. Interest rates hit 20 percent and the interest payments on the debt skyrocketed. Following this period of astronomically high rates, the Bank of Canada launched its high interest rate crusade for zero inflation, which lasted through 1995. The combination meant that by 1992 the accumulating interest payments had outstripped the original debt.

In 1991 the extent to which the revenue crisis and the high interest rates had generated the large federal debt, then at $420 billion, was revealed in a Statistics Canada study that looked at the accumulation of the debt between 1975 and 1991. The study revealed that 50 percent of the debt was due to shortfalls in revenue, that is, a decline in revenues relative to the growth in GDP, much of it a result of tax breaks for corporations and wealthy individuals. Interest charges accounted for 44 percent of the debt by 1991, and just 6 percent was from increases in program spending.[14] The study was so damaging to the carefully constructed neo-con view of the debt that the finance department tried to suppress it.

A 1995 report issued by the Dominion Bond Rating Service confirmed the Statscan study and other analysts' contention that interest charges were a key factor in the burgeoning debt. The study pointed out that the entire accumulated program deficit was created between 1975 and 1985. From 1986 to 1995 there was a $12-billion surplus. All the rest was interest charges.[15]

The Bank of Canada's policy of high interest rates was the principal tool in the government's strategy to create, in the words of the Bank of International Settlements, "more satisfactory levels of unemployment." The cost of "disciplining workers" shows up as a staggering cost to the country.

Examining the figures for 1992 and 1993, economists Diane Bellemare and Lise Poulin-Simon calculated that for each 1 percent of unemployment the costs to the government in lost direct taxes exceeds $3 billion, in indirect taxes just over $2 billion, and $1.2 billion in additional social assistance spending. If the $1.2-billion drain on the unemployment insurance system is added in, the total cost to the treasury of each 1 percent in unemployment weighs in at $7.4 billion a year. The total cost to the economy from unemployment — including lost wages, profits, and tax revenue — amounted to $109 billion in 1993.[16]

The Bank of Canada's high interest policy went far beyond that waged by the Federal Reserve Board in the U.S. and accounts in large measure for the difference in unemployment levels between the two countries. But it wasn't just the high interest rates that flattened the economy. According to a CIBC/Wood Gundy study, the government's spending cuts reduced growth by 3.5 percent from 1994 through 1996.[17] Independent economists such as Pierre Fortin put the cost of the cumulative unemployment at "about $400 billion in forgone national income" between 1990 and 1996, equal to 30 percent of the losses of the Great Depression.[18]

Of the G7 countries only Canada recorded a net loss in GDP between 1990 and 1995. An OECD survey of GDP growth from 1989 to 1996 showed that of thirteen developed countries only Canada had a negative growth.[19]

The question is not whether to bring down the debt but how to bring it down. In a democracy there are always alternatives. The Canadian Centre for Policy Alternatives and the Manitoba social justice coalition Choices have since 1995 produced the Alternative Federal Budget (AFB), a comprehensive proposal for government spending and tax measures presented as a clear and progressive alternative to the federal Liberals' annual budget. Produced with the input of dozens of economists and advocacy groups, it is an unprecedented example of democratic budget making and has garnered endorsements from 164 economists and political economists in universities across the country.

On the spending side, the AFB questions government priorities by focussing on unemployment and poverty. It established National Social Investment Funds in the areas of health care, education, income support, child care, retirement income, unemployment insurance, and housing. It also includes specific measures to increase employment. The AFB would see federal spending gradually increase to levels experienced in the 1980s. Seventy percent of the new revenue would come from increased economic growth, but it would also come from fair tax reform, which would raise an additional $5 billion.[20]

The AFB's plan is deliberately modest. Other critics of the Liberals, including tax policy expert Neil Brooks, have called for more radical measures, including eliminating $3 billion in corporate tax breaks, and an $8-billion decrease in personal tax exemptions. Another method by which the government could raise tax revenue and at the same time cool money speculation, particularly in bond markets, would be to impose a domestic financial transactions tax. This tax, applied to every sale of stocks and bonds, is applied already in a dozen OECD countries, including Britain, France, Germany, Italy, and Hong Kong. Even Singapore, a model of corporate libertarianism, has such a tax. A financial transactions tax set as low as 0.1 percent would reduce speculation and encourage productive investment and would, according to economist Jack Biddell, raise about $27 billion in revenue each year, compared to $17.4 billion raised by the regressive GST.[21]

The question, of course, has never been whether or not the money is there. It has always been a question of corporate domination and the power of the ideology supporting that domination.

Governments of all persuasions, infected by this market ideology, took advantage of the propaganda and have been cutting social spending for ten years. But that cutting was still no easy task. The values of Canadians have proven remarkably resilient. The 1994 Ekos poll "Rethinking Government" proved that, as did many other polls. Neo-cons like Andrew Coyne have ridiculed the idea that a country can define itself by its social programs precisely because they know the idea is so popular. It is also dangerous: it is one of the few aspects of government that people can still identify as fundamentally linked to the notion of a collectivity, a community. So the corporate elite and their neo-con allies had to open a second front: trash public services.

THE ATTACK ON PUBLIC SERVICES

The attack on public services has necessarily been more complicated compared to the debt war, which had the advantage of having a clear and simple target. The motivations for social program cuts were also more complex. There had to be deep cuts to social assistance and unemployment programs because these directly supported workers in their fight with employers for a share of output. Medicare and education were the largest expenditures, but cuts here had additional long-term goals. As we will see below, corporations are eager to invest in these areas because the profits are potentially so enormous. As well, the free marketeers recognize that these programs are the main source of public support for government. Breach this defensive wall, the new right reasons, and the rest will crumble. Other programs, particularly in the area of culture and heritage, were to be cut because they provided a strong national identity, which reinforced people's high expectations of government and their sense of social solidarity.

The attack on public services does not address all these particulars openly. While some programs, like EI and welfare, are criticized for "encouraging unemployment," the attack on public services is broad-based. Various arguments are used to undermine support for social spending. Canadian social programs are "too generous"; government service monopolies are inefficient and need competition; public services are dominated by vested interests and bureaucratic "empires" and are "not accountable" to the public; public services are simply not working any more and therefore not worth defending; universality in social programs should be replaced by two-tiered services, that will "free up" public services while the wealthy pay for their own.

One of the main propaganda pieces in the attack on medicare and education is that we are spending too much, with the mantra that health-care costs are spiralling out of control. This is the classic big lie, since this has never been proven anywhere in Canada. In Saskatchewan, for example, health care took up 31.2 percent of the program budget in 1980 and 32.1 percent in 1990. By 1995, public spending on health care in Canada was at its lowest level since the introduction of medicare in the 1960s.

The attacks on the public education system, like those on medicare, claim we are spending far more than other countries and that we are not getting our money's worth in quality. This too is false. The figures used include post-secondary education, and Canada has had one of the highest levels of post-secondary participation of any country in the OECD.

Of course, these attacks go well beyond costs. The Fraser Institute's bogus study regarding waiting times is typical of the fear-mongering, and critics of public education continue to use false and misleading claims to raise parents' fears. The one-two propaganda punch of excessive spending and poor results is intended to weaken public support for the two most important community services.

The notion that Canadian social programs are too generous was a familiar theme throughout the early 1990s in preparation for the Liberals' huge budget cuts from 1994 on. Yet the argument had no basis in fact, unless the country we were being compared with was the U.S. An OECD comparison of seventeen industrial nations in 1990, when the propaganda campaign began, revealed that Canada was one of the most miserly of any of the developed countries. The study revealed that Canada placed fourteenth, tied with New Zealand, for the percentage of GDP going to social spending. In that year Canada spent 18.8 percent of GDP on social programs, lower only than Australia, the U.S., and Japan (11.6 percent). The U.K., Germany, and France spent more, and Sweden 33.9 percent. Far from being generous, the OECD report indicated that Canada would have had to increase its social spending by $18 billion just to meet the average of the OECD countries.[22]

There are a great deal more data showing how inferior Canada's social programs are compared to other industrial nations. Even in 1991, before the UI program was slashed, Canada ranked sixteenth out of nineteen OECD countries in "generosity."[23] At the time Canada had one of the highest unemployment rates of any OECD nation, putting in serious doubt the claim that our "overly generous" UI was causing unemployment. The program now ranks with Japan's as the stingiest of any OECD country, with 46 percent of the unemployed eligible for benefits, below that of the U.S.

As seen in chapter 6, Canada's social assistance programs rank below many of those in the U.S., in spite of popular notions to the contrary. Canada's maternity-leave provisions are equally tightfisted. Of twenty-four countries (the U.S. and Australia provide no benefits), Canada placed fourth last in the number of weeks covered and dead last for rate of benefits as a percentage of salary.

It is clear now that the Liberal government's objective from the beginning has been not deficit reduction but a dramatically reduced government presence in the economy, a systematic shrinking of the egalitarian state. This was stated explicitly in a 1995 speech by Finance Minister Paul Martin at Jackson Hole, Wyoming, in which he boasted that Canada would be spending significantly less than the U.S. within two years. "Looking to 1996–97, the U.S. budget forecasts a reduction to 16.3 percent of GDP while Canada's ratio will have fallen to just over 13 percent, the lowest level since the 1950s."[24] In fact he beat his own target, lowering spending to 1949 levels and to about 12 percent of GDP. This is what gets Martin his greatest accolades on Bay Street and admiration on Wall Street.

The new-right argument that public services are dominated by bureaucratic "empires" unaccountable to the public is propaganda at its most transparent. The bureaucracy that runs the medicare system in the U.S., the paragon of free-market health care, is so sclerotic and byzantine that it is hard to find anyone who can adequately explain just how it works. We do know that the administrative costs of that system, a good indicator of bureaucratic gridlock, are more than double those of the public system in Canada. As for accountability, that depends on how you define it. For the 40 million Americans who cannot get coverage at all, the American system must seem very unaccountable.

The attack on public services is often couched in terms of attacks on those who provide those services. The number of times that service providers, civil servants, and government employees are referred to as "bureaucrats" in the media reveals something of this campaign to discredit those hired by the community who provide services *to* the community. Government workers are portrayed as selfish, lazy, overpaid, incompetent, resistant to reforms, and, through their unions, protective of an overly generous system of privileges.

The objective in this campaign is to portray public servants as alien to the community, rather than as serving it. Portraying civil servants bargaining for a new contract as selfish individuals who should be grateful that they have a job is intended to distance the public from the people who provide it with services. Characterizing workers in this manner undermines public confidence in public services and therefore makes it easier for government to cut them. It is a divide-and-conquer tactic, encouraging one set of victims to resent another set while the victimizer accomplishes his goals with impunity.

This portrayal of public servants is obviously false. Besides providing crucial services, civil servants, like other working people, are family breadwinners. They spend hundreds of millions of dollars in their local communities, they are volunteers, they are people's neighbours, they have kids in local schools, they pay taxes. They are, in short, citizens like everyone else, with the same concerns for the future and, indeed, in need of the same public services as everyone else.

Everyone has an anecdote about bad service or surly treatment from a civil servant somewhere, but neo-con propaganda often magnifies these to mythic proportions. Meanwhile, in the private sector, fraud, overpricing, price fixing, bad service, shoddy and dangerous goods, and outright theft are commonplace. But this is somehow expected of the private sector, which Canadians instinctively know has lower ethical standards. It is ironic that part of the assault on pubic services, of which people demand high standards, is to suggest that the private sector can do the job better.

The notion that civil servants are overpaid, even at the highest level of the public service, is nonsense. Senior policy analysts and administrators would quickly double or triple their salaries if they went to the private sector. Martin Harts, compensation specialist with the forecasting firm KPMG, points out that a deputy minister's salary and benefits package is about $175,000, compared with the $1.3-million average paid to CEOs with similar responsibilities.[25] Unfortunately, with that kind of lure, many have left. While the federal civil service has declined by 20 percent overall, the number of managers has gone down 33 percent. But thousands of civil servants stay in the public service *despite* working conditions that have steadily worsened throughout the 1990s. Federal civil servants have gone six years without a pay increase, an inexcusable way of treating those who serve the country. The media, content to play their role in discrediting government, miss the irony of business voices attacking "bureaucrats" as incompetent and overpaid and then, when they are laid off, hiring them at many times their former pay.

Social workers face ever increasing caseloads yet stay on the job because they are committed to the work. The same is true of nurses, teachers, childcare workers, employees protecting the environment and ensuring safe food as well as the thousands of people who work in administration. Stevie Cameron's book *On the Take*, on Brian Mulroney, documents many examples of senior civil servants who quietly defended the public interest against an assortment of political crooks and opportunists, often at great cost to their careers.

The assault on public services is, in part, designed to reduce the confidence of people in public services and to drive a wedge between them and government. But a related objective is to open up huge new areas of investment for the billions of dollars roaming the world looking for places to make a profit. Thus the attack on the effectiveness and cost of public services has one ultimate purpose: prepare the ground for privatization. The underlying message: we should hand over our services to the lean, efficient, accountable private sector.

Even a cursory examination of the private sector's record of efficiency, performance, and accountability demonstrates that handing our public services over to corporations would be disaster for citizens and communities. This is true even if we just examine the behaviour, since 1980, of the world's largest corporations. In Canada and elsewhere, they have been characterized by unbridled greed, spectacular examples of fraud, and the longest period of unproductive speculation in decades, possibly in this entire century. Throughout the latter half of the 1980s and continuing to today, enormous sums, far outweighing new productive investment, were put into leveraged buy-outs, hostile takeovers, stock and money market manipulations, and real estate speculation.

Business culture in Canada and the U.S. is full of stories about criminal fraud by people who were lauded as brilliant and heroic before they were caught. The Canadian fraud artists and their takes have included Donald Cormie ($491 million), Robert Vesco ($220 million), Leonard Rosenberg ($131.8 million), and Julius Melnitzer ($75 million), to mention just a few.[26] This kind of outrageous and destructive fraud is nonexistent in the history of the public service. Yet we are being asked to put the most important aspects of our community into the hands of this sick subculture of greed and corruption.

Beyond this sorry record, the structure of North American corporations is more bureaucratic than any government department or "state monopoly." According to researcher David Gordon, corporate bureaucracies in the U.S. and Canada distinguish themselves among developed countries by their extremely top-heavy management. In 1993, as many as 16.6 million people were employed in the U.S. as managers or supervisors. This was almost as many people as worked in the entire public service of the U.S., and the cost in salaries was

$1.3 trillion, almost a quarter of the total national income.[27]

This bloated bureaucracy, according to Gordon, is rooted in the ten-year effort to "discipline" labour out of its share of output. The destruction of unions in the U.S., the downward pressure on wages (real wages are now at 1967 levels), and the trend to contingent labour have resulted in the need for more and more supervisors. The strained labour relations that arise out of this strategy require "the threat of job dismissal as a goad to workers." The alternative model used in Europe emphasizes job security, wage incentives, and employee involvement in decision making.

"In such a hierarchy," Gordon writes, "you need supervisors to supervise the supervisors . . . and superiors above them and managers to watch the higher level supervisors and higher level managers to watch the lower-level managers." You also need to maintain a significant wage differential to keep the hierarchy working. As CEOs demand higher compensation, "they find that they have to allow for upward creep in the salaries of their managerial and supervisory subordinates as well," whether they deserve it or not.[28] The money for high wages for nonproductive staff comes out of workers' earnings and what might otherwise have been retained for new investment.

The corporate drive for privatization has a simple motive: the huge areas covered by public services provide the single largest area for investment remaining in developed countries. Health and education alone would absorb hundreds of billions of dollars in capital investment in Canada. So privatization is a key dimension of the new-right agenda. It not only provides enormous new areas for profitable investment. It also eliminates the example of public-purpose enterprises and the challenge they present to private corporations.

The particular way privatization has been implemented has also has been used to shift the political culture from communitarian values to those of possessive individualism. The privatization program of Margaret Thatcher in Britain, subsequently copied in Australia, New Zealand, and Canada, has often set out to build a social base for the new-right counter-revolution. Madsen Pirie, one of Thatcher's principal consultants on privatization, also advised the Saskatchewan Tory government of Grant Devine.

The strategy is a crude appeal to greed. Make a public share offering to all citizens (offering them the opportunity to "own" shares in an enterprise they already own as citizens) at a cut-rate price. The share price typically rises rapidly to something approaching its real value, whereupon most initial purchasers sell at a tidy profit. The provincially owned potash industry in Saskatchewan was sold off for approximately half its market value. Within a few short years, the majority of the shares were owned by large investors.

But all those "successful" small investors were, or so the theory goes, bitten by the market bug, many for the first time. That would lead, suggested Pirie and others, to more middle-class people getting into the stock market, helping build a base of supporters who would defend later market reforms. The Devine government took the propaganda dimension one step further, setting up a department just for privatization: the Department of Public Participation.

When privatization ideology becomes aggressive, it can simply declare that government should not be in any business, period. This was the case with the Alberta Liquor Control Board. A study by four University of Alberta academics shows that the privatization "was conceived with little thought, and implemented in haste." The study reveals what Albertans already knew: prices were higher, and selection declined The government also lost $400 million in annual revenue, gaining only $40 million from the sale of the board's assets, half what they cost originally. Thirteen hundred full-time jobs, at an average wage of $30,000, were lost. The new system pays workers between 35 and 50 percent less, resulting in less spending in the community and a loss of tax revenue.[29] It was almost as if the Klein government punished itself for ever daring to have a public enterprise.

The experience of privatization in Britain has been a disaster, with the level of services plummeting, CEOs taking home obscene pay packages, companies earning enormous returns on investment, and a failure in most cases to maintain the assets, which in many cases were sold at cut-rate prices. The *Globe and Mail*'s *Report on Business* declared in 1996 that "the British experience with privatization can claim few successes." The article starts out with an account of the newly privatized bus service in North London. "The posted timetable tantalizingly promises a bus every ten minutes . . . But half an hour has elapsed by the time the . . . bus finally trundles into view. There is a near riot when the people realize that they can't get on the crowded vehicle."[30]

The greatest failure of Thatcher's scheme was its inability to make good on the most important free-market promise, that of choice. British citizens, it seems, not only have lost their sovereignty as citizens by losing public assets, but have gained precious little in the way of consumer sovereignty in return. In part, this is because competition, the panacea for everything evil about state services, simply doesn't work for things like water, gas, and electricity. If you get lousy bus service, you don't get to patronize another bus line. There isn't any.

Privatization of water in Britain has become a nightmare. The system breaks down so often that, on average, nearly a third of the water piped into the system leaks out again. Lawsuits abound: there have been 250 successful prosecutions since Thatcher privatized water in 1990; and throughout 1996 there were, on average, successful prosecutions of water companies once every three weeks. The papers are full of stories about people who are forced to make do with a basin of water a day. Low-income people have been cut off for nonpayment. Rate increases have averaged 85 percent and in some areas exceeded 300 percent.[31]

MEDICAL CARE AMERICAN-STYLE

Even if the bloated corporate bureaucracy was pared down, some services just aren't appropriate for the private sector. The record of health care in the U.S. is one dramatic example as horror stories continue to emerge from that profit-driven system.

In the U.S. most people get their medical insurance through their employers as part of their benefits package. In the 1980s medical insurance premiums skyrocketed. Many employers experienced 35 to 40 percent annual premium increases, and they began demanding cost controls. The solution that developed was HMOs, or health maintenance organizations. They are a frightening example of what happens when you apply the ruthlessness of corporate cost control to the health of citizens.

In his book *Health Against Wealth: HMOs and the Breakdown of Medical Trust*, George Anders tells horror story after horror story about the "managed care" system. Like the parents of a baby boy in Georgia who had to race 42 miles to get treatment for their baby's meningococcemia because their managed care plan would not pay for care at a more expensive, but closer hospital. The boy survived, but only after having his feet and hands amputated, something doctors said might not have been necessary had he been treated even a few minutes sooner.[32]

Other evidence was provided by Dr. Linda Pino, a former "physician executive" with managed care plans. Pino explains, "The gate-keeper position . . . limits access to other specialists or other tests because he or she is being paid a lump sum to take care of all their patients and if they don't spend their money they get to keep it when it's left over." Dr. Pino is the kind of doctor who terrifies the HMO industry — a whistle blower. When she testified at a congressional hearing, her opening words brought dead silence to the room. "I am here primarily today to make a public confession. As a physician I denied a man a necessary operation that would have saved his life and thus I caused his death. No person and no group has held me accountable for this because in fact what I did saved a company a half a million dollars . . . If I am an expert here today it is because I know how managed care maims and kills patients so I am here to tell you about the dirty work of managed care."[33] Pino's job as a "physician reviewer" was to maintain a denial rate of at least 10 percent or she would be replaced. Her employer offered the physician with the highest denial rate a Christmas bonus.

Managed care is already coming to Canada. The huge American health corporations, which must like all corporations grow just to survive, are eyeing the Canadian system as an investment opportunity. Hospitals are already applying the gate-keeper principles to health care, according to Pat Armstrong, an authority on Canadian health care. "It has a multitude of forms; it's there already . . . private labs, rehab hospitals, the [corporate] management of the more public hospitals. It's happening in ways that are just so hard for Canadians to see that before we know it there will be nothing to defend unless we defend it right now."[34]

The giant health corporations like Columbia/HCA, Tenet Healthcare Corp., and Kaiser Permanente have plans to set up for-profit hospitals either alone or in partnership with Canadian corporations like MDS. The insurance companies — Sun Life, Manulife, Liberty

Mutual, and Great-West Life — are already taking advantage of governments' continued "delisting" of services from public health coverage. (Virtually none of the new procedures in medicine are now added to the list of covered items.)

Although Canada does not yet have HMOs and for-profit hospitals, the business-oriented philosophy behind them is seeping in. One example of the intrusion of corporate values into public medicare is the software being purchased by Canadian hospitals that includes patient classifications designed according to cost-cutting criteria. "Some call it an evidence-based approach," says Pat Armstrong, "others call it assembly-line health care." It is an insidious process because there are no announcements that the philosophy is changing, no debates in Parliament, and nothing in the way of visible decisions that patients and their families can document. This is commercialization by stealth.

Two parallel developments are setting up Canadian medicare for incremental privatization. Cuts to health-care budgets are forcing more health-care services out of the public realm and into the private. (Nearly 30 percent of money spent on health care was being spent in the private sector in 1997, compared to 23 percent in 1986.) At the same time, medicare is being commercialized, blurring the differences between private and public health care, differences that form the basis of popular support for the public system. As those differences blur, the rationale for defending the public system gradually disappears.

The Canadian government, despite its statements about the importance of medicare, is complicit in this shift. In 1995 representatives of government and health corporations gathered in Singapore to discuss "corporatizing, commercializing and privatizing opportunities [in the] most rapidly expanding market in the world."[35] As Ottawa's power to maintain national standards erodes, most provinces are moving to regionalization of health care, in effect voluntarily weakening provincial authority and empowering regional authorities to pursue private options.

Ever expanding U.S. health corporations are driving the global health-privatization push. According to health researcher Colleen Fuller, U.S. health corporations are amongst the most fierce opponents of so-called trade barriers erected by countries like Canada to protect public health care. Many of these barriers have already fallen with NAFTA, and more will fall if the MAI is successful. The U.S. giants promise to bring a number of trends to Canada. The American focus on high-tech care has resulted in skyrocketing health costs in that country and, says Fuller, "massive increases in profits of drug, hospital and insurance corporations."[36]

In addition, the trend in the U.S. has been to de-skilling, replacing professional staff with unlicensed, unskilled, inexperienced, and cheaper staff. These trends have made the health "industry" the most profitable sector in the States. Dr. David Himmelstein of Harvard Medical School concluded from his comparison of the Canadian and American systems that private providers are "extraordinarily efficient at extracting money from the health-care system. Other than that, there's not an iota of evidence of greater efficiency in the for-profit sector."[37]

TARGETING EDUCATION

Medicare is a hot issue in Canada, and because of its obvious public popularity, the media have provided some coverage of the threat posed by cuts and the growing portion of the system falling into private hands. Yet the ideological attacks on the education system are much more fierce than those against medicare, and the threat of privatization and commercialization of education is very real.

As with other public services, the groundwork for corporate incursions into education begins with a sustained assault on the effectiveness of public education. Corporate-funded think-tanks have led the way, with the Fraser Institute and the C. D. Howe Institute playing key roles. But nearly every corporate voice in the country, from chambers of commerce, boards of trade, the BCNI, and the corporate media have contributed to the campaign to undermine public confidence in public education.

The myth making is now well established: our schools graduate students who are illiterate; students aren't being prepared for work in the new corporate world; their math and science skills are inferior to those of Asian students; there is a 30 percent drop-out rate; teachers and their unions resist change and make it impossible to fire bad teachers; there is no "choice" in education, and public schools and their students remain "prisoners of mediocrity and educational gridlock."

All of these claims are either outright falsehoods or deliberately misleading. According to Statscan, only 3 percent of Canadian-born 16- to 24-year-olds (the most recent graduates and senior students) have any literacy problems, and achieve by far the highest literacy rate of any age group. Figures from a 1988 OECD study show that Canada had the highest rate of post-secondary participation of any of the developed countries and graduated 50 percent more engineers per capita than Japan.

True, Canada placed ninth in math and science out of fifteen countries, but Asian countries devote enormous energy to preparing students for such tests as a matter of national pride, and often hand-pick the schools that take the international tests. The 30 percent drop-out myth is pervasive despite Statscan figures that show the graduation rate to be at least 82 percent and could prove to be as high as 90 percent if surveys tracked people through age 25 rather than 20.[38]

The efforts of corporations to break into public education have taken two separate but related paths in Canada. The first is through the heavy promotion of education "partnerships" between corporations and school boards, schools, universities, and colleges. As budgets are slashed, "good corporate citizens" arrive on the school doorstep offering assistance that inevitably involves hooking the school into product lines or trading materials or cash for the exclusive right to peddle or advertise their products in the school. (Coke, Pepsi, Burger King, and McDonald's are the prime examples of the latter.)

The second track is corporate support (the Royal Bank, Bank of Montreal, Syncrude, the Donner Foundation) for the campaign to promote charter schools, publicly funded private schools first introduced in Britain, the U.S., and New Zealand. The key promoters of the charter concept in Canada are all in the neo-con camp. Charter advocates use the market language that is familiar to other new-right campaigns — parents are "customers," education the "product," and students the "value-added" result. Advocating "choice" promotes the libertarian value of a private benefit over the public-purpose philosophy of the public system. Charters are seen as a halfway step to school vouchers, the ultimate consumer-driven system wherein parents receive a certificate that they can "spend" at any school they choose. Vouchers go hand in hand with contracting out school administration, which is already happening in the U.S. In 1996, Lehman Brothers investment house prepared a detailed study, "Investment Opportunity in the Education Industry," that declares that there is great potential in education privatization. "The health care sector 20 years ago and the education industry today have several similarities that, given the massive private sector growth in the health care sector, make the education sector extremely attractive to investors."[39]

The analysis is explicit about the positive role education "reformers" are playing in preparing the ground for investment opportunities. "Private companies owe much of their success to the ability of the public sector to open the education system to competition. The [for-profit] Edison Project was granted a [contract] from one of Massachusetts' charter schools to manage one grade school, illustrating how a private company can use a public reform movement to grow and build its own credibility."[40] The study goes on to map out strategies for using various reform movements as gateways to investment.

Yet reports out of the U.S. suggest that privatization experiments have virtually all failed to deliver. Education corporations have doctored performance records and provided no cost savings to the states and boards that hired them. So far, contracting out school administration isn't on the Canadian agenda, but corporations are aggressively preparing the ideological ground for direct involvement. Already, dozens of schools in Canada have named classrooms after corporations in return for financial assistance. According to author Tony Clarke, there are twenty thousand business-school partnerships in Canada "with technology and communications corporations such as AT&T, Bell Canada, General Electric, Hewlett Packard, IBM Canada, Northern Telecom, Unitel and YNN."[41]

The corporate message is being injected into the Canadian curriculum through corporate publications. *What!* magazine is targeted at 13- to 19-year-olds, who, says publisher Elliott Ettenburg, "aren't children so much as what I like to call 'evolving consumers.'" The *Globe and Mail* "Classroom Edition" (twelve times a year) is free and accompanied by a teacher's guide and lesson plans. It is promoted as a vehicle for corporate culture to its potential sponsors (fee: $100,000 a year) with the promise that it is "the ideal vehicle to provide information about your company, products or services." In the U.S., curriculum kits and science projects are brazen in their promotion of products. Campbell Soup has a science project designed to prove that its Prego spaghetti sauce is thicker than Unilever's Ragú. McDonald's gives away a kit that shows students how to design a McDonald's restaurant and how to apply for a job at McDonald's.[42]

The direct connection between the control of education and the strategy of transnational corporations to "capture" people at the earliest possible stage is frighteningly apparent in this commercial assault on public education. The transformation of citizens into consumers logically begins with transforming the schools whose original purpose included citizenship training.

The major ingredient missing from the investment picture in Canada is the severe education crisis that has spawned reform movements in the U.S. But business, ever eager to create opportunities where none exist, is working hard on this issue. Hand in hand with such right-wing politicians as former Ontario education minister John Snobelen (famous for talking about the need to "invent a crisis" to kick-start reform), business continues to call for massive cuts to public spending at the same time its think-tanks attack the system for failing.

The unprecedented confrontation between teachers and the Ontario government in 1997 is precisely the "crisis invention" that neo-conservatives require. The government's seizure from school boards and teachers of complete control over education policy making fits the classic mould of taking power away from "vested interests." The billion dollars in cuts to Ontario schools put tremendous pressure on teachers and schools to accept the various freebies and partnerships that are the fifth column in the long-term plan to make profits off education. It also opens the door to a middle-class exodus from the deteriorating public system to the private system, and it increases pressure for charter schools and publicly subsidized private schools.

Universities and colleges have not escaped the encroaching corporate profiteers. Enormous cuts to university budgets are slowly forcing our almost exclusively public system to mimic its private American counterpart. University presidents are being hired on the basis of their ability to get corporate donations. Corporate sponsorships inevitably raise questions about the objectivity of research as well as what research will get done.

In their book *Class Warfare*, Maude Barlow and Heather-jane Robertson state, "Educators who once jealously guarded their autonomy now negotiate curriculum planning with corporate sponsors . . . Professors who once taught are now on company payrolls churning out marketable research while universities pay the cut-rate fee for replacement teaching assistants."[43] Almost every aspect of university funding is commercially tainted. According to Barlow and Robertson, "A professor's ability to attract private investment is now often more important than academic qualifications . . . Provincial and federal funding . . . is also increasingly tied to commercial considerations."[44]

The convergence of corporate and university interests is becoming increasingly formal and explicit. The Corporate-Higher Education Forum is "a national [Canadian] coalition of university presidents and corporate CEOs designed to merge goals and activities . . . [It] promotes corporate-education interaction by placing members on one another's governing bodies."[45] The forum supports lower state funding in order to encourage greater corporate influence.

"GOVERNMENTS CAN'T CREATE JOBS"

By early 1996 the deficit issue was not getting the same play as it had through the first half of the 1990s. Polls consistently showed that unemployment was increasingly the biggest concern of most Canadians, often showing numbers twice as high as those concerned about the deficit and debt. In addition, when asked what the government's priorities should be, Canadians believed that it should be doing something about unemployment and that it *could* do something.

Consequently, the business and neo-con propaganda machine has been working overtime to disabuse citizens of the notion that their governments can do anything useful. The mantra is familiar: "Governments can't create jobs; they can only create the conditions for the private sector to create jobs." The most dramatic example was that given by Jean Chrétien on the famous televised town-hall meeting when he told a Regina woman with three degrees and no job, "Some are lucky, some are not. That's life."

Few areas in public policy demonstrate so clearly the profound shift in the role of the state in society as does job creation. Nothing is more important to an individual or a family than their access to employment, their ability to lead productive lives. Nothing is so central to the health of communities, indeed to their survival, than the employment of their citizens. If democratically elected governments abandon the responsibility of even trying to ensure that people have jobs, they have effectively abandoned everything.

The argument that government can't create jobs rests in part on the assault on public services, and the effort to denigrate public servants and the jobs they do. The Reform Party and the National Citizens' Coalition take this argument the furthest with their declarations that "government jobs are not real jobs." This notion is plain silly. It suggests that someone lobbying on behalf of the tobacco industry has a real job while a nurse attending a cancer patient does not. It also suggests that the person working for the public liquor board in Alberta didn't have a real job until it was privatized and wages cut in half. Indeed, government jobs are so pervasive in the community that a 1996 examination of the résumés of Reform MPs revealed that twenty-three of them used to work at "government jobs."

The work done by teachers, nurses, park-maintenance people, and those providing municipal services and ensuring that our food and water are safe, are real jobs. Without them our communities would collapse. But government job "creation" goes far beyond a direct provision of public services. The government creates hundreds of thousands of jobs in the private sector through government purchases. Road building, the purchasing of supplies by schools and hospitals, the building of coast guard vessels, the purchase of police cars and fire trucks, the provision of electric generators for public utilities, all testify to the fact that most goods and services produced in Canada depend on a complex set of public-sector/private-sector connections. This includes purchases, the salaries of public employees, and transfer payments — family allowance, welfare, and pensions — to individuals.

A study by researcher David Robertson in 1985, before spending cuts started, reveals just how much the private sector relies on government purchases. In the mid-1980s, all levels of government spent some $68 billion on the private sector. Within the service sector, government purchases accounted for 6 percent of the total output; in transportation, communications, and utilities, 14 percent, and in advertising, 18 percent. The percent of purchases consumed by the public sector in publishing and printing was 21.6 percent; ready-mix concrete, 48 percent; pharmaceuticals, 17 percent; shipbuilding and repair, 20.2 percent; petroleum refineries, 17.6 percent.[46]

In total, government purchases of domestically produced goods and services generated or maintained more than one million jobs in Canada, about 12 percent of all private-sector jobs. This was in addition to the half million private-sector jobs created by the spending of $30 billion in government salaries. Transfer payments to individuals accounted for an additional $30 billion spent on private-sector goods and services, creating a further half million private-sector jobs. Robertson's calculations did not even include the multiplier effect of all this spending, nor did they include the expenditure of UI payments.

In 1997 we have chronic unemployment of over 9 percent (add discouraged job hunters and the involuntarily underemployed and it exceeds 17 percent), and it is widely recognized, even in the financial community, that government spending cuts are a major cause. Economist Jeff Rubin of CIBC Wood Gundy stated: "The Canadian economy would be far more closely on a par with the full employment U.S. economy" had governments not tackled their deficits so vigorously.[47]

Government can and does create jobs and has even more tools at its disposal to do more. The government could require that banks, as a condition of their operating licences, devote 0.5 percent of their loans to small community and cooperative enterprises. The banks would then have to encourage and seek out such development, rather than promote leveraged buy-outs.

Government could change labour and tax laws to redistribute work. It could reduce the official work week to thirty-five hours and discourage overtime by raising the ceiling for the maximum insurable earnings under UI/EI. It could increase the annual-leave and educational-leave provisions in labour codes.

The government could implement changes in the tax code that would penalize companies for gratuitous downsizing at times of increasing profitability. It could do the same with respect to the deductibility of interest payments on loans used to engage in mergers, leveraged buy-outs, and currency speculation, thus encouraging productive investment. If companies insist on paying their CEOs outrageous pay packages, the government could limit deductible business expenses to $500,000, or twenty times the average employee's compensation. It could lower the ceiling on the percentage of pension funds and RRSPs that can be invested abroad, making more money available for investment in Canada.

Most important, the government could, as the Alternative Federal Budget and the proposals of other analysts have pointed out, stop the destruction of public services and return to the principle of democratic governance. It could embark on a program to strengthen communities through investments in education, health care, child care, social housing, enhancing the environment, and refurbishing the economic infrastructure in communications and transportation, and it could triple its spending in the area of culture, one of the least expensive and most socially rewarding ways of creating jobs.

The new right's counter-claim that the private sector will create jobs if only the government creates favourable conditions stands thoroughly discredited. The conditions that corporations say are necessary to persuade them to invest are all in place. That is what the FTA and NAFTA were supposed to do; that is what low minimum wages were supposed to do; that is what slashed taxes for high-income Canadians was supposed to do; that was the rationale for wrestling inflation to the ground and for gutting unemployment and social assistance programs.

When critics of the government negatively compare our social programs to those of other Western countries, the answer is always a reference to the U.S., our main trading partner. Yet the notion that the U.S. provides cheaper business costs than Canada has also been thoroughly discredited.

In 1995, a KPMG study, sponsored by the Canadian and U.S. governments, compared business costs in various cities in the two countries. The eight Canadian cities were all found to be less expensive than any of the seven U.S. cities. The study was repeated in 1996 with thirteen Canadian cities. All were cheaper than any of the twelve American cities, on average 6.7 percent cheaper. This was true of every industry examined. Most of the differential was due to Canada's much lower labour costs, in large measure because of our universal medicare. Taxes played a role, too; payroll taxes in Canada were 35 percent lower than in the U.S. Canada would only begin to lose its advantage if the dollar rose to 87 cents U.S.[48]

A further KPMG study, released in 1997, showed that Canada was the least expensive location to invest of seven countries studied, including the U.S., France, Germany, Britain, and Sweden. Most important, given the constant complaint about high business taxes in Canada, the study showed that Canada and Sweden were tied with the lowest corporate taxes of any of the seven countries.[49]

The ten-year decline in real wages and the slashing of social programs was undertaken not to make us more competitive but to re-establish capital's historic "share" of output. The corporate sector has had ten years to show the country that the enormous sacrifices Canadian have made were worth it. Yet the private sector's job creation record is abysmal, despite the best government-created "conditions" in a generation. In 1995, following a pattern set in the late 1980s, net new investment in Canada (after replacement of old plants and equipment costing $100 billion) was $11 billion. In comparison, $78 billion was expended on mergers and acquisitions, and tens of billions more on speculative activity in currency, real estate, the stock market, and other areas, none of which created a single "real" job or any real wealth.[50]

Every few months the headlines in the business press report on the merger mania continuing apace. The *Globe and Mail*'s business section reported in February 1997, "Merger Wave Gathers Momentum"; in July, "Mergers on Target for Record," and in October, "Mergers and Acquisitions Jump in Third Quarter."[51] Every such headline heralds the loss of more jobs and the loss of government revenue — the huge interest charges on the money borrowed for these deals is deductible. In the third quarter of 1997, there were 347 mergers, worth $21.6 billion. Assuming that pace continued, the year-end total was $88 billion, compared to the previous high of $78 billion.[52]

Glenn Bowman, of the investment banking firm Crosbie and Co., expects that the "global drive to consolidation" will continue unabated. Sheer size continues to be seen as an advantage because it allows for efficiency — which means fewer employees. One American example: the merger of Chase Manhattan and Chemical Banking will result in one hundred of the current six hundred branches closing, throwing 12,000 of the 75,000 employees onto the streets.[53]

The corporate record of job creation makes it clear that the conditions demanded by the corporate elite had nothing to do with a commitment to create jobs. It had everything to do with seizing a larger share of output for profits, and of effectively deregulating labour. Indeed, the conditions, both actions and inactions by government, have contributed to the elimination of tens of thousands of the best-paying, highest-skilled jobs in the country. The Canadian Centre for Policy Alternatives, using *Financial Post* figures, has tracked forty-four BCNI corporations since the free-trade deal came into effect. Between 1988 and 1996, thirty-three of those corporations posted a total job destruction of 216,004, on average a cut of 35 percent of their workforces. During this period, these corporations increased their revenues by $40 billion, or 34 percent. Eleven of the companies created 28,073 jobs, with one, Seagram, accounting for half the total.[54]

THE TAX-CUT SCAM

The last piece in the five-part propaganda assault on Canadian expectations of government was tested in the June 1997 federal election. The panacea for economic growth and jobs, everything else having failed to prod business into action, is tax cuts. The argument for tax cuts follows a similar pattern to those for cutting the deficit. Not all proposals are the same, with the Liberals talking about a tax cut for low- and middle-income earners, but only after the deficit is gone; Reform demanding across-the-board tax breaks so that the rich would benefit, and the Tories campaigning on an immediate tax break. The desired effect, as it was with the deficit questions, is a debate about how much and when, not about *whether* to cut taxes.

Because the country is preoccupied with unemployment, the tax cutters must couch their propaganda in terms of job creation. Even Preston Manning, who for the first nine years as leader of the Reform Party never referred to unemployment as a problem, now expresses concern over the issue and promotes a tax cut with the populist rhetorical question: "Who knows better how to spend your money, you or the government?"

Yet most studies demonstrate that tax cuts are a poor way of stimulating job growth. One analysis, done in 1997 by the economic forecasting firm Informetrica, shows that tax cuts, as a strategy for job creation, compare very poorly to direct and indirect government spending. Informetrica looked at how many jobs could be created with a billion dollars spent in various ways. The most effective way: direct hiring. If the government spent a billion dollars hiring back the teachers, nurses, government employees, and cultural workers it laid off in the 1990s, it could create 56,000 jobs. Increased spending of $1 billion on goods and services in the private sector would create 28,000 jobs; in infrastructure spending, 26,000 jobs.

Tax cuts don't even come close. A billion dollars put back into people's pockets through a cut in the GST would produce 17,000 jobs; in corporate taxes, 14,000; in personal income taxes, 12,000, and in payroll taxes, just 9,000 jobs.[55]

As tax-policy professor Neil Brooks points out, the whole notion of a tax cut is based on the false premise that people want to make choices only in the things they purchase in the private sector. But where is the evidence that people really want more private cars and less public transit, more private roads and fewer public ones, more user-pay vacation spots and fewer public campgrounds?

The pressure for tax cuts will not go away. The entire world capitalist system depends on ever expanding consumer spending. With incomes in a nearly permanent state of stagnation in North America, one of the only ways left to increase disposable income is through tax cuts. And although they fare badly as a method of job creation, tax cuts are an effective policy tool for corporate governance. They reduce government finances, making return to egalitarian government programs more difficult, and preserve the debt at high levels so that the debt terror campaign can be revived whenever pressure for government spending arises. Lastly, tax cuts are yet another ideological appeal to individualism, which is part of the neo-liberal campaign to change the political culture.

Despite the barrage of calls for tax cuts — from the same right-wing coalition of forces that campaigned against the deficit — Canadians are resisting the attempted seduction. A *Globe*/Environics poll conducted at the end of 1996 showed that only 9 percent of respondents supported this option. Thirty-one percent wanted money to go to job creation, 25 percent wanted more money spent on health care, and 13 percent preferred spending on benefits for children in poor families.[56] An Environics poll done for the Alberta government in 1997 showed that, even in the province alleged to be the most suspicious of government, citizens want more spent on public services and reject a tax cut. Thirty-seven percent wanted the government's new-found surplus to focus on education, 29 percent chose health care, and 22 percent picked job creation. Just 11 percent cited the debt, and a mere 5 percent, barely more than the margin of error of the poll, wanted tax cuts.[57]

A Vector poll showed that most Canadians are actually willing to pay more taxes for certain public services. For child poverty the support was 74 percent; for training, 68 percent; for free day care so poor parents could work or get training, 57 percent; for education, 58 percent.[58] This is a remarkable endorsement of public services and the principle of equality, given that we are already paying more and getting less for our tax dollars because of debt-servicing charges.

The battle over tax cuts is just beginning, and the new right will be just as relentless on this issue as they were on the deficit. In late 1997, the elite opinion on the issue, however, had not yet reached a consensus. In November, the *Globe and Mail* headlined an Angus Reid poll that claimed Canadians wanted debt reduction and tax cuts by a large margin over new spending. The federal government responded quickly by releasing a poll by Ekos, the same firm conducting its yearly "Rethinking Government" survey, showing that Canadians back increased spending on social programs strongly but not on other items like helicopters and Internet access.[59] The fight for public opinion on tax cuts will be a key indicator of the future of civil society in Canada, and a measure of how resistant Canadians are to appeals to individualism.

❖

Has the counter-revolution of lowered expectations succeeded? It has had dramatic success in the private sector. That is, for those who depend for their livelihood on selling their labour in the private market, expectations are at their lowest in decades, probably since the

Great Depression.

But in the campaign to lower expectations of government, the campaign has largely failed. Despite a plethora of think-tanks, economists, bond raters, and the corporate media to sell the campaign, the collusion of political parties to implement the program of the corporate agenda, and the failure to date of progressive forces to mount an effective opposition to it, the goal of transforming the political culture has not been accomplished. Polls show that people support medicare, think it should continue to be publicly funded, and call for the reinstatement of monies already cut. They are tenacious in their defence of public education. They hold firm to the conviction that governments should do something about unemployment, that areas of high unemployment should get special attention from the federal government, and that the crisis in child poverty should be addressed.

The relative failure of the propaganda campaign should give citizens and those fighting corporate rule some solace, yet overall the news is clearly not good. Obviously, if the vast majority in a democracy opposes virtually every element of a revolutionary change their government is implementing, and the government is able to carry on regardless, there is a crisis in democracy. People's expectations are still high; they still believe in the power of government to improve their lives and to provide them with security. But they are profoundly disillusioned and disappointed that their government seems unwilling to carry out their wishes, and they have very low expectations that governments will change direction.

The old egalitarian state is rapidly transmogrifying into what has been called the market state, which sees as its role promoting not just the accumulation of capital but the commercialization of the functions of government and the commodification of public services. It is the institutional parallel to the transforming of citizens into customers. The state is rapidly restructuring by adopting the corporate ethic as its own, assessing its role by judging how well it adapts to globalization, and by judging its delivery of services as if they were widgets. This is the state that is emerging from ten years of corporate domination and the downsizing of democracy.

10

THE MARKET STATE

Better telephone service. Telecommunications has become very competitive. The range of services is now as great as anywhere in the world. You have to judge [the reforms] by whether the consumer is served or not.
—SIR ROGER DOUGLAS, CREDITED WITH LAUNCHING NEW ZEALAND'S FREE-MARKET REFORMS, ASKED TO DESCRIBE THE COUNTER-REVOLUTION'S GREATEST SUCCESS

The extent to which we have lost democracy in Canada is concealed by the paraphernalia of democracy, its institutions, elections, the sight of politicians from different parties on our TV screens. Yet people obviously take little comfort in these democratic trappings when Ottawa casually dismantles all the things that people say make them proud of the country they have built. The corporate counter-revolution is upon us and our institutions have been hijacked in its service.

Fraser Institute economist Walter Block's musings about a return to the time when only those with property, meaning capital wealth, could vote seem outrageous. Yet formal universal suffrage notwithstanding, this principle has found its way into our "democratic" politics. We now have a political system in which only those with property have "effective" votes, because, when substantive policies and visions for the country are up for debate, only those with property are listened to. The rest of us have been silenced.

Every method of seeking out people's values and desires with regard to government shows that the vast majority believe in a forceful citizen state, that they want more money spent on medicare and education, that they want strong, enforced environmental laws, that they don't trust corporations and don't want them to be involved in social programs, that unemployment should be the main priority for government, that they don't want a tax break, but they do want a well funded state that can do its work effectively.

Yet none of this matters a whit to the governing elite, and they have made it clear that the majority will not rule; they will not be allowed their vision of their country. Paul Martin meets with those who put together the Alternative Federal Budget (AFB) and says, yes, this is a viable document. Polls and other surveys show that most people support the values reflected in the AFB. So what? Property and the propertied rule.

It isn't just the pro-business "policies" of government that have left so many Canadians reeling. Government itself has simply abandoned the ethical and moral ethos of democratic government. It is now no different from Bell Canada, which casually announces the layoff of 10,000 people and then, a couple of years later, with record profits, announces the layoffs of 2,500 more. The federal government just as casually laid off more than 55,000 people. Our governments are becoming corporations in everything but name.

Governments are now infected with the economic rationalism that drives the market. We have prime ministers and premiers who have forgotten their history, their culture, their own values and good sense, and have lost their ethical centre. How can these "leaders," brought up in Canada, display such a stunning indifference to their own communities? Simple: ideology. It allows them to dismiss all facts, all history, all disbelievers. It is government by faith in dogma, and it is immune to the pain and suffering it causes. It is turning our democratic state from an expression of community into a rationale for the market. Just as this new contractual state cannot integrate history or culture into its decisions, it cannot see long-term social consequences, because everything is based on contracts.

If ideology is meaning in the service of power, then since the mid-1970s, but increasingly since the early 1980s, that ideology has been economic rationalism. Understanding this theory of human behaviour makes the actions of our governments more understandable, if no less acceptable. The theory conveniently promotes the liberation of the forces of transnational corporations and international capital, even though it claims to place the "free market" on a philosophical pedestal. Economic rationalism now dominates the thinking of senior policy advisers in nearly every government in Canada.

This abstracted individual is called "economic man" and its innate sexism is the least of its problems. Economic wo/man is always, and only, maximizing his/her satisfactions as if individuals have no other connection to any other human being. By abstracting the individual to this extent, economic rationalism escapes the consequences of its view of human beings as essentially selfish.

The most famous expression of this abstract human-being-with-no-community is Lady Margaret Thatcher's declaration: "There is no such thing as society, only individuals and families each pursuing their own interests." By focussing so exclusively on the individual, economic rationalism ignores the existence of community and all of the human actions, values, attitudes, and principles on which community depends.

Economics has now taken on the status of a religion by which others are judged. Keynes was wary of economics, proposing that it be a matter for specialists and that economists think of themselves as being on a level with dentists. Instead, it has been inflated to imply an entire philosophy, a way of seeing everything. Now, when we choose an economic theory, we choose a particular kind of society.

Before we examine how economic rationalism has been applied to the functions of the state, here is what economic rationalism says about human behaviour and the economy. Economic rationalism states that our economic resources are better allocated through market forces than by government intervention. This proposition is based not on experience or empirical data but on deductive reasoning. It goes something like this: Markets are efficient. Canadian industry is plagued by inefficiencies. Therefore, markets should be freed from all the restrictions that make them inefficient, such as tariffs, financial regulation, environmental and labour regulation, and investment rules.

Economic rationalism today attempts to incorporate empirical evidence to justify its obsession with efficiency. The new right has raised the spectre of "special interests" putting pressure on government for their own narrow benefit. How can an anti-poverty group be a special interest? Creating policies on the basis of people's social condition is inefficient because it puts on the economic model demands that cannot be accommodated. The state's response "is an undesirably complex and incoherent set of policies which obstruct the achievement of the 'national interest.' The prescription: strip away the policies and all will be better off."[1]

Economic rationalism does not recognize the possibility that the so-called special interests are citizens concerned about the quality of life in their communities. The special interests whose pressure has produced the incoherent policies are citizen groups whose sense of social responsibility has led them to take action on child poverty, the degradation of the environment, the erosion of health care and public education, Canada's relationships with poorer countries, and dozens of other aspects of community life.

But economic rationalism does not permit such motivation. Because all individuals are exclusively engaged in a continuous struggle to maximize their satisfactions, all action is, by the deductive reasoning of the economist, self-interested. We are by definition reduced to the status of consumers, on a lifelong quest to quench unquenchable desires. For the economic rationalist, the only sovereignty that matters is consumer sovereignty.

The whole edifice of the counter-revolution is constructed on the single value of efficiency in the name of economic freedom. If we are to make sense of the nonsensical situation in which the economy is doing well but the people are not, we can only do so by returning to this first principle. Everything, including the well-being of citizens, communities, and nations, is to be judged in the court of the market on the basis of whether or not it contributes to efficiency.

Good sense, as opposed to common sense, tells us that efficiency is just one among many values by which to judge an economy. Indeed, a preoccupation with efficiency would make sense in a context of a society that is extremely poor, where people's basic needs are not being met, and where, as a result, it needs to increase its production and consumption. But it makes no sense to be obsessed with efficiency, to the virtual exclusion of all else, when we live in an affluent economy.

In Canada, our goals can and should be much different. The obvious "rational" goal of economics, in the context of community and organized society, is to provide everybody with the necessities of life. The economic rationalists have a lot of explaining to do, for as we become increasingly affluent and productive, we get further and further from that truly rational objective.

We must carefully examine just what "efficiency" means to megacorporations and their ideologues. Efficient at what? As Sam Gindin, economist with the Canadian Auto Workers, argues: "Capitalist competitiveness . . . rules out other forms of 'efficiency' — for instance, the efficient production of high quality goods or the efficient provision of services under conditions controlled by workers, not to maximize the profits for capital but to fulfill social needs."[2] The devastating social consequences of corporate efficiency belie the implied universality of the term.

THE MYTH OF THE MARKET'S INVISIBLE HAND

The Fraser Institute sells Adam Smith ties, and the Adam Smith Institute in Britain promotes free-market policies, including privatization, around the world. Much is done in the name of Adam Smith that the man himself would, and did, object to. Smith's notion of the invisible hand of the market (allocating resources efficiently until an equilibrium was reached to the benefit of everyone) could exist only in a society permanently characterized by thousands of small producers. But the market Smith talked about ceased to exist at around the time he was writing about it. According to Allan Engler, "By the time Smith's vision of the individual in the market had become the foundation of economic theory, the rise of machine industry had made economic independence an unattainable dream for most people. How could factory workers freely exchange the products of their own labour in the market when [their labour] belonged to others?"[3]

Yet Smith had already recognized the changing situation and wrote condemning those who exploited labour. The corporate backers of

Michael Walker's Fraser Institute would recognize themselves in Smith's critique of the great manufacturers and merchants of his day. They were, he declared, "an order of men, whose interest is never exactly the same as that of the public, who have generally an interest to deceive and even to oppress the public and who accordingly have, upon many occasions, both deceived and oppressed it."[4]

Adam Smith had something to say that was directly relevant to today's ruling corporate and political elite and their fondness for permanent recession to encourage "market" forces: "Servants, labourers, and workmen . . . make up the far greater part of . . . political society . . . What improves the circumstances of the greater part can never be regarded as inconvenience to the whole."[5]

The notion that the market can do a better job of allocating resources than an economy regulated by the state assumes that there is still a free market. Yet the kinds of policies promoted by the Fraser Institute, the NCC, and other right-wing think-tanks actually destroy the conditions necessary for entrepreneurial risk taking and innovative investment. As pollster Angus Reid points out in his book *Shakedown*, only in a civil society in which people have trust in their institutions and respect for the political process will people take risks. Yet the market liberals and their ideological sidekicks ridicule the very notion of a civil society.

The Fraser Institute, the C. D. Howe Institute, and the corporate media, for all their talk of the market, have for fifteen years been promoting the accumulation of power and control by transnational corporations, which are notorious for crushing their competitors in the drive for ever greater control of the markets. The corporations' control of capital and their use of an enormous percentage of that capital for nonproductive activity reduce the capital pool that might otherwise be available for creative local investment. Robson Street in Vancouver and Elgin Street in Ottawa, just two examples of once vibrant expressions of local culture and entrepreneurial spirit, are now being wiped clean of any local expression, homogenized by a corporate blitzkrieg.

As giant corporations came to dominate economies and businesses, the invisible hand of the market was replaced by the very visible hand of managers. The multiple units of production and distribution of modern large corporations, and particularly TNCs, saw coordinated administration replace the market in the allocation of resources. Indeed, the evolution of the modern TNC was driven by the effort to gain administrative control over materials, labour, markets, technology, and finance. The more control, the fewer risks, the more profit. Far from taking risk, large corporations, as we saw in chapter 3, devote enormous energy and planning to avoid it.

The constant attacks of business leaders, market liberals, economists, and neo-con politicians on the whole notion of planning the economy and regulating demand are based largely on the myth of "the play of market forces." But what else are large corporations if not huge planning institutions? The largest corporations in the world are larger than most countries and have far more capacity to plan than all but the most developed nations. Furthermore, corporations are not restricted by issues of accountability to broad communities of voters and their varied interests. They are accountable only, and only in the narrowest share-price terms, to shareholders.

Corporations and governments are both hierarchical, bureaucratic planning organizations. The difference isn't in the planning but in the goal of the planning. In governments that goal has been, until recently, to exert some influence over the allocation of national resources to achieve social objectives; for corporations it is to achieve profits and maintain growth. With the increasing application of economic rationalism to the principles of governance, though, the differences are blurring. The egalitarian, public purpose of government planning is being transformed, as market objectives and market-inspired performance assessment seeps into the civil service.

It is revealing of both the objectives of the TNCs and the morality of those who labour in their interests that the two countries often identified as preferred models by the corporate elite and economic rationalists are Chile and New Zealand. Both countries are characterized by almost completely deregulated and privatized economies — and the extreme class divisions that such economies naturally produce. In both cases, the structural adjustment regimes were imposed by coups, a military coup in Chile in 1973 (examined in chapter 3) and a political coup in New Zealand in 1984.

GOVERNMENT BY FUNDAMENTALIST FANATICISM

New Zealand in the 1980s became the darling of the neo-con set as the Little Country That Could. Not only did it completely deregulate its economy, from the financial sector to the ending of tariffs, agriculture supports, and labour deregulation, it privatized virtually every public enterprise, to the tune of $16 billion, and devalued the New Zealand dollar by 20 percent to encourage foreign investment. The result was an orgy of foreign purchasing of the country's assets that was symbolized by foreign corporations buying up every major New Zealand bank. No other developed country has passed policies so favourable to transnational corporations and finance capital.[6]

But the neo-cons' admiration for New Zealand extends to how it has transformed the state in the image of corporation, economic rationalism, and market liberalism. In a manner that has some parallels in Britain, successive New Zealand governments, of both the Labour Party and the conservative National Party, have extended the counter-revolution into the civil service and the services provided by the state, such as education and health care.

What has come to be described as the contract state or market state is not just a deregulated state. It is a state that adopts the competitive principles of corporate management as its model and corporate "values" as its national values. Its assumptions are appropriated from economic rationalism and systematically adapted into operational plans for governance on market principles. It is the corporate takeover of the state in all but name.

Not only has the New Zealand government established a regime of corporate libertarianism, it has superimposed on democratic government a model of management identical to that of corporations. The traditional public service, established and trained to reflect the communitarian values intrinsic to the definition of society, has been all but swept away, replaced by corporate executives, planners, and managers immersed in the ideology of the market.

The application of economic rationalism to democratic governance has spawned a number of faith-based theories, all of them rooted in the notion of what Canadian philosopher C. B. Macpherson called possessive individualism. These abstracted theories about human behaviour have a variety of names, each with special application: public choice theory, agency theory, a whole system of management philosophy applied to the public service called new public management, and the "new contractualism," which reduces all relationships to business contracts. These theories have all found their way into the practice of governance in New Zealand. And they all reduce citizens to customers, and governments to enterprises.

As the University of Waikato's Paul Havemann says, these theories have the effect of "hollowing out the state, that is, cutting out its citizens-focussed core. In New Zealand the contract state is explicitly the undemocratic state . . . The idea that the public service should be run in a 'business-like fashion' has been conflated with running it like a business, i.e. for a profit."[7]

Applying the principle of contractualism transforms the citizen into a customer contracting with the state for a service. Public accountability through the normal democratic process is replaced by contractual responsibility — two otherwise socially disconnected entities fulfil a one-time, binding contract. By this model all social relationships are turned into commodities to be exchanged on the market.

Nowhere do these neo-liberal theories permit the notion that any person would choose to work for government because he was committed to the idea of public service, that any politician would run for office to contribute to her community. "Concepts like 'public spirit,' 'public service,' and 'the public interest' have not figured . . . in the public choice literature . . . because they are thought to lack meaning or relevance."[8]

Their application of such theories starkly demonstrates just how impoverished is the new right's view of human nature. It is ideology gone berserk. As John McMurtry, author of *Unequal Freedoms: The Global Market as an Ethical System*, suggests, the market doctrine smacks of "fundamentalist fanaticism." It is immune to the influence of the everyday experience of even those promoting the theories. Every day we engage in and observe dozens of actions that are not self-interested, at home, in the workplace, on the street. Organized society could not possibly exist if economic rationalism had any basis in reality. Yet this fundamentalist ideology now guides the entire governance structure, social services, and health care and education in New Zealand.

According to Jonathon Boston of Victoria University in Wellington, "Government agencies have been seen as businesses; ministers have been likened to board chairpersons and department heads to chief executives . . . and the taxpayers have been seen as shareholders." The health-care system now carefully separates "purchaser" (the state) from "provider" (private and public health organizations) for fear that conflicts of interest and "bureaucratic capture" will undermine efficiency and accountability.[9]

Government departments have been commercialized and corporatized; commercial operations have been separated from noncommercial. Yet TNCs, one of the most successful organizational forms in history, have grown in size, bureaucratic complexity, and hierarchical structure at the same time as large departmental bureaucracies have been radically decentralized in the name of avoiding bureaucratic capture. This is pure political ideology, a rationalization for deconstructing the democratic state.

No empirical evidence whatsoever suggests that bureaucrats are primarily interested in maximizing their budgets. They are influenced by many factors, not least of which is often a strong belief in community values, but also integrity, reputation, and professional standards. It is presumably these qualities that corporations value when they hire these "bureaucrats" after they have left government.

The wholesale application of this orthodoxy also paints politicians as inevitably opportunistic. Yet politicians repeatedly do things that their constituents don't like and risk losing votes. They are also committed to the ideas their parties espouse. We can presume, for example, that the politicians in New Zealand believed in thier counter-revolution. With laws in place to "reduce the scope for political interference," New Zealand has entered the Orwellian world of laws and constitutional amendments that stop elected representatives from "interfering" in the governing of their country.

Much of the restructuring of the New Zealand government was based on the apprehension of "bureaucratic capture," that is, the manipulation of politicians by "special interests" and by empire-building civil servants. "Special interests," of course, include the people actually providing all the public services. As a result, policy making during the counter-revolutionary period (from 1984 on) was made with the explicit exclusion of those who had knowledge of the policy area in question. Educators and health-care experts were mostly excluded on the assumption that they would resist change in defending their "special interests." In their place were CEOs and business administrators.

Within a few years of the reforms, New Zealand's social statistics gave stunning evidence of what "the market" does to a community when it is unleashed from any controls. New Zealand now has the highest youth suicide rate in the world, skyrocketing prostitution among young women, a child poverty rate of nearly 25 percent, food banks proliferating everywhere in a country that had never seen them, the loss of half the manufacturing jobs in the country, chronic high unemployment (especially among the Maori and as high as 100 percent in some rural communities), and the legislative destruction of unions (from 63 percent of workers organized to 26 percent).

There has been a dramatic reversal of the gains women had made in income and security, and new class divisions are now a permanent feature of New Zealand society. Hundreds of children have been forced to leave their schools because their learning disabilities are "too expensive" to deal with; the diseases of poverty, eradicated in the 1940s, are back with a vengeance; lengthy hospital waiting lists are now "normal," and New Zealand boasts one of the highest violent crime rates in the developed world. Tens of thousands of people have emigrated to escape the destruction of their own country.

No area of policy escaped the zeal of the marketeers, but, of all the areas transformed, the effects in education are perhaps the most revealing. Because public education is such a critical area for creating an egalitarian society, neo-cons and market liberals focus a great deal of attention on reforming it. The free-market revolution is not just indifferent to social equality; it is hostile to it, believing it to be a barrier to rewarding those who excel.

The commercialization of education in New Zealand took the form of charter schools, which marry the principles of private schools, with their emphasis on exclusivity and choice, with the state funding of a public system. The government eliminated school boards and replaced them with local school committees of parents who were mandated to run some aspects of their schools. Super-imposed on this site-based management model was the principle of school "choice," a key concept for the economic rationalist and the libertarian.[10]

Within three years, half the schools in the country were losing students and the other half gaining them, a shift based on the notion of good schools and bad schools in which the privileged rushed headlong to ensure that their children got the "best." What became known as the good school syndrome was a self-fulfilling prophecy.

While student fees weren't compulsory, they were expected, and the disparity between poor and wealthy schools grew even wider. The average student fee paid in the former is $48 a year and in the latter $200, producing an average budgetary difference of over $180,000 a year.

The educational and social consequences were almost immediate. What is termed "white flight" leaves schools of mixed class and race impoverished — they lose model parents, state per-pupil funding, and, increasingly, their best teachers. Parents ended up identifying more with their social class than they did with their neighbourhood.

Segregated by social class and denied the opportunity to work and play with those of other races and classes, children become increasingly vulnerable to the unsympathetic stereotypes of the poor and nonwhites that prevail in New Zealand society. The new school system institutionalizes the class divisions created by other free-market policies.

The fruits of consumer sovereignty are hard to find. Popular schools effectively choose students from those who apply, not the other way around, and to keep their grade-average high (to attract funding), the "good" schools have systematically excluded those students who don't perform well. Parents have little say in the substantive issues of education and spend most of their time on administrative details and fundraising.

As Havemann writes, this "model of development requires the deconstruction of the universalized citizen and promotes the construction of the citizen-customer."[11] In this new world of the citizen-customer, those who do not produce, those who have not earned money to qualify as a customer, are automatically defined as noncitizens. The government's deliberate division of the poor into deserving and undeserving is a reflection of this effort at transformation.

None of the major reforms to New Zealand's economy and

government were ever voted upon, an approach that was begun by chance and developed later into a handbook for the rapid restructuring of Western economies. In the 1984 election, both Labour and the governing National Party ran on their traditional platforms, both of which included continuing tariffs, farm subsidies, currency exchanges, and full-employment policies.

But behind the scenes, some very different thinking had been taking place within a small group in the Labour hierarchy. Led by Roger Douglas, these technopols had seized control of Labour's economic policy making and were determined to bring radical changes to the country's stagnant economy. At the same time, a number of officials in the treasury branch (the finance department) and the Reserve Bank, trained at the monetarist schools in the U.S., were preparing comprehensive plans for restructuring. Beginning in 1982, these two groups and some new, aggressive businessmen in the finance sector met regularly to discuss economic and state restructuring. Immediately after the 1984 election, a currency crisis provided the opportunity for the wholesale introduction of nearly the entire neo-liberal program.

Two other benchmarks of the modern pattern of restructuring featured in the New Zealand counter-revolution. First, the core of new-right radicals moved quickly to expand their base by seizing the initiative in the other main political party, the National Party. At the same time, they systematically neutralized centres of opposition in government or government-funded agencies.

Second, the neo-cons recognized the need to ensure that a significant portion of the population substantially gained income and social status from the changes so that, regardless of obvious failures in other areas, there would be a strong base of support for the reforms. In these two efforts they were largely successful. While poverty increased dramatically, much of the middle class has been seduced by access to foreign luxury goods, lower taxes, access to foreign currency and stock markets, investments in privatized state assets, and the resulting higher incomes and greater social status.

Promoting a comprehensive ideology for corporate rule provides an enormous advantage for those committed to restructuring, for it provides converts with a complete package of reforms that purport to have the answers to everything. This package, including guidelines on how to implement it, presents itself as much more than an economic theory or an approach to governance. It is the equivalent of a religion, and its proponents' faith in its tenets is a powerful incentive to continue the "program" regardless of the consequences; to proselytize to anyone who will listen, and to ostracize those who challenge the faith. Faith is a powerful weapon against all doubters.

The New Zealand model has been marketed around the world by Roger Douglas, now Sir Roger, who has been to Canada nearly a dozen times. One of his most high-profile appearances here was at the 1991 federal Reform Party convention, where he was the keynote speaker. He described ten principles for implementing the counter-revolution. The first was, "Consensus for 'quality decisions' does not arise before they are taken; it develops progressively after they are taken." This principle of deliberate electoral deception must be combined with implementing reforms in "quantum leaps using large packages" because otherwise "interest groups will drag you down." And the reforms must be implemented "at maximum speed."[12]

The New Zealand experiment is now virtually a restructuring handbook for other Western nations. At the colloquium on structural adjustment held by the International Institute of Economics in 1993, a number of key requirements were identified, all of them elements of New Zealand's reform program. There had to be a crisis on which to launch the changes, a systematic program and a core of technocrats and technopols strategically placed to carry it out, support from institutional power, a political leader not concerned about his or her popularity, and "beneficiaries likely to fight to protect the reforms."[13]

STRUCTURAL ADJUSTMENT IN CANADA

Restructuring has been uneven in Canada; ruthless and uncompromising in some jurisdictions, gradual in others, effectively blocked in at least one. But the ideology and parts of the program are present across the country, either explicitly applied by governments or seeping into the civil service and the administration of programs.

The federal government under both the Mulroney Tories and the Chrétien Liberals were not and are not as ideologically consistent or committed as the Labour and National governments of New Zealand.

In part this is true simply because Canada is a federation, and even in national parties competing regional influences interfere with purely ideological programs. But Brian Mulroney was prepared to implement major elements of the corporate economic agenda regardless of the political and personal consequences, sacrificing public popularity to his commitment to corporate objectives.

Mulroney did not achieve everything the corporate elite asked. His failed constitutional initiatives meant that property rights and amendments to the Bank of Canada's mandate (to make inflation fighting its exclusive mandate) did not go through. And his cuts to social spending were nowhere near the levels demanded by the corporate elite. Yet he started all the projects that constitute the program of economic rationalists. Mulroney's greatest success was the free-trade deal, still the centrepiece of corporate rule.

The Liberal Party's Red Book for the 1993 election stated: "We do not believe that the only solution to our economic problems is another five years of cutbacks, job losses and diminished expectations." Yet that solution is what Paul Martin and Jean Chrétien believed in before and after they approved the publication of the Red Book of promises. And it summarizes very well what they did between 1993 and 1997. Their course of action had, in fact, been determined at the "thinkers conference" convened at Chrétien's behest, in Aylmer, Quebec, in the fall of 1991. This conference, and Chrétien himself, set the stage for corporate rule.

To achieve their return to power the Liberals followed Sir Roger Douglas's first principle for implementing structural adjustment: electoral deception and the hope that they could build retroactive consensus in time for the next election. There is no evidence that any of the economic facts changed after the election, and the Liberals do not claim that anything had changed. They simply lied, and with a cavalier indifference that betrayed their eagerness to get down to business. And they did just that. As Maude Barlow and Bruce Campbell write:

> Within two years, the party that came to power on a pledge of jobs, social security, and preserving the nation-state would gut Canada's social programs, break the collective agreement with its public sector workers, privatize its transportation system,

> commercialize its cultural sector, abandon its environmental obligations, endorse world-wide free trade, sever trade from human rights, promote deregulated foreign investment, yield control of the economy to global investment speculators, and become the apologists for the corporate sector it once vilified.[14]

For a party that had staked its ground on opposing free trade and promising to renegotiate NAFTA, the Liberals did more to establish a level playing field for corporate investment than Mulroney ever found the will to do. Taking Michael Walker's advice, Martin cut social spending back to levels (as a percentage of GDP) not seen since 1949 and now below the spending levels of the U.S. Critics of these massive cuts have rightly focussed on their impact on ordinary Canadians, but it is also true that they directly benefit TNCs.

Cuts to social spending constitute a transfer of power from real citizens to corporate citizens, and the rationale for them is clearly revealed in documents, studies, and position papers of international agencies and corporate think-tanks around the world. Social spending is not simply a redistribution of wealth; it is a redistribution of political power, because it provides security and independence otherwise unattainable by those who have to sell their labour to survive. Every time medicare is weakened, or parents have to pay for school materials that used to be paid for by the school board, or labour standards are reduced, or a municipal service is privatized because of the downloading of cuts by higher governments, ordinary people and their communities lose power to corporations.

As we saw in chapter 6, the capacity of people and communities to resist the ever greater demands corporations make on them is very much connected to the role of governments. The federal Liberals' creation of a level playing field in social spending is explicitly a part of the FTA promise made to the U.S. The Americans had tagged as "unfair subsidies" the social programs now being cut. Market liberals saw them as nontariff barriers, a part of the old protectionist past, a market "rigidity" and a barrier to efficiency. Behind all this jargon: the demands of corporations for greater freedom to act.

In fact, the social spending cuts actually promote the growth of TNCs while they undermine the interests of many Canadian corporations. Universal medicare provides an enormous advantage to Canadian manufacturers because they do not have to pay private medicare premiums for their employees, as American corporations do. As Canadian medicare is eroded, employers, in order to keep employees and keep them healthy, will eventually have to start paying their employees' private insurance premiums. The same is true with education. The result is that corporations in Canada will actually be less competitive with their American counterparts, increasing pressures to produce off-shore, externalize costs, and merge into ever larger entities. Despite this contradiction for Canadian corporations, ideology prevails, even, it would seem, at the expense of the shareholder.

One of the Liberal government's most draconian and shameful initiatives has been its savaging of the unemployment insurance system. The special targeting of UI is more evidence of the rapid development of the market state in Canada, a mark of the government's acceptance of the economic rationalists' arguments about human nature. (Economic rationalists assume that no one would work if other income is provided.)

The Liberals' Green Book, its policy paper on UI, is an exact replica of the position of the OECD, the most free market of the multilateral organizations. The Green Book states that UI "may deter some workers from seeking alternate employment, relocating, developing broader skills, or tapping their creativity and potential for small scale enterprise."[15] There is no evidence that generous UI payments cause unemployment except at extremely low levels of joblessness. If that were the case, the unemployment rate should have been coming down steadily, with the UI cutbacks of the 1990s. And the "logic" of forcing workers to relocate is economic rationalism at its most obvious: sacrificing community for abstract economic efficiency.

Another key indicator of how the Canadian state is transforming its role is the Liberal government's changes to its policy on international human rights. Thus, in May 1997, Canada announced that it was considering a request from Indonesia for "military training assistance, police training, regular exercises with the Canadian navy and a full-time military attaché in Ottawa." The fact that the Suharto regime is one of the most ruthless and violent dictatorships in the world and widely condemned for its genocidal policies in East Timor was reduced to a public relations problem. In the words of an unnamed foreign affairs official, "It has to be finessed at a political level. It's the optics of putting out public dollars to train people who are associated with the ABRI [the Armed Forces of the Republic of Indonesia]. That's what makes it so difficult."[16]

In July, the government announced that it would apply sanctions to companies doing business with Burma, another country guilty of gross human rights violations. The reasons for the different approaches to two equally repressive regimes is not difficult to determine. Trade with Indonesia reached $1.5 billion in 1996, almost one hundred times the $16 million with Burma. A spokesman for foreign affairs told a worried mining investor with assets in Burma that the sanctions were, in any case, "largely symbolic."[17]

The most dramatic demonstration of the new role of the Canadian government vis-à-vis international human rights has been the four Team Canada junkets to Asia and Latin America. The prime minister and assorted premiers became cheerleaders for transnational corporations in a whole raft of countries with horrific human rights records, from China with its forced-labour camps, and India and Pakistan, notorious for their child slave-labour factories.

Bill Saunders was an award-winning television documentary maker in New Zealand before the Labour government "commercialized" the public broadcasting system. In doing so, the government eliminated any requirement that the New Zealand Broadcasting Corporation produce any New Zealand content in its programming. Saunders had a meeting with his superior. "The new mandate, I was told, was to deliver the audience to the advertiser. Full stop. I stayed on for a while but I eventually left. All productions had to be geared to the international market. You had to be able to sell them abroad or forget it."[18]

As *Globe and Mail* reporter Doug Saunders pointed out in "Exporting Canadian Culture," the global product imperative is very much in play in Canada. The four-part TV series based on Peter Newman's history of the Hudson's Bay Company was "carefully tailored to appeal to European and U.S. audiences." According to the series producer, Michael Levine, "We could tell it from a Canadian viewpoint in a way that absolutely *nobody* would have bought it." Said Saunders, "This is how culture is created in Canada today."[19] You must prove you have an international audience or you won't get financed.

When Eggleton, then minister of trade, commented on the WTO ruling in favour of split-run magazine editions, he set off a panic. He suggested that protecting culture was a nonstarter, that we should be producing for the global market. This position wasn't stated again so blatantly and may have been a testing of political waters. Heritage Minister Sheila Copps was given the opportunity to lead a charge in favour of doing *something* about the WTO threat. But the reactions all around suggested that the WTO ruling was completely unconnected with anything Canada had done or could do. Yet Canada signed the WTO agreements and when this ruling came down was involved in secret MAI negotiations that would make things far worse.

The door to the attack on Canadian culture was opened by the signing of the FTA and NAFTA and much of the erosion of Canadian culture results directly or indirectly from those deals. Yet most of the actions taken by the federal government were not forced on it by NAFTA. They were voluntarily taken, demonstrating not just that the government knows exactly what NAFTA requires but that it is fully supportive of the ideology.

While the government would no doubt like to look as though it is defending Canadian magazines, its policy decisions in other areas of cultural activity have devastated Canadian production of films, TV documentaries and dramas, and books. Incentives delivered through the tax system for these areas all but disappeared under the Tories between 1990 and 1992, from $1 billion to nearly zero. More than any other single act, this drove media companies to look for foreign funding. The Liberals have done nothing to reverse this policy.

Consistent with that policy are the Liberals' cuts to the CBC, the flagship of Canadian cultural organizations and the one most identified by Canadians as such. The sheer size of the cuts under the Liberals — 29 percent when the average cut to program spending under the Liberals has been 22 percent — suggests that rather than trying to protect the CBC (as promised in the Red Book) they have targeted it for special treatment.[20]

Most of the support programs for Canadian publishers have been eliminated or cut back. Publishers now routinely reject book proposals that would have been viable just a few years ago. While the Canadian film and TV industry is doing well on the international market, many Canadian themes just don't get addressed. According to the CBC's Mark Starowicz, many Canadian stories aren't being told and are not even being brought forward. Writers censor their own ideas. "In documentaries . . . we don't change our stories, we murder them in the crib."[21]

Ironically, the government still spends a great deal on cultural "industries," giving tax breaks to Hollywood studios to encourage investment and funding private production houses. Documentary producer John Kastner asks, "Are [the CBC], Telefilm, the OFDC [Ontario Film Development Corporation], all of these government-funded agencies really subsidizing independent producers who are making American directed stories?"[22]

In 1997 the Senate Subcommittee on Communications tabled its interim report on Canada's international competitive position in communications. With the macho title *Wired to Win*, it is a stunning example of the betrayal of Canadian interests and their capture by the globalization hype. The senators recommend that Canada "move away from policies based on *protection* and towards those that seek to more proactively *promote* Canadian products."[23] The attitude of the senators towards industry spokespeople is reflected in their fawning over IBM's John Warner (with his corporate-rule title of director of government programs); he told the senators that "it is essential to link industrial and cultural policy objectives [which] are best met through an increased reliance on market forces."[24]

The senators as good as endorsed IBM's recommendations that would, "within a period of roughly 10 years, see the complete phasing out of all regulations in the area of licensing, foreign ownership, Canadian content and mandatory contributions to Canadian production."[25] It would be hard to outdo the chamber of sober second thought in handing over parliamentary sovereignty.

The new right's call for decentralization is a piece of the restructuring plan that is less obvious to the casual observer. It is often sold, as it has been by the Reform Party, as a more democratic alternative because it brings services closer to the people who are being served. This is nothing more than deliberate deception.

National programs, such as medicare, social assistance, and post-secondary education, provide national citizenship rights and a national identity, and these become a central part of the political culture. If programs are delivered by a national government, they are then universally available, and opposition to their removal can be more effectively focussed. According to the guru of economic rationalism, Milton Friedman, there are two steps to the elimination of democratic governance. The first is limiting the scope of government. The second is the dispersal of government power. "If government is to exercise power, better in the county than in the state, better in the state than in Washington. If I do not like what my local community does, be it in sewage disposal, or zoning, or schools, I can move to another local community . . . If I do not like what Washington imposes, I have few alternatives in this world of jealous nations."[26]

The BCNI has consistently pushed for decentralization. Its latest initiative, in the summer of 1997, used Quebec nationalism as a stalking horse for the devolution of the national government. The Chrétien Liberals have already taken decisive steps towards decentralization. Building on the policies of the Mulroney Tories, the cuts to social program transfers in themselves are a de facto decentralization, as they reduce the impact of the national government. The Tories cut federal spending on health, education, and welfare to 15 percent of provincial spending (from 20 percent), and the Liberals followed by cutting that to 9 percent.[27]

But the Liberals have taken it further. They placed a cap on federal spending for the Canada Assistance Plan, which established national standards for social assistance, and implemented the Canada Health and Social Transfer by which federal transfers to the provinces for health and education were rolled into one, allowing provinces to spend the money as they saw fit. The government is also increasingly leaving environmental regulation to the provinces, and by its massive privatization of the national transportation system has all but abandoned that field as well. In 1996 it handed over to the provinces employment training and housing.

TRANSFORMING THE PUBLIC SERVICE

The state is far more than just the elected representatives, the provincial legislatures, and the House of Commons. Part of the traditional system of checks and balances built into democratic governance is the professional civil service, whose job it is to devise policy for elected governments and to ensure that those policies are carried out. During the post-war period of the social contract they were a critical element in the creation of the egalitarian democratic state.

That role has now changed dramatically in Ottawa and the provinces. Through attrition and the changing ethos of the bureaucracy, the public servants who created the egalitarian state are rapidly disappearing, replaced by those who are chosen because they reflect the new market ethos or by younger professionals trained in new public management theories or in the ideology of public choice theory.

Duncan Cameron is head of the Canadian Centre for Policy Alternatives and a former civil servant in the finance department. He explains that the transformation of the public service was a complex affair. It started out as a result of how successful the mandarins had been at guiding the policy process. "They had too much success. The mandarins by the time Trudeau came in were dominating the government's agenda. Trudeau wanted the elected ministers to lead the policy debate. It was Trudeau who actually dismantled the mandarin system. As a result, with the growth of government and its complexity, the culture of the policy adviser changed from mandarin as policy thinker to mandarin as manager. Once you're into that management mode you're into the business ethic."[28]

It was in the mid-1970s that the ethos of serving the public began to seriously erode. Cameron compares two deputy finance ministers. Robert Bryce who had studied with Keynes, refused on principle to work for the private sector when he retired. Mickey Cohen, deputy minister through most of the 1980s, after making a tax ruling that put $600 million in the pockets of the Reichmann family, joined the company six months after quitting the department.

Another major change is shown in the way people were recruited. The Public Service Commission used to hire for the entire government, and based its decisions on tests and interviews. But under Simon Reisman's reign as deputy finance minister, the government began hiring straight from university economics departments, dominated by the late seventies by monetarists and economic rationalists. The result was that "they were recruiting the people who had gone through the brainwashing process of academic economics."[29]

Two other trends contributed to the remaking of the public service, completing its transformation into a market-oriented bureaucracy disconnected from the country's history and its social reality. Trudeau undermined the public service ethos, and Mulroney completely politicized the bureaucracy by appointing political operatives like Stanley Hartt into deputy minister positions.

For Cameron, the changes that have taken place since the late 1960s are further symbolized by who in the finance department is in charge of social policy. "When I arrived in 1966 the person on the social policy desk in finance was [the NDP's] Stanley Knowles's son, David. Today the assistant deputy minister for social policy is Tom d'Aquino's wife. She was the one who took over Lloyd Axworthy's social policy review and turned it into a 40 percent funding cut."

The federal public service has by no means been completely purged of those who believe in the idea of government. But the departments that now dominate government, principally finance but also trade and foreign affairs, are captives of this ideology. Treasury Board has studied New Zealand extensively and the economic rationalist theories such as new public management are already being incorporated into the civil service ethos.

In addition, recent years have seen a huge exodus of senior policy people from the federal government. The generous compensation packages in the private sector, combined with a six-year salary freeze for senior public servants, contributed to more than a third of the top officials quitting in 1996. These were mostly people who had joined the service in the 1960s and 1970s and had designed the social programs that are now being slashed. The slow destruction of their accomplishments is also a major factor in their decisions to quit. Many have gone to jobs as lobbyists for large corporations or are with consulting firms.

The transformation of the public service ethic was captured in a 1996 meeting between public-interest groups and officials of the foreign affairs department. The meeting was to discuss funding for the People's Summit on APEC, a forum for representatives of community groups from all the APEC countries, coinciding with the APEC summit in Vancouver in November 1997. In the past these meetings have been friendly affairs, with officials and community representatives sharing a human rights perspective on international relations. The change this time was stunning. Maude Barlow of the Council of Canadians attended.

> We sat down — this was all the big groups, Amnesty International, the CLC, NAC, Canadian Federation of Students, Greenpeace, the churches — and basically it turned into a screaming match. We were coming from different planets. They were coming from a place where they had accepted every single piece of the dogma about trade liberalization . . . At one point MP Raymond Chan, the undersecretary to Lloyd Axworthy, talked about how human rights are improving in China because he had had a meeting with a reporter. The bad news, he said, was that the reporter was arrested a week after this meeting.[30]

It was as if three decades of a democratic ethos had simply never been. When the delegates asked why there had been no consultations on the environmental impact of APEC, an official replied that there had been. Recalls Barlow, "We asked what environmental groups were involved, and he said, 'Well, none.' When we asked why not, he said that the leaders of the economies and the Canadian business people did not want the environment groups involved. He was completely disingenuous when he said this. People were open-mouthed, but he had absolutely no idea what we would be upset about."

The offer on the table to the organizations that have been playing a major role in such consultations for a generation was $100,000, none of which could be used to bring in foreign delegates. Until 1997, CIDA had always insisted that some of the money provided be used for this purpose. And while it was clear that the traditional players were to be cut out of the picture, the government was spending large sums to create its own pro-market, popular sector.

The government spent $700,000 to involve "youth" in the APEC summit. The youth were identified from the subscription list of *Teen Generation* magazine. Similarly, the government sponsored a major women's event called "Stepping Out" for which they hand-picked some Canadian businesswomen and some from the other APEC "economies." Just as the new right ensures that it has a base of support by sharing wealth with a significant section of the middle class, the market state is creating a political base that will support its new approach to international relations.

THE PROVINCES

There is not space in this book to document the commercialization of government at the provincial level. But anyone even casually observing their own provincial government knows well enough that market principles are being expressed in policies affecting everything from medicare and education to municipal services and the arts. The process is uneven, but the influence of monetarist economics, economic rationalism, and the aggressive lobbying of market liberals is everywhere apparent.

Ontario's Mike Harris has followed the restructuring handbook almost to the letter. He has surrounded himself with a core of like-minded ideological true believers whose view of politics really is different from past Tories'. They are prepared to transform the democratic state as a matter of principle, as part of their crusade against equality, regardless of the personal or political consequences. Like Roger Douglas, Margaret Thatcher, and Ronald Reagan, they continue their program with few retreats, and implement it as fast as possible, confident that once the edifice of the egalitarian state is deconstructed it will be impossible to rebuild.

All the elements are there: cuts to health and education; privatization; user fees; slashing of agencies mandated to regulate labour, the environment, and health; elimination of conservation programs; halving of grants to municipalities (seen by many as a direct assault on democracy at the local level); a 22 percent cut to social assistance, the introduction of workfare, and other punitive measures against the poor, and draconian labour legislation that takes workers' rights back to where they were in the 1950s.[31]

In the name of not competing with the private sector, and to advance another ideological principle of the new right, the government eliminated all job creation programs in its first two years. As with the federal Liberals, many of these cuts have no explanation, other than ideology. The government is simply getting out of the business of governing. It is also getting other people out of the business of governing.

The Ontario government's omnibus bill of forty-four pieces of legislation breaks the record previously held by B.C.'s Bill Bennett, whose 1983 Fraser Institute blueprint saw twenty-six pieces of legislation introduced on the first day. Harris's 211-page bill, among other things, gave the government enormous administrative powers to take actions without reference to the legislature, rolled back laws, and gave power to privatize medicare, abolish local governments, raid its employee pension fund, eliminate hospital boards, and amend the Freedom of Information Act to make it harder for citizens to get information about their government.[32]

Later in its mandate, the government cut the number of school boards in half. The resulting Toronto School Board will have 300,000 students, making a mockery of democratic governance. Harris's stubborn confrontation with teachers is classic new-right politics, but it had the effect of identifying teachers as the bulwark for a public education system that is the centrepiece of a democratic society. The government also imposed a megacity structure on the Metro Toronto area, wiping out the democratic governments of half a dozen communities. There was little effort to give these actions a public relations spin. It was being done because there was "too much government." Local politicians were portrayed almost as the enemies of the communities they governed, so self-interested and untrustworthy that Harris put six municipalities under trusteeship until his amalgamation plan was in place.

The Harris Tories reflect another aspect of the state in the new corporate order. The development of the market state necessitates the parallel construction of the security state. As the legitimacy of government breaks down, an inevitable result of the contraction of the egalitarian policies, the coercive function of the state must increase.

To date the opposition to the assault on democracy has been peaceful, with few examples of civil disobedience. Yet ultimately it is only through widespread civil disobedience that the current neo-liberal regimes in Canada will be obliged to change course. And there are already very strong signs that governments will act ruthlessly when civil disobedience begins in earnest. In Ontario the Harris government, early in its mandate, unleashed riot police on a peaceful demonstration at the legislature, producing a scene of unrestrained and highly political police brutality not witnessed in Canada for decades.

In a peaceful demonstration in Ontario, six women were arrested, taken to a high-security prison, and strip-searched. In Harris's own North Bay riding, fifty police were mobilized to break up a demonstration of a thousand teachers, unionists, students, and parents at a fundraising dinner for the Conservative Party at which Harris was speaking. In the background were another sixty tactical police officers in case there was serious resistance.

In Vancouver, during the APEC summit in 1997, the political cleansing of the territory close to the summit site and along the routes leading to it from downtown hotels was reminiscent of the "eyesore free zone" created in Manila for the 1996 summit. A human rights activist, Craig Jones, was held in jail for fourteen hours simply for placing signs reading "Democracy" and "Free Speech" on a roadway where they would be visible to passing APEC motorcades. RCMP spokesmen were aggressive in their defence of these flagrant abuses of democratic rights. Later, students at the University of British Columbia, protesting against the APEC summit being held on their campus, and against the presence of Indonesian dictator Suharto and Chinese president Jiang Zemin, were gratuitously pepper-sprayed in a further demonstration of intimidation. One woman was grabbed from behind by the RCMP, thrown to the ground, and sprayed in the face.

For his part, Prime Minister Chrétien, already infamous for his choking assault on another peaceful demonstrator, and fresh from consorting with dictators, replied to a reporter's question about the police assault with, "For me, pepper, I only put it on my plate. Next [question]." Whether simply a grossly insensitive remark or a signal to police that their actions are approved at the highest level, the meaning for political rights was obvious. Clear, too, was Chrétien's eagerness to comply with Jiang Zemin's request that demonstrators be kept out of earshot and out of sight, a request that became an order to police forces. Several commentators mused about the irony in the federal government's position that dictators like Suharto and Jiang Zemin will moderate their attitudes towards human rights. It seemed as though the result of engagement has been a hardening of attitudes towards human rights in Canada.

These and other recent examples of overwhelming police presence, brutality in the name of security, deliberate intimidation and humiliation are an ominous sign of things to come. So, too, was a memo leaked in November 1997 about a planned $350,000 training program for Atlantic fisheries employees in anticipation of possible "life threatening violence" on the part of fishers at the ending of the TAGS program in May 1998. The message is clear. Accept the downsizing of democracy and the end of social rights or you will get assaulted, or arrested, or both.

❖

Twenty years ago Samuel P. Huntington declared in a Trilateral Commission study that the United States, and by extension other Western countries, was suffering from a crisis of democracy. That crisis was described as an "excess of democracy," a situation in which too many people were engaged in the political process. "Apathy and non-involvement" on the part of a large portion of the population are required for a liberal democracy to be governable, said Huntington. And so we see governments systematically putting in place institutions, bureaucratic regimes, social policies, and a whole new governing ethos, hoping to institutionalize public apathy.

There would seem to be little argument that from the perspective of the ruled in Western societies there is now a severe shortage of democracy. The challenge for those determined to stop the neo-liberal counter-revolution is to renew the struggle for a radical democracy, and to do that we must become intentional citizens.

11

DEMOCRACY IN THE NEW MILLENNIUM

An alternative conception of democracy is that the public must be barred from managing their own affairs, and the means of information must be kept narrowly and rigidly controlled.

— NOAM CHOMSKY

I perceive the divine patience of your people, but where is their divine anger?

— BERTOLT BRECHT

Things are going to get worse before they get better; it doesn't have to be this way. Those two sentiments pretty much sum up the territory on which the struggle for democracy in the new millennium will take place. The so-called new world order is not inevitable, like the weather. It is no more inevitable or "natural" than any other way that we might choose to organize ourselves as human beings. Indeed, it can be forcefully argued that our current path is perhaps the most *unnatural* order of things. For what, if not unnatural, is a way of organizing human existence that virtually guarantees that we will destroy the very basis for that existence? We are now behaving in such a way that an outside observer would conclude that we are a species doomed to die out, incapable of rationally mapping out a sustainable ecological or social future; determined to destroy the means of life.

The demoralization and powerlessness that grip so many Canadians facing the right-wing counter-revolution is its greatest strength. For those who believe that it cannot be stopped, we need to revisit our own fairly recent history. In the Great Depression people faced the ruthless power of corporations and their hired thugs; they experienced incredible deprivation; the organizations they formed to defend their rights were infiltrated by the police, some were declared illegal and their leaders jailed. The unemployed were forced into camps in order to qualify for relief. Activists were murdered. Yet citizens fought back. Indeed, they fought back so effectively, armed with ideas ranging from social democracy to communism, that the most farsighted corporate defenders saw the possibility of a social revolution. And so we got reform.

If people do not resist the onslaught of corporate rule, it not only will get worse; it has to get worse. The logic of the system, driven by a rigid ideology and an elite of true believers, means that there are no natural barriers, no limits to the destructive power of transnational corporations and the international system of finance and mass consumerism they are putting in place. Market liberalism, by its nature unable to value anything but economic efficiency, cannot take into account human suffering, human history, the value of culture or community or the value of the environment.

This system will take everything; it cannot stop. We know what it is capable of because we see it already, in child slave labour abroad, in worsening environmental degradation, in people working in poisonous factories, in advertising aimed at getting children to start smoking cigarettes.

Canadian philosopher John McMurtry suggests that we have entered the "cancer stage" of capitalism. Our social institutions, he argues, acted as an immune system, "recognizing and responding to the vital life needs of social bodies as a whole [which] shielded members from disease, starvation, and disabling morbidities." It is a compelling analogy. The signs of immune system breakdown are unmistakable, says McMurtry, "when money capital lacks any commitment to any life-organization . . . but is free to move in and out of . . . social and environmental life hosts." Our institutional immune system fails to recognize it as a threat and the cancer grows apace, invading every aspect of the host body until it "eventually destroys the life-host in the absence of an effective immune-system recognition and response."[1]

In the absence of effective institutions to protect society and community, the only immune system remaining is the millions of individual citizens who face increasing insecurity and falling standards of living. The challenge will take us back to the roots of democracy, and that is to the struggle of classes over rule by the majority. If we assess our democracy, it is clear that the "problem" identified by the Trilateral Commission more than twenty years ago has been resolved in favour of the corporations that it spoke for.

The "excess" of democracy has been eliminated. For the most part the public no longer questions "the legitimacy of hierarchy, coercion, discipline, secrecy, and deception."[2] We are back to a time when governments can govern simply with the cooperation of a few bankers and Bay Street lawyers, or the modern equivalent, the BCNI. As for the media, the elite no longer has to fear that it is, in Walter Cronkite's words, "inclined to side with humanity rather than with authority and institutions."[3]

Canadians oppose by large majorities nearly every aspect of the new-right agenda and the role of corporations in implementing that agenda. On medicare, education, the environment, job creation, poverty, and the role of government, polls show that Canadian citizens want more money, not less, directed at these issues. Indeed, in most cases we are even prepared to pay more taxes to that end. But the almost universal response of the neo-liberal state is profoundly undemocratic. It is summed up by Brian Mulroney's words, aped by Saskatchewan premier Roy Romanow: When faced with doing the right thing or the popular thing, they always choose the right thing. The willingness to thwart the will of the people has become a sign of the highest level of responsibility that a political leader can aspire to. It is a sign of his or her sense of responsibility to corporations.

The challenge to re-establish majority rule must take place at both the level of ideas and the level of power. At the level of ideas it is really no contest. It would be difficult to imagine a more impoverished set of ideas, principles, assumptions about human nature, and goals for society than those promoted by the new right. They would have us believe that the end point of thousands of years of civilization is a globally homogeneous marketplace of customers (not citizens) whose vision can be summed up in a single value, economic efficiency, and whose ultimate expression of human achievement is the universal availability of Coke and Nike running shoes.

Who are the heroes of this brave new world with whom ordinary citizens have to compete for moral, ethical, and intellectual superiority? Barrick CEO Peter Munk, whose own hero is a fascist general who oversaw the slaughter of thirty thousand unarmed civilians? Conrad Black, who is so contemptuous of his own country that he can't bear to live here? Or does the ultimate challenge come from Bill Gates, a man who with even a minimum of imagination could make fabulous contributions to his community and to humanity with his $54 billion, but instead is satisfied with designing second rate computer software?

Or is the real challenge from the gaggle of new-right columnists and other hangers-on like Andrew Coyne, David Frum, Diane Francis, Barbara Amiel, and *Saturday Night* editor Kenneth Whyte whose generation of new ideas consists of digging through the dust-bins of history to find inspiration in "free markets" that ceased to exist more than a century ago? The collective intellectual leadership of the triumphant new right is the political equivalent of the idiot savant, clever purveyors of ideology but socially retarded, culturally vacuous, and morally indifferent, promoting a world of obscene wealth for the few and insecurity for the many. Is this the vision that is too compelling to challenge?

Clearly it is the sheer power of the corporations and not the ideas of their ideological prostitutes that makes the fight for democracy in the new millennium so formidable. The current institutions of democracy, including the political parties, cannot be expected to lead a radical democratic movement against these forces. That does not mean that we must ignore these institutions or, for example, dismiss the role of the NDP, which is the only voice in Parliament that speaks for social justice and will fight against corporate domination. The rejuvenation of a democratic politics will involve engaging the adversary at every level and in every forum, and supporting others who do so.

But to win against the juggernaut of global corporate rule will require a virtual revolution in citizen consciousness, in people's understanding of what it means to be a citizen. That involves a decision to become deliberate, self-aware citizens who make a commitment of time and resources to rescuing their community from its continued destruction. If those who despair at what is happening to our country are not willing to make sacrifices, and change their lives, it is a certainty that we will lose and end up as observers of ever greater corporate domination.

It means that hundreds of thousands of people who normally pay little attention to politics will have to consciously devote time to learn what is happening, talk with neighbours, join social and political action groups, and stop trading in their rights and obligations as citizens for the right to be a customer in the global marketplace. We cannot be both global customers and global citizens, for to be the former explicitly means the demise of the latter.

Of course, increasing numbers of Canadians don't even get to make that trade. The 4.4 million working Canadians who now slip repeatedly in and out of poverty are disenfranchised in both of the competing democracies, the one person, one vote and the one dollar, one vote varieties. They have no access to the "global products." Imagining a revolution in citizen consciousness has to address the issue of class inequalities.

Citizen participation must be directed at seizing back from corporations power that our "own" governments have ceded. In the vacuum of government deregulation corporations have become the organizational form that is now "regulating" society. It is corporations that increasingly determine what kind of work there will be, who will get it, how much damage will be done to the environment and where, what human needs will be addressed and which will go unfulfilled.

Yet refocussing democratic action on corporations does not mean that we can abandon the state. While the state is now being transformed into the market state from the egalitarian state, it is still the only human organization ultimately powerful enough to challenge corporate rule, whether alone or as part of international institutions for world governance. The struggle for democracy has throughout this century been a struggle for control of the state. That will not change. Creating a genuine democracy means, in the words of political scientist Leo Panitch, building "a different kind of state."

Ironically, we can trace the devolution of the current state and the ease with which corporations are dismantling it in part to the powerful democratic movements of the past. Although they enhanced political democracy by increasing the participation of citizens in the process, they failed to apply that democracy and participation to the social programs and the state itself.

But, says Panitch, "the old welfare state reforms . . . actually had very little to do with reforming the state itself; that is, with reforming the mode of administration in which social policy became embedded. It is a mode of administration that is structured in a fundamentally undemocratic fashion, along strict principles of secrecy and hierarchy that owe much to the principles of organization of the 19th-century British Colonial Office."[4]

The state, even though it responded to the democratic movements of workers and other social groups, was not suddenly transformed into an agency of majority rule. It was still an agency of the market economic system, and the reforms were intended precisely to head off majority rule. The state was still "managing" capitalism for corporations. "It was engaged in the regulation of the people who were its clients, establishing rules that governed their behaviour in many large and small, intimate and public, spheres of life."

The "participatory democracy" programs of the Trudeau Liberals were both a strategy for co-opting radical movements and partly a genuine response to previously marginal groups seeking input into policy making. In both cases, the state was trying to make itself legitimate in the eyes of citizens, both those who were active and those who were influenced by them.

Once movement organizations were recognized by the government, much of their time was spent "dealing" with the state about what sorts of programs would be implemented. Gradually, they moved away from building the grass roots of their own movements in part because they no longer needed active memberships to secure access to government and to policy consultation. In fact, in some cases having an active membership making radical demands was a threat to that access.

One of the reasons that the right has so successfully attacked social programs is that those programs are not integrated into the community but are seen as "delivered" by government agencies that have no real connection with citizens and that citizens had no say in creating. Many poor people are sympathetic to attacks on welfare because their experience with welfare has been punitive, humiliating, and simply inadequate. And it is getting worse. As governments adopt market principles as "performance indicators" to assess public services, there is less and less distinction between those services provided to enhance the well-being of the community and those for sale in the marketplace.

Social reformers of the next millennium will have to take account of the fact that an essential part of democratic reform is the reform of how state services are provided as well as what services are provided. A truly democratic social assistance system would be one in which the front-line workers are mandated to act as advocates for their fellow citizens instead of being obliged by the social control ethic to act as gate-keepers for "clients."

At a 1991 conference on social welfare policy, both policy makers and front-line workers imagined what a democratic policy would look like: "Why could not employees and clients of the ministry elect [assistant deputy ministers] from a panel of choices put forward by the minister? Perhaps there could be referenda among clients on policy — such as among the elderly in extended care institutions who might be encouraged to discuss and vote on a series of policy options being considered inside the ministry . . . Public employees [are] well placed to be facilitators of the collective organization of the poor so they would no longer face the state or the market as powerless and passive individuals."[5]

To have any chance of being successful, social reform must inspire hope. The exciting prospect of creating a whole new democratic ethos involving community and government (they should be the same anyway) could inspire hope. And it integrates both the process of democracy and the outcome of the process, involving an alliance of public servants (we should start to refer to them as community employees) and citizens in political decision making and the service itself. The division between government employee and client would be broken down and reconstructed as a relationship between citizens.

Unless the fight for democracy entails such a radical conception it will fail. Not just because the current relationship between citizen and state has been shown to be vulnerable to attack from the right, but because we are otherwise doomed to remain in our current protective mode, defending an eroding system that we never wanted in the first place. How can we expect to inspire hope by defending a system of social services that we never believed in anyway?

As we keep digging trenches in our rear-guard actions against the encroachment of corporate rule, we find ourselves defending things that are increasingly indefensible. What will we be defending next, a bad two-tier health system against the advent of a terrible one? Will activists ten years from now be demanding better conditions in the camps for the unemployed? There is nothing to suggest that the corporate elite is incapable of such measures.

FIGHTING THE MAI, ABROGATING NAFTA

The most powerful political weapons at the disposal of transnational corporations are the trade and investment agreements that have transformed the powers of the state in the interests of corporate rule. Much of the struggle to regain genuine democracy will of necessity be against these agreements.

There is now in Canada and around the world a vigorous fight against the Multilateral Agreement on Investment (MAI) and against APEC, the Asia Pacific Economic Co-operation regime. This is a common struggle of nations and people who are now alerted to the dangers of such agreements, alerted in part by what has happened in Canada but, even more important, by what has happened in Mexico. The World Trade Organization, which reaches it decisions through the democratic consensus of its member countries, faces serious conflict between the dominant developed countries and a growing number of less developed countries not only over the MAI but over the other agreements coming down the pipe.

NAFTA and the agreements it has helped spawn make the achievement of an economically just world impossible. Those who suggest that we abandon any hope of abrogating NAFTA have, unwittingly, already given up the fight. That is not to say that we should never renegotiate any aspects of NAFTA. We should, but as steps along the road to abrogation. We should start with those sections that prevent us from demanding that investment be productive, not speculative; that take away our sovereignty over energy development; that prevent us from demanding the use of local inputs in exported goods, and that threaten our culture.

The context for abrogating NAFTA is not the current political balance of forces, but that doesn't mean abrogation is unimaginable. The truth is, we have not seriously tried to imagine and plan for the day when it will be possible. When we achieve enough political strength in our social movements to force a Canadian government to renegotiate this deal, then we will be on the way to having the power to abrogate it.

Insisting that NAFTA be abrogated is a statement of faith in the future. Yet any disagreement over whether to maintain the option of cancelling NAFTA may in the end be a question of semantics. For in the longer term creating international institutions for social justice will be necessary to challenge the power of transnationals, and such institutions would change NAFTA and other such agreements so fundamentally that the deals would simply fade away.

INTERNATIONAL DEMOCRACY

There are nation-states who have rejected free trade, rejected structural adjustment, rejected corporate rule and the impoverished ideology that drives it. Norway and Sweden, even after some backward steps, are still egalitarian states and they put the lie to the propaganda that globalization is inevitable and that resistance is futile. Norwegian and Swedish citizens just said no; they made a democratic choice to maintain civilized societies. Their governments remained loyal to the principles of democracy.

But for other countries, like Canada, recapturing that democratic reality will be much more difficult than if we had managed to keep it in the first place. We cannot possibly do it alone. The struggle for economic and social justice will, over the next decade and more, become increasingly international because the nature of the adversary demands it. Transnational corporations and the multilateral deals they are brokering are forcing the labour movement and popular organizations in all countries to cooperate in ways they never have before.

This international movement uniting peoples of the developed nations and the less developed nations was pioneered by the cooperation between activists in Mexico, Canada, and the U.S. in their opposition to NAFTA. It continues to develop in opposition to APEC, as was demonstrated dramatically by the People's Summit against APEC, held in Vancouver in 1997.

As ordinary citizens face the daily crises of insecurity, falling standards of living, and eroding public services it is easy to forget that the TNCs who created the new order will face a crisis, too. If a worldwide financial crisis does not arrive first — and in all likelihood it will — the next big capitalist crisis will be the saturation of the global middle-class market. At that point there will be no more growth strategies and the crisis that capitalism has been avoiding for a century will arrive once more.

No matter what the scenario, the result of the crisis will be a need, as John Dillon points out in his book *Turning the Tide*, for a new United Nations Conference on Money and Finance. It will have to address two main issues, hot money speculation and massive indebtedness of the less developed world, because the world system of finance and production simply cannot continue on these paths forever. Everyone knows this, but it will take a crisis to force a solution. A new U.N. conference would have to return to the intent of the agencies established at Bretton Woods.

A World Bank reformed — or more likely replaced — along the lines of its original goals would make credit available to developing countries with balance-of-payments deficits from deposits by developed countries with balance-of-payments surpluses. An aid program, operating out of the democratically run U.N. and not the financially driven IMF or World Bank, would provide grants and low-cost loans to developing countries without the conditions that now require many countries to destroy their social infrastructure. A World Trade Organization designed to actually work for nations rather than provide guarantees for TNCs would work to stabilize commodity prices and promote the production of key consumer goods, from food to housing, in each member country.

The world has changed dramatically since Keynes lost the battle for the design of international governance in 1944. The principles remain the same, but the mechanisms for such governance will have to address the power of transnational corporations and the ocean of hot money that now drives the global economy. One viable solution to the hot money system is the Tobin tax, a measure even Jean Chrétien briefly flirted with until he was quickly re-educated. The tax, designed by Nobel laureate economist James Tobin, would impose a very low tax, say 0.5 percent or even 0.25 percent, on every currency transaction in the market, worldwide. It would be a universal tax; every country would have to pay it. The money raised, about $500 billion a year at 0.25 percent, would be funnelled into the International Monetary Fund or the World Bank to assist the less developed countries. It would also serve to severely dampen currency speculation because even at such low levels it eliminates the tiny margins on which speculators now make their money.[6]

Besides the Tobin tax, an international agreement setting parameters for corporate taxes could ensure that corporations paid sufficient national taxes to cover the costs of infrastructure and social services (such as the health and education of the workforce) they benefit from in the course of doing business.

International regimes would regulate the flow of capital and control speculative versus productive investment. And instead of the vague and largely unenforceable labour, social, and environmental "side deals," new United Nations charters would make these areas of human concern paramount, or at least equal to concerns about trade and investment. Any country wanting access to loans, grants, or debt restructuring or the benefits of trade regulation through U.N. agencies would be obliged to live up to these charters or face punitive sanctions. In effect, the new institutions, or reformed old ones, would have their functions and their ethos transformed from their current role as agencies for corporate domination to agencies for democratic rule.

CHALLENGING CORPORATE RULE

In 1849, Louis Riel Sr., the father of the great Métis leader, led three hundred armed Métis and surrounded the temporary courthouse in which the Hudson's Bay Company was trying three Métis, independent traders, for violating company laws. The Métis did not fire a shot. They just sat, mounted on their horses, their rifles cradled in their arms. Minutes later the company released the three men. Corporate rule by an occupying power ruling in the interests of Britain and hostile to those it governed had come to an end on the prairies. It's impossible to know at what point the Métis, exploited for decades by the HBC, "imagined" ending its economic monopoly.

Part of the challenge of confronting corporate domination is in overcoming the demoralization of seeing "our leaders" willingly and even eagerly ceding power and sovereignty to transnational corporations. Ursula Franklin captures the appropriate awareness and a useful analogy when she describes us as now being under the rule of an occupying power. Current governments govern with no reference to our needs and desires as citizens, without reference to our history and culture, in complete contradiction of our well-being. Like the Vichy government of wartime France, they have seized the governing structures of our society in the interests of a foreign power. They are dismantling them and setting up new structures of authority that reflect the interests of that foreign power, transnational corporations.

Many groups are still involved in consulting with the government even while the government has absolutely no intention of listening. We should not be engaged in the pretend democracy of consulting with an occupying power, otherwise, as Franklin puts it, we are behaving as "collaborators." By agreeing to take part in a consultation process completely corrupted by the government and without any prospect of success, we provide the occupying power with undeserved legitimacy and ultimately betray the people we are struggling to represent. These are not our governments.

The principal adversaries in our struggle for social and economic justice around the world are the transnational corporations, who are merging their way to ever greater concentrations of power. The task of creating a revolution in citizen consciousness begins here, with building a recognition of the role of corporations and the role of the state in ceding our national sovereignty to them. That entails a shift in focus for every citizen organization in the country away from trying to "influence" currently constituted governments and towards exposing every detail of how corporations rule, who are the people behind the CEO titles, and how they control government and those political parties that are governing in their interests.

There are those challenging corporate power who attempt to force corporations back into what is often called a stakeholder model of corporate management. This model encourages corporations to manage their affairs in the interest of not just shareholders but also employees, communities, consumers, and the environment. Yet this model is outdated. The global economy has passed it by. Transnational corporations dominate the world and their decisions determine the fate and the decisions of lesser corporations. The planners of these corporate behemoths do not think in terms of stakeholders because it is irrelevant to their purpose: profits and global expansion.

This does not mean that we should stop organizing against corporations like Nike for employing labour at slave-level wages or stop boycotting Shell Oil for its complicity in the murder of Ogoni activists in Nigeria. Indeed, we need to multiply those actions across the country as a way of reinforcing people's understanding of the role of corporations. But there are simply too many corporations to expose one by one as bad corporate citizens. By trying to change corporate rule this way, in the words of Henry Demarest Lloyd, an anti-corporate activist of the last century, "we are asking them not to be what we have made them to be. We have put power into their hands and ask them not to use it as power."

Exposing the antisocial and anti-democratic behaviour of corporations like Shell, Nike, Barrick, Bell Canada, the big five banks, and the many other destructive corporate citizens, we can educate and mobilize citizens to take power away from them. We can turn the tables on those who would transform us from citizens into customers and use our power as consumers to act as citizens. But this use of consumer/citizen activism must be aimed at undercutting corporate power. In removing the power we have put in their hands, we put it back into the hands of citizens and begin to create the conditions for a genuinely democratic state.

Removing that corporate power is straightforward when it comes to reforming the democratic institutions they now control or influence. Corporations must be prevented from funding political parties. Legislation severely limiting third-party spending in elections must be forced back on the political agenda. Political lobbying by corporations and by professional lobbyists in their hire must be even more restricted than it is, and the tax deduction allowed for this activity, essentially taxpayers funding the perversion of their own government, must be ended. All contacts between corporate representatives and government officials must be made transparent so that every attempted intervention by corporate interests is out in the open for citizens to see and assess. The growing practice of contracting out policy making to the private sector must be banned.

On the broader political front, we must begin to put faces on the faceless corporations by publicly identifying the people who make the decisions that harm our communities. People need to know the connections between millionaire tax-avoider Paul Martin and the other multimillionaires who make public policy in Canada. How is it in a democracy that the vast majority of citizens do not know the faces or even the names of the people who rule them? How is it that millions of Canadians believe that the Fraser Institute is an independent educational institute when it is nothing more than a front for the largest corporations in the country, pimping "free-market" ideology in the cause of eliminating majority rule?

Some unions and citizen organizations are already shifting their focus to exposing corporate domination of the country. The Canadian Union of Public Employees has a whole research team investigating the large corporations who have targeted our health care, municipal services, and educational institutions. A series of cards, for example, profiles the eight most dangerous corporate vultures currently circling the wounded Canadian medicare system. As we examine below, the labour movement in Ontario confronted corporate rule in its Days of Action city shutdowns. The Ontario teachers' strike raised the question of future privatization of education.

The Council of Canadians has targeted the corporate ownership of the media and the concentration of that ownership in the hands of a very few companies. For several years the council (with its 100,000 members) has been shifting focus from pressuring government to exposing the role of corporations. One of its main themes is the development of a citizens' agenda as a counterpoint to the corporate agenda. Its campaign to expose the Bay Street takeover of the Canada Pension Plan has reached hundreds of thousands of people; in addition, the council has led the way in the fight against the MAI.

The International Forum on Globalization, based in San Francisco but with strong Canadian representation, held a conference on corporate rule in late 1996 and brought ninety anti-corporate activists from twenty-two countries together to develop global strategies against corporate rule. Every APEC summit so far has also featured a People's Summit of social activists from all the APEC countries.

THE CULTURAL REVOLUTION: FROM CONSUMER TO CITIZEN

The long-term struggle for democracy and a democratic state entails not just a power struggle against corporate rule. If that democratic dream is global, then the fight has to be against the very logic of capitalism. To defeat transnational corporations, we clearly have to deny them the source of their power over us, and that power is the promotion of mass consumerism. If we do not resist that global strategy, we cannot win. The only way to defeat the global marketing monster is to starve it into a state of submission and weakness that allows another vision to emerge. The crisis in overproduction hinted at in Asia will be an opportunity for citizens to establish production for human need, not for profit.

The need to choose between dollar-democracy or citizen-democracy, between being customers in the global marketplace or citizens of civil society, entails more than simply devoting more time to being informed and active citizens. "Average" Canadians, who consume hundreds of times their allotted, equal share of the world's resources, are already complicit in the corporate plan for a global consumer society.

International capitalism cannot survive unless it continues to grow, and with the added hot money dimension of finance capitalism it now has to grow even faster. That means that it has to use up the limited resources of the planet even faster, exploit the world's working people even more, and put ever greater pressure on the resources of communities as they try to cope with the erosion of the egalitarian state.

And every time we purchase something we don't really need we feed that process. In the fight against corporate greed we need to do more than organize against Nike because it exploits its workers. We need to organize a culture that refuses to pay $190 for a pair of running shoes.

Out of our growing understanding of the need to consume less has come one key tool for changing our thinking about the economy. It is called the Genuine Progress Indicator and it is intended to replace the use of the Gross Domestic Product, a concept that rationalizes and perpetuates the notion of unlimited growth as a social good.

The Genuine Progress Indicator judges economic activity, products, and the distribution of economic benefits by whether they contribute to the community's well-being. It subtracts the costs of pollution, the depletion of natural resources, air and water degradation, and the costs associated with crime and places a negative on unequal distribution of income. It treats as a plus household and volunteer work, which the group Redefining Progress estimates at $1.3 trillion per year in the U.S., increasing total output by almost a quarter. The GPI also counts other aspects of the quality of life, treating loss of leisure time, for example, as a negative. The GPI counts new productive assets as a plus but consumption through borrowing as a negative. Since 1973 the GPI in the U.S. has seen a steady decline, dropping most rapidly between 1991 and 1995 to 34 percent below its 1950 level.[7]

THE POPULAR SECTOR: CHALLENGING THE POWER OF CAPITAL

The use of consumer power by politically conscious citizens has enormous potential for exposing the role of corporations in society. This can and should go far beyond just the specifically organized boycotts against particularly offensive corporations. It should, as suggested earlier, extend to boycotting mass consumer culture in general and reducing consumption to more closely reflect actual needs. We should, simply as individual citizens, initiate our own ethical boycotts. No union or social activist should have their money in the Toronto Dominion Bank, which in 1997 co-sponsored a Fraser Institute right-to-work conference. Nor should they patronize the Royal Bank, which gave the Fraser Institute $20,100 in 1996. These actions alone should cancel the privileges of these banks to use union members' money. Likewise, we should boycott any corporation that moves jobs out of Canada.

But there is also great potential for challenging capital from the other end of the economic process, production itself. Most public attention on economic issues is focussed on two sectors in the economy, the private and the public. With the private-sector ethos creeping into even what remains of the public sector, we are facing an economy that will be increasingly guided by the principles of private investment. And as the state exits many of the public services we still have, corporations are poised to take them over completely. And that means that many needs for millions of citizens will not be met.

A key area in the fight for a new democracy is the expansion of the third area of the economy largely ignored by the media and government. The popular sector, called the social sector in France, in fact has great potential as the social base for building a genuine democracy. The popular sector, comprising thousands of nonprofit groups that provide countless services to their members and the community at large, is more reflective of genuine democracy and citizen consciousness than the formal political system.

It is precisely people's identification as women, environmentalists, health advocates, educators, cultural workers, peace activists, and aboriginal people in their organizations that makes them intentional, aware citizens. These organizations are not only engaged in democratic political representation; they employ people, meet needs, purchase goods and services, and involve millions of people in activity that is generally seen as noneconomic because it doesn't make a profit. Promoting the popular sector through expanded access to capital has the dual effect of creating economic activity that serves real needs and potentially seizing some of the economic and social territory from the realm of the corporate sector.

One enormous source of capital that could be used to expand the popular sector is the money in union pension funds, now sitting at something over $190 billion in Canada. That is the deferred income of employees, but it is also a huge pool of capital that, if it was controlled by those employees, could break down the distinction between owners and employees in the private sector, a distinction that is at the root of the power imbalance between citizens and corporations.

Unions in Canada have played an important role in opposing corporate rule and are moving even more in that direction as the state devolves into corporate governance. As Duncan Cameron of the Canadian Centre for Policy Alternatives argues, "Broadening the role of unions to include the direct stewardship of investment capital could be the most important step since the right to strike."[8]

At the moment few unions have much say in how their funds are invested because legislated control over that money is in the hands of pension trustees whose singular legislated mandate is to maximize fund income with limited risk. But unions could make it a priority to bargain for greater control over those funds and where they get invested. Resistance from union members should be addressed as part of the unions' efforts to educate and mobilize their members in the fight against corporate rule.

The power of union pension funds has already been demonstrated by the largest pension fund in the U.S., the California Public Employees Retirement System. By 2000 the fund will have $200 billion, and though not directly controlled by the unions, its sympathetic president, Bill Crist, has repeatedly used the fund to pressure corporations to change their behaviour. They hold huge portfolios in some of America's largest corporations and engineered a boardroom coup at General Motors, removing senior management who had refused even to meet with Crist. Crist's approach is to pressure against short-term layoffs and to demand that companies the fund invests in have long-term strategic plans rather than short-term share-price goals.[9]

The American labour movement is now beginning to coordinate the holdings of union pension funds, which now amount to $1.4 trillion, fully 14 percent of outstanding shares in the U.S. In September 1997, the AFL-CIO established the Center for Working Capital. The founders envision many possible actions, from influencing bargaining disputes to demanding that fund managers invest in companies demonstrating a commitment to long-term growth.

The International Association of Machinists and Aerospace Workers is using pension money to help convert a former Cruise missile factory in Seattle into a plant assembling high-speed passenger trains. The first forty employees of the union's company, Pacifica, will be skilled union members who were laid off from the Naval Undersea Warfare Center. The company will use an innovative method of work organization that maximizes worker participation and leadership.[10]

These funds have even more potential for giving greater power to subordinate groups that have a more democratic vision of the economy. Using even a half of 1 percent of the $190 billion in funds to finance cost-recovery projects, social housing, or environmental projects would give the popular sector more clout to force governments to legislate in favour of such groups. One possibility would be to bring together unions, popular groups, and allies in opposition parties to change the Bank Act to force banks to lend a certain minimum percentage of their assets to third-sector community development projects.

There is another potential benefit to such projects. Many popular-sector groups now function in isolation from both the private-sector economy and the popular and union-based political movements against corporate domination. Bringing together popular sector groups and unions in projects that are both political and economic in nature creates the basis for a powerful alternative vision of community. More than any amount of political education or lobbying, such models would be compelling evidence of how society could be organized. Tangible, functioning models of a democratic economy will be increasingly important in a political atmosphere dominated by corporate media, right-wing think-tanks, and the market state.

Equally important, these "real-life" projects create community in ways that anti-corporate organizing simply can't do. Involvement in social movements can be isolating in the sense that the political activity is rarely integrated into people's lives. Engaging in boycotts, getting involved in community groups, demonstrating against corporate headquarters, while inspiring and providing a sense of citizen power, are often transitory events and can seem disconnected from the daily reality of work, family, and popular culture. We still go home to joblessness, debt, or insecurity, and the late-night TV news still assaults us with its right-wing bias. Creating community out of citizen consciousness is a key part of the movement for a new democracy.

The Canadian Auto Workers is one union that has put considerable energy into anti-corporate strategies and social justice unionism. The union's principal economist, Sam Gindin, argues that unions must integrate themselves and their democratic politics into the community. "Consider, for example, the creation of new local union committees that are open to workers' spouses, and teenage sons and daughters . . . Such an initiative . . . organizes workers around other aspects of their lives: air [quality], the safety of the neighborhood, the schools our kids attend."[11]

Gindin further suggests that unions could begin to influence jobs and job security by working at the municipal level to press for elected job development boards. These boards would guarantee everyone either a paid job or training to meet community needs. The board could identify a community's needs and then invite the community's proposals to meet them.

SIGNS OF CRISIS, OPPORTUNITIES FOR CHANGE

The struggle against corporate rule and for democracy is a long-term fight. It is hard to exaggerate the importance of seeing the fight in this way. Popular forces seem to lose battle after battle, and the forces aligned against them are indeed formidable. Developing a long-term view gives three advantages to those who would take on corporate rule. First, it reduces the terrible burden of having to win the next battle when there is a very strong likelihood that it won't be won. We know that things will get worse before they get better, but if the next battle is just one on the road to an imaginable victory it is not so demoralizing.

Taking the long view also allows us, indeed compels us, to take the time to revisit our vision of what a truly democratic, equal, fair, and ecologically sustainable society would look like. This means imagining the future. But it also means imagining the actual struggle for democracy by developing medium-term and long-term strategies of where we want to be, when we want to be there, and how we can get there.

Last, and perhaps most important, the long-term view has the potential to transform the day-to-day fight to save the nation from a perpetually defensive fight to one in which progressive forces begin to take the initiative. A good defence won't win the fight. Until we take back the initiative to change society, we simply cannot win.

There are real signs that the triumphalism of the new right is showing some tarnish. Some of the most famous gurus of unrestrained corporate greed and power are looking over their shoulders and warning their brethren to beware the angry workers. Take the UPS strike of 1997, in which 185,000 members of the supposedly quiescent American working class not only won their struggle but captured the imagination of the American public. By a two-to-one margin

Americans supported the strikers and in particular their demand for more full-time jobs. American pollster Daniel Yankelovich stated that the solid sympathy for the UPS workers could well be a turning point in American political culture. The strike, he said, was "a consciousness-raising event."[12]

Stephen Roach, one of the decamped gurus of downsizing, was quick to offer some advice to the myth makers who talk about the amazing American economy. The great American recovery, scolds Roach, does not come from increased productivity but is built exclusively on the backs of workers squeezed to the breaking point. "The labour crunch recovery is not sustainable. It is a recipe for mounting tensions, in which a raw power struggle occurs between capital and labour. Investors are initially rewarded beyond their wildest dreams but those rewards could eventually be wiped out by a worker backlash."[13] Americans don't talk much about class war, but that is clearly what Roach is talking about.

There are other encouraging signs of resistance to the new world corporate order, ranging from the election of many social democratic and socialist mayors throughout Latin America, to governments backing off their radical market agendas, to hundreds of thousands of citizens demonstrating against the destruction of social programs throughout Europe.

The major reversal of the Alberta government of Ralph Klein, under steady public pressure to reinstate previous levels of funding for medicare and education, is a victory for democracy, coming as it does in a province noted for its individualism and free-market politics. The Chrétien government's pledge to return to its social liberal tradition may well prove to be a ruse, but just that it felt compelled to make the declaration is more evidence that the government fears a loss of legitimacy.

The breakthrough of the NDP in Atlantic Canada in the 1997 federal election is also significant, and not only for what it says about people's attachment to Canada's communitarian tradition and their rejection of the Liberal Party. Although the NDP's social democracy is moderate, it retains commitments to social justice lost in other social democratic parties around the world. The party is prepared to confront corporate rule explicitly and is clearly the parliamentary wing of the anti-corporate fight. It represents a version of majoritarian democracy, and the act of voting for the NDP in the Atlantic provinces was a declaration of citizen consciousness.

The Days of Action organized since 1996 by labour and community groups in Ontario have brought out historically unprecedented numbers of people protesting the Harris government's implementation of the corporate agenda. The way the demonstrations were conceived and organized directly addressed the issue of corporate power and the erosion of the egalitarian state. According to the CAW's Sam Gindin, "The emphasis on workplace shutdowns . . . highlighted — successfully — the employer base behind the government's attacks on working people. The public sector shutdowns of transit, schools and post offices were directed at reminding people of the importance of social services they had taken for granted."[14]

The 1997 strike by 126,000 Ontario teachers is another example of resistance whose impact will be measured over the next decade and longer. Never before have Ontario teachers, historically divided into five often fractious unions, been so united. There is a special irony here. Teachers are the favourite ideological target of neo-cons because public education is seen as a key source of "excess democracy." Yet the Ontario fight actually heightened public consciousness about the issue of democracy because it was in part fought over the issue of who controls education. And teachers will inevitably take their own heightened awareness of democracy back into the classroom, just where the neo-cons don't want it.

Of critical importance to the long-term struggle for democracy is the fact that the teachers' strike and the Days of Action were acts of civil disobedience. The change of political consciousness required to mobilize hundreds of thousands of people in civil disobedience should never be underestimated; neither should the strength of the social solidarity built by engaging in such actions. These are events that, in the words of Michael Walker, "change the ideological fabric of society."

In creating a revolution of intentional citizenship, no other political act is more significant. Each such act builds increasing understanding of the original, radical meaning of democracy — that is, a struggle between social classes over the issue of majority rule. The history of the twentieth century is full of evidence for the claim that fundamental social change cannot be achieved without massive civil disobedience. In fact, of course, much of that evidence is found in our own history, in the class warfare of the 1930s that resulted in the only period in capitalist history during which working people achieved any measure of equality.

There are some other signs that some key projects of corporate rule are in difficulty. In both education and health care the shine is off the great predictions of endless and easy profits. The evidence is pouring in that the business "ethic" and these two fundamental aspects of community just don't mix. Columbia/HCA, the largest U.S. for-profit hospital chain, is facing massive lawsuits and criminal fraud charges arising out of its insatiable greed. The indictments against the firm came down just as the number-two giant, Tenet Healthcare Corp., agreed to pay defrauded former patients $100 million.

In education the story is the same. A 1996 report on public school privatization, by researchers at New York University, demonstrated that privatized schools failed to produce either better academic results or savings for the state government. In Baltimore, students' test scores rose by 11.2 percent in the regular school system while they were falling in the privatized ones. The corporation, Education Alternatives, had its contract terminated.

These are minor skirmishes in the corporate drive to expand its investment territory. More troubling for the transnationals' strategy of ever expanding mass consumerism is word from Asia that the expansion is not going as planned. In August 1997 the World Bank issued two reports raising the alarm about increasing poverty throughout Asia. The reports estimated that "as many as 900 million Asians, from Mongolia to South India, live in dire poverty despite the region's surging economic growth in recent years." It seems the obsessive attention on the region's "four tigers" has obscured the desperation in other countries.

The financial crisis in Asia is just a symptom of more trouble to come. There are also signs that the global middle-class growth strategy for Asia is running into problems, too. The volatility in Asian currency markets is, according to Jeff Uscher, Tokyo-based editor of *Grant's Asia Observer*, easy to explain. "After more than a decade of aggressive investment in new production capacity to make all manner of manufactured goods, the region is saturated. Much of this new capacity was funded with foreign capital with the intention of serving the rapidly growing Asian market. The only problem is that the Asian market is not growing rapidly."[15]

The late-1997 Asian currency and stock market meltdown helped expose the overcapacity/overproduction crisis. Analyst Louis Uchitelle reported on the "worries" amongst the biggest players in the global economy about "the tendency of the unfettered global economy to produce more cars, toys, shoes, airplanes, steel, paper, appliances . . . than people will buy at high enough prices."[16] Quoting GE chair Jack Welch as saying "There is excess capacity in almost every industry," Uchitelle's analysis suggests that the global version of the problem Henry Ford faced in the 1920s is already upon us. If the global middle class doesn't have the money to buy the goods produced by the global working class, the system becomes unsustainable.

In short, stock markets and growth cannot be permanently sustained on the basis of a planned recession, of squeezing more and more out of workers by reducing their wages and reducing the number who receive wages. In Asia, the currency crisis and stock market collapses have caused middle-class customers to pull back. That has left Asia's enormous productive capacity (accounting for half the growth in world output since 1991) intact but in desperate need of other markets. Ironically, the corporate owners of this overcapacity are now looking back to the developed world to sell those goods. These goods are now cheaper. But the developed world's domestic economics and the purchasing power of their consumers have been systematically savaged by corporations and governments in the interests of global markets. Every dollar of additional spending in these economies will require ever greater marketing and advertising costs.

The Asian situation may be just a temporary setback, although the care and feeding of the Asian consumer market was supposed to keep transnational growth going for at least another decade and beyond. What this crisis demonstrates is that the supposedly unstoppable investment and growth juggernaut can very quickly look shaky. And there are signs that mass consumerism in the U.S. and Canada may be in for some fundamental changes. Many middle-class families, forced to cut back their spending because of stagnant or falling incomes, may well have decided that they can live on less. Journalist Alanna Mitchell argues that "there are signs that the punishing shocks they have endured during the nineties have fundamentally changed patterns of consumer behaviour." And according to economist Ruth Berry, "The whole cultural environment guiding the economy is shifting."[17]

The backlash against employers who trash their workers is growing in Canada and the U.S. A 1996 Angus Reid/Southam poll revealed that Canadians are becoming increasingly aware of and critical of corporate behaviour and the close relationship between corporations and government. Forty-five percent of those polled said that they thought corporations had become "less responsible" over the past few years compared with 19 percent saying "more responsible." According to 75 percent, corporations that lay people off should be required to provide them with job search assistance or job training; 77 percent said it was "not acceptable" for large corporations to lay people off while making high profits, and 54 percent said they should be forced to pay a penalty, such as higher taxes, for doing so. And 54 percent believed governments listen to corporations "too much."[18]

Attitudes towards corporations are even more critical in the U.S., where there is an upsurge of support for more "government intervention." Between 70 and 80 percent of all Americans see "serious problems" in how corporations sacrifice the interests of employees and communities for those of CEOs and shareholders, and "69 percent favour government action to promote more responsible citizenship and to penalize bad corporate citizenship."[19]

The seemingly unstoppable transnational corporations may be facing a major contradiction they had not anticipated. On the one hand, the devolution of the welfare state and the pattern of excluding huge numbers of people from the TNCs' economic growth strategy suggests the development of the security state in which the use of coercion increasingly replaces efforts at building legitimacy as an overall strategy. But there is strong evidence that a repressive response to social inequality may not be compatible with the new economic order. The fierce competition unleashed by the growth of TNCs creates demands for stable regimes with no surprises for investors. As we saw in chapter 6, the first democratic election in seventy years in Mexico was, ironically, partly due to the entrenchment of NAFTA.

This need for stability goes beyond the requirement for predictable governance. A French report, "The Enterprise of the XXI Century," prepared by an organization of young managers, warns that the drive for labour flexibility has reached its limits and could backfire by destroying "the autonomy of the workforce necessary in a modern organization." The temptation to "enslave man to the economy" could bring about the collapse of the whole system: "Business is in the process of breaking the social links that it used to build. We are convinced that unregulated capitalism will explode just as communism exploded, if we do not seize the chance to put man back at the centre of society."[20]

That explosion may be closer than anyone could have imagined. Witness the pressures building in Europe as popular opposition grows to the European Community's single currency, the conditions for which require significant cuts in social spending and restrictive monetary policies. The financial crisis in Asia threatens to create social unrest and political instability in two of the largest countries and economies in the region, South Korea and Japan. The IMF deal with South Korea requires, at a minimum, adding 1.5 million to the unemployment rolls and inevitably huge cuts to social spending.

In Japan pressure is building to fundamentally change the historic social contract, one that went far beyond anything Canada or Europe ever established. The profound sense of responsibility to community that has characterized the relationship between Japanese political leaders, industrialists, and senior public servants has its roots in centuries of Japanese culture. Japan Inc. is proving no more immune to the forces of global finance and liberalization than any other developed nation. But the "restructuring" of its economy implies nothing short of a revolution imposed from above. The social unrest it unleashes may well be catastrophic. The supporters of corporate libertarianism insist that stable societies are a necessary condition for investment. Yet it is precisely the ability of governments to control investment decisions through industrial policies that leads to an equitable allocation of wealth and, consequently, political stability.

As Martin Khor of the Third World Network points out, "The MAI will make developing countries politically unstable. But all these industrialists, if you ask them why do you choose a particular country in which to invest, their number-one answer is 'political stability.' So companies that promote the MAI are undermining the conditions under which they can actually function."[21] All the corporations that have made billions in Malaysia would never have been able to do so, says Khor, had there been an MAI in place twenty years ago. What made their profits possible were Malaysian policies that regulated investment in such a way as to foster political stability.

❖

Nothing stays the same for long. That lesson of history gives those committed to a radical new democracy a foothold in their struggle against corporate rule. Not only do things change but they can change with breathtaking speed. It is the apparent speed with which we have lost so much that stuns many people and creates the sense that all of this change is irresistible. Yet in fact it has taken the forces of the New Right and the transnational corporations twenty years of determined effort to accomplish what they have.

Not only are there alternatives to the current trends but there is the real possibility of a genuinely democratic society. There has to be. And the alternative to a world completely dominated by TNCs is not a return to the days of the impoverished notion of the welfare state. It is a fundamentally different kind of society, based on equality, ecological sustainability, and a redefinition of work that serves human society and that actually allows us to create socially and culturally liberating communities.

But just because things change doesn't mean they necessarily change for the better. Twenty years from now we will either be straining our imaginations and efforts creating democratic and sustainable communities or we will be witnessing worldwide social chaos and environmental catastrophe. If we are not well prepared to take advantage of the coming global crisis with imagination, energy, and a willingness to change, and begin creating a truly democratic society, we will almost certainly end up with a truly authoritarian one.

Appendix
What You Can Do

A major hazard of becoming aware of how things work is a feeling of powerlessness. The more we learn about the forces behind the status quo, the more omnipotent they can seem. However, socially conscious citizens of other eras have achieved major victories for human progress in the face of seemingly impossible odds, and you can, too.

One of the first things you can do is recognize and act on the fact that we are subjected daily to a barrage of propaganda aimed at trashing our values and making us apathetic and accepting of the new corporate reality. To believe that we personally are not affected by what we read is like thinking a steady diet of junk food will not affect our health. Cancel your newspaper subscription. With rare exceptions there isn't a paper in the country that is on balance worth reading. The corporate biases, deliberate censorship of stories and contrary points of view, and aggressive neo-liberal editorial policies of newspapers far outweigh the benefits of the information they provide. While you are at it, cancel your cable. Television, perhaps more than any other single factor, is responsible for the dumbing down of Canadian political culture.

The other side of ridding your life of right-wing propaganda is seeking information and analysis that empowers you as a citizen. I can't overstate the positive effect of having solid information to back up the values you hold. People who have put to memory data about how corporations evade and avoid billions in taxes each year and who know the Statscan study showing the real source of the debt are transformed from being passive victims of propaganda into being activists against debt terror.

An excellent source of information and analysis is the publication of the Canadian Centre for Policy Alternatives (CCPA), *The Monitor*, which is published ten times a year. (The CCPA can be reached at 251 Laurier Ave. West, Suite 804, Ottawa, ON K1P 5J6, ph (613) 563-1341, or in Vancouver (604) 801-5121 and Winnipeg (204) 943-9962.)

Many other publications are also helpful. In Canada, *Canadian Forum*, *Canadian Dimension*, *This Magazine*, *Briarpatch*, and *Our Times* are all excellent. In the U.S., *The New Internationalist*, *The Utne Reader*, and *'Z' Magazine* are first-rate. The Internet has a number of very good WEB sites on corporations, progressive organizations, on-line magazines, and much else. You can access these sites by visiting the CCPA WEB site (www.policyalternatives.ca) and clicking on "links."

As suggested in the final chapter of this book, individuals can begin to change their status from global customer to global citizen in a myriad of ways. Besides reducing television and refusing to consume unnecessarily, those who have savings can invest them as ethically as possible. Ethical companies and mutual funds have lots of problems, given the hazards any company faces by pursuing ethical policies, but they are certainly an improvement over those for whom ethics is irrelevant.

Yet even well-informed citizens are extremely limited in what they can do on their own. In terms of working with others, you have three options: join an existing group fighting on social justice/anti-corporate issues; work to address social issues in other organizations you already belong to; or form new groups specifically to address the issue of corporate rule and democracy. There are dozens of advocacy groups defending democracy and becoming involved in fighting corporations. Any grassroots organization worth its salt will welcome the energy and creativity of new members and find ways for you to contribute your own skills.

Advocacy groups are not the only organizations that can address these issues. Whether it is a parent advisory council, a community health clinic, or even a church, the issue of corporate rule will have an impact on their activity. It could be budget cuts, privatization, or corporate incursions into the classroom. The influence of corporations is so pervasive you would have to look hard to find an organization where it was not relevant. Work with others to learn about and deal with the threat of corporations in the sector you are already involved in.

There are a number of organizations that are now explicitly taking on the issue of corporate domination:

* The Council of Canadians has local action groups across the country and focuses on corporate influence in the media and social programs, fighting the Multilateral Agreement on Investment (MAI), and building a Citizens' Agenda. Their address is 251 Laurier Ave. West, Suite 904, Ottawa, ON K1P 5J6; ph 1-800-387-7177 or in Vancouver (604) 688-8846.
* The Canadian Auto Workers, 205 Placer Court, Willowdale, ON M2H 3H9; ph (416) 497-4110.
* Canadian Union of Public Employees (CUPE), 21 Florence St., Ottawa, ON; K2P 0W6 ph (613) 237-1590. Contact: Jim Turk.
* The Social Justice Centre, formerly the Jesuit Centre for Social Faith and Justice, provides resources for social movements on corporate and other issues. 836 Bloor St., Toronto, ON M6G 1M2; ph (416) 516-0009. Contact: David Langille.
* The Sierra Legal Defence Fund takes legal action against corporations, including a major focus on fighting SLAPP suits. 131 Water St., Suite 214, Vancouver, B.C. V6B 4M3; ph (604) 685-6518. Contact: David Boyd. In Toronto, ph (416) 368-7533.
* The Polaris Institute focuses on enhancing citizens' ability to challenge corporations. 4 Jeffrey Ave., Ottawa, ON K1K 0E2; ph (613) 746-8374. Contact: Tony Clarke.
* CODEV targets mining companies. 2929 Commercial Drive, Suite 205, Vancouver, B.C. V5N 4C8; ph (604) 708-1495. Contact: Jim Raider.
* The Task Force on Churches and Corporate Responsibility engages in shareholder campaigns. 129 St. Clair Ave. West, Toronto, ON M4J 4Z2; ph (416) 923-1758.
* Democracy Watch promotes democratic reforms that empower citizens. It has explicitly targeted corporate lobbying and the outrageous profits of the major banks. 1 Nicholas St., Suite 420, Ottawa, ON K1P 5P9; ph (613) 241-5179. Contact: Duff Conacher.
* Greenpeace focuses on corporate assaults on the environment. 185 Spadina Ave., Toronto, ON M5T 2C6; ph (416) 597-8408.

Of course, taking on corporate domination directly is not sufficient to rebuild community and revive democracy. There are groups in virtually every community — involved in connecting market gardeners with urban consumers, creating housing co-ops, running barter systems — that create real examples of democratic "economies." Many of the good things that happen remain hidden from us because of media bias and because many of the groups are small and local. A great source of examples of how people are using their imaginations as citizens is *Get a Life! (How to Make a Good Buck, Dance Around the Dinosaurs and Save the World While You're at It)* by Wayne Roberts and Susan Brandum.

The shape of democracy in the next decade and beyond is impossible to predict and except for broad principles we probably shouldn't try. It will be determined by what intentional citizens, informed, organized, and making social change, do in that time. We need to create hundreds of new organizations beyond the ones mentioned above, organizations that will not only challenge corporate domination directly but will also create the nuclei of democratic politics and democratic economic activity.

If there are no organizations in your area or none that appeal to you, create your own. Set up a study/action group with like-minded friends or colleagues. Fighting the MAI is a good start (contact the Council of Canadians for a kit). Or focus on identifying a corporation in your area that demonstrates corporate irresponsibility and research it (contact CUPE for a booklet on researching corporations). Then develop a public action plan based on your group's newly developed competence and confidence.

Good intentions alone won't make it possible for everyone to get this involved. Many people find it extremely difficult to make time for such commitments given the incredible pressures in their lives. However, the organizations that are fighting for social change are always desperately in need of money, outspent ten or twenty to one by their corporate adversaries. Give until it hurts.

Becoming a conscious citizen requires a different way of thinking and behaving. It means challenging political ignorance whenever you encounter it, incorporating your political convictions into your daily life, and most important, simply seeing the world as a citizen with the responsibilities and the rights that this implies. As we begin to see and interpret the world as citizens the question of what we can do will begin to answer itself.

Notes

Chapter 1
Globalization and the Rise of the Transnational Corporation

1 In David Korten, *When Corporations Rule the World*, Kumarian Press, West Hartford, Conn., 1995, p.123.
2 Sarah Anderson and John Cavanagh, *The Top 200: The Rise of Global Corporate Power.* Report of the Institute for Policy Studies, Washington, D.C., 1996.
3 Tony Clarke, "The Corporate Rule Treaty," Canadian Centre for Policy Alternatives, Ottawa, 1997.
4 "Index on Globalization," *Canadian Forum*, February 1997.
5 Korten, p.223.
6 Allan Engler, *Apostles of Greed*, Fernwood Publishing, Halifax, 1995, p.43.
7 Engler, p.31.
8 Maude Barlow, *Parcel of Rogues*, Key Porter Books, Toronto, 1990, p.21.
9 Korten, p.225.
10 Ibid.
11 Engler, p.37.
12 Korten, p.221.
13 Peter F. Drucker, *Concept of the Corporation*, Mentor, N.Y., 1983, p.49.
14 Korten, p. 224.
15 "A Survey of Multinationals: Everybody's Favorite Monster," *Economist*, March 27, 1993.
16 Richard Barnet, *The Lean Years*, Touchstone Press, N.Y., 1980, p.240.
17 Korten, p.125.
18 Ibid
19 Ibid., p.121.
20 Ibid., p.131.
21 Canadian Auto Workers, "Parliamentary submission regarding the future Canadian manufacturing operations of Northern Telecom Ltd.," 1995.
22 Ibid.
23 *Globe and Mail*, September 9, 1997.
24 CAW, "Parliamentary submission."
25 *Globe and Mail*, September 9, 1997.
26 *Globe and Mail*, September 9, 1997.
27 Ibid.
28 CAW, "Parliamentary submission."
29 Ontario Federation of Labour, "Unfair Shares," 1997, p.33.
30 Canadian Institute for Environmental Law and Policy, brief to the Commons Standing Committee on Natural Resources Regarding Mining and Canada's Environment (#287), April 16, 1996.
31 Project Underground, Corporate Watch, January 21, 1997.
32 Ibid.

33 *Multinational Monitor*, Vol. 17, No. 3 (1996).
34 Mineral Policy Institute, Bondi Junction, Australia, news release.
35 *Globe and Mail*, August 30, 1997.
36 Canadian Environmental Law Association, brief to the House of Commons Standing Committee on Natural Resources Regarding Mining and Canada's Environment, April 16, 1996.

CHAPTER 2
DEMOCRACY, THE STATE, AND THE CORPORATE CITIZEN

1 Gil Yaron, "The Corporation as a Person," research paper, 1997, p.13.
2 Ibid., p.14.
3 In Murray Dobbin, "Democracy and the Politics of Human Rights," CBC *Ideas*, November 30, 1995.
4 Michael Mandel, "Rights, Freedoms, and Market Power," in *The New Era of Global Competition*, ed. D. Drache, McGill-Queen's University Press, Montreal, 1991, pp.130–32.
5 Dobbin, *Ideas*.
6 W. P. M. Kennedy, *Documents of the Canadian Constitution: 1759–1915*, Toronto, 1918, p.34.
7 Donald Creighton, *The Empire of the St. Lawrence*, Toronto, Macmillan, 1956, p.38.
8 Gil Yaron, "The Legal Evolution of the Corporation in Canada," research paper, Vancouver, 1997.
9 Gerry Van Houten, *Corporate Canada: An Historical Outline*, Progress Books, Toronto, 1991, p.52.
10 Richard Grossman and Frank Adams, "Taking Care of Business: Citizenship and Charter of Incorporation," Charter, Inc., Cambridge, Mass., 1933, p.8.
11 Ibid. p.13.
12 Ibid., p.18.
13 Ibid., p.20.
14 Ibid., p.21.
15 Alvin Finkel, "Origins of the Welfare State in Canada," in *The Canadian State: Political Power and Political Economy*, ed. Leo Panitch, University of Toronto Press, Toronto, 1977, p.348.
16 Ibid.
17 Ibid., p.63.
18 Robert Chodos, Rae Murphy, and Eric Hamovitch, *The Unmaking of Canada*, Lorimer, Toronto, 1991, p.15.
19 Quoted in David Lewis, *Louder Voices: The Corporate Welfare Bums*, 1972, p.iv.
20 C.B. Macpherson, *The Real World of Democracy*, House of Anansi, Toronto, 1992, p.5.
21 Ibid., p.6.
22 Ibid., p.10.
23 Stephen McBride and John Shields, *Dismantling a Nation*, Fernwood Publishing, Halifax, 1993, p.17.
24 Interview with the author, August 16, 1997.
25 McBride and Shields, p.47.
26 William Greider, *Who Will Tell the People*, Simon and Schuster, N.Y., 1992, p.332.
27 Ibid.
28 Ibid., p.336.
29 Nick Fillmore, "The Big Oink," *This Magazine*, March/April 1989.
30 Ibid.
31 Barlow, *Parcel of Rogues*, p.8.
32 Fillmore, "The Big Oink."
33 Ibid.

CHAPTER 3
FROM CITIZEN TO CUSTOMER

1 William Leach, *Land of Desire: Merchants, Power and the Rise of a New American Culture*, Pantheon Books, New York, 1993, p.xv.
2 Richard J. Barnet and John Cavanagh, *Global Dreams; Imperial Corporations and the New World Order*, Simon and Schuster, New York, 1994, p.376.
3 Ibid., p.377.
4 Ibid., p.34.
5 Ibid., p.35.
6 Ibid., p.168.
7 Ibid., p.169.
8 Ibid., pp.171–72.

9 David Korten, *When Corporations Rule the World*, Kumarian Press, West Hartford, Conn., 1995, p.127.
10 Akio Morita, "Toward a New World Economic Order," *Atlantic Monthly*, June 1993.
11 Barnet and Cavanagh, p.178.
12 Ibid., p.174.
13 David Leonhardt, "Hey, Kid, Buy This!" *Business Week*, June 30, 1997.
14 Ibid.
15 Barnet and Cavanagh, p.174.

CHAPTER 4
THE TRANSNATIONAL CORPORATE CITIZEN

1 Allan Engler, *Apostles of Greed*, Fernwood Publishing, Halifax, pp.81-82.
2 J. Patrick Wright, *On a Clear Day You Can See General Motors*, 1979, Avon, New York, 1979, p.61.
3 Engler, p.82.
4 Bruce Livesey, "Provide and Conquer," *Globe and Mail Report on Business*, March 1997.
5 Bernard Lietaer, speech to the International Forum on Globalization, Washington, D.C., May 10, 1996.
6 Ibid.
7 David Korten, *When Corporations Rule the World*, Kumarian Press, West Hartford, Conn., 1995, p.193.
8 Lietaer speech.
9 John Dillon, "Turning the Tide: Confronting the Money Changers," Canadian Centre for Policy Alternatives, Ottawa, 1997, p.26.
10 Ibid.
11 Ibid.
12 Korten, p.188.
13 Ibid., p.203.
14 Dillon, p.30.
15 Ibid., p.33.
16 *Globe and Mail*, November 25, 1997.
17 Canadian Press, Web page, November 30, 1997.
18 Korten, p.210.
19 Ibid., p.211.
20 *Globe and Mail Report on Business*, December 1996.
21 Ibid.
22 *Globe and Mail*, November 13, 1996.
23 Korten, pp.213–14.
24 Ibid., p.214.
25 *Globe and Mail*, November 26, 1996.
26 *Ottawa Citizen*, March 29, 1996.
27 Marc Mentzer, Human Resources Management in Canada, 1997.
28 *Saskatoon Star Phoenix*, October 26, 1996.
29 *New York Times*, February 16, 1992.
30 *Wall Street Journal*, in *Globe and Mail*, November 13, 1997.
31 *Calgary Herald*, November 12, 1994.
32 *Globe and Mail*, January 9, 1996.
33 Ibid.
34 *Globe and Mail*, April 12, 1997.
35 Peter F. Drucker, *Concept of the Corporation*, Mentor, N.Y., 1983, pp.35–36.
36 *Globe and Mail*, April 12, 1997.
37 James Petras and Morris Morley, "The United States and Chile: Imperialism and the Overthrow of the Allende Government," Monthly Review Press, New York, 1975, p.31.
38 Ibid., p.33.
39 Ibid., pp.33–34.
40 *Guardian*, January 26, 1997.
41 George Pring and Penelope Canan, "Slapps: Getting Sued for Speaking Out," *Maclean's*, August 26, 1996.
42 Karen Wristen, Sierra Legal Defence Fund, interview with the author, Vancouver, August 1997.
43 Ibid.
44 Mark Margalli and Andy Friedman, *Masks of Deception: Corporate Front Groups in America*, Essential Information, Washington, D.C., 1991.
45 William Groder, *Who Will Tell the People: One Betrayal of American Democracy*, Simon and Shuster, New York, 1992, pp.36–37.

CHAPTER 5
HOW CORPORATIONS RULE THE WORLD

1 John Dillon, "Turning the Tide: Confronting the Money Changers,"

Canadian Centre for Policy Alternatives, Ottawa, 1997, p.94.

2 Jamie Swift, "The Debt Crisis: A Case of Global Usury," in *Conflicts of Interest*, ed. Jamie Swift and Brian Tomlinson, Between the Lines, Toronto, 1991, p.81.

3 In David Korten, *When Corporations Rule the World*, Kumarian Press, West Hartford, Conn., 1995, p.160.

4 Ibid., p.162.

5 Swift, "The Debt Crisis," p.85.

6 Jonathon Cahn, "Challenging the New Imperial Authority: The World Bank and the Democratization of Development," *Harvard Human Rights Journal*, Vol. 6 (1993).

7 Joyce Kolko, *Restructuring the World Economy*, Pantheon Books, New York, 1988, p.225.

8 Duncan Cameron, introduction to *Canada Under Free Trade*, ed. Duncan Cameron and Mel Watkins, Lorimer, Toronto, 1993, p.xi.

9 *The Monitor*, Canadian Centre for Policy Alternatives, October 1997.

10 "Challenging 'Free Trade' in Canada: The Real Story," Canadian Centre for Policy Alternatives, Ottawa, 1996, p.1.

11 Ibid., p.16.

12 Ibid., p.17.

13 Ibid., p.18.

14 Mel Clark, "Canadian State Powers: Comparing the FTA and GATT," in *Canada Under Free Trade*, ed. Cameron and Watkins, p.43.

15 "An Enviromental Guide to the World Trade Organization," Common Front on the World Trade Organization, c/o The Sierra Club, Ottawa, 1997, p.2.

16 Korten, p.174.

17 Ibid., p.178.

18 Elizabeth Smythe, "Your Place or Mine? States, International Organizations and the Negotiation of Investment Rules: The OECD versus the WTO," paper presented at the annual meeting of the International Studies Association, Toronto, March 1997, p.14.

19 Ibid.

20 Ibid., p.19.

21 Bernard Lietaer, speech to the International Forum on Globalization, Washington, D.C., May 10, 1996.

22 Martin Khor, speech to the International Forum on Globalization, Washington, D.C., May 10, 1996.

23 Ibid.

24 Tony Clarke, "The Corporate Rule Treaty," Canadian Centre for Policy Alternatives, Ottawa, 1997.

25 "Multilateral Agreement on Investment, Consolidated Texts and Commentary," OECD, 1997, p.27.

26 Ibid., p.40.

27 Ibid., p.53.

28 Ibid., p.63.

29 Clarke, p.7.

30 Marjorie Griffin Cohen, chair, CCPA-B.C., presentation to Commons Sub-committee on International Trade and the MAI, November 26, 1997.

31 International Chamber of Commerce, Commission on International Trade and Investment Policy, "Multilateral Rules for Investment," 1996.

32 Ibid.

33 *Vancouver Sun*, May 23, 1997.

34 Gary Campbell and Timothy Reid to Art Eggleton, January 26, 1996.

35 "Trading Insults," *Maclean's*, April 28, 1997.

36 Ibid.

37 Jane Kelsey, "APEC Created Solely to Serve Big Business," *The Monitor*, Canadian Centre for Policy Alternatives, July/August 1997.

38 "Canada's Individual Action Plan," 1996.

39 Ibid., p.73.

40 Gord McIntosh, *Calgary Herald*, November 16, 1997.

41 John Dillon, "The Enterprise for the Americas," in *Canada Under Free Trade*, ed. Cameron and Watkins, p.260.

42 Clarke, p.9.

43 Appleton and Associates, "Reservations to the proposed Multilateral Agreement on Investment," November 14, 1997, p.1.

44 Ibid., p.5.
45 Ibid.
46 Ibid., p.23.
47 Ibid., p.16.

Chapter 6
Rulers and Ruled in the New World Order

1 John Cavanagh, "The Challenge of Global Rule," keynote speech at the People's Forum on APEC, Manila, November 1996.
2 Ted Wheelwright, "Economic Controls for Social Ends," in *Beyond the Market: Alternatives to Economic Rationalism*, ed. R. Rees, G. Rodley, and F. Stilwell, Pluto, Sydney, 1993, p.21.
3 *The Monitor*, Canadian Centre for Policy Alternatives, March 1997.
4 *The Monitor*, Canadian Centre for Policy Alternatives, March 1995.
5 *Globe and Mail*, November 11, 1997.
6 Study by Irwin Gillespie, Frank Vermaeten, and Arndt Vermaeten, cited in *The Monitor*, Canadian Centre for Policy Alternatives, September 1994.
7 Robert Reich, "Secession of the Successful," *New York Times Magazine*, January 6, 1991.
8 Ibid.
9 Ibid.
10 Ibid.
11 *The Monitor*, Canadian Centre for Policy Alternatives, September 1996.
12 "Challenging 'Free Trade' in Canada: The Real Story," Canadian Centre for Policy Alternatives, Ottawa, 1996, p.4.
13 Ekos Research Associates, "Rethinking Government, '94," 1995, p.12.
14 Ibid., p.13.
15 *Taipan*, cited in *The Monitor*, Canadian Centre for Policy Alternatives, November 1995.
16 *The Monitor*, Canadian Centre for Policy Alternatives, May 1996.
17 In David Korten, *When Corporations Rule the World*, Kumarian Press, West Hartford, Conn., 1995, p.210.
18 Jack Warnock, letter to the author, July 29, 1997.
19 Jack Warnock, *The Other Mexico*, Black Rose Books, Montreal, 1995, pp.176–77.
20 Ibid., pp.181–83.
21 *The Monitor*, Canadian Centre for Policy Alternatives, November 1995.
22 "Toy Industry Conditions Exposed," in *Together for Human Rights*, Number 1, January 1996. Edited by Kathleen Puff.
23 *The Georgia Straight*, October 16, 1997.
24 *Together for Human Rights*, Number 4, July 1996.
25 Korten, pp.216–17.
26 *Globe and Mail Report on Business*, February 1997.
27 Canadian Press, September 21, 1996. Web page.
28 David Crane, *Toronto Star*, November 11, 1997.
29 Armine Yalnizyan, *Shifting Time: Social Policy and the Future of Work*, Between the Lines, Toronto, 1994, p.47.
30 Jane Jenson, "Some Consequences of Political Restructuring and Readjustment," in *Social Politics*, spring 1996.
31 *The Monitor*, Canadian Centre for Policy Alternatives, November 1996.
32 *Globe and Mail*, January 11, 1997.
33 Ibid.
34 National Council of Welfare statistics. 1995.
35 *The Monitor*, Canadian Centre for Policy Alternatives, September 1996.
36 Jim Stanford, "The Economics of the Debt and the Remaking of Canada," *Studies in Political Economy*, No. 48 (1995).
37 Joyce Kolko, *Restructuring the World Economy*, Pantheon Books, New York, 1988, p.234
38 Peter Gottschalk and Timothy Smeeding, "Cross-national Comparisons of Levels and Trends in Inequality," Luxembourg Income Study Working Paper 126, 1995.
39 World Economic Forum, *World Competitiveness Report*, 1994.
40 Ibid.

41 CBC Radio news, July 18, 1997.
42 Kate Bronfenbrenner, "We'll Close! Plant Closings, Plant Closing Threats, Union Organizing, and NAFTA," *The Multinational Monitor*, January 30, 1997.
43 *Vancouver Sun*, January 30, 1997.
44 Clarence Lochhead and Vivian Shalla, "Delivering the Goods: Income Distribution and the Precarious Middle Class," *Perceptions*, Canadian Council on Social Development, spring 1997.
45 *Globe and Mail*, November 9, 1996.
46 *Toronto Star*, July 29, 1997.

Chapter 7
Creating the Elite Consensus

1 Peter Thompson, "Bilderberg and the West," in *Trilateralism: The Trilateral Commission and Elite Planning for World Management*, ed. Holly Sklar, Black Rose Books, Montreal, 1980, p.161.
2 Ibid., p.171.
3 Ibid., p.177.
4 *Globe and Mail*, May 14, 1997.
5 World Economic Forum Web page, www.weforum.org.
6 Ibid.
7 *Globe and Mail*, February 1, 1997.
8 Elizabeth Smythe, "Your Place or Mine? States, International Organizations and the Negotiation of Investment Rules: The OECD versus the WTO," paper presented at the annual meeting of the International Studies Association, Toronto, March 1997, p.14.
9 Holly Sklar, "Founding the Trilateral Commission, Chronology 1970–77," in *Trilateralism: The Trilateral Commission and Elite Planning for World Management*, ed. Holly Sklar, Black Rose Books, Montreal, 1980, pp.76–83.
10 Ibid., p.69.
11 Richard Falk, "A New Paradigm for International Legal Studies," *Yale Law Review*, Vol. 84, No.5 (1975).
12 Sklar, p.27.
13 Ibid.
14 Ibid., p.87.
15 Joyce Nelson, "The Trilateral Commission," *Canadian Forum*, December 1993.
16 Trilateral Commission, "Membership — Canadian Group of the Trilateral Commission, 1996," Ottawa.
17 Samuel P. Huntington, in M. J. Crozier, S. P. Huntington, and J. Watanuki, *The Crisis of Democracy: Report of the Governability of Democracies to the Trilateral Commission*, New York University Press, 1975, p.113.
18 Ibid., p.98.
19 Ibid., p.93.
20 Ibid., p.75.
21 Ibid., pp.61-62.
22 Ibid., pp.6-7.
23 Ibid., p.106.
24 Daniel Boorstin, "Welcome to the Consumption Community," *Fortune*, September 1, 1967.
25 Huntington, in *The Crisis of Democracy*, pp.203-9.
26 Jim Stanford, "Policy Alternatives for the New World Order," paper presented to the Federal NDP Renewal Conference, Ottawa, September 1994.
27 David Langille, "The Business Council on National Issues and the Canadian State," *Studies in Political Economy*, autumn 1987.
28 Ibid., p.48.
29 Ibid., p.50.
30 Interview with the author, in "Taxes: The Second Certainty," CBC *Ideas*, 1991.
31 Langille, p.59.
32 Ibid., p.56.
33 Ibid., p.65.
34 Murray Dobbin, "Thomas d'Aquino: The De Facto PM," *Canadian Forum*, November 1992.
35 Canadian Institute of Strategic Studies, "Guns and Butter: Defence and the Canadian Economy," Toronto, 1984, p.50.
36 Duncan Cameron, interview with the author, August 1997.
37 Maude Barlow and Bruce Campbell, *Straight Through the Heart*, HarperCollins,

Toronto, 1995, p.49.

38 Ibid., p.94.

39 Ibid., p.93.

40 Jamie Swift, "The Debt Crisis: A Case of Global Usury," in *Conflicts of Interest*, ed Jamie Swift and Brian Tomlinson, Between the Lines, Toronto, 1991, p.97.

41 Business Council on National Issues, "Policy Committee," 1997.

42 A. L. Flood and Tom d'Aquino, "Memorandum for the Right Honourable Jean Chrétien, P.C., M.P., Prime Minister of Canada," June 20, 1997, p.6.

43 Executive Committee, BCNI, "Memorandum for the Honourable Frank McKenna, Premier of New Brunswick and Chairman-Designate, Council of Premiers," July 15, 1997.

44 Ibid., p.4.

45 Ibid., p.6.

46 Cameron interview.

47 Barlow and Campbell, *Straight Through the Heart*, p.52.

48 Cameron interview.

49 Ibid.

CHAPTER 8
CHANGING THE IDEOLOGICAL FABRIC OF CANADA

1 Tony Clarke, *Silent Coup*, Lorimer, Toronto, 1997, p.12.

2 Stephen McBride and John Shields, *Dismantling a Nation: Canada and the New World Order*, Fernwood Publishing, Halifax, 1993, p.37.

3 Beverly Scott, "Against Social Programs: The Campaigns of the Fraser Institute," in *Directions for Social Welfare in Canada: The Public's View,* School of Social Work, University of B.C., p.121.

4 The Fraser Institute, 1996 Annual Report.

5 "Fraser Currents: Special Anniversary Edition 1974–1989" and *Globe and Mail Report on Business*, July 1997.

6 Krishna Rau, "A Million for Your Thoughts," *Canadian Forum*, August 1996.

7 Walter Stewart, *The Charity Game: Greed, Waste and Fraud in Canada's $86-Billion-a-Year Compassion Industry*, Douglas & McIntyre, Vancouver, 1996. p.102.

8 In Allan Garr, *Tough Guy: Bill Bennett and the Taking of British Columbia*, Key Porter Books, Toronto, 1985, pp.91–93.

9 The Fraser Institute, 1996 Annual Report

10 Fraser Institute, "Freedom, Democracy and Economic Welfare," 1988, p.136.

11 In Scott, "Against Social Programs," p.116.

12 Fraser Institute, *Freedom, Democracy and Economic Welfare*, p.140.

13 Ibid., p.76.

14 Ibid., p.44.

15 In Allan Engler, *Apostles of Greed*, Fernwood Publishing, Halifax, 1995, p.101.

16 Ibid., p.xi.

17 Michael Walker, submission to the Standing Committee on Finance, September 17, 1991.

18 Michael Walker, "The Political Problem," notes for a speech to the Fraser Institute Seminar for Members of Parliament, Ottawa, February 1, 1994, p.9.

19 Kathleen Cross, "Off Balance: How the Fraser Institute Slants Its New-Monitoring Studies," *Canadian Forum*, October 1997.

20 Paul Havemann, "Marketing the New Establishment Ideology in Canada", in *Crime and Social Justice*, No. 26 (1986), p.30.

21 Fraser Institute, "Toward the New Millennium: A Five Year Plan for the Fraser Institute."

22 Fraser Institute, 1996 Annual Report, p.4.

23 *Edmonton Journal*, January 16, 1997.

24 Fraser Institute, "Toward the New Millennium."

25 Walter Block and Michael Walker, *On Employment Equity: A Critique of the Abella Royal Commission Report*, Fraser Institute, 1985, note 46, cited in

Michelle Valiquette, unpublished paper.
26 Block and Walker, *On Employment Equity*, note 50.
27 "Beloved Regulations Squeeze Us Badly," *Vancouver Sun*, September 25, 1996.
28 "Bad Work: A Review of Papers from a Fraser Institute Conference on 'Right-to-Work' Laws," Working Paper No. 16, Centre for Research on Work and Society, York University, 1997.
29 "Beloved Regulations Squeeze Us Badly," *Vancouver Sun*, September 25, 1996.
30 David Somerville, address to conference of the International Society for Individual Liberty, Whistler, B.C., August 1996.
31 "The Legacy of Colin M. Brown," in program for the NCC's Tenth Annual Colin M. Brown Freedom Medal Presentation, Toronto, 1996.
32 *Vancouver Sun*, March 5, 1987
33 *Chronicle Herald*, Halifax, December 17, 1985.
34 Program, NCC's Tenth Annual Colin M. Brown Freedom Medal presentation, Toronto, 1996.
35 The NCC's advisory council is not mentioned in its bylaws so it is difficult to know what role advisory council members play. Leitch appears to play a very active role since he was given the responsibility of heading a search committee for Somerville's replacement. Other current members are: John Dobson, Robert Foret, Ian Gray, Neil Harvie, James Kenny, Eric Kipping, William Magyar, Kenneth McDonald, Jack Pirie, Donald Thain, Roger Thompson, and staff members Peter Coleman, Colin T. Brown, and David Somerville.
36 Nick Fillmore, "The Right Stuff: An Inside Look at the National Citizens' Coalition," *This Magazine*, June/July 1986.
37 In Fillmore, "The Right Stuff."
38 *Ottawa Citizen*, June 23, 1997.
39 *Financial Post*, June 25, 1995.
40 National Citizens' Coalition, *Overview*, May 1994.
41 In Fillmore, "The Right Stuff."
42 National Citizens' Coalition, *Consensus*, February 1994.
43 National Citizens' Coalition, "Campaign '87."
44 *Consensus*, June 1994.
45 *Consensus*, December 1985.
46 National Citizens' Coalition, "Who We Are and What We Do 1996"; Ontarians for Responsible Government, news release, May 1, 1995.
47 *Consensus*, June 1991.
48 National Citizens' Coalition, news release, May 13, 1996.
49 Fillmore, "The Right Stuff."
50 Rick Salutin, *Waiting for Democracy*, Viking, Markham, Ont., 1989, p.268.
51 *Financial Post*, February 14, 1993.
52 *This Magazine*, July 1986.
53 National Citizens' Coalition, "Who We Are and What We Do 1996" and a May 30, 1996, *Vancouver Sun* story contain NCC claims that it has 45,000 supporters. The February 1996 issue of *Consensus* and the NCC program for the Tenth Annual Colin M. Brown Freedom Medal Presentation in 1996 claim the NCC has 40,000 supporters.
54 *Consensus*, June 1987. Somerville expressed initial reservations about the Reform Party because he thought it might not be conservative enough.
55 *Consensus*, special edition, June 1994.
56 Ibid.
57 NCC, "Campaign '87."
58 *Consensus*, October 1994.
59 Atlantic Institute for Market Studies, Internet site.
60 Linda McQuaig, *Shooting the Hippo*, Penguin, Markham, Ont., 1995, p.133.
61 Rau, "A Million for Your Thoughts."
62 *Globe and Mail*, December 6, 1996.
63 "Journalists Voice Ownership Concerns," *Edmonton Journal*, May 26, 1997.
64 In Peter C. Newman, *The Establishment Man*, McClelland and Stewart, Toronto,

1982, p.183.
65 Verus Group brochure.
66 In Garr, *Tough Guy*, p.42.
67 Interview with the author, August 1997.

Chapter 9
Propaganda Wars: The Revolution of Falling Expectations

1 Ed Finn, ed., "The Deficit Made Me Do It," Canadian Centre for Policy Alternatives, Ottawa, 1992.
2 Leslie Biggs and Mark Stobbe, eds., *Devine Rule in Saskatchewan: A Decade of Hope and Hardship*, Fifth House, Saskatoon, 1992, Table 1.2, p.298.
3 Jane Kelsey, *The New Zealand Experiment*, Auckland University Press, 1995, p.28.
4 Ibid.
5 Seth Klein, "Good Sense versus Common Sense: Canada's Debt Debate and Competing Hegemonic Projects," master's thesis, Department of Political Science, Simon Fraser University, 1996, p.71.
6 Ibid.
7 Ibid., p.70.
8 Linda McQuaig, *Shooting the Hippo*, Penguin, Markham, Ont., 1995, pp.41–42.
9 *Globe and Mail*, November 10, 1993.
10 *Globe and Mail*, July 25, 1997.
11 Klein, p.60.
12 Jim Stanford, "Rebuilding a Left Economic Alternative," paper presented to Federal NDP Renewal Conference, Ottawa, September 1994.
13 Kirk Falconer, "Corporate Taxation in Canada," *Canadian Review of Social Policy*, No. 26 (November 1990).
14 Murray McIlveen and Hideo Mimoto, "The Federal Government Deficit, 1975 –1988–89," Statistics Canada, unpublished paper, 1990.
15 Dominion Bond Rating Service, "The Massive Federal Debt: How Did It Happen?" February 1995.
16 *The Monitor*, Canadian Centre for Policy Alternatives, May 1994.
17 *Globe and Mail*, August 27, 1996.
18 *Globe and Mail*, September 26, 1997.
19 *Globe and Mail*, September 22, 1997.
20 Canadian Centre for Policy Alternatives/Choices Coalition, Alternative Federal Budget, 1997.
21 John Dillon, "Turning the Tide: Confronting the Money Changers," Canadian Centre for Policy Alternatives, 1997, p.101.
22 OECD, Paris, 1990.
23 OECD, "Unemployment Benefit Replacement Rates in 1991," 1992.
24 Paul Martin, "The Canadian Experience in Reducing Budget Deficits and Debt," notes for a speech to the Federal Reserve Bank of Kansas City, Jackson Hole, Wyoming, September 1, 1995.
25 Martin Harts, compensation specialist with KPMG, interview with the author, December 1, 1997.
26 *The Monitor*, Canadian Centre for Policy Alternatives, May 1996.
27 David Gordon, "Workers Overmanaged by Business Bureaucrats," *The Monitor*, Canadian Centre for Policy Alternatives, October 1996.
28 Ibid.
29 *The Monitor*, Canadian Centre for Policy Alternatives, September 1994.
30 *Globe and Mail Report on Business*, May 1996.
31 Lyndon McIntyre, "The Best Thing Since Water," *the fifth estate*, CBC, January 7, 1997.
32 George Anders, *Health Against Wealth: HMOs and the Breakdown of Medical Trust*, Houghton Mifflin, New York, 1997.
33 *Sunday Morning*, CBC, February 9, 1997.
34 Ibid.
35 Colleen Fuller, "Restructuring Health Care: A Global Enterprise," Council of Canadians discussion paper, 1995.
36 Ibid.
37 Ian Austen, "Relax: This Won't Hurt a Bit," *Elm Street*, September 1997.
38 Nadene Rehnby, "Inventing Crisis: The Erosion of Public Confidence in

Canadian Public Education," B.C. Teachers' Federation, 1996.
39 Lehman Brothers, "Investment Opportunity in the Education Industry," 1996, p.7.
40 Ibid., p.38.
41 Tony Clarke, *Silent Coup*, Lorimer, Toronto, 1997, p.159.
42 *Business Week*, June 30, 1997.
43 Maude Barlow and Heather-jane Robertson, *Class Warfare: The Assault on Canada's Schools*, Key Porter Books, Toronto, 1994, p.105.
44 Ibid.
45 Ibid., p.106.
46 David Robertson, "The Facts," Canadian Union of Public Employees, July-August 1985.
47 *Globe and Mail*, August 27, 1996.
48 KPMG, "A Comparison of Business Costs in Canada and the United States," 1995; *Globe and Mail*, November 13, 1996.
49 *Ottawa Citizen*, October 10, 1997.
50 Jim Stanford, interview with the author.
51 *Globe and Mail*, February 26, July 8, and October 8, 1997.
52 "Mergers and Acquisitions Jump in Third Quarter," *Globe and Mail*, October 8, 1997.
53 "Merger Wave Gathers Momentum," *Globe and Mail*, February 26, 1997.
54 *The Monitor*, Canadian Centre for Policy Alternatives, October 1997.
55 Alternative Federal Budget (short version), p.12.
56 *Globe and Mail*, January 23, 1997.
57 *Edmonton Journal*, July 25, 1997.
58 *The Monitor*, Canadian Centre for Policy Alternatives, May 1995.
59 *Globe and Mail*, November 10, 1997.

Chapter 10
The Market State

1 Frank Stilwell, "Economic Rationalism: Sound Foundations for Policy?" in *Beyond the Market: Alternatives to Economic Rationalism*, ed. R. Rees, G. Rodley, and F. Stilwell, Pluto Press, Sydney, p.33.
2 Sam Gindin, "Notes on Labour at the End of the Century: Starting Over?" *Monthly Review*, July-August 1997.
3 Allan Engler, *Apostles of Greed*, Fernwood Publishing, Halifax, 1995, p.17.
4 Adam Smith, *Wealth of Nations*, Modern Library, New York, 1937, p.250.
5 Ibid., pp.78–79.
6 Analysis and data on New Zealand from Murray Dobbin, "The Remaking of New Zealand," CBC *Ideas*, October 12 and 19, 1994.
7 Paul Havemann, letter to the author, August 21, 1997.
8 Jonathon Boston, "The Ideas and Theories Underpinning the New Zealand Model," in *The Revolution in Public Management*, Oxford University Press, 1996, pp.29–30.
9 Ibid., p.28.
10 For a more detailed description of the charter experience in New Zealand, see Murray Dobbin, "Charter Schools: Charting a Course to Social Division," Canadian Centre for Policy Alternatives, Ottawa, 1997.
11 Havemann letter.
12 Roger Douglas, keynote address to the Reform Party of Canada convention, Saskatoon, 1991, in Murray Dobbin, *Preston Manning and the Reform Party*, Lorimer, Toronto, 1991, p.113.
13 Jane Kelsey, *The New Zealand Experiment*, Auckland University Press, 1995, p.29.
14 Maude Barlow and Bruce Campbell, *Straight Through the Heart*, HarperCollins, Toronto, 1995, p.97.
15 In "The Liberals' Labour Strategy," Canadian Centre for Policy Alternatives, Ottawa, 1995, p.14.
16 *Globe and Mail*, May 3, 1997.
17 *Globe and Mail*, July 31, 1997.
18 Interview with the author, 1992.
19 "Exporting Canadian Culture," *Globe and Mail*, January 25, 1997.
20 Canadian Conference of the Arts, "Reductions in Departmental Spending,

1994–95 – 1998–99," in 1996 Federal Budget, p.39.

21 "Exporting Canadian Culture," *Globe and Mail.*

22 Ibid.

23 Senate Standing Committee on Transport and Communications, Subcommittee on Communications, *Wired to Win: Canada's International Competitive Position in Communications,* interim report, April 1997, p.26.

24 Ibid., p.51.

25 Ibid.

26 In Michael Walker, ed., "Freedom, Democracy and Economic Welfare," Fraser Institute, 1988, p.48.

27 Barlow and Campbell, *Straight Through the Heart*, p.151.

28 Interview with the author, August 1997.

29 Ibid.

30 Interview with the author, September 1997.

31 Ontario Federation of Labour, "The Common Sense Revolution: 210 Days of Destruction," 1996.

32 Ibid.

CHAPTER 11
DEMOCRACY IN THE NEW MILLENNIUM

1 In *The Monitor*, Canadian Centre for Policy Alternatives, July-August 1997.

2 Samuel P. Huntington, in M. J. Crozier, S. P. Huntington, and J. Watanuki, eds., *The Crisis of Democracy: Report of the Governability of Democracies to the Trilateral Commission*, New York University Press, 1975, p.93.

3 Ibid., p.99.

4 Leo Panitch, "Changing Gears: Democratizing the Welfare State," presentation to Fifth Conference on Social Welfare Policy, Bishop's University, Lennoxville, Que., August 1991.

5 Ibid.

6 James Tobin, "The Tobin Tax on International Monetary Transactions," Canadian Centre for Policy Alternatives, Ottawa, 1995.

7 The Genuine Progress Indicator, Redefining Progress, 1995.

8 *The Monitor*, Canadian Centre for Policy Alternatives, February 1996.

9 *The Monitor*, Canadian Centre for Policy Alternatives, November 1996.

10 *The Monitor*, Canadian Centre for Policy Alternatives, December 1996/January 1998.

11 Samuel Gindin, "Notes on Labour at the End of the Century: Starting Over?" *Monthly Review*, July-August 1997.

12 *Toronto Star*, August 12, 1997.

13 *Globe and Mail*, August 27, 1997.

14 Gindin, "Notes on Labour."

15 *Financial Post*, August 23, 1997.

16 "Global Good Times, Meet the Global Glut," *New York Times*, November 16, 1996.

17 *Globe and Mail*, November 9, 1996.

18 Angus Reid home page, March 29, 1996.

19 News release, undates, Preamble Center for Public Policy, home page.

20 Kees van der Pijl, "The History of Class Struggle," *Monthly Review*, May 1997.

21 Martin Khor, keynote address to the Council of Canadians annual general

meeting, October 1997.

Index

INDEX

INDEX